T0336877

MATHEMATICAL METHODS IN ENGINEERING

This text focuses on a variety of topics in mathematics in common usage in graduate engineering programs including vector calculus, linear and nonlinear ordinary differential equations, approximation methods, vector spaces, linear algebra, integral equations, and dynamical systems. The book is designed for engineering graduate students who wonder how much of their basic mathematics will be of use in practice. Following development of the underlying analysis, the book takes students step-by-step through a large number of examples that have been worked in detail. Students can choose to go through each step or to skip ahead if they desire. After seeing all the intermediate steps, they will be in a better position to know what is expected of them when solving homework assignments and examination problems, and when they are on the job. Each chapter concludes with numerous exercises for the student that reinforce the chapter content and help connect the subject matter to a variety of engineering problems. Students today have grown up with computer-based tools including numerical calculations and computer graphics; the worked-out examples as well as the end-of-chapter exercises often use computers for numerical and symbolic computations and for graphical display of the results.

Joseph M. Powers joined the University of Notre Dame in 1989. His research has focused on the dynamics of high-speed reactive fluids and on computational science, especially as it applies to verification and validation of complex multiscale systems. He has held positions at the NASA Lewis Research Center, the Los Alamos National Laboratory, the Air Force Research Laboratory, the Argonne National Laboratory, and the Chinese Academy of Sciences. He is a member of AIAA, APS, ASME, the Combustion Institute, and SIAM. He is the recipient of numerous teaching awards.

Mihir Sen has been active in teaching and in research in thermal-fluids engineering – especially in regard to problems relating to modeling, dynamics, and stability – since obtaining his PhD from MIT. He has worked on reacting flows, natural and forced convection, flow in porous media, falling films, boiling, MEMS, heat exchangers, thermal control, and intelligent systems. He joined the University of Notre Dame in 1986 and received the Kaneb Teaching Award from the College of Engineering in 2001 and the Rev. Edmund P. Joyce, C.S.C., Award for Excellence in Undergraduate Teaching in 2009. He is a Fellow of ASME.

Mathematical Methods in Engineering

Joseph M. Powers
University of Notre Dame

Mihir Sen
University of Notre Dame

Shaftesbury Road, Cambridge CB2 8EA, United Kingdom

One Liberty Plaza, 20th Floor, New York, NY 10006, USA

477 Williamstown Road, Port Melbourne, VIC 3207, Australia

314–321, 3rd Floor, Plot 3, Splendor Forum, Jasola District Centre, New Delhi – 110025, India

103 Penang Road, #05–06/07, Visioncrest Commercial, Singapore 238467

Cambridge University Press is part of Cambridge University Press & Assessment, a department of the University of Cambridge.

We share the University's mission to contribute to society through the pursuit of education, learning and research at the highest international levels of excellence.

www.cambridge.org
Information on this title: www.cambridge.org/9781107037045

© Joseph M. Powers and Mihir Sen 2015

This publication is in copyright. Subject to statutory exception and to the provisions of relevant collective licensing agreements, no reproduction of any part may take place without the written permission of Cambridge University Press & Assessment.

First published 2015

A catalogue record for this publication is available from the British Library

Library of Congress Cataloging-in-Publication data
Powers, Joseph, 1961–
Mathematical methods in engineering / Joseph Powers, University of Notre Dame, Mihir Sen, University of Notre Dame.
pages cm
Includes bibliographical references and index.
ISBN 978-1-107-03704-5 (hardback : alk. paper)
1. Engineering mathematics – Study and teaching (Graduate)   I. Sen, Mihir, 1947–
II. Title.
TA332.5.P69   2015
510–dc23   2014020962

ISBN   978-1-107-03704-5   Hardback

Cambridge University Press & Assessment has no responsibility for the persistence or accuracy of URLs for external or third-party internet websites referred to in this publication and does not guarantee that any content on such websites is, or will remain, accurate or appropriate.

To my parents: Mary Rita, my first reading teacher, and Joseph Leo, my first mathematics teacher – Joseph Michael Powers

To my family: Beatriz, Pradeep, Maya, Yasamin, and Shayan – Mihir Sen

Contents

Preface

Our overarching aim in writing this book is to build a bridge to enable engineers to better traverse the domains of the mathematical and physical worlds. Our focus is on neither the nuances of pure mathematics nor the phenomenology of physical devices but instead is on the mathematical tools used today in many engineering environments. We often compromise strict formalism for the sake of efficient exposition of mathematical tools. Whereas some results are fully derived, others are simply asserted, especially when detailed proofs would significantly lengthen the presentation. Thus, the book emphasizes method and technique over rigor and completeness; readers who require more of the latter can and *should* turn to many of the foundational works cited in the extensive bibliography.

Our specific objective is to survey topics in engineering-relevant applied mathematics, including multivariable calculus, vectors and tensors, ordinary differential equations, approximation methods, linear analysis, linear algebra, linear integral equations, and nonlinear dynamical systems. In short, the text fully explores linear systems and considers some effects of nonlinearity, especially those types that can be treated analytically. Many topics have geometric interpretations, identified throughout the book. Particular attention is paid to the notion of approximation via projection of an entity from a high- or even infinite-dimensional space onto a space of lower dimension. Another goal is to give the student the mathematical background to delve into topics such as dynamics, differential geometry, continuum mechanics, and computational methods; although the material presented is relevant to those fields, specific physical applications are mainly confined to some of the exercises. A final goal is to introduce the engineer to *some* of the notation and rigor of mathematics in a way used in many upper-level graduate engineering and applied mathematics courses.

This book is intended for use in a beginning graduate course in applied mathematics taught to engineers. It arose from a set of notes for such a course taught by the authors for more than 20 years in the Department of Aerospace and Mechanical Engineering at the University of Notre Dame. Students in this course come from a variety of backgrounds, mainly within engineering but also from science. Most enter with some undergraduate-level proficiency in differential and integral multivariable calculus, differential equations, vectors analysis, and linear algebra. This book briefly reviews these subjects but more often builds on an assumed elementary understanding of topics such as continuity, limits, series, and the chain rule.

As such, we often casually introduce subject matter whose full development is deferred to later chapters. For example, although one of the key features of the book is a lengthy discussion of eigensystems in Chapter 6, most engineers will already know what an eigenvalue is. Consequently, we employ eigenvalues in nearly every chapter, starting from Chapter 1. The same can be said for topics such as vector operators, determinants, and linear equations. When such topics are introduced earlier than their formal presentation, we often make a forward reference to the appropriate section, and the student is encouraged to read ahead. In summary, most beginning graduate students and advanced undergraduates will be prepared for the subject matter, though they may find occasion to revisit some trusted undergraduate texts. Although our course is only one semester, we have added some topics to those we usually cover; the instructor of a similar course should be able to omit certain topics and add others.

At a time not very far in the past, mathematics in engineering was largely confined to basic algebra and interpolation of trigonometric and logarithmic tables. Not so today! Much of engineering has come to rely on sophisticated predictive mathematical models to design and control a wide variety of devices, for example, buildings and bridges, air and ground transportation, manufacturing and chemical processes, electrical and electronic devices, and biomedical and robotic equipment. These models may be in the form of algebraic, differential, integral, or other equations. Formulation of such models is often a challenge that calls on the basic sciences of physics, chemistry, and biology. Once they are formulated, the engineer is faced with actually solving the model equations, and for that a variety of tools are of value. Our focus is on the general mathematical tools used for engineering problems but not their formulation or specific physical details. While we sporadically discuss a paradigm problem such as a mass-spring-damper, we focus more on the mathematics. The use of mathematical analysis within engineering has changed greatly over the years. Once it was the sole means to the solution of some problems, but currently engineers rely on it within numerical and experimental approaches. The use of the adjective *numerical* has also changed over the years because of the variety of ways in which computers may be used in engineering. It is in fact becoming more common and necessary for extensive mathematical manipulation to be performed to prepare the computer for efficient and accurate solution generation.

Choices have been made with regard to notation; most of the conventions we adopt are reflected in at least a portion of the literature. In a few cases, we choose to diverge slightly from some of the more common norms. When we do so, explanatory footnotes are usually included. For example, the literature has a number of conventions for the so-called dot product or scalar product between two vectors, $\mathbf{u}$ and $\mathbf{v}$. The most common is probably $\mathbf{u} \cdot \mathbf{v}$. We generally choose the more elaborate $\mathbf{u}^T \cdot \mathbf{v}$, where the T indicates a transpose operation. This emphasizes the fact that vectors are considered to be columns of elements and that to associate a scalar product of two vectors with the ordinary rules of matrix multiplication, one needs to transpose the first vector into a row of elements. Similarly, we generally take the product of a matrix $\mathbf{A}$ and vector $\mathbf{u}$ to be of the form $\mathbf{A} \cdot \mathbf{u}$ rather than the often-used $\mathbf{A}\mathbf{u}$. And we use $\mathbf{u}^T \cdot \mathbf{A}$, while many texts simply write $\mathbf{u}\mathbf{A}$. Unusually, we often apply the transpose notation to the so-called divergence operator, writing, for example, the divergence of a vector field $\mathbf{u}$ as $\operatorname{div} \mathbf{u} = \nabla^T \cdot \mathbf{u}$ rather than as the more common $\nabla \cdot \mathbf{u}$. One could easily infer the nature of the divergence operation without the transpose, but we believe it adds unity to our notational framework. In the text, italicized letters like a will most often be used for scalars, bold lowercase letters like a for vectors, bold uppercase letters like **A** for tensors and operators, and $\mathbb{A}$ for sets and spaces. In general, we use ^{T} for transpose, $^-$ for complex conjugate, * for adjoint, and ^{H} for Hermitian transpose. The student also has to be aware that the same quantities written on a blackboard or paper may appear differently. Whatever the notational choices, the student should be fluent in a variety of usages in the literature; we in fact sometimes deviate from our own conventions to reinforce that our choices are not unique.

Our experience has been that engineering students learn best by exposure to examples, and a hallmark of the text is that much of the material is developed via presentation of a large number of fully worked problems, each of which generally follows a short fundamental development. The solved examples not only illustrate the points made previously but also introduce additional concepts and are thus an integral part of the text. Ultimately, mathematics is learned by doing, and for this reason we have a large number of exercises. Engineers, moreover, have a special purpose for studying mathematics: they need it to solve practical problems; some are included as exercises.

Presentation of many of our specific details has relied on modern software for symbolic computation and plotting. We encourage the reader to utilize these tools as well, as they enable exact solutions and graphics that may otherwise be impossible. The text does not provide details of particular software packages, which often change with each new version; the reader is advised to choose one or more packages, such as Mathematica, Maple, or MATLAB, and to become familiar with its usage. Exercises are included that require the use of such software. The phrase "symbolic computer mathematics" is used to mean tools such as Mathematica or Maple and "discrete computational methods" to connote tools such as MATLAB, Python, Fortran, C, or C++. The problems emphasize plotting to give a geometric overview of the results. The use of visuals or graphics to get a quick appreciation for the quality of an approximation or the behavior of a result cannot be overemphasized.

There are a number of texts on graduate-level mathematics for engineers. Mathematics applied to engineering is a vast discipline; consequently, each book has a unique emphasis. Here we have attempted to include what is actually used by researchers in our field. To be clear, though, because it is for a one-semester introductory course, many topics are left for advanced courses. Among the topics omitted or lightly covered are integral transforms, complex variables, partial differential equations, group theory, probability, statistics, numerical methods, and graph and network theory.

In closing, we express our hope that the readers of this book will find mathematics to be as beautiful and useful a subject as we have over the years. Our appreciation was nurtured by a large number of special people: family members, teachers at all levels, colleagues at home and abroad, authors from many ages, and our own students over the decades. We have learned from all of them and hope that our propagation of their knowledge engenders new discoveries from readers of this book for future generations.

Joseph M. Powers
Mihir Sen
Notre Dame, Indiana, USA

1 Multivariable Calculus

We commence with a presentation of fundamental notions from the *calculus of many variables*, a subject that will be seen to provide a useful language to describe multidimensional geometry, a topic foundational to nearly all of engineering. We focus on presenting its grammar in sufficient richness to enable precise rendering of both simple and more complex geometries. This is of self-evident importance in engineering applications. We begin with a summary of implicitly defined functions and, in what will be a foundation for the book, local linearization. This quickly leads us to consider matrices that arise from linearization. The basic strategies of optimization, both unconstrained and constrained, are considered via identification of maxima and minima as well as the calculus of variations and Lagrange multipliers. The chapter ends with an extensive exposition relating multivariable calculus to the geometry of nonlinear coordinate transformations; in so doing, we develop geometry-based analytical tools that are used throughout the book. Such tools have relevance in fields ranging from computational mechanics widely used in engineering software to the theory of dynamical systems. Understanding the general nonlinear case also prepares one to better frame the more common geometrical methods associated with orthonormal coordinate transformations studied in Chapter 2.

1.1 Implicit Functions

1.1.1 One Independent Variable

We begin with a consideration of implicit functions of the form

$$f(x, y) = 0. \tag{1.1}$$

Here f is allowed to be a fully nonlinear function of x and y. We take the point (x_0, y_0) to satisfy $f(x_0, y_0) = 0$. Near (x_0, y_0), we can locally linearize via the familiar *total differential*,

$$df = \left.\frac{\partial f}{\partial x}\right|_{x_0, y_0} dx + \left.\frac{\partial f}{\partial y}\right|_{x_0, y_0} dy = 0. \tag{1.2}$$

At the point (x_0, y_0), both $\partial f/\partial x$ and $\partial f/\partial y$ are constants, and we consider the local "variables" to be dx and dy, the infinitesimally small deviations of x and y from x_0

and y_0. As long as $\partial f / \partial y \neq 0$, we can scale to say

$$\frac{dy}{dx}\bigg|_{x_0,y_0} = -\frac{\dfrac{\partial f}{\partial x}\bigg|_{x_0,y_0}}{\dfrac{\partial f}{\partial y}\bigg|_{x_0,y_0}}. \tag{1.3}$$

And with the local existence of dy/dx, we can consider y to be a unique local linear function of x and quantify its variation as

$$y - y_0 \approx \frac{dy}{dx}\bigg|_{x_0,y_0} (x - x_0). \tag{1.4}$$

The process of examining the derivative at a point is a fundamental part of the use of linearization to gain local knowledge of a nonlinear entity. From here on we generally do not specify the point at which the evaluation is occurring, as it should be easy to understand from the context of the problem.

These notions can be formalized in the following:

Implicit Function Theorem: For a given $f(x, y)$ with $f = 0$ and $\partial f / \partial y \neq 0$ at the point (x_0, y_0), there corresponds a unique function $y(x)$ in the neighborhood[1] of (x_0, y_0).

When such a condition is satisfied, we can consider the dependent variable y to be locally a unique function of the independent variable x. The theorem says nothing about the case $\partial f / \partial y = 0$. Though such a case explicitly states that at such a point f is not a function of y, it does not speak to $y(x)$.

EXAMPLE 1.1

Consider the implicit function

$$f(x, y) = x^2 - y = 0. \tag{1.5}$$

Determine where y can be considered to be a unique local function of x.

We first recognize that when $f = 0$, we have $y = x^2$, which is a simple parabola. And it is easy to see that for all $x \in (-\infty, \infty)$, y is defined[2] and will be such that $y \in [0, \infty)$. Let us apply the formalism of the implicit function theorem to see how it is exercised on this well-understood curve. Applying Eq. (1.2) to our f, we find

$$df = \frac{\partial f}{\partial x}\, dx + \frac{\partial f}{\partial y}\, dy = 0, \tag{1.6}$$

$$= 2x\, dx - dy = 0. \tag{1.7}$$

Thus,

$$\frac{dy}{dx} = 2x. \tag{1.8}$$

This is defined for all $x \in (-\infty, \infty)$. Thus, a unique linearization $y(x)$ exists for all $x \in (-\infty, \infty)$. In the neighborhood of a point (x_0, y_0), we have the local linearization

$$y - y_0 = 2x_0(x - x_0). \tag{1.9}$$

---

[1] *Neighborhood* is a word whose mathematical meaning is close to its common meaning; nevertheless, it does require a precise definition, which is deferred until Section 6.1.

[2] Here $\in$ stands for "is an element of" and indicates that x could be any number, which we assume to be real. Also, we use the standard notation $(\cdot)$ to describe an open interval on the real axis, that is, one that does not include the endpoints. A closed interval is denoted by $[\cdot]$ and mixed intervals by either $(\cdot]$ or $[\cdot)$.

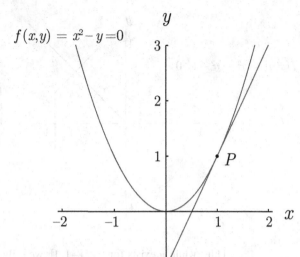

$$f(x,y) = x^2 - y = 0$$

Figure 1.1. The parabola defined implicitly by $f(x,y) = x^2 - y = 0$ along with its local linearization near $P : (1,1)$.

Consider the point $P : (x_0, y_0) = (1,1)$. Obviously at P, $f = 1^2 - 1 = 0$, so P is on the curve. The local linearization at P is

$$y - 1 = 2(x - 1), \tag{1.10}$$

which shows uniquely how y varies with x in the neighborhood of P. The parabola defined by $f(x,y) = x^2 - y = 0$ and the linearization near P are plotted in Figure 1.1.

EXAMPLE 1.2

Consider the implicit function

$$f(x,y) = x^2 + y^2 - 1 = 0. \tag{1.11}$$

Determine where y can be considered to be a unique local function of x.

We first note that f is obviously nonlinear; our analysis will hinge on its local linearization. The function can be rewritten as $x^2 + y^2 = 1$, which is recognized as a circle centered at the origin with radius of unity. By inspection, we will need $x \in [-1, 1]$ to have a real value of y. Let us see how the formalism of the implicit function theorem applies. We have $\partial f / \partial x = 2x$ and $\partial f / \partial y = 2y$. We are thus concerned that wherever $y = 0$, there is no unique function $y(x)$ that can be defined in the neighborhood of such a point. Applying Eq. (1.2) to our f, we find

$$df = \frac{\partial f}{\partial x} dx + \frac{\partial f}{\partial y} dy = 0, \tag{1.12}$$

$$= 2x \, dx + 2y \, dy = 0. \tag{1.13}$$

Thus,

$$\frac{dy}{dx} = -\frac{x}{y}. \tag{1.14}$$

Obviously, whenever $y = 0$, the derivative dy/dx is undefined, and there is no local linearization. In the neighborhood of a point (x_0, y_0), with $y_0 \neq 0$, we have the local linearization

$$y - y_0 = -\frac{x_0}{y_0}(x - x_0). \tag{1.15}$$

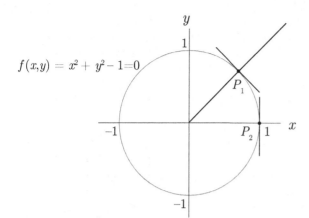

$$f(x,y) = x^2 + y^2 - 1 = 0$$

Figure 1.2. The circle defined implicitly by $f(x,y) = x^2 + y^2 - 1 = 0$ along with its local linearization near $P_1 : (\sqrt{2}/2, \sqrt{2}/2)$ and $P_2 : (1,0)$.

Thus, while y exists for $x \in [-1,1]$, we only have a local linearization for $x \in (-1,1)$. When $x = \pm 1$, $y = 0$, and the local linearization does not exist.

Consider two points on the circle $P_1 : (x_0, y_0) = (\sqrt{2}/2, \sqrt{2}/2)$ and $P_2 : (x_0, y_0) = (1,0)$. Obviously $f = 0$ at both P_1 and P_2, so both are on the circle. At P_1, we have the local linearization

$$y - \frac{\sqrt{2}}{2} = -\left(x - \frac{\sqrt{2}}{2}\right), \tag{1.16}$$

showing uniquely how y varies with x near P_1, the curve being the tangent line to f. At P_2, it is obvious that the tangent line is expressed by $x = 1$, which does not tell us how y varies with x locally. Indeed, we do know that $y = 0$ at P_2, but we lack a unique formula for its variation with x in nearby regions.

The circle defined by $f(x,y) = x^2 + y^2 - 1 = 0$ and the linearizations near P_1 and P_2 are plotted in Figure 1.2. Globally, of course, y does vary with x for this f; in fact, one can see by inspection that

$$y(x) = \pm\sqrt{1 - x^2}. \tag{1.17}$$

This function takes on real values for $x \in [-1,1]$, but within that domain, it is nowhere unique. One might think a Taylor[3] series (see Section 5.1.1) approximation of Eq. (1.17) might be of use near P_2, but that too is nonunique, yielding the nonlinear approximations

$$y(x) \approx \sqrt{2(1-x)}, \tag{1.18}$$

$$y(x) \approx -\sqrt{2(1-x)}, \tag{1.19}$$

for $x \approx 1$. So the impact of the implicit function theorem on this f at $P_2 : (1,0)$ is that there is no *unique* approximating linear function $y(x)$ at P_2. Whenever a linear approximation is available, we can expect uniqueness.

Had our global function been $f(x,y) = x - 1 = 0$, we would have had $\partial f/\partial y = 0$ everywhere and would nowhere have been able to identify $y(x)$; this simply reflects that for $f = x - 1 = 0$, whenever $x = 1$, any value of y will allow satisfaction of $f = 0$.

[3] Brook Taylor, 1685–1731, English mathematician, musician, and painter.

1.1.2 Many Independent Variables

If, instead, we have N independent variables, we can think of a relation such as

$$f(x_1, x_2, \ldots, x_N, y) = 0 \qquad (1.20)$$

in some region as an implicit function of y with respect to the other variables. We cannot have $\partial f / \partial y = 0$, because then f would not depend on y in this region. In principle, we can write

$$y = y(x_1, x_2, \ldots, x_N) \qquad (1.21)$$

if $\partial f / \partial y \neq 0$.

The derivative $\partial y / \partial x_n$ can be determined from $f = 0$ without explicitly solving for y. First, from the definition of the total differential, we have

$$df = \frac{\partial f}{\partial x_1} \, dx_1 + \frac{\partial f}{\partial x_2} \, dx_2 + \cdots + \frac{\partial f}{\partial x_n} \, dx_n + \cdots + \frac{\partial f}{\partial x_N} \, dx_N + \frac{\partial f}{\partial y} \, dy = 0.$$
$$(1.22)$$

Differentiating with respect to x_n while holding all the other $x_m, m \neq n$, constant, we get

$$\frac{\partial f}{\partial x_n} + \frac{\partial f}{\partial y} \frac{\partial y}{\partial x_n} = 0, \qquad (1.23)$$

so that

$$\frac{\partial y}{\partial x_n} = -\frac{\dfrac{\partial f}{\partial x_n}}{\dfrac{\partial f}{\partial y}}, \qquad (1.24)$$

which can be found if $\partial f / \partial y \neq 0$. That is to say, y can be considered a function of x_n if $\partial f / \partial y \neq 0$. Again, local linearization of a nonlinear function provides a key to understanding the behavior of a function in a small neighborhood.

EXAMPLE 1.3

Consider the implicit function

$$f(x_1, x_2, y) = x_1^2 + x_2^2 + y^2 - 1 = 0. \qquad (1.25)$$

Determine where y can be considered to be a unique local function of x_1 and x_2.

This problem is essentially a three-dimensional extension of the previous example. Here the nonlinear $f = 0$ describes a sphere centered at the origin with radius unity: $x_1^2 + x_2^2 + y^2 = 1$. We have $\partial f / \partial x_1 = 2x_1$, $\partial f / \partial x_2 = 2x_2$, and $\partial f / \partial y = 2y$. We thus do not expect a unique $y(x_1, x_2)$ to describe the points on the sphere in the neighborhood of $y = 0$.

Applying Eq. (1.22) to our f, we find

$$df = 2x_1 \, dx_1 + 2x_2 \, dx_2 + 2y \, dy = 0. \qquad (1.26)$$

Holding x_2 constant, we can say

$$\frac{\partial y}{\partial x_1} = -\frac{x_1}{y}. \qquad (1.27)$$

Holding x_1 constant, we can say

$$\frac{\partial y}{\partial x_2} = -\frac{x_2}{y}. \qquad (1.28)$$

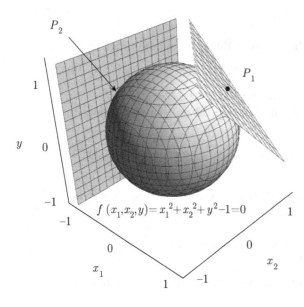

Figure 1.3. The sphere defined implicitly by $f(x_1, x_2, y) = x_1^2 + x_2^2 + y^2 - 1 = 0$ along with its local tangent planes at P_1 : $(1/\sqrt{3}, 1/\sqrt{3}, 1/\sqrt{3})$ and P_2 : $(-1, 0, 0)$.

Near an arbitrary point (x_{1a}, x_{2a}, y_a) on the surface $f = 0$, we get the equation for the *tangent plane*:

$$y - y_a = -\frac{x_{1a}}{y_a}(x_1 - x_{1a}) - \frac{x_{2a}}{y_a}(x_2 - x_{2a}). \tag{1.29}$$

At the point P_1 : $(1/\sqrt{3}, 1/\sqrt{3}, 1/\sqrt{3})$, we can form the unique local linearization

$$y - \frac{1}{\sqrt{3}} = -\left(x_1 - \frac{1}{\sqrt{3}}\right) - \left(x_2 - \frac{1}{\sqrt{3}}\right), \tag{1.30}$$

which describes a tangent plane to $f = 0$ at P_1. At the point P_2 : $(-1, 0, 0)$, the equation

$$x_1 = -1 \tag{1.31}$$

describes the tangent plane. But a unique approximation $y(x_1, x_2)$ for the neighborhood of P_2 does not exist on $f = 0$ at P_2 because $y = 0$ there. Of course, a nonunique approximation does exist, that being $y = \pm\sqrt{1 - x_1^2 - x_2^2}$. The sphere defined by $f(x_1, x_2, y) = x_1^2 + x_2^2 + y^2 - 1 = 0$ along with linearizations near P_1 and P_2 is plotted in Figure 1.3.

1.1.3 Many Dependent Variables

Let us now consider the equations

$$f(x, y, u, v) = 0, \tag{1.32}$$

$$g(x, y, u, v) = 0. \tag{1.33}$$

Under certain circumstances, we can unravel Eqs. (1.32, 1.33), either algebraically or numerically, to form $u = u(x, y)$, $v = v(x, y)$. Those circumstances are given by the multivariable extension of the implicit function theorem. We present it here for two functions of two variables; it can be extended to N functions of N variables.

Theorem: Let $f(x, y, u, v) = 0$ and $g(x, y, u, v) = 0$ be satisfied by (x_0, y_0, u_0, v_0), and suppose that continuous derivatives of f and g exist in the neighborhood of

(x_0, y_0, u_0, v_0) with

$$\begin{vmatrix} \dfrac{\partial f}{\partial u} & \dfrac{\partial f}{\partial v} \\ \dfrac{\partial g}{\partial u} & \dfrac{\partial g}{\partial v} \end{vmatrix}_{x_0, y_0, u_0, v_0} \neq 0. \tag{1.34}$$

Then Eq. (1.34) implies the local existence of $u(x, y)$ and $v(x, y)$.

We will not formally prove this but the following analysis explains its origins. Here we might imagine the two dependent variables u and v both to be functions of the two independent variables x and y. The conditions for the existence of such a functional dependency can be found by local linearized analysis, namely, through differentiation of the original equations. For example, differentiating Eq. (1.32) gives

$$df = \frac{\partial f}{\partial x}\, dx + \frac{\partial f}{\partial y}\, dy + \frac{\partial f}{\partial u}\, du + \frac{\partial f}{\partial v}\, dv = 0. \tag{1.35}$$

Holding y constant and dividing by dx, we get

$$\frac{\partial f}{\partial x} + \frac{\partial f}{\partial u}\frac{\partial u}{\partial x} + \frac{\partial f}{\partial v}\frac{\partial v}{\partial x} = 0. \tag{1.36}$$

Operating on Eq. (1.33) in the same manner, we get

$$\frac{\partial g}{\partial x} + \frac{\partial g}{\partial u}\frac{\partial u}{\partial x} + \frac{\partial g}{\partial v}\frac{\partial v}{\partial x} = 0. \tag{1.37}$$

Similarly, holding x constant and dividing by dy, we get

$$\frac{\partial f}{\partial y} + \frac{\partial f}{\partial u}\frac{\partial u}{\partial y} + \frac{\partial f}{\partial v}\frac{\partial v}{\partial y} = 0, \tag{1.38}$$

$$\frac{\partial g}{\partial y} + \frac{\partial g}{\partial u}\frac{\partial u}{\partial y} + \frac{\partial g}{\partial v}\frac{\partial v}{\partial y} = 0. \tag{1.39}$$

The linear Eqs. (1.36, 1.37) can be solved for $\partial u/\partial x$ and $\partial v/\partial x$, and the linear Eqs. (1.38, 1.39) can be solved for $\partial u/\partial y$ and $\partial v/\partial y$ using the well-known Cramer's[4] rule; see Eq. (7.95) or Section A.2. To solve for $\partial u/\partial x$ and $\partial v/\partial x$, we first write Eqs. (1.36, 1.37) in matrix form:

$$\begin{pmatrix} \dfrac{\partial f}{\partial u} & \dfrac{\partial f}{\partial v} \\ \dfrac{\partial g}{\partial u} & \dfrac{\partial g}{\partial v} \end{pmatrix} \begin{pmatrix} \dfrac{\partial u}{\partial x} \\ \dfrac{\partial v}{\partial x} \end{pmatrix} = \begin{pmatrix} -\dfrac{\partial f}{\partial x} \\ -\dfrac{\partial g}{\partial x} \end{pmatrix}. \tag{1.40}$$

Thus, from Cramer's rule, we have[5]

$$\frac{\partial u}{\partial x} = \frac{\begin{vmatrix} -\dfrac{\partial f}{\partial x} & \dfrac{\partial f}{\partial v} \\ -\dfrac{\partial g}{\partial x} & \dfrac{\partial g}{\partial v} \end{vmatrix}}{\begin{vmatrix} \dfrac{\partial f}{\partial u} & \dfrac{\partial f}{\partial v} \\ \dfrac{\partial g}{\partial u} & \dfrac{\partial g}{\partial v} \end{vmatrix}} \equiv -\frac{\partial(f,g)}{\partial(x,v)}\bigg/\frac{\partial(f,g)}{\partial(u,v)}, \qquad \frac{\partial v}{\partial x} = \frac{\begin{vmatrix} \dfrac{\partial f}{\partial u} & -\dfrac{\partial f}{\partial x} \\ \dfrac{\partial g}{\partial u} & -\dfrac{\partial g}{\partial x} \end{vmatrix}}{\begin{vmatrix} \dfrac{\partial f}{\partial u} & \dfrac{\partial f}{\partial v} \\ \dfrac{\partial g}{\partial u} & \dfrac{\partial g}{\partial v} \end{vmatrix}} \equiv -\frac{\partial(f,g)}{\partial(u,x)}\bigg/\frac{\partial(f,g)}{\partial(u,v)}. \tag{1.41}$$

---

[4] Gabriel Cramer, 1704–1752, Swiss-born mathematician.
[5] Here we are defining notation such as $\partial(f, v)/\partial(u, v)$ in terms of the determinant of quantities, which is easily inferred from the context shown. The notation "$\equiv$" stands for "is defined as."

In a similar fashion, we can form expressions for $\partial u/\partial y$ and $\partial v/\partial y$:

$$\frac{\partial u}{\partial y} = \frac{\begin{vmatrix} -\frac{\partial f}{\partial y} & \frac{\partial f}{\partial v} \\ -\frac{\partial g}{\partial y} & \frac{\partial g}{\partial v} \end{vmatrix}}{\begin{vmatrix} \frac{\partial f}{\partial u} & \frac{\partial f}{\partial v} \\ \frac{\partial g}{\partial u} & \frac{\partial g}{\partial v} \end{vmatrix}} \equiv -\frac{\frac{\partial(f,g)}{\partial(y,v)}}{\frac{\partial(f,g)}{\partial(u,v)}}, \qquad \frac{\partial v}{\partial y} = \frac{\begin{vmatrix} \frac{\partial f}{\partial u} & -\frac{\partial f}{\partial y} \\ \frac{\partial g}{\partial u} & -\frac{\partial g}{\partial y} \end{vmatrix}}{\begin{vmatrix} \frac{\partial f}{\partial u} & \frac{\partial f}{\partial v} \\ \frac{\partial g}{\partial u} & \frac{\partial g}{\partial v} \end{vmatrix}} \equiv -\frac{\frac{\partial(f,g)}{\partial(u,y)}}{\frac{\partial(f,g)}{\partial(u,v)}}. \qquad (1.42)$$

So if these derivatives locally exist, we can form $u(x,y)$ and $v(x,y)$ locally as well. Alternatively, the same procedure has value in simply finding the derivatives themselves.

We take the *Jacobian*[6] *matrix* $\mathbf{J}$ to be defined for this problem as

$$\mathbf{J} = \begin{pmatrix} \frac{\partial f}{\partial u} & \frac{\partial f}{\partial v} \\ \frac{\partial g}{\partial u} & \frac{\partial g}{\partial v} \end{pmatrix}. \qquad (1.43)$$

It is distinguished from the *Jacobian determinant*, J, defined for this problem as

$$J = \det \mathbf{J} = \frac{\partial(f,g)}{\partial(u,v)} = \begin{vmatrix} \frac{\partial f}{\partial u} & \frac{\partial f}{\partial v} \\ \frac{\partial g}{\partial u} & \frac{\partial g}{\partial v} \end{vmatrix} = \frac{\partial f}{\partial u}\frac{\partial g}{\partial v} - \frac{\partial f}{\partial v}\frac{\partial g}{\partial u}. \qquad (1.44)$$

If $J \neq 0$, the derivatives exist, and we indeed can find $u(x,y)$ and $v(x,y)$, at least in a locally linear approximation. This is the condition for existence of implicit to explicit function conversion. If $J = 0$, the analysis is not straightforward; typically one must examine problems on a case-by-case basis to make definitive statements.

Let us now consider a simple globally linear example to see how partial derivatives may be calculated for implicitly defined functions.

EXAMPLE 1.4

If

$$x + y + u + v = 0, \qquad (1.45)$$

$$2x - y + u - 3v = 0, \qquad (1.46)$$

find $\partial u/\partial x$.

We have four unknowns in two equations. Here the problem is sufficiently simple that we can solve for $u(x,y)$ and $v(x,y)$ and then determine all partial derivatives, such as the one desired. Direct solution of the linear equations reveals that

$$u(x,y) = -\frac{5}{4}x - \frac{1}{2}y, \qquad (1.47)$$

$$v(x,y) = \frac{1}{4}x - \frac{1}{2}y. \qquad (1.48)$$

The surface $u(x,y)$ is a plane and is plotted in Figure 1.4. We could generate a similar plot for $v(x,y)$. We can calculate $\partial u/\partial x$ by direct differentiation of Eq. (1.47):

$$\frac{\partial u}{\partial x} = -\frac{5}{4}. \qquad (1.49)$$

---

[6] Carl Gustav Jacob Jacobi, 1804–1851, German/Prussian mathematician.

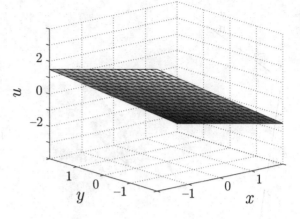

Figure 1.4. Planar surface, $u(x, y) = -5x/4 - y/2$, in the (x, y, u) volume formed by intersection of $x + y + u + v = 0$ and $2x - y + u - 3v = 0$.

Such an approach is generally easiest if $u(x, y)$ is explicitly available. Let us imagine that it is unavailable and show how we can obtain the same result with our more general approach. Equations (1.45) and (1.46) are rewritten as

$$f(x, y, u, v) = x + y + u + v = 0, \tag{1.50}$$

$$g(x, y, u, v) = 2x - y + u - 3v = 0. \tag{1.51}$$

Using the formula from Eq. (1.41) to solve for the desired derivative, we get

$$\frac{\partial u}{\partial x} = \frac{\begin{vmatrix} -\dfrac{\partial f}{\partial x} & \dfrac{\partial f}{\partial v} \\ -\dfrac{\partial g}{\partial x} & \dfrac{\partial g}{\partial v} \end{vmatrix}}{\begin{vmatrix} \dfrac{\partial f}{\partial u} & \dfrac{\partial f}{\partial v} \\ \dfrac{\partial g}{\partial u} & \dfrac{\partial g}{\partial v} \end{vmatrix}}. \tag{1.52}$$

Substituting, we get

$$\frac{\partial u}{\partial x} = \frac{\begin{vmatrix} -1 & 1 \\ -2 & -3 \end{vmatrix}}{\begin{vmatrix} 1 & 1 \\ 1 & -3 \end{vmatrix}} = \frac{3 - (-2)}{-3 - 1} = -\frac{5}{4}, \tag{1.53}$$

as expected. Here $J = -4$ for all real values of x and y (also written as $\forall\, x, y \in \mathbb{R}$), so we can always form $u(x, y)$ and $v(x, y)$.

Let us next consider a similar example, except that it is globally nonlinear.

EXAMPLE 1.5

If

$$x + y + u^6 + u + v = 0, \tag{1.54}$$

$$xy + uv = 1, \tag{1.55}$$

find $\partial u/\partial x$.

We again have four unknowns in two equations. In principle, we could solve for $u(x, y)$ and $v(x, y)$ and then determine all partial derivatives, such as the one desired. In practice, this is not always possible; for example, there is no general solution to sixth-order polynomial equations such as we have here (see Section A.1). The two hypersurfaces defined

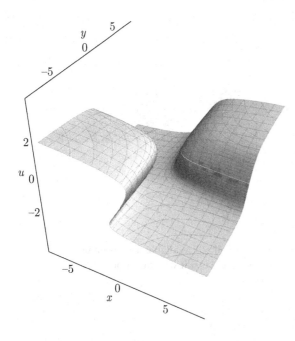

Figure 1.5. Nonplanar surface in the (x, y, u) volume formed by intersection of $x + y + u^6 + u + v = 0$ and $xy + uv = 1$.

by Eqs. (1.54) and (1.55) do in fact intersect to form a surface, $u(x, y)$. The surface can be obtained numerically and is plotted in Figure 1.5. For a given x and y, one can see that u may be nonunique.

To obtain the required partial derivative, the general method of this section is the most convenient. Equations (1.54) and (1.55) are rewritten as

$$f(x, y, u, v) = x + y + u^6 + u + v = 0, \tag{1.56}$$

$$g(x, y, u, v) = xy + uv - 1 = 0. \tag{1.57}$$

Using the formula from Eq. (1.41) to solve for the desired derivative, we get

$$\frac{\partial u}{\partial x} = \frac{\begin{vmatrix} -\frac{\partial f}{\partial x} & \frac{\partial f}{\partial v} \\ -\frac{\partial g}{\partial x} & \frac{\partial g}{\partial v} \end{vmatrix}}{\begin{vmatrix} \frac{\partial f}{\partial u} & \frac{\partial f}{\partial v} \\ \frac{\partial g}{\partial u} & \frac{\partial g}{\partial v} \end{vmatrix}}. \tag{1.58}$$

Substituting, we get

$$\frac{\partial u}{\partial x} = \frac{\begin{vmatrix} -1 & 1 \\ -y & u \end{vmatrix}}{\begin{vmatrix} 6u^5 + 1 & 1 \\ v & u \end{vmatrix}} = \frac{y - u}{u(6u^5 + 1) - v}. \tag{1.59}$$

Although Eq. (1.59) formally solves the problem at hand, we can perform some further analysis to study some of the features of the surface $u(x, y)$. Note if

$$v = 6u^6 + u \tag{1.60}$$

that the relevant Jacobian determinant is zero. At such points, determination of either $\partial u/\partial x$ or $\partial u/\partial y$ may be difficult or impossible; thus, for such points, it may not be possible to form $u(x, y)$. Following lengthy algebra, one can eliminate y and v from Eq. (1.59) to arrive at

$$\frac{\partial u}{\partial x} = \frac{1 + 2u^2 + u^7}{-1 + u^2 + 6u^7 - 2ux - 7u^6x - x^2}. \tag{1.61}$$

From Eq. (1.61), we see a potential singularity[7] in $\partial u / \partial x$ if the denominator is zero. Again, after some algebra, one finds a parametric representation of the curve for which $\partial u / \partial x$ may be singular:

$$x(s) = \frac{1}{2} \left(-2s - 7s^6 \pm \sqrt{-4 + 8s^2 + 52s^7 + 49s^{12}} \right), \tag{1.62}$$

$$y(s) = \frac{-1 - s^2 - s^7 - \frac{1}{2}s \left(-2s - 7s^6 \pm \sqrt{-4 + 8s^2 + 52s^7 + 49s^{12}} \right)}{s + \frac{1}{2} \left(2s + 7s^6 \mp \sqrt{-4 + 8s^2 + 52s^7 + 49s^{12}} \right)}, \tag{1.63}$$

$$u(s) = s. \tag{1.64}$$

Here s is a parameter, introduced for convenience. Certainly, on examination of Figure 1.5, we see there is a set of points for which $\partial u / \partial x$ is infinite.

We also note that $\partial u / \partial x = 0$ if the numerator of Eq. (1.61) is zero (assuming that the corresponding denominator is nonzero):

$$1 + 2u^2 + u^7 = 0. \tag{1.65}$$

This has seven roots, one of which is real, that being found by numerical solution to be $u = -1.21746$. Detailed algebra then reveals the parameterized curve for which $\partial u / \partial x = 0$ to be

$$x(s) = s, \tag{1.66}$$

$$y(s) = \frac{-1.21746 - s}{1 + 0.821383s}, \tag{1.67}$$

$$u(s) = -1.21746. \tag{1.68}$$

This is a curve in (x, y) space confined to a plane of constant u.

In summary, at points where the relevant Jacobian determinant $J = \partial(f, g) / \partial(u, v) \neq 0$ (which for this example includes nearly all of the (x, y) plane), given a local value of (x, y), we can use numerical algebra to find a corresponding u and v, which may be multivalued, and use the formula developed to find the local value of any required partial derivatives.

1.2 Inverse Function Theorem

Consider an explicit general mapping from x and y to u and v:

$$u = u(x, y), \tag{1.69}$$

$$v = v(x, y). \tag{1.70}$$

One can ask under what circumstances one can invert so that given u and v, one might find x and y. Relevant to this question, we have the following.

Inverse Function Theorem: If a mapping $u(x, y), v(x, y)$ is continuously differentiable and its Jacobian is invertible at a point (x_0, y_0), then u and v are uniquely invertible at that point and in its neighborhood.

The theorem can be extended to N functions of N variables.

---

[7] The word *singular* can have many connotations in mathematics, including the one here, which results from a potential division by zero. Other common usages are introduced later, e.g. in Sections 3.11 and 7.9.8.

EXAMPLE 1.6

Apply the inverse function theorem to analyze the unique invertibility of the mapping

$$u(x,y) = x + y, \tag{1.71}$$

$$v(x,y) = x - y. \tag{1.72}$$

The Jacobian **J** of the mapping is given by considering the total differentials, which in general matrix form are

$$\begin{pmatrix} du \\ dv \end{pmatrix} = \underbrace{\begin{pmatrix} \frac{\partial u}{\partial x} & \frac{\partial u}{\partial y} \\ \frac{\partial v}{\partial x} & \frac{\partial v}{\partial y} \end{pmatrix}}_{\mathbf{J}} \begin{pmatrix} dx \\ dy \end{pmatrix}. \tag{1.73}$$

For our problem, we find

$$\begin{pmatrix} du \\ dv \end{pmatrix} = \underbrace{\begin{pmatrix} 1 & 1 \\ 1 & -1 \end{pmatrix}}_{\mathbf{J}} \begin{pmatrix} dx \\ dy \end{pmatrix}. \tag{1.74}$$

The derivative is invertible if $J = \det \mathbf{J} \neq 0$. Here we find $J = -2$. Because $J \neq 0$, the unique inverse always exists and is in fact

$$x = \frac{1}{2}(u + v), \tag{1.75}$$

$$y = \frac{1}{2}(u - v). \tag{1.76}$$

EXAMPLE 1.7

Altering the mapping of the previous example to

$$u = x + y, \tag{1.77}$$

$$v = 2x + 2y, \tag{1.78}$$

we find $J = 0$ everywhere, and the unique inverse mapping never exists. It is straightforward to see that u and v are not independent here as $u = v/2$ for all values of x and y. If u and v are given and $u \neq v/2$, it is impossible to find values for x and y. However, if $u = v/2$, not only can one find values for x and y but there are an infinite number of available solutions. Certainly, then, there is no unique solution for x and y. For example, if $(u,v) = (1,2)$, we find $(x,y) = (1/2, 1/2)$ satisfies, as do an infinite number of other points, such as $(x,y) = (3/2, -1/2)$ or $(x,y) = (1,0)$. In general $(x,y) = (1/2 + t, 1/2 - t)$ for $t \in (-\infty, \infty)$ will satisfy if $(u,v) = (1,2)$. Geometrically, we can say that throughout (x,y) space, lines on which u is constant are parallel to lines on which v is constant. So if lines of constant u intersect lines of constant v, they do so in an infinite number of points. Systems of equations such as this are known as underconstrained and are considered in detail in Section 7.3.2.

EXAMPLE 1.8

Apply the inverse function theorem to analyze the unique invertibility of the mapping

$$u(x,y) = x + y^2, \tag{1.79}$$

$$v(x,y) = y + x^3. \tag{1.80}$$

The Jacobian **J** of the mapping is given by considering the total differentials, which in general matrix form are

$$\begin{pmatrix} du \\ dv \end{pmatrix} = \underbrace{\begin{pmatrix} \frac{\partial u}{\partial x} & \frac{\partial u}{\partial y} \\ \frac{\partial v}{\partial x} & \frac{\partial v}{\partial y} \end{pmatrix}}_{\mathbf{J}} \begin{pmatrix} dx \\ dy \end{pmatrix}. \tag{1.81}$$

For our problem, we find

$$\begin{pmatrix} du \\ dv \end{pmatrix} = \underbrace{\begin{pmatrix} 1 & 2y \\ 3x^2 & 1 \end{pmatrix}}_{\mathbf{J}} \begin{pmatrix} dx \\ dy \end{pmatrix}. \tag{1.82}$$

The derivative is invertible if $J = \det \mathbf{J} \neq 0$. Here we find

$$J = 1 - 6x^2 y. \tag{1.83}$$

So if $y = 1/(6x^2)$, we have $J = 0$, and the unique inverse mapping does not exist for points in the neighborhood of this curve. As we see in the next section, if $J = 0$, we have that u is functionally dependent on v.

Consider the point $P : (x, y) = (1, 1/6)$, where we happen to have $J = 0$. At P, we have $(u, v) = (37/36, 7/6)$. At, and in the neighborhood of, P, we have the linearization

$$u - \frac{37}{36} = (x - 1) + \frac{1}{3}\left(y - \frac{1}{6}\right), \tag{1.84}$$

$$v - \frac{7}{6} = 3(x - 1) + \left(y - \frac{1}{6}\right). \tag{1.85}$$

Let us consider if there are other points than $P : (x, y) = (1, 1/6)$ that bring (u, v) into $(37/36, 7/6)$. Then we seek (x, y) such that

$$0 = (x - 1) + \frac{1}{3}\left(y - \frac{1}{6}\right), \tag{1.86}$$

$$0 = 3(x - 1) + \left(y - \frac{1}{6}\right). \tag{1.87}$$

This can be achieved for $(x, y) = (1 + t, 1/6 - 3t)$ where $t \in (-\infty, \infty)$, at least in the local linearization. Clearly, at P, there is no unique inverse function valid in the neighborhood of P. Once again, the nonunique nature of the solution in the neighborhood of P is a consequence of the linearized system being underconstrained and is considered in detail in Section 7.3.2.

If we multiply Eq. (1.85) by $-1/3$ and add it to Eq. (1.84), we eliminate both x and y and obtain the functional relation between u and v:

$$u = \frac{37}{36} + \frac{1}{3}\left(v - \frac{7}{6}\right). \tag{1.88}$$

This demonstrates how, if $J = 0$, we obtain a functional dependency between the otherwise independent u and v. At P, and at all points where $y = 1/6/x^2$, one finds that locally, lines of constant u are parallel to lines of constant v within (x, y) space.

1.3 Functional Dependence

Here we formalize notions regarding functional dependence. Let $u = u(x, y)$ and $v = v(x, y)$. If we can write $u = g(v)$ or $v = h(u)$, then u and v are said to be *functionally dependent*. If functional dependence between u and v exists, then we can consider $f(u, v) = 0$. So, one can differentiate and expand $df = 0$ to obtain the local linearization

$$\frac{\partial f}{\partial u}\frac{\partial u}{\partial x} + \frac{\partial f}{\partial v}\frac{\partial v}{\partial x} = 0, \tag{1.89}$$

$$\frac{\partial f}{\partial u}\frac{\partial u}{\partial y} + \frac{\partial f}{\partial v}\frac{\partial v}{\partial y} = 0. \tag{1.90}$$

In matrix form, this is

$$\begin{pmatrix} \frac{\partial u}{\partial x} & \frac{\partial v}{\partial x} \\ \frac{\partial u}{\partial y} & \frac{\partial v}{\partial y} \end{pmatrix} \begin{pmatrix} \frac{\partial f}{\partial u} \\ \frac{\partial f}{\partial v} \end{pmatrix} = \begin{pmatrix} 0 \\ 0 \end{pmatrix}. \tag{1.91}$$

Because the right-hand side is zero, and we desire a nontrivial solution, Cramer's rule (see Eq. (7.95) or Section A.2), tells us that the determinant of the coefficient matrix must be zero for functional dependency, that is,

$$\begin{vmatrix} \frac{\partial u}{\partial x} & \frac{\partial v}{\partial x} \\ \frac{\partial u}{\partial y} & \frac{\partial v}{\partial y} \end{vmatrix} = 0. \tag{1.92}$$

Because it can be shown that $\det \mathbf{J} = \det \mathbf{J}^T = J$, this is equivalent to

$$J = \begin{vmatrix} \frac{\partial u}{\partial x} & \frac{\partial u}{\partial y} \\ \frac{\partial v}{\partial x} & \frac{\partial v}{\partial y} \end{vmatrix} = \frac{\partial(u, v)}{\partial(x, y)} = 0. \tag{1.93}$$

That is, the Jacobian determinant J must be zero for functional dependence. As an aside, we note that we employ the common notation of a superscript T, which signifies the transpose operation associated with exchanging nondiagonal terms across the diagonal of the matrix. It will be used from here on, but is first formally defined in Section 2.1.5.

EXAMPLE 1.9

Determine if

$$u = y + z, \tag{1.94}$$

$$v = x + 2z^2, \tag{1.95}$$

$$w = x - 4yz - 2y^2 \tag{1.96}$$

are functionally dependent.

The determinant of the resulting coefficient matrix, by extension to three functions of three variables, is

$$\frac{\partial(u, v, w)}{\partial(x, y, z)} = \begin{vmatrix} \frac{\partial u}{\partial x} & \frac{\partial u}{\partial y} & \frac{\partial u}{\partial z} \\ \frac{\partial v}{\partial x} & \frac{\partial v}{\partial y} & \frac{\partial v}{\partial z} \\ \frac{\partial w}{\partial x} & \frac{\partial w}{\partial y} & \frac{\partial w}{\partial z} \end{vmatrix} = \begin{vmatrix} \frac{\partial u}{\partial x} & \frac{\partial v}{\partial x} & \frac{\partial w}{\partial x} \\ \frac{\partial u}{\partial y} & \frac{\partial v}{\partial y} & \frac{\partial w}{\partial y} \\ \frac{\partial u}{\partial z} & \frac{\partial v}{\partial z} & \frac{\partial w}{\partial z} \end{vmatrix}, \tag{1.97}$$

$$= \begin{vmatrix} 0 & 1 & 1 \\ 1 & 0 & -4(y+z) \\ 1 & 4z & -4y \end{vmatrix}, \tag{1.98}$$

$$= (-1)(-4y - (-4)(y+z)) + (1)(4z), \tag{1.99}$$

$$= 4y - 4y - 4z + 4z, \tag{1.100}$$

$$= 0. \tag{1.101}$$

So, u, v, w are functionally dependent. In fact, $w = v - 2u^2$, as can be verified by direct substitution.

EXAMPLE 1.10

Let

$$x + y + z = 0, \tag{1.102}$$

$$x^2 + y^2 + z^2 + 2xz = 1. \tag{1.103}$$

Can x and y be considered as functions of z?

If $x = x(z)$ and $y = y(z)$, then dx/dz and dy/dz must exist. If we take

$$f(x, y, z) = x + y + z = 0, \tag{1.104}$$

$$g(x, y, z) = x^2 + y^2 + z^2 + 2xz - 1 = 0, \tag{1.105}$$

$$df = \frac{\partial f}{\partial z} dz + \frac{\partial f}{\partial x} dx + \frac{\partial f}{\partial y} dy = 0, \tag{1.106}$$

$$dg = \frac{\partial g}{\partial z} dz + \frac{\partial g}{\partial x} dx + \frac{\partial g}{\partial y} dy = 0, \tag{1.107}$$

$$\frac{\partial f}{\partial z} + \frac{\partial f}{\partial x} \frac{dx}{dz} + \frac{\partial f}{\partial y} \frac{dy}{dz} = 0, \tag{1.108}$$

$$\frac{\partial g}{\partial z} + \frac{\partial g}{\partial x} \frac{dx}{dz} + \frac{\partial g}{\partial y} \frac{dy}{dz} = 0, \tag{1.109}$$

$$\begin{pmatrix} \frac{\partial f}{\partial x} & \frac{\partial f}{\partial y} \\ \frac{\partial g}{\partial x} & \frac{\partial g}{\partial y} \end{pmatrix} \begin{pmatrix} \frac{dx}{dz} \\ \frac{dy}{dz} \end{pmatrix} = \begin{pmatrix} -\frac{\partial f}{\partial z} \\ -\frac{\partial g}{\partial z} \end{pmatrix}, \tag{1.110}$$

then the solution[8] $(dx/dz, dy/dz)^T$ can be obtained by Cramer's rule:

$$\frac{dx}{dz} = \frac{\begin{vmatrix} -\frac{\partial f}{\partial z} & \frac{\partial f}{\partial y} \\ -\frac{\partial g}{\partial z} & \frac{\partial g}{\partial y} \end{vmatrix}}{\begin{vmatrix} \frac{\partial f}{\partial x} & \frac{\partial f}{\partial y} \\ \frac{\partial g}{\partial x} & \frac{\partial g}{\partial y} \end{vmatrix}} = \frac{\begin{vmatrix} -1 & 1 \\ -(2z + 2x) & 2y \end{vmatrix}}{\begin{vmatrix} 1 & 1 \\ 2x + 2z & 2y \end{vmatrix}} = \frac{-2y + 2z + 2x}{2y - 2x - 2z} = -1, \tag{1.111}$$

$$\frac{dy}{dz} = \frac{\begin{vmatrix} \frac{\partial f}{\partial x} & -\frac{\partial f}{\partial z} \\ \frac{\partial g}{\partial x} & -\frac{\partial g}{\partial z} \end{vmatrix}}{\begin{vmatrix} \frac{\partial f}{\partial x} & \frac{\partial f}{\partial y} \\ \frac{\partial g}{\partial x} & \frac{\partial g}{\partial y} \end{vmatrix}} = \frac{\begin{vmatrix} 1 & -1 \\ 2x + 2z & -(2z + 2x) \end{vmatrix}}{\begin{vmatrix} 1 & 1 \\ 2x + 2z & 2y \end{vmatrix}} = \frac{0}{2y - 2x - 2z}, \tag{1.112}$$

In Eq. (1.111), the numerator and denominator cancel; there is no special condition defined by the Jacobian determinant of the denominator being zero. In Eq. (1.112),

---

[8] Note that the transpose operation can also be applied to one-dimensional matrices, sometimes described as vectors. This has the effect of converting a so-called row vector into a column vector, and vice versa. This will be considered in detail in Chapter 2.

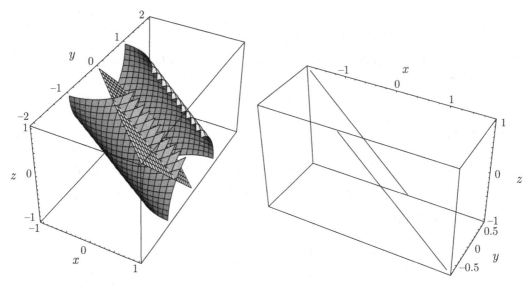

Figure 1.6. Surfaces of $x + y + z = 0$ and $x^2 + y^2 + z^2 + 2xz = 1$, and their loci of intersection.

$dy/dz = 0$ if $y - x - z \neq 0$. If $y - x - z = 0$, Eq. (1.112) is indeterminate for purposes of defining dy/dz, and we cannot be certain that $y(z)$ exists.

Now it is easily shown by algebraic manipulations (which for more general functions are not possible) that $x(z)$ and $y(z)$ do exist and are given by

$$x(z) = -z \pm \frac{\sqrt{2}}{2}, \tag{1.113}$$

$$y(z) = \mp \frac{\sqrt{2}}{2}. \tag{1.114}$$

This forms two distinct lines in (x, y, z) space. On these lines, $J = 2y - 2x - 2z = \mp 2\sqrt{2}$, which is never zero.

The two original functions and their loci of intersection are plotted in Figure 1.6. It is seen that the surface represented by the linear function Eq. (1.102) is a plane, and that represented by the quadratic function Eq. (1.103) is an open cylindrical tube. Note that planes and cylinders may or may not intersect. If they intersect, it is most likely that the intersection will be a closed arc. However, when the plane is parallel to the axis of the cylinder, if an intersection exists, it will be in two parallel lines; such is the case here.

Let us see how slightly altering the equation for the plane removes the degeneracy. Take now

$$5x + y + z = 0, \tag{1.115}$$

$$x^2 + y^2 + z^2 + 2xz = 1. \tag{1.116}$$

Can x and y be considered as functions of z? If $x = x(z)$ and $y = y(z)$, then dx/dz and dy/dz must exist. If we take

$$f(x, y, z) = 5x + y + z = 0, \tag{1.117}$$

$$g(x, y, z) = x^2 + y^2 + z^2 + 2xz - 1 = 0, \tag{1.118}$$

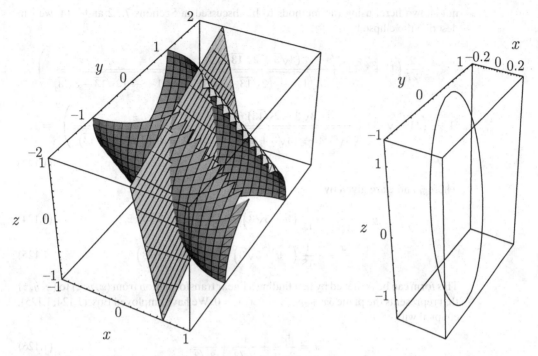

Figure 1.7. Surfaces of $5x + y + z = 0$ and $x^2 + y^2 + z^2 + 2xz = 1$, and their loci of intersection.

then the solution $(dx/dz, dy/dz)^T$ is found as before:

$$\frac{dx}{dz} = \frac{\begin{vmatrix} -\frac{\partial f}{\partial z} & \frac{\partial f}{\partial y} \\ -\frac{\partial g}{\partial z} & \frac{\partial g}{\partial y} \end{vmatrix}}{\begin{vmatrix} \frac{\partial f}{\partial x} & \frac{\partial f}{\partial y} \\ \frac{\partial g}{\partial x} & \frac{\partial g}{\partial y} \end{vmatrix}} = \frac{\begin{vmatrix} -1 & 1 \\ -(2z + 2x) & 2y \end{vmatrix}}{\begin{vmatrix} 5 & 1 \\ 2x + 2z & 2y \end{vmatrix}} = \frac{-2y + 2z + 2x}{10y - 2x - 2z}, \tag{1.119}$$

$$\frac{dy}{dz} = \frac{\begin{vmatrix} \frac{\partial f}{\partial x} & -\frac{\partial f}{\partial z} \\ \frac{\partial g}{\partial x} & -\frac{\partial g}{\partial z} \end{vmatrix}}{\begin{vmatrix} \frac{\partial f}{\partial x} & \frac{\partial f}{\partial y} \\ \frac{\partial g}{\partial x} & \frac{\partial g}{\partial y} \end{vmatrix}} = \frac{\begin{vmatrix} 5 & -1 \\ 2x + 2z & -(2z + 2x) \end{vmatrix}}{\begin{vmatrix} 5 & 1 \\ 2x + 2z & 2y \end{vmatrix}} = \frac{-8x - 8z}{10y - 2x - 2z}. \tag{1.120}$$

The two original functions and their loci of intersection are plotted in Figure 1.7.
Straightforward algebra in this case shows that an explicit dependency exists:

$$x(z) = \frac{-6z \pm \sqrt{2}\sqrt{13 - 8z^2}}{26}, \tag{1.121}$$

$$y(z) = \frac{-4z \mp 5\sqrt{2}\sqrt{13 - 8z^2}}{26}. \tag{1.122}$$

These curves represent the projection of the curve of intersection on the x, z and y, z planes, respectively. In both cases, the projections are ellipses. Considered in three dimensions, the two equations can be thought of as the parameterization of a curve, with z as the parameter. In (x, y, z) space, that curve is also an ellipse. With considerable effort, not shown here, using the methods to be discussed in Sections 7.7.2 and 7.14, we can describe the ellipse by

$$\frac{2}{27}\left(11+\sqrt{13}\right)\left(\frac{\left(3\sqrt{3}-2\sqrt{13}\right)\hat{y}}{5\sqrt{1+\frac{1}{25}\left(2\sqrt{13}-3\sqrt{3}\right)^2}}+\frac{\hat{z}}{\sqrt{1+\frac{1}{25}\left(2\sqrt{13}-3\sqrt{3}\right)^2}}\right)^2$$

$$+\frac{2}{27}\left(11-\sqrt{13}\right)\left(\frac{\left(-3\sqrt{3}-2\sqrt{13}\right)\hat{y}}{5\sqrt{1+\frac{1}{25}\left(3\sqrt{3}+2\sqrt{13}\right)^2}}-\frac{\hat{z}}{\sqrt{1+\frac{1}{25}\left(3\sqrt{3}+2\sqrt{13}\right)^2}}\right)^2=1,$$

$$(1.123)$$

where $\hat{y}$ and $\hat{z}$ are given by

$$\hat{y}=\frac{x}{3\sqrt{3}}+\frac{1}{18}\left(9-5\sqrt{3}\right)y+\frac{1}{18}\left(-9-5\sqrt{3}\right)z, \qquad (1.124)$$

$$\hat{z}=\frac{x}{3\sqrt{3}}+\frac{1}{18}\left(-9-5\sqrt{3}\right)y+\frac{1}{18}\left(9-5\sqrt{3}\right)z. \qquad (1.125)$$

This form can be achieved by first finding a linear transformation from (x,y,z) to $(\hat{x},\hat{y},\hat{z})$ that represents the plane $5x+y+z=0$ as $\hat{x}=0$. We have employed Eqs. (1.124, 1.125), coupled with

$$\hat{x}=\frac{5x}{3\sqrt{3}}+\frac{y}{3\sqrt{3}}+\frac{z}{3\sqrt{3}}, \qquad (1.126)$$

to effect the variable change. This choice is nonunique, and there are others that achieve the same end with a simpler form. Our choice was selected so that, among other things, a particular triple (x,y,z) will have the same magnitude as its transform $(\hat{x},\hat{y},\hat{z})$. Then one studies the cylinder in this transformed space, imposes $\hat{x}=0$ to effect the intersection with the plane, and uses algebraic manipulations to put it into the form of an ellipse.

1.4 Leibniz Rule

Often functions are expressed in terms of integrals. For example,

$$y(x)=\int_{a(x)}^{b(x)}f(x,t)\,dt. \qquad (1.127)$$

Here t is a dummy variable of integration. Although x and t appear in Eq. (1.127), because t is removed by the integration process, it is strictly speaking not multivariable in the same sense as previous sections. However, it is necessary to consider expressions of this type in the upcoming Section 1.5.2, and it is easily extended to truly multivariable scenarios; see Section 2.8.5. We often need to take derivatives of functions in the form expressed in Eq. (1.127), and the *Leibniz*[9] *rule* tells us how to do so. For such functions, the Leibniz rule gives us

$$\frac{dy(x)}{dx}=f(x,b(x))\frac{db(x)}{dx}-f(x,a(x))\frac{da(x)}{dx}+\int_{a(x)}^{b(x)}\frac{\partial f(x,t)}{\partial x}\,dt. \quad (1.128)$$

---

[9] Gottfried Wilhelm von Leibniz, 1646–1716, German mathematician and philosopher; co-inventor with Sir Isaac Newton, 1643–1727, of the calculus.

EXAMPLE 1.11

If

$$y(x) = y(x_0) + \int_{x_0}^{x} f(t) \, dt, \tag{1.129}$$

use the Leibniz rule to find dy/dx.

Here $f(x, t)$ simply reduces to $f(t)$; we also have $b(x) = x$ and $a(x) = x_0$. Applying the Leibniz rule, we find

$$\frac{dy(x)}{dx} = \underbrace{\frac{d}{dx}(y(x_0))}_{=0} + f(x)\underbrace{\frac{d}{dx}(x)}_{=1} - f(x_0)\underbrace{\frac{d}{dx}(x_0)}_{=0} + \int_{x_0}^{x} \underbrace{\frac{\partial}{\partial x}(f(t))}_{=0} \, dt, \tag{1.130}$$

$$= f(x). \tag{1.131}$$

The integral expression naturally includes the initial condition that if $x = x_0$, $y = y(x_0)$. This needs to be expressed separately for the differential version of the equation.

EXAMPLE 1.12

Find dy/dx if

$$y(x) = \int_{x}^{x^2} (x + 1)t^2 \, dt. \tag{1.132}$$

Using the Leibniz rule, we get

$$\frac{dy(x)}{dx} = ((x + 1)x^4)(2x) - ((x + 1)x^2)(1) + \int_{x}^{x^2} t^2 \, dt, \tag{1.133}$$

$$= 2x^6 + 2x^5 - x^3 - x^2 + \left(\frac{t^3}{3}\right)\Big|_{x}^{x^2}, \tag{1.134}$$

$$= 2x^6 + 2x^5 - x^3 - x^2 + \frac{x^6}{3} - \frac{x^3}{3}, \tag{1.135}$$

$$= \frac{7x^6}{3} + 2x^5 - \frac{4x^3}{3} - x^2. \tag{1.136}$$

In this case, it is possible to integrate explicitly to achieve the same result:

$$y(x) = (x + 1)\int_{x}^{x^2} t^2 \, dt, \tag{1.137}$$

$$= (x + 1)\left(\frac{t^3}{3}\right)\Big|_{x}^{x^2}, \tag{1.138}$$

$$= (x + 1)\left(\frac{x^6}{3} - \frac{x^3}{3}\right), \tag{1.139}$$

$$= \frac{x^7}{3} + \frac{x^6}{3} - \frac{x^4}{3} - \frac{x^3}{3}, \tag{1.140}$$

$$\frac{dy(x)}{dx} = \frac{7x^6}{3} + 2x^5 - \frac{4x^3}{3} - x^2. \tag{1.141}$$

So the two methods give identical results.

As an aside, let us take up a common issue of notation, related to our discussion of the Leibniz rule as well as many other topics in the mathematics of engineering systems. Let us imagine we are presented with the differential equation

$$\frac{dy}{dx} = f(x), \quad y(x_0) = y_0. \tag{1.142}$$

One might be tempted to propose the solution to be $y(x) = y_0 + \int_{x_0}^{x} f(x) \, dx$. However, this is in error! The statement is in fact incoherent. Certainly, if $x = x_0$, it predicts the correct value of $y(x_0) = y_0$ because an integral over a domain of width zero has no value. However, if we attempt to evaluate if $x = x_1 \neq x_0$, we see a problem: $y(x_1) = y_0 + \int_{x_0}^{x_1} f(x_1) \, dx_1$ does not make sense in many ways. First, $f(x_1)$ is a constant and could be brought outside of the integral. Second, because x_1 is a constant, $dx_1 = 0$, yielding an incorrect interpretation of $y(x_1) = y_0$. Because the variable of integration must change through the domain of integration, it is important that it have a different nomenclature than the upper limit.

Let us proceed more systematically through the use of dummy variables. First, introduce the dummy variable t into Eq. (1.142) by trading it for x:

$$\frac{dy}{dt} = f(t). \tag{1.143}$$

Now, apply the operator $\int_{x_0}^{x} (\cdot) \, dt$ to both sides:

$$\int_{x_0}^{x} \frac{dy}{dt} \, dt = \int_{x_0}^{x} f(t) \, dt. \tag{1.144}$$

The fundamental theorem of calculus[10] allows evaluation of the integral on the left side to yield

$$y(x) - y(x_0) = \int_{x_0}^{x} f(t) \, dt, \tag{1.145}$$

$$y(x) = y_0 + \int_{x_0}^{x} f(t) \, dt. \tag{1.146}$$

Application of the Leibniz rule to this solution yields the correct $dy/dx = f(x)$.

1.5 Optimization

The basic strategies of optimization have motivated calculus from its inception. Here we consider a variety of approaches for multivariable optimization.[11] We first consider how to select specific points that optimize nonlinear functions of many variables in the absence of external constraints. This is followed by a discussion of how to select specific functions to optimize integrals that take those functions as

---

[10] This well-known theorem has two parts: (1) if $f(x)$ is continuous for $x \in [a, b]$ and the function F can be shown to be the indefinite integral of f in the domain $x \in [a, b]$, then $\int_a^b f(x) \, dx = F(b) - F(a)$, and (2) if on an open interval $F(x) = \int_a^x f(t) \, dt$, then $dF/dx = f(x)$ for a within the interval.

[11] Though the general term *optimization* includes other aspects, we are concerned here only with extremization that is associated with zero derivatives and functions that are not extreme at boundaries.

inputs. The section is completed by a discussion of how to optimize in the presence of external constraints.

1.5.1 Unconstrained Optimization

Consider the real function $f(x)$, where $x \in [a, b]$. Extrema are at $x = x_m$, where $f'(x_m) = 0$, if $x_m \in [a, b]$. It is a local minimum, a local maximum, or an inflection point, according to whether $f''(x_m)$ is positive, negative, or zero, respectively.

Now consider a function of two variables $f(x, y)$, with $x \in [a, b]$, $y \in [c, d]$. A necessary condition for an extremum is

$$\frac{\partial f}{\partial x}(x_m, y_m) = \frac{\partial f}{\partial y}(x_m, y_m) = 0, \tag{1.147}$$

where $x_m \in [a, b]$, $y_m \in [c, d]$. Next, we find the Hessian[12] matrix:

$$\mathbf{H} = \begin{pmatrix} \frac{\partial^2 f}{\partial x^2} & \frac{\partial^2 f}{\partial x \partial y} \\ \frac{\partial^2 f}{\partial x \partial y} & \frac{\partial^2 f}{\partial y^2} \end{pmatrix}. \tag{1.148}$$

We use $\mathbf{H}$ and its elements to determine the character of any local extremum. It can be shown that

- f is a maximum if $\partial^2 f/\partial x^2 < 0$, $\partial^2 f/\partial y^2 < 0$, and $\partial^2 f/\partial x \partial y < \sqrt{(\partial^2 f/\partial x^2)(\partial^2 f/\partial y^2)}$,
- f is a minimum if $\partial^2 f/\partial x^2 > 0$, $\partial^2 f/\partial y^2 > 0$, and $\partial^2 f/\partial x \partial y < \sqrt{(\partial^2 f/\partial x^2)(\partial^2 f/\partial y^2)}$,
- f is a saddle otherwise, as long as $\det \mathbf{H} \neq 0$,
- if $\det \mathbf{H} = 0$, higher order terms need to be considered.

The first two conditions for maximum and minimum require that terms on the diagonal of $\mathbf{H}$ dominate those on the off-diagonal with diagonal terms further required to be of the same sign. For higher dimensional systems, one can show that if all the eigenvalues of $\mathbf{H}$ are negative, f is maximized, and if all the eigenvalues of $\mathbf{H}$ are positive, f is minimized.

One can begin to understand this by considering a Taylor series expansion of $f(x, y)$. Taking $\mathbf{x} = (x, y)^T$, $d\mathbf{x} = (dx, dy)^T$, and $\nabla f = (\partial f/\partial x, \partial f/\partial y)^T$, multivariable Taylor series expansion gives

$$f(\mathbf{x} + d\mathbf{x}) = f(\mathbf{x}) + d\mathbf{x}^T \cdot \underbrace{\nabla f}_{=0} + d\mathbf{x}^T \cdot \mathbf{H} \cdot d\mathbf{x} + \cdots. \tag{1.149}$$

At an extremum, $\nabla f = 0$, so

$$f(\mathbf{x} + d\mathbf{x}) = f(\mathbf{x}) + d\mathbf{x}^T \cdot \mathbf{H} \cdot d\mathbf{x} + \cdots. \tag{1.150}$$

Later (see Section 6.7.1 and Section 7.2.5), we see that, by virtue of the definition of the term *positive definite*, if the Hessian $\mathbf{H}$ is positive definite, then $\forall\, d\mathbf{x} \neq \mathbf{0}$, $d\mathbf{x}^T \cdot \mathbf{H} \cdot d\mathbf{x} > 0$, which corresponds to a minimum. If $\mathbf{H}$ is positive definite within a domain, we say that the surface described by f is *convex*, a topic further considered in Section 6.3. For negative definite $\mathbf{H}$, we have a maximum.

---

[12] Ludwig Otto Hesse, 1811–1874, German mathematician.

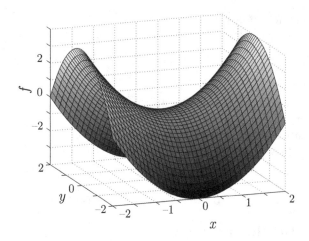

Figure 1.8. Surface of $f = x^2 - y^2$ showing saddle character at the critical point $(x, y) = (0, 0)$.

EXAMPLE 1.13

Find the extrema of

$$f = x^2 - y^2. \tag{1.151}$$

Equating partial derivatives with respect to x and to y to zero, we get

$$\frac{\partial f}{\partial x} = 2x = 0, \tag{1.152}$$

$$\frac{\partial f}{\partial y} = -2y = 0. \tag{1.153}$$

This gives $x = 0, y = 0$. For this f, we find

$$\mathbf{H} = \begin{pmatrix} \frac{\partial^2 f}{\partial x^2} & \frac{\partial^2 f}{\partial x \partial y} \\ \frac{\partial^2 f}{\partial x \partial y} & \frac{\partial^2 f}{\partial y^2} \end{pmatrix}, \tag{1.154}$$

$$= \begin{pmatrix} 2 & 0 \\ 0 & -2 \end{pmatrix}. \tag{1.155}$$

Because (1) $\det \mathbf{H} = -4 \neq 0$ and (2) $\partial^2 f / \partial x^2$ and $\partial^2 f / \partial y^2$ have different signs, the equilibrium is a saddle point. The saddle shape of the surface is visible in the plot of Figure 1.8.

1.5.2 Calculus of Variations

The problem of the *calculus of variations* is to find the function $y(x)$, with $x \in [x_1, x_2]$, and boundary conditions $y(x_1) = y_1, y(x_2) = y_2$, such that

$$I = \int_{x_1}^{x_2} f(x, y, y') \, dx \tag{1.156}$$

is an extremum. For efficiency, we adopt here the common shorthand notation of *prime* denoting differentiation with respect to x, that is, $dy/dx = y'$. In Eq. (1.156), we encounter the operation of mapping a function $y(x)$ into a scalar I, which can be expressed as $I = \mathbf{F}y(x)$. The operator $\mathbf{F}$ that performs this task is known as a *functional*.

If $y(x)$ is the desired solution, let $Y(x) = y(x) + \epsilon h(x)$, where $h(x_1) = h(x_2) = 0$. Thus, $Y(x)$ also satisfies the boundary conditions; also, $Y'(x) = y'(x) + \epsilon h'(x)$. We can write

$$I(\epsilon) = \int_{x_1}^{x_2} f(x, Y, Y') \, dx. \tag{1.157}$$

Taking $dI/d\epsilon$, we get

$$\frac{dI}{d\epsilon} = \int_{x_1}^{x_2} \left(\frac{\partial f}{\partial x} \underbrace{\frac{\partial x}{\partial \epsilon}}_{0} + \frac{\partial f}{\partial Y} \underbrace{\frac{\partial Y}{\partial \epsilon}}_{h(x)} + \frac{\partial f}{\partial Y'} \underbrace{\frac{\partial Y'}{\partial \epsilon}}_{h'(x)} \right) dx. \tag{1.158}$$

Evaluating, we find

$$\frac{dI}{d\epsilon} = \int_{x_1}^{x_2} \left(\frac{\partial f}{\partial x} 0 + \frac{\partial f}{\partial Y} h(x) + \frac{\partial f}{\partial Y'} h'(x) \right) dx. \tag{1.159}$$

Because I is an extremum at $\epsilon = 0$, we have $dI/d\epsilon = 0$ for $\epsilon = 0$. This gives

$$0 = \int_{x_1}^{x_2} \left(\frac{\partial f}{\partial Y} h(x) + \frac{\partial f}{\partial Y'} h'(x) \right) \bigg|_{\epsilon=0} dx. \tag{1.160}$$

When $\epsilon = 0$, we have $Y = y$, $Y' = y'$, so

$$0 = \int_{x_1}^{x_2} \left(\frac{\partial f}{\partial y} h(x) + \frac{\partial f}{\partial y'} h'(x) \right) dx. \tag{1.161}$$

We can look at the second term in this integral. From integration by parts,[13] we get

$$\int_{x_1}^{x_2} \frac{\partial f}{\partial y'} h'(x) \, dx = \int_{x_1}^{x_2} \frac{\partial f}{\partial y'} \frac{dh}{dx} \, dx = \int_{x_1}^{x_2} \frac{\partial f}{\partial y'} \, dh, \tag{1.162}$$

$$= \underbrace{\frac{\partial f}{\partial y'} h(x) \bigg|_{x_1}^{x_2}}_{=0} - \int_{x_1}^{x_2} \frac{d}{dx} \left(\frac{\partial f}{\partial y'} \right) h(x) \, dx, \tag{1.163}$$

$$= - \int_{x_1}^{x_2} \frac{d}{dx} \left(\frac{\partial f}{\partial y'} \right) h(x) \, dx. \tag{1.164}$$

The first term in Eq. (1.163) is zero because of our conditions on $h(x_1)$ and $h(x_2)$. Thus, substituting Eq. (1.164) into the original equation, Eq. (1.161), we find

$$\int_{x_1}^{x_2} \underbrace{\left(\frac{\partial f}{\partial y} - \frac{d}{dx} \left(\frac{\partial f}{\partial y'} \right) \right)}_{0} h(x) \, dx = 0. \tag{1.165}$$

---

[13] Integration by parts tells us $\int u(x)v'(x) \, dx = u(x)v(x) - \int u'(x)v(x) \, dx$, or in short, $\int u \, dv = uv - \int v \, du$.

The equality holds for all $h(x)$, so that we must have

$$\frac{\partial f}{\partial y} - \frac{d}{dx}\left(\frac{\partial f}{\partial y'}\right) = 0. \tag{1.166}$$

Equation (1.166) is called the *Euler[14]-Lagrange[15] equation*; sometimes it is simply called Euler's equation.

While this is, in general, the preferred form of the Euler-Lagrange equation, its explicit dependency on the two end conditions is better displayed by considering a slightly different form. By expanding the term that is differentiated with respect to x, we obtain

$$\frac{d}{dx}\left(\frac{\partial f}{\partial y'}(x,y,y')\right) = \frac{\partial^2 f}{\partial y'\partial x}\underbrace{\frac{dx}{dx}}_{=1} + \frac{\partial^2 f}{\partial y'\partial y}\underbrace{\frac{dy}{dx}}_{y'} + \frac{\partial^2 f}{\partial y'\partial y'}\underbrace{\frac{dy'}{dx}}_{y''}, \tag{1.167}$$

$$= \frac{\partial^2 f}{\partial y'\partial x} + \frac{\partial^2 f}{\partial y'\partial y}y' + \frac{\partial^2 f}{\partial y'\partial y'}y''. \tag{1.168}$$

Thus, the Euler-Lagrange equation, Eq. (1.166), after slight rearrangement, becomes

$$\frac{\partial^2 f}{\partial y'\partial y'}y'' + \frac{\partial^2 f}{\partial y'\partial y}y' + \frac{\partial^2 f}{\partial y'\partial x} - \frac{\partial f}{\partial y} = 0, \tag{1.169}$$

$$f_{y'y'}y'' + f_{y'y}y' + (f_{y'x} - f_y) = 0. \tag{1.170}$$

This is a second-order differential equation for $f_{y'y'} \neq 0$ and, in general, nonlinear. If $f_{y'y'}$ is always nonzero, the problem is said to be *regular*. If $f_{y'y'} = 0$, the equation is no longer second order, and the problem is said to be *singular* at such points. This connotation of the word "singular" is used whenever the highest order term, here y'', is multiplied by zero. Satisfaction of two boundary conditions becomes problematic for equations less than second order.

There are several special cases of the function f.

- $f = f(x,y)$:
  The Euler-Lagrange equation is

$$\frac{\partial f}{\partial y} = 0, \tag{1.171}$$

  which is easily solved:

$$f(x,y) = A(x), \tag{1.172}$$

  which, knowing f, is then solved for $y(x)$.
- $f = f(x,y')$:
  The Euler-Lagrange equation is

$$\frac{d}{dx}\left(\frac{\partial f}{\partial y'}\right) = 0, \tag{1.173}$$

---

[14] Leonhard Euler, 1707–1783, Swiss mathematician.
[15] Joseph-Louis Lagrange, 1736–1813, Italian-born French mathematician.

which yields

$$\frac{\partial f}{\partial y'} = A, \tag{1.174}$$

$$f(x, y') = Ay' + B(x). \tag{1.175}$$

Again, knowing f, the equation is solved for y' and then integrated to find $y(x)$.

- $f = f(y, y')$:

The Euler-Lagrange equation is

$$\frac{\partial f}{\partial y} - \frac{d}{dx}\left(\frac{\partial f}{\partial y'}(y, y')\right) = 0, \tag{1.176}$$

$$\frac{\partial f}{\partial y} - \left(\frac{\partial^2 f}{\partial y \partial y'}\frac{dy}{dx} + \frac{\partial^2 f}{\partial y' \partial y'}\frac{dy'}{dx}\right) = 0, \tag{1.177}$$

$$\frac{\partial f}{\partial y} - \frac{\partial^2 f}{\partial y \partial y'}\frac{dy}{dx} - \frac{\partial^2 f}{\partial y' \partial y'}\frac{d^2 y}{dx^2} = 0. \tag{1.178}$$

Multiply by y' to arrive at

$$y'\left(\frac{\partial f}{\partial y} - \frac{\partial^2 f}{\partial y \partial y'}\frac{dy}{dx} - \frac{\partial^2 f}{\partial y' \partial y'}\frac{d^2 y}{dx^2}\right) = 0. \tag{1.179}$$

Add and subtract $(\partial f/\partial y')y''$ to find

$$y'\left(\frac{\partial f}{\partial y} - \frac{\partial^2 f}{\partial y \partial y'}\frac{dy}{dx} - \frac{\partial^2 f}{\partial y' \partial y'}\frac{d^2 y}{dx^2}\right) + \frac{\partial f}{\partial y'}y'' - \frac{\partial f}{\partial y'}y'' = 0. \tag{1.180}$$

Regroup to get

$$\underbrace{\frac{\partial f}{\partial y}y' + \frac{\partial f}{\partial y'}y''}_{=df/dx} - \underbrace{\left(y'\left(\frac{\partial^2 f}{\partial y \partial y'}\frac{dy}{dx} + \frac{\partial^2 f}{\partial y' \partial y'}\frac{d^2 y}{dx^2}\right) + \frac{\partial f}{\partial y'}y''\right)}_{=d/dx(y'\partial f/\partial y')} = 0. \tag{1.181}$$

Regroup again to show

$$\frac{d}{dx}\left(f - y'\frac{\partial f}{\partial y'}\right) = 0, \tag{1.182}$$

which can be integrated. Thus,

$$f(y, y') - y'\frac{\partial f}{\partial y'} = K, \tag{1.183}$$

where K is an arbitrary constant. What remains is a first-order ordinary differential equation that can be solved. Another integration constant arises. This second constant along with K are determined by the two endpoint conditions.

EXAMPLE 1.14

Find the curve of minimum length between the points (x_1, y_1) and (x_2, y_2).

If $y(x)$ is the curve, then $y(x_1) = y_1$ and $y(x_2) = y_2$. In the most common geometries, the length of the curve can be shown[16] to be

$$\ell = \int_{x_1}^{x_2} \sqrt{1 + (y')^2} \, dx. \tag{1.184}$$

---

[16] If $d\ell^2 = dx^2 + dy^2$, then $d\ell = \sqrt{1 + (dy/dx)^2} \, dx = \sqrt{1 + (y')^2} \, dx$, which can be integrated to find ℓ.

So our f reduces to $f(y') = \sqrt{1 + (y')^2}$. The Euler-Lagrange equation is

$$\frac{d}{dx}\left(\frac{y'}{\sqrt{1 + (y')^2}}\right) = 0, \tag{1.185}$$

which can be integrated to give

$$\frac{y'}{\sqrt{1 + (y')^2}} = K. \tag{1.186}$$

Solving for y', we get

$$y' = \sqrt{\frac{K^2}{1 - K^2}} \equiv A, \tag{1.187}$$

from which

$$y = Ax + B. \tag{1.188}$$

The constants A and B are obtained from the boundary conditions $y(x_1) = y_1$ and $y(x_2) = y_2$. *The shortest distance between two points is a straight line.*

EXAMPLE 1.15

Find the curve through the points (x_1, y_1) and (x_2, y_2), such that the surface area of the body of revolution found by rotating the curve around the x-axis is a minimum.

We wish to minimize the integral that yields the surface area of a body of revolution:

$$I = \int_{x_1}^{x_2} y\sqrt{1 + (y')^2}\, dx. \tag{1.189}$$

Here f reduces to $f(y, y') = y\sqrt{1 + (y')^2}$. So the Euler-Lagrange equation reduces to

$$f(y, y') - y'\frac{\partial f}{\partial y'} = A, \tag{1.190}$$

$$y\sqrt{1 + y'^2} - y'y\frac{y'}{\sqrt{1 + y'^2}} = A, \tag{1.191}$$

$$y(1 + y'^2) - yy'^2 = A\sqrt{1 + y'^2}, \tag{1.192}$$

$$y = A\sqrt{1 + y'^2}, \tag{1.193}$$

$$y' = \sqrt{\left(\frac{y}{A}\right)^2 - 1}, \tag{1.194}$$

$$y(x) = A\cosh\frac{x - B}{A}. \tag{1.195}$$

This is a catenary. The constants A and B are determined from the boundary conditions $y(x_1) = y_1$ and $y(x_2) = y_2$. In general, this requires a trial-and-error solution of simultaneous algebraic equations. If $(x_1, y_1) = (-1, 3.08616)$ and $(x_2, y_2) = (2, 2.25525)$, one finds that solution of the resulting algebraic equations gives $A = 2, B = 1$. For these conditions, the curve $y(x)$ along with the resulting body of revolution of minimum surface area are plotted in Figure 1.9.

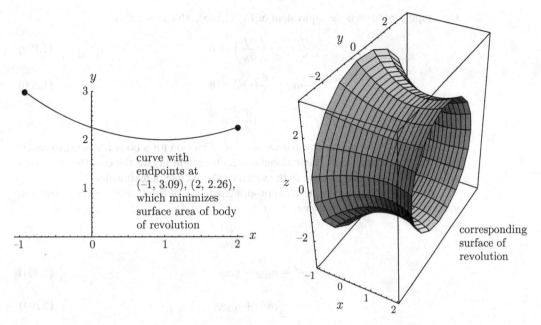

Figure 1.9. Body of revolution of minimum surface area for $(x_1, y_1) = (-1, 3.08616)$ and $(x_2, y_2) = (2, 2.25525)$.

EXAMPLE 1.16

Examine the consequences of the extremization of

$$I = \int_{t_1}^{t_2} \left(\frac{1}{2} m\dot{y}^2 - mgy \right) \, dt, \tag{1.196}$$

where m and g are constants.

Here we have made a trivial change of independent variables from x to t, and simultaneously changed the derivative notation from $'$ to $\dot{}$. The reason for these small changes is that this is a problem that has special significance in mechanics if we consider m to be a mass, g to be a constant gravitational acceleration, y to be position, and t to be time. Within mechanics, a physical principle in which some I is extremized is known as a *variational principle*. And within mechanics, the variational principle we are considering here is known as the principle of least action. Thus, the action that will be least is I. One also often describes $m\dot{y}^2/2$ as the kinetic energy T, mgy as the potential energy V, and their difference as the Lagrangian L. That is, $L = T - V = m\dot{y}^2/2 - mgy$.

So we are considering the equivalent of a limiting case of Eq. (1.156) if it takes the form

$$I = \int_{t_1}^{t_2} f(y, \dot{y}) \, dt, \tag{1.197}$$

where

$$f(y, \dot{y}) = \frac{1}{2} m\dot{y}^2 - mgy. \tag{1.198}$$

If we assert that I is extremized, that is, we assert the variational principle holds, we can immediately state that many forms of the appropriate Euler-Lagrange equation hold.

One important form is the equivalent of Eq. (1.166), which reduces to

$$\frac{\partial f}{\partial y} - \frac{d}{dt}\left(\frac{\partial f}{\partial \dot{y}}\right) = 0, \tag{1.199}$$

$$-mg - \frac{d}{dt}(m\dot{y}) = 0, \tag{1.200}$$

$$m\frac{d^2 y}{dt^2} = -mg. \tag{1.201}$$

This is simply Newton's celebrated second law of motion for a body accelerating under the influence of a constant gravitational force. One might say for this case that Newton's second law is one and the same as the variational principle of least action.

Another form arises from the appropriate version of Eq. (1.183), selecting for convenient physical interpretation $K = -C$:

$$f(y, \dot{y}) - \dot{y}\frac{\partial f}{\partial \dot{y}} = -C, \tag{1.202}$$

$$\frac{1}{2}m\dot{y}^2 - mgy - \dot{y}m\dot{y} = -C, \tag{1.203}$$

$$\frac{1}{2}m\dot{y}^2 + mgy = C. \tag{1.204}$$

This well-known equation is an expression of one type of energy conservation principle, valid in mechanics when no forces of dissipation are present. Here we have the kinetic, $m\dot{y}^2/2$, and potential, mgy, energies combining to form a constant, C.

1.5.3 Constrained Optimization: Lagrange Multipliers

Suppose we have to determine the extremum of $f(x_1, x_2, \ldots, x_M)$ subject to the N constraints

$$g_n(x_1, x_2, \ldots, x_M) = 0, \qquad n = 1, 2, \ldots, N. \tag{1.205}$$

Define

$$f^* = f - \lambda_1 g_1 - \lambda_2 g_2 - \cdots - \lambda_N g_N, \tag{1.206}$$

where the λ_n $(n = 1, 2, \ldots, N)$ are unknown constants called *Lagrange multipliers*. To get the extremum of f^*, we equate to zero its derivative with respect to $x_1, x_2, \ldots, x_M$. Thus, we have the complete set

$$\frac{\partial f^*}{\partial x_m} = 0, \; m = 1, \ldots, M, \tag{1.207}$$

$$g_n = 0, \; n = 1, \ldots, N, \tag{1.208}$$

which are $(M + N)$ equations that can be solved for x_m $(m = 1, 2, \ldots, M)$ and λ_n $(n = 1, 2, \ldots, N)$.

EXAMPLE 1.17

Extremize $f = x^2 + y^2$ subject to the constraint $g = 5x^2 - 6xy + 5y^2 - 8 = 0$.

Here we have $N = 1$ constraint for a function in $M = 2$ variables. Let

$$f^* = x^2 + y^2 - \lambda(5x^2 - 6xy + 5y^2 - 8), \tag{1.209}$$

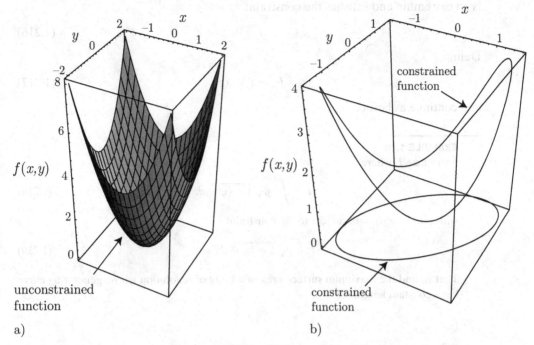

Figure 1.10. (a) Unconstrained function $f(x, y)$ and (b) constrained function and constraint function (image of constrained function.)

from which

$$\frac{\partial f^*}{\partial x} = 2x - 10\lambda x + 6\lambda y = 0, \tag{1.210}$$

$$\frac{\partial f^*}{\partial y} = 2y + 6\lambda x - 10\lambda y = 0, \tag{1.211}$$

$$g = 5x^2 - 6xy + 5y^2 - 8 = 0. \tag{1.212}$$

From Eq. (1.210),

$$\lambda = \frac{2x}{10x - 6y}, \tag{1.213}$$

which, if substituted into Eq. (1.211), gives

$$x = \pm y. \tag{1.214}$$

Equation (1.214), when solved in conjunction with Eq. (1.212), gives the extrema to be at $(x, y) = (\sqrt{2}, \sqrt{2})$, $(-\sqrt{2}, -\sqrt{2})$, $(1/\sqrt{2}, -1/\sqrt{2})$, $(-1/\sqrt{2}, 1/\sqrt{2})$. The first two sets give $f = 4$ (maximum) and the last two $f = 1$ (minimum). The function to be maximized along with the constraint function and its image are plotted in Figure 1.10.

A similar technique can be used for the extremization of a functional with constraints. We wish to find the function $y(x)$, with $x \in [x_1, x_2]$, and $y(x_1) = y_1, y(x_2) = y_2$, such that the integral

$$I = \int_{x_1}^{x_2} f(x, y, y') \, dx \tag{1.215}$$

is an extremum and satisfies the constraint

$$g = 0. \tag{1.216}$$

Define

$$I^* = I - \lambda g, \tag{1.217}$$

and continue as before.

EXAMPLE 1.18

Extremize I, where

$$I = \int_0^a y\sqrt{1 + (y')^2}\, dx, \tag{1.218}$$

with $y(0) = y(a) = 0$, subject to the constraint

$$\int_0^a \sqrt{1 + (y')^2}\, dx = \ell. \tag{1.219}$$

That is, find the maximum surface area of a body of revolution whose generating curve has a constant length.

Let

$$g = \int_0^a \sqrt{1 + (y')^2}\, dx - \ell = 0. \tag{1.220}$$

Then let

$$I^* = I - \lambda g = \int_0^a y\sqrt{1 + (y')^2}\, dx - \lambda \int_0^a \sqrt{1 + (y')^2}\, dx + \lambda\ell, \tag{1.221}$$

$$= \int_0^a (y - \lambda)\sqrt{1 + (y')^2}\, dx + \lambda\ell, \tag{1.222}$$

$$= \int_0^a \left((y - \lambda)\sqrt{1 + (y')^2} + \frac{\lambda\ell}{a} \right) dx. \tag{1.223}$$

With $f^* = (y - \lambda)\sqrt{1 + (y')^2} + \lambda\ell/a$, we have the Euler-Lagrange equation

$$\frac{\partial f^*}{\partial y} - \frac{d}{dx}\left(\frac{\partial f^*}{\partial y'} \right) = 0. \tag{1.224}$$

Integrating from an earlier developed relationship, Eq. (1.183), when $f = f(y, y')$, and absorbing $\lambda\ell/a$ into a constant A, we have

$$(y - \lambda)\sqrt{1 + (y')^2} - y'(y - \lambda)\frac{y'}{\sqrt{1 + (y')^2}} = A, \tag{1.225}$$

from which

$$(y - \lambda)(1 + (y')^2) - (y')^2(y - \lambda) = A\sqrt{1 + (y')^2}, \tag{1.226}$$

$$(y - \lambda)\left(1 + (y')^2 - (y')^2\right) = A\sqrt{1 + (y')^2}, \tag{1.227}$$

$$y - \lambda = A\sqrt{1 + (y')^2}, \tag{1.228}$$

$$y' = \sqrt{\left(\frac{y - \lambda}{A}\right)^2 - 1}, \tag{1.229}$$

$$y = \lambda + A\cosh\frac{x - B}{A}. \tag{1.230}$$

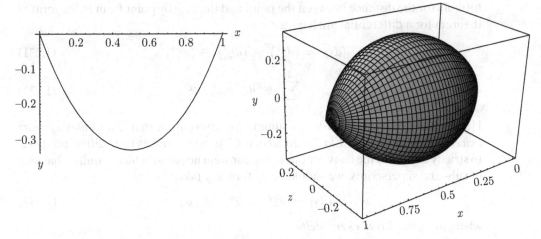

Figure 1.11. Curve of length $\ell = 5/4$ with $y(0) = y(1) = 0$ whose surface area of corresponding body of revolution (also shown) is maximum.

Here A, B, λ have to be numerically determined from the three conditions $y(0) = y(a) = 0, g = 0$. If we take $a = 1, \ell = 5/4$, we find that $A = 0.422752, B = 1/2, \lambda = -0.754549$. For these values, the curve of interest, along with the surface of revolution, is plotted in Figure 1.11.

1.6 Non-Cartesian Coordinate Transformations

Many problems are formulated in ordinary three-dimensional Cartesian[17] space, typically represented by the three variables (x, y, z). However, many of these problems, especially those involving curved geometrical bodies, are more efficiently posed in a non-Cartesian, curvilinear coordinate system. To facilitate analysis involving such geometries, one needs techniques to transform from one coordinate system to another. The transformations that effect this are usually expressed in terms of algebraic relationships between a set of variables, thus rendering the methods of multivariable calculus to be relevant.

In this section, we focus on general nonlinear transformations involving three spatial variables. The analysis can be adapted for an arbitrary number of variables. This general framework is the appropriate one for a variety of tasks in engineering and physics, though it introduces complexities. Coordinate transformations that are Cartesian, the subject of the upcoming Chapter 2, are simpler and often useful, but lack the flexibility afforded by those of this section.

For this section, we utilize the so-called *Einstein*[18] *index notation.* Thus, we take untransformed Cartesian coordinates, most commonly known as (x, y, z), to be represented instead by (ξ^1, ξ^2, ξ^3); that is to say $x \to \xi^1$, $y \to \xi^2$ and $z \to \xi^3$. *Here the superscript is an index and does not represent a power of* ξ. We denote a point by ξ^i, where $i = 1, 2, 3$. A nearby point will be located at $\xi^i + d\xi^i$, where $d\xi^i$ is a differential distance. Because the space is Cartesian, we have the usual Euclidean[19]

---

[17] René Descartes, 1596–1650, French mathematician and philosopher.
[18] Albert Einstein, 1879–1955, German-American physicist and mathematician.
[19] Euclid of Alexandria, c. 325–265 B.C., Greek geometer.

formula for the distance between the point and the nearby point from Pythagoras'[20] theorem for a differential arc length ds:

$$(ds)^2 = \left(d\xi^1\right)^2 + \left(d\xi^2\right)^2 + \left(d\xi^3\right)^2, \tag{1.231}$$

$$= \sum_{i=1}^{3} d\xi^i \, d\xi^i \equiv d\xi^i \, d\xi^i. \tag{1.232}$$

Here we have adopted Einstein's summation convention that *if an index appears twice, a summation from 1 to 3 is understood.* Though it makes little difference here, to strictly adhere to the conventions of the Einstein notation, which require a balance of sub- and superscripts, we should more formally take

$$(ds)^2 = d\xi^j \, \delta_{ji} \, d\xi^i = d\xi_i \, d\xi^i, \tag{1.233}$$

where δ_{ji} is the *Kronecker*[21] *delta,*

$$\delta_{ji} = \delta^{ji} = \delta^i_j = \begin{cases} 1, & i = j, \\ 0, & i \neq j. \end{cases} \tag{1.234}$$

In matrix form, the Kronecker delta is simply the identity matrix $\mathbf{I}$, that is,

$$\delta_{ji} = \delta^{ji} = \delta^i_j = \mathbf{I} = \begin{pmatrix} 1 & 0 & 0 \\ 0 & 1 & 0 \\ 0 & 0 & 1 \end{pmatrix}. \tag{1.235}$$

In matrix form, we can also represent Eq. (1.233) as

$$(ds)^2 = \begin{pmatrix} d\xi_1 & d\xi_2 & d\xi_3 \end{pmatrix} \begin{pmatrix} d\xi^1 \\ d\xi^2 \\ d\xi^3 \end{pmatrix} = d\boldsymbol{\xi}^T \cdot d\boldsymbol{\xi}. \tag{1.236}$$

We associate subscripted triples of scalars with row vectors and superscripted triples of scalars with column vectors. In the final expression of Eq. (1.236), we introduce the familiar bold-faced typography to represent vectors; this is known as *Gibbs*[22] *notation.*[23]

Now let us consider a point P whose representation in Cartesian coordinates is (ξ^1, ξ^2, ξ^3) and transform those coordinates so that P is represented in a more convenient (x^1, x^2, x^3) space. This transformation is achieved by defining the following functional dependencies:

$$x^1 = x^1(\xi^1, \xi^2, \xi^3), \tag{1.237}$$

$$x^2 = x^2(\xi^1, \xi^2, \xi^3), \tag{1.238}$$

$$x^3 = x^3(\xi^1, \xi^2, \xi^3). \tag{1.239}$$

We are thus considering arbitrary nonlinear algebraic functions that take ξ^i into x^j, $i, j = 1, 2, 3$. In this discussion, we make the common presumption that the entity P is invariant and that it has different representations in different coordinate systems.

---

[20] Pythagoras of Samos, c. 570–490 B.C., Ionian Greek mathematician, philosopher, and mystic.

[21] Leopold Kronecker, 1823–1891, German-Prussian mathematician.

[22] Josiah Willard Gibbs, 1839–1903, American mechanical engineer and mathematician.

[23] Here, for the Gibbs notation, we choose to explicitly indicate that the operation requires the first term be transposed to form a row vector. Many notation schemes assume this to be an intrinsic part of the operation and would simply state $(ds)^2 = d\boldsymbol{\xi} \cdot d\boldsymbol{\xi}$.

Thus, the coordinate axes change, but the location of P does not. This is known as an *alias* transformation. This contrasts another common approach in which a point is represented in an original space, and after application of a transformation, it is again represented in the original space in an altered state. This is known as an *alibi* transformation. The alias approach transforms the axes; the alibi approach transforms the elements of the space. Throughout this chapter, as well as in Chapter 2, we are especially concerned with alias transformations, which allow us to represent the same entity in different ways. In other chapters, especially Chapter 7, we focus more on alibi transformations.

Taking derivatives can tell us whether the inverse mapping exists. The derivatives of x^1, x^2, and x^3 are

$$dx^1 = \frac{\partial x^1}{\partial \xi^1} d\xi^1 + \frac{\partial x^1}{\partial \xi^2} d\xi^2 + \frac{\partial x^1}{\partial \xi^3} d\xi^3 = \frac{\partial x^1}{\partial \xi^j} d\xi^j, \tag{1.240}$$

$$dx^2 = \frac{\partial x^2}{\partial \xi^1} d\xi^1 + \frac{\partial x^2}{\partial \xi^2} d\xi^2 + \frac{\partial x^2}{\partial \xi^3} d\xi^3 = \frac{\partial x^2}{\partial \xi^j} d\xi^j, \tag{1.241}$$

$$dx^3 = \frac{\partial x^3}{\partial \xi^1} d\xi^1 + \frac{\partial x^3}{\partial \xi^2} d\xi^2 + \frac{\partial x^3}{\partial \xi^3} d\xi^3 = \frac{\partial x^3}{\partial \xi^j} d\xi^j. \tag{1.242}$$

In matrix form, these reduce to

$$\begin{pmatrix} dx^1 \\ dx^2 \\ dx^3 \end{pmatrix} = \begin{pmatrix} \frac{\partial x^1}{\partial \xi^1} & \frac{\partial x^1}{\partial \xi^2} & \frac{\partial x^1}{\partial \xi^3} \\ \frac{\partial x^2}{\partial \xi^1} & \frac{\partial x^2}{\partial \xi^2} & \frac{\partial x^2}{\partial \xi^3} \\ \frac{\partial x^3}{\partial \xi^1} & \frac{\partial x^3}{\partial \xi^2} & \frac{\partial x^3}{\partial \xi^3} \end{pmatrix} \begin{pmatrix} d\xi^1 \\ d\xi^2 \\ d\xi^3 \end{pmatrix}. \tag{1.243}$$

In Einstein notation, this greatly simplifies to

$$dx^i = \frac{\partial x^i}{\partial \xi^j} d\xi^j. \tag{1.244}$$

For the inverse to exist, we must have a nonzero Jacobian determinant for the transformation, that is,

$$\frac{\partial(x^1, x^2, x^3)}{\partial(\xi^1, \xi^2, \xi^3)} \neq 0. \tag{1.245}$$

As long as Eq. (1.245) is satisfied, the inverse transformation exists:

$$\xi^1 = \xi^1(x^1, x^2, x^3), \tag{1.246}$$

$$\xi^2 = \xi^2(x^1, x^2, x^3), \tag{1.247}$$

$$\xi^3 = \xi^3(x^1, x^2, x^3). \tag{1.248}$$

Likewise, then,

$$d\xi^i = \frac{\partial \xi^i}{\partial x^j} dx^j. \tag{1.249}$$

1.6.1 Jacobian Matrices and Metric Tensors

Defining the Jacobian matrix[24] $\mathbf{J}$ to be associated with the inverse transformation, Eq. (1.249), we take

$$\mathbf{J} = \frac{\partial \xi^i}{\partial x^j} = \begin{pmatrix} \frac{\partial \xi^1}{\partial x^1} & \frac{\partial \xi^1}{\partial x^2} & \frac{\partial \xi^1}{\partial x^3} \\ \frac{\partial \xi^2}{\partial x^1} & \frac{\partial \xi^2}{\partial x^2} & \frac{\partial \xi^2}{\partial x^3} \\ \frac{\partial \xi^3}{\partial x^1} & \frac{\partial \xi^3}{\partial x^2} & \frac{\partial \xi^3}{\partial x^3} \end{pmatrix}. \tag{1.250}$$

We can then rewrite $d\xi^i$ from Eq. (1.249) in Gibbs notation as

$$d\boldsymbol{\xi} = \mathbf{J} \cdot d\mathbf{x}. \tag{1.251}$$

In Gibbs notation, typically lowercase bold is reserved for vectors, and uppercase bold is reserved for matrices.

Now for Euclidean spaces, distances must be independent of coordinate systems, so we require

$$(ds)^2 = d\xi^i d\xi^i = \left(\frac{\partial \xi^i}{\partial x^k} \, dx^k \right) \left(\frac{\partial \xi^i}{\partial x^l} \, dx^l \right) = dx^k \underbrace{\frac{\partial \xi^i}{\partial x^k} \frac{\partial \xi^i}{\partial x^l}}_{g_{kl}} \, dx^l. \tag{1.252}$$

We have introduced new indices k and l. Both are repeated or dummy indices. We need two new indices because in Einstein notation, an index can appear at most twice in any given term. In Gibbs notation, Eq. (1.252) becomes[25]

$$(ds)^2 = d\boldsymbol{\xi}^T \cdot d\boldsymbol{\xi}, \tag{1.253}$$

$$= (\mathbf{J} \cdot d\mathbf{x})^T \cdot (\mathbf{J} \cdot d\mathbf{x}). \tag{1.254}$$

Now, it can be shown that $(\mathbf{J} \cdot d\mathbf{x})^T = d\mathbf{x}^T \cdot \mathbf{J}^T$ (see Section 7.2.5), so

$$(ds)^2 = d\mathbf{x}^T \cdot \underbrace{\mathbf{J}^T \cdot \mathbf{J}}_{\mathbf{G}} \cdot d\mathbf{x}. \tag{1.255}$$

If we define the *metric tensor*, g_{kl} or $\mathbf{G}$, as

$$g_{kl} = \frac{\partial \xi^i}{\partial x^k} \frac{\partial \xi^i}{\partial x^l}, \tag{1.256}$$

$$\mathbf{G} = \mathbf{J}^T \cdot \mathbf{J}, \tag{1.257}$$

---

[24] The definition we adopt influences the form of many of our equations given throughout the remainder of this book. There are three obvious alternates. (1) An argument can be made that a better definition of $\mathbf{J}$ would be the transpose of our Jacobian matrix: $\mathbf{J} \to \mathbf{J}^T$. This is because if one considers that the differential operator acts *first*, the Jacobian matrix is really $\frac{\partial}{\partial x^j} \xi^i$, and the alternative definition is more consistent with traditional matrix notation, which would have the first row as $(\frac{\partial}{\partial x^1} \xi^1, \frac{\partial}{\partial x^1} \xi^2, \frac{\partial}{\partial x^1} \xi^3)$. (2) Many others, (e.g., Kay (1988) adopt as $\mathbf{J}$ the inverse of our Jacobian matrix: $\mathbf{J} \to \mathbf{J}^{-1}$. This Jacobian matrix is thus defined in terms of the forward transformation, $\partial x^i / \partial \xi^j$. (3) One could adopt $\mathbf{J} \to (\mathbf{J}^T)^{-1}$. As long as one remains consistent with one's original choice, the convention adopted ultimately does not matter.

[25] Common alternate formulations of vector mechanics of non-Cartesian spaces view the Jacobian as an *intrinsic* part of the dot product and would say instead that, by definition, $(ds)^2 = d\mathbf{x} \cdot d\mathbf{x}$. Such formulations have no need for the transpose operation, especially because they do not carry forward simply to non-Cartesian systems. The formulation used here has the advantage of *explicitly recognizing* the linear algebra operations necessary to form the scalar ds. These same alternate notations reserve the dot product for that between a vector and a vector and would hold instead that $d\boldsymbol{\xi} = \mathbf{J} \, d\mathbf{x}$. However, this could be confused with raising the dimension of the quantity of interest, whereas we use the dot to lower the dimension.

then we have, equivalently in both Einstein and Gibbs notations,

$$(ds)^2 = dx^k g_{kl} \, dx^l, \tag{1.258}$$

$$(ds)^2 = d\mathbf{x}^T \cdot \mathbf{G} \cdot d\mathbf{x}. \tag{1.259}$$

In Einstein notation, one can loosely imagine superscripted terms in a denominator as being subscripted terms in a corresponding numerator. Now g_{kl} can be represented as a matrix. If we define

$$g = \det g_{kl}, \tag{1.260}$$

it can be shown that the ratio of volumes of differential elements in one space to that of the other is given by

$$d\xi^1 \, d\xi^2 \, d\xi^3 = \sqrt{g} \, dx^1 \, dx^2 \, dx^3. \tag{1.261}$$

Thus, transformations for which $g = 1$ are volume-preserving. Note that $g = J^2$. Volume-preserving transformations also have $J = \det \mathbf{J} = \pm 1$. It can also be shown that if $J > 0$, the transformation is locally orientation-preserving. If $J = \det \mathbf{J} < 0$, the transformation is orientation-reversing and thus involves a reflection. So, if $J = 1$, the transformation is volume- and orientation-preserving.

We also require dependent variables and all derivatives to take on the same values at corresponding points in each space, for example, if $\phi = f(\xi^1, \xi^2, \xi^3) = h(x^1, x^2, x^3)$ is a dependent variable defined at $(\hat{\xi}^1, \hat{\xi}^2, \hat{\xi}^3)$, and $(\hat{\xi}^1, \hat{\xi}^2, \hat{\xi}^3)$ maps into $(\hat{x}^1, \hat{x}^2, \hat{x}^3)$, we require $f(\hat{\xi}^1, \hat{\xi}^2, \hat{\xi}^3) = h(\hat{x}^1, \hat{x}^2, \hat{x}^3)$. The chain rule lets us transform derivatives to other spaces:

$$\begin{pmatrix} \frac{\partial \phi}{\partial x^1} & \frac{\partial \phi}{\partial x^2} & \frac{\partial \phi}{\partial x^3} \end{pmatrix} = \begin{pmatrix} \frac{\partial \phi}{\partial \xi^1} & \frac{\partial \phi}{\partial \xi^2} & \frac{\partial \phi}{\partial \xi^3} \end{pmatrix} \underbrace{\begin{pmatrix} \frac{\partial \xi^1}{\partial x^1} & \frac{\partial \xi^1}{\partial x^2} & \frac{\partial \xi^1}{\partial x^3} \\ \frac{\partial \xi^2}{\partial x^1} & \frac{\partial \xi^2}{\partial x^2} & \frac{\partial \xi^2}{\partial x^3} \\ \frac{\partial \xi^3}{\partial x^1} & \frac{\partial \xi^3}{\partial x^2} & \frac{\partial \xi^3}{\partial x^3} \end{pmatrix}}_{\mathbf{J}}, \tag{1.262}$$

$$\frac{\partial \phi}{\partial x^i} = \frac{\partial \phi}{\partial \xi^j} \frac{\partial \xi^j}{\partial x^i}. \tag{1.263}$$

Equation (1.263) can also be inverted, given that $J \neq 0$, to find $(\partial \phi / \partial \xi^1, \partial \phi / \partial \xi^2, \partial \phi / \partial \xi^3)$.

Employing Gibbs notation,[26] we can write Eq. (1.262) as

$$\nabla_{\mathbf{x}}^T \phi = \nabla_{\boldsymbol{\xi}}^T \phi \cdot \mathbf{J}. \tag{1.264}$$

The fact that the gradient operator required the use of row vectors in conjunction with the Jacobian matrix, while the transformation of distance, earlier in this section, Eq. (1.251), required the use of column vectors, is of fundamental importance and is examined further in Section 1.6.2, where we distinguish between what are known as *covariant* and *contravariant* vectors. We will see that the gradient ∇ is associated with covariance and the differential d is associated with contravariance.

---

[26] In Cartesian coordinates, we take $\nabla_{\boldsymbol{\xi}} \equiv \begin{pmatrix} \frac{\partial}{\partial \xi^1} \\ \frac{\partial}{\partial \xi^2} \\ \frac{\partial}{\partial \xi^3} \end{pmatrix}$. This gives rise to the natural, albeit uncon-

ventional, notation $\nabla_{\boldsymbol{\xi}}^T = \begin{pmatrix} \frac{\partial}{\partial \xi^1} & \frac{\partial}{\partial \xi^2} & \frac{\partial}{\partial \xi^3} \end{pmatrix}$. This notion does not extend easily to non-Cartesian systems, for which index notation is preferred. Here, for convenience, we take $\nabla_{\mathbf{x}}^T \equiv \begin{pmatrix} \frac{\partial}{\partial x^1} & \frac{\partial}{\partial x^2} & \frac{\partial}{\partial x^3} \end{pmatrix}$, and a similar column version for $\nabla_{\mathbf{x}}$.

Transposing both sides of Eq. (1.264), we could also say

$$\nabla_{\mathbf{x}}\phi = \mathbf{J}^T \cdot \nabla_{\boldsymbol{\xi}}\phi. \tag{1.265}$$

Inverting, we then have

$$\nabla_{\boldsymbol{\xi}}\phi = (\mathbf{J}^T)^{-1} \cdot \nabla_{\mathbf{x}}\phi. \tag{1.266}$$

Thus, in general, we could say for the gradient operator

$$\nabla_{\boldsymbol{\xi}} = (\mathbf{J}^T)^{-1} \cdot \nabla_{\mathbf{x}}. \tag{1.267}$$

Contrasting Eq. (1.267) with Eq. (1.251), $d\boldsymbol{\xi} = \mathbf{J} \cdot d\mathbf{x}$, we see the gradient operation ∇ transforms in a fundamentally different way than the differential operation d, unless we restrict attention to an unusual $\mathbf{J}$, one whose transpose is equal to its inverse. We sometimes make this restriction, and sometimes not. When we choose such a special $\mathbf{J}$ and also require that $J = 1$, there are many additional simplifications in the analysis; these are realized because it will be seen for such transformations that all of the original Cartesian character will be retained, albeit in a rotated, but otherwise undeformed, coordinate system. We later identify a matrix whose transpose is equal to its inverse as an orthogonal matrix, $\mathbf{Q}$, which has $\mathbf{Q}^T = \mathbf{Q}^{-1}$, and study it in detail in Section. 2.1.2 and 7.7.

One can also show the relation between $\partial \xi^i / \partial x^j$ and $\partial x^i / \partial \xi^j$ to be

$$\frac{\partial \xi^i}{\partial x^j} = \left(\left(\frac{\partial x^i}{\partial \xi^j} \right)^T \right)^{-1} = \left(\frac{\partial x^j}{\partial \xi^i} \right)^{-1}, \tag{1.268}$$

$$\begin{pmatrix} \frac{\partial \xi^1}{\partial x^1} & \frac{\partial \xi^1}{\partial x^2} & \frac{\partial \xi^1}{\partial x^3} \\ \frac{\partial \xi^2}{\partial x^1} & \frac{\partial \xi^2}{\partial x^2} & \frac{\partial \xi^2}{\partial x^3} \\ \frac{\partial \xi^3}{\partial x^1} & \frac{\partial \xi^3}{\partial x^2} & \frac{\partial \xi^3}{\partial x^3} \end{pmatrix} = \begin{pmatrix} \frac{\partial x^1}{\partial \xi^1} & \frac{\partial x^1}{\partial \xi^2} & \frac{\partial x^1}{\partial \xi^3} \\ \frac{\partial x^2}{\partial \xi^1} & \frac{\partial x^2}{\partial \xi^2} & \frac{\partial x^2}{\partial \xi^3} \\ \frac{\partial x^3}{\partial \xi^1} & \frac{\partial x^3}{\partial \xi^2} & \frac{\partial x^3}{\partial \xi^3} \end{pmatrix}^{-1}. \tag{1.269}$$

Thus, the Jacobian matrix $\mathbf{J}$ of the transformation is simply the inverse of the Jacobian matrix of the inverse transformation. In the special case for which the transpose is the inverse, we have $\partial \xi^i / \partial x^j = \partial x^i / \partial \xi^j$. This allows the i to remain "upstairs" and the j to remain "downstairs." Such a transformation is seen (Section 7.7), to be a pure rotation or reflection.

EXAMPLE 1.19

Transform the Cartesian equation

$$\frac{\partial \phi}{\partial \xi^1} + \frac{\partial \phi}{\partial \xi^2} = \left(\xi^1 \right)^2 + \left(\xi^2 \right)^2 \tag{1.270}$$

under the following.

1. Cartesian to linearly homogeneous affine coordinates.

Consider the following globally linear nonorthogonal transformation:

$$x^1 = \frac{2}{3}\xi^1 + \frac{2}{3}\xi^2, \tag{1.271}$$

$$x^2 = -\frac{2}{3}\xi^1 + \frac{1}{3}\xi^2, \tag{1.272}$$

$$x^3 = \xi^3. \tag{1.273}$$

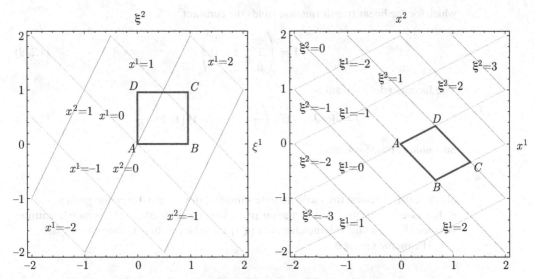

Figure 1.12. Lines of constant x^1 and x^2 in the (ξ^1, ξ^2) plane and lines of constant ξ^1 and ξ^2 in the (x^1, x^2) plane for the homogeneous affine transformation of example problem.

This transformation is of the class of *affine* transformations, which are of the form

$$x^i = A^i_j \xi^j + b^i, \tag{1.274}$$

where A^i_j and b^i are constants. Affine transformations for which $b^i = 0$ are further distinguished as *linear homogeneous* transformations. The transformation of this example is both affine and linear homogeneous.

Equations (1.271–1.273) form a linear system of three equations in three unknowns. Using standard techniques of linear algebra such as Cramer's rule allows us to solve for ξ^1, ξ^2, ξ^3 in terms of x^1, x^2, x^3. Doing so, we find the inverse transformation to be

$$\xi^1 = \frac{1}{2}x^1 - x^2, \tag{1.275}$$

$$\xi^2 = x^1 + x^2, \tag{1.276}$$

$$\xi^3 = x^3. \tag{1.277}$$

Lines of constant x^1 and x^2 in the (ξ^1, ξ^2) plane as well as lines of constant ξ^1 and ξ^2 in the (x^1, x^2) plane are plotted in Figure 1.12. Also shown is a unit square in the first quadrant of the Cartesian (ξ^1, ξ^2) plane, with vertices A, B, C, D. The superposed (x^1, x^2) coordinates in this plane give an aliased representation of the unit square. The alibi image of this square is represented as a parallelogram in the (x^1, x^2) plane. It is seen that the orientation has been preserved in what amounts to a clockwise rotation accompanied by stretching; moreover, the area (and thus the volume in three dimensions) has been slightly decreased.

The appropriate Jacobian matrix for the inverse transformation is

$$\mathbf{J} = \frac{\partial \xi^i}{\partial x^j} = \begin{pmatrix} \frac{\partial \xi^1}{\partial x^1} & \frac{\partial \xi^1}{\partial x^2} & \frac{\partial \xi^1}{\partial x^3} \\ \frac{\partial \xi^2}{\partial x^1} & \frac{\partial \xi^2}{\partial x^2} & \frac{\partial \xi^2}{\partial x^3} \\ \frac{\partial \xi^3}{\partial x^1} & \frac{\partial \xi^3}{\partial x^2} & \frac{\partial \xi^3}{\partial x^3} \end{pmatrix}, \tag{1.278}$$

which for the linear transformation yields the constant

$$\mathbf{J} = \begin{pmatrix} \frac{1}{2} & -1 & 0 \\ 1 & 1 & 0 \\ 0 & 0 & 1 \end{pmatrix}. \tag{1.279}$$

The Jacobian determinant is

$$J = \det \mathbf{J} = (1)\left(\left(\frac{1}{2}\right)(1) - (-1)(1)\right) = \frac{3}{2}. \tag{1.280}$$

So a unique transformation,

$$\boldsymbol{\xi} = \mathbf{J} \cdot \mathbf{x}, \tag{1.281}$$

always exists, because the Jacobian determinant is never zero. Inversion gives $\mathbf{x} = \mathbf{J}^{-1} \cdot \boldsymbol{\xi}$. Because $J > 0$, the transformation preserves the orientation of geometric entities. Because $J > 1$, a unit volume element in $\boldsymbol{\xi}$ space is larger than its image in $\mathbf{x}$ space.

The metric tensor is

$$g_{kl} = \frac{\partial \xi^i}{\partial x^k}\frac{\partial \xi^i}{\partial x^l} = \frac{\partial \xi^1}{\partial x^k}\frac{\partial \xi^1}{\partial x^l} + \frac{\partial \xi^2}{\partial x^k}\frac{\partial \xi^2}{\partial x^l} + \frac{\partial \xi^3}{\partial x^k}\frac{\partial \xi^3}{\partial x^l}. \tag{1.282}$$

For example, for $k = 1, l = 1$, we get

$$g_{11} = \frac{\partial \xi^i}{\partial x^1}\frac{\partial \xi^i}{\partial x^1} = \frac{\partial \xi^1}{\partial x^1}\frac{\partial \xi^1}{\partial x^1} + \frac{\partial \xi^2}{\partial x^1}\frac{\partial \xi^2}{\partial x^1} + \frac{\partial \xi^3}{\partial x^1}\frac{\partial \xi^3}{\partial x^1}, \tag{1.283}$$

$$= \left(\frac{1}{2}\right)\left(\frac{1}{2}\right) + (1)(1) + (0)(0) = \frac{5}{4}. \tag{1.284}$$

Repeating this operation for all terms of g_{kl}, we find that the complete metric tensor is

$$g_{kl} = \begin{pmatrix} \frac{5}{4} & \frac{1}{2} & 0 \\ \frac{1}{2} & 2 & 0 \\ 0 & 0 & 1 \end{pmatrix}. \tag{1.285}$$

Moreover,

$$g = \det g_{kl} = (1)\left(\left(\frac{5}{4}\right)(2) - \left(\frac{1}{2}\right)\left(\frac{1}{2}\right)\right) = \frac{9}{4}. \tag{1.286}$$

One can also calculate the metric tensor through the use of Gibbs notation:

$$\mathbf{G} = \mathbf{J}^T \cdot \mathbf{J}, \tag{1.287}$$

$$= \begin{pmatrix} \frac{1}{2} & 1 & 0 \\ -1 & 1 & 0 \\ 0 & 0 & 1 \end{pmatrix} \begin{pmatrix} \frac{1}{2} & -1 & 0 \\ 1 & 1 & 0 \\ 0 & 0 & 1 \end{pmatrix}, \tag{1.288}$$

$$= \begin{pmatrix} \frac{5}{4} & \frac{1}{2} & 0 \\ \frac{1}{2} & 2 & 0 \\ 0 & 0 & 1 \end{pmatrix}. \tag{1.289}$$

Distance in the transformed system is given by

$$(ds)^2 = dx^k \, g_{kl} \, dx^l, \tag{1.290}$$

$$= d\mathbf{x}^T \cdot \mathbf{G} \cdot d\mathbf{x}, \tag{1.291}$$

$$= \begin{pmatrix} dx^1 & dx^2 & dx^3 \end{pmatrix} \begin{pmatrix} \frac{5}{4} & \frac{1}{2} & 0 \\ \frac{1}{2} & 2 & 0 \\ 0 & 0 & 1 \end{pmatrix} \begin{pmatrix} dx^1 \\ dx^2 \\ dx^3 \end{pmatrix}, \tag{1.292}$$

$$= \underbrace{\left(\left(\tfrac{5}{4} \, dx^1 + \tfrac{1}{2} \, dx^2 \right) \quad \left(\tfrac{1}{2} \, dx^1 + 2 \, dx^2 \right) \quad dx^3 \right)}_{=dx_l = dx^k g_{kl}} \underbrace{\begin{pmatrix} dx^1 \\ dx^2 \\ dx^3 \end{pmatrix}}_{=dx^l} = dx_l \, dx^l, \quad (1.293)$$

$$= \frac{5}{4} \left(dx^1 \right)^2 + 2 \left(dx^2 \right)^2 + \left(dx^3 \right)^2 + dx^1 \, dx^2. \quad (1.294)$$

Detailed algebraic manipulation employing the so-called method of quadratic forms, discussed in Section 7.14, reveals that Eq. (1.294) can be rewritten as follows:

$$(ds)^2 = \frac{9}{20} \left(dx^1 + 2dx^2 \right)^2 + \frac{1}{5} \left(-2dx^1 + dx^2 \right)^2 + \left(dx^3 \right)^2. \quad (1.295)$$

Direct expansion reveals the two forms for $(ds)^2$ to be identical. Note that

- The Jacobian matrix $\mathbf{J}$ is not symmetric.
- The metric tensor $\mathbf{G} = \mathbf{J}^T \cdot \mathbf{J}$ is symmetric.
- The fact that the metric tensor has nonzero off-diagonal elements can be shown to be a consequence of the transformation being nonorthogonal.
- We identify here a new representation of the differential distance vector in the transformed space: $dx_l = dx^k g_{kl}$, whose significance is discussed in Section 1.6.2.
- The distance is guaranteed to be positive. This will be true for all affine transformations in ordinary three-dimensional Euclidean space.

Also, we have the volume ratio of differential elements as

$$d\xi^1 \, d\xi^2 \, d\xi^3 = \sqrt{\frac{9}{4}} \, dx^1 \, dx^2 \, dx^3, \quad (1.296)$$

$$= \frac{3}{2} \, dx^1 \, dx^2 \, dx^3. \quad (1.297)$$

Now we use Eq. (1.266) to find the appropriate derivatives of ϕ. We first note that

$$(\mathbf{J}^T)^{-1} = \begin{pmatrix} \tfrac{1}{2} & 1 & 0 \\ -1 & 1 & 0 \\ 0 & 0 & 1 \end{pmatrix}^{-1} = \begin{pmatrix} \tfrac{2}{3} & -\tfrac{2}{3} & 0 \\ \tfrac{2}{3} & \tfrac{1}{3} & 0 \\ 0 & 0 & 1 \end{pmatrix}. \quad (1.298)$$

So

$$\begin{pmatrix} \frac{\partial \phi}{\partial \xi^1} \\ \frac{\partial \phi}{\partial \xi^2} \\ \frac{\partial \phi}{\partial \xi^3} \end{pmatrix} = \begin{pmatrix} \tfrac{2}{3} & -\tfrac{2}{3} & 0 \\ \tfrac{2}{3} & \tfrac{1}{3} & 0 \\ 0 & 0 & 1 \end{pmatrix} \begin{pmatrix} \frac{\partial \phi}{\partial x^1} \\ \frac{\partial \phi}{\partial x^2} \\ \frac{\partial \phi}{\partial x^3} \end{pmatrix} = \underbrace{\begin{pmatrix} \frac{\partial x^1}{\partial \xi^1} & \frac{\partial x^2}{\partial \xi^1} & \frac{\partial x^3}{\partial \xi^1} \\ \frac{\partial x^1}{\partial \xi^2} & \frac{\partial x^2}{\partial \xi^2} & \frac{\partial x^3}{\partial \xi^2} \\ \frac{\partial x^1}{\partial \xi^3} & \frac{\partial x^2}{\partial \xi^3} & \frac{\partial x^3}{\partial \xi^3} \end{pmatrix}}_{(\mathbf{J}^T)^{-1}} \begin{pmatrix} \frac{\partial \phi}{\partial x^1} \\ \frac{\partial \phi}{\partial x^2} \\ \frac{\partial \phi}{\partial x^3} \end{pmatrix}.$$

$$(1.299)$$

Thus, by inspection,

$$\frac{\partial \phi}{\partial \xi^1} = \frac{2}{3} \frac{\partial \phi}{\partial x^1} - \frac{2}{3} \frac{\partial \phi}{\partial x^2}, \quad (1.300)$$

$$\frac{\partial \phi}{\partial \xi^2} = \frac{2}{3} \frac{\partial \phi}{\partial x^1} + \frac{1}{3} \frac{\partial \phi}{\partial x^2}. \quad (1.301)$$

So the transformed version of Eq. (1.270) becomes

$$\left(\frac{2}{3} \frac{\partial \phi}{\partial x^1} - \frac{2}{3} \frac{\partial \phi}{\partial x^2} \right) + \left(\frac{2}{3} \frac{\partial \phi}{\partial x^1} + \frac{1}{3} \frac{\partial \phi}{\partial x^2} \right) = \left(\frac{1}{2} x^1 - x^2 \right)^2 + \left(x^1 + x^2 \right)^2, \quad (1.302)$$

$$\frac{4}{3} \frac{\partial \phi}{\partial x^1} - \frac{1}{3} \frac{\partial \phi}{\partial x^2} = \frac{5}{4} \left(x^1 \right)^2 + x^1 x^2 + 2 \left(x^2 \right)^2. \quad (1.303)$$

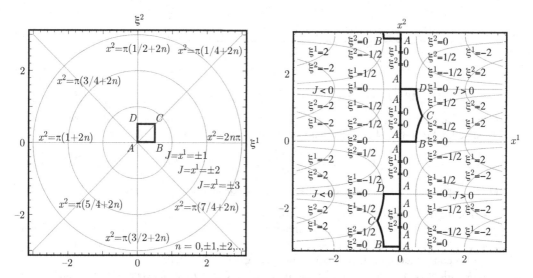

Figure 1.13. Lines of constant x^1 and x^2 in the (ξ^1, ξ^2) plane and lines of constant ξ^1 and ξ^2 in the (x^1, x^2) plane for cylindrical coordinates transformation of example problem.

2. Cartesian to cylindrical coordinates

The globally nonlinear transformations between Cartesian and cylindrical coordinates[27] are

$$x^1 = \pm\sqrt{(\xi^1)^2 + (\xi^2)^2}, \tag{1.304}$$

$$x^2 = \tan^{-1}\left(\frac{\xi^2}{\xi^1}\right), \tag{1.305}$$

$$x^3 = \xi^3. \tag{1.306}$$

Here we have taken the unusual step of admitting negative x^1. This is admissible mathematically but does not reconcile with our geometric intuition as it corresponds to a negative radius. Our analysis will allow for us to interpret this unusual feature. As this system of equations is nonlinear, the transformation can be nonunique, and in fact is. For such systems, we cannot always find an explicit algebraic expression for the inverse transformation. In this case, some straightforward algebraic and trigonometric manipulation reveals that we can find an explicit representation of the inverse transformation, which is

$$\xi^1 = x^1 \cos x^2, \tag{1.307}$$

$$\xi^2 = x^1 \sin x^2, \tag{1.308}$$

$$\xi^3 = x^3. \tag{1.309}$$

Lines of constant x^1 and x^2 in the (ξ^1, ξ^2) plane and lines of constant ξ^1 and ξ^2 in the (x^1, x^2) plane are plotted in Figure 1.13. A square of area $1/2 \times 1/2$ is marked in the (ξ^1, ξ^2) plane. Its unusually shaped alibi image in the (x^1, x^2) plane is also indicated. The nonuniqueness of the mapping from one plane to the other is evident in that (1) A maps from a single point in the (ξ^1, ξ^2) plane to a vertical line in the (x^1, x^2) plane and (2) B, C, and D map from single points in the (ξ^1, ξ^2) plane to an infinite set of discrete points in the (x^1, x^2) plane.

---

[27] The nomenclature "cylindrical coordinates" implies a three-dimensional system; the two-dimensional analog is known as "polar coordinates."

The appropriate Jacobian matrix arising from local linearization for the inverse transformation is

$$\mathbf{J} = \frac{\partial \xi^i}{\partial x^j} = \begin{pmatrix} \frac{\partial \xi^1}{\partial x^1} & \frac{\partial \xi^1}{\partial x^2} & \frac{\partial \xi^1}{\partial x^3} \\ \frac{\partial \xi^2}{\partial x^1} & \frac{\partial \xi^2}{\partial x^2} & \frac{\partial \xi^2}{\partial x^3} \\ \frac{\partial \xi^3}{\partial x^1} & \frac{\partial \xi^3}{\partial x^2} & \frac{\partial \xi^3}{\partial x^3} \end{pmatrix}, \tag{1.310}$$

$$= \begin{pmatrix} \cos x^2 & -x^1 \sin x^2 & 0 \\ \sin x^2 & x^1 \cos x^2 & 0 \\ 0 & 0 & 1 \end{pmatrix}. \tag{1.311}$$

The Jacobian determinant is

$$J = x^1 \cos^2 x^2 + x^1 \sin^2 x^2 = x^1. \tag{1.312}$$

So a unique transformation fails to exist if $x^1 = 0$. For $x^1 > 0$, the transformation is orientation-preserving. For $x^1 = 1$, the transformation is volume-preserving. For $x^1 < 0$, the transformation is orientation-reversing. This is a fundamental mathematical reason why we do not consider negative radius. It fails to preserve the orientation of a mapped element. For $x^1 \in (0,1)$, a unit element in $\boldsymbol{\xi}$-space is smaller than a unit element in **x**-space; the converse holds for $x^1 \in (1,\infty)$.

The metric tensor is

$$g_{kl} = \frac{\partial \xi^i}{\partial x^k} \frac{\partial \xi^i}{\partial x^l} = \frac{\partial \xi^1}{\partial x^k} \frac{\partial \xi^1}{\partial x^l} + \frac{\partial \xi^2}{\partial x^k} \frac{\partial \xi^2}{\partial x^l} + \frac{\partial \xi^3}{\partial x^k} \frac{\partial \xi^3}{\partial x^l}. \tag{1.313}$$

For example, for $k = 1, l = 1$, we get

$$g_{11} = \frac{\partial \xi^i}{\partial x^1} \frac{\partial \xi^i}{\partial x^1} = \frac{\partial \xi^1}{\partial x^1} \frac{\partial \xi^1}{\partial x^1} + \frac{\partial \xi^2}{\partial x^1} \frac{\partial \xi^2}{\partial x^1} + \frac{\partial \xi^3}{\partial x^1} \frac{\partial \xi^3}{\partial x^1}, \tag{1.314}$$

$$= \cos^2 x^2 + \sin^2 x^2 + 0 = 1. \tag{1.315}$$

Repeating this operation, we find the complete metric tensor is

$$g_{kl} = \begin{pmatrix} 1 & 0 & 0 \\ 0 & \left(x^1\right)^2 & 0 \\ 0 & 0 & 1 \end{pmatrix}. \tag{1.316}$$

Moreover,

$$g = \det g_{kl} = \left(x^1\right)^2. \tag{1.317}$$

This is equivalent to the calculation in Gibbs notation:

$$\mathbf{G} = \mathbf{J}^T \cdot \mathbf{J}, \tag{1.318}$$

$$= \begin{pmatrix} \cos x^2 & \sin x^2 & 0 \\ -x^1 \sin x^2 & x^1 \cos x^2 & 0 \\ 0 & 0 & 1 \end{pmatrix} \begin{pmatrix} \cos x^2 & -x^1 \sin x^2 & 0 \\ \sin x^2 & x^1 \cos x^2 & 0 \\ 0 & 0 & 1 \end{pmatrix}, \tag{1.319}$$

$$= \begin{pmatrix} 1 & 0 & 0 \\ 0 & \left(x^1\right)^2 & 0 \\ 0 & 0 & 1 \end{pmatrix}. \tag{1.320}$$

Distance in the transformed system is given by

$$(ds)^2 = dx^k \, g_{kl} \, dx^l, \tag{1.321}$$

$$= d\mathbf{x}^T \cdot \mathbf{G} \cdot d\mathbf{x}, \tag{1.322}$$

$$= (dx^1 \quad dx^2 \quad dx^3) \begin{pmatrix} 1 & 0 & 0 \\ 0 & (x^1)^2 & 0 \\ 0 & 0 & 1 \end{pmatrix} \begin{pmatrix} dx^1 \\ dx^2 \\ dx^3 \end{pmatrix}, \tag{1.323}$$

$$= \underbrace{(dx^1 \quad (x^1)^2 dx^2 \quad dx^3)}_{dx_l = dx^k g_{kl}} \underbrace{\begin{pmatrix} dx^1 \\ dx^2 \\ dx^3 \end{pmatrix}}_{=dx^l} = dx_l \, dx^l, \tag{1.324}$$

$$= (dx^1)^2 + (x^1 dx^2)^2 + (dx^3)^2. \tag{1.325}$$

Note the following:

- The fact that the metric tensor is diagonal can be attributed to the transformation being orthogonal.
- For the nonlinear transformation, **J** and **G** vary with position.
- Because the product of any matrix with its transpose is guaranteed to yield a symmetric matrix, the metric tensor is always symmetric.

Also, we have the volume ratio of differential elements as

$$d\xi^1 \, d\xi^2 \, d\xi^3 = x^1 \, dx^1 \, dx^2 \, dx^3. \tag{1.326}$$

Now we use Eq. (1.266) to find the appropriate derivatives of ϕ. We first note that

$$(\mathbf{J}^T)^{-1} = \begin{pmatrix} \cos x^2 & \sin x^2 & 0 \\ -x^1 \sin x^2 & x^1 \cos x^2 & 0 \\ 0 & 0 & 1 \end{pmatrix}^{-1} = \begin{pmatrix} \cos x^2 & -\frac{\sin x^2}{x^1} & 0 \\ \sin x^2 & \frac{\cos x^2}{x^1} & 0 \\ 0 & 0 & 1 \end{pmatrix}. \tag{1.327}$$

So

$$\begin{pmatrix} \frac{\partial \phi}{\partial \xi^1} \\ \frac{\partial \phi}{\partial \xi^2} \\ \frac{\partial \phi}{\partial \xi^3} \end{pmatrix} = \begin{pmatrix} \cos x^2 & -\frac{\sin x^2}{x^1} & 0 \\ \sin x^2 & \frac{\cos x^2}{x^1} & 0 \\ 0 & 0 & 1 \end{pmatrix} \begin{pmatrix} \frac{\partial \phi}{\partial x^1} \\ \frac{\partial \phi}{\partial x^2} \\ \frac{\partial \phi}{\partial x^3} \end{pmatrix} = \underbrace{\begin{pmatrix} \frac{\partial x^1}{\partial \xi^1} & \frac{\partial x^2}{\partial \xi^1} & \frac{\partial x^3}{\partial \xi^1} \\ \frac{\partial x^1}{\partial \xi^2} & \frac{\partial x^2}{\partial \xi^2} & \frac{\partial x^3}{\partial \xi^2} \\ \frac{\partial x^1}{\partial \xi^3} & \frac{\partial x^2}{\partial \xi^3} & \frac{\partial x^3}{\partial \xi^3} \end{pmatrix}}_{(\mathbf{J}^T)^{-1}} \begin{pmatrix} \frac{\partial \phi}{\partial x^1} \\ \frac{\partial \phi}{\partial x^2} \\ \frac{\partial \phi}{\partial x^3} \end{pmatrix}.$$

$$\tag{1.328}$$

Thus, by inspection,

$$\frac{\partial \phi}{\partial \xi^1} = \cos x^2 \frac{\partial \phi}{\partial x^1} - \frac{\sin x^2}{x^1} \frac{\partial \phi}{\partial x^2}, \tag{1.329}$$

$$\frac{\partial \phi}{\partial \xi^2} = \sin x^2 \frac{\partial \phi}{\partial x^1} + \frac{\cos x^2}{x^1} \frac{\partial \phi}{\partial x^2}. \tag{1.330}$$

So the transformed version of Eq. (1.270) becomes

$$\left(\cos x^2 \frac{\partial \phi}{\partial x^1} - \frac{\sin x^2}{x^1} \frac{\partial \phi}{\partial x^2} \right) + \left(\sin x^2 \frac{\partial \phi}{\partial x^1} + \frac{\cos x^2}{x^1} \frac{\partial \phi}{\partial x^2} \right) = (x^1)^2, \tag{1.331}$$

$$(\cos x^2 + \sin x^2) \frac{\partial \phi}{\partial x^1} + \left(\frac{\cos x^2 - \sin x^2}{x^1} \right) \frac{\partial \phi}{\partial x^2} = (x^1)^2. \tag{1.332}$$

1.6.2 Covariance and Contravariance

Quantities known as *contravariant vectors*[28] transform locally according to

$$\bar{u}^i = \frac{\partial \bar{x}^i}{\partial x^j} u^j. \tag{1.333}$$

We use the restriction implied by "local" to take advantage of the fact that the transformation is locally linear; Eq. (1.333) is not a general recipe for a global transformation rule. Quantities known as *covariant vectors* transform locally according to

$$\bar{u}_i = \frac{\partial x^j}{\partial \bar{x}^i} u_j. \tag{1.334}$$

In both cases, we have considered general transformations from one non-Cartesian coordinate system (x^1, x^2, x^3) to another $(\bar{x}^1, \bar{x}^2, \bar{x}^3)$. Indices associated with contravariant quantities appear as superscripts, and those associated with covariant quantities appear as subscripts. We shall see that the terms *contravariant* and *covariant* can be interpreted as natural alternate alias representations of the same vector entity. The nature of the particular coordinate system dictates the nature of what constitutes contravariance and covariance.

In the special case where the barred coordinate system is Cartesian, we take U to denote the Cartesian vector and say

$$U^i = \frac{\partial \xi^i}{\partial x^j} u^j, \qquad U_i = \frac{\partial x^j}{\partial \xi^i} u_j. \tag{1.335}$$

EXAMPLE 1.20

Returning for the moment to a traditional notation, let us say (x, y, z) is an ordinary Cartesian system and define the globally linear transformation

$$\bar{x} = \lambda x, \qquad \bar{y} = \lambda y, \qquad \bar{z} = \lambda z. \tag{1.336}$$

The transformation represents a uniform stretching of each axis by the factor λ. Now we can assign velocities in both the unbarred and barred systems:

$$u^x = \frac{dx}{dt}, \qquad u^y = \frac{dy}{dt}, \qquad u^z = \frac{dz}{dt}, \tag{1.337}$$

$$\bar{u}^{\bar{x}} = \frac{d\bar{x}}{dt}, \qquad \bar{u}^{\bar{y}} = \frac{d\bar{y}}{dt}, \qquad \bar{u}^{\bar{z}} = \frac{d\bar{z}}{dt}, \tag{1.338}$$

$$\bar{u}^{\bar{x}} = \frac{\partial \bar{x}}{\partial x}\frac{dx}{dt}, \qquad \bar{u}^{\bar{y}} = \frac{\partial \bar{y}}{\partial y}\frac{dy}{dt}, \qquad \bar{u}^{\bar{z}} = \frac{\partial \bar{z}}{\partial z}\frac{dz}{dt}, \tag{1.339}$$

$$\bar{u}^{\bar{x}} = \lambda u^x, \qquad \bar{u}^{\bar{y}} = \lambda u^y, \qquad \bar{u}^{\bar{z}} = \lambda u^z, \tag{1.340}$$

$$\bar{u}^{\bar{x}} = \frac{\partial \bar{x}}{\partial x} u^x, \qquad \bar{u}^{\bar{y}} = \frac{\partial \bar{y}}{\partial y} u^y, \qquad \bar{u}^{\bar{z}} = \frac{\partial \bar{z}}{\partial z} u^z. \tag{1.341}$$

This suggests the velocity vector is contravariant.

---

[28] Many would choose slightly different semantics than we have. Such semantics would hold that vectors (as well as tensors) are intrinsic entities that need no description as contravariant or covariant. These intrinsic entities may, however, be represented in a variety of fashions. For example, one could say a contravariant representation of the vector **u** is one whose scalar components transform according to Eq. (1.333). Such semantics would not identify u^j as a vector but as the scalar contravariant component of a vector. Often this language becomes burdensome, and so we truncate it as earlier.

Now consider a vector that is the gradient of a function $f(x, y, z)$. For example, let

$$f(x, y, z) = x + y^2 + z^3, \tag{1.342}$$

$$u_x = \frac{\partial f}{\partial x}, \qquad u_y = \frac{\partial f}{\partial y}, \qquad u_z = \frac{\partial f}{\partial z}, \tag{1.343}$$

$$u_x = 1, \qquad u_y = 2y, \qquad u_z = 3z^2. \tag{1.344}$$

In the new coordinates,

$$f\left(\frac{\bar{x}}{\lambda}, \frac{\bar{y}}{\lambda}, \frac{\bar{z}}{\lambda}\right) = \frac{\bar{x}}{\lambda} + \frac{\bar{y}^2}{\lambda^2} + \frac{\bar{z}^3}{\lambda^3}, \tag{1.345}$$

so defining a new function $\bar{f}$, we have

$$\bar{f}(\bar{x}, \bar{y}, \bar{z}) = \frac{\bar{x}}{\lambda} + \frac{\bar{y}^2}{\lambda^2} + \frac{\bar{z}^3}{\lambda^3}. \tag{1.346}$$

Now

$$\bar{u}_{\bar{x}} = \frac{\partial \bar{f}}{\partial \bar{x}}, \qquad \bar{u}_{\bar{y}} = \frac{\partial \bar{f}}{\partial \bar{y}}, \qquad \bar{u}_{\bar{z}} = \frac{\partial \bar{f}}{\partial \bar{z}}, \tag{1.347}$$

$$\bar{u}_{\bar{x}} = \frac{1}{\lambda}, \qquad \bar{u}_{\bar{y}} = \frac{2\bar{y}}{\lambda^2}, \qquad \bar{u}_{\bar{z}} = \frac{3\bar{z}^2}{\lambda^3}. \tag{1.348}$$

In terms of x, y, z, we have

$$\bar{u}_{\bar{x}} = \frac{1}{\lambda}, \qquad \bar{u}_{\bar{y}} = \frac{2y}{\lambda}, \qquad \bar{u}_{\bar{z}} = \frac{3z^2}{\lambda}. \tag{1.349}$$

So it is clear here that, in contrast to the velocity vector,

$$\bar{u}_{\bar{x}} = \frac{1}{\lambda} u_x, \qquad \bar{u}_{\bar{y}} = \frac{1}{\lambda} u_y, \qquad \bar{u}_{\bar{z}} = \frac{1}{\lambda} u_z. \tag{1.350}$$

More generally, we find for this case that

$$\bar{u}_{\bar{x}} = \frac{\partial x}{\partial \bar{x}} u_x, \qquad \bar{u}_{\bar{y}} = \frac{\partial y}{\partial \bar{y}} u_y, \qquad \bar{u}_{\bar{z}} = \frac{\partial z}{\partial \bar{z}} u_z, \tag{1.351}$$

which suggests the gradient vector is covariant.

Contravariant tensors transform locally according to

$$\bar{v}^{ij} = \frac{\partial \bar{x}^i}{\partial x^k} \frac{\partial \bar{x}^j}{\partial x^l} v^{kl}. \tag{1.352}$$

Covariant tensors transform locally according to

$$\bar{v}_{ij} = \frac{\partial x^k}{\partial \bar{x}^i} \frac{\partial x^l}{\partial \bar{x}^j} v_{kl}. \tag{1.353}$$

Mixed tensors transform locally according to

$$\bar{v}^i_j = \frac{\partial \bar{x}^i}{\partial x^k} \frac{\partial x^l}{\partial \bar{x}^j} v^k_l. \tag{1.354}$$

Recall that *variance* is another term for *gradient* and that *co*- denotes "with". A vector that is co-variant is aligned with the variance or the gradient. Recalling next that *contra*- denotes "against", a vector that is contra-variant is aligned against the variance or the gradient. This results in a set of contravariant basis vectors being tangent to lines of $x^i = C$, while covariant basis vectors are normal to lines of

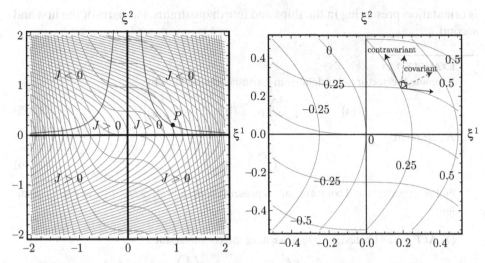

Figure 1.14. Contours for the transformation $x^1 = \xi^1 + (\xi^2)^2$, $x^2 = \xi^2 + (\xi^1)^3$ (left) and a blown-up version (right) including a pair of contravariant basis vectors, which are tangent to the contours, and covariant basis vectors, which are normal to the contours.

$x^i = C$, where in each case C is some constant. A vector in space has two natural representations, one on a contravariant basis and the other on a covariant basis. The contravariant representation seems more natural because it is similar to the familiar $\mathbf{i}, \mathbf{j}$, and $\mathbf{k}$ for Cartesian systems, though both can be used to obtain equivalent results.

For the globally nonlinear transformation, first examined using other variable names in Section 1.2; $x^1 = \xi^1 + (\xi^2)^2$, $x^2 = \xi^2 + (\xi^1)^3$, Figure 1.14 gives a plot of a set of lines of constant x^1 and x^2 in the Cartesian (ξ^1, ξ^2) plane, along with a local set of contravariant and covariant basis vectors. The covariant basis vectors, because they are directly related to the gradient vector, point in the direction of most rapid change of x^1 and x^2 and are orthogonal to contours on which x^1 and x^2 are constant. The contravariant vectors are tangent to the contours. It can be shown that the contravariant vectors are aligned with the columns of $\mathbf{J}$, and the covariant vectors are aligned with the rows of $\mathbf{J}^{-1}$.

This transformation has some special properties. Near the origin, the higher order terms become negligible, and the transformation reduces to the identity mapping $x^1 = \xi^1$, $x^2 = \xi^2$. As such, in the neighborhood of the origin, one has $\mathbf{J} = \mathbf{I}$, and there is no change in area or orientation of an element. Moreover, on each of the coordinate axes, $x^1 = \xi^1$ and $x^2 = \xi^2$; additionally, on each of the coordinate axes, $J = 1$, so in those special locations, the transformation is area- and orientation-preserving. The transformation is singular wherever $J = 0$; this occurs when $\xi^2 = 1/(6(\xi^1)^2)$. As $J \to 0$, the contours of x^1 align more and more with the contours of x^2, and thus the contravariant basis vectors come closer to becoming parallel to each other. When $J = 0$, the two contours of each osculate. At such points, there is only one linearly independent contravariant basis vector, which is not enough to represent an arbitrary vector in a locally linear combination. An analog holds for the covariant basis vectors. The figure shows the point P located on the curve where $J = 0$, near that point that was analyzed earlier in Section 1.2. Clearly, at P, lines of constant x^1 are parallel to lines of constant x^2. In most of the first and second quadrants, the transformation is orientation-reversing. The transformation is orientation-preserving in the third and fourth quadrants and parts of the first and second.

EXAMPLE 1.21

Examine the vector fields defined in Cartesian coordinates by

$$\text{(a)} \quad U^i = \begin{pmatrix} \xi^1 \\ \xi^2 \end{pmatrix}, \qquad \text{(b)} \quad U^i = \begin{pmatrix} \xi^1 \\ 2\xi^2 \end{pmatrix}. \tag{1.355}$$

At the point

$$P : \begin{pmatrix} \xi^1 \\ \xi^2 \end{pmatrix} = \begin{pmatrix} 1 \\ 1 \end{pmatrix}, \tag{1.356}$$

find the covariant and contravariant representations of both cases of U^i in polar coordinates.

(a) At P in the Cartesian system, we have the contravariant

$$U^i = \begin{pmatrix} \xi^1 \\ \xi^2 \end{pmatrix}\Bigg|_{\xi^1=1,\xi^2=1} = \begin{pmatrix} 1 \\ 1 \end{pmatrix}. \tag{1.357}$$

For a Cartesian coordinate system, the metric tensor $g_{ij} = \delta_{ij} = g_{ji} = \delta_{ji}$. Thus, the covariant representation in the Cartesian system is

$$U_j = g_{ji}U^i = \delta_{ji}U^i = \begin{pmatrix} 1 & 0 \\ 0 & 1 \end{pmatrix}\begin{pmatrix} 1 \\ 1 \end{pmatrix} = \begin{pmatrix} 1 \\ 1 \end{pmatrix}. \tag{1.358}$$

In general, we can say that for Cartesian systems, there is no distinction between a vector's contravariant and covariant representations.

Now consider polar coordinates: $\xi^1 = x^1 \cos x^2, \xi^2 = x^1 \sin x^2$. For the inverse transformation, let us insist that $J > 0$, so $x^1 = \sqrt{(\xi^1)^2 + (\xi^2)^2}$, $x^2 = \tan^{-1}(\xi^2/\xi^1)$. Thus, we have the representation of P in polar coordinates of

$$P : \begin{pmatrix} x^1 \\ x^2 \end{pmatrix} = \begin{pmatrix} \sqrt{2} \\ \frac{\pi}{4} \end{pmatrix}. \tag{1.359}$$

For the transformation, we have

$$\mathbf{J} = \begin{pmatrix} \cos x^2 & -x^1 \sin x^2 \\ \sin x^2 & x^1 \cos x^2 \end{pmatrix}, \qquad \mathbf{G} = \mathbf{J}^T \cdot \mathbf{J} = \begin{pmatrix} 1 & 0 \\ 0 & (x^1)^2 \end{pmatrix}. \tag{1.360}$$

At P, we thus have

$$\mathbf{J} = \begin{pmatrix} \frac{\sqrt{2}}{2} & -1 \\ \frac{\sqrt{2}}{2} & 1 \end{pmatrix}, \qquad \mathbf{G} = \mathbf{J}^T \cdot \mathbf{J} = \begin{pmatrix} 1 & 0 \\ 0 & 2 \end{pmatrix}. \tag{1.361}$$

Now, specializing Eq. (1.333) by considering the barred coordinate to be Cartesian, we can say

$$U^i = \frac{\partial \xi^i}{\partial x^j} u^j. \tag{1.362}$$

Locally, we can use the Gibbs notation and say $\mathbf{U} = \mathbf{J} \cdot \mathbf{u}$, and thus get $\mathbf{u} = \mathbf{J}^{-1} \cdot \mathbf{U}$, so that the contravariant representation is

$$\begin{pmatrix} u^1 \\ u^2 \end{pmatrix} = \begin{pmatrix} \frac{\sqrt{2}}{2} & -1 \\ \frac{\sqrt{2}}{2} & 1 \end{pmatrix}^{-1} \begin{pmatrix} 1 \\ 1 \end{pmatrix} = \begin{pmatrix} \frac{1}{\sqrt{2}} & \frac{1}{\sqrt{2}} \\ -\frac{1}{2} & \frac{1}{2} \end{pmatrix}\begin{pmatrix} 1 \\ 1 \end{pmatrix} = \begin{pmatrix} \sqrt{2} \\ 0 \end{pmatrix}. \tag{1.363}$$

In Gibbs notation, one can interpret this as $1\mathbf{i} + 1\mathbf{j} = \sqrt{2}\mathbf{e}_r + 0\mathbf{e}_\theta$, where we have taken $\mathbf{e}_r$ and $\mathbf{e}_\theta$ to be unit vectors associated with the directions of increasing $r = x^1$ and $\theta = x^2$, respectively. This representation is different than the simple polar coordinates of P given by Eq. (1.359).

Let us look closer at the unit vectors $\mathbf{e}_r$ and $\mathbf{e}_\theta$, which, though intuitively obvious, are not yet formally defined. In polar coordinates, the contravariant representations of the unit basis vectors must be $\bar{\mathbf{e}}_r = (1, 0)^T$ and $\bar{\mathbf{e}}_\theta = (0, 1)^T$. So in Cartesian coordinates, those basis vectors are represented as

$$\hat{\mathbf{e}}_r = \mathbf{J} \cdot \bar{\mathbf{e}}_r = \begin{pmatrix} \cos x^2 & -x^1 \sin x^2 \\ \sin x^2 & x^1 \cos x^2 \end{pmatrix} \begin{pmatrix} 1 \\ 0 \end{pmatrix} = \begin{pmatrix} \cos x^2 \\ \sin x^2 \end{pmatrix}, \tag{1.364}$$

$$\hat{\mathbf{e}}_\theta = \mathbf{J} \cdot \bar{\mathbf{e}}_\theta = \begin{pmatrix} \cos x^2 & -x^1 \sin x^2 \\ \sin x^2 & x^1 \cos x^2 \end{pmatrix} \begin{pmatrix} 0 \\ 1 \end{pmatrix} = \begin{pmatrix} -x^1 \sin x^2 \\ x^1 \cos x^2 \end{pmatrix}. \tag{1.365}$$

In general, a unit vector in the transformed space is not a unit vector in the Cartesian space. Note that $\hat{\mathbf{e}}_\theta$ is a unit vector in Cartesian space only if $x^1 = 1$; this is also the condition for $J = 1$. We can recover unit vectors in the original coordinate system by scaling each by their respective magnitudes:

$$\mathbf{e}_r = \frac{\hat{\mathbf{e}}_r}{\sqrt{\hat{\mathbf{e}}_r^T \cdot \hat{\mathbf{e}}_r}} = \frac{1}{\sqrt{\cos^2 x^2 + \sin^2 x^2}} \begin{pmatrix} \cos x^2 \\ \sin x^2 \end{pmatrix} = \begin{pmatrix} \cos x^2 \\ \sin x^2 \end{pmatrix}, \tag{1.366}$$

$$\mathbf{e}_\theta = \frac{\hat{\mathbf{e}}_\theta}{\sqrt{\hat{\mathbf{e}}_\theta^T \cdot \hat{\mathbf{e}}_\theta}} = \frac{1}{\sqrt{(x^1)^2 \sin^2 x^2 + (x^1)^2 \cos^2 x^2}} \begin{pmatrix} -x^1 \sin x^2 \\ x^1 \cos x^2 \end{pmatrix} = \begin{pmatrix} -\sin x^2 \\ \cos x^2 \end{pmatrix}. \tag{1.367}$$

Clearly $\mathbf{e}_r$ and $\mathbf{e}_\theta$ are orthonormal vectors with $\mathbf{e}_r$ pointing in the direction of increasing r and $\mathbf{e}_\theta$ pointing in the direction of increasing θ.[29]

Lastly, we see the covariant representation is given by $u_j = u^i g_{ij}$. Because g_{ij} is symmetric, we can transpose this to get $u_j = g_{ji} u^i$:

$$\begin{pmatrix} u_1 \\ u_2 \end{pmatrix} = \mathbf{G} \cdot \begin{pmatrix} u^1 \\ u^2 \end{pmatrix} = \begin{pmatrix} 1 & 0 \\ 0 & 2 \end{pmatrix} \begin{pmatrix} \sqrt{2} \\ 0 \end{pmatrix} = \begin{pmatrix} \sqrt{2} \\ 0 \end{pmatrix}. \tag{1.368}$$

This simple vector field has an identical contravariant and covariant representation. Appropriate invariant quantities, such as the square of the magnitude of the vector, are independent of the representation:

$$U_i U^i = \begin{pmatrix} 1 & 1 \end{pmatrix} \begin{pmatrix} 1 \\ 1 \end{pmatrix} = 2 \tag{1.369}$$

$$u_i u^i = \begin{pmatrix} \sqrt{2} & 0 \end{pmatrix} \begin{pmatrix} \sqrt{2} \\ 0 \end{pmatrix} = 2. \tag{1.370}$$

Though tempting, we note that there is no invariance at P between $\xi_i \xi^i = 2$ and $x_i x^i = 2 + \pi^2/8$.

(b) This slight change in the vector field introduces many complications. At P in the Cartesian system, we have the contravariant

$$U^i = \begin{pmatrix} \xi^1 \\ 2\xi^2 \end{pmatrix}\Bigg|_{\xi^1 = 1, \xi^2 = 1} = \begin{pmatrix} 1 \\ 2 \end{pmatrix}. \tag{1.371}$$

In the same fashion as demonstrated in part (a), we find the contravariant representation of U^i in polar coordinates at P is

$$\begin{pmatrix} u^1 \\ u^2 \end{pmatrix} = \begin{pmatrix} \frac{\sqrt{2}}{2} & -1 \\ \frac{\sqrt{2}}{2} & 1 \end{pmatrix}^{-1} \begin{pmatrix} 1 \\ 2 \end{pmatrix} = \begin{pmatrix} \frac{1}{\sqrt{2}} & \frac{1}{\sqrt{2}} \\ -\frac{1}{2} & \frac{1}{2} \end{pmatrix} \begin{pmatrix} 1 \\ 2 \end{pmatrix} = \begin{pmatrix} \frac{3}{\sqrt{2}} \\ \frac{1}{2} \end{pmatrix}. \tag{1.372}$$

---

[29] For general orthogonal coordinate systems, we can consider $\mathbf{j}_i$ to be the column vector occupying the ith column of the Jacobian $\mathbf{J}$. Then, we have $\hat{\mathbf{e}}_i = \mathbf{j}_i$, which can be normalized to yield $\mathbf{e}_i = \mathbf{j}_i/\sqrt{\mathbf{j}_i^T \cdot \mathbf{j}_i}$. Here the repeated index does not connote summation.

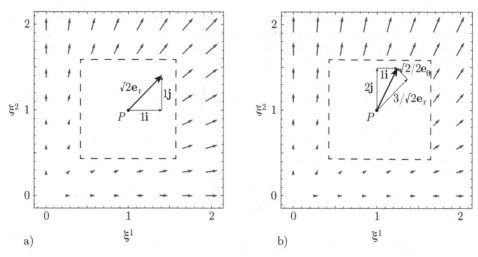

Figure 1.15. Vector fields for (a) $U^i = (\xi^1, \xi^2)^T$ and (b) $U^i = (\xi^1, 2\xi^2)^T$, along with various representations of the vector at P in Cartesian and polar representations.

Let us see how this is interpreted in terms of traditional orthonormal basis vectors. In Gibbs notation, Eq. (1.372) is $\mathbf{u} = \mathbf{J}^{-1} \cdot \mathbf{U}$. Thus, we could also say $\mathbf{U} = \mathbf{J} \cdot \mathbf{u}$:

$$\begin{pmatrix} 1 \\ 2 \end{pmatrix} = \begin{pmatrix} \frac{\sqrt{2}}{2} & -1 \\ \frac{\sqrt{2}}{2} & 1 \end{pmatrix} \begin{pmatrix} \frac{3}{\sqrt{2}} \\ \frac{1}{2} \end{pmatrix} = \frac{3}{\sqrt{2}} \underbrace{\begin{pmatrix} \frac{\sqrt{2}}{2} \\ \frac{\sqrt{2}}{2} \end{pmatrix}}_{\hat{e}_r} + \frac{1}{2} \underbrace{\begin{pmatrix} -1 \\ 1 \end{pmatrix}}_{\hat{e}_\theta}. \tag{1.373}$$

The two column vectors are orthogonal basis vectors, but they are not orthonormal. When we render them to be orthonormal, they become $\mathbf{e}_r$ and $\mathbf{e}_\theta$:

$$\underbrace{\begin{pmatrix} 1 \\ 2 \end{pmatrix}}_{U^i} = \underbrace{\frac{3}{\sqrt{2}}}_{u^1} \underbrace{\begin{pmatrix} \frac{\sqrt{2}}{2} \\ \frac{\sqrt{2}}{2} \end{pmatrix}}_{\mathbf{e}_r} + \underbrace{\frac{1}{2}(\sqrt{2})}_{u^2} \underbrace{\begin{pmatrix} -\frac{\sqrt{2}}{2} \\ \frac{\sqrt{2}}{2} \end{pmatrix}}_{\mathbf{e}_\theta}. \tag{1.374}$$

In Gibbs notation, we could interpret this as $1\mathbf{i} + 2\mathbf{j} = (3/\sqrt{2})\mathbf{e}_r + (\sqrt{2}/2)\mathbf{e}_\theta$.

The covariant representation is given once again by $u_j = g_{ji}u^i$:

$$\begin{pmatrix} u_1 \\ u_2 \end{pmatrix} = \mathbf{G} \cdot \begin{pmatrix} u^1 \\ u^2 \end{pmatrix} = \begin{pmatrix} 1 & 0 \\ 0 & 2 \end{pmatrix} \begin{pmatrix} \frac{3}{\sqrt{2}} \\ \frac{1}{2} \end{pmatrix} = \begin{pmatrix} \frac{3}{\sqrt{2}} \\ 1 \end{pmatrix}. \tag{1.375}$$

This less simple vector field has distinct contravariant and covariant representations. However, appropriate invariant quantities, such as the square of the magnitude of the vector, are independent of the representation:

$$U_i U^i = (1 \quad 2) \begin{pmatrix} 1 \\ 2 \end{pmatrix} = 5, \tag{1.376}$$

$$u_i u^i = (\tfrac{3}{\sqrt{2}} \quad 1) \begin{pmatrix} \frac{3}{\sqrt{2}} \\ \frac{1}{2} \end{pmatrix} = 5. \tag{1.377}$$

The vector fields for both cases, along with cutaway expanded views of the vectors at P and their decompositions, are plotted in Figure 1.15. Figure 1.15b gives a clear depiction of how the vector at P has alias representations as linear combinations of vectors that are natural bases for either (1) Cartesian or (2) polar coordinate systems.

1.6.3 Differentiation and Christoffel Symbols

The idea of covariant and contravariant derivatives plays an important role in engineering disciplines such as continuum mechanics, namely, in that the equations should be formulated such that they are invariant under coordinate transformations. This is not particularly difficult for Cartesian systems, but for nonorthogonal systems, one cannot use differentiation in the ordinary sense but must instead use the notion of covariant and contravariant derivatives, depending on the problem. The role of these terms was especially important in the development of the theory of deformable media in continuum mechanics.

Thus, we next consider how to differentiate vector fields in non-Cartesian coordinates. Consider a contravariant vector u^i defined in x^i which has corresponding components U^i in the Cartesian ξ^i. Take w^i_j and W^i_j to represent the covariant spatial derivatives of u^i and U^i, respectively. Let us use the chain rule and definitions of tensorial quantities to arrive at a formula for covariant differentiation. From the definition of contravariance, Eq. (1.333),

$$U^i = \frac{\partial \xi^i}{\partial x^l} u^l. \tag{1.378}$$

Take the derivative in Cartesian space and then use the chain rule:

$$W^i_j = \frac{\partial U^i}{\partial \xi^j} = \frac{\partial U^i}{\partial x^k}\frac{\partial x^k}{\partial \xi^j}, \tag{1.379}$$

$$= \left(\frac{\partial}{\partial x^k} \underbrace{\left(\frac{\partial \xi^i}{\partial x^l} u^l \right)}_{=U^i} \right) \frac{\partial x^k}{\partial \xi^j}, \tag{1.380}$$

$$= \left(\frac{\partial^2 \xi^i}{\partial x^k \partial x^l} u^l + \frac{\partial \xi^i}{\partial x^l}\frac{\partial u^l}{\partial x^k} \right) \frac{\partial x^k}{\partial \xi^j}, \tag{1.381}$$

$$W^p_q = \left(\frac{\partial^2 \xi^p}{\partial x^k \partial x^l} u^l + \frac{\partial \xi^p}{\partial x^l}\frac{\partial u^l}{\partial x^k} \right) \frac{\partial x^k}{\partial \xi^q}. \tag{1.382}$$

From the definition of a mixed tensor, Eq. (1.354),

$$w^i_j = W^p_q \frac{\partial x^i}{\partial \xi^p}\frac{\partial \xi^q}{\partial x^j}, \tag{1.383}$$

$$= \underbrace{\left(\frac{\partial^2 \xi^p}{\partial x^k \partial x^l} u^l + \frac{\partial \xi^p}{\partial x^l}\frac{\partial u^l}{\partial x^k} \right) \frac{\partial x^k}{\partial \xi^q}}_{=W^p_q} \frac{\partial x^i}{\partial \xi^p}\frac{\partial \xi^q}{\partial x^j}, \tag{1.384}$$

$$= \frac{\partial^2 \xi^p}{\partial x^k \partial x^l}\frac{\partial x^k}{\partial \xi^q}\frac{\partial x^i}{\partial \xi^p}\frac{\partial \xi^q}{\partial x^j} u^l + \frac{\partial \xi^p}{\partial x^l}\frac{\partial x^k}{\partial \xi^q}\frac{\partial x^i}{\partial \xi^p}\frac{\partial \xi^q}{\partial x^j}\frac{\partial u^l}{\partial x^k}, \tag{1.385}$$

$$= \frac{\partial^2 \xi^p}{\partial x^k \partial x^l}\underbrace{\frac{\partial x^k}{\partial x^j}}_{\delta^k_j}\frac{\partial x^i}{\partial \xi^p} u^l + \underbrace{\frac{\partial x^i}{\partial x^l}}_{\delta^i_l}\underbrace{\frac{\partial x^k}{\partial x^j}}_{\delta^k_j}\frac{\partial u^l}{\partial x^k}, \tag{1.386}$$

$$= \frac{\partial^2 \xi^p}{\partial x^k \partial x^l}\delta^k_j\frac{\partial x^i}{\partial \xi^p} u^l + \delta^i_l\delta^k_j\frac{\partial u^l}{\partial x^k}, \tag{1.387}$$

$$= \frac{\partial^2 \xi^p}{\partial x^j \partial x^l}\frac{\partial x^i}{\partial \xi^p} u^l + \frac{\partial u^i}{\partial x^j}. \tag{1.388}$$

Here we have used the identity that

$$\frac{\partial x^i}{\partial x^j} = \delta^i_j, \tag{1.389}$$

where δ^i_j is another form of the Kronecker delta. We define the *Christoffel*[30] *symbols* Γ^i_{jl} as follows:

$$\Gamma^i_{jl} = \frac{\partial^2 \xi^p}{\partial x^j \partial x^l}\frac{\partial x^i}{\partial \xi^p}, \tag{1.390}$$

and we use the symbol Δ_j to represent the covariant derivative. Thus, the covariant derivative of a contravariant vector u^i is as follows:

$$\Delta_j u^i = w^i_j = \frac{\partial u^i}{\partial x^j} + \Gamma^i_{jl} u^l. \tag{1.391}$$

The critical feature, which can be shown not to be the case for Cartesian systems, is that the covariant derivative of u^i contains both differentiated and undifferentiated components of u; moreover, the variation of the coordinate axes is embodied within the Christoffel symbols.

EXAMPLE 1.22

Find $\nabla^T \cdot \mathbf{u}$ in cylindrical coordinates. The orientation-preserving transformations are

$$x^1 = +\sqrt{(\xi^1)^2 + (\xi^2)^2}, \tag{1.392}$$

$$x^2 = \tan^{-1}\left(\frac{\xi^2}{\xi^1}\right), \tag{1.393}$$

$$x^3 = \xi^3. \tag{1.394}$$

Here we restrict attention to positive x^1. The inverse transformation is

$$\xi^1 = x^1 \cos x^2, \tag{1.395}$$

$$\xi^2 = x^1 \sin x^2, \tag{1.396}$$

$$\xi^3 = x^3. \tag{1.397}$$

Finding $\nabla^T \cdot \mathbf{u}$ corresponds to finding

$$\Delta_i u^i = w^i_i = \frac{\partial u^i}{\partial x^i} + \Gamma^i_{il} u^l. \tag{1.398}$$

Now, for $i = j$,

$$\Gamma^i_{il} u^l = \frac{\partial^2 \xi^p}{\partial x^i \partial x^l}\frac{\partial x^i}{\partial \xi^p} u^l, \tag{1.399}$$

$$= \frac{\partial^2 \xi^1}{\partial x^i \partial x^l}\frac{\partial x^i}{\partial \xi^1} u^l + \frac{\partial^2 \xi^2}{\partial x^i \partial x^l}\frac{\partial x^i}{\partial \xi^2} u^l + \underbrace{\frac{\partial^2 \xi^3}{\partial x^i \partial x^l}}_{=0}\frac{\partial x^i}{\partial \xi^3} u^l. \tag{1.400}$$

---

[30] Elwin Bruno Christoffel, 1829–1900, German mathematician.

Noting that all second partials of ξ^3 are zero,

$$\Gamma^i_{il} u^l = \frac{\partial^2 \xi^1}{\partial x^i \partial x^l} \frac{\partial x^i}{\partial \xi^1} u^l + \frac{\partial^2 \xi^2}{\partial x^i \partial x^l} \frac{\partial x^i}{\partial \xi^2} u^l. \tag{1.401}$$

Expanding the i summation,

$$\Gamma^i_{il} u^l = \frac{\partial^2 \xi^1}{\partial x^1 \partial x^l} \frac{\partial x^1}{\partial \xi^1} u^l + \frac{\partial^2 \xi^1}{\partial x^2 \partial x^l} \frac{\partial x^2}{\partial \xi^1} u^l + \underbrace{\frac{\partial^2 \xi^1}{\partial x^3 \partial x^l} \frac{\partial x^3}{\partial \xi^1}}_{=0} u^l$$

$$+ \frac{\partial^2 \xi^2}{\partial x^1 \partial x^l} \frac{\partial x^1}{\partial \xi^2} u^l + \frac{\partial^2 \xi^2}{\partial x^2 \partial x^l} \frac{\partial x^2}{\partial \xi^2} u^l + \underbrace{\frac{\partial^2 \xi^2}{\partial x^3 \partial x^l} \frac{\partial x^3}{\partial \xi^2}}_{=0} u^l. \tag{1.402}$$

Noting that partials of x^3 with respect to ξ^1 and ξ^2 are zero,

$$\Gamma^i_{il} u^l = \frac{\partial^2 \xi^1}{\partial x^1 \partial x^l} \frac{\partial x^1}{\partial \xi^1} u^l + \frac{\partial^2 \xi^1}{\partial x^2 \partial x^l} \frac{\partial x^2}{\partial \xi^1} u^l + \frac{\partial^2 \xi^2}{\partial x^1 \partial x^l} \frac{\partial x^1}{\partial \xi^2} u^l + \frac{\partial^2 \xi^2}{\partial x^2 \partial x^l} \frac{\partial x^2}{\partial \xi^2} u^l. \tag{1.403}$$

Expanding the l summation, we get

$$\Gamma^i_{il} u^l = \frac{\partial^2 \xi^1}{\partial x^1 \partial x^1} \frac{\partial x^1}{\partial \xi^1} u^1 + \frac{\partial^2 \xi^1}{\partial x^1 \partial x^2} \frac{\partial x^1}{\partial \xi^1} u^2 + \underbrace{\frac{\partial^2 \xi^1}{\partial x^1 \partial x^3} \frac{\partial x^1}{\partial \xi^1}}_{=0} u^3$$

$$+ \frac{\partial^2 \xi^1}{\partial x^2 \partial x^1} \frac{\partial x^2}{\partial \xi^1} u^1 + \frac{\partial^2 \xi^1}{\partial x^2 \partial x^2} \frac{\partial x^2}{\partial \xi^1} u^2 + \underbrace{\frac{\partial^2 \xi^1}{\partial x^2 \partial x^3} \frac{\partial x^2}{\partial \xi^1}}_{=0} u^3$$

$$+ \frac{\partial^2 \xi^2}{\partial x^1 \partial x^1} \frac{\partial x^1}{\partial \xi^2} u^1 + \frac{\partial^2 \xi^2}{\partial x^1 \partial x^2} \frac{\partial x^1}{\partial \xi^2} u^2 + \underbrace{\frac{\partial^2 \xi^2}{\partial x^1 \partial x^3} \frac{\partial x^1}{\partial \xi^2}}_{=0} u^3$$

$$+ \frac{\partial^2 \xi^2}{\partial x^2 \partial x^1} \frac{\partial x^2}{\partial \xi^2} u^1 + \frac{\partial^2 \xi^2}{\partial x^2 \partial x^2} \frac{\partial x^2}{\partial \xi^2} u^2 + \underbrace{\frac{\partial^2 \xi^2}{\partial x^2 \partial x^3} \frac{\partial x^2}{\partial \xi^2}}_{=0} u^3. \tag{1.404}$$

Again removing the x^3 variation, we get

$$\Gamma^i_{il} u^l = \frac{\partial^2 \xi^1}{\partial x^1 \partial x^1} \frac{\partial x^1}{\partial \xi^1} u^1 + \frac{\partial^2 \xi^1}{\partial x^1 \partial x^2} \frac{\partial x^1}{\partial \xi^1} u^2 + \frac{\partial^2 \xi^1}{\partial x^2 \partial x^1} \frac{\partial x^2}{\partial \xi^1} u^1 + \frac{\partial^2 \xi^1}{\partial x^2 \partial x^2} \frac{\partial x^2}{\partial \xi^1} u^2$$

$$+ \frac{\partial^2 \xi^2}{\partial x^1 \partial x^1} \frac{\partial x^1}{\partial \xi^2} u^1 + \frac{\partial^2 \xi^2}{\partial x^1 \partial x^2} \frac{\partial x^1}{\partial \xi^2} u^2 + \frac{\partial^2 \xi^2}{\partial x^2 \partial x^1} \frac{\partial x^2}{\partial \xi^2} u^1 + \frac{\partial^2 \xi^2}{\partial x^2 \partial x^2} \frac{\partial x^2}{\partial \xi^2} u^2. \tag{1.405}$$

Substituting for the partial derivatives, we find

$$\Gamma^i_{il} u^l = 0 u^1 - \sin x^2 \cos x^2 u^2 - \sin x^2 \left(\frac{-\sin x^2}{x^1} \right) u^1 - x^1 \cos x^2 \left(\frac{-\sin x^2}{x^1} \right) u^2$$

$$+ 0 u^1 + \cos x^2 \sin x^2 u^2 + \cos x^2 \left(\frac{\cos x^2}{x^1} \right) u^1 - x^1 \sin x^2 \left(\frac{\cos x^2}{x^1} \right) u^2, \tag{1.406}$$

$$= \frac{u^1}{x^1}. \tag{1.407}$$

So, in cylindrical coordinates,

$$\nabla^T \cdot \mathbf{u} = \frac{\partial u^1}{\partial x^1} + \frac{\partial u^2}{\partial x^2} + \frac{\partial u^3}{\partial x^3} + \frac{u^1}{x^1}. \tag{1.408}$$

Note that in standard cylindrical notation, $x^1 = r, x^2 = \theta, x^3 = z$. Considering $\mathbf{u}$ to be a velocity vector, we get

$$\nabla^T \cdot \mathbf{u} = \frac{\partial}{\partial r}\left(\frac{dr}{dt}\right) + \frac{\partial}{\partial \theta}\left(\frac{d\theta}{dt}\right) + \frac{\partial}{\partial z}\left(\frac{dz}{dt}\right) + \frac{1}{r}\left(\frac{dr}{dt}\right), \tag{1.409}$$

$$= \frac{1}{r}\frac{\partial}{\partial r}\left(r\frac{dr}{dt}\right) + \frac{1}{r}\frac{\partial}{\partial \theta}\left(r\frac{d\theta}{dt}\right) + \frac{\partial}{\partial z}\left(\frac{dz}{dt}\right), \tag{1.410}$$

$$= \frac{1}{r}\frac{\partial}{\partial r}\left(ru_r\right) + \frac{1}{r}\frac{\partial u_\theta}{\partial \theta} + \frac{\partial u_z}{\partial z}. \tag{1.411}$$

Here we have also used the more traditional $u_\theta = r(d\theta/dt) = x^1u^2$, along with $u_r = u^1, u_z = u^3$. For practical purposes, this ensures that u_r, u_θ, u_z all have the same dimensions.

EXAMPLE 1.23

Calculate the acceleration vector $d\mathbf{u}/dt$ in cylindrical coordinates.

Here we presume that $\mathbf{u} = \mathbf{u}(\mathbf{x}, t)$. Thus,

$$d\mathbf{u} = \frac{\partial \mathbf{u}}{\partial t}\,dt + (d\mathbf{x}^T \cdot \nabla)\mathbf{u}. \tag{1.412}$$

Scaling by dt and recognizing the velocity is given by $\mathbf{u} = d\mathbf{x}/dt$, we get

$$\frac{d\mathbf{u}}{dt} = \frac{\partial \mathbf{u}}{\partial t} + (\mathbf{u}^T \cdot \nabla)\mathbf{u}. \tag{1.413}$$

Now, we take $\mathbf{u}$ to be a contravariant velocity vector and the gradient operation to be a covariant derivative. Employ index notation to get

$$\frac{d\mathbf{u}}{dt} = \frac{\partial u^i}{\partial t} + u^j \Delta_j u^i, \tag{1.414}$$

$$= \frac{\partial u^i}{\partial t} + u^j\left(\frac{\partial u^i}{\partial x^j} + \Gamma^i_{jl}u^l\right). \tag{1.415}$$

After an extended calculation similar to the previous example, one finds after expanding all terms that

$$\frac{d\mathbf{u}}{dt} = \begin{pmatrix} \dfrac{\partial u^1}{\partial t} \\[2mm] \dfrac{\partial u^2}{\partial t} \\[2mm] \dfrac{\partial u^3}{\partial t} \end{pmatrix} + \begin{pmatrix} u^1\dfrac{\partial u^1}{\partial x^1} + u^2\dfrac{\partial u^1}{\partial x^2} + u^3\dfrac{\partial u^1}{\partial x^3} \\[2mm] u^1\dfrac{\partial u^2}{\partial x^1} + u^2\dfrac{\partial u^2}{\partial x^2} + u^3\dfrac{\partial u^2}{\partial x^3} \\[2mm] u^1\dfrac{\partial u^3}{\partial x^1} + u^2\dfrac{\partial u^3}{\partial x^2} + u^3\dfrac{\partial u^3}{\partial x^3} \end{pmatrix} + \begin{pmatrix} -x^1\left(u^2\right)^2 \\[2mm] 2\dfrac{u^1u^2}{x^1} \\[2mm] 0 \end{pmatrix}. \tag{1.416}$$

The last column matrix is related to the well-known Coriolis[31] and centripetal acceleration terms. However, these are not in the standard form to which most are accustomed. To arrive at that standard form, one must return to a so-called physical representation. Here again take $x^1 = r$, $x^2 = \theta$, and $x^3 = z$. Also, take $u_r = dr/dt = u^1$, $u_\theta = r(d\theta/dt) = x^1u^2$, $u_z = dz/dt = u^3$. Then the r acceleration equation becomes

$$\frac{du_r}{dt} = \frac{\partial u_r}{\partial t} + u_r\frac{\partial u_r}{\partial r} + \frac{u_\theta}{r}\frac{\partial u_r}{\partial \theta} + u_z\frac{\partial u_r}{\partial z} - \underbrace{\frac{u_\theta^2}{r}}_{\text{centripetal}}. \tag{1.417}$$

---

[31] Gaspard-Gustave Coriolis, 1792–1843, French mechanician.

Here the final term is the traditional centripetal acceleration. The θ acceleration is slightly more complicated. First one writes

$$\frac{d}{dt}\left(\frac{d\theta}{dt}\right) = \frac{\partial}{\partial t}\left(\frac{d\theta}{dt}\right) + \frac{dr}{dt}\frac{\partial}{\partial r}\left(\frac{d\theta}{dt}\right) + \frac{d\theta}{dt}\frac{\partial}{\partial\theta}\left(\frac{d\theta}{dt}\right) + \frac{dz}{dt}\frac{\partial}{\partial z}\left(\frac{d\theta}{dt}\right) + \frac{2}{r}\frac{dr}{dt}\frac{d\theta}{dt}. \tag{1.418}$$

Now, here one is actually interested in du_θ/dt, so both sides are multiplied by r, and then one operates to get

$$\frac{du_\theta}{dt} = r\frac{\partial}{\partial t}\left(\frac{d\theta}{dt}\right) + r\frac{dr}{dt}\frac{\partial}{\partial r}\left(\frac{d\theta}{dt}\right) + r\frac{d\theta}{dt}\frac{\partial}{\partial\theta}\left(\frac{d\theta}{dt}\right) + r\frac{dz}{dt}\frac{\partial}{\partial z}\left(\frac{d\theta}{dt}\right) + 2\frac{dr}{dt}\frac{d\theta}{dt}, \tag{1.419}$$

$$= \frac{\partial}{\partial t}\left(r\frac{d\theta}{dt}\right) + \frac{dr}{dt}\left(\frac{\partial}{\partial r}\left(r\frac{d\theta}{dt}\right) - \frac{d\theta}{dt}\right) + \frac{r\frac{d\theta}{dt}}{r}\frac{\partial}{\partial\theta}\left(r\frac{d\theta}{dt}\right) + \frac{dz}{dt}\frac{\partial}{\partial z}\left(r\frac{d\theta}{dt}\right) + 2\frac{dr}{dt}\frac{\left(r\frac{d\theta}{dt}\right)}{r}, \tag{1.420}$$

$$= \frac{\partial u_\theta}{\partial t} + u_r\frac{\partial u_\theta}{\partial r} + \frac{u_\theta}{r}\frac{\partial u_\theta}{\partial\theta} + u_z\frac{\partial u_\theta}{\partial z} + \underbrace{\frac{u_r u_\theta}{r}}_{\text{Coriolis}}. \tag{1.421}$$

The final term here is the Coriolis acceleration. The z acceleration then is easily seen to be

$$\frac{du_z}{dt} = \frac{\partial u_z}{\partial t} + u_r\frac{\partial u_z}{\partial r} + \frac{u_\theta}{r}\frac{\partial u_z}{\partial\theta} + u_z\frac{\partial u_z}{\partial z}. \tag{1.422}$$

1.6.4 Summary of Identities

We summarize some useful identities, all of which can be proved, as well as some other common notation, as follows:

$$g_{kl} = \frac{\partial\xi^i}{\partial x^k}\frac{\partial\xi^i}{\partial x^l}, \tag{1.423}$$

$$g = \det g_{ij}, \tag{1.424}$$

$$g_{ik}g^{kj} = g_i^j = g_j^i = \delta_i^j = \delta_j^i = \delta_{ij} = \delta^{ij}, \tag{1.425}$$

$$u_j = u^i g_{ij}, \tag{1.426}$$

$$u^i = g^{ij}u_j, \tag{1.427}$$

$$\mathbf{u}^T \cdot \mathbf{v} = u_i v^i = u^i v_i = u^i g_{ij} v^j = u_i g^{ij} v_j, \tag{1.428}$$

$$\mathbf{u} \times \mathbf{v} = \epsilon^{ijk} g_{jm} g_{kn} u^m v^n = \epsilon^{ijk} u_j v_k, \tag{1.429}$$

$$\Gamma^i_{jk} = \frac{\partial^2\xi^p}{\partial x^j \partial x^k}\frac{\partial x^i}{\partial\xi^p} = \frac{1}{2}g^{ip}\left(\frac{\partial g_{pj}}{\partial x^k} + \frac{\partial g_{pk}}{\partial x^j} - \frac{\partial g_{jk}}{\partial x^p}\right), \tag{1.430}$$

$$\nabla\mathbf{u} = \Delta_j u^i = u^i_{,j} = \frac{\partial u^i}{\partial x^j} + \Gamma^i_{jl}u^l, \tag{1.431}$$

$$\text{div}\,\mathbf{u} = \nabla^T \cdot \mathbf{u} = \Delta_i u^i = u^i_{,i} = \frac{\partial u^i}{\partial x^i} + \Gamma^i_{il}u^l = \frac{1}{\sqrt{g}}\frac{\partial}{\partial x^i}\left(\sqrt{g}\,u^i\right), \tag{1.432}$$

$$\text{curl}\,\mathbf{u} = \nabla \times \mathbf{u} = \epsilon^{ijk}u_{k,j} = \epsilon^{ijk}g_{kp}u^p_{,j} = \epsilon^{ijk}g_{kp}\left(\frac{\partial u^p}{\partial x^j} + \Gamma^p_{jl}u^l\right), \tag{1.433}$$

$$\frac{d\mathbf{u}}{dt} = \frac{\partial \mathbf{u}}{\partial t} + \mathbf{u}^T \cdot \nabla \mathbf{u} = \frac{\partial u^i}{\partial t} + u^j \frac{\partial u^i}{\partial x^j} + \Gamma^i_{jl} u^l u^j, \tag{1.434}$$

$$\operatorname{grad}\phi = \nabla\phi = \phi_{,i} = \frac{\partial \phi}{\partial x^i}, \tag{1.435}$$

$$\operatorname{div}\operatorname{grad}\phi = \nabla^2\phi = \nabla^T \cdot \nabla\phi = g^{ij}\phi_{,ij} = \frac{\partial}{\partial x^j}\left(g^{ij}\frac{\partial \phi}{\partial x^i}\right) + \Gamma^j_{jk} g^{ik}\frac{\partial \phi}{\partial x^i}, \tag{1.436}$$

$$= \frac{1}{\sqrt{g}}\frac{\partial}{\partial x^j}\left(\sqrt{g}\, g^{ij}\frac{\partial \phi}{\partial x^i}\right), \tag{1.437}$$

$$\nabla \mathbf{T} = T^{ij}_{,k} = \frac{\partial T^{ij}}{\partial x^k} + \Gamma^i_{lk} T^{lj} + \Gamma^j_{lk} T^{il}, \tag{1.438}$$

$$\operatorname{div}\mathbf{T} = \nabla^T \cdot \mathbf{T} = T^{ij}_{,j} = \frac{\partial T^{ij}}{\partial x^j} + \Gamma^i_{lj} T^{lj} + \Gamma^j_{lj} T^{il}, \tag{1.439}$$

$$= \frac{1}{\sqrt{g}}\frac{\partial}{\partial x^j}\left(\sqrt{g}\, T^{ij}\right) + \Gamma^i_{jk} T^{jk} = \frac{1}{\sqrt{g}}\frac{\partial}{\partial x^j}\left(\sqrt{g}\, T^{kj}\frac{\partial \xi^i}{\partial x^k}\right). \tag{1.440}$$

1.6.5 Nonorthogonal Coordinates: Alternate Approach

Some of the information in previous sections regarding non-Cartesian systems can be better visualized with an alternate approach focusing on nonorthogonal basis vectors. The only requirement we will place on the basis vectors is linear independence: they must point in different directions. They need not be unit vectors, and their lengths may differ from one another. Consider the nonorthogonal basis vectors $\mathbf{e}_1, \mathbf{e}_2$, aligned with the x^1 and x^2 directions shown in Figure 1.16a. The vector $\mathbf{u}$ can then be written as

$$\mathbf{u} = u^1 \mathbf{e}_1 + u^2 \mathbf{e}_2. \tag{1.441}$$

Here u^1 and u^2 are the contravariant components of $\mathbf{u}$, and the vectors $\mathbf{e}_1$ and $\mathbf{e}_2$ are the contravariant basis vectors, even though they are subscripted. The entity $\mathbf{u}$ is best thought of as either an entity unto itself or perhaps as a column vector whose

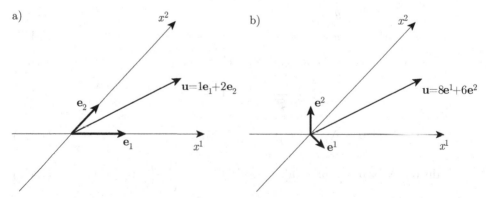

Figure 1.16. Example of a vector $\mathbf{u}$ represented in a nonorthogonal coordinate system along with its (a) basis vectors and (b) dual basis vectors.

components are Cartesian. In matrix form, we can think of $\mathbf{u}$ as

$$\mathbf{u} = u^1 \begin{pmatrix} \vdots \\ \mathbf{e}_1 \\ \vdots \end{pmatrix} + u^2 \begin{pmatrix} \vdots \\ \mathbf{e}_2 \\ \vdots \end{pmatrix} = \underbrace{\begin{pmatrix} \vdots & \vdots \\ \mathbf{e}_1 & \mathbf{e}_2 \\ \vdots & \vdots \end{pmatrix}}_{\mathbf{J}} \begin{pmatrix} u^1 \\ u^2 \end{pmatrix}. \tag{1.442}$$

The matrix of basis vectors really acts as a local Jacobian matrix, $\mathbf{J}$, which relates the Cartesian and nonorthogonal representations of $\mathbf{u}$. Although one might be tempted to think of Eq. (1.442) as an alibi mapping from non-Cartesian to Cartesian coordinates, we instead consider it as part of an alias mapping in which $\mathbf{u}$ is represented in the same plane in different fashions.

Vectors that compose a *dual* or *reciprocal* basis have two characteristics: they are orthogonal to all the original basis vectors with different indices, and the dot product of each dual vector with respect to the original vector of the same index must be unity. We expand on this topic in Section 6.5.4 in context of linear analysis. The covariant basis vectors $\mathbf{e}^1, \mathbf{e}^2$ are dual to $\mathbf{e}_1, \mathbf{e}_2$, as shown in Figure 1.16b. Specifically, we have $\mathbf{e}^{1^T} \cdot \mathbf{e}_2 = 0$, $\mathbf{e}^{2^T} \cdot \mathbf{e}_1 = 0$, $\mathbf{e}^{1^T} \cdot \mathbf{e}_1 = 1$, and $\mathbf{e}^{2^T} \cdot \mathbf{e}_2 = 1$. In matrix form, this is

$$\underbrace{\begin{pmatrix} \cdots & \mathbf{e}^{1^T} & \cdots \\ \cdots & \mathbf{e}^{2^T} & \cdots \end{pmatrix}}_{\mathbf{J}^{-1}} \underbrace{\begin{pmatrix} \vdots & \vdots \\ \mathbf{e}_1 & \mathbf{e}_2 \\ \vdots & \vdots \end{pmatrix}}_{\mathbf{J}} = \mathbf{I}. \tag{1.443}$$

Obviously, the matrix of dual vectors can be formed by inverting the matrix of the original basis vectors. We can also represent $\mathbf{u}$ as

$$\mathbf{u} = u_1 \mathbf{e}^1 + u_2 \mathbf{e}^2. \tag{1.444}$$

In matrix form, we can think of $\mathbf{u}$ as

$$\mathbf{u} = u_1 \begin{pmatrix} \vdots \\ \mathbf{e}^1 \\ \vdots \end{pmatrix} + u_2 \begin{pmatrix} \vdots \\ \mathbf{e}^2 \\ \vdots \end{pmatrix} = \underbrace{\begin{pmatrix} \vdots & \vdots \\ \mathbf{e}^1 & \mathbf{e}^2 \\ \vdots & \vdots \end{pmatrix}}_{\mathbf{J}^{-1^T}} \begin{pmatrix} u_1 \\ u_2 \end{pmatrix}. \tag{1.445}$$

We might also say

$$\mathbf{u}^T = \begin{pmatrix} u_1 & u_2 \end{pmatrix} \underbrace{\begin{pmatrix} \cdots & \mathbf{e}^{1^T} & \cdots \\ \cdots & \mathbf{e}^{2^T} & \cdots \end{pmatrix}}_{\mathbf{J}^{-1}}. \tag{1.446}$$

Because the magnitude of $\mathbf{u}$ is independent of its coordinate system, we can say

$$\mathbf{u}^T \cdot \mathbf{u} = \underbrace{\begin{pmatrix} u_1 & u_2 \end{pmatrix} \underbrace{\begin{pmatrix} \cdots & \mathbf{e}^{1^T} & \cdots \\ \cdots & \mathbf{e}^{2^T} & \cdots \end{pmatrix}}_{\mathbf{J}^{-1}} \underbrace{\begin{pmatrix} \vdots & \vdots \\ \mathbf{e}_1 & \mathbf{e}_2 \\ \vdots & \vdots \end{pmatrix}}_{\mathbf{J}} \begin{pmatrix} u^1 \\ u^2 \end{pmatrix}}_{\mathbf{I}}, \tag{1.447}$$

and thus

$$\mathbf{u}^T \cdot \mathbf{u} = (\, u_1 \quad u_2 \,) \begin{pmatrix} u^1 \\ u^2 \end{pmatrix} = u_i u^i. \tag{1.448}$$

Now we can also transpose Eq. (1.442) to obtain

$$\mathbf{u}^T = (\, u^1 \quad u^2 \,) \underbrace{\begin{pmatrix} \cdots & \mathbf{e}_1^T & \cdots \\ \cdots & \mathbf{e}_2^T & \cdots \end{pmatrix}}_{\mathbf{J}^T}. \tag{1.449}$$

Now combining this with Eq. (1.442) to form $\mathbf{u}^T \cdot \mathbf{u}$, we also see

$$\mathbf{u}^T \cdot \mathbf{u} = (\, u^1 \quad u^2 \,) \underbrace{\begin{pmatrix} \cdots & \mathbf{e}_1^T & \cdots \\ \cdots & \mathbf{e}_2^T & \cdots \end{pmatrix}}_{\mathbf{J}^T} \underbrace{\begin{pmatrix} \vdots & \vdots \\ \mathbf{e}_1 & \mathbf{e}_2 \\ \vdots & \vdots \end{pmatrix}}_{\mathbf{J}} \begin{pmatrix} u^1 \\ u^2 \end{pmatrix}. \tag{1.450}$$

Now with the metric tensor $g_{ij} = \mathbf{G} = \mathbf{J}^T \cdot \mathbf{J}$, we have

$$\mathbf{u}^T \cdot \mathbf{u} = (\, u^1 \quad u^2 \,) \cdot \mathbf{G} \cdot \begin{pmatrix} u^1 \\ u^2 \end{pmatrix}. \tag{1.451}$$

One can now compare Eq. (1.451) with Eq. (1.448) to infer the covariant components u_1 and u_2. For the same vector $\mathbf{u}$, the covariant components are different than the contravariant components. Thus, for example,

$$u_i = (\, u^1 \quad u^2 \,) \cdot \mathbf{G} = u^j g_{ij} = g_{ij} u^j. \tag{1.452}$$

Remembering from Eq. (1.425) that $g^{ij} = \mathbf{G}^{-1}$, we also see

$$g^{ij} u_i = u^j. \tag{1.453}$$

In Cartesian coordinates, a basis and its dual are the same, and so also are the contravariant and covariant components of a vector. For this reason, Cartesian vectors and tensors are usually written with only subscripts.

EXAMPLE 1.24

Consider the vector $\mathbf{u}$ whose Cartesian representation is

$$\mathbf{u} = \begin{pmatrix} 4 \\ 2 \end{pmatrix}. \tag{1.454}$$

Consider also the set of nonorthogonal basis vectors

$$\mathbf{e}_1 = \begin{pmatrix} 2 \\ 0 \end{pmatrix}, \qquad \mathbf{e}_2 = \begin{pmatrix} 1 \\ 1 \end{pmatrix}. \tag{1.455}$$

Find the contravariant and covariant components, u^i and u_i, of $\mathbf{u}$.

Here the Jacobian matrix is

$$\mathbf{J} = \begin{pmatrix} \vdots & \vdots \\ \mathbf{e}_1 & \mathbf{e}_2 \\ \vdots & \vdots \end{pmatrix} = \begin{pmatrix} 2 & 1 \\ 0 & 1 \end{pmatrix}. \tag{1.456}$$

The contravariant components u^i are given by solving

$$\mathbf{u} = \mathbf{J} \cdot \begin{pmatrix} u^1 \\ u^2 \end{pmatrix}, \tag{1.457}$$

$$\begin{pmatrix} 4 \\ 2 \end{pmatrix} = \begin{pmatrix} 2 & 1 \\ 0 & 1 \end{pmatrix} \begin{pmatrix} u^1 \\ u^2 \end{pmatrix}. \tag{1.458}$$

Inverting, we find

$$\begin{pmatrix} u^1 \\ u^2 \end{pmatrix} = \begin{pmatrix} 1 \\ 2 \end{pmatrix}. \tag{1.459}$$

Thus, we have

$$\mathbf{u} = 1\mathbf{e}_1 + 2\mathbf{e}_2, \tag{1.460}$$

$$\begin{pmatrix} 4 \\ 2 \end{pmatrix} = 1 \begin{pmatrix} 2 \\ 0 \end{pmatrix} + 2 \begin{pmatrix} 1 \\ 1 \end{pmatrix}. \tag{1.461}$$

The covariant basis vectors are given by

$$\mathbf{J}^{-1} = \begin{pmatrix} \cdots & \mathbf{e}^{1\,T} & \cdots \\ \cdots & \mathbf{e}^{2\,T} & \cdots \end{pmatrix} = \begin{pmatrix} \frac{1}{2} & -\frac{1}{2} \\ 0 & 1 \end{pmatrix}. \tag{1.462}$$

Thus, we have

$$\mathbf{e}^1 = \begin{pmatrix} \frac{1}{2} \\ -\frac{1}{2} \end{pmatrix}, \qquad \mathbf{e}^2 = \begin{pmatrix} 0 \\ 1 \end{pmatrix}. \tag{1.463}$$

The metric tensor is given by

$$g_{ij} = \mathbf{G} = \mathbf{J}^T \cdot \mathbf{J} = \begin{pmatrix} 2 & 0 \\ 1 & 1 \end{pmatrix} \begin{pmatrix} 2 & 1 \\ 0 & 1 \end{pmatrix} = \begin{pmatrix} 4 & 2 \\ 2 & 2 \end{pmatrix}. \tag{1.464}$$

We can get the covariant components in many ways. Let us choose

$$u_i = g_{ij} u^j = \begin{pmatrix} 4 & 2 \\ 2 & 2 \end{pmatrix} \begin{pmatrix} 1 \\ 2 \end{pmatrix} = \begin{pmatrix} 8 \\ 6 \end{pmatrix}. \tag{1.465}$$

Thus, we have the covariant representation of $\mathbf{u}$ as

$$\mathbf{u} = 8\mathbf{e}^1 + 6\mathbf{e}^2, \tag{1.466}$$

$$\begin{pmatrix} 4 \\ 2 \end{pmatrix} = 8 \begin{pmatrix} \frac{1}{2} \\ -\frac{1}{2} \end{pmatrix} + 6 \begin{pmatrix} 0 \\ 1 \end{pmatrix}. \tag{1.467}$$

In Cartesian coordinates, we have

$$\mathbf{u}^T \cdot \mathbf{u} = \begin{pmatrix} 4 & 2 \end{pmatrix} \begin{pmatrix} 4 \\ 2 \end{pmatrix} = 20. \tag{1.468}$$

This is invariant under coordinate transformation, as in our nonorthogonal coordinate system, we have

$$u_i u^i = \begin{pmatrix} 8 & 6 \end{pmatrix} \begin{pmatrix} 1 \\ 2 \end{pmatrix} = 20. \tag{1.469}$$

The vectors represented in Figure 1.16 are proportional to those of this problem.

1.6.6 Orthogonal Curvilinear Coordinates

In this section we specialize our discussion to widely used orthogonal curvilinear coordinate transformations. Such transformations admit nonconstant diagonal metric tensors. Because of the diagonal nature of the metric tensor, many simplifications arise. For such systems, subscripts alone suffice. Here we simply summarize the results.

For an orthogonal curvilinear coordinate system (q_1, q_2, q_3), we have

$$ds^2 = (h_1 \, dq_1)^2 + (h_2 \, dq_2)^2 + (h_3 \, dq_3)^2, \tag{1.470}$$

where the *scale factors* h_i are

$$h_i = \sqrt{\left(\frac{\partial x_1}{\partial q_i}\right)^2 + \left(\frac{\partial x_2}{\partial q_i}\right)^2 + \left(\frac{\partial x_3}{\partial q_i}\right)^2}. \tag{1.471}$$

We can show that

$$\text{grad } \phi = \nabla \phi = \frac{1}{h_1}\frac{\partial \phi}{\partial q_1}\mathbf{e}_1 + \frac{1}{h_2}\frac{\partial \phi}{\partial q_2}\mathbf{e}_2 + \frac{1}{h_3}\frac{\partial \phi}{\partial q_3}\mathbf{e}_3, \tag{1.472}$$

$$\text{div } \mathbf{u} = \nabla^T \cdot \mathbf{u} = \frac{1}{h_1 h_2 h_3}\left(\frac{\partial}{\partial q_1}(u_1 h_2 h_3) + \frac{\partial}{\partial q_2}(u_2 h_3 h_1) + \frac{\partial}{\partial q_3}(u_3 h_1 h_2)\right), \tag{1.473}$$

$$\text{curl } \mathbf{u} = \nabla \times \mathbf{u} = \frac{1}{h_1 h_2 h_3}\begin{vmatrix} h_1 \mathbf{e}_1 & h_2 \mathbf{e}_2 & h_3 \mathbf{e}_3, \\ \frac{\partial}{\partial q_1} & \frac{\partial}{\partial q_2} & \frac{\partial}{\partial q_3} \\ u_1 h_1 & u_2 h_2 & u_3 h_3 \end{vmatrix}, \tag{1.474}$$

$$\text{div grad } \phi = \nabla^2 \phi =$$

$$\frac{1}{h_1 h_2 h_3}\left(\frac{\partial}{\partial q_1}\left(\frac{h_2 h_3}{h_1}\frac{\partial \phi}{\partial q_1}\right) + \frac{\partial}{\partial q_2}\left(\frac{h_3 h_1}{h_2}\frac{\partial \phi}{\partial q_2}\right) + \frac{\partial}{\partial q_3}\left(\frac{h_1 h_2}{h_3}\frac{\partial \phi}{\partial q_3}\right)\right). \tag{1.475}$$

Here $\mathbf{e}_i$ is the orthonormal vector associated with the i coordinate.

EXAMPLE 1.25

Find expressions for the gradient, divergence, and curl in cylindrical coordinates (r, θ, z) where

$$x_1 = r \cos \theta, \tag{1.476}$$

$$x_2 = r \sin \theta, \tag{1.477}$$

$$x_3 = z. \tag{1.478}$$

The 1, 2, and 3 directions are associated with r, θ, and z, respectively. From Eq. (1.471), the scale factors are

$$h_r = \sqrt{\left(\frac{\partial x_1}{\partial r}\right)^2 + \left(\frac{\partial x_2}{\partial r}\right)^2 + \left(\frac{\partial x_3}{\partial r}\right)^2}, \tag{1.479}$$

$$= \sqrt{\cos^2 \theta + \sin^2 \theta}, \tag{1.480}$$

$$= 1, \tag{1.481}$$

$$h_\theta = \sqrt{\left(\frac{\partial x_1}{\partial \theta}\right)^2 + \left(\frac{\partial x_2}{\partial \theta}\right)^2 + \left(\frac{\partial x_3}{\partial \theta}\right)^2}, \tag{1.482}$$

$$= \sqrt{r^2 \sin^2 \theta + r^2 \cos^2 \theta}, \tag{1.483}$$

$$= r, \tag{1.484}$$

$$h_z = \sqrt{\left(\frac{\partial x_1}{\partial z}\right)^2 + \left(\frac{\partial x_2}{\partial z}\right)^2 + \left(\frac{\partial x_3}{\partial z}\right)^2}, \tag{1.485}$$

$$= 1, \tag{1.486}$$

so that

$$\text{grad } \phi = \frac{\partial \phi}{\partial r}\mathbf{e}_r + \frac{1}{r}\frac{\partial \phi}{\partial \theta}\mathbf{e}_\theta + \frac{\partial \phi}{\partial z}\mathbf{e}_z, \tag{1.487}$$

$$\text{div } \mathbf{u} = \frac{1}{r}\left(\frac{\partial}{\partial r}(u_r r) + \frac{\partial}{\partial \theta}(u_\theta) + \frac{\partial}{\partial z}(u_z r)\right) = \frac{\partial u_r}{\partial r} + \frac{u_r}{r} + \frac{1}{r}\frac{\partial u_\theta}{\partial \theta} + \frac{\partial u_z}{\partial z}, \tag{1.488}$$

$$\text{curl } \mathbf{u} = \frac{1}{r}\begin{vmatrix} \mathbf{e}_r & r\mathbf{e}_\theta & \mathbf{e}_z \\ \frac{\partial}{\partial r} & \frac{\partial}{\partial \theta} & \frac{\partial}{\partial z} \\ u_r & u_\theta r & u_z \end{vmatrix}. \tag{1.489}$$

EXERCISES

1. Defining

$$u(x,y) = \frac{x^3 - y^3}{x^2 + y^2},$$

   except at $x = y = 0$, where $u = 0$, show that $\partial u/\partial x$ exists at $x = y = 0$ but is not continuous there.

2. Consider $f(x,y) = y^2 - x = 0$. Show that it does not satisfy all the requirements of the implicit function theorem at $x = 0$, $y = 0$. Indicate in a sketch why $y(x)$ is not unique in a neighborhood of this point.

3. If $z^3 + zx + x^4 y = 2y^3$, find a general expression for $\partial z/\partial x|_y$ and $\partial z/\partial y|_x$. Evaluate $\partial z/\partial x|_y$ and $\partial z/\partial y|_x$ at $(x,y) = (1,2)$, considering only real values of x, y, z. Give a computer-generated plot of the surface $z(x,y)$ for $x \in [-2,2]$, $y \in [-2,2]$, $z \in [-2,2]$.

4. A particle is constrained to a path that is defined by the function $s(x,y) = x^2 + y - 5 = 0$. The velocity component in the x-direction is $dx/dt = 2y$. What are the position and velocity components in the y-direction when $x = 4$?

5. Consider the function $y = f(x)$. Show that, if the inverse $x = g(y)$ exists, $f'(x)\, g'(y) = 1$.

6. Assuming that the mapping $u(x,y)$, $v(x,y)$ is invertible, find the Jacobian of the inverse in terms of that of the forward map.

7. If $u(x,y) = e^x \sin y$, $v(x,y) = e^x \cos y$, show that the map $(x,y) \mapsto (u,v)$ is invertible.[32]

8. Show that

$$u(x,y) = \frac{x+y}{x-y},$$

$$v(x,y) = \frac{xy}{(x-y)^2}$$

   are functionally dependent.

9. Show that $u(x,y) = f(x,y)$ and $v(x,y) = g(x,y)$ are functionally dependent if $g = g(f)$.

10. Prove the Leibniz rule using the chain rule.

---

[32] Here $\mapsto$ is to be interpreted as "maps to."

11. Find $d\phi/dx$ in two different ways, where

$$\phi = \int_{x^2}^{x^4} x\sqrt{y}\, dy.$$

12. Given

$$y(x, z) = \int_{x^2}^{x^3} f(s, z)\, ds,$$

find $\partial y/\partial x$ and $\partial y/\partial z$. If $f(s, z) = s^2 - z^2$, find the derivatives in two different ways.

13. Calculate $\partial f/\partial s$ and $\partial f/\partial y$ if

$$f(s, y) = \int_0^s (x^2 + y^2)\, dx.$$

14. Find the extremum of the functional

$$I = \int_0^1 (x^2 y'^2 + 40x^4 y)\, dx,$$

with $y(0) = 0$ and $y(1) = 1$. Plot $y(x)$ which renders I to be at an extremum.

15. Find the point on the plane $ax + by + cz = d$ that is nearest to the origin.

16. Choose $y(x)$ to extremize the integral

$$\int_0^1 y'^2\, dx,$$

subject to the end conditions $y(0) = 0$, $y(1) = 0$, as well as the constraint

$$\int_0^1 y\, dx = 1.$$

Plot this function $y(x)$.

17. Find the point on the curve of intersection of $z - xy = 10$ and $x + y + z = 1$ that is closest to the origin.

18. Find a function $y(x)$ with $y(0) = 1$, $y(1) = 0$ that extremizes the integral

$$I = \int_0^1 \frac{1}{y}\sqrt{1 + \left(\frac{dy}{dx}\right)^2}\, dx.$$

Plot $y(x)$ for this function.

19. Determine the curve $y(x)$, with $x \in [x_1, x_2]$, of total length L with endpoints $y(x_1) = y_1$ and $y(x_2) = y_2$ fixed, for which the area under the curve,

$$\int_{x_1}^{x_2} y\, dx,$$

is a maximum. Show that if $(x_1, y_1) = (0, 0)$; $(x_2, y_2) = (1, 1)$; $L = 3/2$, that the curve that maximizes the area and satisfies all constraints is the circle $(y + 0.254272)^2 + (x - 1.2453)^2 = (1.26920)^2$. Plot this curve, find its area, and verify that each constraint is satisfied. What function $y(x)$ minimizes the area and satisfies all constraints? Plot this curve, find its area, and verify that each constraint is satisfied.

20. Find the form of a quadrilateral with perimeter P such that its area is a maximum. What is this area?

21. Determine the shape of a parallelogram with a given area that has the least perimeter.

22. Consider the integral

$$I = \int_0^1 (y' - y + e^x)^2 \, dx.$$

What kind of extremum does this integral have (maximum or minimum)? What is $y(x)$ for this extremum? What does the solution of the Euler-Lagrange equation give if $y(0) = 0$ and $y(1) = -e$? Find the value of the extremum. Plot $y(x)$ for the extremum. If $y_0(x)$ is the solution of the Euler-Lagrange equation, compute I for $y_1(x) = y_0(x) + h(x)$, where you can take any $h(x)$ you like, but with $h(0) = h(1) = 0$.

23. Find the length of the shortest curve between two points with cylindrical coordinates $(r, \theta, z) = (a, 0, 0)$ and $(r, \theta, z) = (a, \Theta, Z)$ along the surface of the cylinder $r = a$.

24. For the standard transformation from Cartesian to polar coordinates, $x = r \cos \theta$, $y = r \sin \theta$,

(a) identify all points in the (x, y) plane for which curves of constant r are orthogonal to curves of constant θ

(b) identify all points in the (r, θ) plane for which curves of constant x are orthogonal to curves of constant y

25. Find the covariant derivative of the contravariant velocity vector in cylindrical coordinates.

26. Determine the expressions for the gradient, divergence, and curl in spherical coordinates (x^1, x^2, x^3) which are related to the Cartesian coordinates (ξ^1, ξ^2, ξ^3) by

$$\xi^1 = x^1 \sin x^2 \, \cos x^3,$$

$$\xi^2 = x^1 \sin x^2 \, \sin x^3,$$

$$\xi^3 = x^1 \cos x^2,$$

with $x^1 \geq 0, 0 \leq x^2 \leq \pi, 0 \leq x^3 \leq 2\pi$.

27. Transform the Cartesian equation

$$\frac{\partial \phi}{\partial \xi^1} + \frac{\partial \phi}{\partial \xi^2} = (\xi_1)^2 + (\xi_2)^2$$

to spherical coordinates.

28. For elliptic cylindrical coordinates

$$\xi^1 = \cosh x^1 \cos x^2,$$

$$\xi^2 = \sinh x^1 \sin x^2,$$

$$\xi^3 = x^3,$$

find the Jacobian matrix $\mathbf{J}$ and the metric tensor $\mathbf{G}$. Find the transformation $x^i = x^i(\xi^j)$. Plot lines of constant x^1 and x^2 in the ξ^1 and ξ^2 planes. Determine

$\nabla f, \nabla^T \cdot \mathbf{u}, \nabla \times \mathbf{u}$, and $\nabla^2 f$ in this system, where f is a scalar field and $\mathbf{u}$ is a vector field. See Section A.6 for definitions of the hyperbolic trigonometric functions.

29. For orthogonal parabolic coordinates

$$\xi^1 = x^1 x^2 \cos x^3,$$

$$\xi^2 = x^1 x^2 \sin x^3,$$

$$\xi^3 = \frac{1}{2}\left((x^2)^2 - (x^1)^2\right),$$

find the Jacobian matrix $\mathbf{J}$ and the metric tensor $\mathbf{G}$. Find the transformation $x^i = x^i(\xi^j)$. Plot lines of constant x^1 and x^2 in the ξ^1 and ξ^2 planes. Derive an expression for the gradient, divergence, curl, and Laplacian[33] operators, and determine the scale factors.

30. Orthogonal bipolar coordinates (x^1, x^2, x^3) are defined by

$$\xi^1 = \frac{\alpha \sinh x^2}{\cosh x^2 - \cos x^1},$$

$$\xi^2 = \frac{\alpha \sin x^1}{\cosh x^2 - \cos x^1},$$

$$\xi^3 = x^3.$$

For $\alpha = 1$, plot some of the surfaces of constant ξ^1 and ξ^2 in the x^1-x^2 plane.

31. Given the equations, appropriate for the thermodynamics of an ideal gas; $P = \rho R T, e = c_v T, c_p - c_v = R, \gamma = c_p/c_v, v = 1/\rho$, and $de = T ds - P dv$, show that $\partial P/\partial \rho|_T = RT$ and $\partial P/\partial \rho|_s = \gamma RT$.

32. A fluid of variable density $\rho(x,t)$, where x is the axial coordinate and t is time, flows in a pipe of constant area A. Show that the mass of fluid between $x = L_1(t)$ and $L_2(t)$ is

$$m(t) = A \int_{L_1}^{L_2} \rho(x,s)\, ds.$$

Show from this that

$$\frac{dm}{dt} = A\left(\rho(L_2,t)\dot{L}_2(t) - \rho(L_1,t)\dot{L}_1(t)\right).$$

33. A body slides due to gravity from point A to point B along the curve $y = f(x)$. There is no friction, and the initial velocity is zero. If points A and B are fixed, find $f(x)$ for which the time taken will be the least. What is this time? If A is $(1,2)$ and B is $(0,0)$, where distances are in meters, plot the minimum-time curve, and find the minimum time if the gravitational acceleration is $\mathbf{g} = -9.81$ m/s$^2\mathbf{j}$ (brachistochrone problem).

34. Show that if a ray of light is reflected from a mirror, the shortest distance of travel is when the angle of incidence on the mirror is equal to the angle of reflection.

35. The speeds of light in different media separated by a planar interface are c_1 and c_2. Show that if the time taken for light to go from a fixed point in one medium

---

[33] Pierre-Simon Laplace, 1749–1827, French mathematician.

to another in the second is a minimum, the angle of incidence, α_i, and the angle of refraction, α_r, are related by

$$\frac{\sin \alpha_i}{\sin \alpha_r} = \frac{c_1}{c_2}.$$

36. The stress tensor σ in a material at a given point is

$$\sigma = \begin{pmatrix} 1 & 2 & 3 \\ 2 & 4 & 1 \\ 3 & 1 & 5 \end{pmatrix},$$

in units of force per unit area. Find the principal axes of σ. Find also the traction (i.e., the force) $d\mathbf{F}$ exerted on a small surface of area dA, which has a normal in the direction $\mathbf{i} + 2\mathbf{j} + \mathbf{k}$.

37. Find the extrema of $f(x, y) = (x - 1)^2 + y^2$ subject to the constraint $x^2 + xy + y^2 = 3$ using symbolic computer mathematics.

38. Show that the coordinate transformation

$$x = X,$$
$$y = Y - X,$$

implies

$$\frac{\partial}{\partial x} = \frac{\partial}{\partial X} + \frac{\partial}{\partial Y},$$
$$\frac{\partial}{\partial y} = \frac{\partial}{\partial Y}.$$

2 Vectors and Tensors in Cartesian Coordinates

This chapter reviews much of traditional *vector calculus* and includes an introduction to differential geometry. It focuses on coordinate systems that are *Cartesian* and thus specializes the analysis introduced in Section 1.6. The topics discussed in this chapter are associated with the most commonly used language for both geometry and mechanics and thus are essential for understanding the wide spectrum of engineering applications for which spatial description is required. In contrast to Section 1.6, which considered general coordinate systems and transformations, if we restrict to either Cartesian systems or those that have been rotated about the origin, many simplifications result. More generally, it should be recognized that vectors and tensors are intrinsically characterized by linearity. This was first discussed in Section 1.6.2 and is further developed throughout the book.

2.1 Preliminaries

We begin with a review of the notation and concepts associated with vectors and tensors. This includes several familiar topics, and some less familiar, especially the Cartesian index notation, to which we turn next.

2.1.1 Cartesian Index Notation

We specialize the index notation of Section 1.6 for rotation transformations. For such transformations, the distinction between contravariance and covariance disappears, as does the necessity for Christoffel symbols and also the need for an "upstairs-downstairs" index notation. Thus, for this chapter, we are mainly confined to indices in the form of subscripts, which are used to distinguish coordinate axes.

Many vector relations can be written in a compact form by using Cartesian index notation. For this chapter, let x_1, x_2, x_3 represent the three coordinate directions and $\mathbf{e}_1, \mathbf{e}_2, \mathbf{e}_3$ the orthonormal vectors in those directions. Then a vector $\mathbf{u}$ may be written in equivalent forms as

$$\mathbf{u} = \begin{pmatrix} u_1 \\ u_2 \\ u_3 \end{pmatrix} = u_1 \mathbf{e}_1 + u_2 \mathbf{e}_2 + u_3 \mathbf{e}_3 = \sum_{i=1}^{3} u_i \mathbf{e}_i = u_i \mathbf{e}_i = u_i, \tag{2.1}$$

where u_1, u_2, and u_3 are the three Cartesian components of **u**. Here we take the entity of the vector to be embodied within **u**. The vector **u** has a representation as a column vector with three scalar entries, u_1, u_2, and u_3. This column vector is considered to represent a linear combination of the product of the three scalars with the orthonormal basis vectors $\mathbf{e}_1$, $\mathbf{e}_2$, and $\mathbf{e}_3$. We do not need to use the summation sign every time if we use the Einstein convention to sum from 1 to 3 if an index is repeated. The single free index on the right side of Eq. (2.1) indicates that a product with $\mathbf{e}_i$ is assumed. We consider u_i as a vector, but it may be more precise to consider it to be the ith Cartesian scalar component of the vector **u**.

Two additional symbols are needed for later use. They are the Kronecker delta, as specialized from Eq. (1.234),

$$\delta_{ij} \equiv \begin{cases} 0, & \text{if } i \neq j, \\ 1, & \text{if } i = j, \end{cases} \tag{2.2}$$

and the alternating symbol (or permutation symbol or Levi-Civita[1] symbol):

$$\epsilon_{ijk} \equiv \begin{cases} 1, & \text{if indices are not repeated and in cyclical order: 1,2,3, or 2,3,1, or 3,1,2,} \\ -1, & \text{if indices are not repeated and not in cyclical order,} \\ 0, & \text{if two or more indices are the same.} \end{cases} \tag{2.3}$$

The identity

$$\epsilon_{ijk}\epsilon_{lmn} = \delta_{il}\delta_{jm}\delta_{kn} + \delta_{im}\delta_{jn}\delta_{kl} + \delta_{in}\delta_{jl}\delta_{km} - \delta_{il}\delta_{jn}\delta_{km} - \delta_{im}\delta_{jl}\delta_{kn} - \delta_{in}\delta_{jm}\delta_{kl} \tag{2.4}$$

relates the two and can be verified by direct substitution from the definitions of ϵ_{ijk} and δ_{ij}. The following identities are also easily shown by direct substitution:

$$\delta_{ii} = 3, \tag{2.5}$$

$$\delta_{ij} = \delta_{ji}, \tag{2.6}$$

$$\delta_{ij}\delta_{jk} = \delta_{ik}, \tag{2.7}$$

$$\epsilon_{ijk}\epsilon_{ilm} = \delta_{jl}\delta_{km} - \delta_{jm}\delta_{kl}, \tag{2.8}$$

$$\epsilon_{ijk}\epsilon_{ljk} = 2\delta_{il}, \tag{2.9}$$

$$\epsilon_{ijk}\epsilon_{ijk} = 6, \tag{2.10}$$

$$\epsilon_{ijk} = -\epsilon_{ikj}, \tag{2.11}$$

$$\epsilon_{ijk} = -\epsilon_{jik}, \tag{2.12}$$

$$\epsilon_{ijk} = -\epsilon_{kji}, \tag{2.13}$$

$$\epsilon_{ijk} = \epsilon_{kij} = \epsilon_{jki}. \tag{2.14}$$

Regarding index notation, one can say

- a repeated index indicates summation on that index,
- a nonrepeated index is known as a free index,

---

[1] Tullio Levi-Civita, 1883–1941, Italian mathematician.

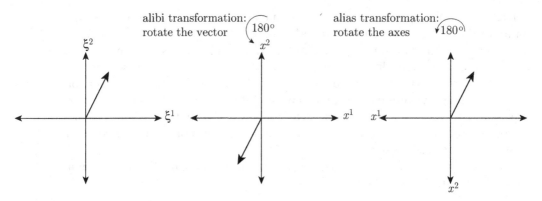

Figure 2.1. Alibi and alias rotation transformations of the vector $(1, 2, 0)^T$ through the angle of $180°$.

- the number of free indices gives the order of the tensor:
  - u, uv, $u_i v_i w$, u_{ii}, $u_{ij} v_{ij}$: zeroth-order tensor, or a scalar,
  - u_i, $u_i v_{ij}$: first-order tensor, or a vector,
  - u_{ij}, $u_{ij} v_{jk}$, $u_i v_j$: second-order tensor,
  - u_{ijk}, $u_i v_j w_k$, $u_{ij} v_{km} w_m$: third-order tensor,
  - u_{ijkl}, $u_{ij} v_{kl}$: fourth-order tensor.
- indices cannot be repeated more than once:
  - u_{iik}, u_{ij}, u_{iijj}, $v_i u_{jk}$ are correct.
  - $u_i v_i w_i$, u_{iiij}, $u_{ij} v_{ii}$ are incorrect!
- Cartesian components commute: $u_{ij} v_i w_{klm} = v_i w_{klm} u_{ij}$,
- Cartesian indices *do not* commute: $u_{ijkl} \neq u_{jlik}$.

EXAMPLE 2.1

Analyze, using generalized coordinates described in Section 1.6, a simple rotation transformation from the Cartesian ξ^i coordinates to the transformed coordinates x^i:

$$x^1 = -\xi^1, \qquad x^2 = -\xi^2, \qquad x^3 = \xi^3. \tag{2.15}$$

Here, we are returning to the more general upstairs-downstairs index notation of Section 1.6. It is easy to envision that the transformation rotates vectors about the ξ^3-axis by $180°$. Figure 2.1 displays the action of the transformation, in both an alibi and alias sense, on the vector $(\xi^1, \xi^2, \xi^3)^T = (1, 2, 0)^T$; the new representation is $(x^1, x^2, x^3)^T = (-1, -2, 0)^T$.

Recalling Eq. (1.250), the Jacobian of the transformation is

$$\mathbf{J} = \frac{\partial \xi^i}{\partial x^j} = \begin{pmatrix} -1 & 0 & 0 \\ 0 & -1 & 0 \\ 0 & 0 & 1 \end{pmatrix}. \tag{2.16}$$

By inspection, $J = \det \mathbf{J} = 1$, so the transformation is orientation- and volume-preserving. From Eq. (1.257), the metric tensor then is

$$g_{ij} = \mathbf{G} = \mathbf{J}^T \cdot \mathbf{J} = \begin{pmatrix} -1 & 0 & 0 \\ 0 & -1 & 0 \\ 0 & 0 & 1 \end{pmatrix} \begin{pmatrix} -1 & 0 & 0 \\ 0 & -1 & 0 \\ 0 & 0 & 1 \end{pmatrix} = \begin{pmatrix} 1 & 0 & 0 \\ 0 & 1 & 0 \\ 0 & 0 & 1 \end{pmatrix} = \mathbf{I} = \delta_{ij}. \tag{2.17}$$

We find for this transformation that the covariant and contravariant representations of a general vector $\mathbf{u}$ are one and the same:

$$u_i = g_{ij} u^j = \delta_{ij} u^j = \delta_j^i u^j = u^i. \tag{2.18}$$

This result extends to all transformations that are rotations. Consequently, for Cartesian vectors, there is no need to use a notation that distinguishes covariant and contravariant representations. We hereafter write all Cartesian vectors with only a subscript notation.

2.1.2 Direction Cosines

Consider the alias transformation of the (x_1, x_2) Cartesian coordinate system by rotation of each coordinate axis by angle α to the rotated Cartesian coordinate system $(\overline{x}_1, \overline{x}_2)$ as depicted in Figure 2.2. Relative to our earlier notation for general non-Cartesian systems (Section 1.6), x plays the role of the earlier ξ, and $\overline{x}$ plays the role of the earlier x.

We define the angle between the x_1 and $\overline{x}_1$ axes as α:

$$\alpha \equiv [x_1, \overline{x}_1], \tag{2.19}$$

where $[\cdot, \cdot]$ is notation to identify the axes for which the angle α is being determined. With $\beta = \pi/2 - \alpha$, the angle between the $\overline{x}_1$ and x_2 axes is

$$\beta \equiv [x_2, \overline{x}_1]. \tag{2.20}$$

The point P can be represented in both coordinate systems. In the unrotated system, P is represented by

$$P : (x_1^*, x_2^*). \tag{2.21}$$

In the rotated coordinate system, P is represented by

$$P : (\overline{x}_1^*, \overline{x}_2^*). \tag{2.22}$$

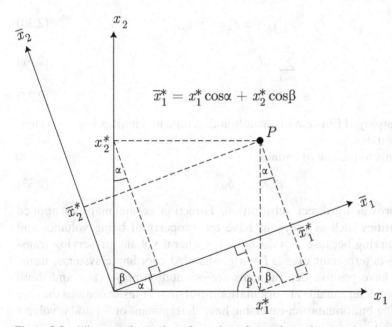

Figure 2.2. Alias transformation of rotation of axes through an angle α in a two-dimensional Cartesian system.

Trigonometry shows us that

$$\overline{x}_1^* = x_1^* \cos\alpha + x_2^* \cos\beta, \tag{2.23}$$

$$= x_1^* \cos[x_1, \overline{x}_1] + x_2^* \cos[x_2, \overline{x}_1]. \tag{2.24}$$

Dropping the asterisks, and extending to three dimensions, we find that

$$\overline{x}_1 = x_1 \cos[x_1, \overline{x}_1] + x_2 \cos[x_2, \overline{x}_1] + x_3 \cos[x_3, \overline{x}_1]. \tag{2.25}$$

Extending to expressions for $\overline{x}_2$ and $\overline{x}_3$ and writing in matrix form, we get

$$\underbrace{(\,\overline{x}_1 \quad \overline{x}_2 \quad \overline{x}_3\,)}_{=\overline{x}_j = \overline{\mathbf{x}}^T} = \underbrace{(\,x_1 \quad x_2 \quad x_3\,)}_{=x_i = \mathbf{x}^T} \underbrace{\begin{pmatrix} \cos[x_1, \overline{x}_1] & \cos[x_1, \overline{x}_2] & \cos[x_1, \overline{x}_3] \\ \cos[x_2, \overline{x}_1] & \cos[x_2, \overline{x}_2] & \cos[x_2, \overline{x}_3] \\ \cos[x_3, \overline{x}_1] & \cos[x_3, \overline{x}_2] & \cos[x_3, \overline{x}_3] \end{pmatrix}}_{=\ell_{ij} = \mathbf{Q}}.$$

$$\tag{2.26}$$

Using the notation

$$\ell_{ij} = \cos[x_i, \overline{x}_j] \equiv \mathbf{Q}, \tag{2.27}$$

Eq. (2.26) is written as

$$\underbrace{(\,\overline{x}_1 \quad \overline{x}_2 \quad \overline{x}_3\,)}_{=\overline{x}_j = \overline{\mathbf{x}}^T} = \underbrace{(\,x_1 \quad x_2 \quad x_3\,)}_{=x_i = \mathbf{x}^T} \underbrace{\begin{pmatrix} \ell_{11} & \ell_{12} & \ell_{13} \\ \ell_{21} & \ell_{22} & \ell_{23} \\ \ell_{31} & \ell_{32} & \ell_{33} \end{pmatrix}}_{=\mathbf{Q}}. \tag{2.28}$$

Here ℓ_{ij} are known as the *direction cosines*. Expanding the first term, we find

$$\overline{x}_1 = x_1 \ell_{11} + x_2 \ell_{21} + x_3 \ell_{31}. \tag{2.29}$$

More generally, we have

$$\overline{x}_j = x_1 \ell_{1j} + x_2 \ell_{2j} + x_3 \ell_{3j}, \tag{2.30}$$

$$= \sum_{i=1}^{3} x_i \ell_{ij}, \tag{2.31}$$

$$= x_i \ell_{ij}. \tag{2.32}$$

Here we have employed Einstein's convention that repeated indices implies a summation over that index.

What amounts to the *law of cosines*,

$$\ell_{ij} \ell_{kj} = \delta_{ik}, \tag{2.33}$$

can easily be proven by direct substitution. Direction cosine matrices applied to geometric entities such as polygons have the property of being volume- and orientation-preserving because $J = \det \ell_{ij} = 1$. General volume-preserving transformations have determinant of ± 1. For right-handed coordinate systems, transformations that have positive determinants are orientation-preserving, and those that have negative determinants are orientation-reversing. Transformations that are volume-preserving but orientation-reversing have determinant of -1 and involve a reflection.

EXAMPLE 2.2

Show for the two-dimensional system described in Figure 2.2 that $\ell_{ij}\ell_{kj} = \delta_{ik}$ holds.

Expanding for the two-dimensional system, we get

$$\ell_{i1}\ell_{k1} + \ell_{i2}\ell_{k2} = \delta_{ik}. \tag{2.34}$$

First, take $i = 1, k = 1$. We get then

$$\ell_{11}\ell_{11} + \ell_{12}\ell_{12} = \delta_{11} = 1, \tag{2.35}$$

$$\cos\alpha\cos\alpha + \cos(\alpha + \pi/2)\cos(\alpha + \pi/2) = 1, \tag{2.36}$$

$$\cos\alpha\cos\alpha + (-\sin(\alpha))(-\sin(\alpha)) = 1, \tag{2.37}$$

$$\cos^2\alpha + \sin^2\alpha = 1. \tag{2.38}$$

This is obviously true. Next, take $i = 1, k = 2$. We get then

$$\ell_{11}\ell_{21} + \ell_{12}\ell_{22} = \delta_{12} = 0, \tag{2.39}$$

$$\cos\alpha\cos(\pi/2 - \alpha) + \cos(\alpha + \pi/2)\cos(\alpha) = 0, \tag{2.40}$$

$$\cos\alpha\sin\alpha - \sin\alpha\cos\alpha = 0. \tag{2.41}$$

Next, take $i = 2, k = 1$. We get then

$$\ell_{21}\ell_{11} + \ell_{22}\ell_{12} = \delta_{21} = 0, \tag{2.42}$$

$$\cos(\pi/2 - \alpha)\cos\alpha + \cos\alpha\cos(\pi/2 + \alpha) = 0, \tag{2.43}$$

$$\sin\alpha\cos\alpha + \cos\alpha(-\sin\alpha) = 0. \tag{2.44}$$

Next, take $i = 2, k = 2$. We get then

$$\ell_{21}\ell_{21} + \ell_{22}\ell_{22} = \delta_{22} = 1, \tag{2.45}$$

$$\cos(\pi/2 - \alpha)\cos(\pi/2 - \alpha) + \cos\alpha\cos\alpha = 1, \tag{2.46}$$

$$\sin\alpha\sin\alpha + \cos\alpha\cos\alpha = 1. \tag{2.47}$$

Thus, the contention holds for all cases in two dimensions.

Using the law of cosines, Eq. (2.33), we can easily find the inverse transformation back to the unbarred coordinates via the following operations. First operate on Eq. (2.32) with ℓ_{kj}:

$$\ell_{kj}\overline{x}_j = \ell_{kj}x_i\ell_{ij}, \tag{2.48}$$

$$= \ell_{ij}\ell_{kj}x_i, \tag{2.49}$$

$$= \delta_{ik}x_i, \tag{2.50}$$

$$= x_k, \tag{2.51}$$

$$\ell_{ij}\overline{x}_j = x_i, \tag{2.52}$$

$$x_i = \ell_{ij}\overline{x}_j. \tag{2.53}$$

The Jacobian matrix of the transformation is $\mathbf{J} = \partial x_i/\partial\overline{x}_j = \ell_{ij}$. It can be shown that the metric tensor is $\mathbf{G} = \mathbf{J}^T \cdot \mathbf{J} = \ell_{ji}\ell_{ki} = \delta_{jk} = \mathbf{I}$, so $g = 1$, and the transformation is volume-preserving. Moreover, because $\mathbf{J}^T \cdot \mathbf{J} = \mathbf{I}$, we see that $\mathbf{J}^T = \mathbf{J}^{-1}$. As such, it is precisely the type of matrix for which the gradient takes on the same form in original and transformed coordinates, as presented in the discussion surrounding

Eq. (1.267). As is discussed in detail in Section 7.7, matrices that have these properties are known as *orthogonal* and are often denoted by $\mathbf{Q}$. So for this class of transformations, $\mathbf{J} = \mathbf{Q} = \partial x_i / \partial \overline{x}_j = \ell_{ij}$. Note that $\mathbf{Q}^T \cdot \mathbf{Q} = \mathbf{I}$ and that $\mathbf{Q}^T = \mathbf{Q}^{-1}$. The matrix $\mathbf{Q}$ is a rotation matrix when its elements are composed of the direction cosines ℓ_{ij}. Note then that $\mathbf{Q}^T = \ell_{ji}$. For a coordinate system that obeys the right-hand rule, we require $\det \mathbf{Q} = 1$ so that it is also orientation-preserving.

EXAMPLE 2.3

Examine in detail a matrix that rotates the coordinate axes through an angle α, such as in Figure 2.2, using matrix methods.

We have

$$\mathbf{J} = \frac{\partial x_i}{\partial \overline{x}_j} = \ell_{ij} = \mathbf{Q} = \begin{pmatrix} \cos\alpha & \cos\left(\alpha + \frac{\pi}{2}\right) \\ \cos\left(\frac{\pi}{2} - \alpha\right) & \cos\alpha \end{pmatrix} = \begin{pmatrix} \cos\alpha & -\sin\alpha \\ \sin\alpha & \cos\alpha \end{pmatrix}. \tag{2.54}$$

We get the rotated coordinates via Eq. (2.26):

$$\overline{\mathbf{x}}^T = \mathbf{x}^T \cdot \mathbf{Q}, \tag{2.55}$$

$$(\overline{x}_1 \quad \overline{x}_2) = (x_1 \quad x_2) \begin{pmatrix} \cos\alpha & -\sin\alpha \\ \sin\alpha & \cos\alpha \end{pmatrix}, \tag{2.56}$$

$$= (x_1 \cos\alpha + x_2 \sin\alpha \quad -x_1 \sin\alpha + x_2 \cos\alpha), \tag{2.57}$$

$$\begin{pmatrix} \overline{x}_1 \\ \overline{x}_2 \end{pmatrix} = \begin{pmatrix} x_1 \cos\alpha + x_2 \sin\alpha \\ -x_1 \sin\alpha + x_2 \cos\alpha \end{pmatrix}. \tag{2.58}$$

We can also rearrange to say

$$\overline{\mathbf{x}} = \mathbf{Q}^T \cdot \mathbf{x}, \tag{2.59}$$

$$\mathbf{Q} \cdot \overline{\mathbf{x}} = \underbrace{\mathbf{Q} \cdot \mathbf{Q}^T}_{\mathbf{I}} \cdot \mathbf{x}, \tag{2.60}$$

$$\mathbf{Q} \cdot \overline{\mathbf{x}} = \mathbf{I} \cdot \mathbf{x}, \tag{2.61}$$

$$\mathbf{x} = \mathbf{Q} \cdot \overline{\mathbf{x}}. \tag{2.62}$$

The law of cosines holds because

$$\mathbf{Q} \cdot \mathbf{Q}^T = \begin{pmatrix} \cos\alpha & -\sin\alpha \\ \sin\alpha & \cos\alpha \end{pmatrix} \begin{pmatrix} \cos\alpha & \sin\alpha \\ -\sin\alpha & \cos\alpha \end{pmatrix}, \tag{2.63}$$

$$= \begin{pmatrix} \cos^2\alpha + \sin^2\alpha & 0 \\ 0 & \sin^2\alpha + \cos^2\alpha \end{pmatrix}, \tag{2.64}$$

$$= \begin{pmatrix} 1 & 0 \\ 0 & 1 \end{pmatrix}, \tag{2.65}$$

$$= \mathbf{I} = \delta_{ij}. \tag{2.66}$$

Consider the determinant of $\mathbf{Q}$:

$$\det \mathbf{Q} = \cos^2\alpha - (-\sin^2\alpha) = \cos^2\alpha + \sin^2\alpha = 1. \tag{2.67}$$

Thus, the transformation is volume- and orientation-preserving; hence, it is a rotation. The rotation is through an angle α.

For instance, consider $\alpha = \tan^{-1}(4/3) = 53.13°$ and the vector $\mathbf{x} = (1,1)^T$. The rotation matrix is then

$$\mathbf{Q} = \begin{pmatrix} \frac{3}{5} & -\frac{4}{5} \\ \frac{4}{5} & \frac{3}{5} \end{pmatrix}, \tag{2.68}$$

alibi transformation: alias transformation:
rotate the vector   rotate the axes

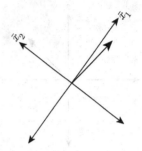

Figure 2.3. Alibi and alias rotation transformations of the vector $(1, 1)^T$ through the angle $\alpha = \tan^{-1}(4/3) = 53.13°$.

and its action on **x** yields

$$\bar{\mathbf{x}} = \mathbf{Q}^T \cdot \mathbf{x} = \begin{pmatrix} \frac{3}{5} & \frac{4}{5} \\ -\frac{4}{5} & \frac{3}{5} \end{pmatrix} \begin{pmatrix} 1 \\ 1 \end{pmatrix} = \begin{pmatrix} \frac{7}{5} \\ -\frac{1}{5} \end{pmatrix}. \tag{2.69}$$

Both **x** and $\bar{\mathbf{x}}$ have magnitudes of $\sqrt{2}$. We sketch the various relevant configurations for this problem in Figure 2.3.

EXAMPLE 2.4

Examine the properties of the so-called *reflection matrix* in two dimensions:

$$\mathbf{Q} = \begin{pmatrix} \cos\alpha & \sin\alpha \\ \sin\alpha & -\cos\alpha \end{pmatrix}. \tag{2.70}$$

The reflection matrix is obtained by multiplying the second column of the rotation matrix of Eq. (2.54) by -1. We see that

$$\mathbf{Q} \cdot \mathbf{Q}^T = \begin{pmatrix} \cos\alpha & \sin\alpha \\ \sin\alpha & -\cos\alpha \end{pmatrix} \begin{pmatrix} \cos\alpha & \sin\alpha \\ \sin\alpha & -\cos\alpha \end{pmatrix}, \tag{2.71}$$

$$= \begin{pmatrix} \cos^2\alpha + \sin^2\alpha & 0 \\ 0 & \sin^2\alpha + \cos^2\alpha \end{pmatrix}, \tag{2.72}$$

$$= \begin{pmatrix} 1 & 0 \\ 0 & 1 \end{pmatrix} = \mathbf{I} = \delta_{ij}. \tag{2.73}$$

The determinant of the reflection matrix is

$$\det \mathbf{Q} = -\cos^2\alpha - \sin^2\alpha = -1. \tag{2.74}$$

Thus, the transformation is volume-preserving but not orientation-preserving. Also, note that for the reflection matrix, $\mathbf{Q} = \mathbf{Q}^T$ and thus $\mathbf{Q} = \mathbf{Q}^{-1}$. Therefore $\mathbf{Q} \cdot \mathbf{Q} = \mathbf{I}$. That can be understood by realizing that two reflections of a vector should return to the original vector. A matrix which, when applied twice to a vector, returns to the original vector is known is *involutary* (see Section 7.2.5). One can show by considering the action of the reflection matrix on coordinate axes to represent vectors **x** that, in an alias sense, the action is a reflection of the coordinate axes about a line passing through the origin inclined at an angle of $\alpha/2$ to the horizontal. In an alibi sense, the vector is reflected about the same angle.

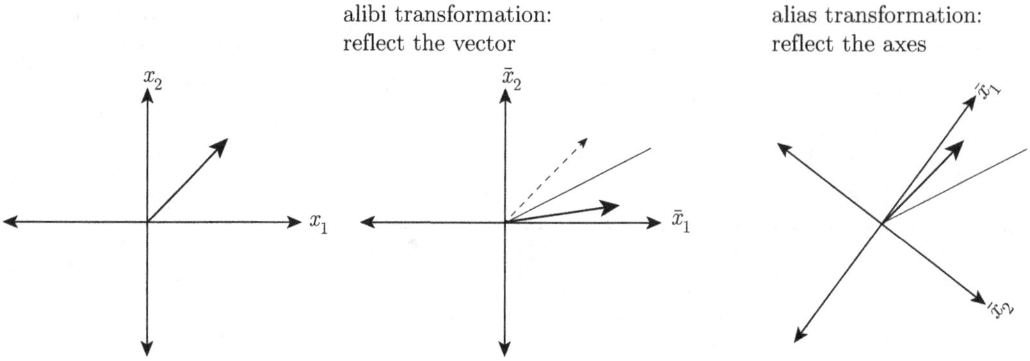

Figure 2.4. Alibi and alias reflection transformations of the vector $(1,1)^T$ about an axis inclined at $\alpha/2 = (1/2)\tan^{-1}(4/3) = 26.57°$.

For instance, consider $\alpha = \tan^{-1}(4/3) = 53.13°$ and the vector $\mathbf{x} = (1,1)^T$. The axis of reflection is then inclined at $\alpha/2 = (1/2)\tan^{-1}(4/3) = 26.57°$ to the horizontal. The reflection matrix is then

$$\mathbf{Q} = \begin{pmatrix} \frac{3}{5} & \frac{4}{5} \\ \frac{4}{5} & -\frac{3}{5} \end{pmatrix}, \tag{2.75}$$

and its action on $\mathbf{x}$ yields

$$\overline{\mathbf{x}} = \mathbf{Q}^T \cdot \mathbf{x} = \mathbf{Q} \cdot \mathbf{x} = \begin{pmatrix} \frac{3}{5} & \frac{4}{5} \\ \frac{4}{5} & -\frac{3}{5} \end{pmatrix} \begin{pmatrix} 1 \\ 1 \end{pmatrix} = \begin{pmatrix} \frac{7}{5} \\ \frac{1}{5} \end{pmatrix}. \tag{2.76}$$

Both $\mathbf{x}$ and $\overline{\mathbf{x}}$ have magnitudes of $\sqrt{2}$. We sketch the various relevant configurations for this problem in Figure 2.4.

2.1.3 Scalars

An entity ϕ is a scalar if it is invariant under a rotation of coordinate axes.

2.1.4 Vectors

A set of three scalars $(v_1, v_2, v_3)^T$ is defined as a *vector* if, under a rotation of coordinate axes, the triple also transforms according to[2]

$$\overline{v}_j = v_i \ell_{ij}, \qquad \overline{\mathbf{v}}^T = \mathbf{v}^T \cdot \mathbf{Q}. \tag{2.77}$$

We could also transpose both sides and have

$$\overline{\mathbf{v}} = \mathbf{Q}^T \cdot \mathbf{v}. \tag{2.78}$$

A vector associates a scalar with a chosen direction in space.

---

[2] In this and many other places, we will repeat equations in different notations to accustom the reader to some of the variety that is possible.

EXAMPLE 2.5

Returning to generalized coordinate notation, show the equivalence between covariant and contravariant representations for pure rotations of a vector **v**.

Consider then a transformation from a Cartesian space ξ^j to a transformed space x^i via a pure rotation:

$$\xi^i = \ell^i_j x^j. \tag{2.79}$$

Here ℓ^i_j is simply a matrix of direction cosines as we have previously defined; we employ the upstairs-downstairs index notation for consistency. The Jacobian is

$$\frac{\partial \xi^i}{\partial x^j} = \ell^i_j. \tag{2.80}$$

From Eq. (1.257), the metric tensor is

$$g_{kl} = \frac{\partial \xi^i}{\partial x^k} \frac{\partial \xi^i}{\partial x^l} = \ell^i_k \ell^i_l = \delta_{kl}. \tag{2.81}$$

Here we have employed the law of cosines, which is easily extensible to the "upstairs-downstairs" notation. So a vector **v** has the same covariant and contravariant components because

$$v_i = g_{ij} v^j = \delta_{ij} v^j = \delta^i_j v^j = v^i. \tag{2.82}$$

The vector itself has components that transform under rotation:

$$v^i = \ell^i_j V^j. \tag{2.83}$$

Here V^j is the contravariant representation of the vector **v** in the unrotated coordinate system. One could also show that $V_j = V^j$, as always for a Cartesian system.

2.1.5 Tensors

A set of nine scalars is defined as a second-order *tensor* if, under a rotation of coordinate axes they transform as

$$\overline{T}_{ij} = \ell_{ki} \ell_{lj} T_{kl}, \qquad \overline{\mathbf{T}} = \mathbf{Q}^T \cdot \mathbf{T} \cdot \mathbf{Q}. \tag{2.84}$$

A tensor associates a vector with each direction in space. It will be seen that

- the first subscript gives the associated direction of a face (hence, first face, and so on),
- the second subscript gives the vector components for that face.

Graphically, one can use the sketch in Figure 2.5 to visualize a second-order tensor. In Figure 2.5, $\mathbf{q}^{(1)}$, $\mathbf{q}^{(2)}$, and $\mathbf{q}^{(3)}$ are the vectors associated with the 1, 2, and 3 faces, respectively.

Matrix Representation

Tensors can be represented as matrices (but all matrices are not tensors!). The following is useful to understand how certain vectors compose a tensor:

$$T_{ij} = \begin{pmatrix} T_{11} & T_{12} & T_{13} \\ T_{21} & T_{22} & T_{23} \\ T_{31} & T_{32} & T_{33} \end{pmatrix} \quad \begin{matrix} \text{vector associated with first face,} \\ \text{vector associated with second face,} \\ \text{vector associated with third face.} \end{matrix} \tag{2.85}$$

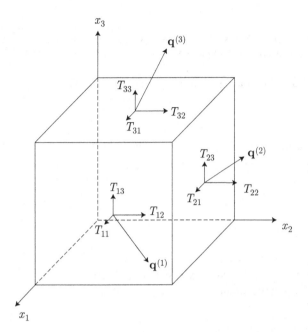

Figure 2.5. Tensor visualization.

A simple way to choose a vector q_j associated with a plane of arbitrary orientation is to form the product of the unit normal associated with the plane n_i and the tensor T_{ij}:

$$q_j = n_i T_{ij}, \qquad \mathbf{q}^T = \mathbf{n}^T \cdot \mathbf{T}. \tag{2.86}$$

Here n_i has three components, which are the direction cosines of the chosen direction. For example, to determine the vector associated with face 2, we choose

$$n_i = \begin{pmatrix} 0 \\ 1 \\ 0 \end{pmatrix}. \tag{2.87}$$

Thus, in Gibbs notation, we have

$$\mathbf{n}^T \cdot \mathbf{T} = (0, 1, 0) \begin{pmatrix} T_{11} & T_{12} & T_{13} \\ T_{21} & T_{22} & T_{23} \\ T_{31} & T_{32} & T_{33} \end{pmatrix} = (T_{21}, T_{22}, T_{23}). \tag{2.88}$$

In Einstein notation, we arrive at the same conclusion via

$$n_i T_{ij} = n_1 T_{1j} + n_2 T_{2j} + n_3 T_{3j}, \tag{2.89}$$

$$= (0) T_{1j} + (1) T_{2j} + (0) T_{3j}, \tag{2.90}$$

$$= (T_{21}, T_{22}, T_{23}). \tag{2.91}$$

Transpose of a Tensor, Symmetric and Antisymmetric Tensors

The *transpose* T_{ij}^T of a tensor T_{ij} is found by trading elements across the diagonal

$$T_{ij}^T \equiv T_{ji}, \tag{2.92}$$

so

$$T_{ij}^T = \begin{pmatrix} T_{11} & T_{21} & T_{31} \\ T_{12} & T_{22} & T_{32} \\ T_{13} & T_{23} & T_{33} \end{pmatrix}. \tag{2.93}$$

A tensor is *symmetric* if it is equal to its transpose, that is,

$$T_{ij} = T_{ji}, \qquad \mathbf{T} = \mathbf{T}^T. \tag{2.94}$$

A tensor is *antisymmetric* if it is equal to the additive inverse of its transpose, that is,

$$T_{ij} = -T_{ji}, \qquad \mathbf{T} = -\mathbf{T}^T. \tag{2.95}$$

A tensor is *asymmetric* if it is neither symmetric nor antisymmetric.

The tensor inner product of a symmetric tensor S_{ij} and antisymmetric tensor A_{ij} can be shown to be 0:

$$S_{ij} A_{ij} = 0, \qquad \mathbf{S} : \mathbf{A} = 0. \tag{2.96}$$

Here the ":" notation indicates a double summation, also called a double contraction. It is a double contraction because it renders objects that are two-dimensional matrices into zero-dimensional scalars.

EXAMPLE 2.6

Show $S_{ij} A_{ij} = 0$ for two-dimensional tensors.

Take a general symmetric tensor to be

$$S_{ij} = \begin{pmatrix} a & b \\ b & c \end{pmatrix}. \tag{2.97}$$

Take a general antisymmetric tensor to be

$$A_{ij} = \begin{pmatrix} 0 & d \\ -d & 0 \end{pmatrix}. \tag{2.98}$$

So

$$S_{ij} A_{ij} = S_{11} A_{11} + S_{12} A_{12} + S_{21} A_{21} + S_{22} A_{22}, \tag{2.99}$$

$$= a(0) + bd - bd + c(0), \tag{2.100}$$

$$= 0. \tag{2.101}$$

An arbitrary tensor can be represented as the sum of symmetric and antisymmetric tensors:

$$T_{ij} = \underbrace{\frac{1}{2}T_{ij} + \frac{1}{2}T_{ij}}_{=T_{ij}} + \underbrace{\frac{1}{2}T_{ji} - \frac{1}{2}T_{ji}}_{=0}, \tag{2.102}$$

$$= \underbrace{\frac{1}{2}\left(T_{ij} + T_{ji}\right)}_{\equiv T_{(ij)}} + \underbrace{\frac{1}{2}\left(T_{ij} - T_{ji}\right)}_{\equiv T_{[ij]}}. \tag{2.103}$$

So with

$$T_{(ij)} \equiv \frac{1}{2}\left(T_{ij} + T_{ji}\right), \tag{2.104}$$

$$T_{[ij]} \equiv \frac{1}{2}\left(T_{ij} - T_{ji}\right), \tag{2.105}$$

we arrive at

$$T_{ij} = \underbrace{T_{(ij)}}_{\text{symmetric}} + \underbrace{T_{[ij]}}_{\text{antisymmetric}} . \tag{2.106}$$

The first term, $T_{(ij)}$, is called the symmetric part of T_{ij}; the second term, $T_{[ij]}$, is called the antisymmetric part of T_{ij}.

Dual Vector of an Antisymmetric Tensor

As the antisymmetric part of a three by three tensor has only three independent components, we might expect that a three-component vector can be associated with it. Let us define the *dual vector* to be

$$d_i \equiv \frac{1}{2}\epsilon_{ijk}T_{jk} = \frac{1}{2}\underbrace{\epsilon_{ijk}T_{(jk)}}_{=0} + \frac{1}{2}\epsilon_{ijk}T_{[jk]}. \tag{2.107}$$

For fixed i, ϵ_{ijk} is antisymmetric. So the first term is zero, being for fixed i the tensor inner product of an antisymmetric and symmetric tensor. Thus,

$$d_i = \frac{1}{2}\epsilon_{ijk}T_{[jk]}. \tag{2.108}$$

Let us find the inverse. Apply ϵ_{ilm} to both sides of Eq. (2.107) to get

$$\epsilon_{ilm}d_i = \frac{1}{2}\epsilon_{ilm}\epsilon_{ijk}T_{jk}, \tag{2.109}$$

$$= \frac{1}{2}(\delta_{lj}\delta_{mk} - \delta_{lk}\delta_{mj})T_{jk}, \tag{2.110}$$

$$= \frac{1}{2}(T_{lm} - T_{ml}), \tag{2.111}$$

$$= T_{[lm]}, \tag{2.112}$$

$$T_{[lm]} = \epsilon_{ilm}d_i, \tag{2.113}$$

$$T_{[ij]} = \epsilon_{kij}d_k, \tag{2.114}$$

$$= \epsilon_{ijk}d_k. \tag{2.115}$$

Expanding, we can see that

$$T_{[ij]} = \epsilon_{ijk}d_k = \epsilon_{ij1}d_1 + \epsilon_{ij2}d_2 + \epsilon_{ij3}d_3 = \begin{pmatrix} 0 & d_3 & -d_2 \\ -d_3 & 0 & d_1 \\ d_2 & -d_1 & 0 \end{pmatrix}. \tag{2.116}$$

The matrix form realized is obvious when one considers that an individual term, such as $\epsilon_{ij1}d_1$, only has a value when $i, j = 2, 3$ or $i, j = 3, 2$, and takes on values of $\pm d_1$ in those cases. In summary, the general dimension 3 tensor can be written as

$$T_{ij} = T_{(ij)} + \epsilon_{ijk}d_k. \tag{2.117}$$

Principal Axes and Tensor Invariants

Given a tensor T_{ij}, we would often like to find the associated direction such that the vector components in this associated direction are parallel to the direction itself. So we want

$$n_i T_{ij} = \lambda n_j. \tag{2.118}$$

This defines an eigenvalue problem; this will be discussed further in Section 6.8. Linear algebra gives us the eigenvalues and associated eigenvectors:

$$n_i T_{ij} = \lambda n_i \delta_{ij}, \tag{2.119}$$

$$n_i (T_{ij} - \lambda \delta_{ij}) = 0, \tag{2.120}$$

$$(n_1, n_2, n_3) \begin{pmatrix} T_{11} - \lambda & T_{12} & T_{13} \\ T_{21} & T_{22} - \lambda & T_{23} \\ T_{31} & T_{32} & T_{33} - \lambda \end{pmatrix} = (0, 0, 0). \tag{2.121}$$

This is equivalent to $\mathbf{n}^T \cdot (\mathbf{T} - \lambda \mathbf{I}) = \mathbf{0}^T$ or $(\mathbf{T} - \lambda \mathbf{I})^T \cdot \mathbf{n} = \mathbf{0}$. We get nontrivial solutions if

$$\begin{vmatrix} T_{11} - \lambda & T_{12} & T_{13} \\ T_{21} & T_{22} - \lambda & T_{23} \\ T_{31} & T_{32} & T_{33} - \lambda \end{vmatrix} = 0. \tag{2.122}$$

We are actually finding the so-called *left* eigenvectors of T_{ij}. These arise with less frequency than the *right* eigenvectors, which are defined by $T_{ij} u_j = \lambda \delta_{ij} u_j$. Right and left eigenvalue problems are discussed later in Section 6.8.

We know from linear algebra that such an equation for a third-order matrix gives rise to a characteristic equation for λ of the form

$$\lambda^3 - I_T^{(1)} \lambda^2 + I_T^{(2)} \lambda - I_T^{(3)} = 0, \tag{2.123}$$

where $I_T^{(1)}, I_T^{(2)}, I_T^{(3)}$ are scalars that are functions of all the scalars T_{ij}. The I_Ts are known as the *invariants* of the tensor T_{ij}. The invariants will not change if the coordinate axes are transformed; in contrast, the scalar components T_{ij} will change under transformation. The invariants can be shown to be given by

$$I_T^{(1)} = T_{ii} = T_{11} + T_{22} + T_{33} = \text{tr}\,\mathbf{T}, \tag{2.124}$$

$$I_T^{(2)} = \frac{1}{2}(T_{ii} T_{jj} - T_{ij} T_{ji}) = \frac{1}{2}\left((\text{tr}\,\mathbf{T})^2 - \text{tr}(\mathbf{T} \cdot \mathbf{T})\right) = (\det \mathbf{T})(\text{tr}\,\mathbf{T}^{-1}), \tag{2.125}$$

$$= \frac{1}{2}\left(T_{(ii)} T_{(jj)} + T_{[ij]} T_{[ij]} - T_{(ij)} T_{(ij)}\right), \tag{2.126}$$

$$I_T^{(3)} = \epsilon_{ijk} T_{1i} T_{2j} T_{3k} = \det \mathbf{T}. \tag{2.127}$$

Here "tr" denotes the trace, that is, the sum of the diagonal terms of the matrix as defined in Section 7.4.1. It can also be shown that if $\lambda^{(1)}, \lambda^{(2)}, \lambda^{(3)}$ are the three eigenvalues, then the invariants can also be expressed as

$$I_T^{(1)} = \lambda^{(1)} + \lambda^{(2)} + \lambda^{(3)}, \tag{2.128}$$

$$I_T^{(2)} = \lambda^{(1)} \lambda^{(2)} + \lambda^{(2)} \lambda^{(3)} + \lambda^{(3)} \lambda^{(1)}, \tag{2.129}$$

$$I_T^{(3)} = \lambda^{(1)} \lambda^{(2)} \lambda^{(3)}. \tag{2.130}$$

If T_{ij} is real and symmetric, then, as shown in Section 6.8,

- the eigenvalues are real,
- the eigenvectors corresponding to distinct eigenvalues are real and orthogonal,
- the left and right eigenvectors can always be rendered to be identical.

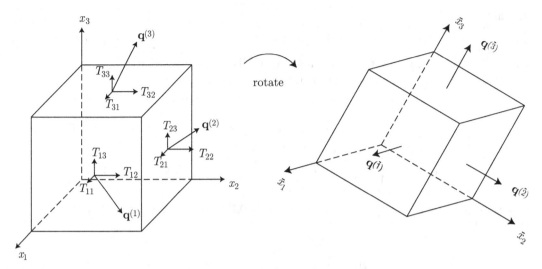

Figure 2.6. Sketch depicting rotation of volume element to be aligned with principal axes. Tensor T_{ij} must be symmetric to guarantee existence of orthogonal principal directions.

A sketch of a volume element rotated in an alias sense to be aligned with a set of orthogonal principal axes is shown in Figure 2.6.

If the matrix is asymmetric, the eigenvalues could be complex, and the eigenvectors are not orthogonal. It is often most physically relevant to decompose a tensor into symmetric and antisymmetric parts and find the orthonormal basis vectors and real eigenvalues associated with the symmetric part and the dual vector associated with the antisymmetric part.

In continuum mechanics,

- the symmetric part of a tensor can be associated with extensional deformation along principal axes,
- the antisymmetric part of a tensor can be associated with rotation of an element.

EXAMPLE 2.7

For the tensor T_{ij},

$$T_{ij} = \begin{pmatrix} 1 & 1 & -2 \\ 3 & 2 & -3 \\ -4 & 1 & 1 \end{pmatrix}, \tag{2.131}$$

find the associated dual vector and orthonormal basis vectors.

First, determine the symmetric and antisymmetric parts of T_{ij}:

$$T_{(ij)} = \frac{1}{2}(T_{ij} + T_{ji}) = \begin{pmatrix} 1 & 2 & -3 \\ 2 & 2 & -1 \\ -3 & -1 & 1 \end{pmatrix}, \tag{2.132}$$

$$T_{[ij]} = \frac{1}{2}(T_{ij} - T_{ji}) = \begin{pmatrix} 0 & -1 & 1 \\ 1 & 0 & -2 \\ -1 & 2 & 0 \end{pmatrix}. \tag{2.133}$$

Next, get the dual vector d_i:

$$d_i = \frac{1}{2}\epsilon_{ijk}T_{[jk]}, \tag{2.134}$$

$$d_1 = \frac{1}{2}\epsilon_{1jk}T_{[jk]} = \frac{1}{2}(\epsilon_{123}T_{[23]} + \epsilon_{132}T_{[32]}) = \frac{1}{2}((1)(-2) + (-1)(2)) = -2, \tag{2.135}$$

$$d_2 = \frac{1}{2}\epsilon_{2jk}T_{[jk]} = \frac{1}{2}(\epsilon_{213}T_{[13]} + \epsilon_{231}T_{[31]}) = \frac{1}{2}((-1)(1) + (1)(-1)) = -1, \tag{2.136}$$

$$d_3 = \frac{1}{2}\epsilon_{3jk}T_{[jk]} = \frac{1}{2}(\epsilon_{312}T_{[12]} + \epsilon_{321}T_{[21]}) = \frac{1}{2}((1)(-1) + (-1)(1)) = -1, \tag{2.137}$$

$$d_i = (-2, -1, -1)^T. \tag{2.138}$$

Note that Eq. (2.116) is satisfied.

Now find the eigenvalues and eigenvectors for the symmetric part:

$$\begin{vmatrix} 1-\lambda & 2 & -3 \\ 2 & 2-\lambda & -1 \\ -3 & -1 & 1-\lambda \end{vmatrix} = 0. \tag{2.139}$$

We get the characteristic equation

$$\lambda^3 - 4\lambda^2 - 9\lambda + 9 = 0. \tag{2.140}$$

The eigenvalue and associated normalized eigenvector for each root is

$$\lambda^{(1)} = 5.36488, \qquad n_i^{(1)} = (-0.630537, -0.540358, 0.557168)^T, \tag{2.141}$$

$$\lambda^{(2)} = -2.14644, \qquad n_i^{(2)} = (-0.740094, 0.202303, -0.641353)^T, \tag{2.142}$$

$$\lambda^{(3)} = 0.781562, \qquad n_i^{(3)} = (-0.233844, 0.816754, 0.527476)^T. \tag{2.143}$$

It is easily verified that each eigenvector is orthonormal. When the coordinates are transformed to be aligned with the principal axes, the magnitude of the vector associated with each face is the eigenvalue; this vector points in the same direction of the unit normal associated with the face.

EXAMPLE 2.8

For a given tensor, which we will take to be symmetric, though the theory applies to asymmetric tensors as well,

$$T_{ij} = \mathbf{T} = \begin{pmatrix} 1 & 2 & 4 \\ 2 & 3 & -1 \\ 4 & -1 & 1 \end{pmatrix}, \tag{2.144}$$

find the three basic tensor invariants $I_T^{(1)}$, $I_T^{(2)}$, and $I_T^{(3)}$, and show they are actually invariant when the tensor is subjected to a rotation with direction cosine matrix of

$$\ell_{ij} = \mathbf{Q} = \begin{pmatrix} \frac{1}{\sqrt{6}} & \sqrt{\frac{2}{3}} & \frac{1}{\sqrt{6}} \\ \frac{1}{\sqrt{3}} & -\frac{1}{\sqrt{3}} & \frac{1}{\sqrt{3}} \\ \frac{1}{\sqrt{2}} & 0 & -\frac{1}{\sqrt{2}} \end{pmatrix}. \tag{2.145}$$

Calculation shows that $\det \mathbf{Q} = 1$, and $\mathbf{Q} \cdot \mathbf{Q}^T = \mathbf{I}$, so the matrix $\mathbf{Q}$ is volume- and orientation-preserving and thus a rotation matrix. As an aside, construction of an arbitrary rotation matrix is nontrivial. One method of construction involves determining a set of orthogonal vectors via a process to be described later (see Section 6.4).

The eigenvalues of $\mathbf{T}$, which are the principal values, are easily calculated to be

$$\lambda^{(1)} = 5.28675, \qquad \lambda^{(2)} = -3.67956, \qquad \lambda^{(3)} = 3.39281. \tag{2.146}$$

The three invariants of T_{ij} from Eqs. (2.124, 2.125, 2.127) are

$$I_T^{(1)} = \mathrm{tr}\,(\mathbf{T}) = \mathrm{tr} \begin{pmatrix} 1 & 2 & 4 \\ 2 & 3 & -1 \\ 4 & -1 & 1 \end{pmatrix} = 1 + 3 + 1 = 5, \tag{2.147}$$

$$I_T^{(2)} = \frac{1}{2}\left((\mathrm{tr}\,(\mathbf{T}))^2 - \mathrm{tr}\,(\mathbf{T} \cdot \mathbf{T}) \right),$$

$$= \frac{1}{2}\left(\left(\mathrm{tr} \begin{pmatrix} 1 & 2 & 4 \\ 2 & 3 & -1 \\ 4 & -1 & 1 \end{pmatrix} \right)^2 - \mathrm{tr} \left(\begin{pmatrix} 1 & 2 & 4 \\ 2 & 3 & -1 \\ 4 & -1 & 1 \end{pmatrix} \begin{pmatrix} 1 & 2 & 4 \\ 2 & 3 & -1 \\ 4 & -1 & 1 \end{pmatrix} \right) \right),$$

$$= \frac{1}{2}\left(5^2 - \mathrm{tr} \begin{pmatrix} 21 & 4 & 6 \\ 4 & 14 & 4 \\ 6 & 4 & 18 \end{pmatrix} \right),$$

$$= \frac{1}{2}(25 - 21 - 14 - 18),$$

$$= -14, \tag{2.148}$$

$$I_T^{(3)} = \det \mathbf{T} = \det \begin{pmatrix} 1 & 2 & 4 \\ 2 & 3 & -1 \\ 4 & -1 & 1 \end{pmatrix} = -66. \tag{2.149}$$

Now, when we rotate the axes, we get a transformed tensor representation given by

$$\overline{\mathbf{T}} = \mathbf{Q}^T \cdot \mathbf{T} \cdot \mathbf{Q} = \begin{pmatrix} \frac{1}{\sqrt{6}} & \frac{1}{\sqrt{3}} & \frac{1}{\sqrt{2}} \\ \sqrt{\frac{2}{3}} & -\frac{1}{\sqrt{3}} & 0 \\ \frac{1}{\sqrt{6}} & \frac{1}{\sqrt{3}} & -\frac{1}{\sqrt{2}} \end{pmatrix} \begin{pmatrix} 1 & 2 & 4 \\ 2 & 3 & -1 \\ 4 & -1 & 1 \end{pmatrix} \begin{pmatrix} \frac{1}{\sqrt{6}} & \sqrt{\frac{2}{3}} & \frac{1}{\sqrt{6}} \\ \frac{1}{\sqrt{3}} & -\frac{1}{\sqrt{3}} & \frac{1}{\sqrt{3}} \\ \frac{1}{\sqrt{2}} & 0 & -\frac{1}{\sqrt{2}} \end{pmatrix}, \tag{2.150}$$

$$= \begin{pmatrix} 4.10238 & 2.52239 & 1.60948 \\ 2.52239 & -0.218951 & -2.91291 \\ 1.60948 & -2.91291 & 1.11657 \end{pmatrix}. \tag{2.151}$$

We then seek the tensor invariants of $\overline{\mathbf{T}}$. Leaving out the details, which are the same as those for calculating the invariants of $\mathbf{T}$, we find that they are indeed invariant:

$$I_T^{(1)} = 4.10238 - 0.218951 + 1.11657 = 5, \tag{2.152}$$

$$I_T^{(2)} = \frac{1}{2}(5^2 - 53) = -14, \tag{2.153}$$

$$I_T^{(3)} = -66. \tag{2.154}$$

Finally, we verify that the tensor invariants are related to the principal values (the eigenvalues of the tensor) as follows:

$$I_T^{(1)} = \lambda^{(1)} + \lambda^{(2)} + \lambda^{(3)} = 5.28675 - 3.67956 + 3.39281 = 5, \tag{2.155}$$

$$I_T^{(2)} = \lambda^{(1)}\lambda^{(2)} + \lambda^{(2)}\lambda^{(3)} + \lambda^{(3)}\lambda^{(1)},$$

$$= (5.28675)(-3.67956) + (-3.67956)(3.39281) + (3.39281)(5.28675) = -14, \tag{2.156}$$

$$I_T^{(3)} = \lambda^{(1)}\lambda^{(2)}\lambda^{(3)} = (5.28675)(-3.67956)(3.39281) = -66. \tag{2.157}$$

Quotient Rule

All tensors are matrices, but not all matrices are tensors, because tensors must also obey the transformation rule, Eq. (2.84). Let us look at one test for whether a matrix is in fact a tensor. Let us imagine that $\mathbf{A}$, $\mathbf{B}$, and $\mathbf{C}$ are matrices that follow the standard rules of matrix algebra. Let us also imagine that $\mathbf{B}$ and $\mathbf{C}$ are known to be tensors and that we have $\mathbf{A}$ such that

$$\mathbf{A} \cdot \mathbf{B} = \mathbf{C}. \tag{2.158}$$

The *quotient rule* allows us to assert that $\mathbf{A}$ is also a tensor. This is easy to prove. Let us assume that $\mathbf{A}, \mathbf{B}$ and $\mathbf{C}$ are all represented in a new coordinate system, and show that $\mathbf{A}$ must then transform as a tensor. In the new coordinate system, we have then

$$\overline{\mathbf{A}} \cdot \overline{\mathbf{B}} = \overline{\mathbf{C}}. \tag{2.159}$$

Because $\mathbf{C}$ is a tensor, Eq. (2.84) allows us to rewrite $\overline{\mathbf{C}}$, thus yielding for a given rotation $\mathbf{Q}$

$$\overline{\mathbf{A}} \cdot \overline{\mathbf{B}} = \mathbf{Q}^T \cdot \mathbf{C} \cdot \mathbf{Q}. \tag{2.160}$$

Now use Eq. (2.158) to eliminate $\mathbf{C}$, giving

$$\overline{\mathbf{A}} \cdot \overline{\mathbf{B}} = \mathbf{Q}^T \cdot \mathbf{A} \cdot \mathbf{B} \cdot \mathbf{Q}. \tag{2.161}$$

Because $\mathbf{B}$ is known to be a tensor, Eq. (2.84) allows us to rewrite as

$$\overline{\mathbf{A}} \cdot \overline{\mathbf{B}} = \mathbf{Q}^T \cdot \mathbf{A} \cdot \underbrace{\mathbf{Q} \cdot \overline{\mathbf{B}} \cdot \mathbf{Q}^T}_{\mathbf{B}} \cdot \mathbf{Q}, \tag{2.162}$$

$$= \mathbf{Q}^T \cdot \mathbf{A} \cdot \mathbf{Q} \cdot \overline{\mathbf{B}}, \tag{2.163}$$

$$(\overline{\mathbf{A}} - \mathbf{Q}^T \cdot \mathbf{A} \cdot \mathbf{Q}) \cdot \overline{\mathbf{B}} = 0. \tag{2.164}$$

Because this must hold for *all* $\overline{\mathbf{B}}$, we must have

$$\overline{\mathbf{A}} = \mathbf{Q}^T \cdot \mathbf{A} \cdot \mathbf{Q}. \tag{2.165}$$

Because $\mathbf{A}$ is seen to satisfy the requirement for a tensor, Eq. (2.84), it too must be a tensor.

2.2 Algebra of Vectors

Let us now consider more formally the algebra of vectors. We introduce some concepts and notation on which we expand later in Chapter 6.

2.2.1 Definitions and Properties

The following definitions and properties are useful.

- Null vector: a vector with zero magnitude,
- Multiplication by a scalar α: $\alpha \mathbf{u} = \alpha u_1 \mathbf{e}_1 + \alpha u_2 \mathbf{e}_2 + \alpha u_3 \mathbf{e}_3 = \alpha u_i$,
- Sum of vectors: $\mathbf{u} + \mathbf{v} = (u_1 + v_1)\mathbf{e}_1 + (u_2 + v_2)\mathbf{e}_2 + (u_3 + v_3)\mathbf{e}_3 = (u_i + v_i)$,
- Magnitude, length, or norm of a vector: $||\mathbf{u}||_2 = \sqrt{u_1^2 + u_2^2 + u_3^2} = \sqrt{u_i u_i}$,
- Triangle inequality: $||\mathbf{u} + \mathbf{v}||_2 \leq ||\mathbf{u}||_2 + ||\mathbf{v}||_2$.

As discussed in detail in Section 6.3.1, the subscript 2 in $|| \cdot ||_2$ indicates that we are considering a so-called Euclidean norm, also called a 2-norm. In many sources in the literature, this subscript is omitted, and the norm is understood to be the Euclidean norm. It is correct to think of the Euclidean norm of a distance vector to be its ordinary length as given by Pythagoras' theorem. In a more general sense, we can still retain the property of a norm for a more general Schatten[3] p-norm for a three-dimensional vector:

$$||\mathbf{u}||_p = (|u_1|^p + |u_2|^p + |u_3|^p)^{1/p}, \qquad 1 \le p < \infty. \tag{2.166}$$

For example, the 1-norm of a vector is the sum of the absolute values of its components:

$$||\mathbf{u}||_1 = (|u_1| + |u_2| + |u_3|). \tag{2.167}$$

The ∞-norm selects the component with the largest magnitude:

$$||\mathbf{u}||_\infty = \lim_{p \to \infty} (|u_1|^p + |u_2|^p + |u_3|^p)^{1/p} = \max_{i=1,2,3} |u_i|. \tag{2.168}$$

2.2.2 Scalar Product

The *scalar product*, also known as the *dot product* or *inner product*, of $\mathbf{u}$ and $\mathbf{v}$ is defined for vectors with real components as

$$\langle \mathbf{u}, \mathbf{v} \rangle = \mathbf{u}^T \cdot \mathbf{v} = \begin{pmatrix} u_1 & u_2 & u_3 \end{pmatrix} \begin{pmatrix} v_1 \\ v_2 \\ v_3 \end{pmatrix} = u_1 v_1 + u_2 v_2 + u_3 v_3 = u_i v_i. \tag{2.169}$$

The term $u_i v_i$ is a scalar, which explains the nomenclature *scalar product*. Because it reduces vectors into a scalar, it is sometimes classified as a contraction. Because scalars are invariant under coordinate transformation, the scalar product cannot depend on the coordinate system chosen. The transpose operation renders a column vector into a row vector.

The vectors $\mathbf{u}$ and $\mathbf{v}$ are said to be *orthogonal* if $\mathbf{u}^T \cdot \mathbf{v} = 0$. Also

$$\langle \mathbf{u}, \mathbf{u} \rangle = \mathbf{u}^T \cdot \mathbf{u} = \begin{pmatrix} u_1 & u_2 & u_3 \end{pmatrix} \begin{pmatrix} u_1 \\ u_2 \\ u_3 \end{pmatrix} = u_1^2 + u_2^2 + u_3^2 = u_i u_i = (||\mathbf{u}||_2)^2. \tag{2.170}$$

We consider important modifications for vectors with complex components later in Section 6.3.2. In the same section, we consider the generalized notion of an inner product, denoted here by $\langle \cdot, \cdot \rangle$.

2.2.3 Cross Product

The cross product of $\mathbf{u}$ and $\mathbf{v}$ is defined as

$$\mathbf{u} \times \mathbf{v} = \begin{vmatrix} \mathbf{e}_1 & \mathbf{e}_2 & \mathbf{e}_3 \\ u_1 & u_2 & u_3 \\ v_1 & v_2 & v_3 \end{vmatrix} = \epsilon_{ijk} u_j v_k. \tag{2.171}$$

The cross product of two vectors is a vector.

---

[3] Robert Schatten, 1911–1977, Polish and American mathematician.

Property: $\mathbf{u} \times \alpha\mathbf{u} = \mathbf{0}$. Let us use Cartesian index notation to prove this

$$\mathbf{u} \times \alpha\mathbf{u} = \epsilon_{ijk}u_j\alpha u_k, \tag{2.172}$$

$$= \alpha\epsilon_{ijk}u_j u_k, \tag{2.173}$$

$$= \alpha(\epsilon_{i11}u_1 u_1 + \epsilon_{i12}u_1 u_2 + \epsilon_{i13}u_1 u_3, \tag{2.174}$$

$$+ \epsilon_{i21}u_2 u_1 + \epsilon_{i22}u_2 u_2 + \epsilon_{i23}u_2 u_3 \tag{2.175}$$

$$+ \epsilon_{i31}u_3 u_1 + \epsilon_{i32}u_3 u_2 + \epsilon_{i33}u_3 u_3), \tag{2.176}$$

$$= 0, \quad \text{for } i = 1, 2, 3, \tag{2.177}$$

because $\epsilon_{i11} = \epsilon_{i22} = \epsilon_{i33} = 0$ and $\epsilon_{i12} = -\epsilon_{i21}, \epsilon_{i13} = -\epsilon_{i31}$, and $\epsilon_{i23} = -\epsilon_{i32}$.

2.2.4 Scalar Triple Product

The scalar triple product of three vectors $\mathbf{u}$, $\mathbf{v}$, and $\mathbf{w}$ is defined by

$$[\mathbf{u}, \mathbf{v}, \mathbf{w}] = \mathbf{u}^T \cdot (\mathbf{v} \times \mathbf{w}), \tag{2.178}$$

$$= \epsilon_{ijk}u_i v_j w_k. \tag{2.179}$$

The scalar triple product is a scalar. Geometrically, it represents the volume of the parallelepiped with edges parallel to the three vectors.

2.2.5 Identities

The following identities can be proved:

$$[\mathbf{u}, \mathbf{v}, \mathbf{w}] = -[\mathbf{u}, \mathbf{w}, \mathbf{v}], \tag{2.180}$$

$$\mathbf{u} \times (\mathbf{v} \times \mathbf{w}) = (\mathbf{u}^T \cdot \mathbf{w})\mathbf{v} - (\mathbf{u}^T \cdot \mathbf{v})\mathbf{w}, \tag{2.181}$$

$$(\mathbf{u} \times \mathbf{v}) \times (\mathbf{w} \times \mathbf{x}) = [\mathbf{u}, \mathbf{w}, \mathbf{x}]\mathbf{v} - [\mathbf{v}, \mathbf{w}, \mathbf{x}]\mathbf{u}, \tag{2.182}$$

$$(\mathbf{u} \times \mathbf{v})^T \cdot (\mathbf{w} \times \mathbf{x}) = (\mathbf{u}^T \cdot \mathbf{w})(\mathbf{v}^T \cdot \mathbf{x}) - (\mathbf{u}^T \cdot \mathbf{x})(\mathbf{v}^T \cdot \mathbf{w}). \tag{2.183}$$

EXAMPLE 2.9

Prove Eq. (2.181) using Cartesian index notation:

$$\mathbf{u} \times (\mathbf{v} \times \mathbf{w}) = \epsilon_{ijk}u_j \left(\epsilon_{klm}v_l w_m\right), \tag{2.184}$$

$$= \epsilon_{ijk}\epsilon_{klm}u_j v_l w_m, \tag{2.185}$$

$$= \epsilon_{kij}\epsilon_{klm}u_j v_l w_m, \tag{2.186}$$

$$= (\delta_{il}\delta_{jm} - \delta_{im}\delta_{jl})\, u_j v_l w_m, \tag{2.187}$$

$$= u_j v_i w_j - u_j v_j w_i, \tag{2.188}$$

$$= u_j w_j v_i - u_j v_j w_i, \tag{2.189}$$

$$= (\mathbf{u}^T \cdot \mathbf{w})\mathbf{v} - (\mathbf{u}^T \cdot \mathbf{v})\mathbf{w}. \tag{2.190}$$

2.3 Calculus of Vectors

Next, one can consider the calculus of vectors.

2.3.1 Vector Functions

If we have the scalar function $\phi(\tau)$ and vector functions $\mathbf{u}(\tau)$ and $\mathbf{v}(\tau)$, each of a single scalar variable τ, some useful identities, based on the product rule, which can be proved include

$$\frac{d}{d\tau}(\phi\mathbf{u}) = \phi\frac{d\mathbf{u}}{d\tau} + \frac{d\phi}{d\tau}\mathbf{u}, \qquad \frac{d}{d\tau}(\phi u_i) = \phi\frac{du_i}{d\tau} + \frac{d\phi}{d\tau}u_i, \qquad (2.191)$$

$$\frac{d}{d\tau}(\mathbf{u}^T\cdot\mathbf{v}) = \mathbf{u}^T\cdot\frac{d\mathbf{v}}{d\tau} + \frac{d\mathbf{u}^T}{d\tau}\cdot\mathbf{v}, \qquad \frac{d}{d\tau}(u_iv_i) = u_i\frac{dv_i}{d\tau} + \frac{du_i}{d\tau}v_i, \qquad (2.192)$$

$$\frac{d}{d\tau}(\mathbf{u}\times\mathbf{v}) = \mathbf{u}\times\frac{d\mathbf{v}}{d\tau} + \frac{d\mathbf{u}}{d\tau}\times\mathbf{v}, \qquad \frac{d}{d\tau}(\epsilon_{ijk}u_jv_k) = \epsilon_{ijk}u_j\frac{dv_k}{d\tau} + \epsilon_{ijk}v_k\frac{du_j}{d\tau}. \qquad (2.193)$$

Here τ is a general scalar parameter, which may or may not have a simple physical interpretation.

2.3.2 Differential Geometry of Curves

Now let us begin a general discussion of curves in space, which in fact provides an introduction to the broader field known as *differential geometry*.

If

$$\mathbf{r}(\tau) = x_i(\tau)\mathbf{e}_i = x_i(\tau), \qquad (2.194)$$

then $\mathbf{r}(\tau)$ describes a curve in three-dimensional space. If we require that the basis vectors be constants (this will not be the case in most general coordinate systems, but is for ordinary Cartesian systems), the derivative of Eq. (2.194) is

$$\frac{d\mathbf{r}(\tau)}{d\tau} = \mathbf{r}'(\tau) = x_i'(\tau)\mathbf{e}_i = x_i'(\tau). \qquad (2.195)$$

Now $\mathbf{r}'(\tau)$ is a vector that is tangent to the curve. A unit vector in this direction is

$$\mathbf{t} = \frac{\mathbf{r}'(\tau)}{||\mathbf{r}'(\tau)||_2}, \qquad (2.196)$$

where

$$||\mathbf{r}'(\tau)||_2 = \sqrt{x_i'x_i'}. \qquad (2.197)$$

In the special case in which τ is time t, we denote the derivative by dot ($\dot{}$) notation rather than prime ($'$) notation; $\dot{\mathbf{r}}$ is the velocity vector, $\dot{x}_i$ is its ith component, and $||\dot{\mathbf{r}}||_2$ is the velocity magnitude. The unit tangent vector $\mathbf{t}$ *is not related* to scalar parameter for time t. Also, we occasionally use the scalar components of $\mathbf{t}$; t_i, which again are not related to time t.

Take $s(t)$ to be the time-varying distance along the curve. Pythagoras' theorem tells us for differential distances that

$$ds^2 = dx_1^2 + dx_2^2 + dx_3^2, \qquad (2.198)$$

$$ds = \sqrt{dx_1^2 + dx_2^2 + dx_3^2}, \qquad (2.199)$$

$$= ||dx_i||_2, \tag{2.200}$$

$$\frac{ds}{dt} = \left\|\frac{dx_i}{dt}\right\|_2, \tag{2.201}$$

$$= ||\dot{\mathbf{r}}(t)||_2, \tag{2.202}$$

so that

$$\mathbf{t} = \frac{\dot{\mathbf{r}}}{||\dot{\mathbf{r}}||_2} = \frac{\frac{d\mathbf{r}}{dt}}{\frac{ds}{dt}} = \frac{d\mathbf{r}}{ds}, \qquad t_i = \frac{dr_i}{ds}. \tag{2.203}$$

Also, integrating Eq. (2.202) with respect to t gives

$$s = \int_a^b ||\dot{\mathbf{r}}(t)||_2\, dt = \int_a^b \sqrt{\frac{dx_i}{dt}\frac{dx_i}{dt}}\, dt = \int_a^b \sqrt{\frac{dx_1}{dt}\frac{dx_1}{dt} + \frac{dx_2}{dt}\frac{dx_2}{dt} + \frac{dx_3}{dt}\frac{dx_3}{dt}}\, dt, \tag{2.204}$$

to be the distance along the curve between $t = a$ and $t = b$.

EXAMPLE 2.10

If

$$\mathbf{r}(t) = 2t^2\mathbf{i} + t^3\mathbf{j}, \tag{2.205}$$

find the unit tangent at $t = 1$ and the length of the curve from $t = 0$ to $t = 1$.

Assuming t is time, the time derivative is

$$\dot{\mathbf{r}}(t) = 4t\mathbf{i} + 3t^2\mathbf{j}. \tag{2.206}$$

At $t = 1$,

$$\dot{\mathbf{r}}(t = 1) = 4\mathbf{i} + 3\mathbf{j}, \tag{2.207}$$

so that the unit vector in this direction is

$$\mathbf{t} = \frac{4}{5}\mathbf{i} + \frac{3}{5}\mathbf{j}. \tag{2.208}$$

The length of the curve from $t = 0$ to $t = 1$ is

$$s = \int_0^1 \sqrt{16t^2 + 9t^4}\, dt, \tag{2.209}$$

$$= \frac{1}{27}(16 + 9t^2)^{3/2}\,|_0^1, \tag{2.210}$$

$$= \frac{61}{27}. \tag{2.211}$$

In Figure 2.7, $\mathbf{r}(t)$ describes a circle. Two unit tangents, $\mathbf{t}$ and $\hat{\mathbf{t}}$, are drawn at times t and $t + \Delta t$. At time t, we have

$$\mathbf{t} = -\sin\theta\mathbf{i} + \cos\theta\mathbf{j}. \tag{2.212}$$

At time $t + \Delta t$, we have

$$\hat{\mathbf{t}} = -\sin(\theta + \Delta\theta)\mathbf{i} + \cos(\theta + \Delta\theta)\mathbf{j}. \tag{2.213}$$

Expanding Eq. (2.213) in a Taylor series (see Section 5.1.1) about $\Delta\theta = 0$, we get

$$\hat{\mathbf{t}} = \left(-\sin\theta - \Delta\theta\cos\theta + O(\Delta\theta)^2\right)\mathbf{i} + \left(\cos\theta - \Delta\theta\sin\theta + O(\Delta\theta)^2\right)\mathbf{j}, \tag{2.214}$$

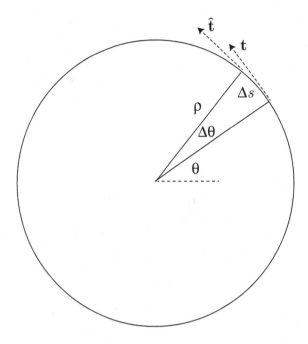

Figure 2.7. Sketch for determination of radius of curvature.

so as $\Delta\theta \to 0$,

$$\hat{\mathbf{t}} - \mathbf{t} = -\Delta\theta \cos\theta \mathbf{i} - \Delta\theta \sin\theta \mathbf{j}, \tag{2.215}$$

$$\Delta\mathbf{t} = \Delta\theta \underbrace{(-\cos\theta \mathbf{i} - \sin\theta \mathbf{j})}_{\text{unit vector}}. \tag{2.216}$$

It is easily verified that $\Delta\mathbf{t}^T \cdot \mathbf{t} = 0$, so $\Delta\mathbf{t}$ is normal to $\mathbf{t}$. Furthermore, because $-\cos\theta \mathbf{i} - \sin\theta \mathbf{j}$ is a unit vector,

$$||\Delta\mathbf{t}||_2 = \Delta\theta. \tag{2.217}$$

Now, for small $\Delta\theta$,

$$\Delta s = \rho \Delta\theta, \tag{2.218}$$

where ρ is the *radius of curvature*. So

$$||\Delta\mathbf{t}||_2 = \frac{\Delta s}{\rho}. \tag{2.219}$$

Thus,

$$\left|\left|\frac{\Delta\mathbf{t}}{\Delta s}\right|\right|_2 = \frac{1}{\rho}. \tag{2.220}$$

Taking all limits to zero, we get

$$\left|\left|\frac{d\mathbf{t}}{ds}\right|\right|_2 = \frac{1}{\rho}. \tag{2.221}$$

The term on the right side of Eq. (2.221) is defined as the *curvature*, κ:

$$\kappa = \frac{1}{\rho}. \tag{2.222}$$

Thus, the curvature κ is the magnitude of dt/ds; it gives a measure of how the unit tangent changes as one moves along the curve.

Curves on a Plane

The plane curve $y = f(x)$ in the (x, y) plane can be represented as

$$\mathbf{r}(t) = x(t)\mathbf{i} + y(t)\mathbf{j}, \tag{2.223}$$

where $x(t) = t$ and $y(t) = f(t)$. Differentiating, we have

$$\dot{\mathbf{r}}(t) = \dot{x}(t)\mathbf{i} + \dot{y}(t)\mathbf{j}. \tag{2.224}$$

The unit vector from Eq. (2.196) is

$$\mathbf{t} = \frac{\dot{x}\mathbf{i} + \dot{y}\mathbf{j}}{(\dot{x}^2 + \dot{y}^2)^{1/2}}, \tag{2.225}$$

$$= \frac{\mathbf{i} + y'\mathbf{j}}{(1 + (y')^2)^{1/2}}. \tag{2.226}$$

Because

$$ds^2 = dx^2 + dy^2, \tag{2.227}$$

$$ds = \left(dx^2 + dy^2\right)^{1/2}, \tag{2.228}$$

$$\frac{ds}{dx} = \frac{1}{dx}\left(dx^2 + dy^2\right)^{1/2}, \tag{2.229}$$

$$= (1 + (y')^2)^{1/2}, \tag{2.230}$$

we have, by first expanding dt/ds with the chain rule, then applying the quotient rule for derivatives to expand the derivative of Eq. (2.226) along with the use of Eq. (2.230),

$$\frac{dt}{ds} = \frac{\frac{dt}{dx}}{\frac{ds}{dx}}, \tag{2.231}$$

$$= \underbrace{\frac{(1 + (y')^2)^{1/2}y''\mathbf{j} - (\mathbf{i} + y'\mathbf{j})(1 + (y')^2)^{-1/2}y'y''}{1 + (y')^2}}_{dt/dx} \underbrace{\frac{1}{(1 + (y')^2)^{1/2}}}_{1/(ds/dx)}, \tag{2.232}$$

$$= \underbrace{\frac{y''}{(1 + (y')^2)^{3/2}}}_{=\kappa} \underbrace{\frac{-y'\mathbf{i} + \mathbf{j}}{(1 + (y')^2)^{1/2}}}_{\mathbf{n}}. \tag{2.233}$$

As the second factor of Eq. (2.233) is a unit vector, the leading scalar factor must be the magnitude of dt/ds. We define this unit vector to be $\mathbf{n}$ and note that it is orthogonal to the unit tangent vector $\mathbf{t}$:

$$\mathbf{n}^T \cdot \mathbf{t} = \frac{-y'\mathbf{i} + \mathbf{j}}{(1 + (y')^2)^{1/2}} \cdot \frac{\mathbf{i} + y'\mathbf{j}}{(1 + (y')^2)^{1/2}}, \tag{2.234}$$

$$= \frac{-y' + y'}{1 + (y')^2}, \tag{2.235}$$

$$= 0. \tag{2.236}$$

Expanding our notion of curvature and radius of curvature, we define dt/ds such that

$$\frac{d\mathbf{t}}{ds} = \kappa\mathbf{n}, \tag{2.237}$$

$$\left\|\frac{d\mathbf{t}}{ds}\right\|_2 = \kappa = \frac{1}{\rho}. \tag{2.238}$$

Thus,

$$\kappa = \frac{y''}{(1 + (y')^2)^{3/2}}, \tag{2.239}$$

$$\rho = \frac{(1 + (y')^2)^{3/2}}{y''}, \tag{2.240}$$

for curves on a plane.

Curves in Three-Dimensional Space

We next expand these notions to three-dimensional space. A set of local, right-handed, orthogonal coordinates can be defined at a point on a curve $\mathbf{r}(t)$. The unit vectors at this point are the tangent $\mathbf{t}$, the principal normal $\mathbf{n}$, and the binormal $\mathbf{b}$, where

$$\mathbf{t} = \frac{d\mathbf{r}}{ds}, \tag{2.241}$$

$$\mathbf{n} = \frac{1}{\kappa}\frac{d\mathbf{t}}{ds}, \tag{2.242}$$

$$\mathbf{b} = \mathbf{t} \times \mathbf{n}. \tag{2.243}$$

We first show that $\mathbf{t}$, $\mathbf{n}$, and $\mathbf{b}$ form an orthogonal system of unit vectors. We have already seen that $\mathbf{t}$ is a unit vector tangent to the curve. By the product rule for vector differentiation, we have the identity

$$\mathbf{t}^T \cdot \frac{d\mathbf{t}}{ds} = \frac{1}{2}\frac{d}{ds}(\underbrace{\mathbf{t}^T \cdot \mathbf{t}}_{=1}). \tag{2.244}$$

Because $\mathbf{t}^T \cdot \mathbf{t} = \|\mathbf{t}\|_2^2 = 1$, we recover

$$\mathbf{t}^T \cdot \frac{d\mathbf{t}}{ds} = 0. \tag{2.245}$$

Thus, $\mathbf{t}$ is orthogonal to $d\mathbf{t}/ds$. Because $\mathbf{n}$ is parallel to $d\mathbf{t}/ds$, it is orthogonal to $\mathbf{t}$ also. From Eqs. (2.221) and (2.242), we see that $\mathbf{n}$ is a unit vector. Furthermore, $\mathbf{b}$ is a unit vector orthogonal to both $\mathbf{t}$ and $\mathbf{n}$ because of its definition in terms of a cross product of those vectors in Eq. (2.243).

Next, we derive some basic relations involving the unit vectors and the characteristics of the curve. Take d/ds of Eq. (2.243):

$$\frac{d\mathbf{b}}{ds} = \frac{d}{ds}(\mathbf{t} \times \mathbf{n}), \tag{2.246}$$

$$= \frac{d\mathbf{t}}{ds} \times \underbrace{\mathbf{n}}_{(1/\kappa)\,d\mathbf{t}/ds} + \mathbf{t} \times \frac{d\mathbf{n}}{ds}, \tag{2.247}$$

$$= \frac{dt}{ds} \times \frac{1}{\kappa}\frac{dt}{ds} + t \times \frac{dn}{ds}, \qquad (2.248)$$

$$= \frac{1}{\kappa}\underbrace{\frac{dt}{ds} \times \frac{dt}{ds}}_{=0} + t \times \frac{dn}{ds}, \qquad (2.249)$$

$$= t \times \frac{dn}{ds}. \qquad (2.250)$$

So we see that db/ds is orthogonal to t. In addition, because $||b||_2 = 1$,

$$b^T \cdot \frac{db}{ds} = \frac{1}{2}\frac{d}{ds}(b^T \cdot b), \qquad (2.251)$$

$$= \frac{1}{2}\frac{d}{ds}(||b||_2^2), \qquad (2.252)$$

$$= \frac{1}{2}\frac{d}{ds}(1^2), \qquad (2.253)$$

$$= 0. \qquad (2.254)$$

So db/ds is orthogonal to b also. Because db/ds is orthogonal to both t and b, it must be aligned with the only remaining direction, n. So, we can write

$$\frac{db}{ds} = \tau n, \qquad (2.255)$$

where τ is the magnitude of db/ds, which we call the *torsion* of the curve.

From Eq. (2.243), it is easily deduced that $n = b \times t$. Differentiating this with respect to s, we get

$$\frac{dn}{ds} = \frac{db}{ds} \times t + b \times \frac{dt}{ds}, \qquad (2.256)$$

$$= \tau n \times t + b \times \kappa n, \qquad (2.257)$$

$$= -\tau b - \kappa t. \qquad (2.258)$$

Summarizing,

$$\frac{dt}{ds} = \kappa n, \qquad (2.259)$$

$$\frac{dn}{ds} = -\kappa t - \tau b, \qquad (2.260)$$

$$\frac{db}{ds} = \tau n. \qquad (2.261)$$

These are the Frenet-Serret[4] relations. In matrix form, we can say that

$$\frac{d}{ds}\begin{pmatrix} t \\ n \\ b \end{pmatrix} = \begin{pmatrix} 0 & \kappa & 0 \\ -\kappa & 0 & -\tau \\ 0 & \tau & 0 \end{pmatrix}\begin{pmatrix} t \\ n \\ b \end{pmatrix}. \qquad (2.262)$$

The coefficient matrix is antisymmetric.

---

[4] Jean Frédéric Frenet, 1816–1900, French mathematician, and Joseph Alfred Serret, 1819–1885, French mathematician.

EXAMPLE 2.11

Find the local coordinates, the curvature, and the torsion for the helix

$$\mathbf{r}(t) = a\cos t\mathbf{i} + a\sin t\mathbf{j} + bt\mathbf{k}. \tag{2.263}$$

Taking the derivative and finding its magnitude, we get

$$\frac{d\mathbf{r}(t)}{dt} = -a\sin t\mathbf{i} + a\cos t\mathbf{j} + b\mathbf{k}, \tag{2.264}$$

$$\left\| \frac{d\mathbf{r}(t)}{dt} \right\|_2 = \sqrt{a^2\sin^2 t + a^2\cos^2 t + b^2}, \tag{2.265}$$

$$= \sqrt{a^2 + b^2}. \tag{2.266}$$

This gives us the unit tangent vector $\mathbf{t}$:

$$\mathbf{t} = \frac{\frac{d\mathbf{r}}{dt}}{\left\|\frac{d\mathbf{r}}{dt}\right\|_2} = \frac{-a\sin t\mathbf{i} + a\cos t\mathbf{j} + b\mathbf{k}}{\sqrt{a^2 + b^2}}. \tag{2.267}$$

We also have

$$\frac{ds}{dt} = \sqrt{\left(\frac{dx}{dt}\right)^2 + \left(\frac{dy}{dt}\right)^2 + \left(\frac{dz}{dt}\right)^2}, \tag{2.268}$$

$$= \sqrt{a^2\sin^2 t + a^2\cos^2 t + b^2}, \tag{2.269}$$

$$= \sqrt{a^2 + b^2}. \tag{2.270}$$

Continuing, we have

$$\frac{d\mathbf{t}}{ds} = \frac{\frac{d\mathbf{t}}{dt}}{\frac{ds}{dt}}, \tag{2.271}$$

$$= -a\frac{\cos t\mathbf{i} + \sin t\mathbf{j}}{\sqrt{a^2 + b^2}}\frac{1}{\sqrt{a^2 + b^2}}, \tag{2.272}$$

$$= \underbrace{\frac{a}{a^2 + b^2}}_{\kappa}\underbrace{(-\cos t\mathbf{i} - \sin t\mathbf{j})}_{\mathbf{n}}, \tag{2.273}$$

$$= \kappa\mathbf{n}. \tag{2.274}$$

Thus, the unit principal normal is

$$\mathbf{n} = -(\cos t\mathbf{i} + \sin t\mathbf{j}). \tag{2.275}$$

The curvature is

$$\kappa = \frac{a}{a^2 + b^2}. \tag{2.276}$$

The radius of curvature is

$$\rho = \frac{a^2 + b^2}{a}. \tag{2.277}$$

We also find the unit binormal

$$\mathbf{b} = \mathbf{t} \times \mathbf{n}, \tag{2.278}$$

$$= \frac{1}{\sqrt{a^2 + b^2}}\begin{vmatrix} \mathbf{i} & \mathbf{j} & \mathbf{k} \\ -a\sin t & a\cos t & b \\ -\cos t & -\sin t & 0 \end{vmatrix}, \tag{2.279}$$

$$= \frac{b\sin t\,\mathbf{i} - b\cos t\,\mathbf{j} + a\,\mathbf{k}}{\sqrt{a^2 + b^2}}. \tag{2.280}$$

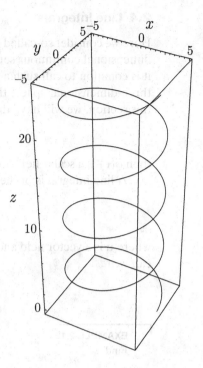

Figure 2.8. Three-dimensional curve parameterized by $x(t) = a \cos t$, $y(t) = a \sin t$, $z(t) = bt$, with $a = 5$, $b = 1$, for $t \in [0, 25]$.

The torsion is determined from

$$\tau \mathbf{n} = \frac{\frac{d\mathbf{b}}{dt}}{\frac{ds}{dt}}, \tag{2.281}$$

$$= b \frac{\cos t \, \mathbf{i} + \sin t \, \mathbf{j}}{a^2 + b^2}, \tag{2.282}$$

$$= \underbrace{\frac{-b}{a^2 + b^2}}_{\tau} \underbrace{(- \cos t \, \mathbf{i} - \sin t \, \mathbf{j})}_{\mathbf{n}}, \tag{2.283}$$

from which

$$\tau = -\frac{b}{a^2 + b^2}. \tag{2.284}$$

Further identities that can be proved relate directly to the time parameterization of $\mathbf{r}$:

$$\frac{d\mathbf{r}}{dt} \times \frac{d^2\mathbf{r}}{dt^2} = \kappa v^3 \mathbf{b}, \tag{2.285}$$

$$\left(\frac{d\mathbf{r}}{dt} \times \frac{d^2\mathbf{r}}{dt^2} \right)^T \cdot \frac{d^3\mathbf{r}}{dt^3} = -\kappa^2 v^6 \tau, \tag{2.286}$$

$$\frac{\sqrt{\|\ddot{\mathbf{r}}\|_2^2 \, \|\dot{\mathbf{r}}\|_2^2 - (\dot{\mathbf{r}}^T \cdot \ddot{\mathbf{r}})^2}}{\|\dot{\mathbf{r}}\|_2^3} = \kappa, \tag{2.287}$$

where $v = ds/dt$.

2.4 Line Integrals

Here we consider so-called line integrals. For line integrals, we integrate along a one-dimensional continuous set of points; they may or may not form a straight line, but it is common to call such a process line integration. Although we focus on ordinary three-dimensional space, the analysis can be extended for higher dimensions. For this section, we will have that if $\mathbf{r}$ is a position vector,

$$\mathbf{r} = x_i \mathbf{e}_i, \tag{2.288}$$

then $\phi(\mathbf{r})$ is a scalar field, and $\mathbf{u}(\mathbf{r})$ is a vector field.

A line integral is of the form

$$I = \int_C \mathbf{u}^T \cdot d\mathbf{r}, \tag{2.289}$$

where $\mathbf{u}$ is a vector field and dr is an element of curve C. If $\mathbf{u} = u_i$, and $d\mathbf{r} = dx_i$, we can write

$$I = \int_C u_i \, dx_i. \tag{2.290}$$

EXAMPLE 2.12

Find

$$I = \int_C \mathbf{u}^T \cdot d\mathbf{r}, \tag{2.291}$$

if

$$\mathbf{u} = yz\mathbf{i} + xy\mathbf{j} + xz\mathbf{k}, \tag{2.292}$$

and C goes from $(0,0,0)$ to $(1,1,1)$ along

(a) the curve $x = y^2 = z$
(b) the straight line $x = y = z$.

The vector field and two paths are sketched in Figure 2.9. We have

$$\int_C \mathbf{u}^T \cdot d\mathbf{r} = \int_C (yz \, dx + xy \, dy + xz \, dz). \tag{2.293}$$

(a) Substituting $x = y^2 = z$, and thus $dx = 2y \, dy$, $dx = dz$, we get

$$I = \int_0^1 y^3 (2y \, dy) + y^3 \, dy + y^4 (2y \, dy), \tag{2.294}$$

$$= \int_0^1 (2y^4 + y^3 + 2y^5) \, dy, \tag{2.295}$$

$$= \frac{2y^5}{5} + \frac{y^4}{4} + \frac{y^6}{3} \Big|_0^1, \tag{2.296}$$

$$= \frac{59}{60}. \tag{2.297}$$

We can achieve the same result in an alternative way that is often more useful for curves whose representation is more complicated. Let us parameterize C by taking $x = t$,

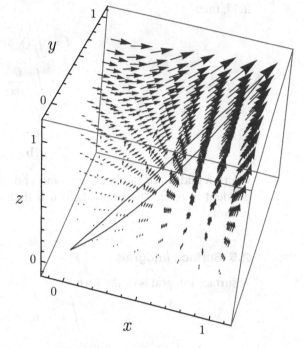

Figure 2.9. The vector field $\mathbf{u} = yz\mathbf{i} + xy\mathbf{j} + xz\mathbf{k}$ and the curves (a) $x = y^2 = z$ and (b) $x = y = z$.

$y = t^2$, $z = t$. Thus $dx = dt$, $dy = 2t\,dt$, $dz = dt$. The endpoints of C are at $t = 0$ and $t = 1$. So the integral is

$$I = \int_0^1 (t^2 t\,dt + tt^2(2t)\,dt + tt\,dt), \tag{2.298}$$

$$= \int_0^1 (t^3 + 2t^4 + t^2)\,dt, \tag{2.299}$$

$$= \frac{t^4}{4} + \frac{2t^5}{5} + \frac{t^3}{3}\bigg|_0^1, \tag{2.300}$$

$$= \frac{59}{60}. \tag{2.301}$$

(b) Substituting $x = y = z$, and thus $dx = dy = dz$, we get

$$I = \int_0^1 (x^2\,dx + x^2\,dx + x^2\,dx) = \int_0^1 3x^2\,dx = x^3|_0^1 = 1. \tag{2.302}$$

A different value for I was obtained on path (b) relative to that found on path (a); thus, the line integral here is path-dependent.

In general the value of a line integral depends on the path. If, however, we have the special case in which we can form $\mathbf{u} = \nabla\phi$ in Eq. (2.289),[5] where ϕ is a scalar

---

[5] Some sources choose an alternate definition, $\mathbf{u} = -\nabla\phi$, usually with a physical motivation. As long as one is consistent, the sign is not critical.

field, then

$$I = \int_C (\nabla\phi)^T \cdot d\mathbf{r}, \tag{2.303}$$

$$= \int_C \frac{\partial\phi}{\partial x_i} \, dx_i, \tag{2.304}$$

$$= \int_C d\phi, \tag{2.305}$$

$$= \phi(\mathbf{b}) - \phi(\mathbf{a}), \tag{2.306}$$

where $\mathbf{a}$ and $\mathbf{b}$ are the beginning and end of curve C. The integral I is then independent of path. The vector field $\mathbf{u}$ is then called a *conservative* field, and ϕ is its *potential*.

2.5 Surface Integrals

A surface integral is of the form

$$I = \int_S \mathbf{u}^T \cdot \mathbf{n} \, dS = \int_S u_i n_i \, dS, \tag{2.307}$$

where $\mathbf{u}$ (or u_i) is a vector field, S is an open or closed surface, dS is an element of this surface, and $\mathbf{n}$ (or n_i) is a unit vector normal to the surface element.

2.6 Differential Operators

Surface integrals can be used for coordinate-independent definitions of differential operators. Beginning with some well-known theorems; the divergence theorem for a scalar, the divergence theorem, and a third theorem with no common name, which are possible to demonstrate, we have, where S is a surface enclosing volume V,

$$\int_V \nabla\phi \, dV = \int_S \mathbf{n}\phi \, dS, \tag{2.308}$$

$$\int_V \nabla^T \cdot \mathbf{u} \, dV = \int_S \mathbf{n}^T \cdot \mathbf{u} \, dS, \tag{2.309}$$

$$\int_V (\nabla \times \mathbf{u}) \, dV = \int_S \mathbf{n} \times \mathbf{u} \, dS. \tag{2.310}$$

Now, we invoke the mean value theorem, which asserts that somewhere within the limits of integration, the integrand takes on its mean value, which we denote with an overline, so that, for example, $\int_V \alpha \, dV = \overline{\alpha}V$. Thus, we get

$$\overline{(\nabla\phi)} \, V = \int_S \mathbf{n}\phi \, dS, \tag{2.311}$$

$$\overline{(\nabla^T \cdot \mathbf{u})} \, V = \int_S \mathbf{n}^T \cdot \mathbf{u} \, dS, \tag{2.312}$$

$$\overline{(\nabla \times \mathbf{u})} \, V = \int_S \mathbf{n} \times \mathbf{u} \, dS. \tag{2.313}$$

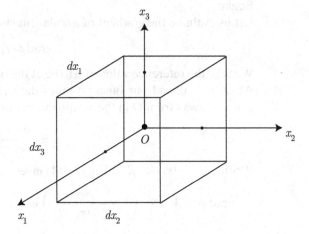

Figure 2.10. Element of volume.

As we let $V \to 0$, mean values approach local values, so we get

$$\nabla \phi \equiv \text{grad } \phi = \lim_{V \to 0} \frac{1}{V} \int_S \mathbf{n} \phi \, dS, \qquad (2.314)$$

$$\nabla^T \cdot \mathbf{u} \equiv \text{div } \mathbf{u} = \lim_{V \to 0} \frac{1}{V} \int_S \mathbf{n}^T \cdot \mathbf{u} \, dS, \qquad (2.315)$$

$$\nabla \times \mathbf{u} \equiv \text{curl } \mathbf{u} = \lim_{V \to 0} \frac{1}{V} \int_S \mathbf{n} \times \mathbf{u} \, dS, \qquad (2.316)$$

where $\phi(\mathbf{r})$ is a scalar field and $\mathbf{u}(\mathbf{r})$ is a vector field. The volume V is the region enclosed within a closed surface S, and $\mathbf{n}$ is the unit normal to an element of the surface dS. Here "grad" is the gradient operator, "div" is the divergence operator, and "curl" is the curl operator.

Consider the element of volume in Cartesian coordinates shown in Figure 2.10. The differential operations in this coordinate system can be deduced from the definitions and written in terms of the vector operator ∇:

$$\nabla = \mathbf{e}_1 \frac{\partial}{\partial x_1} + \mathbf{e}_2 \frac{\partial}{\partial x_2} + \mathbf{e}_3 \frac{\partial}{\partial x_3} = \begin{pmatrix} \frac{\partial}{\partial x_1} \\ \frac{\partial}{\partial x_2} \\ \frac{\partial}{\partial x_3} \end{pmatrix}, \qquad (2.317)$$

$$= \frac{\partial}{\partial x_i}. \qquad (2.318)$$

We also adopt the unconventional, but pedagogically useful, row vector operator

$$\nabla^T = \begin{pmatrix} \frac{\partial}{\partial x_1} & \frac{\partial}{\partial x_2} & \frac{\partial}{\partial x_3} \end{pmatrix}. \qquad (2.319)$$

The operator ∇^T is well defined for Cartesian coordinate systems but does not extend easily to nonorthogonal systems.

2.6.1 Gradient

Let us consider the gradient operator, which is commonly applied to scalar and vector fields.

Scalar

Let us evaluate the gradient of a scalar function of a vector

$$\text{grad}\,\phi(x_i). \tag{2.320}$$

We take the reference value of ϕ to be at the origin O. Consider first the x_1 variation. At O, $x_1 = 0$, and our function takes the value of ϕ. At the faces a distance $x_1 = \pm\, dx_1/2$ away from O in the x_1-direction, our function takes values of

$$\phi \pm \frac{\partial\phi}{\partial x_1}\frac{dx_1}{2}. \tag{2.321}$$

Writing $V = dx_1\,dx_2\,dx_3$, Eq. (2.314) gives

$$\text{grad}\,\phi = \lim_{V\to 0}\frac{1}{V}\left(\left(\phi + \frac{\partial\phi}{\partial x_1}\frac{dx_1}{2}\right)\mathbf{e}_1\,dx_2\,dx_3 - \left(\phi - \frac{\partial\phi}{\partial x_1}\frac{dx_1}{2}\right)\mathbf{e}_1\,dx_2\,dx_3\right.$$

$$\left. +\;\text{similar terms from the }x_2\text{ and }x_3\text{ faces}\right), \tag{2.322}$$

$$= \frac{\partial\phi}{\partial x_1}\mathbf{e}_1 + \frac{\partial\phi}{\partial x_2}\mathbf{e}_2 + \frac{\partial\phi}{\partial x_3}\mathbf{e}_3, \tag{2.323}$$

$$= \frac{\partial\phi}{\partial x_i}\mathbf{e}_i = \frac{\partial\phi}{\partial x_i}, \tag{2.324}$$

$$= \nabla\phi. \tag{2.325}$$

The derivative of ϕ on a particular path is called the directional derivative. If the path has a unit tangent $\mathbf{t}$, the derivative in this direction is

$$(\nabla\phi)^T\cdot\mathbf{t} = t_i\frac{\partial\phi}{\partial x_i}. \tag{2.326}$$

If $\phi(x,y,z) = $ constant is a surface, then $d\phi = 0$ on this surface. Also

$$d\phi = \frac{\partial\phi}{\partial x_i}\,dx_i, \tag{2.327}$$

$$= (\nabla\phi)^T\cdot d\mathbf{r}. \tag{2.328}$$

Because $d\mathbf{r}$ is tangent to the surface, $\nabla\phi$ must be normal to it. The tangent plane at $\mathbf{r} = \mathbf{r}_0$ is defined by the position vector $\mathbf{r}$ such that

$$(\nabla\phi)^T\cdot(\mathbf{r} - \mathbf{r}_0) = 0. \tag{2.329}$$

EXAMPLE 2.13

At the point (1,1,1), find the unit normal to the surface

$$z^3 + xz = x^2 + y^2. \tag{2.330}$$

Define

$$\phi(x,y,z) = z^3 + xz - x^2 - y^2 = 0. \tag{2.331}$$

A normal vector at (1,1,1) is

$$\nabla\phi = (z - 2x)\mathbf{i} - 2y\mathbf{j} + (3z^2 + x)\mathbf{k}, \tag{2.332}$$

$$= -1\mathbf{i} - 2\mathbf{j} + 4\mathbf{k}. \tag{2.333}$$

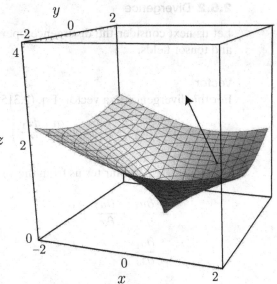

Figure 2.11. Plot of surface $z^3 + xz = x^2 + y^2$ and normal vector at $(1, 1, 1)$.

A unit normal vector is

$$\mathbf{n} = \frac{\nabla \phi}{\|\nabla \phi\|_2}, \qquad (2.334)$$

$$= \frac{1}{\sqrt{21}}(-1\mathbf{i} - 2\mathbf{j} + 4\mathbf{k}). \qquad (2.335)$$

Vector

We can take the gradient of a vector field $\mathbf{u}(\mathbf{x}) = u_i(x_j)$ in the following manner. Consider each scalar component of $\mathbf{u}$ individually and apply the results of the previous section. Thus, for example,

$$\text{grad}\, u_1 = \frac{\partial u_1}{\partial x_1}\mathbf{e}_1 + \frac{\partial u_1}{\partial x_2}\mathbf{e}_2 + \frac{\partial u_1}{\partial x_3}\mathbf{e}_3. \qquad (2.336)$$

In general, we could expect

$$\text{grad}\, u_i = \frac{\partial u_i}{\partial x_1}\mathbf{e}_1 + \frac{\partial u_i}{\partial x_2}\mathbf{e}_2 + \frac{\partial u_i}{\partial x_3}\mathbf{e}_3. \qquad (2.337)$$

This suggests the matrix form of

$$\text{grad}\, \mathbf{u} = \nabla \mathbf{u}^T = \begin{pmatrix} \frac{\partial}{\partial x_1} \\ \frac{\partial}{\partial x_2} \\ \frac{\partial}{\partial x_3} \end{pmatrix} (u_1 \quad u_2 \quad u_3) = \frac{\partial u_i}{\partial x_j} = \begin{pmatrix} \frac{\partial u_1}{\partial x_1} & \frac{\partial u_1}{\partial x_2} & \frac{\partial u_1}{\partial x_3} \\ \frac{\partial u_2}{\partial x_1} & \frac{\partial u_2}{\partial x_2} & \frac{\partial u_2}{\partial x_3} \\ \frac{\partial u_3}{\partial x_1} & \frac{\partial u_3}{\partial x_2} & \frac{\partial u_3}{\partial x_3} \end{pmatrix}. \qquad (2.338)$$

If one considers $\mathbf{u}$ to be a new set of coordinates transformed from the original $\mathbf{x}$, the gradient of the vector field is the Jacobian of the transformation, similar to that discussed in Section 1.6.1.

2.6.2 Divergence

Let us next consider the divergence operator, which can be applied to both vector and tensor fields.

Vector

For the divergence of a vector, Eq. (2.315) becomes

$$\text{div } \mathbf{u} = \lim_{V \to 0} \frac{1}{V} \left(\left(u_1 + \frac{\partial u_1}{\partial x_1} \frac{dx_1}{2} \right) dx_2 \, dx_3 - \left(u_1 - \frac{\partial u_1}{\partial x_1} \frac{dx_1}{2} \right) dx_2 \, dx_3 \right.$$

$$\left. + \text{ similar terms from the } x_2 \text{ and } x_3 \text{ faces} \right), \tag{2.339}$$

$$= \frac{\partial u_1}{\partial x_1} + \frac{\partial u_2}{\partial x_2} + \frac{\partial u_3}{\partial x_3}, \tag{2.340}$$

$$= \frac{\partial u_i}{\partial x_i}, \tag{2.341}$$

$$= \nabla^T \cdot \mathbf{u} = \begin{pmatrix} \frac{\partial}{\partial x_1} & \frac{\partial}{\partial x_2} & \frac{\partial}{\partial x_3} \end{pmatrix} \begin{pmatrix} u_1 \\ u_2 \\ u_3 \end{pmatrix}. \tag{2.342}$$

Note the difference between Eq. (2.338) and Eq. (2.342).

Tensor

The extension for divergence of a tensor is straightforward:

$$\text{div } \mathbf{T} = \nabla^T \cdot \mathbf{T}, \tag{2.343}$$

$$= \frac{\partial T_{ij}}{\partial x_i}. \tag{2.344}$$

Notice that this yields a vector quantity. The divergence of tensor fields often arises in continuum mechanics, especially when considering the effect of surface forces.

2.6.3 Curl

The application of Eq. (2.316) is not obvious. Consider just one of the faces: the face whose outer normal is $\mathbf{e}_1$. For that face, one needs to evaluate

$$\int_S \mathbf{n} \times \mathbf{u} \, dS. \tag{2.345}$$

On this face, one has $\mathbf{n} = \mathbf{e}_1$, and

$$\mathbf{u} = \left(u_1 + \frac{\partial u_1}{\partial x_1} dx_1 \right) \mathbf{e}_1 + \left(u_2 + \frac{\partial u_2}{\partial x_1} dx_1 \right) \mathbf{e}_2 + \left(u_3 + \frac{\partial u_3}{\partial x_1} dx_1 \right) \mathbf{e}_3. \tag{2.346}$$

So, on this face, the integrand is

$$\mathbf{n} \times \mathbf{u} = \begin{vmatrix} \mathbf{e}_1 & \mathbf{e}_2 & \mathbf{e}_3 \\ 1 & 0 & 0 \\ \left(u_1 + \frac{\partial u_1}{\partial x_1} dx_1 \right) & \left(u_2 + \frac{\partial u_2}{\partial x_1} dx_1 \right) & \left(u_3 + \frac{\partial u_3}{\partial x_1} dx_1 \right) \end{vmatrix}, \tag{2.347}$$

$$= \left(u_2 + \frac{\partial u_2}{\partial x_1} dx_1 \right) \mathbf{e}_3 - \left(u_3 + \frac{\partial u_3}{\partial x_1} dx_1 \right) \mathbf{e}_2. \tag{2.348}$$

Two similar terms appear on the opposite face, whose unit vector points in the $-\mathbf{e}_1$ direction.

Carrying out the integration, then, for Eq. (2.316), one gets

$$\operatorname{curl} \mathbf{u} = \lim_{V \to 0} \frac{1}{V} \left(\left(u_2 + \frac{\partial u_2}{\partial x_1} \frac{dx_1}{2} \right) \mathbf{e}_3 dx_2 dx_3 - \left(u_3 + \frac{\partial u_3}{\partial x_1} \frac{dx_1}{2} \right) \mathbf{e}_2 dx_2 dx_3 \right.$$

(2.349)

$$- \left(u_2 - \frac{\partial u_2}{\partial x_1} \frac{dx_1}{2} \right) \mathbf{e}_3 dx_2 dx_3 + \left(u_3 - \frac{\partial u_3}{\partial x_1} \frac{dx_1}{2} \right) \mathbf{e}_2 dx_2 dx_3$$

$$\left. + \text{ similar terms from the } x_2 \text{ and } x_3 \text{ faces} \right),$$

$$= \begin{vmatrix} \mathbf{e}_1 & \mathbf{e}_2 & \mathbf{e}_3 \\ \frac{\partial}{\partial x_1} & \frac{\partial}{\partial x_2} & \frac{\partial}{\partial x_3} \\ u_1 & u_2 & u_3 \end{vmatrix},$$

(2.350)

$$= \epsilon_{ijk} \frac{\partial u_k}{\partial x_j},$$

(2.351)

$$= \nabla \times \mathbf{u}.$$

(2.352)

The curl of a tensor does not arise often in practice.

2.6.4 Laplacian

The Laplacian operator plays a key role throughout mathematical modeling of engineering systems. It is especially prevalent in models describing diffusion processes.

Scalar

The Laplacian is simply "div grad" and can be written, when operating on ϕ, as

$$\operatorname{div} \operatorname{grad} \phi = \nabla^T \cdot (\nabla \phi) = \nabla^2 \phi = \frac{\partial^2 \phi}{\partial x_i \partial x_i}.$$

(2.353)

The notation ∇^2 is commonly used but is better expanded as $\nabla^T \cdot \nabla$.

Vector

Equation (2.362) is used to evaluate the Laplacian of a vector:

$$\nabla^2 \mathbf{u} = \nabla^T \cdot \nabla \mathbf{u} = \nabla(\nabla^T \cdot \mathbf{u}) - \nabla \times (\nabla \times \mathbf{u}).$$

(2.354)

2.6.5 Identities

The following identities can be proven:

$$\nabla \times \nabla \phi = \mathbf{0},$$

(2.355)

$$\nabla^T \cdot (\nabla \times \mathbf{u}) = 0,$$

(2.356)

$$\nabla^T \cdot (\phi \mathbf{u}) = \phi \nabla^T \cdot \mathbf{u} + (\nabla \phi)^T \cdot \mathbf{u},$$

(2.357)

$$\nabla \times (\phi \mathbf{u}) = \phi \nabla \times \mathbf{u} + \nabla \phi \times \mathbf{u},$$

(2.358)

$$\nabla^T \cdot (\mathbf{u} \times \mathbf{v}) = \mathbf{v}^T \cdot (\nabla \times \mathbf{u}) - \mathbf{u}^T \cdot (\nabla \times \mathbf{v}),$$

(2.359)

$$\nabla \times (\mathbf{u} \times \mathbf{v}) = (\mathbf{v}^T \cdot \nabla)\mathbf{u} - (\mathbf{u}^T \cdot \nabla)\mathbf{v} + \mathbf{u}(\nabla^T \cdot \mathbf{v}) - \mathbf{v}(\nabla^T \cdot \mathbf{u}), \tag{2.360}$$

$$\nabla(\mathbf{u}^T \cdot \mathbf{v}) = (\mathbf{u}^T \cdot \nabla)\mathbf{v} + (\mathbf{v}^T \cdot \nabla)\mathbf{u} + \mathbf{u} \times (\nabla \times \mathbf{v}) + \mathbf{v} \times (\nabla \times \mathbf{u}), \tag{2.361}$$

$$\nabla \cdot \nabla^T \mathbf{u} = \nabla(\nabla^T \cdot \mathbf{u}) - \nabla \times (\nabla \times \mathbf{u}). \tag{2.362}$$

EXAMPLE 2.14

Show that Eq. (2.362),

$$\nabla \cdot \nabla^T \mathbf{u} = \nabla(\nabla^T \cdot \mathbf{u}) - \nabla \times (\nabla \times \mathbf{u}), \tag{2.363}$$

is true.

Going from right to left,

$$\nabla(\nabla^T \cdot \mathbf{u}) - \nabla \times (\nabla \times \mathbf{u}) = \frac{\partial}{\partial x_i} \frac{\partial u_j}{\partial x_j} - \epsilon_{ijk} \frac{\partial}{\partial x_j} \left(\epsilon_{klm} \frac{\partial u_m}{\partial x_l} \right), \tag{2.364}$$

$$= \frac{\partial}{\partial x_i} \frac{\partial u_j}{\partial x_j} - \epsilon_{kij} \epsilon_{klm} \frac{\partial}{\partial x_j} \left(\frac{\partial u_m}{\partial x_l} \right), \tag{2.365}$$

$$= \frac{\partial^2 u_j}{\partial x_i \partial x_j} - (\delta_{il}\delta_{jm} - \delta_{im}\delta_{jl}) \frac{\partial^2 u_m}{\partial x_j \partial x_l}, \tag{2.366}$$

$$= \frac{\partial^2 u_j}{\partial x_i \partial x_j} - \frac{\partial^2 u_j}{\partial x_j \partial x_i} + \frac{\partial^2 u_i}{\partial x_j \partial x_j}, \tag{2.367}$$

$$= \frac{\partial}{\partial x_j} \left(\frac{\partial u_i}{\partial x_j} \right), \tag{2.368}$$

$$= \nabla^T \cdot \nabla \mathbf{u}. \tag{2.369}$$

2.7 Curvature Revisited

Let us consider again the notion of curvature, first for curves, then for surfaces, in light of our knowledge of differential vector operators.

2.7.1 Trajectory

Let us first consider one-dimensional curves embedded within a high-dimensional space. The curves may be considered to be defined as solutions to the differential equations of the form

$$\frac{d\mathbf{x}}{dt} = \mathbf{v}(\mathbf{x}). \tag{2.370}$$

This form is equivalent to a form we shall consider in more detail in Section 9.3.1. We can consider $\mathbf{v}(\mathbf{x})$ to be a velocity field that is dependent on position $\mathbf{x}$ but independent of time. A particle with a known initial condition will move through the field, acquiring a new velocity at each new spatial point it encounters, thus tracing a nontrivial trajectory. We now take the velocity gradient tensor to be $\mathbf{L}$, with

$$\mathbf{L} = \nabla \mathbf{v}^T. \tag{2.371}$$

With this, it can then be shown after detailed analysis that the curvature of the trajectory is given by

$$\kappa = \frac{\sqrt{(\mathbf{v}^T \cdot \mathbf{L} \cdot \mathbf{L}^T \cdot \mathbf{v})(\mathbf{v}^T \cdot \mathbf{v}) - (\mathbf{v}^T \cdot \mathbf{L}^T \cdot \mathbf{v})^2}}{(\mathbf{v}^T \cdot \mathbf{v})^{3/2}}. \tag{2.372}$$

In terms of the unit tangent vector, $\mathbf{t} = \mathbf{v}/\|\mathbf{v}\|_2$, Eq. (2.372) reduces to

$$\kappa = \frac{\sqrt{(\mathbf{t}^T \cdot \mathbf{L} \cdot \mathbf{L}^T \cdot \mathbf{t}) - (\mathbf{t}^T \cdot \mathbf{L}^T \cdot \mathbf{t})^2}}{\|\mathbf{v}\|_2}. \tag{2.373}$$

EXAMPLE 2.15

Find the curvature of the curve given by

$$\frac{dx}{dt} = -y, \qquad x(0) = 0, \tag{2.374}$$

$$\frac{dy}{dt} = x, \qquad y(0) = 2. \tag{2.375}$$

We can of course solve this exactly by first dividing one equation by the other to get

$$\frac{dy}{dx} = -\frac{x}{y}, \qquad y(x = 0) = 2. \tag{2.376}$$

Separating variables, we get

$$y \, dy = -x \, dx, \tag{2.377}$$

$$\frac{y^2}{2} = -\frac{x^2}{2} + C, \tag{2.378}$$

$$\frac{2^2}{2} = -\frac{0^2}{2} + C, \tag{2.379}$$

$$C = 2. \tag{2.380}$$

Thus,

$$x^2 + y^2 = 4 \tag{2.381}$$

is the curve of interest. It is a circle whose radius is 2 and thus whose radius of curvature $\rho = 2$; thus, its curvature $\kappa = 1/\rho = 1/2$.

Let us reproduce this result using Eq. (2.372). We can think of the two-dimensional velocity vector as

$$\mathbf{v} = \begin{pmatrix} u(x,y) \\ v(x,y) \end{pmatrix} = \begin{pmatrix} -y \\ x \end{pmatrix}. \tag{2.382}$$

The velocity gradient is then

$$\mathbf{L} = \nabla \mathbf{v}^T = \begin{pmatrix} \frac{\partial}{\partial x} \\ \frac{\partial}{\partial y} \end{pmatrix} \begin{pmatrix} u(x,y) & v(x,y) \end{pmatrix} = \begin{pmatrix} \frac{\partial u}{\partial x} & \frac{\partial v}{\partial x} \\ \frac{\partial u}{\partial y} & \frac{\partial v}{\partial y} \end{pmatrix} = \begin{pmatrix} 0 & 1 \\ -1 & 0 \end{pmatrix}. \tag{2.383}$$

Now, let us use Eq. (2.372) to directly compute the curvature. The simple nature of our velocity field induces several simplifications. First, because the velocity gradient tensor here is antisymmetric, we have

$$\mathbf{v}^T \cdot \mathbf{L}^T \cdot \mathbf{v} = \begin{pmatrix} -y & x \end{pmatrix} \begin{pmatrix} 0 & -1 \\ 1 & 0 \end{pmatrix} \begin{pmatrix} -y \\ x \end{pmatrix} = \begin{pmatrix} -y & x \end{pmatrix} \begin{pmatrix} -x \\ -y \end{pmatrix} = xy - xy = 0. \tag{2.384}$$

Second, we see that

$$\mathbf{L} \cdot \mathbf{L}^T = \begin{pmatrix} 0 & 1 \\ -1 & 0 \end{pmatrix} \begin{pmatrix} 0 & -1 \\ 1 & 0 \end{pmatrix} = \begin{pmatrix} 1 & 0 \\ 0 & 1 \end{pmatrix} = \mathbf{I}. \tag{2.385}$$

So for this problem, Eq. (2.372) reduces to

$$\kappa = \frac{\sqrt{(\mathbf{v}^T \cdot \underbrace{\mathbf{L} \cdot \mathbf{L}^T}_{\mathbf{I}} \cdot \mathbf{v})(\mathbf{v}^T \cdot \mathbf{v}) - \underbrace{(\mathbf{v}^T \cdot \mathbf{L}^T \cdot \mathbf{v})^2}_{=0}}}{(\mathbf{v}^T \cdot \mathbf{v})^{3/2}}, \tag{2.386}$$

$$= \frac{\sqrt{(\mathbf{v}^T \cdot \mathbf{v})(\mathbf{v}^T \cdot \mathbf{v})}}{(\mathbf{v}^T \cdot \mathbf{v})^{3/2}}, \tag{2.387}$$

$$= \frac{(\mathbf{v}^T \cdot \mathbf{v})}{(\mathbf{v}^T \cdot \mathbf{v})^{3/2}}, \tag{2.388}$$

$$= \frac{1}{\sqrt{\mathbf{v}^T \cdot \mathbf{v}}}, \tag{2.389}$$

$$= \frac{1}{||\mathbf{v}||_2}, \tag{2.390}$$

$$= \frac{1}{\sqrt{x^2 + y^2}}, \tag{2.391}$$

$$= \frac{1}{\sqrt{4}}, \tag{2.392}$$

$$= \frac{1}{2}. \tag{2.393}$$

If a curve in two-dimensional space is given implicitly by the function

$$\phi(x, y) = 0, \tag{2.394}$$

it can be shown that the curvature is given by the formula

$$\kappa = \nabla \cdot \left(\frac{\nabla \phi}{||\nabla \phi||_2} \right), \tag{2.395}$$

provided one takes precautions to preserve the sign, as demonstrated in the following example. Note that $\nabla \phi$ is a gradient vector that must be normal to any so-called *level set* curve for which ϕ is constant; moreover, it points in the direction of most rapid change of ϕ. The corresponding vector $\nabla \phi / ||\nabla \phi||_2$ must be a unit normal vector to level sets of ϕ.

EXAMPLE 2.16

Show Eq. (2.395) is equivalent to Eq. (2.239) if $y = f(x)$.

Let us take

$$\phi(x, y) = f(x) - y = 0. \tag{2.396}$$

Then, we get

$$\nabla \phi = \frac{\partial \phi}{\partial x} \mathbf{i} + \frac{\partial \phi}{\partial y} \mathbf{j} \tag{2.397}$$

$$= f'(x) \mathbf{i} - \mathbf{j}. \tag{2.398}$$

We then see that

$$||\nabla \phi||_2 = \sqrt{f'(x)^2 + 1}, \tag{2.399}$$

so that

$$\frac{\nabla\phi}{\|\nabla\phi\|_2} = \frac{f'(x)\mathbf{i} - \mathbf{j}}{\sqrt{1 + f'(x)^2}}. \tag{2.400}$$

Then, applying Eq. (2.395), we get

$$\kappa = \nabla \cdot \left(\frac{\nabla\phi}{\|\nabla\phi\|_2} \right), \tag{2.401}$$

$$= \nabla \cdot \left(\frac{f'(x)\mathbf{i} - \mathbf{j}}{\sqrt{1 + f'(x)^2}} \right), \tag{2.402}$$

$$= \frac{\partial}{\partial x} \left(\frac{f'(x)}{\sqrt{1 + f'(x)^2}} \right) + \underbrace{\frac{\partial}{\partial y} \left(\frac{-1}{\sqrt{1 + f'(x)^2}} \right)}_{=0}, \tag{2.403}$$

$$= \frac{\sqrt{1 + f'(x)^2}f''(x) - f'(x)f'(x)f''(x)\left(1 + f'(x)^2\right)^{-1/2}}{1 + f'(x)^2}, \tag{2.404}$$

$$= \frac{\left(1 + f'(x)^2\right)f''(x) - f'(x)f'(x)f''(x)}{\left(1 + f'(x)^2\right)^{3/2}}, \tag{2.405}$$

$$= \frac{f''(x)}{\left(1 + f'(x)^2\right)^{3/2}}. \tag{2.406}$$

Equation (2.406) is fully equivalent to the earlier developed Eq. (2.239). Note, however, that if we had chosen $\phi(x, y) = y - f(x) = 0$, we would have recovered a formula for curvature with the opposite sign.

2.7.2 Principal

In a three-dimensional space, a point P on a surface S will have a tangent plane. It will also have an infinite number of normal planes. Each of these normal planes has an intersection with S which forms a curve, and each of these curves has a curvature at P. The *principal curvatures* κ_1 and κ_2 are the maximum and minimum values, respectively, of the curvature of such curves.

2.7.3 Gaussian

The *Gaussian curvature* K in three-dimensional space is the product of the two principal curvatures:

$$K = \kappa_1 \kappa_2. \tag{2.407}$$

One can also show that

$$K = \det \mathbf{H}, \tag{2.408}$$

where $\mathbf{H}$ is the Hessian matrix evaluated at P.

2.7.4 Mean

The *mean curvature H* is the average of the two principal curvatures:

$$H = \frac{\kappa_1 + \kappa_2}{2}. \tag{2.409}$$

It can be shown that $H = (\kappa_a + \kappa_b)/2$ also, where κ_a and κ_b are the curvatures in any two orthogonal planes. Alternatively, if the surface in three-space is defined implicitly by

$$\phi(x, y, z) = 0, \tag{2.410}$$

it can be shown that the mean curvature is given by the extension of Eq. (2.395):

$$H = \nabla \cdot \left(\frac{\nabla \phi}{\|\nabla \phi\|_2} \right). \tag{2.411}$$

2.8 Special Theorems

We present here some well-known theorems.

2.8.1 Green's Theorem

Let $\mathbf{u} = u_x \mathbf{i} + u_y \mathbf{j}$ be a vector field, C a closed curve, and D the region enclosed by C, all in the (x, y) plane. Then *Green's theorem*[6] is

$$\oint_C \mathbf{u}^T \cdot d\mathbf{r} = \iint_D \left(\frac{\partial u_y}{\partial x} - \frac{\partial u_x}{\partial y} \right) dx\, dy. \tag{2.412}$$

Here $\oint$ indicates a line integral about a closed contour.

EXAMPLE 2.17

Show that Green's theorem is valid if $\mathbf{u} = y\mathbf{i} + 2xy\mathbf{j}$, and C consists of the straight lines (0,0) to (1,0) to (1,1) to (0,0):

$$\oint_C \mathbf{u}^T \cdot d\mathbf{r} = \int_{C_1} \mathbf{u}^T \cdot d\mathbf{r} + \int_{C_2} \mathbf{u}^T \cdot d\mathbf{r} + \int_{C_3} \mathbf{u}^T \cdot d\mathbf{r}, \tag{2.413}$$

where C_1, C_2, and C_3 are the straight lines (0,0) to (1,0), (1,0) to (1,1), and (1,1) to (0,0), respectively. This is sketched in Figure 2.12.

For this problem, we have

$$C_1: \qquad\qquad y = 0, \qquad dy = 0, \qquad x \in [0,1], \qquad \mathbf{u} = 0\mathbf{i} + 0\mathbf{j}, \tag{2.414}$$

$$C_2: \qquad\qquad x = 1, \qquad dx = 0, \qquad y \in [0,1], \qquad \mathbf{u} = y\mathbf{i} + 2y\mathbf{j}, \tag{2.415}$$

$$C_3: \quad x = y, \quad dx = dy, \quad x \in [1,0], \qquad y \in [1,0], \qquad \mathbf{u} = x\mathbf{i} + 2x^2\mathbf{j}. \tag{2.416}$$

Thus,

$$\oint_C \mathbf{u} \cdot d\mathbf{r} = \underbrace{\int_0^1 (0\mathbf{i} + 0\mathbf{j}) \cdot (dx\, \mathbf{i})}_{C_1} + \underbrace{\int_0^1 (y\mathbf{i} + 2y\mathbf{j}) \cdot (dy\, \mathbf{j})}_{C_2} + \underbrace{\int_1^0 (x\mathbf{i} + 2x^2\mathbf{j}) \cdot (dx\, \mathbf{i} + dx\, \mathbf{j})}_{C_3},$$
$$\tag{2.417}$$

$$= \int_0^1 2y\, dy + \int_1^0 (x + 2x^2)\, dx, \tag{2.418}$$

---

[6] George Green, 1793–1841, English corn miller and mathematician.

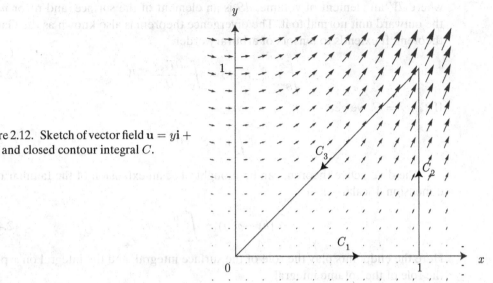

Figure 2.12. Sketch of vector field $\mathbf{u} = y\mathbf{i} + 2xy\mathbf{j}$ and closed contour integral C.

$$= y^2\big|_0^1 + \left(\frac{1}{2}x^2 + \frac{2}{3}x^3\right)\bigg|_1^0 = 1 - \frac{1}{2} - \frac{2}{3}, \tag{2.419}$$

$$= -\frac{1}{6}. \tag{2.420}$$

On the other hand,

$$\iint_D \left(\frac{\partial u_y}{\partial x} - \frac{\partial u_x}{\partial y}\right) dx\, dy = \int_0^1 \int_0^x (2y - 1)\, dy\, dx, \tag{2.421}$$

$$= \int_0^1 \left((y^2 - y)\big|_0^x\right) dx, \tag{2.422}$$

$$= \int_0^1 (x^2 - x)\, dx, \tag{2.423}$$

$$= \left(\frac{x^3}{3} - \frac{x^2}{2}\right)\bigg|_0^1, \tag{2.424}$$

$$= \frac{1}{3} - \frac{1}{2}, \tag{2.425}$$

$$= -\frac{1}{6}. \tag{2.426}$$

2.8.2 Divergence Theorem

Let us consider Eq. (2.312) in more detail. Let S be a closed surface and V the region enclosed within it; then the divergence theorem is

$$\int_S \mathbf{u}^T \cdot \mathbf{n}\, dS = \int_V \nabla^T \cdot \mathbf{u}\, dV, \tag{2.427}$$

$$\int_S u_i n_i\, dS = \int_V \frac{\partial u_i}{\partial x_i}\, dV, \tag{2.428}$$

where dV an element of volume, dS is an element of the surface, and $\mathbf{n}$ (or n_i) is the outward unit normal to it. The divergence theorem is also known as the Gauss[7] theorem. It extends to tensors of arbitrary order:

$$\int_S T_{ijk...} n_i \, dS = \int_V \frac{\partial T_{ijk...}}{\partial x_i} \, dV. \tag{2.429}$$

If $T_{ijk...} = C$, we get

$$\int_S n_i \, dS = 0. \tag{2.430}$$

The divergence theorem can be thought of as an extension of the familiar one-dimensional scalar result:

$$\phi(b) - \phi(a) = \int_a^b \frac{d\phi}{dx} \, dx. \tag{2.431}$$

Here the endpoints play the role of the surface integral, and the integral on x plays the role of the volume integral.

EXAMPLE 2.18

Show that the divergence theorem is valid if

$$\mathbf{u} = x\mathbf{i} + y\mathbf{j} + 0\mathbf{k}, \tag{2.432}$$

and S is a closed surface that consists of a circular base and the hemisphere of unit radius with center at the origin and $z \geq 0$, that is,

$$x^2 + y^2 + z^2 = 1. \tag{2.433}$$

In spherical coordinates, defined by

$$x = r \sin\theta \cos\phi, \tag{2.434}$$

$$y = r \sin\theta \sin\phi, \tag{2.435}$$

$$z = r \cos\theta, \tag{2.436}$$

the hemispherical surface is described by

$$r = 1. \tag{2.437}$$

A sketch of the surface of interest along with the vector field is shown in Figure 2.13. We split the surface integral into two parts:

$$\int_S \mathbf{u}^T \cdot \mathbf{n} \, dS = \int_B \mathbf{u}^T \cdot \mathbf{n} \, dS + \int_H \mathbf{u}^T \cdot \mathbf{n} \, dS, \tag{2.438}$$

where B is the base and H is the curved surface of the hemisphere. The first term on the right is zero because $\mathbf{n} = -\mathbf{k}$, and $\mathbf{u}^T \cdot \mathbf{n} = 0$ on B. In general, the unit normal pointing in the r direction can be shown to be

$$\mathbf{e}_r = \mathbf{n} = \sin\theta \cos\phi \mathbf{i} + \sin\theta \sin\phi \mathbf{j} + \cos\theta \mathbf{k}. \tag{2.439}$$

---

[7] Johann Carl Friedrich Gauss, 1777–1855, Brunswick-born German mathematician.

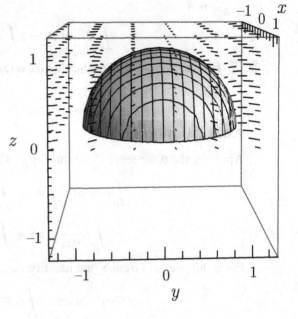

Figure 2.13. Plot depicting $x^2 + y^2 + z^2 = 1$, $z \geq 0$ and vector field $\mathbf{u} = x\mathbf{i} + y\mathbf{j} + 0\mathbf{k}$.

This is in fact the unit normal on H. Thus, on H, where $r = 1$, we have

$$\mathbf{u}^T \cdot \mathbf{n} = (x\mathbf{i} + y\mathbf{j} + 0\mathbf{k})^T \cdot (\sin\theta\cos\phi\mathbf{i} + \sin\theta\sin\phi\mathbf{j} + \cos\theta\mathbf{k}), \tag{2.440}$$

$$= (r\sin\theta\cos\phi\mathbf{i} + r\sin\theta\sin\phi\mathbf{j} + 0\mathbf{k})^T \cdot (\sin\theta\cos\phi\mathbf{i} + \sin\theta\sin\phi\mathbf{j} + \cos\theta\mathbf{k}), \tag{2.441}$$

$$= \underbrace{r}_{1}\sin^2\theta\cos^2\phi + \underbrace{r}_{1}\sin^2\theta\sin^2\phi, \tag{2.442}$$

$$= \sin^2\theta\cos^2\phi + \sin^2\theta\sin^2\phi, \tag{2.443}$$

$$= \sin^2\theta, \tag{2.444}$$

$$\int_H \mathbf{u}^T \cdot \mathbf{n}\,dS = \int_0^{2\pi}\int_0^{\pi/2} \underbrace{\sin^2\theta}_{\mathbf{u}^T\cdot\mathbf{n}} \underbrace{(\sin\theta\,d\theta\,d\phi)}_{dS}, \tag{2.445}$$

$$= \int_0^{2\pi}\int_0^{\pi/2} \sin^3\theta\,d\theta\,d\phi, \tag{2.446}$$

$$= \int_0^{2\pi}\int_0^{\pi/2} \left(\frac{3}{4}\sin\theta - \frac{1}{4}\sin 3\theta\right)d\theta\,d\phi, \tag{2.447}$$

$$= 2\pi\int_0^{\pi/2}\left(\frac{3}{4}\sin\theta - \frac{1}{4}\sin 3\theta\right)d\theta, \tag{2.448}$$

$$= 2\pi\left(\frac{3}{4} - \frac{1}{12}\right), \tag{2.449}$$

$$= \frac{4}{3}\pi. \tag{2.450}$$

On the other hand, if we use the divergence theorem, we find that

$$\nabla^T \cdot \mathbf{u} = \frac{\partial}{\partial x}(x) + \frac{\partial}{\partial y}(y) + \frac{\partial}{\partial z}(0) = 2, \tag{2.451}$$

so that

$$\int_V \nabla^T \cdot \mathbf{u}\, dV = 2\int_V dV = 2\frac{2}{3}\pi = \frac{4}{3}\pi, \tag{2.452}$$

because the volume of the hemisphere is $(2/3)\pi$.

2.8.3 Green's Identities

Applying the divergence theorem, Eq. (2.428), to the vector $\mathbf{u} = \phi\nabla\psi$, we get

$$\int_S \phi(\nabla\psi)^T \cdot \mathbf{n}\, dS = \int_V \nabla^T \cdot (\phi\nabla\psi)\, dV, \tag{2.453}$$

$$\int_S \phi\frac{\partial\psi}{\partial x_i} n_i\, dS = \int_V \frac{\partial}{\partial x_i}\left(\phi\frac{\partial\psi}{\partial x_i}\right) dV. \tag{2.454}$$

From this, we get Green's first identity:

$$\int_S \phi(\nabla\psi)^T \cdot \mathbf{n}\, dS = \int_V (\phi\nabla^2\psi + (\nabla\phi)^T \cdot \nabla\psi)\, dV, \tag{2.455}$$

$$\int_S \phi\frac{\partial\psi}{\partial x_i} n_i\, dS = \int_V \left(\phi\frac{\partial^2\psi}{\partial x_i\partial x_i} + \frac{\partial\phi}{\partial x_i}\frac{\partial\psi}{\partial x_i}\right) dV. \tag{2.456}$$

Interchanging ϕ and ψ in Eq. (2.455), we get

$$\int_S \psi(\nabla\phi)^T \cdot \mathbf{n}\, dS = \int_V (\psi\nabla^2\phi + (\nabla\psi)^T \cdot \nabla\phi)\, dV, \tag{2.457}$$

$$\int_S \psi\frac{\partial\phi}{\partial x_i} n_i\, dS = \int_V \left(\psi\frac{\partial^2\phi}{\partial x_i\partial x_i} + \frac{\partial\psi}{\partial x_i}\frac{\partial\phi}{\partial x_i}\right) dV. \tag{2.458}$$

Subtracting Eq. (2.457) from Eq. (2.455), we get Green's second identity:

$$\int_S (\phi\nabla\psi - \psi\nabla\phi)^T \cdot \mathbf{n}\, dS = \int_V (\phi\nabla^2\psi - \psi\nabla^2\phi)\, dV, \tag{2.459}$$

$$\int_S \left(\phi\frac{\partial\psi}{\partial x_i} - \psi\frac{\partial\phi}{\partial x_i}\right) n_i\, dS = \int_V \left(\phi\frac{\partial^2\psi}{\partial x_i\partial x_i} - \psi\frac{\partial^2\phi}{\partial x_i\partial x_i}\right) dV. \tag{2.460}$$

2.8.4 Stokes' Theorem

Consider Stokes'[8] theorem. Let S be an open surface and the curve C its boundary. Then

$$\int_S (\nabla \times \mathbf{u})^T \cdot \mathbf{n}\, dS = \oint_C \mathbf{u}^T \cdot d\mathbf{r}, \tag{2.461}$$

$$\int_S \epsilon_{ijk}\frac{\partial u_k}{\partial x_j} n_i\, dS = \oint_C u_i\, dr_i, \tag{2.462}$$

where $\mathbf{n}$ is the unit vector normal to the element dS and $d\mathbf{r}$ is an element of curve C.

---

[8] George Gabriel Stokes, 1819–1903, Irish-born English mathematician.

EXAMPLE 2.19

Evaluate

$$I = \int_S (\nabla \times \mathbf{u})^T \cdot \mathbf{n}\, dS, \tag{2.463}$$

using Stokes' theorem, where

$$\mathbf{u} = x^3 \mathbf{j} - (z+1)\mathbf{k}, \tag{2.464}$$

and S is the surface $z = 4 - 4x^2 - y^2$ for $z \geq 0$.

Using Stokes' theorem, the surface integral can be converted to a line integral along the boundary C, which is the curve $4 - 4x^2 - y^2 = 0$:

$$I = \oint_C \mathbf{u}^T \cdot d\mathbf{r}, \tag{2.465}$$

$$= \oint_C \underbrace{(x^3 \mathbf{j} - (z+1)\mathbf{k})}_{\mathbf{u}^T} \cdot \underbrace{(dx\mathbf{i} + dy\mathbf{j})}_{d\mathbf{r}}, \tag{2.466}$$

$$= \int_C x^3\, dy. \tag{2.467}$$

The curve C can be represented by the parametric equations $x = \cos t$, $y = 2\sin t$. This is easily seen by direct substitution on C:

$$4 - 4x^2 - y^2 = 4 - 4\cos^2 t - (2\sin t)^2 = 4 - 4(\cos^2 t + \sin^2 t) = 4 - 4 = 0. \tag{2.468}$$

Thus, $dy = 2\cos t\, dt$, so that

$$I = \int_0^{2\pi} \underbrace{\cos^3 t}_{x^3}\, \underbrace{(2\cos t\, dt)}_{dy}, \tag{2.469}$$

$$= 2\int_0^{2\pi} \cos^4 t\, dt, \tag{2.470}$$

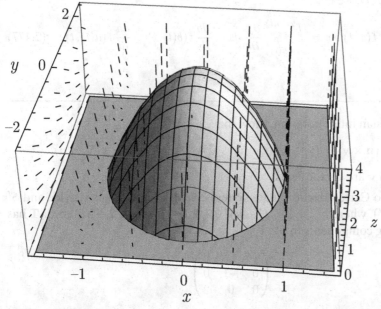

Figure 2.14. Plot depicting $z = 4 - 4x^2 - y^2$ and vector field $\mathbf{u} = x^3\mathbf{j} - (z+1)\mathbf{k}$.

$$= 2 \int_0^{2\pi} \left(\frac{1}{8} \cos 4t + \frac{1}{2} \cos 2t + \frac{3}{8} \right) dt, \qquad (2.471)$$

$$= 2 \left. \left(\frac{1}{32} \sin 4t + \frac{1}{4} \sin 2t + \frac{3}{8} t \right) \right|_0^{2\pi}, \qquad (2.472)$$

$$= \frac{3}{2} \pi. \qquad (2.473)$$

A sketch of the surface of interest along with the vector field is shown in Figure 2.14. The curve C is on the boundary $z = 0$.

2.8.5 Extended Leibniz Rule

If we consider an arbitrary moving volume $V(t)$ with a corresponding surface area $S(t)$ with surface volume elements moving at velocity w_k, the Leibniz rule, extended from the earlier Eq. (1.128), gives us a means to calculate the time derivatives of integrated quantities. For a tensor of arbitrary order, it is

$$\frac{d}{dt} \int_{V(t)} T_{jk\ldots}(x_i, t) \, dV = \int_{V(t)} \frac{\partial T_{jk\ldots}(x_i, t)}{\partial t} \, dV + \int_{S(t)} n_m w_m T_{jk\ldots}(x_i, t) \, dS.$$
$$(2.474)$$

If $T_{jk\ldots}(x_i, t) = 1$, we get

$$\frac{d}{dt} \int_{V(t)} (1) \, dV = \int_{V(t)} \frac{\partial}{\partial t} (1) \, dV + \int_{S(t)} n_m w_m (1) \, dS, \qquad (2.475)$$

$$\frac{dV}{dt} = \int_{S(t)} n_m w_m \, dS. \qquad (2.476)$$

Here the volume changes due to the net surface motion. In one dimension, where $T_{jk\ldots}(x_i, t) = f(x, t)$, we get

$$\frac{d}{dt} \int_{x=a(t)}^{x=b(t)} f(x, t) \, dx = \int_{x=a(t)}^{x=b(t)} \frac{\partial f}{\partial t} \, dx + \frac{db}{dt} f(b(t), t) - \frac{da}{dt} f(a(t), t). \quad (2.477)$$

EXERCISES

1. Using Cartesian index notation, show that

$$\nabla \times (\mathbf{u} \times \mathbf{v}) = (\mathbf{v}^T \cdot \nabla)\mathbf{u} - (\mathbf{u}^T \cdot \nabla)\mathbf{v} + \mathbf{u}(\nabla^T \cdot \mathbf{v}) - \mathbf{v}(\nabla^T \cdot \mathbf{u}),$$

where $\mathbf{u}$ and $\mathbf{v}$ are vector fields.

2. Consider two Cartesian coordinate systems: S with unit vectors $(\mathbf{i}, \mathbf{j}, \mathbf{k})$, and S' with $(\mathbf{i}', \mathbf{j}', \mathbf{k}')$, where $\mathbf{i}' = \mathbf{i}$, $\mathbf{j}' = (\mathbf{j} - \mathbf{k})/\sqrt{2}$, $\mathbf{k}' = (\mathbf{j} + \mathbf{k})/\sqrt{2}$. The tensor $\mathbf{T}$ has the following components in S:

$$\begin{pmatrix} 1 & 0 & 0 \\ 0 & -1 & 0 \\ 0 & 0 & 2 \end{pmatrix}.$$

Find its components in S'.

3. Prove the following identities using Cartesian index notation:
   (a) $(\mathbf{a} \times \mathbf{b})^T \cdot \mathbf{c} = \mathbf{a}^T \cdot (\mathbf{b} \times \mathbf{c})$,
   (b) $\mathbf{a} \times (\mathbf{b} \times \mathbf{c}) = \mathbf{b}(\mathbf{a}^T \cdot \mathbf{c}) - \mathbf{c}(\mathbf{a}^T \cdot \mathbf{b})$,
   (c) $(\mathbf{a} \times \mathbf{b})^T \cdot (\mathbf{c} \times \mathbf{d}) = ((\mathbf{a} \times \mathbf{b}) \times \mathbf{c})^T \cdot \mathbf{d}$.

4. Show using Cartesian index notation that
   (a) $\nabla \times \nabla\phi = 0$,
   (b) $\nabla^T \cdot \nabla \times \mathbf{u} = 0$,
   (c) $\nabla(\mathbf{u}^T \cdot \mathbf{v}) = (\mathbf{u}^T \cdot \nabla)\mathbf{v} + (\mathbf{v}^T \cdot \nabla)\mathbf{u} + \mathbf{u} \times (\nabla \times \mathbf{v}) + \mathbf{v} \times (\nabla \times \mathbf{u})$,
   (d) $\frac{1}{2}\nabla(\mathbf{u}^T \cdot \mathbf{u}) = (\mathbf{u}^T \cdot \nabla)\mathbf{u} + \mathbf{u} \times (\nabla \times \mathbf{u})$,
   (e) $\nabla^T \cdot (\mathbf{u} \times \mathbf{v}) = \mathbf{v}^T \cdot \nabla \times \mathbf{u} - \mathbf{u}^T \cdot \nabla \times \mathbf{v}$,
   (f) $\nabla \times (\nabla \times \mathbf{u}) = \nabla(\nabla^T \cdot \mathbf{u}) - \nabla^2\mathbf{u}$,
   (g) $\nabla \times (\mathbf{u} \times \mathbf{v}) = (\mathbf{v}^T \cdot \nabla)\mathbf{u} - (\mathbf{u}^T \cdot \nabla)\mathbf{v} + \mathbf{u}(\nabla^T \cdot \mathbf{v}) - \mathbf{v}(\nabla^T \cdot \mathbf{u})$.

5. Show that the Laplacian operator $\partial^2/\partial x_i \partial x_i$ has the same form in S and a rotated frame S'.

6. If
$$T_{ij} = \begin{pmatrix} x_1 x_2^2 & 3x_3 & x_1 - x_2 \\ x_2 x_1 & x_1 x_3 & x_3^2 + 1 \\ 0 & 4 & 2x_2 - x_3 \end{pmatrix},$$

   (a) evaluate T_{ij} at $P = (3, 1, 2)$; (b) find $T_{(ij)}$ and $T_{[ij]}$ at P; (c) find the associated dual vector d_i; (d) find the principal values and the orientations of each associated normal vector for the symmetric part of T_{ij} evaluated at P; (e) evaluate the divergence of T_{ij} at P; (f) evaluate the curl of the divergence of T_{ij} at P.

7. Consider the tensor
$$T_{ij} = \begin{pmatrix} 2 & -1 & 2 \\ 3 & 1 & 0 \\ 0 & 1 & 4 \end{pmatrix},$$

   defined in a Cartesian coordinate system. Consider the vector associated with the plane whose normal points in the direction $(2, 5, -1)$. What is the magnitude of the component of the associated vector that is aligned with the normal to the plane?

8. Find the invariants of the tensors
   (a)
$$T_{ij} = \begin{pmatrix} 1 & 2 \\ 2 & 2 \end{pmatrix},$$

   (b)
$$T_{ij} = \begin{pmatrix} 0 & 1 \\ -1 & -2 \end{pmatrix}.$$

   Select a rotation matrix, represent each tensor in the rotated coordinate system, and demonstrate that the invariants remain unchanged in the new coordinate system.

9. A tensor has a representation of

$$T_{ij} = \begin{pmatrix} 1 & 2 & 4 \\ 2 & 3 & -1 \\ 4 & -1 & 1 \end{pmatrix}$$

in an original Cartesian coordinate system. When the coordinate axes are rotated via the action of

$$\ell_{ij} = \begin{pmatrix} \frac{1}{\sqrt{6}} & \sqrt{\frac{2}{3}} & \frac{1}{\sqrt{6}} \\ \frac{1}{\sqrt{3}} & -\frac{1}{\sqrt{3}} & \frac{1}{\sqrt{3}} \\ \frac{1}{\sqrt{2}} & 0 & -\frac{1}{\sqrt{2}} \end{pmatrix},$$

find the representation in the rotated coordinates, $\overline{T}_{ij} = \ell_{ki}\ell_{lj}T_{ij}$. Find also the transformation that takes the original tensor to a diagonal form, and show that the three invariants remain the same.

10. A cube with corners initially at (0,0,0), (0,1,0), (1,1,0), (1,0,0), (0,0,1), (0,1,1), (1,1,1), (1,0,1) is rotated at a constant angular velocity ω around an axis passing through the origin and pointing in the direction of the vector $\mathbf{i} + 2\mathbf{j} + 3\mathbf{k}$. Find the initial velocity of each of the corners and their location with respect to time.

11. The position of a point is given by $\mathbf{r} = \mathbf{i}a\cos\omega t + \mathbf{j}b\sin\omega t$, where t is time. Show that the path of the point is an ellipse. Find its velocity $\mathbf{v}$ and show that $\mathbf{r} \times \mathbf{v}$ = constant. Show also that the acceleration of the point is directed toward the origin, and its magnitude is proportional to the distance from the origin.

12. A Cartesian system S is defined by the unit vectors $\mathbf{e}_1$, $\mathbf{e}_2$, and $\mathbf{e}_3$. Another Cartesian system S' is defined by unit vectors $\mathbf{e}'_1$, $\mathbf{e}'_2$, and $\mathbf{e}'_3$ in directions $\mathbf{a}$, $\mathbf{b}$, and $\mathbf{c}$, where

$$\mathbf{a} = \mathbf{e}_1,$$

$$\mathbf{b} = \mathbf{e}_2 - \mathbf{e}_3,$$

$$\mathbf{c} = \mathbf{e}_2 + \mathbf{e}_3.$$

(a) Find $\mathbf{e}'_1$, $\mathbf{e}'_2$, $\mathbf{e}'_3$, (b) find the transformation array A_{ij}, (c) show that $\delta_{ij} = A_{ki}A_{kj}$ is satisfied, and (d) find the components of the vector $\mathbf{e}_1 + \mathbf{e}_2 + \mathbf{e}_3$ in S'.

13. Find the unit vector normal to the plane passing through the points (1,0,0), (0,1,0), and (0,0,2).

14. Determine the unit vector normal to the surface $x^3 - 2xyz + z^3 = 0$ at the point (1,1,1).

15. Find the angle between the planes

$$3x - y + 2z = 2$$

$$x - 2y = 1.$$

16. Determine a unit vector in the plane of the vectors $\mathbf{i} - \mathbf{j}$ and $\mathbf{j} + \mathbf{k}$ and perpendicular to the vector $\mathbf{i} - \mathbf{j} + \mathbf{k}$.

17. Determine a unit vector perpendicular to the plane that includes the vectors $\mathbf{a} = \mathbf{i} + 2\mathbf{j} - \mathbf{k}$ and $\mathbf{b} = 2\mathbf{i} + \mathbf{j}$.

18. Find the curvature and torsion of the curve defined parametrically by $x = t^2$, $y = t^3$, $z = t^4$ at the point $(1, 1, 1)$.

19. Find the curvature and the radius of curvature of $y = a \sin x$ at its peaks and valleys.

20. Find the largest and smallest radii of curvature of the ellipse

$$\frac{x^2}{a^2} + \frac{y^2}{b^2} = 1.$$

21. Find the curve of intersection of the cylinders $x^2 + y^2 = 1$ and $y^2 + z^2 = 1$. Determine also the radius of curvature of this curve at the points $(0,1,0)$ and $(1,0,1)$.

22. Show that

$$\mathbf{t}^T \cdot \frac{d\mathbf{t}}{ds} \times \frac{d^2\mathbf{t}}{ds^2} = \kappa^2 \tau$$

$$\frac{\frac{d\mathbf{r}^T}{ds} \cdot \frac{d^2\mathbf{r}}{ds^2} \times \frac{d^3\mathbf{r}}{ds^3}}{\frac{d^2\mathbf{r}^T}{ds^2} \cdot \frac{d^2\mathbf{r}}{ds^2}} = \tau$$

for a curve $\mathbf{r}(t)$, where $\mathbf{t}$ is the unit tangent, s is the length along the curve, κ is the curvature, and τ is the torsion.

23. Find the equation for the tangent to the curve of intersection of $x = 2$ and $y = 1 + xz \sin y^2 z$ at the point $(2, 1, \pi)$.

24. Find the curvature and torsion of the curve $\mathbf{r}(t) = 2t\mathbf{i} + t^2\mathbf{j} + 2t^3\mathbf{k}$ at the point $(2, 1, 2)$.

25. Find the tangent to the curve of intersection of the surfaces $y^2 = x$ and $y = xy$ at $(x, y, z) = (1, 1, 1)$.

26. Evaluate

$$I = \oint_C \left(\left(e^{-x^2} - yz \right) dx + \left(e^{-y^2} - xz + 2x \right) dy + e^{-z^2} dz \right),$$

where C is a circle written in cylindrical coordinates as $x = \cos \theta, y = \sin \theta, z = 2$.

27. Find the vector field $\mathbf{u}$ corresponding to the two-dimensional conservative potential $\phi = x^2 + y^2$. Thus find $I = \int_A^B \mathbf{u} \cdot ds$, where $A = (x_A, y_A)$ and $B = (x_B, y_B)$ in two different ways.

28. Calculate

$$\int_0^1 \int_0^1 (x^2 + y^2) \, dx \, dy.$$

29. Calculate the surface integral $\int_S \mathbf{F} \cdot \mathbf{n} \, dS$, where $\mathbf{F} = x^2\mathbf{i} + y^2\mathbf{j} + z^2\mathbf{k}$ and $\mathbf{n}$ is the unit normal over the surface S defined by $z = x - y$ with $x \in [0, 1], y \in [0, 1]$.

30. For $\mathbf{f}(x, y) = x\mathbf{i} + y\mathbf{j} + z\mathbf{k}$, calculate

   (a)

$$\int_S \mathbf{f} \, dx \, dy,$$

   (b)

$$\int_S \mathbf{f} \cdot \mathbf{n} \, dx \, dy,$$

where S is the square with corners $(0,0,0)$, $(0,1,0)$, $(1,1,0)$, and $(0,1,0)$ and $\mathbf{n}$ is the unit outer normal to it.

31. Find the gradient of the scalar field $\phi = x^2 + y^2 + z^2$.

32. Find the gradient, divergence, and curl of the vector field $\mathbf{u} = (x - y)\mathbf{i} + (-x + 2y + z)\mathbf{j} + (y + 3z)\mathbf{j}$.

33. For the curve traced by $\mathbf{r} = \mathbf{i}a\cos\omega t + \mathbf{j}b\sin\omega t$, find the minimum and maximum curvatures.

34. Apply Stokes' theorem to the plane vector field $\mathbf{u}(x, y) = u_x\mathbf{i} + u_y\mathbf{j}$ and a closed curve enclosing a plane region. Use the result to find $\oint_C \mathbf{u}^T \cdot d\mathbf{r}$, where $\mathbf{u} = -y\mathbf{i} + x\mathbf{j}$ and the integration is counterclockwise along the sides C of the trapezoid with corners at $(0,0)$, $(2,0)$, $(2,1)$, and $(1,1)$.

35. Use Green's theorem to calculate $\oint_C \mathbf{u}^T \cdot d\mathbf{r}$, where $\mathbf{u} = x^2\mathbf{i} + 2xy\mathbf{j}$, and C is the counterclockwise path around a rectangle with vertices at $(0,0)$, $(2,0)$, $(0,4)$, and $(2,4)$.

36. For the vector field $\mathbf{u} = x\mathbf{i} + y\mathbf{j} + z\mathbf{k}$ and S being a sphere centered on the origin, show by calculation that the divergence theorem holds.

37. In Cartesian coordinates, the moment of inertia tensor of a body is defined as

$$\mathbf{I} = \int_V \rho(x, y, z) \begin{pmatrix} y^2 + z^2 & -xy & -xz \\ -xy & z^2 + x^2 & -yz \\ -xz & -yz & x^2 + y^2 \end{pmatrix} dx\, dy\, dz.$$

Find the tensor for (a) a sphere and, (b) a cylinder. Assume that the density ρ is uniform, and use a suitable orientation of the coordinates.

38. Heat conduction in an anisotropic solid is described by Duhamel's[9] extension of Fourier's[10] law as $q_i = k_{ij}\partial T/\partial x_j$, where q_i is the heat flux vector, T is the temperature, and k_{ij} is the constant conductivity tensor. Additionally, the first law of thermodynamics requires that $\rho c\, dT/dt = -\partial q_i/\partial x_i$, where ρ is the density and c is the specific heat, both constant. The second law of thermodynamics for such a system requires that $-(1/T^2)q_i\partial T/\partial x_i \geq 0$. (a) Show that when Fourier's law is implemented within the first law of thermodynamics, the antisymmetric part of k_{ij} cannot influence the evolution of T. (b) Show that when Fourier's law is implemented within the second law of thermodynamics for a three-dimensional geometry, one finds that three principal invariants of k_{ij} are required to be positive semidefinite.

39. Calculate the vorticity of the velocity field $\mathbf{u}(x, y, z) = x^2\mathbf{i} + y^2\mathbf{j} + z^2\mathbf{k}$, where the vorticity is defined as the curl of the velocity field.

40. The work done by a force is $W = \int_C \mathbf{F} \cdot d\mathbf{s}$, where $\mathbf{F}$ is the force and $d\mathbf{s}$ is an element of the curve C over which the point of application moves. Compute W if $\mathbf{F} = x\mathbf{i} + y\mathbf{j} + z\mathbf{k}$ and C is the circle of radius R centered on the origin.

41. It is common in fluid mechanics to assume an irrotational velocity field $\mathbf{u}$ for which (a) $\nabla \times \mathbf{u} = 0$, (b) $\oint_C \mathbf{u} \cdot \mathbf{t}\, ds = 0$, for any closed curve C, or (c) $\mathbf{u} = \nabla\phi$. Show that all three are equivalent.

42. It is common in fluid mechanics to work with a solenoidal velocity field $\mathbf{u}$ for which (a) $\nabla \cdot \mathbf{u} = 0$, (b) $\iint_S \mathbf{u} \cdot \mathbf{n}\, dS = 0$, for any closed surface S, or (c) $\mathbf{u} = \nabla \times \mathbf{v}$, where $\mathbf{v}$ is also solenoidal. Show that all three are equivalent.

[9] Jean-Marie Constant Duhamel, 1797–1872, French mathematician.
[10] Jean Baptiste Joseph Fourier, 1768–1830, French mathematician.

3 First-Order Ordinary Differential Equations

We consider here the solution of so-called *first-order ordinary differential equations*. We thus depart from direct consideration of geometry to introduce another major theme of this book: how to solve differential equations. Differential equations are, of course, one of the major mathematical tools in engineering and have great utility in describing the nature of many physical phenomena. This particular chapter is confined to first-order equations, which must be understood before consideration of higher order systems. Their first-order nature will allow us to obtain general solutions to *all* linear problems and solve analytically a remarkably large number of fully nonlinear problems. Moreover, it will be seen in Chapter 9 that a high-order equation can be recast as a coupled system of first-order equations, thus rendering the study of these foundational systems even more critical. When such higher order systems are considered, the strong connection of differential equations to differential geometry will be manifestly obvious. Lastly, although we delay formal definition of linearity to Chapter 4, most readers will be sufficiently familiar with the concept to understand its usage in this chapter.

3.1 Paradigm Problem

A paradigm problem in general form for first-order ordinary differential equations is

$$F(x, y, y') = 0, \tag{3.1}$$

where $y' = dy/dx$. This is fully nonlinear. Solution of a first-order equation typically requires the dependent variable to be specified at one point, although for nonlinear equations, this does not guarantee uniqueness.

EXAMPLE 3.1

Probably the most important linear example, with wide application in engineering, is

$$ay + \frac{dy}{dx} = 0, \qquad y(0) = y_0. \tag{3.2}$$

It can be verified by direct substitution that the unique solution is

$$y(x) = y_0 e^{-ax}. \tag{3.3}$$

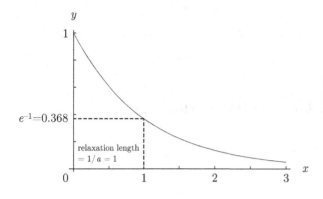

Figure 3.1. Solution $y(x)$, which satisfies the first-order linear equation $y' = -y$, $y(0) = 1$.

We take $a > 0$. When $x = 0$, $y = y_0$, and as $x \to \infty$, $y \to 0$. The solution is monotonically decaying as x increases. We can get a useful estimate of the value of x for which the decay is significant. When $x = 1/a = \ell$, we have $y = y_0 e^{-1} \approx 0.368 y_0$. If x is a distance, ℓ is often called the *relaxation length*. If the independent variable had been time t, we would have defined a so-called *time constant* as $\tau = 1/a$ and considered it to be the time necessary for the initial condition to decay to 0.368 of its initial value. For $y_0 = 1$, $a = 1$, we plot $y(x)$ in Figure 3.1. Here the relaxation length $\ell = 1/a = 1$ is shown in the figure.

EXAMPLE 3.2

A nonlinear example, which we will not try to solve in full analytically, is

$$\left(xy^2 \left(\frac{dy}{dx} \right)^3 + 2\frac{dy}{dx} + \ln\left(\sin xy\right) \right)^2 - 1 = 0, \qquad y(1) = 1. \tag{3.4}$$

The solution can be obtained numerically. Numerical methods give indication that there are two distinct real solutions that satisfy the differential equation and condition at $x = 1$. A plot of $y(x)$ for both solutions for $x \in [1, 5]$ is given in Figure 3.2. One branch is entirely real for $x \in [1, 5]$; the other is entirely real for $x \in [1, 2.060]$. Even though we do not have an exact solution, one can consider an approximate solution, a topic that will formally be considered in Chapter 5. Informally, we see from Figure 3.2 that the solution with the smaller y value seems to be flattening as $x \to \infty$. We can use this fact as a motivation

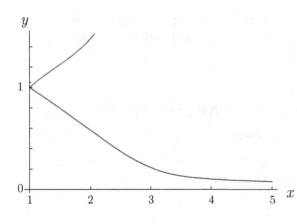

Figure 3.2. Numerical solutions for real $y(x)$ that satisfy the first-order nonlinear Eq. (3.4) for $x \in [1, 5]$.

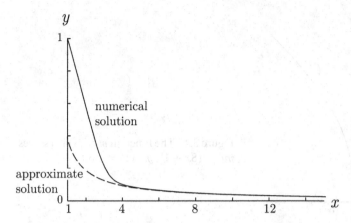

Figure 3.3. Numerical solution for real $y(x)$ that satisfies the first-order nonlinear Eq. (3.4) for $x \in [1, 15]$ along with an approximate solution valid for large x.

for simplifying Eq. (3.4). Let us consider that equation in the limit as $dy/dx \to 0$, which yields

$$(\ln(\sin xy))^2 \approx 1, \tag{3.5}$$

$$\ln(\sin xy) \approx \pm 1, \tag{3.6}$$

$$\sin xy \approx \begin{cases} e, \\ 1/e. \end{cases} \tag{3.7}$$

The $\sin^{-1}$ function has many branches. Trial and error reveals that the relevant branch that matches our numerical solution to the full equation is the hyperbola

$$y \approx \frac{\sin^{-1}(1/e)}{x}. \tag{3.8}$$

A plot of $y(x)$ for the smaller real numerical solution along with the approximation of Eq. (3.8) for $x \in [1, 15]$ is given in Figure 3.3. The approximation is not good for $x \in [1, 2]$; however, it is remarkably accurate for $x > 4$. The availability of a numerical solution motivated us to consider a closed form approximation. Without the numerical solution, we would have had little guidance to generate the approximation.

In summary, despite that Eq. (3.4) has daunting complexity, the solutions are well behaved. In fact, one of them can be approximated accurately if x is large. Thus, one can recognize that seemingly complicated equations may in fact yield solutions whose structure is relatively simple in certain domains. In later chapters, we shall find examples in which seemingly simple equations yield solutions with remarkable complexity.

Returning to the present topic, an exact solution exists for all linear first-order equations and many nonlinear ones; we turn to those with exact solution for the remainder of the chapter.

3.2 Separation of Variables

Equation (3.1) is separable if it can be written in the form

$$P(x)\,dx = Q(y)\,dy, \tag{3.9}$$

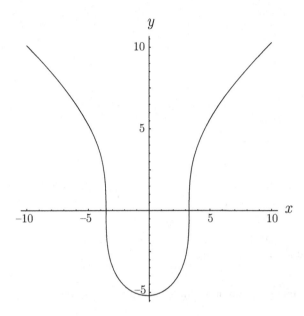

Figure 3.4. The function $y(x)$, which solves $yy' = (8x + 1)/y$, $y(1) = -5$.

which can then be integrated. This simple method is appropriate for both linear and nonlinear equations.

EXAMPLE 3.3

Solve the nonlinear equation

$$yy' = \frac{8x + 1}{y}, \qquad y(1) = -5. \tag{3.10}$$

Separating variables, we get

$$y^2 \, dy = 8x \, dx + dx. \tag{3.11}$$

Integrating, we have

$$\frac{y^3}{3} = 4x^2 + x + C. \tag{3.12}$$

The initial condition gives $C = -140/3$, so that the solution is

$$y^3 = 12x^2 + 3x - 140. \tag{3.13}$$

The solution is plotted in Figure 3.4.

3.3 Homogeneous Equations

A first-order differential equation is defined by many as *homogeneous*[1] if it can be written in the form

$$y' = f\left(\frac{y}{x}\right). \tag{3.14}$$

---

[1] The word *homogeneous* has two distinct interpretations in differential equations. In the present section, the word actually refers to the function f, which is better considered as a so-called homogeneous function of degree zero, which implies $f(tx, ty) = t^0 f(x, y) = f(x, y)$. Obviously, $f(y/x)$ satisfies this criterion. A more common interpretation is that an equation of the form $\mathbf{L}y = f$ is homogeneous iff (i.e., if and only if) $f = 0$.

Defining

$$u = \frac{y}{x}, \tag{3.15}$$

we get

$$y = ux, \tag{3.16}$$

from which

$$y' = u + xu'. \tag{3.17}$$

Substituting in Eq. (3.14) and separating variables, we have

$$u + xu' = f(u), \tag{3.18}$$

$$u + x\frac{du}{dx} = f(u), \tag{3.19}$$

$$x\frac{du}{dx} = f(u) - u, \tag{3.20}$$

$$\frac{du}{f(u) - u} = \frac{dx}{x}, \tag{3.21}$$

which can be integrated. Equations of the form

$$y' = f\left(\frac{a_1 x + a_2 y + a_3}{a_4 x + a_5 y + a_6}\right) \tag{3.22}$$

can be similarly integrated. For equations of this type, there is no restriction of linearity.

EXAMPLE 3.4

Solve

$$xy' = 3y + \frac{y^2}{x}, \qquad y(1) = 4. \tag{3.23}$$

This can be written as

$$y' = 3\left(\frac{y}{x}\right) + \left(\frac{y}{x}\right)^2. \tag{3.24}$$

Let $u = y/x$. Then

$$f(u) = 3u + u^2. \tag{3.25}$$

Using Eq. (3.21), we get

$$\frac{du}{2u + u^2} = \frac{dx}{x}. \tag{3.26}$$

Because by partial fraction expansion we have

$$\frac{1}{2u + u^2} = \frac{1}{2u} - \frac{1}{4 + 2u}, \tag{3.27}$$

Eq. (3.26) can be rewritten as

$$\frac{du}{2u} - \frac{du}{4 + 2u} = \frac{dx}{x}. \tag{3.28}$$

Both sides can be integrated to give

$$\frac{1}{2}\left(\ln|u| - \ln|2 + u|\right) = \ln|x| + C. \tag{3.29}$$

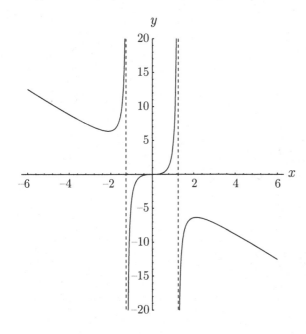

Figure 3.5. The function $y(x)$, which solves $xy' = 3y + y^2/x$, $y(1) = 4$.

The initial condition gives $C = (1/2)\ln(2/3)$, so that the solution can be reduced to

$$\left|\frac{y}{2x + y}\right| = \frac{2}{3}x^2.$$

This can be solved explicitly for $y(x)$ for each case of the absolute value. The first case,

$$y(x) = \frac{\frac{4}{3}x^3}{1 - \frac{2}{3}x^2}, \tag{3.30}$$

is seen to satisfy the condition at $x = 1$. The second case is discarded as it does not satisfy the condition at $x = 1$. The solution is plotted in Figure 3.5.

3.4 Exact Equations

A differential equation is exact if it can be written in the form

$$dF(x, y) = 0, \tag{3.31}$$

where $F(x, y) = 0$ is a solution to the differential equation. Such problems may be fully nonlinear. The total differential of $F(x, y)$ is

$$dF = \frac{\partial F}{\partial x}\, dx + \frac{\partial F}{\partial y}\, dy = 0. \tag{3.32}$$

So, for an equation of the form

$$P(x, y)\, dx + Q(x, y)\, dy = 0, \tag{3.33}$$

we have an exact differential if

$$\frac{\partial F}{\partial x} = P(x, y), \qquad \frac{\partial F}{\partial y} = Q(x, y), \tag{3.34}$$

$$\frac{\partial^2 F}{\partial x \partial y} = \frac{\partial P}{\partial y}, \qquad \frac{\partial^2 F}{\partial y \partial x} = \frac{\partial Q}{\partial x}. \tag{3.35}$$

As long as $F(x, y)$ is continuous and differentiable, the mixed second partials are equal, thus

$$\frac{\partial P}{\partial y} = \frac{\partial Q}{\partial x} \tag{3.36}$$

must hold if $F(x, y)$ is to exist and render the original differential equation to be exact.

EXAMPLE 3.5

Solve the nonlinear problem

$$\frac{dy}{dx} = \frac{e^{x-y}}{e^{x-y} - 1}, \tag{3.37}$$

$$\underbrace{\left(e^{x-y}\right)}_{=P} dx + \underbrace{\left(1 - e^{x-y}\right)}_{=Q} dy = 0, \tag{3.38}$$

$$\frac{\partial P}{\partial y} = -e^{x-y}, \tag{3.39}$$

$$\frac{\partial Q}{\partial x} = -e^{x-y}. \tag{3.40}$$

Because $\partial P/\partial y = \partial Q/\partial x$, the equation is exact. Thus,

$$\frac{\partial F}{\partial x} = P(x, y), \tag{3.41}$$

$$\frac{\partial F}{\partial x} = e^{x-y}, \tag{3.42}$$

$$F(x, y) = e^{x-y} + A(y), \tag{3.43}$$

$$\frac{\partial F}{\partial y} = -e^{x-y} + \frac{dA}{dy} = Q(x, y) = 1 - e^{x-y}, \tag{3.44}$$

$$\frac{dA}{dy} = 1, \tag{3.45}$$

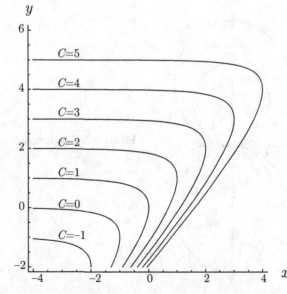

Figure 3.6. The function $y(x)$, which solves $y' = \exp(x - y)/(\exp(x - y) - 1)$.

$$A(y) = y - C, \tag{3.46}$$

$$F(x, y) = e^{x-y} + y - C = 0, \tag{3.47}$$

$$e^{x-y} + y = C. \tag{3.48}$$

The solution for various values of C is plotted in Figure 3.6.

3.5 Integrating Factors

Sometimes an equation of the form of Eq. (3.33) is not exact but can be made so by multiplication by a function $u(x, y)$, where u is called the *integrating factor*. It is not always obvious that integrating factors exist; sometimes they do not. When one exists, it may not be unique. Unfortunately, there is no general procedure to identify integrating factors for arbitrary first-order differential equations, even with forms as simple as Eq. (3.33).

EXAMPLE 3.6

Solve the nonlinear problem

$$\frac{dy}{dx} = \frac{2xy}{x^2 - y^2}. \tag{3.49}$$

Separating variables, we get

$$(x^2 - y^2)\, dy = 2xy\, dx. \tag{3.50}$$

This is not exact according to Eq. (3.36). It turns out that the integrating factor is y^{-2}, so that on multiplication, we get

$$\frac{2x}{y}\, dx - \left(\frac{x^2}{y^2} - 1\right) dy = 0. \tag{3.51}$$

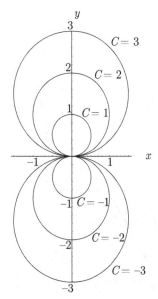

Figure 3.7. The function $y(x)$, which solves $y'(x) = 2xy/(x^2 - y^2)$.

This can be written as

$$d\left(\frac{x^2}{y} + y\right) = 0,$$ (3.52)

which gives

$$\frac{x^2}{y} + y = C,$$ (3.53)

$$x^2 + y^2 = Cy.$$ (3.54)

The solution for various values of C is plotted in Figure 3.7.

3.6 General Linear Solution

Thus far we have considered special methods that could be applied to some nonlinear problems. For *all* linear first-order ordinary differential equations, unique solutions exist and can be easily found, as is shown here. The general first-order linear equation

$$\frac{dy(x)}{dx} + P(x)y(x) = Q(x), \qquad y(0) = y_0,$$ (3.55)

can be solved by applying the integrating factor

$$e^{\int_a^x P(s)\,ds} = e^{F(x) - F(a)}.$$ (3.56)

Here, depending on the form of P, F may or may not be available in explicit analytical form. We choose a such that

$$F(a) = 0.$$ (3.57)

Multiply Eq. (3.55) by the integrating factor and proceed:

$$\left(e^{\int_a^x P(s)\,ds}\right)\frac{dy(x)}{dx} + \left(e^{\int_a^x P(s)\,ds}\right) P(x)\,y(x) = \left(e^{\int_a^x P(s)\,ds}\right) Q(x).$$ (3.58)

Now use the product rule to combine terms on the left side:

$$\frac{d}{dx}\left(e^{\int_a^x P(s)\,ds}y(x)\right) = \left(e^{\int_a^x P(s)\,ds}\right) Q(x).$$ (3.59)

Next, replace x by t:

$$\frac{d}{dt}\left(e^{\int_a^t P(s)\,ds}y(t)\right) = \left(e^{\int_a^t P(s)\,ds}\right) Q(t).$$ (3.60)

Now apply the operator $\int_{x_0}^x (\cdot)\,dt$ to both sides:

$$\int_{x_0}^x \frac{d}{dt}\left(e^{\int_a^t P(s)\,ds}y(t)\right)\,dt = \int_{x_0}^x \left(e^{\int_a^t P(s)\,ds}\right) Q(t)\,dt.$$ (3.61)

The fundamental theorem of calculus allows evaluation of the left side so that

$$e^{\int_a^x P(s)\,ds}y(x) - e^{\int_a^{x_0} P(s)\,ds}y(x_0) = \int_{x_0}^x \left(e^{\int_a^t P(s)\,ds}\right) Q(t)\,dt,$$ (3.62)

which yields an exact solution for $y(x)$ in terms of arbitrary $P(x)$, $Q(x)$, x_0, and y_0:

$$y(x) = e^{-\int_a^x P(s)\, ds} \left(e^{\int_a^{x_0} P(s)\, ds} y_0 + \int_{x_0}^x \left(e^{\int_a^t P(s)\, ds} \right) Q(t)\, dt \right). \quad (3.63)$$

In terms of F, we can say

$$y(x) = e^{-F(x)} \left(e^{F(x_0)} y_0 + \int_{x_0}^x e^{F(t)} Q(t)\, dt \right), \quad (3.64)$$

$$= e^{F(x_0)-F(x)} y_0 + \int_{x_0}^x e^{F(t)-F(x)} Q(t)\, dt. \quad (3.65)$$

EXAMPLE 3.7

Solve the linear problem

$$y' - y = e^{2x}, \qquad y(0) = y_0. \quad (3.66)$$

Here

$$P(x) = -1, \quad (3.67)$$

or

$$P(s) = -1, \quad (3.68)$$

$$\int_a^x P(s)\, ds = \int_a^x (-1)\, ds, \quad (3.69)$$

$$= -s\big|_a^x, \quad (3.70)$$

$$= -x - (-a). \quad (3.71)$$

So

$$F(x) = -x. \quad (3.72)$$

For $F(a) = 0$, take $a = 0$. So the integrating factor is

$$e^{\int_0^x (-1)\, ds} = e^{-x}. \quad (3.73)$$

Multiplying and rearranging, we get

$$e^{-x} \frac{dy(x)}{dx} - e^{-x} y(x) = e^x, \quad (3.74)$$

$$\frac{d}{dx} \left(e^{-x} y(x) \right) = e^x, \quad (3.75)$$

$$\frac{d}{dt} \left(e^{-t} y(t) \right) = e^t, \quad (3.76)$$

$$\int_{x_0=0}^x \frac{d}{dt} \left(e^{-t} y(t) \right) dt = \int_{x_0=0}^x e^t\, dt, \quad (3.77)$$

$$e^{-x} y(x) - e^{-0} y(0) = e^x - e^0, \quad (3.78)$$

$$e^{-x} y(x) - y_0 = e^x - 1, \quad (3.79)$$

$$y(x) = e^x \left(y_0 + e^x - 1 \right), \quad (3.80)$$

$$= e^{2x} + (y_0 - 1) e^x. \quad (3.81)$$

The solution for various values of y_0 is plotted in Figure 3.8.

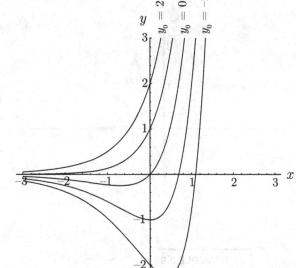

Figure 3.8. The function $y(x)$, which solves $y' - y = e^{2x}$, $y(0) = y_0$.

3.7 Bernoulli Equation

Some first-order nonlinear equations can be transformed into linear problems with exact solutions. An example is the *Bernoulli*[2] equation,

$$y' + P(x)y = Q(x)y^n, \tag{3.82}$$

where $n \neq 1$. Let us define a new dependent variable u such that

$$u = y^{1-n}, \tag{3.83}$$

so that

$$y = u^{\frac{1}{1-n}}. \tag{3.84}$$

The derivative is

$$y' = \frac{1}{1-n}\left(u^{\frac{n}{1-n}}\right)u'. \tag{3.85}$$

Substituting in Eq. (3.82), we get

$$\frac{1}{1-n}\left(u^{\frac{n}{1-n}}\right)u' + P(x)u^{\frac{1}{1-n}} = Q(x)u^{\frac{n}{1-n}}. \tag{3.86}$$

This can be written as

$$u' + (1-n)P(x)u = (1-n)Q(x), \tag{3.87}$$

which is a first-order linear equation of the form of Eq. (3.55) and can be solved.

---

[2] Jacob Bernoulli, 1654–1705, Swiss-born mathematician.

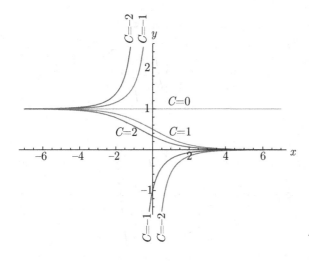

Figure 3.9. The function $y(x)$, which solves $y' + y = y^2$ for various C.

EXAMPLE 3.8

Solve the nonlinear equation

$$y' + y = y^2. \tag{3.88}$$

This is a Bernoulli equation with $P(x) = 1$, $Q(x) = 1$, and $n = 2$. We let $u = y^{-1}$. Therefore $y = u^{-1}$ and $y' = -u^{-2}u'$. Equation (3.88) becomes

$$-\frac{1}{u^2}u' + \frac{1}{u} = \frac{1}{u^2}. \tag{3.89}$$

Assuming $u \neq 0$, we multiply by $-u^2$ to get the linear equation

$$u' - u = -1. \tag{3.90}$$

Multiplying by the integrating factor e^{-x}, we get

$$e^{-x}u' - e^{-x}u = -e^{-x}. \tag{3.91}$$

Analyzing, we find

$$(e^{-x}u)' = -e^{-x}, \tag{3.92}$$

$$e^{-x}u = e^{-x} + C, \tag{3.93}$$

$$u = 1 + Ce^x, \tag{3.94}$$

$$y = \frac{1}{1 + Ce^x}. \tag{3.95}$$

The solution for various values of C is plotted in Figure 3.9.

3.8 Riccati Equation

A *Riccati*[3] equation is of the form

$$\frac{dy}{dx} = P(x)y^2 + Q(x)y + R(x). \tag{3.96}$$

---

[3] Jacopo Riccati, 1676–1754, Venetian mathematician.

The presence of y^2 renders it nonlinear. Studied by several Bernoullis and two Riccatis, it was solved by Euler. If we know a specific solution $y = S(x)$ of this equation, the general solution can then be found. There is no general procedure to identify $S(x)$ for an arbitrary Riccati equation.

Let

$$y = S(x) + \frac{1}{z(x)}. \tag{3.97}$$

Thus

$$\frac{dy}{dx} = \frac{dS}{dx} - \frac{1}{z^2}\frac{dz}{dx}. \tag{3.98}$$

Substituting into Eq. (3.96), we get

$$\frac{dS}{dx} - \frac{1}{z^2}\frac{dz}{dx} = P\left(S + \frac{1}{z}\right)^2 + Q\left(S + \frac{1}{z}\right) + R, \tag{3.99}$$

$$= P\left(S^2 + \frac{2S}{z} + \frac{1}{z^2}\right) + Q\left(S + \frac{1}{z}\right) + R, \tag{3.100}$$

$$\underbrace{\frac{dS}{dx} - (PS^2 + QS + R)}_{=0} - \frac{1}{z^2}\frac{dz}{dx} = P\left(\frac{2S}{z} + \frac{1}{z^2}\right) + Q\left(\frac{1}{z}\right), \tag{3.101}$$

$$-\frac{dz}{dx} = P(2Sz + 1) + Qz, \tag{3.102}$$

$$\frac{dz}{dx} + (2P(x)S(x) + Q(x))z = -P(x). \tag{3.103}$$

Again this is a first-order linear equation in z and x of the form of Eq. (3.55) and can be solved.

EXAMPLE 3.9

Solve

$$y' = \frac{e^{-3x}}{x}y^2 - \frac{1}{x}y + 3e^{3x}. \tag{3.104}$$

One solution is

$$y = S(x) = e^{3x}. \tag{3.105}$$

Verify

$$3e^{3x} = \frac{e^{-3x}}{x}e^{6x} - \frac{1}{x}e^{3x} + 3e^{3x}, \tag{3.106}$$

$$= \frac{e^{3x}}{x} - \frac{e^{3x}}{x} + 3e^{3x}, \tag{3.107}$$

$$= 3e^{3x}, \tag{3.108}$$

so let

$$y = e^{3x} + \frac{1}{z}. \tag{3.109}$$

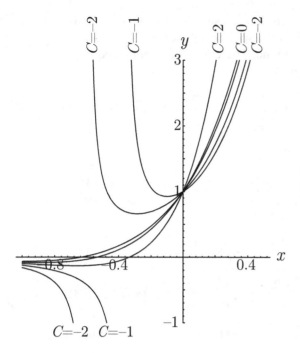

Figure 3.10. The function $y(x)$, which solves $y' = \exp(-3x)/x - y/x + 3\exp(3x)$.

Also we have

$$P(x) = \frac{e^{-3x}}{x}, \tag{3.110}$$

$$Q(x) = -\frac{1}{x}, \tag{3.111}$$

$$R(x) = 3e^{3x}. \tag{3.112}$$

Substituting into Eq. (3.103), we get

$$\frac{dz}{dx} + \left(2\frac{e^{-3x}}{x}e^{3x} - \frac{1}{x}\right)z = -\frac{e^{-3x}}{x}, \tag{3.113}$$

$$\frac{dz}{dx} + \frac{z}{x} = -\frac{e^{-3x}}{x}. \tag{3.114}$$

The integrating factor here is

$$e^{\int \frac{dx}{x}} = e^{\ln x} = x. \tag{3.115}$$

Multiplying by the integrating factor, we get

$$x\frac{dz}{dx} + z = -e^{-3x}, \tag{3.116}$$

$$\frac{d(xz)}{dx} = -e^{-3x}, \tag{3.117}$$

which can be integrated to yield

$$z = \frac{e^{-3x}}{3x} + \frac{C}{x} = \frac{e^{-3x} + 3C}{3x}. \tag{3.118}$$

Because $y = S(x) + 1/z$, the solution is

$$y = e^{3x} + \frac{3x}{e^{-3x} + 3C}. \tag{3.119}$$

The solution for various values of C is plotted in Figure 3.10.

3.9 Reduction of Order

There are higher order equations that can be reduced to first-order equations and then solved.

3.9.1 Dependent Variable *y* Absent

If

$$f(x, y', y'') = 0, \tag{3.120}$$

then let $u(x) = y'$. Thus, $u'(x) = y''$, and the equation reduces to

$$f\left(x, u, \frac{du}{dx}\right) = 0, \tag{3.121}$$

which is an equation of first order.

EXAMPLE 3.10

Solve

$$xy'' + 2y' = 4x^3. \tag{3.122}$$

Let $u = y'$, so that

$$x\frac{du}{dx} + 2u = 4x^3. \tag{3.123}$$

Multiplying by x,

$$x^2\frac{du}{dx} + 2xu = 4x^4, \tag{3.124}$$

$$\frac{d}{dx}(x^2 u) = 4x^4. \tag{3.125}$$

This can be integrated to give

$$u = \frac{4}{5}x^3 + \frac{C_1}{x^2}, \tag{3.126}$$

from which

$$y = \frac{1}{5}x^4 - \frac{C_1}{x} + C_2, \tag{3.127}$$

for $x \neq 0$.

3.9.2 Independent Variable *x* Absent

If

$$f(y, y', y'') = 0, \tag{3.128}$$

let $u(x) = y'$, so that

$$y'' = \frac{dy'}{dx} = \frac{dy'}{dy}\frac{dy}{dx} = \frac{du}{dy}u. \tag{3.129}$$

Equation (3.128) becomes

$$f\left(y, u, u\frac{du}{dy}\right) = 0, \qquad (3.130)$$

which is also an equation of first order. However, the independent variable is now y, while the dependent variable is u.

EXAMPLE 3.11

Solve

$$y'' - 2yy' = 0, \qquad y(0) = y_0, \ y'(0) = y_0'. \qquad (3.131)$$

Let $u = y'$, so that $y'' = du/dx = (dy/dx)(du/dy) = u(du/dy)$. Equation (3.131) becomes

$$u\frac{du}{dy} - 2yu = 0. \qquad (3.132)$$

Now

$$u = 0, \qquad (3.133)$$

satisfies Eq. (3.132). Thus,

$$\frac{dy}{dx} = 0, \qquad (3.134)$$

$$y = C. \qquad (3.135)$$

Applying one initial condition, we get

$$y = y_0. \qquad (3.136)$$

This satisfies the initial conditions only under special circumstances, that is, $y_0' = 0$. For $u \neq 0$,

$$\frac{du}{dy} = 2y, \qquad (3.137)$$

$$u = y^2 + C_1. \qquad (3.138)$$

Now, apply the initial conditions to get

$$y_0' = y_0^2 + C_1, \qquad (3.139)$$

$$C_1 = y_0' - y_0^2, \qquad (3.140)$$

$$\frac{dy}{dx} = y^2 + y_0' - y_0^2, \qquad (3.141)$$

$$\frac{dy}{y^2 + y_0' - y_0^2} = dx, \qquad (3.142)$$

from which, for $y_0' - y_0^2 > 0$,

$$\frac{1}{\sqrt{y_0' - y_0^2}} \tan^{-1}\left(\frac{y}{\sqrt{y_0' - y_0^2}}\right) = x + C_2, \qquad (3.143)$$

$$\frac{1}{\sqrt{y_0' - y_0^2}} \tan^{-1}\left(\frac{y_0}{\sqrt{y_0' - y_0^2}}\right) = C_2, \qquad (3.144)$$

$$y(x) = \sqrt{y_0' - y_0^2} \tan\left(x\sqrt{y_0' - y_0^2} + \tan^{-1}\left(\frac{y_0}{\sqrt{y_0' - y_0^2}}\right)\right). \qquad (3.145)$$

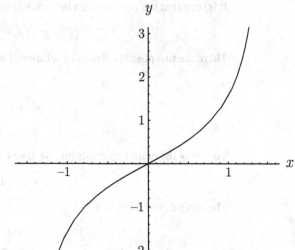

Figure 3.11. The function $y(x)$, which solves $y'' - 2yy' = 0$, $y(0) = 0$, $y'(0) = 1$.

The solution for $y_0 = 0$, $y_0' = 1$ is plotted in Figure 3.11.

For $y_0' - y_0^2 = 0$,

$$\frac{dy}{dx} = y^2, \tag{3.146}$$

$$\frac{dy}{y^2} = dx, \tag{3.147}$$

$$-\frac{1}{y} = x + C_2, \tag{3.148}$$

$$-\frac{1}{y_0} = C_2, \tag{3.149}$$

$$-\frac{1}{y} = x - \frac{1}{y_0}, \tag{3.150}$$

$$y = \frac{1}{\dfrac{1}{y_0} - x}. \tag{3.151}$$

For $y_0' - y_0^2 < 0$, one would obtain solutions in terms of hyperbolic trigonometric functions (see Section A.6).

3.10 Factorable Equations

Some nonlinear first-order differential equations can be factored and solved.

EXAMPLE 3.12

Find all solutions to

$$y'^2 - y' - y^2 y' + y^2 = 0, \qquad y(0) = 1. \tag{3.152}$$

It is seen that the equation can be factored into

$$\left(y' - y^2\right)\left(y' - 1\right) = 0. \tag{3.153}$$

There are two possible classes of solution. The first possibility is

$$y' - 1 = 0, \tag{3.154}$$

$$y' = 1, \tag{3.155}$$

$$y = x + C_1. \tag{3.156}$$

To satisfy the boundary condition, we take $C_1 = 1$, so

$$y = x + 1. \tag{3.157}$$

The second possibility is

$$y' - y^2 = 0, \tag{3.158}$$

$$y' = y^2, \tag{3.159}$$

$$\frac{dy}{y^2} = dx, \tag{3.160}$$

$$-\frac{1}{y} = x + C_2, \tag{3.161}$$

$$y = \frac{-1}{x + C_2}. \tag{3.162}$$

To satisfy the boundary condition, we take $C_2 = -1$, so

$$y = \frac{1}{1 - x}. \tag{3.163}$$

The nonlinear equation obviously has two solutions, both of which satisfy the governing equation and initial condition.

3.11 Uniqueness and Singular Solutions

Not all differential equations have solutions, as can be seen by considering

$$y' = \frac{y}{x}\ln y, \qquad y(0) = 2. \tag{3.164}$$

The general solution is $y = e^{Cx}$, but no finite value of C allows the boundary condition to be satisfied. Let us check this by direct substitution:

$$y = e^{Cx}, \tag{3.165}$$

$$y' = Ce^{Cx}, \tag{3.166}$$

$$\frac{y}{x}\ln y = \frac{e^{Cx}}{x}\ln e^{Cx}, \tag{3.167}$$

$$= \frac{e^{Cx}}{x}Cx, \tag{3.168}$$

$$= Ce^{Cx}, \tag{3.169}$$

$$= y'. \tag{3.170}$$

So the differential equation is satisfied for all values of C. Now to satisfy the boundary condition, we must have

$$2 = e^{C(0)}. \tag{3.171}$$

However, for all finite C, $e^{C(0)} = 1$; thus, there is no finite value of C that allows satisfaction of the boundary condition.

The original differential equation can be rewritten as $xy' = y \ln y$. The point $x = 0$ is *singular* because at that point, the highest derivative is multiplied by 0, leaving only $0 = y \ln y$ at $x = 0$. For the special boundary condition $y(0) = 1$, the solution $y = e^{Cx}$ is valid for *all* values of C. Thus, for this singular equation, for most boundary conditions, no solution exists. For one special boundary condition, a solution exists, but it is not unique. One can prove the following.

Theorem: Let $f(x, y)$ be continuous and satisfy $|f(x, y)| \le m$ and the *Lipschitz*[4] *condition* $|f(x, y) - f(x, y_0)| \le k|y - y_0|$ in a bounded region. Then the equation $y' = f(x, y)$ has one and only one solution containing the point (x_0, y_0).

A stronger condition is that if $f(x, y)$ and $\partial f/\partial y$ are finite and continuous at (x_0, y_0), then a solution of $y' = f(x, y)$ exists and is unique in the neighborhood of this point.

EXAMPLE 3.13

Analyze the uniqueness of the solution of

$$\frac{dy}{dt} = -K\sqrt{y}, \qquad y(t_0) = 0. \tag{3.172}$$

Here, t is the independent variable instead of x. Taking

$$f(t, y) = -K\sqrt{y}, \tag{3.173}$$

we have

$$\frac{\partial f}{\partial y} = -\frac{K}{2\sqrt{y}}, \tag{3.174}$$

which is not finite at $y = 0$. So the solution cannot be guaranteed to be unique. In fact, one solution is

$$y(t) = \frac{1}{4}K^2(t - t_0)^2. \tag{3.175}$$

Another solution that satisfies the initial condition and differential equation is

$$y(t) = 0. \tag{3.176}$$

Obviously the solution is not unique.

EXAMPLE 3.14

Solve the differential equation and boundary condition

$$\frac{dy}{dx} = 3y^{2/3}, \qquad y(2) = 0. \tag{3.177}$$

---

[4] Rudolf Otto Sigismund Lipschitz, 1832–1903, German mathematician.

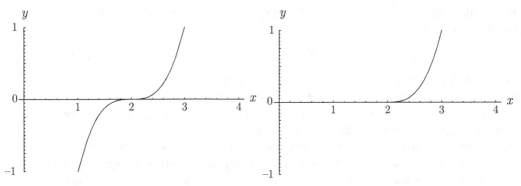

Figure 3.12. Two solutions $y(x)$ that satisfy $y' = 3y^{2/3}$, $y(2) = 0$.

On separating variables and integrating, we get

$$3y^{1/3} = 3x + 3C,$$ (3.178)

so that the *general* solution is

$$y = (x + C)^3.$$ (3.179)

Applying the boundary condition, we find

$$y = (x - 2)^3.$$ (3.180)

However,

$$y = 0,$$ (3.181)

and

$$y = \begin{cases} (x - 2)^3, & x \geq 2, \\ 0, & x < 2, \end{cases}$$ (3.182)

are also solutions. These *singular* solutions cannot be obtained from the general solution. However, values of y' and y are the same at intersections. Both satisfy the differential equation. The two solutions are plotted in Figure 3.12.

3.12 Clairaut Equation

The solution of a Clairaut[5] equation,

$$y = xy' + f(y'),$$ (3.183)

can be obtained by letting $y' = u(x)$, so that

$$y = xu + f(u).$$ (3.184)

Clairaut equations are generally nonlinear, depending on f.
    Differentiating Eq. (3.184) with respect to x, we get

$$y' = xu' + u + \frac{df}{du}u',$$ (3.185)

$$u = xu' + u + \frac{df}{du}u',$$ (3.186)

$$\left(x + \frac{df}{du}\right)u' = 0.$$ (3.187)

---

[5] Alexis Claude Clairaut, 1713–1765, French mathematician.

There are two possible solutions: $u' = 0$ or $x + df/du = 0$. If we consider the first and take

$$u' = \frac{du}{dx} = 0, \tag{3.188}$$

we can integrate to get

$$u = C, \tag{3.189}$$

where C is a constant. Then, from Eq. (3.184), we get the general solution

$$y = Cx + f(C). \tag{3.190}$$

Applying a boundary condition $y(x_0) = y_0$ gives what we will call the *regular* solution.
    But if we take the second,

$$x + \frac{df}{du} = 0, \tag{3.191}$$

and rearrange to get

$$x = -\frac{df}{du}, \tag{3.192}$$

then Eq. (3.192) along with the rearranged Eq. (3.184),

$$y = -u\frac{df}{du} + f(u), \tag{3.193}$$

form a set of parametric equations for what we call the *singular* solution. It is singular because the coefficient on the highest derivative in Eq. (3.187) is itself 0.

EXAMPLE 3.15

Solve

$$y = xy' + (y')^3, \qquad y(0) = y_0. \tag{3.194}$$

Take

$$u = y', \tag{3.195}$$

Then

$$f(u) = u^3, \tag{3.196}$$

$$\frac{df}{du} = 3u^2, \tag{3.197}$$

so specializing Eq. (3.190) gives

$$y = Cx + C^3 \tag{3.198}$$

as the general solution. Use the boundary condition to evaluate C and get the regular solution:

$$y_0 = C(0) + C^3, \tag{3.199}$$

$$C = y_0^{1/3}, \tag{3.200}$$

$$y = y_0^{1/3}x + y_0. \tag{3.201}$$

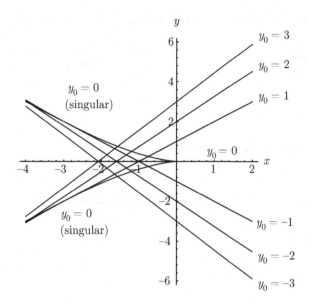

Figure 3.13. Solutions $y(x)$ that satisfy $y = xy' + (y')^3, y(0) = y_0$ for various values of y_0.

If $y_0 \in \mathbb{R}^1$, there are actually three roots $C = y_0^{1/3}, (-1/2 \pm i\sqrt{3}/2)y_0^{1/3}$ (see Section A.1). So the solution is nonunique. However, if we confine our attention to real-valued solutions, there is a unique real solution, with $C = y_0^{1/3}$.

The parametric form of the singular solution is

$$y = -2u^3, \tag{3.202}$$

$$x = -3u^2. \tag{3.203}$$

Eliminating the parameter u, we obtain

$$y = \pm 2\left(-\frac{x}{3}\right)^{3/2} \tag{3.204}$$

as the explicit form of the singular solution.

The regular solutions and singular solution are plotted in Figure 3.13. Note the following:

- In contrast to solutions for equations linear in y', the trajectories $y(x; y_0)$ cross at numerous locations in the (x, y) plane. This is a consequence of the differential equation's *nonlinearity*.
- Although the singular solution satisfies the differential equation, it satisfies this boundary condition only when $y_0 = 0$.
- For real-valued x and y, the singular solution is only valid for $x \leq 0$.
- Because of nonlinearity, addition of the regular and singular solutions does not yield a solution to the differential equation.

3.13 Picard Iteration

The existence and uniqueness of the solutions to first-order differential equations have proofs, not given here, that rely on so-called *Picard*[6] *iteration*. The iterative procedure is as follows. Consider a potentially nonlinear first-order differential equation

---

[6] Émile Picard, 1856–1941, French mathematician.

of the general form

$$\frac{dy}{dx} = f(x, y), \qquad y(x_0) = y_0. \tag{3.205}$$

By inspection, one can write the solution in the integral form

$$y = y_0 + \int_{x_0}^{x} f(t, y(t)) \, dt. \tag{3.206}$$

Application of the Leibniz rule, Eq. (1.128), easily verifies this is a solution. However, it is of little use because we have no explicit representation of $y(x)$. It is easy to invoke an iterative scheme to estimate the solution $y(x)$; we outline that approach here.

- Estimate the solution to be

$$y(x) \approx y_0. \tag{3.207}$$

  This will obviously satisfy the boundary condition at $x = x_0$, but little else.
- Employ y_0 within Eq. (3.206) to generate the next estimate, $y_1(x)$:

$$y(x) \approx y_1(x) = y_0 + \int_{x_0}^{x} f(t, y_0) \, dt. \tag{3.208}$$

  The estimate y_1 is likely to have nontrivial variation with x.
- Employ $y_1(x)$ to obtain a related estimate, $y_2(x)$:

$$y(x) \approx y_2(x) = y_0 + \int_{x_0}^{x} f(t, y_1(t)) \, dt. \tag{3.209}$$

- Employ $y_2(x)$ to obtain a related estimate, $y_3(x)$:

$$y(x) \approx y_3(x) = y_0 + \int_{x_0}^{x} f(t, y_2(t)) \, dt. \tag{3.210}$$

- Continue as necessary.

Sometimes the integrals in Picard iteration can be evaluated analytically. Sometimes they require numerical quadrature. If one is using Picard iteration to generate a solution and numerical quadrature is required, one may wish to reconsider and simply use a numerical method on the original differential equation.

EXAMPLE 3.16

Use Picard iteration to estimate the solution to

$$y' = y, \qquad y(0) = 1. \tag{3.211}$$

Here $f(x, y) = y$, $x_0 = 0$, and $y_0 = 1$. Our first guess is $y_0 = 1$. Using this, we get

$$y_1(x) = 1 + \int_{0}^{x} (1) \, dt, \tag{3.212}$$

$$= 1 + x. \tag{3.213}$$

Our second guess is

$$y_2(x) = 1 + \int_0^x (1+t)\, dt, \tag{3.214}$$

$$= 1 + \left(t + \frac{t^2}{2} \right) \Big|_0^x, \tag{3.215}$$

$$= 1 + x + \frac{x^2}{2}. \tag{3.216}$$

We can carry on as far as we like and induce that

$$y_N = \sum_{n=0}^{N} \frac{x^n}{n!}, \tag{3.217}$$

which is the well-known Taylor series for e^x. It is also the well-known exact solution to our original differential equation.

EXAMPLE 3.17

Use Picard iteration to estimate the solution to

$$y' = x^2 - y^3, \qquad y(0) = 1. \tag{3.218}$$

There is no obvious method to solve this problem in closed form. Here $f(x, y) = x^2 - y^3$, $x_0 = 0$, and $y_0 = 1$. Our first guess is $y_0 = 1$. Using this, we get

$$y_1(x) = 1 + \int_0^x (t^2 - 1^3)\, dt, \tag{3.219}$$

$$= 1 - x + \frac{x^3}{3}. \tag{3.220}$$

Our second guess is

$$y_2(x) = 1 - x + \frac{x^3}{3} + \int_0^x \left(t^2 - \left(1 - t + \frac{t^3}{3} \right)^3 \right) dt, \tag{3.221}$$

$$= 1 - x + \frac{3}{2}x^2 - \frac{2}{3}x^3 + \frac{2}{5}x^5 - \frac{1}{6}x^6 - \frac{1}{21}x^7 + \frac{1}{24}x^8 - \frac{1}{270}x^{10}. \tag{3.222}$$

We could obtain more terms, but we stop here. We compare our estimate of $y_2(x)$ with a high-precision numerical solution in Figure 3.14.

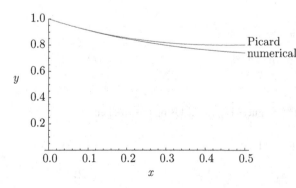

Figure 3.14. High-precision numerical estimate along with the Picard iteration estimate for $y(x)$ that satisfies $y' = x^2 - y^3$, $y(0) = 1$.

3.14 Solution by Taylor Series

One can also solve general first-order equations with the assistance of Taylor series. Consider again Eq. (3.205):

$$\frac{dy}{dx} = f(x, y), \qquad y(x_0) = y_0. \tag{3.223}$$

We can take as many derivatives as we need, using the chain rule to assist. For example, the second derivative is

$$\frac{d^2 y}{dx^2} = \frac{\partial f}{\partial x} + \frac{\partial f}{\partial y}\frac{dy}{dx}, \tag{3.224}$$

$$= \frac{\partial f}{\partial x} + f\frac{\partial f}{\partial y}. \tag{3.225}$$

This can be evaluated at (x_0, y_0). We can continue this process to obtain expressions for $d^3 y/dx^3$, $d^4 y/dx^4, \ldots$, with the application of the chain rule becoming more complicated at higher orders. For example, one can verify that

$$\frac{d^3 y}{dx^3} = \frac{\partial^2 f}{\partial x^2} + \frac{\partial f}{\partial y}\frac{\partial f}{\partial x} + 2f\frac{\partial^2 f}{\partial x \partial y} + f\left(\frac{\partial f}{\partial y}\right)^2 + f^2\frac{\partial^2 f}{\partial y^2}. \tag{3.226}$$

Each derivative will be able to be evaluated at (x_0, y_0). Then we can use these values in the Taylor series for y about $x = x_0$:

$$y(x) = y(x_0) + \left.\frac{dy}{dx}\right|_{x_0, y_0}(x - x_0) + \frac{1}{2}\left.\frac{d^2 y}{dx^2}\right|_{x_0, y_0}(x - x_0)^2 + \cdots \tag{3.227}$$

The method can be extended to higher order differential equations.

EXAMPLE 3.18

Use the Taylor series method to develop an estimate for the same problem addressed in an earlier example by Picard iteration:

$$\frac{dy}{dx} = x^2 - y^3, \qquad y(0) = 1. \tag{3.228}$$

We have $y(x_0) = 1$. By inspection, we also have

$$\left.\frac{dy}{dx}\right|_{x_0, y_0} = 0^2 - 1^3 = -1. \tag{3.229}$$

Now we can differentiate the original equation to obtain

$$\frac{d^2 y}{dx^2} = 2x - 3y^2\frac{dy}{dx}, \tag{3.230}$$

$$= 2x - 3y^2(x^2 - y^3), \tag{3.231}$$

$$= 2x - 3x^2 y^2 + 3y^5. \tag{3.232}$$

We can differentiate again to get

$$\frac{d^3 y}{dx^3} = 2 - 6x^2 y\frac{dy}{dx} - 6xy^2 + 15y^4\frac{dy}{dx}, \tag{3.233}$$

$$= 2 - 6x^2 y(x^2 - y^3) - 6xy^2 + 15y^4(x^2 - y^3). \tag{3.234}$$

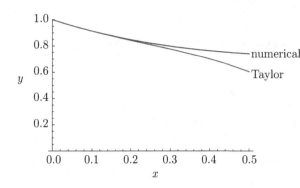

Figure 3.15. High-precision numerical estimate along with Taylor series estimate for $y(x)$ that satisfies $y' = x^2 - y^3$, $y(0) = 1$.

At the boundary, we thus have

$$\left.\frac{d^2y}{dx^2}\right|_{x_0,y_0} = 2(0) - 3(0)^2(1)^2 + 3(1)^5 = 3, \tag{3.235}$$

$$\left.\frac{d^3y}{dx^3}\right|_{x_0,y_0} = 2 - 6(0)^2(1)((0)^2 - (1)^3) - 6(0)(1)^2 + 15(1)^4((0)^2 - (1)^3) = -13. \tag{3.236}$$

Thus, our Taylor series about $x = 0$ is

$$y(x) = 1 - x + \frac{3}{2}x^2 - \frac{13}{6}x^3 + \cdots \tag{3.237}$$

The first three terms are identical to those obtained by Picard iteration. If we had performed another Picard iteration, all four terms would have been identical. We compare our four-term Taylor series estimate of $y(x)$ with a high-precision numerical solution in Figure 3.15.

3.15 Delay Differential Equations

Occasionally a physical system's evolution may not depend on its local value but on previous values. Such systems can often be modeled by so-called *delay differential equations*. The theory of delay differential equations is rich, and we shall only give a brief introduction here. Consider a first-order delay differential equation of the form

$$\frac{d}{dx}(y(x)) = f(x, y(x), y(x - x_1)), \qquad x \geq x_0, \tag{3.238}$$

$$y(x) = Y(x), \qquad (x_0 - x_1) \leq x \leq x_0. \tag{3.239}$$

Here f, Y, x_1, and x_0 are considered known, and we seek $y(x)$ for $x \geq x_0$. We note a significant deviation from ordinary first-order differential equations in that instead of a boundary condition at $x = x_0$, we have specified a function for $(x_0 - x_1) \leq x \leq x_0$. This is necessary because as x advances, the local evolution of y will depend on a continuum of states of y for $x < x_0$.

EXAMPLE 3.19

Determine $y(x)$ for $x_1 = 1/4, 1/2, 1, 2$ if

$$\frac{d}{dx}(y(x)) = -y(x - x_1), \qquad x \geq 0, \tag{3.240}$$

$$y(x) = 1, \qquad -x_1 \leq x \leq 0. \tag{3.241}$$

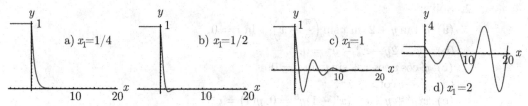

Figure 3.16. Solution to delay differential equation $dy(x)/dx = -y(x - x_1), x \geq 0; y(x) = 1, x \leq 0$: (a) $x_1 = 1/4$, (b) $x_1 = 1/2$, (c) $x_1 = 1$, (d) $x_1 = 2$.

Here we have taken $x_0 = 0$, $f = -y(x - x_1)$, and $Y(x) = 1$. Also, recognize that $y(x - x_1)$ means that the function y is evaluated at $(x - x_1)$. We first present the solutions for the various x_1 values for $y(x)$ for $x \in [-5, 20]$ in Figure 3.16. For $x_1 = 1/4$, the solution closely resembles the solution of its nondelay analog, $x_1 = 0$, which yields $y(x) = e^{-x}$. It also shares identical behavior for small positive x. The nondelay analog has a relaxation scale of $x_r \approx 1$. For $x_1 = 1/4$, this relaxation scale is greater than the delay scale, which induces both solutions to be similar for $x > 0$. For $x_1 = 1/2$, the delay is approaching x_r, and we notice a phenomenon similar to that seen in critically damped second-order differential systems (see Section 4.8); there is a small overshoot where $y < 0$, followed by relaxation to $y = 0$. For $x_1 = 1 = x_r$, we see stable oscillatory behavior. When $x_1 = 2 > x_r$, we see unstable oscillatory behavior.

The actual solution algorithm is known as the *method of steps*. In this method, we break the domain into appropriate intervals and solve for y on as many intervals as we need. We present results for $x_1 = 1$. For this problem, we know for $x < 1$ that $y(x - 1) = 1$. So for $x \in [0, 1]$, we can solve

$$\frac{dy}{dx} = -1, \qquad y(0) = 1. \tag{3.242}$$

Solution yields

$$y(x) = 1 - x, \qquad x \in [0, 1]. \tag{3.243}$$

For $x \in [1, 2]$, we have $y(x - 1) = 1 - (x - 1) = 2 - x$. So in this domain, our differential equation is

$$\frac{dy}{dx} = -(2 - x), \qquad y(1) = 0, \tag{3.244}$$

which has solution

$$y(x) = \frac{x^2}{2} - 2x + \frac{3}{2}, \qquad x \in [1, 2]. \tag{3.245}$$

This can be continued as far as desired. We see there is a lack of continuity of many of the higher order derivatives of the solutions of delay equations.

EXERCISES

1. Show that

$$y = \int_0^\infty e^{-t^2} \cos 2tx \, dt$$

satisfies the differential equation

$$\frac{dy}{dx} + 2xy = 0.$$

2. Solve:

(a) $y' \tan y + 2 \sin x \sin \left(\dfrac{\pi}{2} + x \right) + \ln x = 0.$

(b) $xy' - 2y - x^4 - y^2 = 0.$

(c) $y' \cos y \cos x + \sin y \sin x = 0.$

(d) $y' + y \cot x = e^x.$

(e) $x^5 y' + y + e^{x^2}(x^6 - 1)y^3 = 0, y(1) = e^{-1/2}.$

(f) $y' + y^2 - xy - 1 = 0.$

(g) $y'(x + y^2) - y = 0.$

(h) $y' = \dfrac{x + 2y - 5}{-2x - y + 4}.$

(i) $y' + xy = y.$

Plot solutions, when possible, for $y(0) = -1, 0, 1.$

3. Solve

$$\frac{dy}{dx} = \frac{x - y}{x + y}.$$

4. Show that $y'' - 2yy' = 0$ can be integrated to give a first-order equation, which can then be solved. Plot a solution for $y(0) = 0, y'(0) = 3.$

5. Solve

$$(y')^2 + 2yy' + y^2 = 1.$$

6. Determine $y(x)$ if

$$(y - x)(e^y y' + x) = 0$$

and $y(1) = 1.$

7. Solve:

(a) $y' = \dfrac{y}{x} + \left(\dfrac{y}{x} \right)^2.$

(b) $y' = \dfrac{x + y}{x + y + 1}.$

(c) $y' - \dfrac{1}{x^2}y^2 + \dfrac{1}{x}y = 1.$

8. Show that

(a) $2xy\,dx + (1 + x^2)\,dy = 0,$

(b) $3x^2 y^2\,dx + 2x^3 y\,dy = 0,$

(c) $2y' \left(\dfrac{1}{y} + xye^x \right) + y^2 e^x (x + 1) = 0,$

are exact, and find their general solutions.

9. Find the integrating factor and solve:

(a) $2\dfrac{y}{x}\,dx + dy = 0.$

(b) $x\,dy + y\,dx = x^2 y^2\,dx.$

(c) $y \ln y\,dx + x\,dy = 0.$

10. Find the general solution of

$$\frac{dy}{dt} + t^3 y = f(t)$$

for $t \geq 0$ with $y(0) = 0.$

11. Solve

$$y' + x^2 y = x,$$

with $y(0) = 1$.

12. Solve

$$\dot{x} = 2tx + te^{-t^2} x^2.$$

Plot a solution for $x(0) = 1$.

13. Find the Bernoulli equation corresponding to the solution

$$y(x) = \frac{1}{x^4(c - \ln x)},$$

for $x > 0$, where c is a constant.

14. Show that $y = 0$ is one solution of

$$xy' - 2y = -x^3 y^2.$$

Find the general solution.

15. Think of one simple solution of the differential equation

$$y' + x^2 y(1 + y) = 1 + x^3(1 + x),$$

and from this find the general solution. Plot solutions for $y(0) = -2, 0$, and 2.

16. The Abel[7] equation of the first kind is

$$\frac{dy}{dx} = f_0(x) + f_1(x)y + f_2(x)y^2 + f_3(x)y^3.$$

(a) Find four different choices of the fs for which the equation becomes separable, linear, Riccati, or Bernoulli. (b) Show that by assuming

$$y = u(z)v(x) + F(x), \quad v(x) = \exp\left(\int (f_1 + f_2 F)\, dx \right),$$

$$F(x) = -\frac{f_2}{3f_3},$$

and

$$z = \int f_3 v^2(x)\, dx,$$

the equation reduces to

$$u'(z) = U(x) + u^3,$$

where

$$f_3 v^3 U = f_0 - F' + f_1 F + \frac{2}{3} f_2 F^2.$$

17. Solve

$$y'' + y' = 1.$$

---

[7] Niels Henrick Abel, 1802–1829, Norwegian mathematician.

18. Solve

$$xy'' + 2y' = x.$$

Plot a solution for $y(1) = 1, y'(1) = 1$.

19. Solve

$$y'' + (y')^2 = 0.$$

20. Solve the boundary value problem

$$\frac{d^2y}{dx^2} + y\frac{dy}{dx} = 0,$$

with boundary conditions $y(0) = 0$ and $y(\pi/2) = -1$. Plot your result.

21. Solve:
  (a) $(y' - 1)(y' + 1) = 0$.
  (b) $(y' + a)\,(y' + x) = 0$.
  (c) $(y' + xy)(y' + y) = 0$.
  (d) $(y')^2 + 2yy' + y^2 = 1$.

22. Find an a for which a unique real solution of

$$(y')^4 + 8(y')^3 + (3a + 16)(y')^2 + 12ay' + 2a^2 = 0, \qquad y(1) = -2,$$

exists. Find the solution.

23. Solve

$$y = xy' + 2\sqrt{1 + (y')^2}.$$

24. Solve

$$\frac{y}{(y')^2} - \frac{x}{y'} - (y')^2 = 0,$$

with $y(0) = y_0$.

25. Using Picard iteration, obtain a series solution of

$$y' = xy,$$

with $y(0) = 1$. Compare with the series expansion of the exact solution.

26. Approximately solve:
  (a) $y' = -2xy^2, y(0) = 1$,
  (b) $y' = x - y^2, y(0) = 1$,
  (c) $y' = x + y + 1, y(0) = 0$,
  with the help of Taylor series.

27. Find $x(t)$ in the interval $\tau \le t \le 2\tau$ for

$$\dot{x} = ax(t - \tau),$$

with $x(t) = 1$ for $-\tau \le t \le 0$.

28. Show that solutions of the type $x = e^{\lambda t}$ for the delay equation

$$\dot{x} = a_0 x + a_1 x(t - \tau_1) + \cdots + a_n x(t - \tau_n)$$

require that λ satisfy the characteristic equation

$$a_0 + a_1 e^{-\lambda \tau_1} + \cdots + a_n e^{-\lambda \tau_n} - \lambda = 0.$$

29. Show that oscillatory solutions of the delay equation

$$\frac{dx}{dt}(t) + x(t) + bx(t-1) = 0$$

are possible only when $b = 2.2617$, and find the corresponding frequency.

30. Show that the equation

$$\dot{x} = x + \int_{-\infty}^{0} x(t+\tau)\, e^{\tau}\, d\tau$$

can be reduced to the form $\dot{\mathbf{x}} = \mathbf{A} \cdot \mathbf{x}$. Hint: Take $y = \int_{-\infty}^{0} x(t+\tau)\, e^{\tau}\, d\tau$, and show that $\dot{y} = x - y$.

31. The water level in a bucket with a hole in the bottom is governed by $\dot{y} = a\sqrt{y}$. Show that it is possible to find the time it will take for a full bucket to become empty, but given an empty bucket, that it is not possible to uniquely say when it was full.

32. A surge of wind with velocity $u(t)$, where

$$u(t) = \begin{cases} 0, & t \le 0 \\ V, & t > 0 \end{cases}$$

hits a stationary sphere of large mass. Assuming that the drag force on the sphere is proportional to the square of the velocity difference between the wind and the sphere, show that the differential equation governing the displacement $x(t)$ of the sphere in the direction of the wind is of the form

$$\ddot{x} = \epsilon \left(\dot{x} - u \right)^2.$$

What is the starting acceleration of the sphere if ϵ is small?

33. The heat loss by radiation from a mass m of specific heat c and surface area A is governed by

$$mc\frac{dT}{dt} + \sigma A(T^4 - T_\infty^4) = 0,$$

where t is time, σ is the Stefan[8]-Boltzmann[9] constant, $T(t)$ is the temperature of the mass, and T_∞ is that of the ambient. Find an implicit solution for $T(t)$ for $T(0) = T_0$.

34. For a step change in valve position, the flow rate $Q(t)$ in a pipe is given by

$$\frac{dQ}{dt} + \alpha(t)Q^n = 1,$$

where $\alpha(t)$ is a step function at $t = 0$. Solve numerically, and plot $Q(t)$ going from one steady state to another. Take $n = 1.75$ and α going from 1 to 2.

---

[8] Jožef Stefan, 1835–1893, Carinthian Slovene Austrian physicist.
[9] Ludwig Eduard Boltzmann, 1844–1906, Austrian physicist and philosopher.

4   Linear Ordinary Differential Equations

We consider in this chapter *linear ordinary differential equations*. We have already introduced first-order linear differential equations in Chapter 3. Here we are mainly concerned with equations that are second order or higher in a single dependent variable. We review several topics that are commonly covered in undergraduate mathematics, including complementary functions, particular solutions, the superposition principle, Sturm-Liouville equations, and resonance of a sinusoidally forced linear oscillator. We close with a discussion of linear difference equations. Strictly speaking, these are not differential equations, but they certainly arise in many discretized forms of linear differential equations, and their solution has analog to the solution of differential equations. Intrinsic in much of our discussion will be the notion of oscillation at a variety of frequencies. This lays the foundation of the exercise of seeking repetitive patterns, a topic of relevance in engineering. The chapter also introduces the important concept of representation of a function by infinite trigonometric and nontrigonometric Fourier series, as well as projection of a function onto a basis composed of a finite Fourier series. This motivates important abstractions that will be considered in detail in the later Chapter 6. Advanced topics such as Green's functions for particular solutions and discrete/continuous spectra are included as well. All of these topics have relevance in the wide assortment of engineering systems that are well modeled by linear systems. We will provide some focus on linear oscillators, such as found in mass-spring systems. Analogs abound and are too numerous to be delineated.

4.1 Linearity and Linear Independence

An ordinary differential equation can be written in the form

$$\mathbf{L}y = f(x), \tag{4.1}$$

where $\mathbf{L}$ is a known operator, y is an unknown function, and $f(x)$ is a known function of the independent variable x. The equation is said to be *homogeneous* if $f(x) = 0$, giving, then,

$$\mathbf{L}y = 0. \tag{4.2}$$

This is the most common usage for the term *homogeneous*. The operator $\mathbf{L}$ can involve a combination of derivatives $d/dx, d^2/dx^2$, and so on. The operator $\mathbf{L}$ is linear if

$$\mathbf{L}(y_1 + y_2) = \mathbf{L}y_1 + \mathbf{L}y_2 \tag{4.3}$$

and

$$\mathbf{L}(\alpha y) = \alpha \mathbf{L}y, \tag{4.4}$$

where α is a scalar. The first condition is known as the *superposition principle*. The second condition is known as a type of *homogeneity* condition, more specifically, it describes homogeneity of degree one.[1] We can contrast this definition of linearity with the definition of the more general term *affine* given by Eq. (1.274), which, while similar, admits a constant inhomogeneity.

For the remainder of this chapter, we will take $\mathbf{L}$ to be a linear differential operator. The general form of $\mathbf{L}$ is

$$\mathbf{L} = P_N(x)\frac{d^N}{dx^N} + P_{N-1}(x)\frac{d^{N-1}}{dx^{N-1}} + \cdots + P_1(x)\frac{d}{dx} + P_0(x). \tag{4.5}$$

The ordinary differential equation, Eq. (4.1), is then linear when $\mathbf{L}$ has the form of Eq. (4.5).

Definition: The functions $y_1(x), y_2(x), \ldots, y_N(x)$ are said to be *linearly independent* when $C_1 y_1(x) + C_2 y_2(x) + \cdots + C_N y_N(x) = 0$ is true only when $C_1 = C_2 = \cdots = C_N = 0$ for all x in some interval.

A homogeneous equation of order N can be shown to have N linearly independent solutions. These are called *complementary functions*. If y_n $(n = 1, \ldots, N)$ are the complementary functions of Eq. (4.2), then

$$y(x) = \sum_{n=1}^{N} C_n y_n(x) \tag{4.6}$$

is the general solution of the homogeneous Eq. (4.2). In language to be defined (see Section 6.3), we can say the complementary functions are linearly independent and span the space of solutions of the homogeneous Eq. (4.2); they are the bases of the *null space* of the differential operator $\mathbf{L}$. The null space is the set of y for which $\mathbf{L}y = 0$; that is, it is the space in which $\mathbf{L}$ maps y to zero. If $y_p(x)$ is any *particular solution* of Eq. (4.1), the general solution to Eq. (4.2) is then

$$y(x) = y_p(x) + \sum_{n=1}^{N} C_n y_n(x). \tag{4.7}$$

Now we would like to show that any solution $\phi(x)$ to the homogeneous equation $\mathbf{L}y = 0$ can be written as a linear combination of the N complementary functions $y_n(x)$:

$$C_1 y_1(x) + C_2 y_2(x) + \cdots + C_N y_N(x) = \phi(x). \tag{4.8}$$

---

[1] In this sense, it is of the same family of the homogeneity of degree zero described in Section 3.3. Generalizing, we can say a function f is homogeneous of degree k if $f(\alpha y) = \alpha^k f(y)$. Recognize again the unfortunate use of the same term for entirely different concepts. This homogeneity is not that of the more common Eq. (4.2)!

We can form additional equations by taking a series of derivatives up to $N - 1$:

$$C_1 y_1'(x) + C_2 y_2'(x) + \cdots + C_N y_N'(x) = \phi'(x), \qquad (4.9)$$

$$\vdots$$

$$C_1 y_1^{(N-1)}(x) + C_2 y_2^{(N-1)}(x) + \cdots + C_N y_N^{(N-1)}(x) = \phi^{(N-1)}(x). \qquad (4.10)$$

This is a linear system of algebraic equations:

$$\begin{pmatrix} y_1 & y_2 & \cdots & y_N \\ y_1' & y_2' & \cdots & y_N' \\ \vdots & \vdots & \cdots & \vdots \\ y_1^{(N-1)} & y_2^{(N-1)} & \cdots & y_N^{(N-1)} \end{pmatrix} \begin{pmatrix} C_1 \\ C_2 \\ \vdots \\ C_N \end{pmatrix} = \begin{pmatrix} \phi(x) \\ \phi'(x) \\ \vdots \\ \phi^{(N-1)}(x) \end{pmatrix}. \qquad (4.11)$$

We could solve Eq. (4.11) by Cramer's rule, which requires the use of determinants. For a unique solution, we need the determinant of the coefficient matrix of Eq. (4.11) to be nonzero. This particular determinant is known as the *Wronskian*[2] W of $y_1(x), y_2(x), \ldots, y_N(x)$ and is defined as

$$W = \begin{vmatrix} y_1 & y_2 & \cdots & y_N \\ y_1' & y_2' & \cdots & y_N' \\ \vdots & \vdots & \cdots & \vdots \\ y_1^{(N-1)} & y_2^{(N-1)} & \cdots & y_N^{(N-1)} \end{vmatrix}. \qquad (4.12)$$

The condition $W \neq 0$ indicates linear independence of the functions $y_1(x), y_2(x), \ldots, y_N(x)$, because if $\phi(x) \equiv 0$, the only solution is $C_n = 0, n = 1, \ldots, N$. Unfortunately, the converse is not always true; that is, if $W = 0$, the complementary functions may or may not be linearly dependent, though in most cases $W = 0$ indeed implies linear dependence.

EXAMPLE 4.1

Determine the linear independence of (a) $y_1 = x$ and $y_2 = 2x$, (b) $y_1 = x$ and $y_2 = x^2$, and (c) $y_1 = x^2$ and $y_2 = x|x|$ for $x \in (-1, 1)$:

(a) $W = \begin{vmatrix} x & 2x \\ 1 & 2 \end{vmatrix} = 0$, linearly dependent.

(b) $W = \begin{vmatrix} x & x^2 \\ 1 & 2x \end{vmatrix} = x^2 \neq 0$, linearly independent, except at $x = 0$.

(c) We can restate y_2 as

$$y_2(x) = -x^2, \quad x \in (-1, 0], \qquad (4.13)$$

$$y_2(x) = x^2, \quad x \in (0, 1), \qquad (4.14)$$

so that

$$W = \begin{vmatrix} x^2 & -x^2 \\ 2x & -2x \end{vmatrix} = -2x^3 + 2x^3 = 0, \quad x \in (-1, 0], \qquad (4.15)$$

$$W = \begin{vmatrix} x^2 & x^2 \\ 2x & 2x \end{vmatrix} = 2x^3 - 2x^3 = 0, \quad x \in (0, 1). \qquad (4.16)$$

Thus, $W = 0$ for $x \in (-1, 1)$, which suggests the functions may be linearly dependent. However, when we seek C_1 and C_2 such that $C_1 y_1 + C_2 y_2 = 0$, we find the only solution

---

[2] Józef Maria Hoene-Wroński , 1778–1853, Polish-born French mathematician.

is $C_1 = 0, C_2 = 0$; therefore, the functions are in fact linearly independent, despite the fact that $W = 0$! Let us check this. For $x \in (-1, 0]$,

$$C_1 x^2 + C_2(-x^2) = 0, \tag{4.17}$$

so we will at least need $C_1 = C_2$. For $x \in (0, 1)$,

$$C_1 x^2 + C_2 x^2 = 0, \tag{4.18}$$

which gives the requirement that $C_1 = -C_2$. Substituting the first condition into the second gives $C_2 = -C_2$, which is only satisfied if $C_2 = 0$, thus requiring that $C_1 = 0$; hence, the functions are indeed linearly independent.

EXAMPLE 4.2

Determine the linear independence of the set of polynomials

$$y_n(x) = \left\{ 1, x, \frac{x^2}{2}, \frac{x^3}{6}, \dots, \frac{x^{N-1}}{(N-1)!} \right\}. \tag{4.19}$$

The Wronskian is

$$W = \begin{vmatrix} 1 & x & \frac{1}{2}x^2 & \frac{1}{6}x^3 & \cdots & \frac{1}{(N-1)!}x^{N-1} \\ 0 & 1 & x & \frac{1}{2}x^2 & \cdots & \frac{1}{(N-2)!}x^{N-2} \\ 0 & 0 & 1 & x & \cdots & \frac{1}{(N-3)!}x^{N-3} \\ 0 & 0 & 0 & 1 & \cdots & \frac{1}{(N-4)!}x^{N-4} \\ \vdots & \vdots & \vdots & \vdots & \cdots & \vdots \\ 0 & 0 & 0 & 0 & \cdots & 1 \end{vmatrix} = 1. \tag{4.20}$$

The determinant is unity $\forall\ N$. As such, the polynomials are linearly independent.

4.2 Complementary Functions

This section will consider solutions to the homogeneous part of the differential equation, that is, Eq. (4.2).

4.2.1 Constant Coefficients

First consider differential equations with constant coefficients.

Arbitrary Order

Consider the homogeneous equation of order N with constant coefficients

$$a_N y^{(N)} + a_{N-1} y^{(N-1)} + \cdots + a_1 y' + a_0 y = 0, \tag{4.21}$$

where a_n, $n = 0, \dots, N$, are constants. To find the solution of Eq. (4.21), we make the assumption

$$y = e^{rx}, \tag{4.22}$$

where r is a constant. Indeed, solving differential equations often has many aspects of a trial and error process. Substituting, we get

$$a_N r^N e^{rx} + a_{N-1} r^{(N-1)} e^{rx} + \cdots + a_1 r^1 e^{rx} + a_0 e^{rx} = 0. \tag{4.23}$$

Eliminating the nonzero common factor e^{rx}, we get

$$a_N r^N + a_{N-1} r^{(N-1)} + \cdots + a_1 r^1 + a_0 r^0 = 0, \qquad (4.24)$$

$$\sum_{n=0}^{N} a_n r^n = 0. \qquad (4.25)$$

This is called the *characteristic* equation.[3] It is an Nth-order polynomial which has N roots (some of which could be repeated, some of which could be complex; see Section A.1), $r_n, n = 1, \ldots, N$, from which N linearly independent complementary functions $y_n(x), n = 1, \ldots, N$, have to be obtained. The general solution is then given by Eq. (4.6).

If all roots are real and distinct, the complementary functions are simply $e^{r_n x}$, $n = 1, \ldots, N$. If, however, k of these roots are repeated, that is, $r_1 = r_2 = \cdots = r_k = r$, the linearly independent complementary functions are obtained by multiplying e^{rx} by $1, x, x^2, \ldots, x^{k-1}$. For a pair of complex conjugate roots $p \pm qi$, one can use de Moivre's[4] formula (see Eq. (A.136)) to show that the complementary functions are $e^{px} \cos qx$ and $e^{px} \sin qx$.

EXAMPLE 4.3
Solve

$$\frac{d^4 y}{dx^4} - 2\frac{d^3 y}{dx^3} + \frac{d^2 y}{dx^2} + 2\frac{dy}{dx} - 2y = 0. \qquad (4.26)$$

Substituting $y = e^{rx}$, we get a characteristic equation

$$r^4 - 2r^3 + r^2 + 2r - 2 = 0, \qquad (4.27)$$

which can be factored as

$$(r + 1)(r - 1)(r^2 - 2r + 2) = 0, \qquad (4.28)$$

from which

$$r_1 = -1, \qquad r_2 = 1 \qquad r_3 = 1 + i \qquad r_4 = 1 - i. \qquad (4.29)$$

The general solution is

$$y(x) = C_1 e^{-x} + C_2 e^x + C_3' e^{(1+i)x} + C_4' e^{(1-i)x}, \qquad (4.30)$$

$$= C_1 e^{-x} + C_2 e^x + C_3' e^x e^{ix} + C_4' e^x e^{-ix}, \qquad (4.31)$$

$$= C_1 e^{-x} + C_2 e^x + e^x (C_3' e^{ix} + C_4' e^{-ix}), \qquad (4.32)$$

$$= C_1 e^{-x} + C_2 e^x + e^x \left(C_3' \left(\cos x + i \sin x \right) + C_4' \left(\cos(-x) + i \sin(-x) \right) \right), \qquad (4.33)$$

$$= C_1 e^{-x} + C_2 e^x + e^x \left((C_3' + C_4') \cos x + i(C_3' - C_4') \sin x \right), \qquad (4.34)$$

$$= C_1 e^{-x} + C_2 e^x + e^x (C_3 \cos x + C_4 \sin x), \qquad (4.35)$$

where $C_3 = C_3' + C_4'$ and $C_4 = i(C_3' - C_4')$.

---

[3] It is sometimes called the secular equation.
[4] Abraham de Moivre, 1667–1754, French mathematician.

First-Order

The characteristic equation of the first-order equation

$$ay' + by = 0 \tag{4.36}$$

is

$$ar + b = 0. \tag{4.37}$$

So

$$r = -\frac{b}{a}. \tag{4.38}$$

Thus, the complementary function for Eq. (4.36) is $e^{-bx/a}$, and the solution is simply

$$y = Ce^{-bx/a}. \tag{4.39}$$

Second-Order

The characteristic equation of the second-order equation

$$a\frac{d^2y}{dx^2} + b\frac{dy}{dx} + cy = 0 \tag{4.40}$$

is

$$ar^2 + br + c = 0. \tag{4.41}$$

Depending on the coefficients of this quadratic equation, there are three cases to be considered:

- $b^2 - 4ac > 0$: two distinct real roots r_1 and r_2; the complementary functions are $y_1 = e^{r_1 x}$ and $y_2 = e^{r_2 x}$,
- $b^2 - 4ac = 0$: one real root; the complementary functions are $y_1 = e^{rx}$ and $y_2 = xe^{rx}$,
- $b^2 - 4ac < 0$: two complex conjugate roots $p \pm qi$; the complementary functions are $y_1 = e^{px} \cos qx$ and $y_2 = e^{px} \sin qx$.

EXAMPLE 4.4

Solve

$$\frac{d^2y}{dx^2} - 3\frac{dy}{dx} + 2y = 0. \tag{4.42}$$

The characteristic equation is

$$r^2 - 3r + 2 = 0, \tag{4.43}$$

with solutions

$$r_1 = 1, \qquad r_2 = 2. \tag{4.44}$$

The general solution is then

$$y = C_1 e^x + C_2 e^{2x}. \tag{4.45}$$

EXAMPLE 4.5

Solve

$$\frac{d^2y}{dx^2} - 2\frac{dy}{dx} + y = 0. \tag{4.46}$$

The characteristic equation is

$$r^2 - 2r + 1 = 0, \tag{4.47}$$

with repeated roots

$$r_1 = 1, \qquad r_2 = 1. \tag{4.48}$$

The general solution is then

$$y = C_1 e^x + C_2 x e^x. \tag{4.49}$$

EXAMPLE 4.6

Solve

$$\frac{d^2y}{dx^2} - 2\frac{dy}{dx} + 10y = 0. \tag{4.50}$$

The characteristic equation is

$$r^2 - 2r + 10 = 0, \tag{4.51}$$

with solutions

$$r_1 = 1 + 3i, \qquad r_2 = 1 - 3i. \tag{4.52}$$

The general solution is then

$$y = e^x(C_1 \cos 3x + C_2 \sin 3x). \tag{4.53}$$

Systems of Equations

To this point, we have focused on Eq. (4.2), where y is a single function of a dependent variable, for example, $y(x)$. The method is easily extended to the case where y is a vector of functions, for example, $y(x) = (y_1(x), y_2(x), \ldots)^T$. We defer a full discussion of linear systems, both homogeneous and inhomogeneous, to Section 9.5. Here we give a brief example of how homogeneous linear systems may be addressed. Typically, methods of linear algebra are employed, and we shall do so here.

EXAMPLE 4.7

Find the general solution to

$$\frac{dy_1}{dx} = y_1 + 2y_2, \tag{4.54}$$

$$\frac{dy_2}{dx} = y_1 - y_2. \tag{4.55}$$

We can rewrite this in the form of Eq. (4.2) as follows:

$$\left(\begin{pmatrix} 1 & 0 \\ 0 & 1 \end{pmatrix} \frac{d}{dx} + \begin{pmatrix} -1 & -2 \\ -1 & 1 \end{pmatrix} \right) \begin{pmatrix} y_1 \\ y_2 \end{pmatrix} = \begin{pmatrix} 0 \\ 0 \end{pmatrix}, \tag{4.56}$$

$$\underbrace{\begin{pmatrix} -1 + \frac{d}{dx} & -2 \\ -1 & 1 + \frac{d}{dx} \end{pmatrix}}_{\text{L}} \underbrace{\begin{pmatrix} y_1 \\ y_2 \end{pmatrix}}_{y} = \begin{pmatrix} 0 \\ 0 \end{pmatrix}. \tag{4.57}$$

Although this formulation illustrates the applicability of our general theory, solution of the problem is better achieved by using the original form. To solve, we can make an assumption that is an extension of our earlier Eq. (4.22), namely,

$$y_1 = a_1 e^{rx}, \tag{4.58}$$

$$y_2 = a_2 e^{rx}. \tag{4.59}$$

Substituting into our original system, we find

$$a_1 r e^{rx} = a_1 e^{rx} + 2a_2 e^{rx} \tag{4.60}$$

$$a_2 r e^{rx} = a_1 e^{rx} - a_2 e^{rx}. \tag{4.61}$$

Significantly, and as a consequence of the system being linear with constant coefficients, one can scale by the never-zero e^{rx} to obtain the linear algebraic system

$$a_1 r = a_1 + 2a_2, \tag{4.62}$$

$$a_2 r = a_1 - a_2. \tag{4.63}$$

We cast this in matrix form as

$$r \begin{pmatrix} a_1 \\ a_2 \end{pmatrix} = \begin{pmatrix} 1 & 2 \\ 1 & -1 \end{pmatrix} \begin{pmatrix} a_1 \\ a_2 \end{pmatrix}, \tag{4.64}$$

$$\begin{pmatrix} 0 \\ 0 \end{pmatrix} = \begin{pmatrix} 1-r & 2 \\ 1 & -1-r \end{pmatrix} \begin{pmatrix} a_1 \\ a_2 \end{pmatrix}. \tag{4.65}$$

This is an eigenvalue problem, where the eigenvalues are r and the eigenvectors are $(a_1, a_2)^T$. For nontrivial solution, we need to select r so that the determinant of the coefficient matrix is zero:

$$(1-r)(-1-r) - 2 = 0, \tag{4.66}$$

$$r^2 - 3 = 0, \tag{4.67}$$

$$r = \pm\sqrt{3}. \tag{4.68}$$

It is easily shown that the corresponding eigenvectors are

$$\begin{pmatrix} a_1 \\ a_2 \end{pmatrix} = \begin{pmatrix} 1 \pm \sqrt{3} \\ 1 \end{pmatrix}. \tag{4.69}$$

Now multiplying either of the eigenvectors by any nonzero constant still yields a nontrivial solution, so we can take the general solution to be

$$\begin{pmatrix} y_1 \\ y_2 \end{pmatrix} = C_1 \begin{pmatrix} 1 + \sqrt{3} \\ 1 \end{pmatrix} e^{\sqrt{3}x} + C_2 \begin{pmatrix} 1 - \sqrt{3} \\ 1 \end{pmatrix} e^{-\sqrt{3}x}. \tag{4.70}$$

Or one might say that

$$y_1(x) = C_1(1 + \sqrt{3})e^{\sqrt{3}x} + C_2(1 - \sqrt{3})e^{-\sqrt{3}x}, \tag{4.71}$$

$$y_2(x) = C_1 e^{\sqrt{3}x} + C_2 e^{-\sqrt{3}x}. \tag{4.72}$$

The general approach can be extended to larger systems as well as to inhomogeneous systems. A variety of special cases must be addressed for systems such as those with repeated eigenvalues. Full consideration is given in Section 9.5.

4.2.2 Variable Coefficients

Next we consider methods to solve differential equations whose nonconstant coefficients are functions of the independent variable. We consider a few common equations, but one must realize there is no absolutely general technique.

One Solution to Find Another

If $y_1(x)$ is a known solution of

$$y'' + P(x)y' + Q(x)y = 0, \tag{4.73}$$

let the other solution be $y_2(x) = u(x)y_1(x)$, where $u(x)$ is as of yet unknown. We then form derivatives of y_2 and substitute into the original differential equation. First compute the derivatives:

$$y_2' = uy_1' + u'y_1, \tag{4.74}$$

$$y_2'' = uy_1'' + u'y_1' + u'y_1' + u''y_1, \tag{4.75}$$

$$= uy_1'' + 2u'y_1' + u''y_1. \tag{4.76}$$

Substituting into Eq. (4.73), we get

$$\underbrace{(uy_1'' + 2u'y_1' + u''y_1)}_{y_2''} + P(x)\underbrace{(uy_1' + u'y_1)}_{y_2'} + Q(x)\underbrace{uy_1}_{y_2} = 0, \tag{4.77}$$

$$u''y_1 + u'(2y_1' + P(x)y_1) + u\underbrace{(y_1'' + P(x)y_1' + Q(x)y_1)}_{=0} = 0. \tag{4.78}$$

We can cancel the term multiplying u because y_1 is a known solution, yielding

$$u''y_1 + u'(2y_1' + P(x)y_1) = 0. \tag{4.79}$$

This can be written as a first-order equation in v, where $v = u'$:

$$v'y_1 + v(2y_1' + P(x)y_1) = 0, \tag{4.80}$$

which is solved for $v(x)$ using known methods for first-order equations. Knowing v, u can be found, and knowing u, y_2 can be found.

Euler-Cauchy Equation

An equation of the type

$$x^2 \frac{d^2y}{dx^2} + Ax\frac{dy}{dx} + By = 0, \tag{4.81}$$

where A and B are constants, is known a *Euler-Cauchy*[5] *equation* and can be solved by a change of independent variables. The Euler-Cauchy equation is singular at $x = 0$, as at that point the term multiplying the highest order derivative vanishes. Let

$$z = \ln x, \tag{4.82}$$

so that

$$x = e^z. \tag{4.83}$$

---

[5] Augustin-Louis Cauchy, 1789–1857, French mathematician and physicist.

Then

$$\frac{dz}{dx} = \frac{1}{x} = e^{-z}, \tag{4.84}$$

$$\frac{dy}{dx} = \frac{dy}{dz}\frac{dz}{dx} = e^{-z}\frac{dy}{dz}. \tag{4.85}$$

Thus, we can write the differential operator in the transformed space as $d/dx = e^{-z}d/dz$. We find

$$\frac{d^2y}{dx^2} = \frac{d}{dx}\left(\frac{dy}{dx}\right), \tag{4.86}$$

$$= e^{-z}\frac{d}{dz}\left(e^{-z}\frac{dy}{dz}\right), \tag{4.87}$$

$$= e^{-2z}\left(\frac{d^2y}{dz^2} - \frac{dy}{dz}\right). \tag{4.88}$$

Substituting into Eq. (4.81), we get

$$\frac{d^2y}{dz^2} + (A-1)\frac{dy}{dz} + By = 0, \tag{4.89}$$

which is an equation with constant coefficients.

In what amounts to the same approach, one can alternatively assume a solution of the form $y = Cx^r$. This leads to a characteristic equation for r:

$$r(r-1) + Ar + B = 0. \tag{4.90}$$

The two roots for r induce two linearly independent complementary functions.

EXAMPLE 4.8
Solve

$$x^2y'' - 2xy' + 2y = 0, \qquad x > 0. \tag{4.91}$$

Here $A = -2$ and $B = 2$ in Eq. (4.81). Using this, along with $x = e^z$, we get Eq. (4.89) to reduce to

$$\frac{d^2y}{dz^2} - 3\frac{dy}{dz} + 2y = 0. \tag{4.92}$$

The solution is

$$y = C_1e^z + C_2e^{2z} = C_1x + C_2x^2. \tag{4.93}$$

This equation can also be solved by letting $y = Cx^r$. Substituting, we get $r^2 - 3r + 2 = 0$, so that $r_1 = 1$ and $r_2 = 2$. The solution is then obtained as a linear combination of x^{r_1} and x^{r_2}.

EXAMPLE 4.9
Solve

$$x^2\frac{d^2y}{dx^2} + 3x\frac{dy}{dx} + 15y = 0. \tag{4.94}$$

Let us assume here that $y = Cx^r$. Substituting this assumption into Eq. (4.94) yields

$$x^2Cr(r-1)x^{r-2} + 3xCrx^{r-1} + 15Cx^r = 0. \tag{4.95}$$

For $x \neq 0$, $C \neq 0$, we divide by Cx^r to get

$$r(r-1) + 3r + 15 = 0, \tag{4.96}$$

$$r^2 + 2r + 15 = 0. \tag{4.97}$$

Solving gives

$$r = -1 \pm i\sqrt{14}. \tag{4.98}$$

Thus, there are two linearly independent complementary functions, and

$$y(x) = C_1 x^{-1+i\sqrt{14}} + C_2 x^{-1-i\sqrt{14}}. \tag{4.99}$$

Factoring gives

$$y(x) = \frac{1}{x}\left(C_1 x^{i\sqrt{14}} + C_2 x^{-i\sqrt{14}}\right). \tag{4.100}$$

Expanding in terms of exponentials and logarithms gives

$$y(x) = \frac{1}{x}\left(C_1(\exp(\ln x))^{i\sqrt{14}} + C_2(\exp(\ln x))^{-i\sqrt{14}}\right), \tag{4.101}$$

$$= \frac{1}{x}\left(C_1 \exp(i\sqrt{14}\ln x) + C_2 \exp(i\sqrt{14}\ln x)\right), \tag{4.102}$$

$$= \frac{1}{x}\left(\hat{C}_1 \cos(\sqrt{14}\ln x) + \hat{C}_2 \sin(\sqrt{14}\ln x)\right). \tag{4.103}$$

4.3 Particular Solutions

We now consider particular solutions of the inhomogeneous Eq. (4.1) when **L** is linear.

4.3.1 Undetermined Coefficients

To implement the *method of undetermined coefficients*, we guess a solution with unknown coefficients and then substitute in the equation to determine these coefficients. Our guess is often motivated by the form of the inhomogeneity. The number of undetermined coefficients has no relation to the order of the differential equation. The method has the advantage of being straightforward but the disadvantage of not being guaranteed to work.

EXAMPLE 4.10

Find the most general solution to

$$y'' + 4y' + 4y = 169\sin 3x. \tag{4.104}$$

First find the complementary functions via

$$r^2 + 4r + 4 = 0, \tag{4.105}$$

$$(r+2)(r+2) = 0, \tag{4.106}$$

$$r_1 = -2, \qquad r_2 = -2. \tag{4.107}$$

Because the roots are repeated, the complementary functions are

$$y_1 = e^{-2x}, \qquad y_2 = xe^{-2x}. \tag{4.108}$$

For the particular function, guess a linear combination of trigonometric functions with the same frequency as the inhomogeneous term:

$$y_p = a \sin 3x + b \cos 3x, \tag{4.109}$$

so

$$y_p' = 3a \cos 3x - 3b \sin 3x, \tag{4.110}$$

$$y_p'' = -9a \sin 3x - 9b \cos 3x. \tag{4.111}$$

Substituting into Eq. (4.104), we get

$$\underbrace{(-9a \sin 3x - 9b \cos 3x)}_{y_p''} + 4 \underbrace{(3a \cos 3x - 3b \sin 3x)}_{y_p'}$$

$$+ 4 \underbrace{(a \sin 3x + b \cos 3x)}_{y_p} = 169 \sin 3x, \tag{4.112}$$

$$(-5a - 12b) \sin 3x + (12a - 5b) \cos 3x = 169 \sin 3x, \tag{4.113}$$

$$\underbrace{(-5a - 12b - 169)}_{=0} \sin 3x + \underbrace{(12a - 5b)}_{=0} \cos 3x = 0. \tag{4.114}$$

Now sine and cosine can be shown to be linearly independent. Because of this, as the right-hand side of Eq. (4.114) is zero, the constants on the sine and cosine functions must also be zero. This yields the simple system of linear algebraic equations

$$\begin{pmatrix} -5 & -12 \\ 12 & -5 \end{pmatrix} \begin{pmatrix} a \\ b \end{pmatrix} = \begin{pmatrix} 169 \\ 0 \end{pmatrix}, \tag{4.115}$$

from which we find $a = -5$ and $b = -12$. The solution is then

$$y(x) = (C_1 + C_2 x)e^{-2x} - 5 \sin 3x - 12 \cos 3x. \tag{4.116}$$

EXAMPLE 4.11
Solve

$$y'''' - 2y''' + y'' + 2y' - 2y = x^2 + x + 1. \tag{4.117}$$

Let the particular solution be of the same general form as the inhomogeneous forcing function: $y_p = ax^2 + bx + c$. Substituting and reducing, we get

$$\underbrace{-(2a + 1)}_{=0} x^2 + \underbrace{(4a - 2b - 1)}_{=0} x + \underbrace{(2a + 2b - 2c - 1)}_{=0} = 0. \tag{4.118}$$

Because x^2, x^1, and x^0 are linearly independent, their coefficients in Eq. (4.118) must be zero, from which $a = -1/2$, $b = -3/2$, and $c = -5/2$. Thus,

$$y_p = -\frac{1}{2}(x^2 + 3x + 5). \tag{4.119}$$

The solution of the homogeneous equation was found in a previous example (see Eq. (4.35)), so that the general solution is

$$y = C_1 e^{-x} + C_2 e^x + e^x (C_3 \cos x + C_4 \sin x) - \frac{1}{2}(x^2 + 3x + 5). \tag{4.120}$$

A variant must be attempted if any term of $f(x)$ is a complementary function.

EXAMPLE 4.12
Solve

$$y'' + 4y = 6\sin 2x. \tag{4.121}$$

Because $\sin 2x$ is a complementary function, we will try

$$y_p = x(a\sin 2x + b\cos 2x), \tag{4.122}$$

from which

$$y_p' = 2x(a\cos 2x - b\sin 2x) + (a\sin 2x + b\cos 2x), \tag{4.123}$$

$$y_p'' = -4x(a\sin 2x + b\cos 2x) + 4(a\cos 2x - b\sin 2x). \tag{4.124}$$

Substituting into Eq. (4.121), we compare coefficients and get $a = 0$, $b = -3/2$. The general solution is then

$$y = C_1\sin 2x + C_2\cos 2x - \frac{3}{2}x\cos 2x. \tag{4.125}$$

EXAMPLE 4.13
Solve

$$y'' + 2y' + y = xe^{-x}. \tag{4.126}$$

The complementary functions are e^{-x} and xe^{-x}. To get the particular solution, we have to choose a function of the kind $y_p = ax^3 e^{-x}$. On substitution, we find that $a = 1/6$. Thus, the general solution is

$$y = C_1 e^{-x} + C_2 xe^{-x} + \frac{1}{6}x^3 e^{-x}. \tag{4.127}$$

4.3.2 Variation of Parameters

The method of undetermined coefficients is straightforward but may not provide a solution. A more robust method to find particular solutions is given by the method of *variation of parameters*. For an equation of the class

$$P_N(x)y^{(N)} + P_{N-1}(x)y^{(N-1)} + \cdots + P_1(x)y' + P_0(x)y = f(x), \tag{4.128}$$

we propose

$$y_p = \sum_{n=1}^{N} u_n(x)y_n(x), \tag{4.129}$$

where $y_n(x)$, $n = 1, \ldots, N$, are complementary functions of the equation, and $u_n(x)$, $n = 1, \ldots, N$, are N unknown functions. Differentiating Eq. (4.129), we find

$$y_p' = \underbrace{\sum_{n=1}^{N} u_n' y_n}_{\text{choose to be 0}} + \sum_{n=1}^{N} u_n y_n'. \tag{4.130}$$

We set $\sum_{n=1}^{N} u'_n y_n$ to zero as a first condition. Differentiating the rest of Eq. (4.130), we obtain

$$y''_p = \underbrace{\sum_{n=1}^{N} u'_n y'_n}_{\text{choose to be } 0} + \sum_{n=1}^{N} u_n y''_n. \tag{4.131}$$

Again we set the first term on the right side of Eq. (4.131) to zero as a second condition. Following this procedure repeatedly, we arrive at

$$y_p^{(N-1)} = \underbrace{\sum_{n=1}^{N} u'_n y_n^{(N-2)}}_{\text{choose to be } 0} + \sum_{n=1}^{N} u_n y_n^{(N-1)}. \tag{4.132}$$

The vanishing of the first term on the right gives us the $(N-1)^{th}$ condition. Substituting these into Eq. (4.128), the last condition,

$$P_N(x) \sum_{n=1}^{N} u'_n y_n^{(N-1)} + \sum_{n=1}^{N} u_n \underbrace{\left(P_N y_n^{(N)} + P_{N-1} y_n^{(N-1)} + \cdots + P_1 y'_n + P_0 y_n \right)}_{=0} = f(x), \tag{4.133}$$

is obtained. Because each of the functions y_n is a complementary function, the term within parentheses is zero.

To summarize, we have the following N equations in the N unknowns u'_n, $n = 1, \ldots, N$:

$$\sum_{n=1}^{N} u'_n y_n = 0,$$

$$\sum_{n=1}^{N} u'_n y'_n = 0,$$

$$\vdots \tag{4.134}$$

$$\sum_{n=1}^{N} u'_n y_n^{(N-2)} = 0,$$

$$P_N(x) \sum_{n=1}^{N} u'_n y_n^{(N-1)} = f(x).$$

These can be solved for u'_n and then integrated to give the u_ns.

EXAMPLE 4.14

Solve

$$y'' + y = \tan x. \tag{4.135}$$

The complementary functions are

$$y_1 = \cos x, \qquad y_2 = \sin x. \tag{4.136}$$

The equations for $u_1(x)$ and $u_2(x)$ are

$$u_1' y_1 + u_2' y_2 = 0, \tag{4.137}$$

$$u_1' y_1' + u_2' y_2' = \tan x. \tag{4.138}$$

Solving this system, which is linear in u_1' and u_2', we get

$$u_1' = -\sin x \tan x, \tag{4.139}$$

$$u_2' = \cos x \tan x. \tag{4.140}$$

Integrating, we get

$$u_1 = \int -\sin x \tan x \, dx = \sin x - \ln|\sec x + \tan x|, \tag{4.141}$$

$$u_2 = \int \cos x \tan x \, dx = -\cos x. \tag{4.142}$$

The particular solution is

$$y_p = u_1 y_1 + u_2 y_2, \tag{4.143}$$

$$= (\sin x - \ln|\sec x + \tan x|)\cos x - \cos x \sin x, \tag{4.144}$$

$$= -\cos x \ln|\sec x + \tan x|. \tag{4.145}$$

The complete solution, obtained by adding the complementary and particular, is

$$y = C_1 \cos x + C_2 \sin x - \cos x \ln|\sec x + \tan x|. \tag{4.146}$$

It would be difficult to use the method of undetermined coefficients to guess the proper form of the particular solution. In contrast, variation of parameters provided the answer without difficulty.

4.3.3 Green's Functions

We take up the *Green's function* here for use in generating particular solutions for linear differential equations.

Boundary Value Problems

Determining a particular solution can be achieved for boundary value problems involving a more general linear operator $\mathbf{L}$, where $\mathbf{L}$ is given by Eq. (4.5). Let us say that on the closed interval $x \in [a, b]$, we have a *two-point boundary value problem* for a general linear differential equation of the form from Eq. (4.1):

$$\mathbf{L}y = f(x), \tag{4.147}$$

where the highest derivative in $\mathbf{L}$ is order N and with general homogeneous boundary conditions at $x = a$ and $x = b$ on linear combinations of y and $N - 1$ of its derivatives:

$$\mathbf{A} \cdot (y(a), y'(a), \ldots, y^{(N-1)}(a))^T + \mathbf{B} \cdot (y(b), y'(b), \ldots, y^{(N-1)}(b))^T = \mathbf{0}, \tag{4.148}$$

where $\mathbf{A}$ and $\mathbf{B}$ are $N \times N$ constant coefficient matrices. Then, knowing $\mathbf{L}$, $\mathbf{A}$, and $\mathbf{B}$, we can generate a solution of the form

$$y(x) = \int_a^b f(s) g(x, s) \, ds. \tag{4.149}$$

This is desirable as

- once $g(x, s)$ is known, the solution is defined for *all* f, including
  - forms of f for which no simple explicit integrals can be written,
  - piecewise continuous forms of f,
- numerical solution of the quadrature problem is more robust than direct numerical solution of the original differential equation,
- the solution will automatically satisfy all boundary conditions,
- the solution is useful in experiments in which the system dynamics are well characterized (e.g., mass-spring-damper) but the forcing may be erratic (perhaps digitally specified).

If the boundary conditions are inhomogeneous, a simple transformation of the dependent variables can be effected to render the boundary conditions to be homogeneous. Understanding of Eq. (4.149) will be enriched by study of the related topic of integral equations, which is considered in detail in Chapter 8.

We now define the Green's function, $g(x, s)$, and proceed to show that with our definition, we are guaranteed to achieve the solution to the differential equation in the desired form as shown at the beginning of the section. We take $g(x, s)$ to be the Green's function for the linear differential operator **L**, as defined by Eq. (4.5), if it satisfies the following conditions:

- $\mathbf{L}g(x, s) = \delta(x - s)$,
- $g(x, s)$ satisfies all boundary conditions given on x,
- $g(x, s)$ is a solution of $\mathbf{L}g = 0$ on $x \in [a, s)$ and on $x \in (s, b]$,
- $g(x, s), g'(x, s), \ldots, g^{(N-2)}(x, s)$ are continuous for $x \in [a, b]$,
- $g^{(N-1)}(x, s)$ is continuous for $x \in [a, b]$, except at $x = s$, where it has a jump of $1/P_N(s)$; the jump is defined from left to right.

Also, *for purposes of these conditions*, s is thought of as a constant parameter. In the actual Green's function representation of the solution, s is a dummy variable. The well-known *Dirac*[6] *delta function* $\delta(x - s)$ is discussed in Section A.7.6.

These conditions are not all independent, nor is the dependence obvious. Consider, for example,

$$\mathbf{L} = P_2(x)\frac{d^2}{dx^2} + P_1(x)\frac{d}{dx} + P_0(x). \qquad (4.150)$$

Then we have

$$P_2(x)\frac{d^2g}{dx^2} + P_1(x)\frac{dg}{dx} + P_0(x)g = \delta(x - s), \qquad (4.151)$$

$$\frac{d^2g}{dx^2} + \frac{P_1(x)}{P_2(x)}\frac{dg}{dx} + \frac{P_0(x)}{P_2(x)}g = \frac{\delta(x - s)}{P_2(x)}. \qquad (4.152)$$

Now integrate both sides with respect to x in a neighborhood enveloping $x = s$:

$$\int_{s-\epsilon}^{s+\epsilon} \frac{d^2g}{dx^2}\,dx + \int_{s-\epsilon}^{s+\epsilon} \frac{P_1(x)}{P_2(x)}\frac{dg}{dx}\,dx + \int_{s-\epsilon}^{s+\epsilon} \frac{P_0(x)}{P_2(x)}g\,dx = \int_{s-\epsilon}^{s+\epsilon} \frac{\delta(x - s)}{P_2(x)}\,dx. \qquad (4.153)$$

---

[6] Paul Adrien Maurice Dirac, 1902–1984, English physicist.

Because P's are continuous, as we let $\epsilon \to 0$, we get

$$\int_{s-\epsilon}^{s+\epsilon} \frac{d^2g}{dx^2}\, dx + \frac{P_1(s)}{P_2(s)} \int_{s-\epsilon}^{s+\epsilon} \frac{dg}{dx}\, dx + \frac{P_0(s)}{P_2(s)} \int_{s-\epsilon}^{s+\epsilon} g\, dx = \frac{1}{P_2(s)} \int_{s-\epsilon}^{s+\epsilon} \delta(x-s)\, dx.$$

(4.154)

Integrating, we find

$$\frac{dg}{dx}\bigg|_{s+\epsilon} - \frac{dg}{dx}\bigg|_{s-\epsilon} + \frac{P_1(s)}{P_2(s)} \underbrace{(g|_{s+\epsilon} - g|_{s-\epsilon})}_{\to 0} + \frac{P_0(s)}{P_2(s)} \underbrace{\int_{s-\epsilon}^{s+\epsilon} g\, dx}_{\to 0} = \frac{1}{P_2(s)} \underbrace{H(x-s)|_{s-\epsilon}^{s+\epsilon}}_{\to 1}.$$

(4.155)

Here $H(x-s)$ is the *Heaviside*[7] unit step function (see Section A.7.6 for details). Because g is continuous, Eq. (4.155) reduces to

$$\frac{dg}{dx}\bigg|_{s+\epsilon} - \frac{dg}{dx}\bigg|_{s-\epsilon} = \frac{1}{P_2(s)}.$$

(4.156)

This is consistent with the final point, that the second highest derivative of g suffers a jump at $x = s$.

Next, we show that applying this definition of $g(x, s)$ to our desired result lets us recover the original differential equation, rendering $g(x, s)$ to be appropriately defined. This can be easily shown by direct substitution:

$$y(x) = \int_a^b f(s)g(x, s)\, ds,$$

(4.157)

$$\mathbf{L}y = \mathbf{L} \int_a^b f(s)g(x, s)\, ds.$$

(4.158)

Now $\mathbf{L}$ behaves as linear combinations of d^n/dx^n, so via the Leibniz rule, Eq. (1.128), we get

$$\mathbf{L}y = \int_a^b f(s) \underbrace{\mathbf{L}g(x, s)}_{\delta(x-s)}\, ds,$$

(4.159)

$$= \int_a^b f(s)\delta(x-s)\, ds,$$

(4.160)

$$= f(x).$$

(4.161)

While Green's functions are most commonly employed in second-order systems, it is straightforward to use them on higher order systems. For example, consider the differential equation

$$P_4(x)y'''' + P_3(x)y''' + P_2(x)y'' + P_1(x)y' + P_0(x)y = f(x).$$

(4.162)

The Green's function $g(x, s)$ is defined by the solution of the equation

$$P_4(x)g'''' + P_3(x)g''' + P_2(x)g'' + P_1(x)g' + P_0(x)g = \delta(x-s).$$

(4.163)

---

[7] Oliver Heaviside, 1850–1925, English mathematician.

Assuming $P_4 \neq 0$, divide by it to get

$$g'''' + \frac{P_3(x)}{P_4(x)}g''' + \frac{P_2(x)}{P_4(x)}g'' + \frac{P_1(x)}{P_4(x)}g' + \frac{P_0(x)}{P_4(x)}g = \frac{\delta(x-s)}{P_4(x)}. \quad (4.164)$$

Integrating in the neighborhood of the discontinuity, that is, from $x = s - \epsilon$ to $x = s + \epsilon$, where $0 < \epsilon \ll 1$, gives

$$g'''\Big|_{s-\epsilon}^{s+\epsilon} + \int_{s-\epsilon}^{s+\epsilon}\left(\frac{P_3(x)}{P_4(x)}g''' + \frac{P_2(x)}{P_4(x)}g'' + \frac{P_1(x)}{P_4(x)}g' + \frac{P_0(x)}{P_4(x)}g\right)dx = \int_{s-\epsilon}^{s+\epsilon}\frac{\delta(x-s)}{P_4(x)}dx,$$

$$(4.165)$$

$$= \frac{1}{P_4(s)}. \quad (4.166)$$

Taking g, g', and g'' to be continuous with respect to x, but with g''' having a discontinuity at $x = s$, the second term from the left vanishes as $\epsilon \to 0$. Thus, the jump in g''' at $x = s$ is given by

$$g'''\Big|_{s-\epsilon}^{s+\epsilon} = \frac{1}{P_4(s)}. \quad (4.167)$$

EXAMPLE 4.15

Find the Green's function and the corresponding solution of the differential equation

$$\frac{d^2y}{dx^2} = f(x), \quad (4.168)$$

subject to boundary conditions

$$y(0) = 0, \quad y(1) = 0. \quad (4.169)$$

Verify the solution if $f(x) = 6x$.

Here

$$\mathbf{L} = \frac{d^2}{dx^2}. \quad (4.170)$$

Let us take an alternate approach to obtain the solution. Let us (1) break the problem into two domains: (a) $x < s$, (b) $x > s$; (2) solve $\mathbf{L}g = 0$ in both domains – four constants arise; (3) use boundary conditions to obtain two constants; (4) use conditions at $x = s$: continuity of g and a jump of dg/dx, to obtain the other two constants.

a). $x < s$

$$\frac{d^2g}{dx^2} = 0, \quad (4.171)$$

$$\frac{dg}{dx} = C_1, \quad (4.172)$$

$$g = C_1x + C_2, \quad (4.173)$$

$$g(0) = 0 = C_1(0) + C_2, \quad (4.174)$$

$$C_2 = 0, \quad (4.175)$$

$$g(x,s) = C_1x, \quad x < s \quad (4.176)$$

b). $x > s$

$$\frac{d^2 g}{dx^2} = 0,$$ (4.177)

$$\frac{dg}{dx} = C_3,$$ (4.178)

$$g = C_3 x + C_4,$$ (4.179)

$$g(1) = 0 = C_3(1) + C_4,$$ (4.180)

$$C_4 = -C_3,$$ (4.181)

$$g(x,s) = C_3(x - 1), \qquad x > s$$ (4.182)

Now invoke continuity of $g(x, s)$ when $x = s$:

$$C_1 s = C_3(s - 1),$$ (4.183)

$$C_1 = C_3 \frac{s - 1}{s},$$ (4.184)

$$g(x,s) = C_3 \frac{s - 1}{s} x, \qquad x \le s,$$ (4.185)

$$g(x,s) = C_3(x - 1), \qquad x \ge s.$$ (4.186)

Now enforce the jump in dg/dx at $x = s$ (note $P_2(x) = 1$):

$$\left. \frac{dg}{dx} \right|_{s+\epsilon} - \left. \frac{dg}{dx} \right|_{s-\epsilon} = 1,$$ (4.187)

$$C_3 - C_3 \frac{s - 1}{s} = 1,$$ (4.188)

$$C_3 = s.$$ (4.189)

Thus our Green's function is

$$g(x,s) = x(s - 1), \qquad x \le s,$$ (4.190)

$$g(x,s) = s(x - 1), \qquad x \ge s.$$ (4.191)

Note some properties of $g(x, s)$ that are common in such problems:

- it is broken into two domains,
- it is continuous in and through both domains,
- its $N - 1$ (here $N = 2$, so first) derivative is discontinuous at $x = s$,
- it is symmetric in s and x across the two domains (this will not be true for all systems),
- it is seen by inspection to satisfy both boundary conditions.

The general solution in integral form can be written by breaking the integral into two pieces as

$$y(x) = \int_0^x f(s)\, s(x - 1)\, ds + \int_x^1 f(s)\, x(s - 1)\, ds,$$ (4.192)

$$= (x - 1) \int_0^x f(s)\, s\, ds + x \int_x^1 f(s)\, (s - 1)\, ds.$$ (4.193)

Now evaluate the integral if $f(x) = 6x$; thus, $f(s) = 6s$:

$$y(x) = (x - 1) \int_0^x (6s)\, s\, ds + x \int_x^1 (6s)\, (s - 1)\, ds,$$ (4.194)

$$= (x - 1) \int_0^x 6s^2\, ds + x \int_x^1 \left(6s^2 - 6s\right)\, ds,$$ (4.195)

$$= (x - 1) \left(2s^3\right)\big|_0^x + x \left(2s^3 - 3s^2\right)\big|_x^1,$$ (4.196)

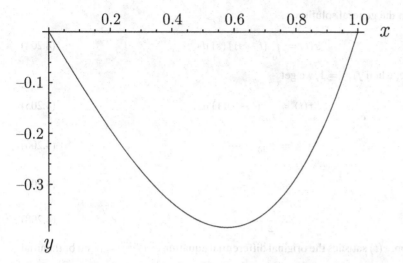

Figure 4.1. Plot of function $y(x) = x^3 - x$ that satisfies $d^2y/dx^2 = 6x, y(0) = y(1) = 0$.

$$= (x - 1)(2x^3 - 0) + x((2 - 3) - (2x^3 - 3x^2)), \qquad (4.197)$$

$$= 2x^4 - 2x^3 - x - 2x^4 + 3x^3, \qquad (4.198)$$

$$= x^3 - x. \qquad (4.199)$$

The original differential equation and both boundary conditions are automatically satisfied by the solution, plotted in Figure 4.1.

Initial Value Problems

Green's functions are not restricted to boundary value problems; they can also be developed for so-called *initial value problems*. For such problems, the function and a sufficient number of its derivatives are specified at a single point. If in the general boundary conditions of Eq. (4.148) we take either $\mathbf{A} = \mathbf{0}$ or $\mathbf{B} = \mathbf{0}$, our problem reduces to an initial value problem.

In engineering, initial value problems are usually associated with time-dependent calculations, and so we often label the independent variable as t. We take here the dependent variable to be x. Thus, we are considering Eq. (4.1) when it takes the form

$$\mathbf{L}x = f(t). \qquad (4.200)$$

EXAMPLE 4.16

Find the general solution in terms of the Green's function for the initial value problem

$$\frac{d^2x}{dt^2} = f(t), \qquad x(0) = 0, \qquad \left.\frac{dx}{dt}\right|_{t=0} = 0. \qquad (4.201)$$

Details of the analysis are similar to those given in previous examples. It can be easily verified that

$$g(t, s) = 0, \qquad t \le s, \qquad (4.202)$$

$$g(t, s) = t - s, \qquad t \ge s. \qquad (4.203)$$

This leads to the general solution

$$x(t) = \int_0^t (t - s)f(s)\, ds. \tag{4.204}$$

For example, when $f(s) = 1$, we get

$$x(t) = \int_0^t (t - s)(1)\, ds, \tag{4.205}$$

$$= ts\big|_0^t - \frac{s^2}{2}\bigg|_0^t, \tag{4.206}$$

$$= t^2 - \frac{t^2}{2}, \tag{4.207}$$

$$= \frac{t^2}{2}. \tag{4.208}$$

By inspection, $x(t)$ satisfies the original differential equation $d^2x/dt^2 = 1$ and both initial conditions.

4.3.4 Operator D

Here we consider another strategy of identifying particular solutions to linear differential equations of the form $\mathbf{L}y = f(x)$ using the more rudimentary linear differential operator $\mathbf{D}$. The linear operator $\mathbf{D}$ is defined by

$$\mathbf{D}y = \frac{dy}{dx}, \tag{4.209}$$

or, in terms of the operator alone,

$$\mathbf{D} = \frac{d}{dx}. \tag{4.210}$$

The operator can be repeatedly applied, so that

$$\mathbf{D}^n y = \frac{d^n y}{dx^n}. \tag{4.211}$$

Another example of its use is

$$(\mathbf{D} - a)(\mathbf{D} - b)f(x) = (\mathbf{D} - a)((\mathbf{D} - b)f(x)), \tag{4.212}$$

$$= (\mathbf{D} - a)\left(\frac{df}{dx} - bf\right), \tag{4.213}$$

$$= \frac{d^2 f}{dx^2} - (a + b)\frac{df}{dx} + abf. \tag{4.214}$$

Negative powers of $\mathbf{D}$ are related to integrals. This comes from

$$\frac{dy(x)}{dx} = f(x) \qquad y(x_0) = y_0, \tag{4.215}$$

$$y(x) = y_0 + \int_{x_0}^x f(s)\, ds; \tag{4.216}$$

then

$$\mathbf{D}y(x) = f(x), \tag{4.217}$$

$$\mathbf{D}^{-1}\mathbf{D}y(x) = \mathbf{D}^{-1}f(x), \tag{4.218}$$

$$y(x) = \mathbf{D}^{-1}f(x), \tag{4.219}$$

$$= y_0 + \int_{x_0}^{x} f(s)\, ds, \tag{4.220}$$

$$\mathbf{D}^{-1}(\cdot) = y_0 + \int_{x_0}^{x} (\cdot)\, ds. \tag{4.221}$$

We can evaluate $h(x)$, where

$$h(x) = \frac{1}{\mathbf{D} - a} f(x), \tag{4.222}$$

in the following way:

$$(\mathbf{D} - a)\, h(x) = (\mathbf{D} - a)\left(\frac{1}{\mathbf{D} - a} f(x)\right), \tag{4.223}$$

$$= f(x), \tag{4.224}$$

$$\frac{dh(x)}{dx} - ah(x) = f(x), \tag{4.225}$$

$$e^{-ax}\frac{dh(x)}{dx} - ae^{-ax}h(x) = f(x)e^{-ax}, \tag{4.226}$$

$$\frac{d}{dx}\left(e^{-ax}h(x)\right) = f(x)e^{-ax}, \tag{4.227}$$

$$\frac{d}{ds}\left(e^{-as}h(s)\right) = f(s)e^{-as}, \tag{4.228}$$

$$\int_{x_0}^{x} \frac{d}{ds}\left(e^{-as}h(s)\right)\, ds = \int_{x_0}^{x} f(s)e^{-as}\, ds, \tag{4.229}$$

$$e^{-ax}h(x) - e^{-ax_0}h(x_0) = \int_{x_0}^{x} f(s)e^{-as}\, ds, \tag{4.230}$$

$$h(x) = e^{a(x-x_0)}h(x_0) + e^{ax}\int_{x_0}^{x} f(s)e^{-as}\, ds, \tag{4.231}$$

$$\frac{1}{\mathbf{D} - a} f(x) = e^{a(x-x_0)}h(x_0) + e^{ax}\int_{x_0}^{x} f(s)e^{-as}\, ds. \tag{4.232}$$

This gives us $h(x)$ explicitly in terms of the known function f such that h satisfies $Dh - ah = f$.

We can find the solution to higher order equations such as

$$(\mathbf{D} - a)(\mathbf{D} - b)y(x) = f(x), \qquad y(x_0) = y_0,\ y'(x_0) = y_0', \tag{4.233}$$

$$(\mathbf{D} - b)y(x) = \frac{1}{\mathbf{D} - a} f(x), \tag{4.234}$$

$$= h(x), \tag{4.235}$$

$$y(x) = y_0 e^{b(x-x_0)} + e^{bx}\int_{x_0}^{x} h(s)e^{-bs}\, ds. \tag{4.236}$$

We see that

$$\frac{dy}{dx} = y_0 b e^{b(x-x_0)} + h(x) + b e^{bx} \int_{x_0}^x h(s) e^{-bs}\, ds, \qquad (4.237)$$

$$\left.\frac{dy}{dx}\right|_{x=x_0} = y_0' = y_0 b + h(x_0), \qquad (4.238)$$

which can be rewritten as

$$((\mathbf{D} - b)y)|_{x=x_0} = h(x_0), \qquad (4.239)$$

which is what one would expect.

Returning to the problem at hand, we take our expression for $h(x)$, evaluate it at $x = s$, and substitute into the expression for $y(x)$ to get

$$y(x) = y_0 e^{b(x-x_0)} + e^{bx} \int_{x_0}^x \left(h(x_0) e^{a(s-x_0)} + e^{as} \int_{x_0}^s f(t) e^{-at}\, dt \right) e^{-bs}\, ds,$$
$$(4.240)$$

$$= y_0 e^{b(x-x_0)} + e^{bx} \int_{x_0}^x \left((y_0' - y_0 b) e^{a(s-x_0)} + e^{as} \int_{x_0}^s f(t) e^{-at}\, dt \right) e^{-bs}\, ds,$$
$$(4.241)$$

$$= y_0 e^{b(x-x_0)} + e^{bx} \int_{x_0}^x \left((y_0' - y_0 b) e^{(a-b)s - ax_0} + e^{(a-b)s} \int_{x_0}^s f(t) e^{-at}\, dt \right) ds,$$
$$(4.242)$$

$$= y_0 e^{b(x-x_0)} + e^{bx} (y_0' - y_0 b) \int_{x_0}^x e^{(a-b)s - ax_0}\, ds$$
$$+ e^{bx} \int_{x_0}^x e^{(a-b)s} \left(\int_{x_0}^s f(t) e^{-at}\, dt \right) ds, \qquad (4.243)$$

$$= y_0 e^{b(x-x_0)} + e^{bx} (y_0' - y_0 b) \frac{e^{a(x-x_0) - xb} - e^{-bx_0}}{a - b}$$
$$+ e^{bx} \int_{x_0}^x e^{(a-b)s} \left(\int_{x_0}^s f(t) e^{-at}\, dt \right) ds, \qquad (4.244)$$

$$= y_0 e^{b(x-x_0)} + (y_0' - y_0 b) \frac{e^{a(x-x_0)} - e^{b(x-x_0)}}{a - b}$$
$$+ e^{bx} \int_{x_0}^x e^{(a-b)s} \left(\int_{x_0}^s f(t) e^{-at}\, dt \right) ds, \qquad (4.245)$$

$$= y_0 e^{b(x-x_0)} + (y_0' - y_0 b) \frac{e^{a(x-x_0)} - e^{b(x-x_0)}}{a - b}$$
$$+ e^{bx} \int_{x_0}^x \int_{x_0}^s e^{(a-b)s} f(t) e^{-at}\, dt\, ds. \qquad (4.246)$$

Changing the order of integration and integrating on s, we get

$$y(x) = y_0 e^{b(x-x_0)} + (y_0' - y_0 b) \frac{e^{a(x-x_0)} - e^{b(x-x_0)}}{a - b}$$

$$+ e^{bx} \int_{x_0}^x \int_t^x e^{(a-b)s} f(t) e^{-at} \, ds \, dt, \tag{4.247}$$

$$= y_0 e^{b(x-x_0)} + (y_0' - y_0 b) \frac{e^{a(x-x_0)} - e^{b(x-x_0)}}{a - b}$$

$$+ e^{bx} \int_{x_0}^x f(t) e^{-at} \left(\int_t^x e^{(a-b)s} \, ds \right) dt, \tag{4.248}$$

$$= y_0 e^{b(x-x_0)} + (y_0' - y_0 b) \frac{e^{a(x-x_0)} - e^{b(x-x_0)}}{a - b}$$

$$+ \int_{x_0}^x \frac{f(t)}{a - b} \left(e^{a(x-t)} - e^{b(x-t)} \right) dt. \tag{4.249}$$

Thus, we have a solution to the second-order linear differential equation with constant coefficients and arbitrary forcing expressed in integral form. A similar alternate expression can be developed when $a = b$.

4.4 Sturm-Liouville Analysis

Within engineering, second-order linear differential equations arise in a variety of scenarios, so much so that many relevant analysis tools have been developed. Solution of homogeneous versions of these equations gives rise to complementary functions, which are the key to unraveling many intrinsic patterns and harmonies observed in nature. For example, the well-known trigonometric functions sine and cosine are complementary functions to the differential equation modeling a linear mass-spring system; the harmonies of such a system realized in their natural frequencies are a prime subject of mathematical analysis. When the constant coefficients of the equations that induce sinusoidal oscillations are allowed to vary, one predicts modulated patterns and harmonies. It is this ability to allow description of widely varying patterns that gives this analysis great utility in describing the natural world.

This section considers trigonometric and other functions that arise from the solution of a variety of linear homogeneous second-order differential equations with constant and nonconstant coefficients. The notions of *eigenvalues*, *eigenfunctions*, and *orthogonal* and *orthonormal* functions are introduced in a natural fashion, which will be familiar to many readers. These notions are formalized in a general sense in the upcoming Chapter 6 on linear analysis.

A key result of this and the next section will be to show how one can expand an arbitrary function in terms of infinite sums of the product of scalar amplitudes with orthogonal basis functions. Such a summation is known as a *Fourier series*. When the infinite sum is truncated, we have an approximation of the original function, which can be thought of geometrically as a projection of a high- or infinite-dimensional entity onto a lower dimensional space. Projection is discussed in more generality in Section 6.5.

4.4.1 General Formulation

Consider on the domain $x \in [x_0, x_1]$ the following general linear homogeneous second-order differential equation with general homogeneous boundary conditions:

$$a(x)\frac{d^2y}{dx^2} + b(x)\frac{dy}{dx} + c(x)y + \lambda y = 0, \tag{4.250}$$

$$\alpha_1 y(x_0) + \alpha_2 y'(x_0) = 0, \tag{4.251}$$

$$\beta_1 y(x_1) + \beta_2 y'(x_1) = 0. \tag{4.252}$$

Define the following functions:

$$p(x) = \exp\left(\int_{x_0}^{x} \frac{b(s)}{a(s)} \, ds\right), \tag{4.253}$$

$$r(x) = \frac{1}{a(x)} \exp\left(\int_{x_0}^{x} \frac{b(s)}{a(s)} \, ds\right), \tag{4.254}$$

$$q(x) = \frac{c(x)}{a(x)} \exp\left(\int_{x_0}^{x} \frac{b(s)}{a(s)} \, ds\right). \tag{4.255}$$

With these definitions, Eq. (4.250) is transformed to the type known as a *Sturm-Liouville*[8] equation:

$$\frac{d}{dx}\left(p(x)\frac{dy}{dx}\right) + (q(x) + \lambda r(x))\, y(x) = 0, \tag{4.256}$$

$$-\left(\frac{1}{r(x)}\left(\frac{d}{dx}\left(p(x)\frac{d}{dx}\right) + q(x)\right)\right) y(x) = \lambda y(x). \tag{4.257}$$

$$\underbrace{\qquad\qquad\qquad\qquad\qquad\qquad}_{\mathbf{L}_s}$$

Here the Sturm-Liouville linear operator $\mathbf{L}_s$ is

$$\mathbf{L}_s = -\frac{1}{r(x)}\left(\frac{d}{dx}\left(p(x)\frac{d}{dx}\right) + q(x)\right), \tag{4.258}$$

so we have Eq. (4.257) compactly stated as

$$\mathbf{L}_s y(x) = \lambda y(x). \tag{4.259}$$

Now the trivial solution $y(x) = 0$ will satisfy the differential equation and boundary conditions, Eqs. (4.250–4.252). In addition, for special real values of λ, known as *eigenvalues*, there are special nontrivial functions, known as *eigenfunctions*, that also satisfy Eqs. (4.250–4.252). Eigenvalues and eigenfunctions are discussed in more general terms in Section 6.8.

Now it can be shown that if we have, for $x \in [x_0, x_1]$,

$$p(x) > 0, \tag{4.260}$$

$$r(x) > 0, \tag{4.261}$$

$$q(x) \geq 0, \tag{4.262}$$

that an infinite number of real positive eigenvalues λ and corresponding eigenfunctions $y_n(x)$ exist for which Eqs. (4.250–4.252) are satisfied. Moreover, it can also be

---

[8] Jacques Charles François Sturm, 1803–1855, Swiss-born French mathematician, and Joseph Liouville, 1809–1882, French mathematician.

shown that a consequence of the homogeneous boundary conditions is the *orthogonality condition*:

$$\langle y_n, y_m \rangle = \int_{x_0}^{x_1} r(x) y_n(x) y_m(x) \, dx = 0, \qquad n \neq m, \tag{4.263}$$

$$\langle y_n, y_n \rangle = \int_{x_0}^{x_1} r(x) y_n(x) y_n(x) \, dx = K^2. \tag{4.264}$$

Consequently, in the same way that in ordinary vector mechanics $(2\mathbf{i}) \cdot \mathbf{j} = 0$, $(2\mathbf{i}) \cdot (3\mathbf{k}) = 0$, $(2\mathbf{i}) \cdot (2\mathbf{i}) = 4$ implies $2\mathbf{i}$ is orthogonal to $\mathbf{j}$ and $3\mathbf{k}$ but not to itself, the eigenfunctions of a Sturm-Liouville operator $\mathbf{L}_s$ are said to be orthogonal to each other. The so-called inner product notation, $\langle \cdot, \cdot \rangle$, is explained in detail in Section 6.3.2. Here $K \in \mathbb{R}^1$ is a real constant. Equations (4.264,4.263) can be written compactly using the Kronecker delta, δ_{nm}, as

$$\int_{x_0}^{x_1} r(x) y_n(x) y_m(x) \, dx = K^2 \delta_{nm}. \tag{4.265}$$

Sturm-Liouville theory shares more analogies with vector algebra. In the same sense that the dot product of a vector with itself is guaranteed positive, we have defined an inner product for the eigenfunctions in which the inner product of a Sturm-Liouville eigenfunction with itself is guaranteed positive.

Motivated by Eq. (4.265), we can define functions $\varphi_n(x)$;

$$\varphi_n(x) = \frac{\sqrt{r(x)}}{K} y_n(x), \tag{4.266}$$

so that

$$\langle \varphi_n, \varphi_m \rangle = \int_{x_0}^{x_1} \varphi_n(x) \varphi_m(x) \, dx = \delta_{nm}. \tag{4.267}$$

Such functions are said to be *orthonormal*, in the same way that $\mathbf{i}$, $\mathbf{j}$, and $\mathbf{k}$ are orthonormal, for example, $\mathbf{i} \cdot \mathbf{j} = 0$, $\mathbf{i} \cdot \mathbf{k} = 0$, $\mathbf{i} \cdot \mathbf{i} = 1$. While orthonormal functions have great utility, in the context of our Sturm-Liouville nomenclature, $\varphi_n(x)$ does not in general satisfy the Sturm-Liouville equation: $\mathbf{L}_s \varphi_n(x) \neq \lambda_n \varphi_n(x)$.

4.4.2 Adjoint of Differential Operators

In this section, we briefly characterize what are known as the *adjoint operators* of linear differential systems. We also consider adjoints for the Sturm-Liouville operator but do not present a detailed exposition of all of the various important implications. For full details of adjoints for broader classes of linear systems, see the upcoming Section 6.7.2.

For the Nth-order linear differential operator $\mathbf{L}$ defined by

$$\mathbf{L} y = \sum_{n=0}^{N} a_n(x) \frac{d^n y}{dx^n}, \tag{4.268}$$

the adjoint can be shown to be $\mathbf{L}^*$, where[9]

$$\mathbf{L}^*y = \sum_{n=0}^{N}(-1)^n\frac{d^n}{dx^n}\left(a_n(x)y\right).\tag{4.269}$$

As applied to linear differential equations, the adjoint operator is such that

$$\langle\mathbf{L}z, y\rangle = \langle z, \mathbf{L}^*y\rangle.\tag{4.270}$$

For real-valued functions, this becomes

$$\int_{x_0}^{x_1}y\mathbf{L}z\,dx = \int_{x_0}^{x_1}z\mathbf{L}^*y\,dx.\tag{4.271}$$

Here $y = y(x)$, $z = z(x)$. This implies that

$$\int_{x_0}^{x_1}\left(y\mathbf{L}z - z\mathbf{L}^*y\right)dx = 0.\tag{4.272}$$

The Sturm-Liouville operator is sometimes alternatively defined such that $\mathbf{L}_sy(x) = \lambda r(x)y(x)$. Then, with $\mathbf{L}_s$ defined by

$$\mathbf{L}_sy = -\frac{d}{dx}\left(p(x)\frac{dy}{dx}\right) - q(x)y,\tag{4.273}$$

$$= -p\frac{d^2y}{dx^2} - \frac{dp}{dx}\frac{dy}{dx} - qy,\tag{4.274}$$

$\mathbf{L}_s$ has the adjoint

$$\mathbf{L}_s^*y = -\frac{d^2}{dx^2}\left(py\right) + \frac{d}{dx}\left(\frac{dp}{dx}y\right) - qy,\tag{4.275}$$

$$= -\frac{d}{dx}\left(\frac{dp}{dx}y + p\frac{dy}{dx}\right) + \frac{d}{dx}\left(\frac{dp}{dx}y\right) - qy,\tag{4.276}$$

$$= -\frac{d}{dx}\left(p\frac{dy}{dx}\right) - qy.\tag{4.277}$$

Because $\mathbf{L}_s = \mathbf{L}_s^*$, $\mathbf{L}_s$ is known as *self-adjoint*.

Any linear, homogeneous second-order differential equation

$$a(x)\frac{d^2y}{dx^2} + b(x)\frac{dy}{dx} + c(x)y = 0,\tag{4.278}$$

which may not be self-adjoint, may be made self-adjoint by multiplying by an appropriate integrating factor

$$\frac{1}{a(x)}\exp\left(\int_{x_0}^{x}\frac{b(s)}{a(s)}\,ds\right).\tag{4.279}$$

Thus, for example, the physicists' Hermite equation,

$$\frac{d^2y}{dx^2} - 2x\frac{dy}{dx} + \lambda y = 0,\tag{4.280}$$

---

[9] In this adjoint, we are ignoring the boundary conditions. This is sometimes known as the "formal adjoint." For a more completely defined adjoint, boundary conditions must be considered, and we have to define an inner product space, which will be done in Chapter 6. The adjoint will be defined in Section 6.7.2

is not self-adjoint. However, multiplying it by

$$\exp\left(\int_{-\infty}^{x}(-2s)\,ds\right) = e^{-x^2} \tag{4.281}$$

converts it into

$$\frac{d}{dx}\left(e^{-x^2}\frac{dy}{dx}\right) + \lambda e^{-x^2}y = 0, \tag{4.282}$$

which is in Sturm-Liouville form and thus is self-adjoint.

The operator studied in Example 4.15, $\mathbf{L} = d^2/dx^2$ with homogeneous boundary conditions at $x = 0$ and $x = 1$, is a Sturm-Liouville operator and thus self-adjoint with $\mathbf{L}^* = \mathbf{L} = d^2/dx^2$. As a result of this self-adjoint property, the associated Green's function $g(x, s)$ was found to be symmetric: $g(x, s) = g(s, x)$. Let us next study an example that is not self-adjoint and see that, for such a system, $g(x, s) \neq g(s, x)$.

EXAMPLE 4.17

Consider $\mathbf{L}y = f(x)$, with $y(0) = y(1) = 0$ and $\mathbf{L} = d^2/dx^2 + d/dx$. Find the adjoint operator $\mathbf{L}^*$ and the Green's function $g(x, s)$. Verify the solution when $f(x) = 6x$.

Our problem is

$$\frac{d^2y}{dx^2} + \frac{dy}{dx} = f(x), \qquad y(0) = y(1) = 0. \tag{4.283}$$

By Eq. (4.269), the adjoint operator is

$$\mathbf{L}^* = \frac{d^2}{dx^2} - \frac{d}{dx}. \tag{4.284}$$

Because $\mathbf{L} \neq \mathbf{L}^*$, the operator $\mathbf{L}$ is not self-adjoint. We can verify that $\mathbf{L}^*$ is the adjoint by examining its application to Eq. (4.272):

$$\int_0^1 \left(z\left(\frac{d^2y}{dx^2} + \frac{dy}{dx}\right) - y\left(\frac{d^2z}{dx^2} - \frac{dz}{dx}\right)\right) dx = 0, \tag{4.285}$$

$$\int_0^1 \left(z\frac{d^2y}{dx^2} - y\frac{d^2z}{dx^2} + z\frac{dy}{dx} + y\frac{dz}{dx}\right) dx = 0, \tag{4.286}$$

$$\int_0^1 z\frac{d^2y}{dx^2}\,dx - \int_0^1 y\frac{d^2z}{dx^2}\,dx + \int_0^1 \frac{d}{dx}(yz)\,dx = 0. \tag{4.287}$$

Now integrate by parts twice on the first term and integrate the last term to get

$$z\frac{dy}{dx}\Big|_0^1 - \int_0^1 \frac{dz}{dx}\frac{dy}{dx}\,dx - \int_0^1 y\frac{d^2z}{dx^2}\,dx + (yz)|_0^1 = 0, \tag{4.288}$$

$$z\frac{dy}{dx}\Big|_0^1 - \frac{dz}{dx}y\Big|_0^1 + \int_0^1 y\frac{d^2z}{dx^2}\,dx - \int_0^1 y\frac{d^2z}{dx^2}\,dx + (yz)|_0^1 = 0. \tag{4.289}$$

Now let us insist that $z(0) = z(1) = 0$. Using this and the fact that $y(0) = y(1) = 0$ eliminates all remaining terms. Thus, $\mathbf{L}^* = d^2/dx^2 - d/dx$, when accompanied by homogeneous boundary values at $x = 0$ and $x = 1$, is in fact the adjoint operator.

For the Green's function, we have for $x < s$

$$\frac{d^2g}{dx^2} + \frac{dg}{dx} = 0, \tag{4.290}$$

$$\frac{dg}{dx} + g = C_1, \tag{4.291}$$

$$g = C_2 e^{-x} + C_1, \tag{4.292}$$

$$g(0) = = 0 = C_2 e^0 + C_1. \tag{4.293}$$

Thus, $C_2 = -C_1$ and

$$g(x, s) = C_1(1 - e^{-x}), \qquad x < s. \tag{4.294}$$

Now for $x > s$, we have

$$\frac{d^2 g}{dx^2} + \frac{dg}{dx} = 0, \tag{4.295}$$

$$\frac{dg}{dx} + g = C_3, \tag{4.296}$$

$$g = C_4 e^{-x} + C_3, \tag{4.297}$$

$$g(1) = 0 = C_4 e^{-1} + C_3. \tag{4.298}$$

So $C_4 = -C_3 e$ and

$$g(x, s) = C_3(1 - e^{1-x}), \qquad x > s. \tag{4.299}$$

For continuity of $g(x, s)$ at $x = s$, we have

$$C_1(1 - e^{-s}) = C_3(1 - e^{1-s}). \tag{4.300}$$

This gives

$$C_1 = \frac{1 - e^{1-s}}{1 - e^{-s}} C_3. \tag{4.301}$$

Thus, we have

$$g(x, s) = C_3 \frac{\left(1 - e^{1-s}\right)\left(1 - e^{-x}\right)}{1 - e^{-s}}, \qquad x \leq s, \tag{4.302}$$

$$g(x, s) = C_3(1 - e^{1-x}), \qquad x \geq s. \tag{4.303}$$

Now, we enforce a jump of unity in dg/dx at $x = s$:

$$\left.\frac{dg}{dx}\right|_{x=s+\epsilon} - \left.\frac{dg}{dx}\right|_{x=s-\epsilon} = 1. \tag{4.304}$$

Leaving out details, this calculation yields

$$C_3 = \frac{e^s - 1}{e - 1}. \tag{4.305}$$

This yields the Green's function

$$g(x, s) = \frac{e\left(e^{s-1} - 1\right)\left(1 - e^{-x}\right)}{e - 1}, \qquad x \leq s, \tag{4.306}$$

$$g(x, s) = \frac{\left(1 - e^s\right)\left(e^{1-x} - 1\right)}{e - 1}, \qquad x \geq s. \tag{4.307}$$

We see that $g(x, s) \neq g(s, x)$. This is a consequence of $\mathbf{L}$ not being self-adjoint. Nevertheless, the Green's function is still valid for obtaining the solution for arbitrary $f(x)$:

$$y(x) = \int_0^x \frac{\left(1 - e^s\right)\left(e^{1-x} - 1\right)}{e - 1} f(s)\, ds + \int_x^1 \frac{e\left(e^{s-1} - 1\right)\left(1 - e^{-x}\right)}{e - 1} f(s)\, ds. \tag{4.308}$$

When $f(x) = 6x$, detailed calculation reveals the solution to be

$$y(x) = \int_0^x \frac{\left(1 - e^s\right)\left(e^{1-x} - 1\right)}{e - 1}(6s)\, ds + \int_x^1 \frac{e\left(e^{s-1} - 1\right)\left(1 - e^{-x}\right)}{e - 1}(6s)\, ds, \tag{4.309}$$

$$= \frac{3\left(e\left(x^2 - 2x + 1\right) - x^2 + 2x - e^{1-x}\right)}{e - 1}. \tag{4.310}$$

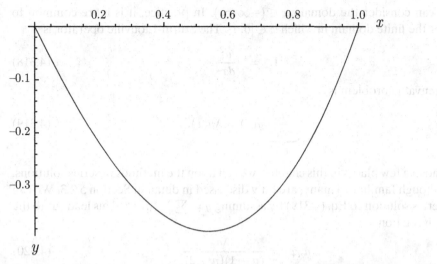

Figure 4.2. Plot of function that satisfies $d^2y/dx^2 + dy/dx = 6x, y(0) = y(1) = 0$.

While complicated, the plot of the result (see Figure 4.2) is remarkably similar to that obtained earlier in the self-adjoint problem $d^2y/dx^2 = 6x$, $y(0) = y(1) = 0$ (see Figure 4.1).

4.4.3 Linear Oscillator

A linear oscillator gives perhaps the simplest example of a Sturm-Liouville problem. We will consider the domain $x \in [0, 1]$. For other domains, we could easily transform coordinates; for example, if $x \in [x_0, x_1]$, then the linear mapping $\tilde{x} = (x - x_0)/(x_1 - x_0)$ lets us consider $\tilde{x} \in [0, 1]$.

The equations governing a linear oscillator with general homogeneous boundary conditions are

$$\frac{d^2y}{dx^2} + \lambda y = 0, \qquad \alpha_1 y(0) + \alpha_2 \frac{dy}{dx}(0) = 0, \qquad \beta_1 y(1) + \beta_2 \frac{dy}{dx}(1) = 0. \quad (4.311)$$

Here we have

$$a(x) = 1, \qquad (4.312)$$

$$b(x) = 0, \qquad (4.313)$$

$$c(x) = 0, \qquad (4.314)$$

so

$$p(x) = \exp\left(\int_{x_0}^{x} \frac{0}{1} \, ds\right) = e^0 = 1, \qquad (4.315)$$

$$r(x) = \frac{1}{1} \exp\left(\int_{x_0}^{x} \frac{0}{1} \, ds\right) = e^0 = 1, \qquad (4.316)$$

$$q(x) = \frac{0}{1} \exp\left(\int_{x_0}^{x} \frac{0}{1} \, ds\right) = 0. \qquad (4.317)$$

So, we can consider the domain $x \in (-\infty, \infty)$. In practice, it is more common to consider the finite domain in which $x \in [0, 1]$. The Sturm-Liouville operator is

$$\mathbf{L}_s = -\frac{d^2}{dx^2}. \tag{4.318}$$

The eigenvalue problem is

$$-\underbrace{\frac{d^2}{dx^2}}_{\mathbf{L}_s} y(x) = \lambda y(x). \tag{4.319}$$

Here and in a few places in this chapter, we call upon the methods of series solutions, which, though familiar to many, are only discussed in detail in Section 5.2.3. We can find a series solution to Eq. (4.319) by assuming $y = \sum_{n=0}^{\infty} a_n x^n$. This leads us to the recursion relation

$$a_{n+2} = \frac{-\lambda a_n}{(n+1)(n+2)}. \tag{4.320}$$

So, given two seed values, a_0 and a_1, detailed but familiar analysis reveals that the solution can be expressed as the infinite series

$$y(x) = a_0 \underbrace{\left(1 - \frac{(\sqrt{\lambda}x)^2}{2!} + \frac{(\sqrt{\lambda}x)^4}{4!} - \cdots\right)}_{\cos(\sqrt{\lambda}x)} + a_1 \underbrace{\left(\sqrt{\lambda}x - \frac{(\sqrt{\lambda}x)^3}{3!} + \frac{(\sqrt{\lambda}x)^5}{5!} - \cdots\right)}_{\sin(\sqrt{\lambda}x)}. \tag{4.321}$$

The series is recognized as being composed of linear combinations of the Taylor series for $\cos(\sqrt{\lambda}x)$ and $\sin(\sqrt{\lambda}x)$ about $x = 0$. Letting $a_0 = C_1$ and $a_1 = C_2$, we can express the general solution in terms of these two complementary functions as

$$y(x) = C_1 \cos(\sqrt{\lambda}x) + C_2 \sin(\sqrt{\lambda}x). \tag{4.322}$$

Applying the general homogeneous boundary conditions from Eq. (4.311) leads to a challenging problem for determining admissible eigenvalues λ. To apply the boundary conditions, we need dy/dx, which is

$$\frac{dy}{dx} = -C_1\sqrt{\lambda}\sin(\sqrt{\lambda}x) + C_2\sqrt{\lambda}\cos(\sqrt{\lambda}x). \tag{4.323}$$

Enforcing the boundary conditions at $x = 0$ and $x = 1$ leads us to two equations:

$$\alpha_1 C_1 + \alpha_2\sqrt{\lambda}C_2 = 0, \tag{4.324}$$

$$C_1\left(\beta_1\cos\sqrt{\lambda} - \beta_2\sqrt{\lambda}\sin\sqrt{\lambda}\right) + C_2\left(\beta_1\sin\sqrt{\lambda} + \beta_2\sqrt{\lambda}\cos\sqrt{\lambda}\right) = 0. \tag{4.325}$$

This can be posed as the linear system

$$\begin{pmatrix} \alpha_1 & \alpha_2\sqrt{\lambda} \\ \beta_1\cos\sqrt{\lambda} - \beta_2\sqrt{\lambda}\sin\sqrt{\lambda} & \beta_1\sin\sqrt{\lambda} + \beta_2\sqrt{\lambda}\cos\sqrt{\lambda} \end{pmatrix}\begin{pmatrix} C_1 \\ C_2 \end{pmatrix} = \begin{pmatrix} 0 \\ 0 \end{pmatrix}. \tag{4.326}$$

For nontrivial solutions, the determinant of the coefficient matrix must be zero, which leads to the transcendental equation

$$\alpha_1\left(\beta_1\sin\sqrt{\lambda} + \beta_2\sqrt{\lambda}\cos\sqrt{\lambda}\right) - \alpha_2\sqrt{\lambda}\left(\beta_1\cos\sqrt{\lambda} - \beta_2\sqrt{\lambda}\sin\sqrt{\lambda}\right) = 0. \tag{4.327}$$

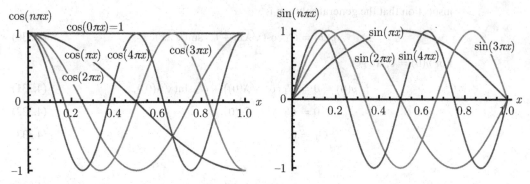

Figure 4.3. Solutions to the linear oscillator equation, Eq. (4.311), in terms of two sets of complementary functions, $\cos(n\pi x)$ and $\sin(n\pi x)$.

Assuming $\alpha_1, \beta_1 \neq 0$, we can scale by $\alpha_1\beta_1$ to get the equivalent

$$\left(\sin\sqrt{\lambda} + \frac{\beta_2}{\beta_1}\sqrt{\lambda}\cos\sqrt{\lambda}\right) - \frac{\alpha_2}{\alpha_1}\sqrt{\lambda}\left(\cos\sqrt{\lambda} - \frac{\beta_2}{\beta_1}\sqrt{\lambda}\sin\sqrt{\lambda}\right) = 0. \quad (4.328)$$

For known values of α_2/α_1, β_2/β_1, one seeks values of λ that satisfy Eq. (4.328). This is a solution that in general must be obtained numerically, except for the simplest of cases.

One important case has $\alpha_2 = \beta_2 = 0$. This gives the boundary conditions to be $y(0) = y(1) = 0$. Boundary conditions for which the function values are specified are known as *Dirichlet*[10] conditions. In this case, Eq. (4.328) reduces to $\sin\sqrt{\lambda} = 0$, which is easily solved as $\sqrt{\lambda} = n\pi$, with $n = 0, \pm1, \pm2, \ldots$. We also get $C_1 = 0$; consequently, $y = C_2\sin(n\pi x)$. For $n = 0$, the solution is the trivial $y = 0$.

Another set of conditions also leads to a similarly simple result. Taking $\alpha_1 = \beta_1 = 0$, the boundary conditions become $y'(0) = y'(1) = 0$. Boundary conditions where the function's derivative values are specified are known as *Neumann*[11] conditions. In this case, Eq. (4.328) reduces to $-\lambda\sin\sqrt{\lambda} = 0$, which is easily solved as $\sqrt{\lambda} = n\pi$, with $n = 0, \pm1, \pm2, \ldots$. We also get $C_2 = 0$; consequently, $y = C_1\cos(n\pi x)$. Here, for $n = 0$, the solution is the nontrivial $y = C_1$.

Some of the eigenfunctions for pure Neumann and pure Dirichlet boundary conditions are plotted in Figure 4.3. These two families form the linearly independent complementary functions of Eq. (4.311). Also, as n increases, the number of zero-crossings within the domain increases. This will be seen to be characteristic of all sets of eigenfunctions for Sturm-Liouville equations.

EXAMPLE 4.18

Find the eigenvalues and eigenfunctions for a linear oscillator equation with Dirichlet boundary conditions:

$$\frac{d^2y}{dx^2} + \lambda y = 0, \qquad y(0) = y(\ell) = 0. \quad (4.329)$$

We could transform the domain via $\tilde{x} = x/\ell$ so that $\tilde{x} \in [0, 1]$, but this problem is sufficiently straightforward to allow us to deal with the original domain. We know by

---

[10] Johann Peter Gustav Lejeune Dirichlet, 1805–1859, German mathematician.
[11] Carl Gottfried Neumann, 1832–1925, German mathematician.

inspection that the general solution is

$$y(x) = C_1 \cos(\sqrt{\lambda}x) + C_2 \sin(\sqrt{\lambda}x). \tag{4.330}$$

For $y(0) = 0$, we get

$$y(0) = 0 = C_1 \cos(\sqrt{\lambda}(0)) + C_2 \sin(\sqrt{\lambda}(0)), \tag{4.331}$$

$$0 = C_1(1) + C_2(0), \tag{4.332}$$

$$C_1 = 0. \tag{4.333}$$

So

$$y(x) = C_2 \sin(\sqrt{\lambda}x). \tag{4.334}$$

At the boundary at $x = \ell$, we have

$$y(\ell) = 0 = C_2 \sin(\sqrt{\lambda}\,\ell). \tag{4.335}$$

For nontrivial solutions, we need $C_2 \neq 0$, which then requires that

$$\sqrt{\lambda}\ell = n\pi \qquad n = \pm 1, \pm 2, \pm 3, \ldots, \tag{4.336}$$

so

$$\lambda = \left(\frac{n\pi}{\ell}\right)^2. \tag{4.337}$$

The eigenvalues and eigenfunctions are

$$\lambda_n = \frac{n^2\pi^2}{\ell^2}, \tag{4.338}$$

and

$$y_n(x) = \sin\left(\frac{n\pi x}{\ell}\right), \tag{4.339}$$

respectively.

Check orthogonality for $y_2(x)$ and $y_3(x)$:

$$I = \int_0^\ell \sin\left(\frac{2\pi x}{\ell}\right) \sin\left(\frac{3\pi x}{\ell}\right) dx, \tag{4.340}$$

$$= \frac{\ell}{2\pi} \left(\sin\left(\frac{\pi x}{\ell}\right) - \frac{1}{5}\sin\left(\frac{5\pi x}{\ell}\right)\right)\Bigg|_0^\ell, \tag{4.341}$$

$$= 0. \tag{4.342}$$

Check orthogonality for $y_4(x)$ and $y_4(x)$:

$$I = \int_0^\ell \sin\left(\frac{4\pi x}{\ell}\right) \sin\left(\frac{4\pi x}{\ell}\right) dx, \tag{4.343}$$

$$= \left(\frac{x}{2} - \frac{\ell}{16\pi}\sin\left(\frac{8\pi x}{\ell}\right)\right)\Bigg|_0^\ell, \tag{4.344}$$

$$= \frac{\ell}{2}. \tag{4.345}$$

In fact,

$$\int_0^\ell \sin\left(\frac{n\pi x}{\ell}\right) \sin\left(\frac{n\pi x}{\ell}\right) dx = \frac{\ell}{2}, \tag{4.346}$$

so the orthonormal functions $\varphi_n(x)$ for this problem are

$$\varphi_n(x) = \sqrt{\frac{2}{\ell}} \sin\left(\frac{n\pi x}{\ell}\right). \tag{4.347}$$

With this choice, we recover the orthonormality condition:

$$\int_0^\ell \varphi_n(x)\varphi_m(x)\, dx = \delta_{nm}, \tag{4.348}$$

$$\frac{2}{\ell}\int_0^\ell \sin\left(\frac{n\pi x}{\ell}\right)\sin\left(\frac{m\pi x}{\ell}\right)\, dx = \delta_{nm}. \tag{4.349}$$

4.4.4 Legendre Differential Equation

The Legendre[12] differential equation is given next. Here it is convenient to let the term $n(n+1)$ play the role of λ:

$$(1-x^2)\frac{d^2 y}{dx^2} - 2x\frac{dy}{dx} + \underbrace{n(n+1)}_{\lambda} y = 0. \tag{4.350}$$

The Legendre differential equation is singular at $x = \pm 1$. We have

$$a(x) = 1 - x^2, \tag{4.351}$$

$$b(x) = -2x, \tag{4.352}$$

$$c(x) = 0. \tag{4.353}$$

Then, taking $x_0 = -1$, we have

$$p(x) = \exp\int_{-1}^x \frac{-2s}{1-s^2}\, ds, \tag{4.354}$$

$$= \exp\left(\ln\left(1-s^2\right)\right)\big|_{-1}^x, \tag{4.355}$$

$$= (1-s^2)\big|_{-1}^x, \tag{4.356}$$

$$= 1 - x^2. \tag{4.357}$$

We find then that

$$r(x) = 1, \tag{4.358}$$

$$q(x) = 0. \tag{4.359}$$

Thus, we require $x \in (-1, 1)$. In Sturm-Liouville form, Eq. (4.350) reduces to

$$\frac{d}{dx}\left((1-x^2)\frac{dy}{dx}\right) + n(n+1)y = 0, \tag{4.360}$$

$$\underbrace{-\frac{d}{dx}\left((1-x^2)\frac{d}{dx}\right)}_{L_s} y(x) = n(n+1)y(x). \tag{4.361}$$

[12] Adrien-Marie Legendre, 1752–1833, French mathematician.

So

$$\mathbf{L}_s = -\frac{d}{dx}\left((1-x^2)\frac{d}{dx}\right). \tag{4.362}$$

Again consider an infinite series solution. Let

$$y = \sum_{m=0}^{\infty} a_m x^m. \tag{4.363}$$

Substituting into Eq. (4.350), we find after detailed analysis the recursion relation

$$a_{m+2} = a_m \frac{(m+n+1)(m-n)}{(m+1)(m+2)}. \tag{4.364}$$

With a_0 and a_1 as given seeds, we can thus generate all values of a_m for $m \geq 2$. We find

$$y(x) = a_0 \underbrace{\left(1 - n(n+1)\frac{x^2}{2!} + n(n+1)(n-2)(n+3)\frac{x^4}{4!} - \cdots\right)}_{y_1(x)}$$

$$+ a_1 \underbrace{\left(x - (n-1)(n+2)\frac{x^3}{3!} + (n-1)(n+2)(n-3)(n+4)\frac{x^5}{5!} - \cdots\right)}_{y_2(x)}.$$

$$\tag{4.365}$$

Thus, the general solution takes the form

$$y(x) = a_0 y_1(x) + a_1 y_2(x), \tag{4.366}$$

with complementary functions $y_1(x)$ and $y_2(x)$ defined as

$$y_1(x) = 1 - n(n+1)\frac{x^2}{2!} + n(n+1)(n-2)(n+3)\frac{x^4}{4!} - \cdots, \tag{4.367}$$

$$y_2(x) = x - (n-1)(n+2)\frac{x^3}{3!} + (n-1)(n+2)(n-3)(n+4)\frac{x^5}{5!} - \cdots. \tag{4.368}$$

This solution holds for arbitrary real values of n. However, for $n = 0, 2, 4, \ldots, y_1(x)$ is a finite polynomial, while $y_2(x)$ is an infinite series that diverges at $|x| = 1$. For $n = 1, 3, 5, \ldots$, it is the other way around. Thus, for integer, nonnegative n, either (1) y_1 is a polynomial of degree n, and y_2 is a polynomial of infinite degree, or (2) y_1 is a polynomial of infinite degree, and y_2 is a polynomial of degree n.

We could in fact treat y_1 and y_2 as the complementary functions for Eq. (4.350). However, the existence of finite degree polynomials in special cases has led to an alternate definition of the standard complementary functions for Eq. (4.350). The finite polynomials (y_1 for even n and y_2 for odd n) can be normalized by dividing through by their values at $x = 1$ to give the *Legendre polynomials*, $P_n(x)$:

$$P_n(x) = \begin{cases} \dfrac{y_1(x)}{y_1(1)}, & n \text{ even}, \\[2ex] \dfrac{y_2(x)}{y_2(1)}, & n \text{ odd}. \end{cases} \tag{4.369}$$

The Legendre polynomials are thus

$$n = 0, \qquad P_0(x) = 1, \tag{4.370}$$

$$n = 1, \qquad P_1(x) = x, \tag{4.371}$$

$$n = 2, \qquad P_2(x) = \frac{1}{2}(3x^2 - 1), \tag{4.372}$$

$$n = 3, \qquad P_3(x) = \frac{1}{2}(5x^3 - 3x), \tag{4.373}$$

$$n = 4, \qquad P_4(x) = \frac{1}{8}(35x^4 - 30x^2 + 3), \tag{4.374}$$

$$\vdots$$

$$n, \qquad P_n(x) = \frac{1}{2^n n!} \frac{d^n}{dx^n}(x^2 - 1)^n, \qquad \text{Rodrigues formula.} \tag{4.375}$$

This gives a generating formula for general n, also known as a type of Rodrigues[13] formula.

The orthogonality condition is

$$\int_{-1}^{1} P_n(x) P_m(x) \, dx = \frac{2}{2n + 1} \delta_{nm}. \tag{4.376}$$

Direct substitution shows that $P_n(x)$ satisfies both the differential equation, Eq. (4.350), and the orthogonality condition. It is then easily shown that the following functions are orthonormal on the interval $x \in (-1, 1)$:

$$\varphi_n(x) = \sqrt{n + \frac{1}{2}} P_n(x), \tag{4.377}$$

giving

$$\int_{-1}^{1} \varphi_n(x) \varphi_m(x) \, dx = \delta_{nm}. \tag{4.378}$$

The total solution, Eq. (4.366), can be recast as the sum of the finite sum of polynomials $P_n(x)$ (Legendre functions of the first kind and degree n) and the infinite sum of polynomials $Q_n(x)$ (Legendre functions of the second kind and degree n):

$$y(x) = C_1 P_n(x) + C_2 Q_n(x). \tag{4.379}$$

Here $Q_n(x)$, the infinite series portion of the solution, is obtained by

$$Q_n(x) = \begin{cases} y_1(1)y_2(x), & n \text{ even,} \\ -y_2(1)y_1(x), & n \text{ odd.} \end{cases} \tag{4.380}$$

One can also show the Legendre functions of the second kind, $Q_n(x)$, satisfy a similar orthogonality condition. Additionally, $Q_n(\pm 1)$ is singular. One can further show that the infinite series of polynomials that form $Q_n(x)$ can be recast as a finite series of polynomials along with a logarithmic function. The first few values of $Q_n(x)$ are in fact

$$n = 0, \qquad Q_0(x) = \frac{1}{2} \ln\left(\frac{1+x}{1-x}\right), \tag{4.381}$$

$$n = 1, \qquad Q_1(x) = \frac{x}{2} \ln\left(\frac{1+x}{1-x}\right) - 1, \tag{4.382}$$

---

[13] Benjamin Olinde Rodrigues, 1794–1851, French mathematician.

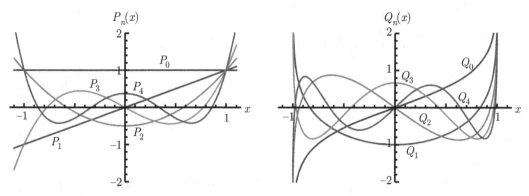

Figure 4.4. Solutions to the Legendre equation, Eq. (4.350), in terms of two sets of complementary functions, $P_n(x)$ and $Q_n(x)$.

$$n = 2, \qquad Q_2(x) = \frac{3x^2 - 1}{4} \ln\left(\frac{1+x}{1-x}\right) - \frac{3}{2}x, \qquad (4.383)$$

$$n = 3, \qquad Q_3(x) = \frac{5x^3 - 3x}{4} \ln\left(\frac{1+x}{1-x}\right) - \frac{5}{2}x^2 + \frac{2}{3}, \qquad (4.384)$$

$$\vdots$$

The first few eigenfunctions of Eq. (4.350) for the two families of complementary functions are plotted in Figure 4.4.

4.4.5 Chebyshev Equation

The Chebyshev[14] equation is

$$(1 - x^2)\frac{d^2y}{dx^2} - x\frac{dy}{dx} + \lambda y = 0. \qquad (4.385)$$

It is singular at $x = \pm 1$. Let us get this into Sturm-Liouville form:

$$a(x) = 1 - x^2, \qquad (4.386)$$

$$b(x) = -x, \qquad (4.387)$$

$$c(x) = 0. \qquad (4.388)$$

Now, taking $x_0 = -1$,

$$p(x) = \exp\left(\int_{-1}^{x} \frac{b(s)}{a(s)}\, ds\right), \qquad (4.389)$$

$$= \exp\left(\int_{-1}^{x} \frac{-s}{1 - s^2}\, ds\right), \qquad (4.390)$$

$$= \exp\left(\frac{1}{2}\ln(1 - s^2)\right)\Big|_{-1}^{x}, \qquad (4.391)$$

$$= \sqrt{1 - s^2}\,\Big|_{-1}^{x}, \qquad (4.392)$$

---

[14] Pafnuty Lvovich Chebyshev, 1821–1894, Russian mathematician.

$$= \sqrt{1 - x^2}, \tag{4.393}$$

$$r(x) = \frac{\exp\left(\int_{-1}^{x} \frac{b(s)}{a(s)} \, ds\right)}{a(x)} = \frac{1}{\sqrt{1 - x^2}}, \tag{4.394}$$

$$q(x) = 0. \tag{4.395}$$

Thus, for $p(x) > 0$, we require $x \in (-1, 1)$. The Chebyshev equation, Eq. (4.385), in Sturm-Liouville form is

$$\frac{d}{dx}\left(\sqrt{1 - x^2}\,\frac{dy}{dx}\right) + \frac{\lambda}{\sqrt{1 - x^2}}y = 0, \tag{4.396}$$

$$\underbrace{-\sqrt{1 - x^2}\,\frac{d}{dx}\left(\sqrt{1 - x^2}\,\frac{d}{dx}\right)}_{\mathbf{L}_s} y(x) = \lambda y(x). \tag{4.397}$$

Thus,

$$\mathbf{L}_s = -\sqrt{1 - x^2}\,\frac{d}{dx}\left(\sqrt{1 - x^2}\,\frac{d}{dx}\right). \tag{4.398}$$

That the two forms are equivalent can be easily checked by direct expansion.

Series solution techniques reveal for eigenvalues of λ one family of complementary functions of Eq. (4.385) can be written in terms of the so-called *Chebyshev polynomials*, $T_n(x)$. These are also known as Chebyshev polynomials of the first kind. These polynomials can be obtained by a regular series expansion of the original differential equation. These eigenvalues and eigenfunctions are listed next:

$$\lambda = 0, \qquad T_0(x) = 1, \tag{4.399}$$

$$\lambda = 1, \qquad T_1(x) = x, \tag{4.400}$$

$$\lambda = 4, \qquad T_2(x) = -1 + 2x^2, \tag{4.401}$$

$$\lambda = 9, \qquad T_3(x) = -3x + 4x^3, \tag{4.402}$$

$$\lambda = 16, \qquad T_4(x) = 1 - 8x^2 + 8x^4, \tag{4.403}$$

$$\vdots$$

$$\lambda = n^2, \qquad T_n(x) = \cos(n \cos^{-1} x), \qquad \text{Rodrigues formula.} \tag{4.404}$$

The orthogonality condition is

$$\int_{-1}^{1} \frac{T_n(x)T_m(x)}{\sqrt{1 - x^2}} \, dx = \begin{cases} \pi \delta_{nm}, & n = 0, \\ \frac{\pi}{2}\delta_{nm}, & n = 1, 2, \ldots. \end{cases} \tag{4.405}$$

Direct substitution shows that $T_n(x)$ satisfies both the differential equation, Eq. (4.385), and the orthogonality condition. We can deduce, then, that the functions

$$\varphi_n(x) = \begin{cases} \sqrt{\dfrac{1}{\pi\sqrt{1 - x^2}}}\, T_n(x), & n = 0, \\[4mm] \sqrt{\dfrac{2}{\pi\sqrt{1 - x^2}}}\, T_n(x), & n = 1, 2, \ldots \end{cases} \tag{4.406}$$

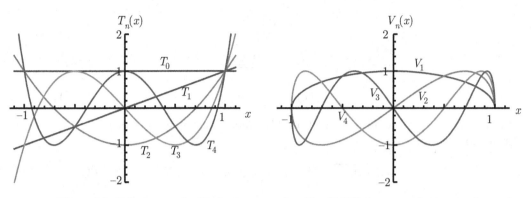

Figure 4.5. Solutions to the Chebyshev equation, Eq. (4.385), in terms of two sets of complementary functions, $T_n(x)$ and $V_n(x)$.

are an orthonormal set of functions on the interval $x \in (-1, 1)$; that is,

$$\int_{-1}^{1} \varphi_n(x)\varphi_m(x) \, dx = \delta_{nm}. \tag{4.407}$$

The Chebyshev polynomials of the first kind, $T_n(x)$, form one set of complementary functions which satisfy Eq. (4.385). The other set of complementary functions are $V_n(x)$ and can be shown to be

$$\lambda = 0, \quad V_0(x) = 0, \tag{4.408}$$

$$\lambda = 1, \quad V_1(x) = \sqrt{1 - x^2}, \tag{4.409}$$

$$\lambda = 4, \quad V_2(x) = \sqrt{1 - x^2}(2x), \tag{4.410}$$

$$\lambda = 9, \quad V_3(x) = \sqrt{1 - x^2}(-1 + 4x^2), \tag{4.411}$$

$$\lambda = 16, \quad V_4(x) = \sqrt{1 - x^2}(-4x^2 + 8x^3), \tag{4.412}$$

$$\vdots$$

$$\lambda = n^2, \quad V_n(x) = \sin(n \cos^{-1} x), \qquad \text{Rodrigues formula.} \tag{4.413}$$

The general solution to Eq. (4.503) is a linear combination of the two complementary functions:

$$y(x) = C_1 T_n(x) + C_2 V_n(x). \tag{4.414}$$

One can also show that $V_n(x)$ satisfies an orthogonality condition:

$$\int_{-1}^{1} \frac{V_n(x)V_m(x)}{\sqrt{1 - x^2}} \, dx = \frac{\pi}{2}\delta_{nm}. \tag{4.415}$$

The first few eigenfunctions of Eq. (4.385) for the two families of complementary functions are plotted in Figure 4.5.

4.4.6 Hermite Equation

The Hermite[15] equation is discussed next. There are two common formulations, the physicists' and the probabilists'. We focus on the first and briefly discuss the second.

Physicists'

The physicists' Hermite equation is

$$\frac{d^2y}{dx^2} - 2x\frac{dy}{dx} + \lambda y = 0. \tag{4.416}$$

We find that

$$p(x) = e^{-x^2}, \tag{4.417}$$

$$r(x) = e^{-x^2}, \tag{4.418}$$

$$q(x) = 0. \tag{4.419}$$

Thus, we allow $x \in (-\infty, \infty)$. In Sturm-Liouville form, Eq. (4.416) becomes

$$\frac{d}{dx}\left(e^{-x^2}\frac{dy}{dx}\right) + \lambda e^{-x^2} y = 0, \tag{4.420}$$

$$\underbrace{-e^{x^2}\frac{d}{dx}\left(e^{-x^2}\frac{d}{dx}\right)}_{\mathbf{L}_s} y(x) = \lambda y(x). \tag{4.421}$$

So

$$\mathbf{L}_s = -e^{x^2}\frac{d}{dx}\left(e^{-x^2}\frac{d}{dx}\right). \tag{4.422}$$

One set of complementary functions can be expressed in terms of polynomials known as the *Hermite polynomials*, $H_n(x)$. These polynomials can be obtained by a regular series expansion of the original differential equation. The eigenvalues and eigenfunctions corresponding to the physicists' Hermite polynomials are listed next:

$$\lambda = 0, \qquad H_0(x) = 1, \tag{4.423}$$

$$\lambda = 2, \qquad H_1(x) = 2x, \tag{4.424}$$

$$\lambda = 4, \qquad H_2(x) = -2 + 4x^2, \tag{4.425}$$

$$\lambda = 6, \qquad H_3(x) = -12x + 8x^3, \tag{4.426}$$

$$\lambda = 8, \qquad H_4(x) = 12 - 48x^2 + 16x^4, \tag{4.427}$$

$$\vdots$$

$$\lambda = 2n, \qquad H_n(x) = (-1)^n e^{x^2}\frac{d^n e^{-x^2}}{dx^n}, \qquad \text{Rodrigues formula.} \tag{4.428}$$

The orthogonality condition is

$$\int_{-\infty}^{\infty} e^{-x^2} H_n(x)H_m(x)\,dx = 2^n n!\sqrt{\pi}\delta_{nm}. \tag{4.429}$$

---

[15] Charles Hermite, 1822–1901, French mathematician.

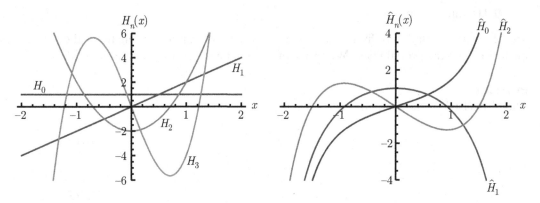

Figure 4.6. Solutions to the physicists' Hermite equation, Eq. (4.416), in terms of two sets of complementary functions $H_n(x)$ and $\hat{H}_n(x)$.

Direct substitution shows that $H_n(x)$ satisfies both the differential equation, Eq. (4.416), and the orthogonality condition. It is then easily shown that the following functions are orthonormal on the interval $x \in (-\infty, \infty)$:

$$\varphi_n(x) = \frac{e^{-x^2/2} H_n(x)}{\sqrt{\sqrt{\pi} 2^n n!}}, \tag{4.430}$$

giving

$$\int_{-\infty}^{\infty} \varphi_n(x) \varphi_m(x) \, dx = \delta_{mn}. \tag{4.431}$$

The general solution to Eq. (4.416) is

$$y(x) = C_1 H_n(x) + C_2 \hat{H}_n(x), \tag{4.432}$$

where the other set of complementary functions is $\hat{H}_n(x)$. For general n, $\hat{H}_n(x)$ is a version of the so-called Kummer[16] confluent hypergeometric function of the first kind, $\hat{H}_n(x) = {}_1F_1\left(-n/2; 1/2; x^2\right)$ (see Section A.7.4). This general solution should be treated carefully, especially as the second complementary function, $\hat{H}_n(x)$, is rarely discussed in the literature and notation is often nonstandard. For our eigenvalues of n, somewhat simpler results can be obtained in terms of the *imaginary error function*, $\mathrm{erfi}(x)$; see Section A.7.2. The first few of these functions are

$$\lambda = 0, \quad n = 0, \qquad \hat{H}_0(x) = \frac{\sqrt{\pi}}{2} \mathrm{erfi}(x), \tag{4.433}$$

$$\lambda = 2, \quad n = 1, \qquad \hat{H}_1(x) = e^{x^2} - \sqrt{\pi x^2}\, \mathrm{erfi}(\sqrt{x^2}), \tag{4.434}$$

$$\lambda = 4, \quad n = 2, \qquad \hat{H}_2(x) = -x e^{x^2} + \sqrt{\pi}\, \mathrm{erfi}(x) \left(x^2 - \frac{1}{2} \right), \tag{4.435}$$

$$\lambda = 6, \quad n = 3, \qquad \hat{H}_3(x) = e^{x^2}\left(1 - x^2\right) + \sqrt{\pi x^2}\, \mathrm{erfi}(x) \left(x^2 - \frac{3}{2} \right). \tag{4.436}$$

The first few eigenfunctions of the Hermite equation, Eq. (4.416), for the two families of complementary functions are plotted in Figure 4.6.

---

[16] Ernst Kummer, 1810–1893, German mathematician.

Probabilists'

The probabilists' Hermite equation is

$$\frac{d^2y}{dx^2} - x\frac{dy}{dx} + \lambda y = 0. \tag{4.437}$$

We find that

$$p(x) = e^{-x^2/2}, \tag{4.438}$$

$$r(x) = e^{-x^2/2}, \tag{4.439}$$

$$q(x) = 0. \tag{4.440}$$

Thus, we allow $x \in (-\infty, \infty)$. In Sturm-Liouville form, Eq. (4.437) becomes

$$\frac{d}{dx}\left(e^{-x^2/2}\frac{dy}{dx}\right) + \lambda e^{-x^2/2}y = 0, \tag{4.441}$$

$$\underbrace{-e^{x^2/2}\frac{d}{dx}\left(e^{-x^2/2}\frac{d}{dx}\right)}_{\mathbf{L}_s}y(x) = \lambda y(x), \tag{4.442}$$

So

$$\mathbf{L}_s = -e^{x^2/2}\frac{d}{dx}\left(e^{-x^2/2}\frac{d}{dx}\right). \tag{4.443}$$

One set of complementary functions can be expressed in terms of polynomials known as the probabilists' Hermite polynomials, $He_n(x)$. These polynomials can be obtained by a regular series expansion of the original differential equation. The eigenvalues and eigenfunctions corresponding to the probabilists' Hermite polynomials are listed next:

$$\lambda = 0, \quad He_0(x) = 1, \tag{4.444}$$

$$\lambda = 1, \quad He_1(x) = x, \tag{4.445}$$

$$\lambda = 2, \quad He_2(x) = -1 + x^2, \tag{4.446}$$

$$\lambda = 3, \quad He_3(x) = -3x + x^3, \tag{4.447}$$

$$\lambda = 4, \quad He_4(x) = 3 - 6x^2 + x^4, \tag{4.448}$$

$$\vdots$$

$$\lambda = n, \quad He_n(x) = (-1)^n e^{x^2/2}\frac{d^n e^{-x^2/2}}{dx^n}, \quad \text{Rodrigues formula.} \tag{4.449}$$

The orthogonality condition is

$$\int_{-\infty}^{\infty} e^{-x^2/2}He_n(x)He_m(x)\, dx = n!\sqrt{2\pi}\delta_{nm}. \tag{4.450}$$

Direct substitution shows that $He_n(x)$ satisfies both the differential equation, Eq. (4.437), and the orthogonality condition. It is then easily shown that the following functions are orthonormal on the interval $x \in (-\infty, \infty)$:

$$\varphi_n(x) = \frac{e^{-x^2/4}He_n(x)}{\sqrt{\sqrt{2\pi}n!}}, \tag{4.451}$$

giving

$$\int_{-\infty}^{\infty} \varphi_n(x)\varphi_m(x)\, dx = \delta_{mn}. \tag{4.452}$$

Plots and the second set of complementary functions for the probabilists' Hermite equation are obtained in a similar manner to those for the physicists'. One can easily show the relation between the two to be

$$He_n(x) = 2^{-n/2} H_n\left(\frac{x}{\sqrt{2}}\right). \tag{4.453}$$

4.4.7 Laguerre Equation

The Laguerre[17] equation is

$$x\frac{d^2y}{dx^2} + (1-x)\frac{dy}{dx} + \lambda y = 0. \tag{4.454}$$

It is singular when $x = 0$.

We find that

$$p(x) = xe^{-x}, \tag{4.455}$$

$$r(x) = e^{-x}, \tag{4.456}$$

$$q(x) = 0. \tag{4.457}$$

Thus, we require $x \in (0, \infty)$.

In Sturm-Liouville form, Eq. (4.454) becomes

$$\frac{d}{dx}\left(xe^{-x}\frac{dy}{dx}\right) + \lambda e^{-x}y = 0, \tag{4.458}$$

$$\underbrace{-e^x\frac{d}{dx}\left(xe^{-x}\frac{d}{dx}\right)}_{\mathbf{L}_s}y(x) = \lambda y(x). \tag{4.459}$$

So

$$\mathbf{L}_s = -e^x\frac{d}{dx}\left(xe^{-x}\frac{d}{dx}\right). \tag{4.460}$$

One set of the complementary functions can be expressed in terms of polynomials of finite order known as the *Laguerre polynomials*, $L_n(x)$. These polynomials can be obtained by a regular series expansion of Eq. (4.454). Eigenvalues and eigenfunctions corresponding to the Laguerre polynomials are listed next:

$$\lambda = 0, \qquad L_0(x) = 1, \tag{4.461}$$

$$\lambda = 1, \qquad L_1(x) = 1 - x, \tag{4.462}$$

$$\lambda = 2, \qquad L_2(x) = 1 - 2x + \frac{1}{2}x^2, \tag{4.463}$$

$$\lambda = 3, \qquad L_3(x) = 1 - 3x + \frac{3}{2}x^2 - \frac{1}{6}x^3, \tag{4.464}$$

---

[17] Edmond Nicolas Laguerre, 1834–1886, French mathematician.

$$\lambda = 4, \qquad L_4(x) = 1 - 4x + 3x^2 - \frac{2}{3}x^3 + \frac{1}{24}x^4, \tag{4.465}$$

$$\vdots$$

$$\lambda = n, \qquad L_n(x) = \frac{1}{n!}e^x \frac{d^n \left(x^n e^{-x} \right)}{dx^n}, \qquad \text{Rodrigues formula.} \tag{4.466}$$

The orthogonality condition reduces to

$$\int_0^\infty e^{-x} L_n(x) L_m(x) \, dx = \delta_{nm}. \tag{4.467}$$

Direct substitution shows that $L_n(x)$ satisfies both the differential equation, Eq. (4.454), and the orthogonality condition. It is then easily shown that the following functions are orthonormal on the interval $x \in (0, \infty)$:

$$\varphi_n(x) = e^{-x/2} L_n(x), \tag{4.468}$$

so that

$$\int_0^\infty \varphi_n(x) \varphi_m(x) \, dx = \delta_{mn}. \tag{4.469}$$

The general solution to Eq. (4.454) is

$$y(x) = C_1 L_n(x) + C_2 \hat{L}_n(x), \tag{4.470}$$

where the other set of complementary functions is $\hat{L}_n(x)$. For general n, $\hat{L}_n(x) = U(-n, 1, x)$, one of the so-called Tricomi[18] confluent hypergeometric functions. Again the literature is not extensive on these functions. For integer eigenvalues n, $\hat{L}_n(x)$ reduces somewhat and can be expressed in terms of the exponential integral function, $\text{Ei}(x)$; see Section A.7.3. The first few of these functions are

$$\lambda = n = 0, \qquad \hat{L}_0(x) = \text{Ei}(x), \tag{4.471}$$

$$\lambda = n = 1, \qquad \hat{L}_1(x) = -e^x - \text{Ei}(x)(1 - x), \tag{4.472}$$

$$\lambda = n = 2, \qquad \hat{L}_2(x) = \frac{1}{4} \left(e^x(3 - x) + \text{Ei}(x) \left(2 - 4x + x^2 \right) \right), \tag{4.473}$$

$$\lambda = n = 3, \qquad \hat{L}_3(x) = \frac{1}{36} \left(e^x \left(-11 + 8x - x^2 \right) + \text{Ei}(x) \left(-6 + 18x - 9x^2 + x^3 \right) \right). \tag{4.474}$$

The first few eigenfunctions of the Laguerre equation, Eq. (4.454), for the two families of complementary functions are plotted in Figure 4.7.

4.4.8 Bessel Differential Equation

One of the more important and common equations of engineering is known as the *Bessel[19] differential equation*. There are actually many types of such equations, and we shall survey them here.

---

[18] Francesco Giacomo Tricomi, 1897–1978, Italian mathematician.
[19] Friedrich Wilhelm Bessel, 1784–1846, Westphalia-born German mathematician.

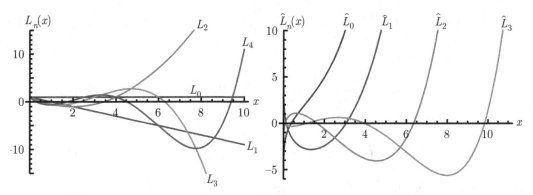

Figure 4.7. Solutions to the Laguerre equation, Eq. (4.454), in terms of two sets of complementary functions, $L_n(x)$ and $\hat{L}_n(x)$.

First and Second Kind

The Bessel differential equation is as follows, with it being convenient to define $\lambda = -\nu^2$:

$$x^2 \frac{d^2y}{dx^2} + x\frac{dy}{dx} + (\mu^2 x^2 - \nu^2)y = 0. \tag{4.475}$$

It is singular when $x = 0$. We find that

$$p(x) = x, \tag{4.476}$$

$$r(x) = \frac{1}{x}, \tag{4.477}$$

$$q(x) = \mu^2 x. \tag{4.478}$$

We thus require $x \in (0, \infty)$, though in practice, it is more common to employ a finite domain such as $x \in (0, \ell)$. In Sturm-Liouville form, we have

$$\frac{d}{dx}\left(x\frac{dy}{dx}\right) + \left(\mu^2 x - \frac{\nu^2}{x}\right)y = 0, \tag{4.479}$$

$$\underbrace{-\left(x\left(\frac{d}{dx}\left(x\frac{d}{dx}\right) + \mu^2 x\right)\right)}_{\mathbf{L}_s}y(x) = -\nu^2 y(x). \tag{4.480}$$

The Sturm-Liouville operator is

$$\mathbf{L}_s = -x\left(\frac{d}{dx}\left(x\frac{d}{dx}\right) + \mu^2 x\right). \tag{4.481}$$

In some other cases, it is more convenient to take $\lambda = \mu^2$, in which case we get

$$p(x) = x, \tag{4.482}$$

$$r(x) = x, \tag{4.483}$$

$$q(x) = -\frac{\nu^2}{x}, \tag{4.484}$$

and the Sturm-Liouville form and operator are

$$\underbrace{-\left(\frac{1}{x}\left(\frac{d}{dx}\left(x\frac{d}{dx}\right) - \frac{\nu^2}{x}\right)\right)}_{\mathbf{L}_s} y(x) = \mu^2 y(x), \tag{4.485}$$

$$\mathbf{L}_s = -\frac{1}{x}\left(\frac{d}{dx}\left(x\frac{d}{dx}\right) - \frac{\nu^2}{x}\right). \tag{4.486}$$

The general solution is

$$y(x) = C_1 J_\nu(\mu x) + C_2 Y_\nu(\mu x), \qquad \text{if } \nu \text{ is an integer,} \tag{4.487}$$

$$y(x) = C_1 J_\nu(\mu x) + C_2 J_{-\nu}(\mu x), \qquad \text{if } \nu \text{ is not an integer,} \tag{4.488}$$

where $J_\nu(\mu x)$ and $Y_\nu(\mu x)$ are called the Bessel and Neumann functions of order ν. Often $J_\nu(\mu x)$ is known as a *Bessel function of the first kind* and $Y_\nu(\mu x)$ is known as a *Bessel function of the second kind*. Both J_ν and Y_ν are represented by infinite series rather than finite series such as the series for Legendre polynomials.

The Bessel function of the first kind of order ν, $J_\nu(\mu x)$, is represented by

$$J_\nu(\mu x) = \left(\frac{1}{2}\mu x\right)^\nu \sum_{k=0}^\infty \frac{\left(-\frac{1}{4}\mu^2 x^2\right)^k}{k!\,\Gamma(\nu + k + 1)}. \tag{4.489}$$

The Neumann function $Y_\nu(\mu x)$ has a complicated series representation. The representations for $J_0(\mu x)$ and $Y_0(\mu x)$ are

$$J_0(\mu x) = 1 - \frac{\left(\frac{1}{4}\mu^2 x^2\right)^1}{(1!)^2} + \frac{\left(\frac{1}{4}\mu^2 x^2\right)^2}{(2!)^2} + \cdots + \frac{\left(-\frac{1}{4}\mu^2 x^2\right)^n}{(n!)^2}, \tag{4.490}$$

$$Y_0(\mu x) = \frac{2}{\pi}\left(\ln\left(\frac{1}{2}\mu x\right) + \gamma\right) J_0(\mu x)$$

$$+ \frac{2}{\pi}\left(\frac{\left(\frac{1}{4}\mu^2 x^2\right)^1}{(1!)^2} - \left(1 + \frac{1}{2}\right)\frac{\left(\frac{1}{4}\mu^2 x^2\right)^2}{(2!)^2}\cdots\right). \tag{4.491}$$

It can be shown using term-by-term differentiation that

$$\frac{dJ_\nu(\mu x)}{dx} = \mu\frac{J_{\nu-1}(\mu x) - J_{\nu+1}(\mu x)}{2}, \qquad \frac{dY_\nu(\mu x)}{dx} = \mu\frac{Y_{\nu-1}(\mu x) - Y_{\nu+1}(\mu x)}{2},$$

$$\tag{4.492}$$

$$\frac{d}{dx}\left(x^\nu J_\nu(\mu x)\right) = \mu x^\nu J_{\nu-1}(\mu x), \qquad \frac{d}{dx}\left(x^\nu Y_\nu(\mu x)\right) = \mu x^\nu Y_{\nu-1}(\mu x). \tag{4.493}$$

The Bessel functions $J_0(\mu_0 x)$, $J_0(\mu_1 x)$, $J_0(\mu_2 x)$, $J_0(\mu_3 x)$ are plotted in Figure 4.8. Here the eigenvalues μ_n can be determined from trial and error. The first four are found to be $\mu_0 = 2.40483$, $\mu_1 = 5.52008$, $\mu_2 = 8.65373$, and $\mu_3 = 11.7915$. In general, one can say

$$\lim_{n\to\infty} \mu_n = n\pi + O(1). \tag{4.494}$$

The Bessel functions $J_0(x), J_1(x), J_2(x), J_3(x)$, and $J_4(x)$ along with the Neumann functions $Y_0(x), Y_1(x), Y_2(x), Y_3(x)$, and $Y_4(x)$ are plotted in Figure 4.9 (so here $\mu = 1$).

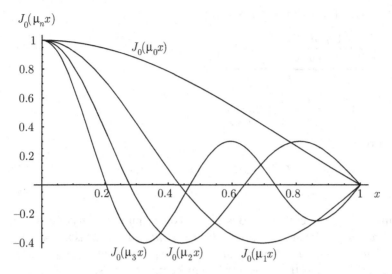

Figure 4.8. Bessel functions $J_0(\mu_0 x)$, $J_0(\mu_1 x)$, $J_0(\mu_2 x)$, $J_0(\mu_3 x)$.

The orthogonality condition for a domain $x \in [0, 1]$, taken here for the case in which the eigenvalue is μ_n, can be shown to be

$$\int_0^1 x J_\nu(\mu_n x) J_\nu(\mu_m x)\, dx = \frac{1}{2} \left(J_{\nu+1}(\mu_n)\right)^2 \delta_{nm}. \tag{4.495}$$

Here we must choose μ_n such that $J_\nu(\mu_n) = 0$, which corresponds to a vanishing of the function at the outer limit $x = 1$. So the orthonormal Bessel function is

$$\varphi_n(x) = \frac{\sqrt{2x} J_\nu(\mu_n x)}{|J_{\nu+1}(\mu_n)|}. \tag{4.496}$$

Third Kind

Hankel[20] functions, also known as *Bessel functions of the third kind*, are defined by

$$H_\nu^{(1)}(x) = J_\nu(x) + i Y_\nu(x), \tag{4.497}$$

$$H_\nu^{(2)}(x) = J_\nu(x) - i Y_\nu(x). \tag{4.498}$$

Modified Bessel Functions

The modified Bessel equation is

$$x^2 \frac{d^2 y}{dx^2} + x \frac{dy}{dx} - (x^2 + \nu^2)y = 0, \tag{4.499}$$

the solutions of which are the modified Bessel functions. The *modified Bessel function of the first kind of order* ν is

$$I_\nu(x) = i^{-\nu} J_\nu(ix). \tag{4.500}$$

The *modified Bessel function of the second kind of order* ν is

$$K_\nu(x) = \frac{\pi}{2} i^{\nu+1} H_n^{(1)}(ix). \tag{4.501}$$

---

[20] Hermann Hankel, 1839–1873, German mathematician.

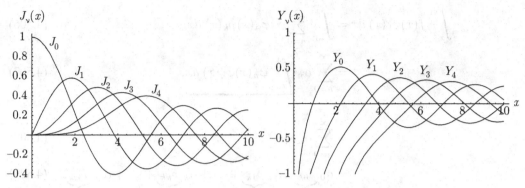

Figure 4.9. Bessel functions $J_0(x)$, $J_1(x)$, $J_2(x)$, $J_3(x)$, $J_4(x)$ and Neumann functions $Y_0(x)$, $Y_1(x)$, $Y_2(x)$, $Y_3(x)$, $Y_4(x)$.

Ber and Bei Functions

The real and imaginary parts of the solutions of

$$x^2 \frac{d^2 y}{dx^2} + x \frac{dy}{dx} - (p^2 + ix^2)y = 0, \tag{4.502}$$

where p is a real constant, are called the *ber* and *bei functions*. They are also known as *Kelvin*[21] *functions*.

4.5 Fourier Series Representation

It is often useful, especially when solving partial differential equations, to be able to represent an arbitrary function $f(x)$ in the domain $x \in [x_0, x_1]$ with an appropriately weighted sum of orthonormal functions $\varphi_n(x)$:

$$f(x) = \sum_{n=0}^{\infty} \alpha_n \varphi_n(x). \tag{4.503}$$

We generally truncate the infinite series to a finite number of N terms so that $f(x)$ is approximated by

$$f(x) \approx \sum_{n=0}^{N} \alpha_n \varphi_n(x). \tag{4.504}$$

We can better label an N-term approximation of a function as a *projection* of the function from an infinite-dimensional function space onto an N-dimensional function space. This is discussed further in Section 6.5. The projection is useful only if the infinite series converges so that the error incurred in neglecting terms past N is small relative to the terms included.

The problem is to determine what the coefficients α_n must be. They can be found in the following manner. We first assume the expansion exists, multiply both sides by $\varphi_k(x)$, and integrate from x_0 to x_1, analyzing as follows:

$$f(x)\varphi_k(x) = \sum_{n=0}^{\infty} \alpha_n \varphi_n(x)\varphi_k(x), \tag{4.505}$$

---

[21] William Thomson (Lord Kelvin), 1824–1907, Irish-born British mathematical physicist.

$$\int_{x_0}^{x_1} f(x)\varphi_k(x)\,dx = \int_{x_0}^{x_1} \sum_{n=0}^{\infty} \alpha_n \varphi_n(x)\varphi_k(x)\,dx, \tag{4.506}$$

$$= \sum_{n=0}^{\infty} \alpha_n \underbrace{\int_{x_0}^{x_1} \varphi_n(x)\varphi_k(x)\,dx}_{\delta_{nk}}, \tag{4.507}$$

$$= \sum_{n=0}^{\infty} \alpha_n \delta_{nk}, \tag{4.508}$$

$$= \alpha_0 \underbrace{\delta_{0k}}_{=0} + \alpha_1 \underbrace{\delta_{1k}}_{=0} + \cdots + \alpha_k \underbrace{\delta_{kk}}_{=1} + \cdots + \alpha_\infty \underbrace{\delta_{\infty k}}_{=0}, \tag{4.509}$$

$$= \alpha_k. \tag{4.510}$$

So, trading k and n,

$$\alpha_n = \int_{x_0}^{x_1} f(x)\varphi_n(x)\,dx. \tag{4.511}$$

The series of Eq. (4.503) with coefficients given by Eq. (4.511) is known as a *Fourier series*. Depending on the expansion functions, the series is often specialized as Fourier-sine, Fourier-cosine, Fourier-Legendre, Fourier-Bessel, and so on. We have inverted Eq. (4.503) to solve for the unknown α_n. The inversion was aided greatly by the fact that the basis functions were orthonormal. For nonorthonormal as well as nonorthogonal bases, more general techniques exist for the determination of α_n; these are given in detail in the upcoming Section 6.5.

EXAMPLE 4.19

Represent

$$f(x) = x^2, \qquad \text{on} \qquad x \in [0, 3], \tag{4.512}$$

with a finite series of

- trigonometric functions,
- Legendre polynomials,
- Chebyshev polynomials,
- Bessel functions.

Trigonometric Series

For the trigonometric series, let us try a five-term Fourier-sine series. The orthonormal functions in this case are, from Eq. (4.347),

$$\varphi_n(x) = \sqrt{\frac{2}{3}} \sin\left(\frac{n\pi x}{3}\right). \tag{4.513}$$

The coefficients from Eq. (4.511) are thus

$$\alpha_n = \int_0^3 \underbrace{x^2}_{f(x)} \underbrace{\left(\sqrt{\frac{2}{3}} \sin\left(\frac{n\pi x}{3}\right)\right)}_{\varphi_n(x)}\,dx, \tag{4.514}$$

so the first five terms are

$$\alpha_0 = 0, \tag{4.515}$$

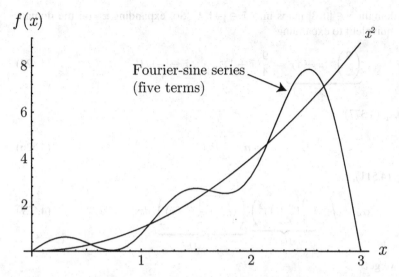

Figure 4.10. Five-term Fourier-sine series approximation to $f(x) = x^2$.

$$\alpha_1 = 4.17328, \tag{4.516}$$

$$\alpha_2 = -3.50864, \tag{4.517}$$

$$\alpha_3 = 2.23376, \tag{4.518}$$

$$\alpha_4 = -1.75432, \tag{4.519}$$

$$\alpha_5 = 1.3807. \tag{4.520}$$

The magnitude of the coefficient on the orthonormal function, α_n, decreases as n increases. From this, one can loosely infer that the higher frequency modes contain less "energy." While one can find analogies between this mathematical construct of energy and more physical concepts, it should be recognized that it is not a traditional energy from engineering or physics:

$$f(x) = \sqrt{\frac{2}{3}} \left(4.17328 \sin\left(\frac{\pi x}{3}\right) - 3.50864 \sin\left(\frac{2\pi x}{3}\right) \right. \tag{4.521}$$

$$\left. + 2.23376 \sin\left(\frac{3\pi x}{3}\right) - 1.75432 \sin\left(\frac{4\pi x}{3}\right) + 1.3807 \sin\left(\frac{5\pi x}{3}\right) + \cdots \right). \tag{4.522}$$

The function $f(x) = x^2$ and five terms of the approximation are plotted in Figure 4.10.

Legendre Polynomials

Next, let us try the Legendre polynomials. The Legendre polynomials are orthogonal on $x \in [-1, 1]$, but we have $x \in [0, 3]$, so let us define

$$\tilde{x} = \frac{2}{3}x - 1, \tag{4.523}$$

$$x = \frac{3}{2}(\tilde{x} + 1), \tag{4.524}$$

so that the domain $x \in [0, 3]$ maps into $\tilde{x} \in [-1, 1]$. So, expanding x^2 on the domain $x \in [0, 3]$ is equivalent to expanding

$$\underbrace{\left(\frac{3}{2}\right)^2 (\tilde{x} + 1)^2}_{x^2} = \frac{9}{4}(\tilde{x} + 1)^2, \qquad \tilde{x} \in [-1, 1]. \tag{4.525}$$

Now, from Eq. (4.377),

$$\varphi_n(\tilde{x}) = \sqrt{n + \frac{1}{2}} P_n(\tilde{x}). \tag{4.526}$$

So, from Eq. (4.511),

$$\alpha_n = \int_{-1}^{1} \underbrace{\left(\frac{9}{4}(\tilde{x} + 1)^2\right)}_{f(\tilde{x})} \underbrace{\left(\sqrt{n + \frac{1}{2}} P_n(\tilde{x})\right)}_{\varphi_n(\tilde{x})} d\tilde{x}. \tag{4.527}$$

Evaluating, we get

$$\alpha_0 = 3\sqrt{2} = 4.24264, \tag{4.528}$$

$$\alpha_1 = 3\sqrt{\frac{3}{2}} = 3.67423, \tag{4.529}$$

$$\alpha_2 = \frac{3}{\sqrt{10}} = 0.948683, \tag{4.530}$$

$$\alpha_3 = 0, \tag{4.531}$$

$$\vdots$$

$$\alpha_n = 0, \qquad n \geq 3. \tag{4.532}$$

Once again, the fact that $\alpha_0 > \alpha_1 > \alpha_2$ indicates that the bulk of the "energy" is contained in the lower frequency modes. Interestingly for $n \geq 3$, $\alpha_n = 0$; thus the series that gives an exact representation is finite. Carrying out the multiplication and returning to x-space gives the series, which can be expressed in a variety of forms:

$$x^2 = \alpha_0 \varphi_0(\tilde{x}) + \alpha_1 \varphi_1(\tilde{x}) + \alpha_2 \varphi_2(\tilde{x}), \tag{4.533}$$

$$= 3\sqrt{2} \underbrace{\left(\sqrt{\frac{1}{2}} P_0\left(\frac{2}{3}x - 1\right)\right)}_{=\varphi_0(\tilde{x})} + 3\sqrt{\frac{3}{2}} \underbrace{\left(\sqrt{\frac{3}{2}} P_1\left(\frac{2}{3}x - 1\right)\right)}_{=\varphi_1(\tilde{x})}$$

$$+ \frac{3}{\sqrt{10}} \underbrace{\left(\sqrt{\frac{5}{2}} P_2\left(\frac{2}{3}x - 1\right)\right)}_{=\varphi_2(\tilde{x})}, \tag{4.534}$$

$$= 3 P_0\left(\frac{2}{3}x - 1\right) + \frac{9}{2} P_1\left(\frac{2}{3}x - 1\right) + \frac{3}{2} P_2\left(\frac{2}{3}x - 1\right), \tag{4.535}$$

$$= 3(1) + \frac{9}{2}\left(\frac{2}{3}x - 1\right) + \frac{3}{2}\left(-\frac{1}{2} + \frac{3}{2}\left(\frac{2}{3}x - 1\right)^2\right), \tag{4.536}$$

$$= 3 + \left(-\frac{9}{2} + 3x\right) + \left(\frac{3}{2} - 3x + x^2\right), \tag{4.537}$$

$$= x^2. \tag{4.538}$$

Thus, the Fourier-Legendre representation is *exact* over the entire domain. This is because the function that is being expanded has the same general functional form as the Legendre polynomials; both are polynomials.

Chebyshev Polynomials

Let us now try the Chebyshev polynomials. These are orthogonal on the same domain as the Legendre polynomials, so let us use the same transformation as before. Now, from Eq. (4.406),

$$\varphi_0(\tilde{x}) = \sqrt{\frac{1}{\pi\sqrt{1 - \tilde{x}^2}}} T_0(\tilde{x}), \tag{4.539}$$

$$\varphi_n(\tilde{x}) = \sqrt{\frac{2}{\pi\sqrt{1 - \tilde{x}^2}}} T_n(\tilde{x}), \qquad n > 0. \tag{4.540}$$

So

$$\alpha_0 = \int_{-1}^{1} \underbrace{\frac{9}{4}(\tilde{x}+1)^2}_{f(\tilde{x})} \underbrace{\sqrt{\frac{1}{\pi\sqrt{1 - \tilde{x}^2}}} T_0(\tilde{x})}_{\varphi_0(\tilde{x})} \ d\tilde{x} \tag{4.541}$$

$$\alpha_n = \int_{-1}^{1} \underbrace{\frac{9}{4}(\tilde{x}+1)^2}_{f(\tilde{x})} \underbrace{\sqrt{\frac{2}{\pi\sqrt{1 - \tilde{x}^2}}} T_n(\tilde{x})}_{\varphi_n(\tilde{x})} \ d\tilde{x} \tag{4.542}$$

Evaluating, we get

$$\alpha_0 = 4.2587, \tag{4.543}$$

$$\alpha_1 = 3.4415, \tag{4.544}$$

$$\alpha_2 = -0.28679, \tag{4.545}$$

$$\alpha_3 = -1.1472, \tag{4.546}$$

$$\vdots$$

With this representation, we see that $|\alpha_3| > |\alpha_2|$, so it is not yet clear that the "energy" is concentrated in the high frequency modes. Consideration of more terms would verify that in fact it is the case that the "energy" of higher modes is decaying; in fact, $\alpha_4 = -0.683$, $\alpha_5 = -0.441$, $\alpha_6 = -0.328$, $\alpha_7 = -0.254$. So

$$f(x) = x^2 = \sqrt{\frac{2}{\pi\sqrt{1 - \left(\frac{2}{3}x - 1\right)^2}}} \left(\frac{4.2587}{\sqrt{2}} T_0\left(\frac{2}{3}x - 1\right) + 3.4415 \, T_1\left(\frac{2}{3}x - 1\right) \right.$$

$$\tag{4.547}$$

$$\left. -0.28679 \, T_2\left(\frac{2}{3}x - 1\right) - 1.1472 \, T_3\left(\frac{2}{3}x - 1\right) + \cdots \right). \tag{4.548}$$

The function $f(x) = x^2$ and four terms of the approximation are plotted in Figure 4.11.

This expansion based on the orthonormalized Chebyshev polynomials yielded an infinite series. In fact, because the function x^2 is itself a polynomial, it can easily be shown that it has a finite series representation in terms of the original Chebyshev polynomials, that being

$$f(x) = x^2 = \frac{27}{8} T_0\left(\frac{2}{3}x - 1\right) + \frac{9}{2} T_1\left(\frac{2}{3}x - 1\right) + \frac{9}{8} T_2\left(\frac{2}{3}x - 1\right), \tag{4.549}$$

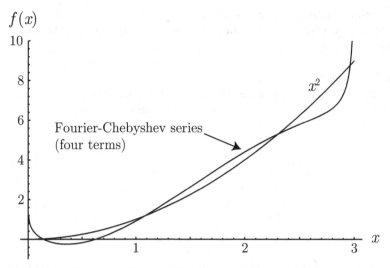

Figure 4.11. Four-term Fourier-Chebyshev series approximation to $f(x) = x^2$.

$$= \frac{27}{8}(1) + \frac{9}{2}\left(\frac{2}{3}x - 1\right) + \frac{9}{8}\left(-1 + 2\left(\frac{2}{3}x - 1\right)\right)^2, \tag{4.550}$$

$$= x^2. \tag{4.551}$$

Bessel Functions

Now let us expand in terms of Bessel functions. The Bessel functions have been defined such that they are orthogonal on a domain between zero and unity when the eigenvalues are the zeros of the Bessel function. To achieve this, we adopt the transformation (and inverse)

$$\tilde{x} = \frac{x}{3}, \qquad x = 3\tilde{x}. \tag{4.552}$$

With this, our domain transforms as follows:

$$x \in [0, 3] \longrightarrow \tilde{x} \in [0, 1]. \tag{4.553}$$

Also, $x^2 = 9\tilde{x}^2$. So in the transformed space, we seek an expansion

$$\underbrace{9\tilde{x}^2}_{f(\tilde{x})} = \sum_{n=0}^{\infty} \alpha_n J_\nu(\mu_n \tilde{x}). \tag{4.554}$$

Let us choose to expand on J_0, so we take

$$9\tilde{x}^2 = \sum_{n=0}^{\infty} \alpha_n J_0(\mu_n \tilde{x}). \tag{4.555}$$

Now, the eigenvalues μ_n are such that $J_0(\mu_n) = 0$. We find using trial- and-error methods that solutions for all the zeros can be found:

$$\mu_0 = 2.4048, \tag{4.556}$$

$$\mu_1 = 5.5201, \tag{4.557}$$

$$\mu_2 = 8.6537, \tag{4.558}$$

$$\vdots$$

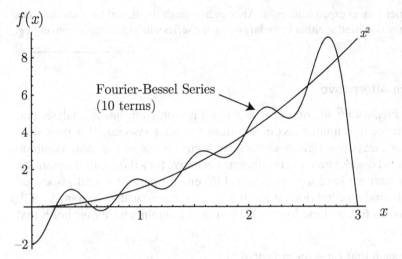

Figure 4.12. Ten-term Fourier-Bessel series approximation to $f(x) = x^2$.

Similar to the other functions, we could expand in terms of the orthonormalized Bessel functions, $\varphi_n(x)$. Instead, for variety, let us directly operate on Eq. (4.555) to determine the values for α_n:

$$9\tilde{x}^2 \tilde{x} J_0(\mu_k \tilde{x}) = \sum_{n=0}^{\infty} \alpha_n \tilde{x} J_0(\mu_n \tilde{x}) J_0(\mu_k \tilde{x}), \tag{4.559}$$

$$\int_0^1 9\tilde{x}^3 J_0(\mu_k \tilde{x})\, d\tilde{x} = \int_0^1 \sum_{n=0}^{\infty} \alpha_n \tilde{x} J_0(\mu_n \tilde{x}) J_0(\mu_k \tilde{x})\, d\tilde{x}, \tag{4.560}$$

$$9\int_0^1 \tilde{x}^3 J_0(\mu_k \tilde{x})\, d\tilde{x} = \sum_{n=0}^{\infty} \alpha_n \int_0^1 \tilde{x} J_0(\mu_n \tilde{x}) J_0(\mu_k \tilde{x})\, d\tilde{x}, \tag{4.561}$$

$$= \alpha_k \int_0^1 \tilde{x} J_0(\mu_k \tilde{x}) J_0(\mu_k \tilde{x})\, d\tilde{x}. \tag{4.562}$$

So replacing k by n and dividing, we get

$$\alpha_n = \frac{9 \int_0^1 \tilde{x}^3 J_0(\mu_n \tilde{x})\, d\tilde{x}}{\int_0^1 \tilde{x} J_0(\mu_n \tilde{x}) J_0(\mu_n \tilde{x})\, d\tilde{x}}. \tag{4.563}$$

Evaluating the first three terms, we get

$$\alpha_0 = 4.446, \tag{4.564}$$

$$\alpha_1 = -8.325, \tag{4.565}$$

$$\alpha_2 = 7.253, \tag{4.566}$$

$$\vdots$$

Because the basis functions are not normalized, it is difficult to infer how the amplitude is decaying by looking at α_n alone. The function $f(x) = x^2$ and 10 terms of the Fourier-Bessel series approximation are plotted in Figure 4.12 The Fourier-Bessel approximation is

$$f(x) = x^2 = 4.446\, J_0\left(2.4048\left(\frac{x}{3}\right)\right) - 8.325\, J_0\left(5.5201\left(\frac{x}{3}\right)\right)$$

$$+7.253\, J_0\left(8.6537\left(\frac{x}{3}\right)\right) + \cdots. \tag{4.567}$$

Other Fourier-Bessel expansions exist. Also, even though the Bessel function does not match the function itself at either boundary point, the series still appears to be converging.

4.6 Fredholm Alternative

The so-called *Fredholm*[22] *alternative* is a general notion from linear analysis that speaks to existence and uniqueness of solutions to linear systems. It is presented in a few different ways in different sources; we focus on one of the more common definitions. Let us consider here the Fredholm alternative for a differential equations. Let us imagine that we have a given linear differential operator $\mathbf{L}$ and associated homogeneous boundary conditions at $x = 0, x = 1$, a nonzero scalar constant $\lambda \in \mathbb{C}^1$, along with a known forcing function $f(x)$. Then the Fredholm alternative holds that there exists either

- a unique y such that $\mathbf{L}y - \lambda y = f$, or
- a nonunique y such that $\mathbf{L}y - \lambda y = 0$.

EXAMPLE 4.20

Identify the Fredholm alternative as it applies to $\mathbf{L} = -d^2/dx^2$ with $y(0) = y(1) = 0$ and $f(x) = -x$.

Consider, then,

$$\frac{d^2y}{dx^2} + \lambda y = x, \qquad y(0) = y(1) = 0. \tag{4.568}$$

For $\lambda \neq n^2\pi^2, n = \pm 1, \pm 2, \ldots$, the unique solution is

$$y(x) = \frac{1}{\lambda}\left(x - \frac{\sin\left(\sqrt{\lambda}x\right)}{\sin\sqrt{\lambda}}\right). \tag{4.569}$$

If $\lambda = n^2\pi^2$ with $n = \pm 1, \pm 2, \ldots$, there is no solution to $\mathbf{L}y - \lambda y = f$, but there is a solution to $\mathbf{L}y - \lambda y = 0$. That problem is

$$\frac{d^2y}{dx^2} + n^2\pi^2 y = 0, \qquad y(0) = y(1) = 0. \tag{4.570}$$

The nonunique solution is

$$y(x) = C\sin(n\pi x), \qquad C \in \mathbb{R}^1. \tag{4.571}$$

EXAMPLE 4.21

Identify the Fredholm alternative as it applies to $\mathbf{L} = -d^2/dx^2$ with $y(0) = y(1) = 0$, $\lambda = \pi^2$, and (a) $f(x) = -\sin(\pi x)$, (b) $f(x) = -\sin(2\pi x)$.

a). For $f(x) = -\sin(\pi x)$, we have

$$\frac{d^2y}{dx^2} + \pi^2 y = \sin(\pi x), \qquad y(0) = y(1) = 0. \tag{4.572}$$

---

[22] Erik Ivar Fredholm, 1866–1927, Swedish mathematician.

The complementary functions are $\sin(\pi x)$ and $\cos(\pi x)$. Because the forcing function is itself a complementary function, we need to multiply by x for the particular solution. Omitting details, we find a general solution that satisfies the differential equation to be

$$y(x) = C_1 \sin(\pi x) + C_2 \cos(\pi x) - \frac{x}{2\pi} \cos(\pi x). \qquad (4.573)$$

For $y(0) = 0$, we need $C_2 = 0$, so

$$y(x) = C_1 \sin(\pi x) - \frac{x}{2\pi} \cos(\pi x). \qquad (4.574)$$

But there is no finite C_1 that allows us to satisfy $y(1) = 0$. Because we cannot satisfy $\mathbf{L}y - \lambda y = f$, we must instead take the Fredholm alternative of satisfying $\mathbf{L}y - \lambda y = 0$, which, for this problem, gives

$$y(x) = C_1 \sin(\pi x). \qquad (4.575)$$

b). For $f(x) = -\sin(2\pi x)$, we have

$$\frac{d^2 y}{dx^2} + \pi^2 y = \sin(2\pi x), \qquad y(0) = y(1) = 0. \qquad (4.576)$$

There is a nonunique solution that satisfies $\mathbf{L}y - \lambda y = f$ and both boundary conditions:

$$y(x) = C_1 \sin(\pi x) - \frac{1}{3\pi^2} \sin(2\pi x), \qquad C_1 \in \mathbb{R}^1. \qquad (4.577)$$

So for this problem, there are either no solutions or an infinite number of solutions to $\mathbf{L}y - \lambda y = f$.

4.7 Discrete and Continuous Spectra

In the Sturm-Liouville equations we have considered, we have generally obtained an infinite number of real eigenvalues; however, each eigenvalue has been isolated on the real axis. We say these eigenvalues form a *spectrum*. An eigenvalue spectrum whose elements are isolated from one another is known as a *discrete spectrum*. In many cases, discrete spectra are associated with nonsingular Sturm-Liouville problems cast within a finite domain. In contrast, there are also problems for which the set of eigenvalues form a *continuous spectrum*. Such spectra are typically associated with infinite or semi-infinite domains or singular Sturm-Liouville operators. For the many physical problems that are posed on infinite or semi-infinite domains, the notion of a continuous spectrum is of critical importance.

EXAMPLE 4.22

Examine a modification of the Sturm-Liouville problem considered first in Example 4.18:

$$-\frac{d^2 y}{dx^2} = \lambda y, \qquad \begin{matrix} y(0) = 0,\ y(\ell) = 0, \\ y(\ell) = 0,\ y(\infty)\text{ is bounded.} \end{matrix} \qquad (4.578)$$

One might envision y to be pinned at $x = 0$ and $x = \ell$ for $x \in [0, \ell]$. And for $x \in [\ell, \infty)$, it is pinned only at $x = \ell$. For $x \in [0, \ell]$, we know the eigenvalues and eigenfunctions are given by

$$\lambda_n = \frac{n^2 \pi^2}{\ell^2}, \qquad n = 1, 2, \dots, \qquad (4.579)$$

$$y_n(x) = \sin\left(\frac{n\pi x}{\ell}\right), \qquad n = 1, 2, \dots. \qquad (4.580)$$

Clearly there is a discrete spectrum for $x \in [0, \ell]$.

For $x \in [\ell, \infty)$, let us first translate the axes taking $\overline{x} = x - \ell$; thus, we are considering

$$-\frac{d^2y}{d\overline{x}^2} = \lambda y, \qquad y(0) = 0, \quad y(\infty) \text{ is bounded.} \tag{4.581}$$

The differential equation and boundary condition at $\overline{x} = 0$ are satisfied by

$$y = C\sin(\sqrt{\lambda}\overline{x}), \qquad \forall \lambda \geq 0. \tag{4.582}$$

There is no need to isolate values of λ; a continuum of values is available. However, here it is impossible to satisfy a boundary condition at infinity, other than y remaining bounded. Thus, arbitrary values for both C and λ are allowed.

To see how this problem is actually singular, we can transform the semi-infinite domain $\overline{x} \in [0, \infty)$ to a finite domain $z \in [0, 1]$ via

$$z = \frac{\overline{x}}{1 + \overline{x}}. \tag{4.583}$$

Application of the chain rule reveals that

$$\frac{d}{d\overline{x}} = \frac{dz}{d\overline{x}}\frac{d}{dz} = (1 - z)^2 \frac{d}{dz}. \tag{4.584}$$

Thus, Eq. (4.578) transforms to

$$-(1 - z)^2 \frac{d}{dz}\left((1 - z)^2 \frac{dy}{dz}\right) = \lambda y, \qquad y(z = 0) = 0, \quad y(z = 1) = 0. \tag{4.585}$$

This expands to

$$(1 - z)^4 \frac{d^2y}{dz^2} + 2(1 - z)^3 \frac{dy}{dz} + \lambda y = 0, \qquad y(0) = 0, \quad y(1) = 0. \tag{4.586}$$

Clearly the domain of interest is now finite, which would give one hope for a finite spectrum. However, the equation itself is singular at $z = 1$, because at that point, the highest order derivative is multiplied by zero.

_____

4.8 Resonance

One of the more important topics of engineering is that of *resonance*, especially in the context of solutions to Eq. (4.1). Such problems can either be boundary value problems or initial value problems; here we briefly consider those of the second type. Thus, as in Section 4.3.3, let us consider x to be the dependent variable and t to be the independent variable so that Eq. (4.1) takes the form

$$\mathbf{L}x = f(t). \tag{4.587}$$

Many operators $\mathbf{L}$ yield complementary functions that have behavior in which the dependent variable oscillates at one or more *natural frequencies*, and such cases are considered here. One can obtain the response x at large values of t, which arise because of the imposed forcing $f(t)$. Of most interest is a forcing function that is closely related to the complementary functions of $\mathbf{L}$. We shall see that when f oscillates with a so-called *resonant frequency* similar to the natural frequencies of $\mathbf{L}$, the response $x(t)$ may be significantly amplified relative to forcing at frequencies far from resonant. It is this feature of tunable amplification that renders the phenomenon to be of engineering interest. Let us consider an important example that contains some additional physical motivation.

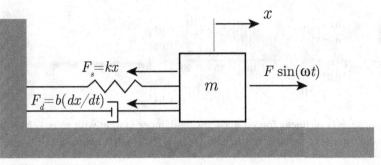

Figure 4.13. Initial configuration for the linear forced mass-spring-damper problem.

EXAMPLE 4.23

Consider

$$m\frac{d^2x}{dt^2} + b\frac{dx}{dt} + kx = F\sin(\omega t), \qquad x(0) = 0, \quad \left.\frac{dx}{dt}\right|_{t=0} = 0. \tag{4.588}$$

Find the response $x(t)$ for large values of t for various values of ω when $m = 1, b = 2, k = 101$, and $F = 1$. This models a physical system, and as such the variables and parameters must have appropriate units selected, omitted here.

Let us first provide a physical motivation. Consider a body of mass m moving in the x direction (see Figure 4.13). Initially the body has displacement $x(0) = 0$ and velocity $dx/dt(0) = 0$. The body is subjected to (1) a linear spring force F_s oriented to be negative when $x > 0$, (2) a linear damping force F_d that is negative when $dx/dt > 0$, and (3) an imposed forcing function $F\sin(\omega t)$. Newton's second law of motion gives us

$$m\frac{d^2x}{dt^2} = -F_s - F_d + F\sin(\omega t). \tag{4.589}$$

We take the spring and damping forces, F_s and F_d, respectively, to be

$$F_s = kx, \tag{4.590}$$

$$F_d = b\frac{dx}{dt}. \tag{4.591}$$

Here k is the spring constant and b is the constant damping coefficient. With this, Newton's second law reduces to

$$m\frac{d^2x}{dt^2} = -kx - b\frac{dx}{dt} + F\sin(\omega t), \tag{4.592}$$

$$m\frac{d^2x}{dt^2} + b\frac{dx}{dt} + kx = F\sin(\omega t). \tag{4.593}$$

As an aside, one sees for m and b sufficiently small to counteract any finite d^2x/dt^2 and dx/dt that $x(t) \sim (F/k)\sin(\omega t)$. In this limit, the system's dynamics are dictated by a balance of the spring force and the imposed forcing function.

For the parameters of our problem, $m = 1, b = 2, k = 101, F = 1$. Examining first $\omega = 1$, we can consider $\mathbf{L}x = f(t)$ where

$$\mathbf{L} = \frac{d^2}{dt^2} + 2\frac{d}{dt} + 101, \qquad f(t) = \sin t. \tag{4.594}$$

The characteristic equation associated with $\mathbf{L}$ is

$$r^2 + 2r + 101 = 0, \tag{4.595}$$

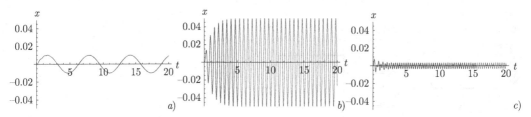

Figure 4.14. Response $x(t)$ of linear forced mass-spring-damper system with $m = 1$, $b = 2$, $k = 101$ to sinusoidal forcing of variable frequency: a) $\omega = 1$, b) $\omega = 10$, c) $\omega = 20$.

which has solutions $r = -1 \pm 10i$. Such solutions with a nonzero imaginary part are known as *underdamped*. Had the roots been real, negative, and distinct, the solution would be *overdamped*. Had the roots been real, negative, and repeated, the solution would have been *critically damped*. In our case, the complementary functions take the form $e^{-t} \sin(10t)$, $e^{-t} \cos(10t)$. They decay for large t and have a natural frequency of $\omega_n = 10$. Solving, we obtain

$$x(t) = \underbrace{-\frac{49e^{-t} \sin(10t)}{50020} + \frac{e^{-t} \cos(10t)}{5002}}_{\text{decaying transient}} + \underbrace{\frac{25 \sin(t)}{2501} - \frac{\cos(t)}{5002}}_{\text{long time solution}}. \tag{4.596}$$

Direct substitution reveals that this solution satisfies the differential equation and both initial conditions. More importantly, we see that there are dynamics at two distinct frequencies present in the solution. The first has a frequency of 10, which is the natural frequency of **L**. These terms are modulated by an exponentially decreasing amplitude associated with the damping coefficient $b = 1$. The other dynamics have the same frequency as the forcing function, that is, unity. Taking the long time limit and applying standard trigonometric reduction reveals

$$\lim_{t \to \infty} x(t) = \frac{\sin\left(t + \text{Tan}^{-1}(-2, 100)\right)}{2\sqrt{2501}}. \tag{4.597}$$

See Eq. (A.96) for a description of the function Tan^{-1}, also known as atan2, which is a simple variant of $\tan^{-1}$. For the parameters chosen here, the phase shift, $\text{Tan}^{-1}(-2, 100) = \tan^{-1}(-2/100) = -0.02$, is small. The long time amplitude is $|x(t)| = 1/2/\sqrt{2501} \sim 0.009998$. This is essentially identical to our long time estimate of $F/k = 1/101 = 0.00990099$. For these parameters, the spring force is effectively balanced by the imposed force.

We show the full solutions, including decaying transients, for $\omega = 1, 10, 20$ in Figure 4.14. Figure 4.14a gives our result for $\omega = 1$. One notes diminishing amplitude oscillations at ω_n for small t.

When we force the system at $\omega = \omega_n = 10$, we notice a remarkable phenomenon. At long time, the system again oscillates with the frequency of the imposed force; however, the amplitude of the output $x(t)$ is much increased relative to that for $\omega = 1$. The complete solution for $\omega = 10$, displayed in Figure 4.14b, is

$$x(t) = \underbrace{\frac{1}{401} e^{-t} \sin(10t) + \frac{20}{401} e^{-t} \cos(10t)}_{\text{decaying transient}} + \underbrace{\frac{1}{401} \sin(10t) - \frac{20}{401} \cos(10t)}_{\text{long time solution}}. \tag{4.598}$$

Taking the long time limit and applying standard trigonometric reduction reveals

$$\lim_{t \to \infty} x(t) = \frac{\sin\left(10t + \text{Tan}^{-1}(-20, 1)\right)}{\sqrt{401}}. \tag{4.599}$$

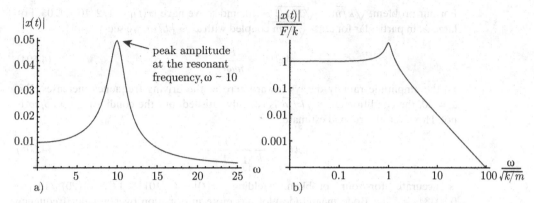

Figure 4.15. Long time amplitude of $x(t)$ for linear forced mass-spring-damper system with $m = 1$, $b = 2$, $k = 101$ as a function of forcing frequency ω: (a) linear plot, (b) log-log Bode magnitude plot.

The long time amplitude $|x(t)|$ is $1/\sqrt{401} \sim 0.0499376$. The phase shift, $\text{Tan}^{-1}(-20, 1) = \tan^{-1}(-20/1)$, which roughly evaluates to $-\pi/2$, is moderate.

Further increase of the forcing frequency increases the frequency of the response $x(t)$ and decreases its amplitude. For $\omega = 20$, solving, taking the long time limit, and applying standard trigonometric reduction reveals

$$\lim_{t \to \infty} x(t) = \frac{\sin\left(20t + \text{Tan}^{-1}(-40, -299)\right)}{\sqrt{91001}}. \qquad (4.600)$$

The long time amplitude is $|x(t)| = 1/\sqrt{91001} \sim 0.00331495$. The phase shift, $\text{Tan}^{-1}(-40, -299) = \tan^{-1}(-40/(-299)) - \pi$, which roughly evaluates to $-\pi$, has doubled from the previous case. The full dynamics, including the initial transients, are displayed in Figure 4.14c.

Additional analysis, not shown here, reveals that the amplitude at long time is given by the relation

$$|x(t)| = \frac{F}{k} \frac{1}{\sqrt{\left(1 - \frac{m\omega^2}{k}\right)^2 + \left(\frac{b\omega}{k}\right)^2}}. \qquad (4.601)$$

A plot of the long time amplitude $|x(t)|$ as a function of forcing frequency ω is given in Figure 4.15. Figure 4.15a gives an unscaled linear plot, and Figure 4.15b gives a scaled log-log plot known as a Bode[23] magnitude plot.

For small ω, Eq. (4.601) simplifies to

$$|x(t)| = \frac{F}{k}. \qquad (4.602)$$

One can actually quantify just how small ω needs to be for Eq. (4.602) to be accurate by examining Eq. (4.601), where one sees that one must have both $\omega \ll \sqrt{k/m}$ and $\omega \ll k/b$. For $\omega = 1$, these conditions are satisfied for our problem's parameters, and this estimate gives $|x(t)| = F/k = 1/101 = 0.0099099$, which matches well the actual value of 0.009998.

For $\omega = \sqrt{k/m}$, we see

$$|x(t)| = \frac{F}{b\omega}. \qquad (4.603)$$

---

[23] Hendrik Wade Bode, 1905–1982, American mathematician and engineer.

For our problem, $\sqrt{k/m} = \sqrt{101/1} \sim 10$, and so we have $|x(t)| \sim 1/2/10 = 0.05$. For large ω, in particular for $\omega \gg \sqrt{k/m}$ coupled with $\omega \gg kb/m$, we see

$$|x(t)| \sim \frac{F}{m\omega^2}. \tag{4.604}$$

So the amplitude rapidly decays toward zero as the driving frequency increases. For $\omega = 20$, the condition $\omega \gg \sqrt{k/m}$ is roughly satisfied, but the condition $\omega \gg kb/m$ is not. However, the refined estimate,

$$|x(t)| \sim \frac{F}{k} \frac{1}{|1 - \frac{m\omega^2}{k}|}, \tag{4.605}$$

is accurate for our problem, yielding $|x(t)| \sim (1/101) \times 1/|1 - 1(20)^2/101| = 0.00334448$. The Bode magnitude plot has more information over a wider frequency range. Clearly for low frequency input, the output is insensitive. And for large frequency input, the output amplitude drops two orders of magnitude for every order of magnitude increase in input frequency. Such behavior is characteristic of what is known as a *low pass filter*. For low frequency input, the output does not vary significantly with frequency; the signal passes through, so to say. But high frequency signals are severely attenuated. One might say they do not pass or that they are filtered.

One can find the actual resonant frequency and amplitude at the resonant frequency by applying standard optimization methods to Eq. (4.601). Taking its derivative with respect to ω and solving for ω yields three solutions. The relevant physical solution is

$$\omega_r = \sqrt{\frac{k}{m}\left(1 - \frac{b^2}{2km}\right)}. \tag{4.606}$$

For the parameters of our problem, this yields $\omega_r = 3\sqrt{11} = 9.9499$. This is close to the natural frequency of $\mathbf{L}$, $\omega_n = 10$. The peak amplitude, found for $\omega = \omega_r$, is then seen to be

$$|x(t)|_{max} = \frac{F}{b\sqrt{\frac{k}{m}}} \frac{1}{\sqrt{1 - \frac{b^2}{4km}}}. \tag{4.607}$$

For the parameters of our problem, we find $|x(t)|_{max} = 1/20 = 0.05$.

Lastly, we note that our solutions as $t \to \infty$ take the form

$$x(t) = |x(t)| \sin(\omega t + \phi), \tag{4.608}$$

where $|x(t)|$ is given by Eq. (4.601), and the so-called *phase angle* ϕ can be shown to be given by

$$\phi = \mathrm{Tan}^{-1}\left(-b\omega, k - m\omega^2\right) = \begin{cases} \tan^{-1}\left(\frac{-b\omega}{k - m\omega^2}\right), & \omega < \sqrt{\frac{k}{m}}, \\ -\frac{\pi}{2}, & \omega = \sqrt{\frac{k}{m}}, \\ \tan^{-1}\left(\frac{-b\omega}{k - m\omega^2}\right) - \pi, & \omega > \sqrt{\frac{k}{m}}. \end{cases} \tag{4.609}$$

When we plot the phase angle as a function of driving frequency with the frequency given on a logarithmic scale, we get what is known as a Bode phase plot, given in Figure 4.16. We see for $\omega \ll \omega_r$ that the phase shift $\phi \sim 0$; that is, the output is *in phase* with the input. There is a transition near the resonant frequency, and for $\omega \gg \omega_r$, we have a phase shift of $\phi \sim -\pi$; in common parlance, one might say the output is 180° *out of phase* with the input.

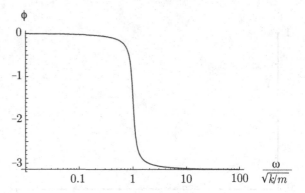

Figure 4.16. Bode phase plot for phase ϕ of $x(t)$ for linear forced mass-spring-damper system with $m = 1, b = 2, k = 101$ as a function of scaled forcing frequency $\omega/\sqrt{k/m}$.

4.9 Linear Difference Equations

We briefly consider the subject of so-called *difference equations,* whose theory is similar to that of differential equations. Difference equations, however, lack the continuity properties associated with most differential equations. A general nonlinear difference equation of order p takes the form

$$x_{n+p} = f(n, x_n, x_{n+1}, \ldots, x_{n+p-1}), \qquad n = 0, 1, 2, \ldots. \qquad (4.610)$$

That is to say, a new value of x is taken to be a function of an iteration counter n and p of the old values of x. Here both n and p are integers. The integer n plays the role of the continuous analog t from differential equations. Generally one iterates to find x as a function of n. However, in special cases, we can actually form an explicit relationship between x and n; one might consider such a solution $x(n)$ to be the analog to the continuous $x(t)$. We will not consider nonlinear difference equations.

For linear difference equations, one could appeal to matrix methods. We will not perform such a formulation. However, an inhomogeneous linear equation with variable coefficients can easily be given by the form

$$x_{n+p} = a_0(n)x_n + \cdots + a_{p-1}(n)x_{n+p-1} + a_p(n), \qquad n = 0, 1, 2, \cdots \qquad (4.611)$$

A homogeneous linear difference equation with constant coefficients is given by the form

$$x_{n+p} = a_0 x_n + \cdots + a_{p-1}x_{n+p-1}, \qquad n = 0, 1, 2, \ldots. \qquad (4.612)$$

A homogeneous second-order ($p = 2$) linear difference equation with constant coefficients is given by the form

$$x_{n+2} = a_0 x_n + a_1 x_{n+1}, \qquad n = 0, 1, 2, \ldots. \qquad (4.613)$$

Given seed values of x_0 and x_1, we may predict x_2 and as many values of x as desired.

Difference equations can be studied in their own right without reference to any underlying continuous system. In fact, we shall do so later for a nonlinear system in Section 9.1. Here, however, we shall study linear difference equations in the context of finite difference discretization of differential equations, as shown in the following example problems.

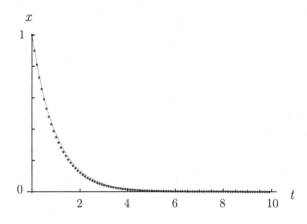

Figure 4.17. Exact and finite difference solutions $x(t)$ for exponential decay.

EXAMPLE 4.24

Examine a finite difference approximation to the linear differential equation and initial condition

$$\frac{dx}{dt} + x = 0, \qquad x(0) = 1, \tag{4.614}$$

in the context of the theory of linear difference equations.

We first note that the exact solution is easily seen to be $x(t) = e^{-t}$. Let us, however, discretize the equation. We choose to use the common approximations of

$$t_n = n\Delta t, \tag{4.615}$$

$$\underbrace{\frac{x_{n+1} - x_n}{\Delta t}}_{\sim dx/dt} + x_n = 0, \qquad n = 0, 1, 2, \dots. \tag{4.616}$$

The approximation is an application of what is known as a *forward Euler method*. It is a *first-order method*, in that the error can be shown to have the same order of magnitude as Δt; in contrast, a second-order method would have error related to $(\Delta t)^2$. It is *explicit* in that it is possible to solve for x_{n+1} directly. We simulate the initial conditions by requiring

$$x_0 = 1. \tag{4.617}$$

We rewrite Eq. (4.616) in the form of Eq. (4.613):

$$x_{n+1} = (1 - \Delta t)x_n. \tag{4.618}$$

We shall analyze this further to find analogs with the theory of linear differential equations, but note simply that Eq. (4.618) coupled with the seed value from Eq. (4.617) and advancement of t from Eq. (4.615) is sufficient, when iterated, to approximate the solution. We show the exact solution and its finite difference approximation in Figure 4.17. Here we have utilized $\Delta t = 1/10$. The approximation captures the essence of the exact solution. Had the exact solution had faster relaxation, $\Delta t = 1/10$ may have been too large.

Let us now obtain an exact solution in a single formula to the linear difference equation, Eq. (4.618), using an analogous set of methods we use for differential equations. In practice, this is rarely done, but when it can be achieved, it is instructive. Much as we assume linear differential equations have solutions of the form Ce^{rt}, we assume linear difference equations to have solutions of the form

$$x_n = C\alpha^n. \tag{4.619}$$

Let us examine the consequences of this assumption by substituting into Eq. (4.618) to obtain

$$C\alpha^{n+1} = (1 - \Delta t)C\alpha^n. \tag{4.620}$$

In the same fashion by which Ce^{rt} is factored from linear differential equations, we can remove the common factor $C\alpha^n$ to obtain the solution

$$\alpha = 1 - \Delta t. \tag{4.621}$$

This is also the characteristic equation for the linear difference equation. We also see immediately that the general solution is

$$x_n = C(1 - \Delta t)^n, \qquad n = 0, 1, 2, \ldots. \tag{4.622}$$

Analogous to differential equations, we can solve for the constant by applying the initial seed values for $x_0 = 1$, giving $C = 1$. Thus,

$$x_n = (1 - \Delta t)^n, \qquad n = 0, 1, 2, \ldots. \tag{4.623}$$

Using Eq. (4.615), we can replace n in favor of t_n to get

$$x_n = (1 - \Delta t)^{t_n/\Delta t}. \tag{4.624}$$

Considering t_n as a continuous variable, and examining x_n for small Δt, we see after a detailed Taylor series expansion that

$$x_n(t_n) = e^{-t_n} \left(1 - \frac{1}{2} t_n \Delta t + \frac{1}{24} t_n \left(3 t_n - 8 \right) \Delta t^2 + O \left(\Delta t^3 \right) \right). \tag{4.625}$$

So when $\Delta t = 0$, it is clear that the difference formula yields the same solution as the continuous analog. And the approximation is good when $\Delta t \ll 1/t_n$.

EXAMPLE 4.25

Examine a discrete approximation to the linear differential equation and initial conditions

$$\frac{d^2 x}{dt^2} + x = 0, \qquad x(0) = 1, \quad \dot{x}(0) = 0, \tag{4.626}$$

in the context of the theory of linear difference equations.

We first note that the exact solution is easily seen to be $x(t) = \cos t$. Let us, however, discretize the equation. We choose to use the common approximations of

$$t_n = n\Delta t, \tag{4.627}$$

$$\underbrace{\frac{x_{n+1} - 2x_n + x_{n-1}}{\Delta t^2}}_{\sim d^2 x/dt^2} + x_n = 0, \qquad n = 0, 1, 2, \ldots. \tag{4.628}$$

We simulate the initial conditions by requiring

$$x_0 = 1, \qquad x_1 = 1. \tag{4.629}$$

We rewrite Eq. (4.628) as

$$x_{n+1} - (2 - \Delta t^2)x_n + x_{n-1} = 0. \tag{4.630}$$

We then increment by 1 and solve for x_{n+2} giving the linear difference equation in the form of Eq. (4.613):

$$x_{n+2} = -x_n + (2 - \Delta t^2)x_{n+1}. \tag{4.631}$$

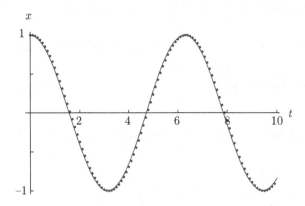

Figure 4.18. Exact and finite difference solutions $x(t)$ for a linear oscillator.

We shall again analyze this further to find analogs with the theory of linear differential equations, but note simply that Eq. (4.631) coupled with seed values from Eq. (4.629) and advancement of t from Eq. (4.627) is sufficient, when iterated, to approximate the solution. We show the exact solution and its finite difference approximation in Figure 4.18. Here we have utilized $\Delta t = 1/10$. The approximation captures the essence of the exact solution. If the exact solution had higher frequency oscillations, $\Delta t = 1/10$ may have been too large.

Let us now obtain an exact solution in a single formula to the linear difference equation, Eq. (4.631), using an analogous set of methods we use for differential equations. Again, we assume solutions to have the form

$$x_n = C\alpha^n. \tag{4.632}$$

We substitute into Eq. (4.631) to obtain

$$C\alpha^{n+2} = -C\alpha^n + (2 - \Delta t^2)C\alpha^{n+1}. \tag{4.633}$$

We can then remove the common factor $C\alpha^n$ to obtain the quadratic equation

$$\alpha^2 = -1 + (2 - \Delta t^2)\alpha. \tag{4.634}$$

This is the characteristic equation for the linear difference equation. It has two roots:

$$\alpha_{1,2} = 1 - \frac{\Delta t^2}{2} \pm i\Delta t \sqrt{1 - \left(\frac{\Delta t}{2}\right)^2}. \tag{4.635}$$

For $\Delta t < 2$, the roots are complex, which indicates an oscillatory behavior. Because our difference operator is linear, we can superpose solutions and take the general solution to be of the form

$$x_n = C_1\alpha_1^n + C_2\alpha_2^n, \qquad n = 0, 1, 2, \ldots. \tag{4.636}$$

Analogous to differential equations, we can solve for the two constants by applying the two initial seed values for $x_0 = 1$ and $x_1 = 1$, giving two linear equations:

$$1 = C_1 + C_2, \tag{4.637}$$

$$1 = C_1\alpha_1 + C_2\alpha_2. \tag{4.638}$$

Solution yields

$$C_1 = \frac{1 - \alpha_2}{\alpha_1 - \alpha_2}, \tag{4.639}$$

$$C_2 = 1 + \frac{\alpha_2 - 1}{\alpha_1 - \alpha_2}. \tag{4.640}$$

Because the algebra is lengthy, we proceed from here on with $\Delta t = 1/10$. This choice yields

$$\alpha_{1,2} = \frac{199}{200} \pm i\frac{\sqrt{399}}{200}. \tag{4.641}$$

One has, then, the solution for x_n for arbitrary n as

$$x_n = \left(\frac{1}{2} - \frac{i}{2\sqrt{399}}\right)\left(\frac{199}{200} + i\frac{\sqrt{399}}{200}\right)^n$$

$$+ \left(\frac{1}{2} + \frac{i}{2\sqrt{399}}\right)\left(\frac{199}{200} - \frac{i\sqrt{399}}{200}\right)^n, \quad n = 0, 1, 2, \ldots. \tag{4.642}$$

Detailed analysis reveals that the imaginary components cancel and that one can rewrite the approximation exactly as

$$x_n = \frac{20\cos\left(\tan^{-1}\left(\frac{1}{\sqrt{399}}\right) - n\tan^{-1}\left(\frac{\sqrt{399}}{199}\right)\right)}{\sqrt{399}}, \quad n = 0, 1, 2, \ldots. \tag{4.643}$$

So once again, we have actually found a general solution for arbitrary n in closed form. Moreover, the solution is in the form of a cosine with an amplitude near unity, as suggested by the solution to the continuous analog. In terms of $t_n = n\Delta t = n/10$, we have

$$x_n(t_n) = \frac{20\cos\left(\tan^{-1}\left(\frac{1}{\sqrt{399}}\right) - 10t_n\tan^{-1}\left(\frac{\sqrt{399}}{199}\right)\right)}{\sqrt{399}}, \tag{4.644}$$

which direct evaluation reveals closely approximates $x(t) = \cos t$.

EXERCISES

1. If $y_1(x)$ and $y_2(x)$ are two solutions of

$$y'' + p(x)y' + q(x)y = 0,$$

   show that the Wronskian of y_1 and y_2 satisfies

$$W' + pW = 0.$$

   From this show Abel's formula $W(x) = W(x_0)\exp\left(-\int_{x_0}^{x} p(s)\,ds\right)$.

2. Solve

$$\frac{d^3y}{dx^3} - 3\frac{d^2y}{dx^2} + 4y = 0.$$

3. Find the general solution of

$$y'''' - 2y''' - 7y'' + 20y' - 12y = 0.$$

4. Solve

$$\frac{d^2y}{dt^2} + C\frac{dy}{dt} + 4y = 0$$

   for (a) $C = 6$, (b) $C = 4$, and (c) $C = 3$, with $y(0) = 1$ and $y'(0) = -3$. Plot your results.

5. Find a linear, homogeneous differential equation that has the two complementary functions e^x and $\sin x$.

6. Show that the functions $y_1 = \sin x$, $y_2 = x \cos x$, and $y_3 = x$ are linearly independent for $x \in (-\infty, \infty)$. Find the lowest order differential equation of which they are the complementary functions.

7. If $y_1(x)$ is one complementary function of $ay'' + by' + cy = 0$, with $b^2 = 4ac$, that comes from the characteristic equation, then show that $y_2 = xy_1$ is the second.

8. Given that $y_1 = x^{-1}$ is one solution of $y'' + (3/x)y' + (1/x^2)y = 0$, find the other solution.

9. Find the exact solution of

$$y'' + \frac{2}{x}y' + \alpha y = 0$$

for positive, negative, and zero values of α. Hint: Let $u = xy$.

10. Find the general solution of the Euler-Cauchy equation,

$$x^2 y'' - 5xy' + 8y = 0.$$

11. Solve the nonlinear equation $(y' - x)y'' + 2y' = 2x$ with the aid of factoring.

12. Solve

$$(y'' + 2y' + y)(y' + y) = 0.$$

13. Show that the transformation

$$u(x) = \exp\left(\int_0^x y(s)\, ds\right),$$

applied to

$$y'' + 3yy' + y^3 = 0,$$

gives a third-order linear equation. Hint: Differentiate $u(x)$ three times.

14. Solve the Lane[24]-Emden[25] equation,

$$xy'' + 2y' + xy^n = 0,$$

for $n = 0$ and $n = 1$. Hint: For $n = 1$, take $y = u/x$. For $n = 5$, show that $y = (1 + x^2/3)^{-1/2}$ is a solution.

15. Solve:
   (a) $y'' + 2y = 6e^x + \cos 2x$.
   (b) $y'' + 2y' + y = 2e^{-x}$.
   (c) $x^2 y'' + xy' - 4y = 6x$.
   (d) $x^2 y'' - 3xy' - 5y = x^2 \ln x$.
   (e) $y'' + y = 2e^x \cos x + (e^x - 2)\sin x$.
   (f) $y''' - 2y'' - y' + 2y = \sin^2 x$.
   (g) $y''' + 6y'' + 12y' + 8y = e^x - 3\sin x - 8e^{-2x}$.
   (h) $x^4 y'''' + 7x^3 y''' + 8x^2 y'' = 4x^{-3}$.
   (i) $y'''' - 5y''' + 11y'' - 7y' = 12$.

16. Find the general solution of
   (a) $y'' + 2y' + y = xe^{-x}$.
   (b) $y'' - y' - 6y = 6x^3 + 26\sin 2x$.

---

[24] Jonathan Homer Lane, 1819–1880, American astrophysicist.
[25] Jacob Robert Emden, 1862–1940, Swiss astrophysicist.

17. Find the complementary function of $y' + f(x)y = g(x)$, and then use the method of variation of parameters to find the particular solution.

18. Solve the equation

$$2y'' - 4y' + 2y = \frac{e^x}{x},$$

where $x > 0$.

19. Solve

$$y'' - 2y' + y = \frac{e^x}{x} + 1$$

using variation of parameters.

20. Show that x^{-1} and x^5 are solutions of the equation

$$x^2 y'' - 3xy' - 5y = 0.$$

Then, find the general solution of

$$x^2 y'' - 3xy' - 5y = x^2.$$

21. Find a particular solution to the following equation using (a) variation of parameters and (b) undetermined coefficients:

$$\frac{d^2 y}{dx^2} - 4y = \cosh 2x.$$

22. Solve

$$2x^2 \frac{d^3 y}{dx^3} + 2x \frac{d^2 y}{dx^2} - 8 \frac{dy}{dx} = 1,$$

with $y(1) = 4$, $y'(1) = 8$, $y(2) = 11$. Plot your result.

23. Solve

$$y'' - 2y' + y = \frac{e^x}{x} + 1,$$

using the **D** operator.

24. Solve

$$(\mathbf{D} - 1)(\mathbf{D} - 2)y = x.$$

25. Find the Green's function for the equation

$$u'''' = f(x),$$

with $u(0) = u''(0) = 0$, $u'(0) = 2u'(1)$, $u(1) = a$.

26. Use the appropriate Green's function to present the solution of

$$\frac{d^2 x}{dt^2} + k^2 x = f(t), \qquad x(0) = a, \ x(\pi) = b.$$

27. Find the Green's function solution of

$$y'' + 4y = f(x),$$

with $y(0) = y(1)$, $y'(0) = 0$. Verify that this is the correct solution when $f(x) = x^2$. Plot your result.

28. Find the Green's function solution of

$$y'' + y' - 2y = f(x),$$

with $y(0) = 0$, $y'(1) = 0$. Determine $y(x)$ if $f(x) = 3 \sin x$. Plot your result.

29. For the differential operator

$$L_n(y) = a_0(x)\frac{d^n y}{dx^n} + a_1(x)\frac{d^{n-1}y}{dx^{n-1}} + \cdots + a_{n-1}(x)\frac{dy}{dx} + a_n(x)y,$$

the formal adjoint operator is

$$L_n^*(y) = (-1)^n \frac{d^n}{dx^n}(a_0(x)y) + (-1)^{n-1}\frac{d^{n-1}}{dx^{n-1}}(a_1(x)y) + \cdots$$

$$- \frac{d}{dx}(a_{n-1}(x)y) + a_n(x)y.$$

See Section 4.4.2. For $n = 2, 3$, and 4, show that

$$vL_n(u) - uL_n^*(v) = \frac{d}{dx}P(u, v),$$

where

$$P(u, v) = \sum_{k=1}^{n}\sum_{j=1}^{k}(-1)^{j-1}\frac{d^{k-j}u}{dx^{k-j}}\frac{d^{j-1}v}{dx^{j-1}}.$$

30. Find the eigenvalues and eigenfunctions of

$$x^2 y'' + xy' + \lambda y = 0,$$

with $y(1) = 0$, $y(a) = 0$.

31. Show that all $y(x)$ that extremize

$$\int_a^b \left(p(x)(y')^2 - q(x)y^2\right)\, dx,$$

subject to the constraint

$$\int_a^b r(x)y^2\, dx = 1,$$

are solutions of the Sturm-Liouville equation

$$(py')' + (q + \lambda r)y = 0$$

for some constants λ.

32. Show that

$$\int_{-1}^{1} xP_n(x)P_{n-1}(x)\, dx = \frac{2n}{4n^2 - 1},$$

where the Ps are Legendre polynomials.

33. Show that $x^a J_\nu(bx^c)$ is a solution of

$$y'' - \frac{2a-1}{x}y' + \left(b^2 c^2 x^{2c-2} + \frac{a^2 - \nu^2 c^2}{x^2}\right)y = 0.$$

With this, solve in terms of Bessel functions

(a) $\dfrac{d^2y}{dx^2} + k^2xy = 0$,

(b) $\dfrac{d^2y}{dx^2} + x^4y = 0$.

34. Let $y = y_1(x)$ and $y = y_2(x)$ for $\lambda = \lambda_1$ and $\lambda = \lambda_2$, respectively, be two solutions to the Sturm-Liouville equation $\mathbf{L}_s y = \lambda y$ with homogeneous boundary conditions at $x = x_0$ and $x = x_1$. Defining the inner product of $u(x)$ and $v(x)$ as

$$\langle u, v \rangle = \int_{x_0}^{x_1} r(x)u(x)v(x)\, dx,$$

where $r > 0$, show that $\mathbf{L}_s$ is self-adjoint by proving that

$$\langle \mathbf{L}_s y_1, y_2 \rangle = \langle y_1, \mathbf{L}_s y_2 \rangle$$

as long as $\lambda_1 \neq \lambda_2$.

35. Find the adjoint of the differential equation

$$\frac{d^2y}{dx^2} + x\frac{dy}{dx} + y = 0,$$

where $y(0) = y(1) = 0$.

36. Consider the function $y(x) = 0, x \in [0,1)$, $y(x) = 2x - 2, x \in [1,2]$. Find an eight-term Fourier-Legendre expansion of this function. Plot the function and the eight-term expansion for $x \in [0,2]$.

37. Find the first three terms of the Fourier-Legendre series for

$$f(x) = \begin{cases} -2, & -1 \leq x < 0 \\ 1, & 0 \leq x \leq 1. \end{cases}$$

Graph $f(x)$ and its approximation.

38. Expand the function $f(x) = x$ in $-1 \leq x \leq 3$ in a five-term Fourier series using (a) sines, (b) Legendre polynomials, (c) Chebyshev polynomials, and (d) Bessel functions. Plot each approximation and compare with the exact function.

39. Show graphically that the Fourier trigonometric series representation of the function

$$f(t) = \begin{cases} -1, & -\pi \leq t < 0 \\ 1, & 0 \leq t \leq \pi \end{cases}$$

always has an overshoot near $x = 0$, however many terms one takes (Gibbs phenomenon). Estimate the overshoot.

40. Expand $f(x) = e^{-2x}$ in $0 \leq x < \infty$ in a five-term Fourier series using Laguerre polynomials. Plot the approximation and compare with the exact function.

41. Expand $f(x) = x(1 - x)$ in $-\infty < x < \infty$ in a five-term Fourier series using Hermite polynomials (physicists'). Plot the approximation and compare with the exact function.

42. Find the first three terms of the Fourier-Legendre series for $f(x) = \cos(\pi x/2)$ for $x \in [-1,1]$. Compare graphically with exact function.

43. Find the first three terms of the Fourier-Legendre series for

$$f(x) = \begin{cases} 0, & -1 \leq x < 0 \\ x, & 0 \leq x \leq 1. \end{cases}$$

44. Consider the function $y(x) = x^2 - 2x + 1$ defined for $x \in [0, 4]$. Find eight-term expansions in terms of (a) Fourier-sine, (b) Fourier-Legendre, (c) Fourier-Hermite (physicists'), and (d) Fourier-Bessel series and plot your results on a single graph.

45. Consider the function $y(x) = 2x, x \in [0, 6]$. Find an eight-term (a) Fourier-Chebyshev and (b) Fourier-sine expansion of this function. Plot the function and the eight-term expansions for $x \in [0, 6]$. Which expansion minimizes the error in representation of the function?

46. Consider the function $y(x) = \cos^2(x^2)$. Find an eight-term (a) Fourier-Laguerre in $x \in [0, \infty)$, and (b) Fourier-sine in $x \in [0, 10]$ expansion of this function. Plot the function and the eight-term expansions for $x \in [0, 10]$. Which expansion minimizes the error in representation of the function?

47. Show that one of the two problems $y'' + \lambda^2 y = 0$ with (a) $y(0) = 0, y(1) = 0$ and (b) $y(0) = 0, y'(0) = 1$ has solutions for discrete values of λ and the other for continuous. In each case, find the values of λ for which solutions are possible.

48. Solve:

(a) $\ddot{y} + y = \sin t$.

(b) $\ddot{y} + y = \sin 2t \, \sin 3t$.

49. Find the solution of

$$\ddot{y} + y = f(t),$$

where $f(t)$ is a periodic square wave pulse (for half of the period $f = 0$, and for the other half $f = 1$) at the same frequency as the unforced equation.

50. Find the first five terms of the sequence $\{x_1, x_2, \ldots\}$ if

(a) $x_n = 2x_{n-1} + 1, x_1 = 0$.

(b) $x_n = 3x_{n-1} + 2x_{n-2} + 4, x_1 = 0, x_2 = 2$.

51. Determine x_2, x_3 and x_4, if $x_1 = 1$, $x_5 = 2$, and $x_{n-1} - 2x_n + x_{n+1} = 0$ for $n = 2, 3, 4$.

52. Mass 1 is connected by a spring A to a wall, and by another spring B to mass 2; mass 2, in addition to being connected as well as identical to mass 1, is also connected by a spring C to a wall. The in-line motion of each mass is governed by

$$m\ddot{x}_1 = k(x_2 - 2x_1),$$

$$m\ddot{x}_2 = k(x_1 - 2x_2),$$

where the masses are m and the spring constants are k. Find the natural frequencies of the system. The total mechanical energy of the system, E, is composed of the kinetic energy of the three masses and the potential energy of the three springs. Show from the differential equations (without solving them) that $dE/dt = 0$.

53. The following differential equation is obtained in a problem in forced convection in porous media:

$$T'' + \frac{x}{2}T' = 0,$$

with $T(0) = 0, T(\infty) = 1$. Find the solution, $T(x)$, in terms of the error function (see Section A.7.2),

$$\mathrm{erf}(x) = \frac{1}{\sqrt{\pi}} \int_0^x e^{-s^2} \, ds,$$

where T is the temperature.

54. The steady state temperature $T(x)$ along a fin due to axial conduction, convection from the side, and internal heat generation $g(x)$ is

$$\frac{d^2 T}{dx^2} - m^2 (T - T_\infty) = \frac{g(x)}{k},$$

where m is a constant, T_∞ is the ambient temperature, and k is the thermal conductivity of the material. With the boundary conditions $T(0) = T(L) = T_\infty$, find the general solution $T(x)$.

55. The transverse deflection of a column $y(x)$ is given by

$$y'' + \frac{P}{EI}y = 0,$$

with $y(0) = y(L) = 0$, where P is the axial load and EI is a constant. What is the smallest load at which the column buckles?

56. The steady state form of a string, pinned at its two extremes, under tension and with small deflection due to an external load per unit length, is governed by an equation of the form

$$y'' = f(x),$$

with $y(0) = y(1) = 0$. (a) Defend the statement that the Green's function $g(x, s)$ represents the deflection at one point due to a point load at another. (b) Show that the deflection at location A due to a point load at location B is the same as the deflection at B due to the same load at A (Maxwell's[26] reciprocity theorem). (c) Show by direct substitution that

$$y = \int_0^1 f(s)g(x, s) \, ds$$

is a particular integral of the differential equation.

57. Consider the differential equation

$$y'' + y' + y = 0,$$

with boundary conditions $y(0) = 0$, $y'(1) = 1$. (a) Solve the equation symbolically using computer mathematics. (b) Solve the equation numerically using discrete methods. (d) Compare the analytical and numerical solutions on a graph.

---

[26] James Clerk Maxwell, 1831–1879, Scottish mathematical physicist.

58. Solve

$$(1 + x^2)y' + 2xy = 0$$

for $y(x)$ using symbolic computer mathematics.

59. Find a second-order ordinary differential differential equation with appropriate initial conditions that can be solved by symbolic computer mathematics and another that cannot.

60. Write a symbolic computer mathematics script that solves $x'' + \omega^2 x = 0$, with $x(0) = a$, $x'(0) = 0$.

61. Use symbolic computer mathematics (for analytical integration) and discrete computational methods (for numerical integration) to show that the first five Chebyshev polynomials are orthogonal with respect to each other.

62. Work out the following shooting method problem using discrete computational methods. (a) Solve $y'' = 6x$ many times in the interval $0 \leq x \leq 1$ with $y(0) = 0$, $y'(0) = c$, where c is varied from -2 to 0 in steps of 0.01. Plot $y(1)$ versus c. (b) From the graph, find c for which $y(1) = 0$. (c) Using this value of c, plot the solution of $y'' = 6x$ that satisfies both $y(0) = 0$ and $y(1) = 0$. Compare with the exact solution.

63. Work out the following shooting method problem using discrete computational methods. (a) Solve $y'' - 2y' + y = e^x/x + 1$ many times in the interval $1 \leq x \leq 2$ with $y(1) = 0$, $y'(1) = c$, where c is varied. Plot $y(2)$ versus c. (b) From the graph, find c for which $y(2) = 1$. (c) Using this value of c, plot the solution of the equation that satisfies both $y(1) = 0$ and $y(2) = 1$. Compare with the exact solution.

5 Approximation Methods

This chapter deals with *approximation methods*, mainly through the use of series. After a short discussion of approximation of known functions, we focus on approximately solving equations for unknown functions. One might wonder why anyone should bother with an approximate solution in favor of an exact solution. There are many justifications. Often physical systems are described by complicated equations with detailed exact solutions; the details of the solution may in fact obscure easy interpretation of results, rendering the solution to be of small aid in discerning trends or identifying the most important causal agents. A carefully crafted approximate solution will often yield a result that exposes the important driving physics and filters away extraneous features of the solution. Colloquially, one hopes for an approximate solution that segregates the so-called signal from the noise. This can aid the engineer greatly in building or reinforcing intuition and sometimes lead to a more efficient design and control strategy. In other cases, including those with practical importance, exact solutions are not available. In such cases, engineers often resort to numerically based approximation methods. Indeed, these methods have been established as an essential design tool; however, short of exhaustive parametric studies, it can be difficult to induce significant general insight from numerics alone. Numerical approximation is a broad topic and is not is studied here in any real detail; instead, we focus on analysis-based approximation methods. They do not work for all problems, but in those cases where they do, they are potent aids to the engineer as a predictive tool for design.

Often, though not always, approximation methods rely on some form of linearization to capture the behavior of some local nonlinearity. Such methods are useful in solving algebraic, differential, and integral equations. We begin with a consideration of Taylor series and the closely related Padé[1] approximant. The class of methods we next consider, power series, employed already in Section 4.4 for solutions of ordinary differential equations, is formally exact in that an infinite number of terms can be obtained. Moreover, many such series can be shown to have absolute and uniform convergence properties as well as analytical estimates of errors incurred by truncation at a finite number of terms. One common use for such series is to truncate at a small finite number of terms for an approximate solution, and this often leads to physical insights in engineering problems. A second method, asymptotic series solution,

---

[1] Henri Eugène Padé, 1863–1953, French mathematician.

is less rigorous in that convergence is not always guaranteed; in fact, convergence is rarely examined because that aspect of many problems tends to be intractable. Still, asymptotic methods will be seen to be useful in interpreting the results of highly nonlinear equations in local domains. We consider asymptotic methods for algebraic and differential equations as well as evaluation of integrals. Especially because approximations are widely used in engineering and the physical sciences, often with little explanation, it is useful to explore their mathematical foundations.

5.1 Function Approximation

Here we discuss some commonly used methods to approximate functions. As with all approximations, these are nonunique, and each has its own advantages and its own error.

5.1.1 Taylor Series

Functions $y(x)$ are often approximated by the well-known Taylor series expansion of $y(x)$ about the point $x = x_0$. Such an expansion is

$$y(x) = y(x_0) + \frac{dy}{dx}\bigg|_{x=x_0} (x - x_0) + \frac{1}{2}\frac{d^2y}{dx^2}\bigg|_{x=x_0} (x - x_0)^2 + \frac{1}{6}\frac{d^3y}{dx^3}\bigg|_{x=x_0} (x - x_0)^3 + \cdots$$

$$+ \frac{1}{n!}\frac{d^ny}{dx^n}\bigg|_{x=x_0} (x - x_0)^n + \cdots. \tag{5.1}$$

EXAMPLE 5.1

Find a Taylor series of $y(x)$ about $x = 0$ if

$$y(x) = \frac{1}{(1+x)^n}. \tag{5.2}$$

Direct substitution into Eq. (5.1) reveals that the answer is

$$y(x) = 1 - nx + \frac{(-n)(-n-1)}{2!}x^2 + \frac{(-n)(-n-1)(-n-2)}{3!}x^3 + \cdots. \tag{5.3}$$

So if $n = 3$, we have

$$y(x) = \frac{1}{(1+x)^3} = 1 - 3x + 6x^2 - 10x^3 + \cdots. \tag{5.4}$$

The function along with its four-term Taylor series approximation is shown Figure 5.1. The four-term approximation is not good for $x > 1/2$.

Generalizing the Taylor series to multiple dimensions gives the equivalent of Eq. (1.149):

$$y(\mathbf{x}) = y(\mathbf{x}_0) + (\mathbf{x} - \mathbf{x}_0)^T \cdot \nabla y + (\mathbf{x} - \mathbf{x}_0)^T \cdot \mathbf{H} \cdot (\mathbf{x} - \mathbf{x}_0) + \cdots, \tag{5.5}$$

where $\mathbf{x} \in \mathbb{R}^N$ and $\mathbf{H}$ is the Hessian matrix as defined in Eq. (1.148).

It is useful to keep in mind an important definition. A function $f(x)$ is said to be *analytic* if it is an infinitely differentiable function such that the Taylor series, $\sum_{n=0}^{\infty} f^{(n)}(x_0)(x - x_0)^n/n!$, at any point $x = x_0$ in its domain converges to $f(x)$ in

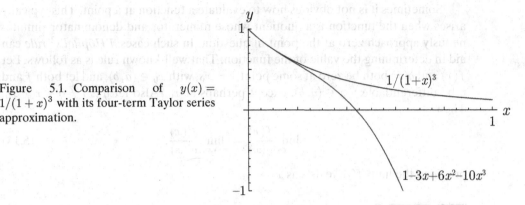

Figure 5.1. Comparison of $y(x) = 1/(1+x)^3$ with its four-term Taylor series approximation.

a neighborhood of $x = x_0$. In the complex plane, the *radius of convergence* is the radius of the largest disk within which the power series converges.

EXAMPLE 5.2

Find the radius of convergence of the Taylor series expansion of the complex function

$$f(z) = \frac{1}{z^2 + 1}, \tag{5.6}$$

about $z = 0$.

The Taylor series expansion of f about $z = 0$ is

$$f(z) = 1 - z^2 + z^4 - \cdots = \sum_{n=0}^{\infty} (-1)^n z^{2n}. \tag{5.7}$$

Cauchy's root test tells us about the convergence of this series. To apply this test, we recall that $z = re^{i\theta}$ (see Section A.8) and consider

$$C = \lim_{n \to \infty} \sup \sqrt[n]{|(-1)^n z^{2n}|}, \tag{5.8}$$

$$= \lim_{n \to \infty} \sup \sqrt[n]{|r^{2n} e^{2ni\theta}|}, \tag{5.9}$$

$$= \lim_{n \to \infty} \sup \sqrt[n]{r^{2n}}, \tag{5.10}$$

$$= r^2. \tag{5.11}$$

For absolute convergence, we require $C < 1$, so we must have $r < 1$. If we confine to the real axis, this forces $x \in (-1, 1)$ for convergence. Let us check with a six-term estimate:

$$f(x) = \frac{1}{1 + x^2} \sim 1 - x^2 + x^4 - x^6 + x^8 - x^{10}. \tag{5.12}$$

If $x = 1/2$, $f(x) = 4/5$, and the six-term estimate yields 0.799805. So it seems the finite series estimate is good, which might be expected when x is within the radius of convergence. If $x = 3/2$, $f(x) = 4/13$, but the six-term approximation is -39.6143. More importantly, for x within the radius of convergence, the estimate converges to the exact solution as the number of terms goes to infinity. For x outside of the radius of convergence, adding more terms does not improve the estimate, even for a large number of terms.

Sometimes it is not obvious how to evaluate a function at a point. This typically arises when the function is a quotient whose numerator and denominator simultaneously approach zero at the point in question. In such cases, *l'Hôpital's*[2] *rule* can aid in determining the value of the function. That well-known rule is as follows. Let $f(x)$ and $g(x)$ both be zero at some point $x = x_0$, with $x_0 \in (a, b)$, and let both f and g be differentiable $\forall\ x \in (a, b)$, except perhaps at x_0. Also let $g'(x) \neq 0 \ \forall\ x \neq x_0$. Then

$$\lim_{x \to x_0} \frac{f(x)}{g(x)} = \lim_{x \to x_0} \frac{f'(x)}{g'(x)}, \tag{5.13}$$

provided the limit f'/g' exists as $x \to x_0$.

EXAMPLE 5.3
Evaluate

$$\lim_{x \to 0} \frac{\sin x}{x}. \tag{5.14}$$

Here $f(x) = \sin x$, $g(x) = x$, and $f(0) = g(0) = 0$, so the value of the function $f(0)/g(0)$ is indeterminate. We can use l'Hôpital's rule, however, to evaluate the limit

$$\lim_{x \to 0} \frac{\sin x}{x} = \lim_{x \to 0} \frac{\cos x}{1} = \frac{\cos 0}{1} = 1. \tag{5.15}$$

Some common useful notation is defined as follows. Consider two functions $f(x)$ and $g(x)$. We write that

- $f(x) \sim g(x)$, if $\displaystyle\lim_{x \to a} \frac{f(x)}{g(x)} = 1$; here f is said to be *asymptotic* to g.

- $f(x) = o(g(x))$, if $\displaystyle\lim_{x \to a} \frac{f(x)}{g(x)} = 0$; here f is said to be "little o" g.

- $f(x) = O(g(x))$, if $\displaystyle\lim_{x \to a} \left| \frac{f(x)}{g(x)} \right| = $ constant; here f is said to be "big O" g.

5.1.2 Padé Approximants

Although Taylor series are widely used to approximate $y(x)$ near a point, for many functions, they may not provide the most accurate approximation. One alternative is known as a *Padé approximant*. This is a rational function that can provide an improvement over a Taylor series, though improvement is not guaranteed. We take the rational function to be

$$y_P(x) = \frac{\sum_{n=0}^{M} a_n x^n}{1 + \sum_{n=1}^{N} b_n x^n}. \tag{5.16}$$

For $y_P(x)$ to be an order $M + N + 1$ Padé approximant to $y(x)$, we must have the error in approximation such that

$$y(x) - y_P(x) = O(x^{N+M+1}). \tag{5.17}$$

---

[2] Guillaume François Antoine, Marquis de l'Hôpital, 1661–1704, French mathematician.

Now we have available a Taylor series of $y(x)$ for as many terms as we desire. This takes the form

$$y(x) = \sum_{n=0}^{\infty} c_n x^n. \tag{5.18}$$

We take the c_n to be known quantities. We choose to retain terms up to $N + M$ and expand Eq. (5.17) as

$$\sum_{n=0}^{N+M} c_n x^n - \frac{\sum_{n=0}^{M} a_n x^n}{1 + \sum_{n=1}^{N} b_n x^n} = 0. \tag{5.19}$$

We have neglected the $O(x^{N+M+1})$ term as small. There are many ways to proceed from here. One first notes that when $x = 0$, we have $a_0 = c_0$, so a_0 is no longer an unknown. One of the most straightforward following steps is to choose M and N, perform synthetic division on the quotient that forms $y_P(x)$, group terms in common powers of x^n, and solve $N + M$ equations for the M values of b_n and the N remaining values of a_n. This is illustrated in the following example.

EXAMPLE 5.4

Find a five-term Padé approximant to

$$y(x) = \sqrt{1 + x}, \tag{5.20}$$

valid near $x = 0$.

Let us take $N = M = 2$, so we need to find the five terms a_0, a_1, a_2, b_1, and b_2. The five-term Taylor series is

$$y(x) = \sqrt{1 + x} = 1 + \frac{x}{2} - \frac{x^2}{8} + \frac{x^3}{16} - \frac{5x^4}{128} + \cdots. \tag{5.21}$$

We thus specialize Eq. (5.19) as

$$1 + \frac{x}{2} - \frac{x^2}{8} + \frac{x^3}{16} - \frac{5x^4}{128} - \frac{a_0 + a_1 x + a_2 x^2}{1 + b_1 x + b_2 x^2} = 0. \tag{5.22}$$

We first note that when $x = 0$, it must be that $a_0 = 1$. Performing synthetic division on the quotient and retaining terms at or below $O(x^4)$ gives

$$1 + \frac{x}{2} - \frac{x^2}{8} + \frac{x^3}{16} - \frac{5x^4}{128} = x^4 \left(-a_1 b_1^3 + a_2 b_1^2 + 2a_1 b_2 b_1 - a_2 b_2 + b_1^4 - 3b_2 b_1^2 + b_2^2 \right)$$

$$+ x^3 \left(a_1 b_1^2 - a_2 b_1 - a_1 b_2 - b_1^3 + 2b_2 b_1 \right)$$

$$+ x^2 \left(-a_1 b_1 + a_2 + b_1^2 - b_2 \right)$$

$$+ x(a_1 - b_1)$$

$$+ 1. \tag{5.23}$$

Grouping terms involving common powers of x and setting the coefficients to zero leads to the following four equations for a_1, a_2, b_1, and b_2:

$$-a_1 + b_1 + \frac{1}{2} = 0, \tag{5.24}$$

$$a_1 b_1 - a_2 - b_1^2 + b_2 - \frac{1}{8} = 0, \tag{5.25}$$

$$-a_1 b_1^2 + a_2 b_1 + a_1 b_2 + b_1^3 - 2b_2 b_1 + \frac{1}{16} = 0, \tag{5.26}$$

$$a_1 b_1^3 - a_2 b_1^2 - 2a_1 b_2 b_1 + a_2 b_2 - b_1^4 + 3b_2 b_1^2 - b_2^2 - \frac{5}{128} = 0. \tag{5.27}$$

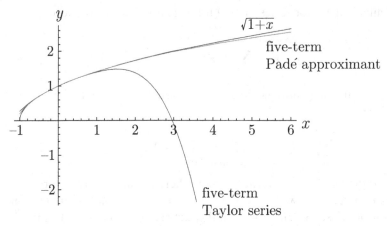

Figure 5.2. Comparison of $y(x) = \sqrt{1+x}$ with its five-term Taylor and Padé approximants.

These have the unique solution $a_1 = 5/4$, $a_2 = 5/16$, $b_1 = 3/4$, $b_2 = 1/16$. Thus, our five-term Padé approximant is

$$y(x) \sim y_P(x) = \frac{1 + \frac{5x}{4} + \frac{5x^2}{16}}{1 + \frac{3x}{4} + \frac{x^2}{16}}. \tag{5.28}$$

The function along with five-term Taylor and Padé approximants are shown in Figure 5.2. Clearly the Padé approximant is superior, especially for $x \in [1, 6]$.

5.2 Power Series

The previous section considered approximations of functions. Here we turn to approximations to solutions of equations. Solutions to many equations cannot be found in a closed form expressed, for instance, in terms of polynomials and transcendental functions such as sine and cosine. Often, instead, the solutions can be expressed as an infinite series of polynomials. It is desirable to get a complete expression for the nth term of the series so that one can make statements regarding absolute and uniform convergence of the series. Such solutions are approximate in that if one uses a finite number of terms to represent the solution, there is a truncation error. Formally, though, for series that converge, an infinite number of terms gives a true representation of the actual solution, and hence the method is exact. In this section, we first briefly consider power series solutions to so-called *functional equations*. We then turn attention to power series solution of ordinary differential equations.

5.2.1 Functional Equations

In a functional equation, the functional form is unknown, but its values for different arguments are related. In general, these are difficult to solve. Occasionally power series combined with Taylor series can be useful, as demonstrated in the following example.

EXAMPLE 5.5

Find $y(x)$ if

$$y(x) + y(2x) = e^x. \tag{5.29}$$

The power series strategy is to expand every term in a Taylor series, equate terms of common powers of x, and form a general series solution for the form of $y(x)$. We first assume the existence of a solution with the general form $y(t)$ with as many continuous derivatives as we require. With the existence of $y(t)$ comes the existence of its Taylor series expansion about $t = 0$:

$$y(t) = y(0) + y'(0)t + \frac{1}{2}y''(0)t^2 + \cdots + \frac{1}{n!}y^{(n)}(0)t^n + \cdots. \tag{5.30}$$

Evaluating $y(t)$ at $t = x$ gives, trivially,

$$y(x) = y(0) + y'(0)x + \frac{1}{2}y''(0)x^2 + \cdots + \frac{1}{n!}y^{(n)}(0)x^n + \cdots, \tag{5.31}$$

Evaluating $y(t)$ at $t = 2x$ gives

$$y(2x) = y(0) + y'(0)(2x) + \frac{1}{2}y''(0)(2x)^2 + \cdots + \frac{1}{n!}y^{(n)}(0)(2x)^n + \cdots, \tag{5.32}$$

$$= y(0) + 2y'(0)x + 2y''(0)x^2 + \cdots + \frac{2^n}{n!}y^{(n)}(0)x^n + \cdots. \tag{5.33}$$

We also have the Taylor series of e^x as

$$e^x = 1 + x + \frac{1}{2}x^2 + \cdots + \frac{1}{n!}x^n + \cdots. \tag{5.34}$$

Our functional equation is then

$$\underbrace{\left(y(0) + y'(0)x + \frac{1}{2}y''(0)x^2 + \cdots + \frac{1}{n!}y^{(n)}(0)x^n + \cdots \right)}_{y(x)}$$

$$+ \underbrace{\left(y(0) + 2y'(0)x + 2y''(0)x^2 + \cdots + \frac{2^n}{n!}y^{(n)}(0)x^n + \cdots \right)}_{y(2x)}$$

$$= \underbrace{1 + x + \frac{1}{2}x^2 + \cdots + \frac{1}{n!}x^n + \cdots}_{e^x}. \tag{5.35}$$

Grouping terms on common factors of x, we get the system of equations

$$2y(0) = 1, \tag{5.36}$$

$$3y'(0) = 1, \tag{5.37}$$

$$5y''(0) = 1, \tag{5.38}$$

$$\vdots \tag{5.39}$$

$$(1 + 2^n)y^{(n)} = 1. \tag{5.40}$$

This yields $y(0) = 1/2$, $y'(0) = 1/3$, $y''(0) = 1/5, \ldots, y^{(n)} = 1/(1 + 2^n)$. We thus have all the Taylor series coefficients for our function $y(x)$, which is

$$y(x) = \frac{1}{2} + \frac{1}{3}x + \frac{1}{10}x^2 + \cdots + \frac{1}{n!(1 + 2^n)}x^n + \cdots. \tag{5.41}$$

The function $y(x)$ is shown in Figure 5.3.

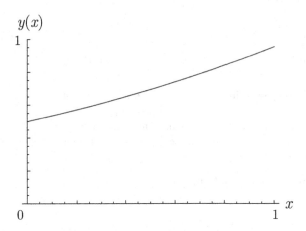

Figure 5.3. Power series solution for $y(x)$ to functional equation $y(x) + y(2x) = e^x$.

5.2.2 First-Order Differential Equations

An equation of the form

$$\frac{dy}{dx} + P(x)y = Q(x), \tag{5.42}$$

where $P(x)$ and $Q(x)$ are analytic at $x = a$, has a power series solution

$$y(x) = \sum_{n=0}^{\infty} a_n (x - a)^n \tag{5.43}$$

around this point.

EXAMPLE 5.6

Find the power series solution of

$$\frac{dy}{dx} = y, \qquad y(0) = y_0, \tag{5.44}$$

around $x = 0$.

Let

$$y = a_0 + a_1 x + a_2 x^2 + a_3 x^3 + \cdots, \tag{5.45}$$

so that

$$\frac{dy}{dx} = a_1 + 2a_2 x + 3a_3 x^2 + 4a_4 x^3 + \cdots. \tag{5.46}$$

Substituting into Eq. (5.44), we have

$$\underbrace{a_1 + 2a_2 x + 3a_3 x^2 + 4a_4 x^3 + \cdots}_{dy/dx} = \underbrace{a_0 + a_1 x + a_2 x^2 + a_3 x^3 + \cdots}_{y}, \tag{5.47}$$

$$\underbrace{(a_1 - a_0)}_{=0} + \underbrace{(2a_2 - a_1)}_{=0} x + \underbrace{(3a_3 - a_2)}_{=0} x^2 + \underbrace{(4a_4 - a_3)}_{=0} x^3 + \cdots = 0. \tag{5.48}$$

Because the polynomials $x^0, x^1, x^2, \ldots$ are linearly independent, the coefficients must be all zero. Thus,

$$a_1 = a_0, \tag{5.49}$$

$$a_2 = \frac{1}{2} a_1 = \frac{1}{2} a_0, \tag{5.50}$$

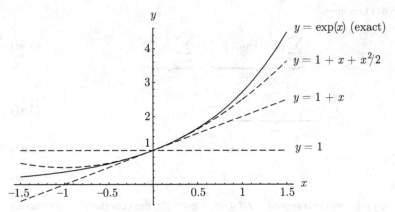

Figure 5.4. Comparison of truncated series and exact solutions to $dy/dx = y$, $y(0) = 1$.

$$a_3 = \frac{1}{3}a_2 = \frac{1}{3!}a_0, \tag{5.51}$$

$$a_4 = \frac{1}{4}a_3 = \frac{1}{4!}a_0, \tag{5.52}$$

$$\vdots$$

so that

$$y(x) = a_0\left(1 + x + \frac{x^2}{2!} + \frac{x^3}{3!} + \frac{x^4}{4!} + \cdots\right). \tag{5.53}$$

Applying the initial condition at $x = 0$ gives $a_0 = y_0$, so

$$y(x) = y_0\left(1 + x + \frac{x^2}{2!} + \frac{x^3}{3!} + \frac{x^4}{4!} + \cdots\right). \tag{5.54}$$

Of course this power series is the Taylor series expansion (see Section 5.1.1), of the closed form solution $y = y_0 e^x$ about $x = 0$. The power series solution about a different point will give a different solution.

For $y_0 = 1$, the exact solution and three approximations to the exact solution are shown in Figure 5.4. Alternatively, one can use a compact summation notation. Thus,

$$y = \sum_{n=0}^{\infty} a_n x^n, \tag{5.55}$$

$$\frac{dy}{dx} = \sum_{n=0}^{\infty} n a_n x^{n-1}, \tag{5.56}$$

$$= \sum_{n=1}^{\infty} n a_n x^{n-1}. \tag{5.57}$$

Now take $m = n - 1$ and get

$$\frac{dy}{dx} = \sum_{m=0}^{\infty} (m+1) a_{m+1} x^m, \tag{5.58}$$

$$= \sum_{n=0}^{\infty} (n+1) a_{n+1} x^n. \tag{5.59}$$

Thus, Eq. (5.44) becomes

$$\underbrace{\sum_{n=0}^{\infty}(n+1)a_{n+1}x^n}_{dy/dx} = \underbrace{\sum_{n=0}^{\infty}a_n x^n}_{y}, \tag{5.60}$$

$$\sum_{n=0}^{\infty}\underbrace{\left((n+1)a_{n+1} - a_n\right)}_{=0}x^n = 0, \tag{5.61}$$

$$(n+1)a_{n+1} = a_n, \tag{5.62}$$

$$a_{n+1} = \frac{1}{n+1}a_n. \tag{5.63}$$

Equation (5.63) is known as a *recursion relation*; it gives the next coefficient of the series in terms of previous coefficients. In this case, the recursion relation can be generalized to give a_n for arbitrary n, yielding

$$a_n = \frac{a_0}{n!}, \tag{5.64}$$

$$y = a_0 \sum_{n=0}^{\infty} \frac{x^n}{n!}, \tag{5.65}$$

$$= y_0 \sum_{n=0}^{\infty} \frac{x^n}{n!}. \tag{5.66}$$

The ratio test tells us that

$$\lim_{n\to\infty}\left|\frac{a_{n+1}}{a_n}\right| = \frac{1}{n+1} \to 0, \tag{5.67}$$

so the series converges absolutely.

If a series is *uniformly* convergent in a domain, it converges at the same rate for all x in that domain. We can use the Weierstrass[3] M-test for uniform convergence. That is, for a series

$$\sum_{n=0}^{\infty} u_n(x), \tag{5.68}$$

to be convergent, we need a convergent series of constants M_n to exist such that

$$|u_n(x)| \le M_n, \tag{5.69}$$

for all x in the domain. For our problem, we take $x \in [-A, A]$, where $A > 0$.

So for *uniform* convergence, we must have

$$\left|\frac{x^n}{n!}\right| \le M_n. \tag{5.70}$$

So take

$$M_n = \frac{A^n}{n!}. \tag{5.71}$$

Note M_n is thus strictly positive. So

$$\sum_{n=0}^{\infty} M_n = \sum_{n=0}^{\infty} \frac{A^n}{n!}. \tag{5.72}$$

---

[3] Karl Theodor Wilhelm Weierstrass, 1815–1897, Westphalia-born German mathematician.

By the ratio test, this is convergent if

$$\lim_{n \to \infty} \left| \frac{\frac{A^{n+1}}{(n+1)!}}{\frac{A^n}{(n)!}} \right| \leq 1, \tag{5.73}$$

$$\lim_{n \to \infty} \left| \frac{A}{n+1} \right| \leq 1. \tag{5.74}$$

This holds for all A, so for $x \in (-\infty, \infty)$, the series converges absolutely and uniformly.

EXAMPLE 5.7

Find a power series solution of

$$\frac{dy}{dx} = -\sqrt{x}y, \qquad y(0) = y_0, \tag{5.75}$$

for $x \geq 0$.

Separation of variables reveals the exact solution to be

$$y = y_0 \exp\left(-\frac{2}{3}x^{3/2}\right), \tag{5.76}$$

which has a Taylor series expansion of

$$y = y_0 \left(1 - \frac{2}{3}x^{3/2} + \frac{2}{9}x^3 - \frac{4}{81}x^{9/2} + \frac{2}{243}x^6 - \frac{4}{3645}x^{15/2} + \cdots\right). \tag{5.77}$$

Let us see if we can recover this by a power series method. Assume a power series solution of the form

$$y = \sum_{n=0}^{\infty} a_n x^{\alpha n}. \tag{5.78}$$

Here we are allowing a more general form than for the previous example. Detailed analysis, not given here, would reveal that $\alpha = 1/2$ would provide an acceptable solution. We shall proceed from that point. So we are now considering

$$y = \sum_{n=0}^{\infty} a_n x^{n/2}. \tag{5.79}$$

The derivative is thus

$$\frac{dy}{dx} = \sum_{n=1}^{\infty} \frac{n}{2} a_n x^{\frac{n-2}{2}}. \tag{5.80}$$

The differential equation expands to

$$\underbrace{\sum_{n=1}^{\infty} \frac{n}{2} a_n x^{\frac{n-2}{2}}}_{dy/dx} + \sqrt{x} \underbrace{\sum_{n=0}^{\infty} a_n x^{n/2}}_{y} = 0, \tag{5.81}$$

$$\sum_{n=1}^{\infty} \frac{n}{2} a_n x^{\frac{n-2}{2}} + \sum_{n=0}^{\infty} a_n x^{\frac{n+1}{2}} = 0. \tag{5.82}$$

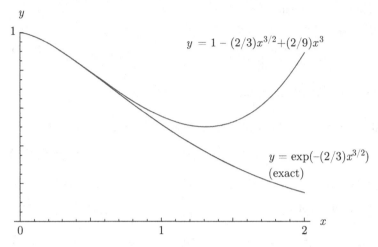

Figure 5.5. Comparison of truncated series and exact solutions to $dy/dx = \sqrt{x}y$, $y(0) = 1$.

Now take $n = m + 3$ and alter the first sum to get

$$\sum_{m=-2}^{\infty} \frac{m+3}{2}a_{m+3}x^{\frac{m+1}{2}} + \sum_{n=0}^{\infty} a_n x^{\frac{n+1}{2}} = 0, \qquad (5.83)$$

$$\sum_{n=-2}^{\infty} \frac{n+3}{2}a_{n+3}x^{\frac{n+1}{2}} + \sum_{n=0}^{\infty} a_n x^{\frac{n+1}{2}} = 0, \qquad (5.84)$$

$$\frac{1}{2}a_1 x^{-1/2} + a_2 + \sum_{n=0}^{\infty} \left(\frac{n+3}{2}a_{n+3} + a_n \right) x^{\frac{n+1}{2}} = 0. \qquad (5.85)$$

From this we deduce that $a_1 = a_2 = 0$ and

$$a_{n+3} = -\frac{2}{n+3}a_n. \qquad (5.86)$$

From the initial condition, we must have

$$a_0 = y_0. \qquad (5.87)$$

We then see that $a_3 = -(2/3)y_0$, $a_4 = a_5 = 0$, $a_6 = (2/9)y_0, \ldots$, consistent with the Taylor series of Eq. (5.77). With $y(0) = y_0 = 1$, we plot the exact solution and a three-term power series approximation in Figure 5.5.

5.2.3 Second-Order Differential Equations

We consider series solutions of

$$P(x)\frac{d^2y}{dx^2} + Q(x)\frac{dy}{dx} + R(x)y = 0, \qquad (5.88)$$

around $x = a$. There are three different cases, depending on the behavior of $P(a)$, $Q(a)$, and $R(a)$, in which $x = a$ is classified as an ordinary point, a regular singular point, or an irregular singular point. These are described next.

Ordinary Point

If $P(a) \neq 0$ and Q/P, R/P are analytic at $x = a$, this point is called an *ordinary* point. The general solution is $y = C_1 y_1(x) + C_2 y_2(x)$ where y_1 and y_2 are of the form $\sum_{n=0}^{\infty} a_n (x - a)^n$. The radius of convergence of the series is the distance to the nearest complex singularity, that is, the distance between $x = a$ and the closest point on the complex plane at which Q/P or R/P is not analytic.

EXAMPLE 5.8

Find the series solution of

$$y'' + xy' + y = 0, \qquad y(0) = y_0, \qquad y'(0) = y_0' \tag{5.89}$$

around $x = 0$.

The point $x = 0$ is an ordinary point, so that we have

$$y = \sum_{n=0}^{\infty} a_n x^n, \tag{5.90}$$

$$y' = \sum_{n=1}^{\infty} n a_n x^{n-1}, \tag{5.91}$$

$$xy' = \sum_{n=1}^{\infty} n a_n x^n, \tag{5.92}$$

$$= \sum_{n=0}^{\infty} n a_n x^n, \tag{5.93}$$

$$y'' = \sum_{n=2}^{\infty} n(n-1) a_n x^{n-2}. \tag{5.94}$$

Take $m = n - 2$, so that

$$y'' = \sum_{m=0}^{\infty} (m+1)(m+2) a_{m+2} x^m, \tag{5.95}$$

$$= \sum_{n=0}^{\infty} (n+1)(n+2) a_{n+2} x^n. \tag{5.96}$$

Substituting into Eq. (5.89), we get

$$\sum_{n=0}^{\infty} \underbrace{\left((n+1)(n+2) a_{n+2} + n a_n + a_n \right)}_{=0} x^n = 0. \tag{5.97}$$

Equating the coefficients to zero, we get the recursion relation

$$a_{n+2} = -\frac{1}{n+2} a_n, \tag{5.98}$$

so that

$$y = a_0 \left(1 - \frac{x^2}{2} + \frac{x^4}{4 \cdot 2} - \frac{x^6}{6 \cdot 4 \cdot 2} + \cdots \right) + a_1 \left(x - \frac{x^3}{3} + \frac{x^5}{5 \cdot 3} - \frac{x^7}{7 \cdot 5 \cdot 3} + \cdots \right), \tag{5.99}$$

$$= y_0 \left(1 - \frac{x^2}{2} + \frac{x^4}{4 \cdot 2} - \frac{x^6}{6 \cdot 4 \cdot 2} + \cdots \right) + y_0' \left(x - \frac{x^3}{3} + \frac{x^5}{5 \cdot 3} - \frac{x^7}{7 \cdot 5 \cdot 3} + \cdots \right), \tag{5.100}$$

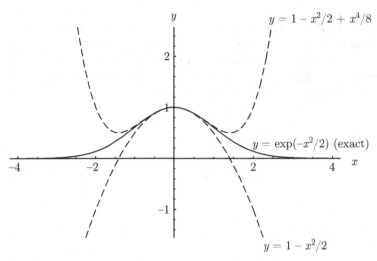

Figure 5.6. Comparison of truncated series and exact solutions to $y'' + xy' + y = 0, y(0) = 1$, $y'(0) = 0$.

$$= y_0 \sum_{n=0}^{\infty} \frac{(-1)^n}{2^n n!} x^{2n} + y_0' \sum_{n=1}^{\infty} \frac{(-1)^{n-1} 2^n n!}{(2n)!} x^{2n-1}, \tag{5.101}$$

$$= y_0 \sum_{n=0}^{\infty} \frac{1}{n!} \left(\frac{-x^2}{2} \right)^n - \frac{y_0'}{x} \sum_{n=1}^{\infty} \frac{n!}{(2n)!} \left(-2x^2 \right)^n. \tag{5.102}$$

The series converges for all x. For $y_0 = 1, y_0' = 0$ the exact solution, which can be shown to be

$$y = \exp\left(-\frac{x^2}{2} \right), \tag{5.103}$$

and two approximations to the exact solution are shown in Figure 5.6. For arbitrary y_0 and y_0', the solution can be shown to be

$$y = \exp\left(-\frac{x^2}{2} \right) \left(y_0 + \sqrt{\frac{\pi}{2}} y_0' \text{erfi} \left(\frac{x}{\sqrt{2}} \right) \right). \tag{5.104}$$

Here again "erfi" is the imaginary error function; see Section A.7.2.

Regular Singular Point

If $P(a) = 0$, then $x = a$ is a *singular* point. Furthermore, if $(x - a)Q/P$ and $(x - a)^2 R/P$ are both analytic at $x = a$, this point is called a *regular* singular point. Then there exists at least *one* solution of the form

$$y(x) = (x - a)^r \sum_{n=0}^{\infty} a_n (x - a)^n = \sum_{n=0}^{\infty} a_n (x - a)^{n+r}. \tag{5.105}$$

This is known as the *Frobenius*[4] method. The radius of convergence of the series is again the distance to the nearest complex singularity.

----

[4] Ferdinand Georg Frobenius, 1849–1917, Prussian-German mathematician.

An equation for r is called the *indicial* equation. The following are the different kinds of solutions of the indicial equation:

- $r_1 \neq r_2$, and $r_1 - r_2$ not an integer; then

$$y_1 = (x-a)^{r_1} \sum_{n=0}^{\infty} a_n(x-a)^n = \sum_{n=0}^{\infty} a_n(x-a)^{n+r_1}, \qquad (5.106)$$

$$y_2 = (x-a)^{r_2} \sum_{n=0}^{\infty} b_n(x-a)^n = \sum_{n=0}^{\infty} a_n(x-a)^{n+r_2}. \qquad (5.107)$$

- $r_1 = r_2 = r$ then

$$y_1 = (x-a)^{r} \sum_{n=0}^{\infty} a_n(x-a)^n = \sum_{n=0}^{\infty} a_n(x-a)^{n+r}, \qquad (5.108)$$

$$y_2 = y_1 \ln x + (x-a)^{r} \sum_{n=0}^{\infty} b_n(x-a)^n = y_1 \ln x + \sum_{n=0}^{\infty} b_n(x-a)^{n+r}. \qquad (5.109)$$

- $r_1 \neq r_2$, and $r_1 - r_2$ is a positive integer; then

$$y_1 = (x-a)^{r_1} \sum_{n=0}^{\infty} a_n(x-a)^n = \sum_{n=0}^{\infty} a_n(x-a)^{n+r_1}, \qquad (5.110)$$

$$y_2 = k y_1 \ln x + (x-a)^{r_2} \sum_{n=0}^{\infty} b_n(x-a)^n = k y_1 \ln x + \sum_{n=0}^{\infty} b_n(x-a)^{n+r_2}. \qquad (5.111)$$

The constants a_n and k are determined by the differential equation. The general solution is

$$y(x) = C_1 y_1(x) + C_2 y_2(x). \qquad (5.112)$$

EXAMPLE 5.9

Find the series solution of

$$4xy'' + 2y' + y = 0, \qquad (5.113)$$

around $x = 0$.

The point $x = 0$ is a regular singular point. So we have $a = 0$ and take

$$y = x^{r} \sum_{n=0}^{\infty} a_n x^n, \qquad (5.114)$$

$$= \sum_{n=0}^{\infty} a_n x^{n+r}, \qquad (5.115)$$

$$y' = \sum_{n=0}^{\infty} a_n(n+r)x^{n+r-1}, \qquad (5.116)$$

$$y'' = \sum_{n=0}^{\infty} a_n(n+r)(n+r-1)x^{n+r-2}, \qquad (5.117)$$

$$\underbrace{4 \sum_{n=0}^{\infty} a_n(n+r)(n+r-1)x^{n+r-1}}_{=4xy''} + \underbrace{2 \sum_{n=0}^{\infty} a_n(n+r)x^{n+r-1}}_{=2y'} + \underbrace{\sum_{n=0}^{\infty} a_n x^{n+r}}_{=y} = 0, \qquad (5.118)$$

$$2 \sum_{n=0}^{\infty} a_n(n+r)(2n+2r-1)x^{n+r-1} + \sum_{n=0}^{\infty} a_n x^{n+r} = 0, \tag{5.119}$$

$$m = n-1 \qquad 2 \sum_{m=-1}^{\infty} a_{m+1}(m+1+r)(2(m+1)+2r-1)x^{m+r} + \sum_{n=0}^{\infty} a_n x^{n+r} = 0, \tag{5.120}$$

$$2 \sum_{n=-1}^{\infty} a_{n+1}(n+1+r)(2(n+1)+2r-1)x^{n+r} + \sum_{n=0}^{\infty} a_n x^{n+r} = 0, \tag{5.121}$$

$$2a_0 r(2r-1)x^{-1+r} + 2 \sum_{n=0}^{\infty} a_{n+1}(n+1+r)(2(n+1)+2r-1)x^{n+r} + \sum_{n=0}^{\infty} a_n x^{n+r} = 0. \tag{5.122}$$

The first term ($n = -1$) gives the indicial equation:

$$r(2r-1) = 0, \tag{5.123}$$

from which $r = 0, 1/2$. We then have

$$2 \sum_{n=0}^{\infty} a_{n+1}(n+r+1)(2n+2r+1)x^{n+r} + \sum_{n=0}^{\infty} a_n x^{n+r} = 0, \tag{5.124}$$

$$\sum_{n=0}^{\infty} \underbrace{(2a_{n+1}(n+r+1)(2n+2r+1) + a_n)}_{=0} x^{n+r} = 0. \tag{5.125}$$

For $r = 0$,

$$a_{n+1} = -a_n \frac{1}{(2n+2)(2n+1)}, \tag{5.126}$$

$$y_1 = a_0 \left(1 - \frac{x}{2!} + \frac{x^2}{4!} - \frac{x^3}{6!} + \cdots \right). \tag{5.127}$$

For $r = 1/2$,

$$a_{n+1} = -a_n \frac{1}{2(2n+3)(n+1)}, \tag{5.128}$$

$$y_2 = a_0 x^{1/2} \left(1 - \frac{x}{3!} + \frac{x^2}{5!} - \frac{x^3}{7!} + \cdots \right). \tag{5.129}$$

The series converges for all x to $y_1 = \cos \sqrt{x}$ and $y_2 = \sin \sqrt{x}$. The general solution is

$$y = C_1 y_1 + C_2 y_2, \tag{5.130}$$

or

$$y(x) = C_1 \cos \sqrt{x} + C_2 \sin \sqrt{x}. \tag{5.131}$$

Note that $y(x)$ is real and nonsingular for $x \in [0, \infty)$. However, the first derivative,

$$y'(x) = -C_1 \frac{\sin \sqrt{x}}{2\sqrt{x}} + C_2 \frac{\cos \sqrt{x}}{2\sqrt{x}}, \tag{5.132}$$

is singular at $x = 0$. The nature of the singularity is seen from a Taylor series expansion of $y'(x)$ about $x = 0$, which gives

$$y'(x) \sim C_1 \left(-\frac{1}{2} + \frac{x}{12} + \cdots \right) + C_2 \left(\frac{1}{2\sqrt{x}} - \frac{\sqrt{x}}{4} + \cdots \right). \tag{5.133}$$

So there is a weak $1/\sqrt{x}$ singularity in $y'(x)$ at $x = 0$.

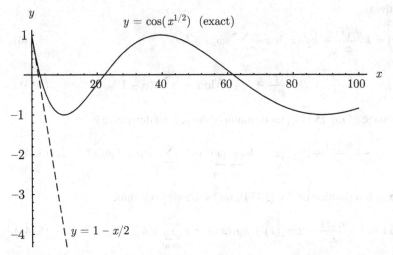

Figure 5.7. Comparison of truncated series and exact solutions to $4xy'' + 2y' + y = 0, y(0) = 1, y'(0) < \infty$.

For $y(0) = 1, y'(0) < \infty$, the exact solution and the linear approximation to the exact solution are shown in Figure 5.7. For this case, one has $C_1 = 1$ to satisfy the condition on $y(0)$, and one must have $C_2 = 0$ to satisfy the nonsingular condition on $y'(0)$.

EXAMPLE 5.10

Find the series solution of

$$xy'' - y = 0, \tag{5.134}$$

around $x = 0$.

Let $y = \sum_{n=0}^{\infty} a_n x^{n+r}$. Then, from Eq. (5.134),

$$r(r-1)a_0 x^{r-1} + \sum_{n=1}^{\infty} \left((n+r)(n+r-1)a_n - a_{n-1} \right) x^{n+r-1} = 0. \tag{5.135}$$

The indicial equation is $r(r-1) = 0$, from which $r = 0, 1$.
   Consider the larger of the two, that is, $r = 1$. For this we get

$$a_n = \frac{1}{n(n+1)} a_{n-1}, \tag{5.136}$$

$$= \frac{1}{n!(n+1)!} a_0. \tag{5.137}$$

Thus,

$$y_1(x) = x + \frac{1}{2}x^2 + \frac{1}{12}x^3 + \frac{1}{144}x^4 + \cdots. \tag{5.138}$$

From Eq. (5.111), the second solution is

$$y_2(x) = ky_1(x) \ln x + \sum_{n=0}^{\infty} b_n x^n. \tag{5.139}$$

It has derivatives

$$y_2'(x) = k\frac{y_1(x)}{x} + ky_1'(x)\ln x + \sum_{n=0}^{\infty} nb_n x^{n-1}, \tag{5.140}$$

$$y_2''(x) = -k\frac{y_1(x)}{x^2} + 2k\frac{y_1'(x)}{x} + ky_1''(x)\ln x + \sum_{n=0}^{\infty} n(n-1)b_n x^{n-2}. \tag{5.141}$$

To take advantage of Eq. (5.134), let us multiply the second derivative by x:

$$xy_2''(x) = -k\frac{y_1(x)}{x} + 2ky_1'(x) + k\underbrace{xy_1''(x)}_{=y_1(x)}\ln x + \sum_{n=0}^{\infty} n(n-1)b_n x^{n-1}. \tag{5.142}$$

Now because y_1 is a solution of Eq. (5.134), we have $xy_1'' = y_1$; thus,

$$xy_2''(x) = -k\frac{y_1(x)}{x} + 2ky_1'(x) + ky_1(x)\ln x + \sum_{n=0}^{\infty} n(n-1)b_n x^{n-1}. \tag{5.143}$$

Now subtract Eq. (5.139) from both sides and then enforce Eq. (5.134) to get

$$0 = xy_2''(x) - y_2(x) = -k\frac{y_1(x)}{x} + 2ky_1'(x) + ky_1(x)\ln x + \sum_{n=0}^{\infty} n(n-1)b_n x^{n-1}$$

$$- \left(ky_1(x)\ln x + \sum_{n=1}^{\infty} b_n x^n\right). \tag{5.144}$$

Simplifying and rearranging, we get

$$-\frac{ky_1(x)}{x} + 2ky_1'(x) + \sum_{n=0}^{\infty} n(n-1)b_n x^{n-1} - \sum_{n=0}^{\infty} b_n x^n = 0. \tag{5.145}$$

Substituting the solution $y_1(x)$ already obtained, we get

$$0 = -k\left(1 + \frac{1}{2}x + \frac{1}{12}x^2 + \cdots\right) + 2k\left(1 + x + \frac{1}{2}x^2 + \cdots\right)$$

$$+ \left(2b_2 x + 6b_3 x^2 + \cdots\right) - \left(b_0 + b_1 x + b_2 x^2 + \cdots\right). \tag{5.146}$$

Collecting terms, we have

$$k = b_0, \tag{5.147}$$

$$b_{n+1} = \frac{1}{n(n+1)}\left(b_n - \frac{k(2n+1)}{n!(n+1)!}\right), \qquad n = 1, 2, \ldots. \tag{5.148}$$

Thus,

$$y_2(x) = b_0 y_1 \ln x + b_0\left(1 - \frac{3}{4}x^2 - \frac{7}{36}x^3 - \frac{35}{1728}x^4 - \cdots\right)$$

$$+ b_1 \underbrace{\left(x + \frac{1}{2}x^2 + \frac{1}{12}x^3 + \frac{1}{144}x^4 + \cdots\right)}_{=y_1}. \tag{5.149}$$

Because the last part of the series, shown in an underbraced term, is actually $y_1(x)$, and we already have $C_1 y_1$ as part of the solution, we choose $b_1 = 0$. Because we also allow for a C_2, we can then set $b_0 = 1$. Thus, we take

$$y_2(x) = y_1 \ln x + \left(1 - \frac{3}{4}x^2 - \frac{7}{36}x^3 - \frac{35}{1728}x^4 - \cdots\right). \tag{5.150}$$

The general solution, $y = C_1 y_1 + C_2 y_2$, is

$$y(x) = C_1 \underbrace{\left(x + \frac{1}{2}x^2 + \frac{1}{12}x^3 + \frac{1}{144}x^4 + \cdots \right)}_{y_1}$$

$$+ C_2 \underbrace{\left(\left(x + \frac{1}{2}x^2 + \frac{1}{12}x^3 + \frac{1}{144}x^4 + \cdots \right) \ln x + \left(1 - \frac{3}{4}x^2 - \frac{7}{36}x^3 - \frac{35}{1728}x^4 - \cdots \right) \right)}_{y_2}.$$

$$\text{(5.151)}$$

It can also be shown that the solution can be represented compactly as

$$y(x) = \sqrt{x} \left(C_1 I_1(2\sqrt{x}) + C_2 K_1(2\sqrt{x}) \right), \tag{5.152}$$

where I_1 and K_1 are *modified Bessel functions of the first and second kinds, respectively, both of order 1* (see Section 4.4.8). The function $I_1(s)$ is nonsingular, while $K_1(s)$ is singular at $s = 0$.

Irregular Singular Point

If $P(a) = 0$ and in addition either $(x - a)Q/P$ or $(x - a)^2 R/P$ is not analytic at $x = a$, this point is an *irregular* singular point. In this case, a series solution cannot be guaranteed.

5.2.4 Higher-Order Differential Equations

Similar techniques can sometimes be used for equations of higher order.

EXAMPLE 5.11

Solve

$$y''' - xy = 0, \tag{5.153}$$

around $x = 0$.

Let

$$y = \sum_{n=0}^{\infty} a_n x^n, \tag{5.154}$$

from which

$$xy = \sum_{n=1}^{\infty} a_{n-1} x^n, \tag{5.155}$$

$$y''' = 6a_3 + \sum_{n=1}^{\infty} (n+1)(n+2)(n+3)a_{n+3} x^n. \tag{5.156}$$

Substituting into Eq. (5.153), we find that

$$a_3 = 0, \tag{5.157}$$

$$a_{n+3} = \frac{1}{(n+1)(n+2)(n+3)} a_{n-1}, \tag{5.158}$$

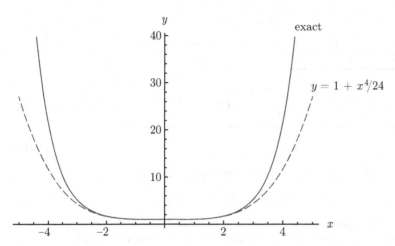

Figure 5.8. Comparison of truncated series and exact solutions to $y''' - xy = 0$, $y(0) = 1$, $y'(0) = 0$, $y''(0) = 0$.

which gives the general solution

$$y(x) = a_0 \left(1 + \frac{1}{24}x^4 + \frac{1}{8064}x^8 + \cdots \right)$$

$$+ a_1 x \left(1 + \frac{1}{60}x^4 + \frac{1}{30240}x^8 + \cdots \right)$$

$$+ a_2 x^2 \left(1 + \frac{1}{120}x^4 + \frac{1}{86400}x^8 + \cdots \right). \tag{5.159}$$

For $y(0) = 1$, $y'(0) = 0$, $y''(0) = 0$, we get $a_0 = 1$, $a_1 = 0$, and $a_2 = 0$. The exact solution and the linear approximation to the exact solution, $y \sim 1 + x^4/24$, are shown in Figure 5.8. The exact solution is expressed in terms of one of the hypergeometric functions (see Section A.7.4) and is

$$y = {}_0F_2\left(\{\} ; \left\{ \frac{1}{2}, \frac{3}{4} \right\} ; \frac{x^4}{64} \right). \tag{5.160}$$

5.3 Taylor Series Solution

In some differential equations, we can directly use Taylor series to generate approximate solutions.

EXAMPLE 5.12

Solve

$$y' = e^{xy} \tag{5.161}$$

near the origin with $y(0) = 0$.

Differentiating

$$y'' = e^{xy}\left(xy' + y \right), \tag{5.162}$$

$$y''' = e^{xy}\left(xy'' + 2y' \right) + e^{xy}\left(xy' + y \right)^2, \tag{5.163}$$

$\vdots$

so that

$$y(0) = 0, \tag{5.164}$$

$$y'(0) = 1, \tag{5.165}$$

$$y''(0) = 0, \tag{5.166}$$

$$y'''(0) = 2, \tag{5.167}$$

$$\vdots$$

the Taylor series of $y(x)$ around $x = 0$ is

$$y = y(0) + xy'(0) + \frac{x^2}{2!}y''(0) + \frac{x^3}{3!}y'''(0) + \cdots \tag{5.168}$$

$$= x + \frac{x^3}{3} \cdots. \tag{5.169}$$

EXAMPLE 5.13

Solve

$$y'' = e^{xy} \tag{5.170}$$

near the origin with $y(0) = 0, y'(0) = 1$.

Differentiating

$$y''' = e^{xy}\left(xy' + y\right), \tag{5.171}$$

$$y^{iv} = e^{xy}\left(xy'' + 2y'\right) + e^{xy}\left(xy' + y\right)^2, \tag{5.172}$$

$$\vdots$$

so that

$$y(0) = 0, \tag{5.173}$$

$$y'(0) = 1, \tag{5.174}$$

$$y''(0) = 1, \tag{5.175}$$

$$y'''(0) = 0, \tag{5.176}$$

$$y^{iv} = 2, \tag{5.177}$$

$$\vdots$$

the Taylor series of $y(x)$ around $x = 0$ is

$$y = y(0) + xy'(0) + \frac{x^2}{2!}y''(0) + \frac{x^3}{3!}y'''(0) + \cdots \tag{5.178}$$

$$= x + \frac{x^2}{2} + \frac{x^4}{24} + \cdots. \tag{5.179}$$

5.4 Perturbation Methods

Perturbation methods, also known as linearization or asymptotic techniques, are not as rigorous as infinite series methods in that usually it is impossible to make a statement regarding convergence. Nevertheless, the methods have proven to be powerful in many areas of applied mathematics, science, and engineering.

The method hinges on the identification of a small parameter ϵ, $0 < \epsilon \ll 1$. Typically there is an easily obtained solution when $\epsilon = 0$. One then uses this solution as a seed to construct a linear theory about it. The resulting set of linear equations is then solved giving a solution that is valid in a region near $\epsilon = 0$.

5.4.1 Polynomial and Transcendental Equations

To illustrate the method of solution, we begin with quadratic algebraic equations for which exact solutions are available. We can then easily see the advantages and limitations of the method.

EXAMPLE 5.14

For $0 < \epsilon \ll 1$, solve

$$x^2 + \epsilon x - 1 = 0. \tag{5.180}$$

Let

$$x = x_0 + \epsilon x_1 + \epsilon^2 x_2 + \cdots. \tag{5.181}$$

Substituting into Eq. (5.180),

$$\underbrace{\left(x_0 + \epsilon x_1 + \epsilon^2 x_2 + \cdots\right)^2}_{=x^2} + \epsilon \underbrace{\left(x_0 + \epsilon x_1 + \epsilon^2 x_2 + \cdots\right)}_{=x} - 1 = 0, \tag{5.182}$$

expanding the square by polynomial multiplication,

$$\left(x_0^2 + 2x_1x_0\epsilon + \left(x_1^2 + 2x_2x_0\right)\epsilon^2 + \cdots\right) + \left(x_0\epsilon + x_1\epsilon^2 + \cdots\right) - 1 = 0. \tag{5.183}$$

Regrouping, we get

$$\underbrace{(x_0^2 - 1)}_{=0}\epsilon^0 + \underbrace{(2x_1x_0 + x_0)}_{=0}\epsilon^1 + \underbrace{(x_1^2 + 2x_0x_2 + x_1)}_{=0}\epsilon^2 + \cdots = 0. \tag{5.184}$$

Because the equality holds for any small, positive ϵ, the coefficients in Eq. (5.184) must each equal zero. This amounts to collecting terms of common order in ϵ. Thus, we get

$$O(\epsilon^0): \qquad x_0^2 - 1 = 0 \Rightarrow x_0 = 1, \quad -1,$$
$$O(\epsilon^1): \qquad 2x_0x_1 + x_0 = 0 \Rightarrow x_1 = -\tfrac{1}{2}, \ -\tfrac{1}{2}, \tag{5.185}$$
$$O(\epsilon^2): x_1^2 + 2x_0x_2 + x_1 = 0 \Rightarrow x_2 = \tfrac{1}{8}, \quad -\tfrac{1}{8},$$
$$\vdots$$

The solutions are

$$x = 1 - \frac{\epsilon}{2} + \frac{\epsilon^2}{8} + \cdots, \tag{5.186}$$

and

$$x = -1 - \frac{\epsilon}{2} - \frac{\epsilon^2}{8} + \cdots. \tag{5.187}$$

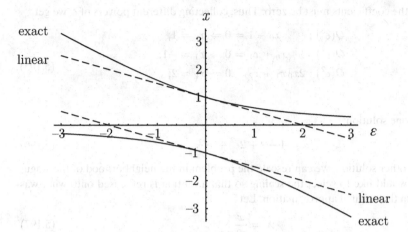

Figure 5.9. Comparison of asymptotic and exact solutions to $x^2 + \epsilon x - 1 = 0$ as a function of ϵ.

The exact solutions can also be expanded,

$$x = \frac{1}{2} \left(-\epsilon \pm \sqrt{\epsilon^2 + 4} \right), \tag{5.188}$$

$$= \pm 1 - \frac{\epsilon}{2} \pm \frac{\epsilon^2}{8} + \cdots, \tag{5.189}$$

to give the same results. The exact solution and the linear approximation are shown in Figure 5.9.

EXAMPLE 5.15

For $0 < \epsilon \ll 1$ solve

$$\epsilon x^2 + x - 1 = 0. \tag{5.190}$$

As $\epsilon \to 0$, the equation becomes singular in the sense that the coefficient modulating the highest order term is approaching zero. Let

$$x = x_0 + \epsilon x_1 + \epsilon^2 x_2 + \cdots. \tag{5.191}$$

Substituting into Eq. (5.190), we get

$$\epsilon \underbrace{\left(x_0 + \epsilon x_1 + \epsilon^2 x_2 + \cdots \right)^2}_{x^2} + \underbrace{\left(x_0 + \epsilon x_1 + \epsilon^2 x_2 + \cdots \right)}_{x} - 1 = 0. \tag{5.192}$$

Expanding the quadratic term gives

$$\epsilon \left(x_0^2 + 2\epsilon x_0 x_1 + \cdots \right) + \left(x_0 + \epsilon x_1 + \epsilon^2 x_2 + \cdots \right) - 1 = 0, \tag{5.193}$$

$$\underbrace{(x_0 - 1)}_{=0} \epsilon^0 + \underbrace{(x_0^2 + x_1)}_{=0} \epsilon^1 + \underbrace{(2x_0 x_1 + x_2)}_{=0} \epsilon^2 + \cdots = 0. \tag{5.194}$$

As before the coefficients must be zero. Thus, collecting different powers of ϵ, we get

$$
\begin{aligned}
O(\epsilon^0): & \quad x_0 - 1 = 0 \Rightarrow x_0 = 1, \\
O(\epsilon^1): & \quad x_0^2 + x_1 = 0 \Rightarrow x_1 = -1, \\
O(\epsilon^2): & \quad 2x_0 x_1 + x_2 = 0 \Rightarrow x_2 = 2, \\
& \quad \vdots
\end{aligned}
\tag{5.195}
$$

This gives one solution:

$$
x = 1 - \epsilon + 2\epsilon^2 + \cdots.
\tag{5.196}
$$

To get the other solution, we can rescale the problem in the neighborhood of the singularity. We would like to select the scaling so that attention is refocused onto what was neglected in the original approximation. Let

$$
X = \frac{x}{\epsilon^\alpha}.
\tag{5.197}
$$

Equation (5.190) becomes

$$
\epsilon^{2\alpha+1} X^2 + \epsilon^\alpha X - 1 = 0.
\tag{5.198}
$$

The first two terms are of the same order if $2\alpha + 1 = \alpha$. This demands $\alpha = -1$. With this,

$$
X = x\epsilon, \qquad \epsilon^{-1} X^2 + \epsilon^{-1} X - 1 = 0.
\tag{5.199}
$$

This gives

$$
X^2 + X - \epsilon = 0.
\tag{5.200}
$$

With this scaling, the quadratic term remains, even when $\epsilon \to 0$. We expand

$$
X = X_0 + \epsilon X_1 + \epsilon^2 X_2 + \cdots,
\tag{5.201}
$$

so

$$
\underbrace{\left(X_0 + \epsilon X_1 + \epsilon^2 X_2 + \cdots\right)^2}_{X^2} + \underbrace{\left(X_0 + \epsilon X_1 + \epsilon^2 X_2 + \cdots\right)}_{X} - \epsilon = 0,
\tag{5.202}
$$

$$
\left(X_0^2 + 2\epsilon X_0 X_1 + \epsilon^2(X_1^2 + 2X_0 X_2) + \cdots\right) + \left(X_0 + \epsilon X_1 + \epsilon^2 X_2 + \cdots\right) - \epsilon = 0.
\tag{5.203}
$$

Collecting terms of the same order,

$$
\begin{aligned}
O(\epsilon^0): & \quad X_0^2 + X_0 = 0 \Rightarrow X_0 = -1, \quad 0, \\
O(\epsilon^1): & \quad 2X_0 X_1 + X_1 = 1 \Rightarrow X_1 = -1, \quad 1, \\
O(\epsilon^2): & \quad X_1^2 + 2X_0 X_2 + X_2 = 0 \Rightarrow X_2 = 1, \quad -1, \\
& \quad \vdots
\end{aligned}
\tag{5.204}
$$

gives the two solutions

$$
X = -1 - \epsilon + \epsilon^2 + \cdots,
\tag{5.205}
$$

$$
X = \epsilon - \epsilon^2 + \cdots,
\tag{5.206}
$$

or, with $X = x\epsilon$,

$$
x = \frac{1}{\epsilon}\left(-1 - \epsilon + \epsilon^2 + \cdots\right),
\tag{5.207}
$$

$$
x = 1 - \epsilon + \cdots.
\tag{5.208}
$$

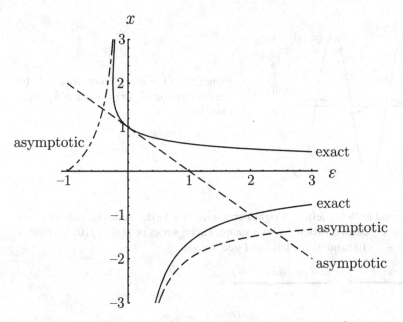

Figure 5.10. Comparison of asymptotic and exact solutions to $\epsilon x^2 + x - 1 = 0$ as a function of ϵ.

Expansion of the exact solutions,

$$x = \frac{1}{2\epsilon}\left(-1 \pm \sqrt{1 + 4\epsilon}\right), \tag{5.209}$$

$$= \frac{1}{2\epsilon}\left(-1 \pm \left(1 + 2\epsilon - 2\epsilon^2 + 4\epsilon^4 + \cdots\right)\right), \tag{5.210}$$

gives the same results. The exact solution and the linear approximation are shown in Figure 5.10.

EXAMPLE 5.16
Solve

$$\cos x = \epsilon \sin(x + \epsilon), \tag{5.211}$$

for x near $\pi/2$.

Figure 5.11 shows a plot of $f(x) = \cos x$ and $f(x) = \epsilon \sin(x + \epsilon)$ for $\epsilon = 1/10$. It is seen that there are multiple intersections near $x = \left(n + \frac{1}{2}\pi\right)$, where $n = 0, \pm 1, \pm 2, \ldots$. We seek only one of these. When we substitute

$$x = x_0 + \epsilon x_1 + \epsilon^2 x_2 + \cdots, \tag{5.212}$$

into Eq. (5.211), we find

$$\cos(\underbrace{x_0 + \epsilon x_1 + \epsilon^2 x_2 + \cdots}_{x}) = \epsilon \sin(\underbrace{x_0 + \epsilon x_1 + \epsilon^2 x_2 + \cdots}_{x} + \epsilon). \tag{5.213}$$

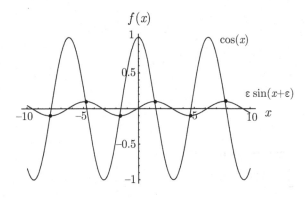

Figure 5.11. Location of roots for $\epsilon = 1/10$: the intersection of $f(x) = \cos x$ and $f(x) = \epsilon \sin(x + \epsilon)$.

Now we expand both the left- and right-hand sides in a Taylor series in ϵ about $\epsilon = 0$. We note that a general function $f(\epsilon)$ has such a Taylor series of $f(\epsilon) \sim f(0) + \epsilon f'(0) + (\epsilon^2/2)f''(0) + \cdots$ Expanding the left hand side, we get

$$\underbrace{\cos(x_0 + \epsilon x_1 + \cdots)}_{=\cos x} = \underbrace{\cos(x_0 + \epsilon x_1 + \cdots)|_{\epsilon=0}}_{=\cos x|_{\epsilon=0}}$$

$$+ \epsilon \overbrace{\underbrace{(-\sin(x_0 + \epsilon x_1 + \cdots))}_{=d/dx(\cos x)|_{\epsilon=0}} \underbrace{(x_1 + 2\epsilon x_2 + \cdots)}_{= dx/d\epsilon|_{\epsilon=0}}}^{= d/d\epsilon(\cos x)|_{\epsilon=0}}\Bigg|_{\epsilon=0} + \cdots, \quad (5.214)$$

$$\cos(x_0 + \epsilon x_1 + \cdots) = \cos x_0 - \epsilon x_1 \sin x_0 + \cdots. \quad (5.215)$$

The right-hand side is similar. We then arrive at Eq. (5.211) being expressed as

$$\cos x_0 - \epsilon x_1 \sin x_0 + \cdots = \epsilon(\sin x_0 + \cdots). \quad (5.216)$$

Collecting terms

$$O(\epsilon^0): \qquad \cos x_0 = 0 \Rightarrow x_0 = \frac{\pi}{2},$$
$$O(\epsilon^1): -x_1 \sin x_0 - \sin x_0 = 0 \Rightarrow x_1 = -1, \quad (5.217)$$
$$\vdots$$

the solution is

$$x = \frac{\pi}{2} - \epsilon + \cdots. \quad (5.218)$$

5.4.2 Regular Perturbations

Differential equations can also be solved using perturbation techniques.

EXAMPLE 5.17

For $0 < \epsilon \ll 1$ solve the nonlinear ordinary differential equation

$$y'' + \epsilon y^2 = 0, \quad (5.219)$$

$$y(0) = 1, \qquad y'(0) = 0. \quad (5.220)$$

Let

$$y(x) = y_0(x) + \epsilon y_1(x) + \epsilon^2 y_2(x) + \cdots, \tag{5.221}$$

$$y'(x) = y_0'(x) + \epsilon y_1'(x) + \epsilon^2 y_2'(x) + \cdots, \tag{5.222}$$

$$y''(x) = y_0''(x) + \epsilon y_1''(x) + \epsilon^2 y_2''(x) + \cdots. \tag{5.223}$$

Substituting into Eq. (5.219),

$$\underbrace{\left(y_0''(x) + \epsilon y_1''(x) + \epsilon^2 y_2''(x) + \cdots\right)}_{y''} + \epsilon \underbrace{\left(y_0(x) + \epsilon y_1(x) + \epsilon^2 y_2(x) + \cdots\right)^2}_{y^2} = 0, \tag{5.224}$$

$$\left(y_0''(x) + \epsilon y_1''(x) + \epsilon^2 y_2''(x) + \cdots\right) + \epsilon \left(y_0^2(x) + 2\epsilon y_1(x)y_0(x) + \cdots\right) = 0. \tag{5.225}$$

Substituting into the boundary conditions, Eq. (5.220):

$$y_0(0) + \epsilon y_1(0) + \epsilon^2 y_2(0) + \cdots = 1, \tag{5.226}$$

$$y_0'(0) + \epsilon y_1'(0) + \epsilon^2 y_2'(0) + \cdots = 0. \tag{5.227}$$

Collecting terms

$$\begin{aligned}
O(\epsilon^0) &: y_0'' = 0, & y_0(0) = 1, \; y_0'(0) = 0 \Rightarrow y_0 = 1, \\
O(\epsilon^1) &: y_1'' = -y_0^2, & y_1(0) = 0, \; y_1'(0) = 0 \Rightarrow y_1 = -\frac{x^2}{2}, \\
O(\epsilon^2) &: y_2'' = -2y_0 y_1, & y_2(0) = 0, \; y_2'(0) = 0 \Rightarrow y_2 = \frac{x^4}{12}, \\
&\vdots
\end{aligned} \tag{5.228}$$

the solution is

$$y = 1 - \epsilon \frac{x^2}{2} + \epsilon^2 \frac{x^4}{12} + \cdots. \tag{5.229}$$

For validity of the asymptotic solution, we must have

$$1 \gg \epsilon \frac{x^2}{2}. \tag{5.230}$$

This solution becomes invalid when the first term is as large as or larger than the second:

$$1 \leq \epsilon \frac{x^2}{2}, \tag{5.231}$$

$$|x| \geq \sqrt{\frac{2}{\epsilon}}. \tag{5.232}$$

Using the techniques of Section 3.9.2, it is seen that Eqs. (5.219, 5.220) possess an exact solution. With

$$u = \frac{dy}{dx}, \qquad \frac{d^2y}{dx^2} = \frac{dy'}{dy}\frac{dy}{dx} = \frac{du}{dy}u, \tag{5.233}$$

Eq. (5.219) becomes

$$u\frac{du}{dy} + \epsilon y^2 = 0, \tag{5.234}$$

$$u \, du = -\epsilon y^2 \, dy, \tag{5.235}$$

$$\frac{u^2}{2} = -\frac{\epsilon}{3}y^3 + C_1. \tag{5.236}$$

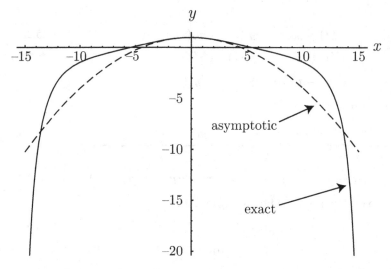

Figure 5.12. Comparison of asymptotic and exact solutions for $y'' + \epsilon y^2 = 0$, $y(0) = 1$, $y'(0) = 0$, $\epsilon = 1/10$.

We have $u = 0$ when $y = 1$, so $C = \epsilon/3$. Thus

$$u = \pm\sqrt{\frac{2\epsilon}{3}(1-y^3)}, \tag{5.237}$$

$$\frac{dy}{dx} = \pm\sqrt{\frac{2\epsilon}{3}(1-y^3)}, \tag{5.238}$$

$$dx = \pm\frac{dy}{\sqrt{\frac{2\epsilon}{3}(1-y^3)}}, \tag{5.239}$$

$$x = \pm\int_1^y \frac{ds}{\sqrt{\frac{2\epsilon}{3}(1-s^3)}}. \tag{5.240}$$

It can be shown that this integral can be represented in terms of (1) the gamma function Γ (see Section A.7.1), and (2) the Gauss hypergeometric function, $_2F_1(a,b,c,z)$ (see Section A.7.4), as follows:

$$x = \mp\sqrt{\frac{\pi}{6\epsilon}}\frac{\Gamma\left(\frac{1}{3}\right)}{\Gamma\left(\frac{5}{6}\right)} \pm \sqrt{\frac{3}{2\epsilon}}\, y\left(\,_2F_1\left(\frac{1}{3}, \frac{1}{2}, \frac{4}{3}, y^3\right)\right). \tag{5.241}$$

It is likely difficult to invert either Eq. (5.240) or (5.241) to get $y(x)$ explicitly. For small ϵ, the essence of the solution is better conveyed by the asymptotic solution. Portions of the asymptotic and exact solutions for $\epsilon = 1/10$ are shown in Figure 5.12. For this value, the asymptotic solution is expected to be invalid for $|x| \geq \sqrt{2/\epsilon} = 4.47$, and Figure 5.12 verifies this.

EXAMPLE 5.18
Solve

$$y'' + \epsilon y^2 = 0, \qquad y(0) = 1, \qquad y'(0) = \epsilon. \tag{5.242}$$

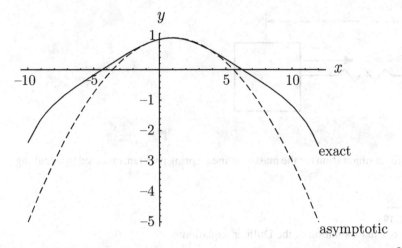

Figure 5.13. Comparison of asymptotic and exact solutions for $y'' + \epsilon y^2 = 0$, $y(0) = 1$, $y'(0) = \epsilon$, $\epsilon = 1/10$.

Let

$$y(x) = y_0(x) + \epsilon y_1(x) + \epsilon^2 y_2(x) + \cdots. \tag{5.243}$$

Substituting into Eq. (5.242) and collecting terms

$$O(\epsilon^0) : y_0'' = 0, \qquad y_0(0) = 1,\ y_0'(0) = 0 \Rightarrow y_0 = 1,$$

$$O(\epsilon^1) : y_1'' = -y_0^2, \qquad y_1(0) = 0,\ y_1'(0) = 1 \Rightarrow y_1 = -\frac{x^2}{2} + x,$$

$$O(\epsilon^2) : y_2'' = -2y_0 y_1, \qquad y_2(0) = 0,\ y_2'(0) = 0 \Rightarrow y_2 = \frac{x^4}{12} - \frac{x^3}{3}, \tag{5.244}$$

$$\vdots$$

the solution is

$$y = 1 - \epsilon \left(\frac{x^2}{2} - x \right) + \epsilon^2 \left(\frac{x^4}{12} - \frac{x^3}{3} \right) + \cdots. \tag{5.245}$$

A portion of the asymptotic and exact solutions for $\epsilon = 1/10$ are shown in Figure 5.13. Compared to the previous example, there is a slight shift in the direction of increasing x.

5.4.3 Strained Coordinates

The regular perturbation expansion may not be valid over the complete domain of interest. The method of *strained coordinates*, also known as the Poincaré[5]-Lindstedt[6] method, is designed to address this. In a slightly different context, this method is known as Lighthill's[7] method. Let us apply this method to a problem of sufficiently broad interest in engineering that we will give it a full physical description: a nonlinear oscillator for which inertia is balanced by nonlinear spring forces.

---

[5] Henri Poincaré, 1854–1912, French polymath.
[6] Anders Lindstedt, 1854–1939, Swedish mathematician, astronomer, and actuarial scientist.
[7] Sir Michael James Lighthill, 1924–1998, British applied mathematician.

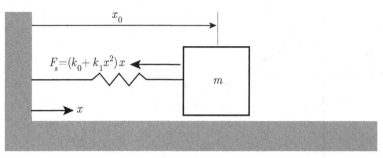

Figure 5.14. Initial configuration for the mass-nonlinear spring problem modeled by a Duffing equation.

EXAMPLE 5.19

Find an approximate solution of the Duffing[8] equation:

$$\ddot{x} + x + \epsilon x^3 = 0, \qquad x(0) = 1, \qquad \dot{x}(0) = 0. \tag{5.246}$$

Let us give some physical motivation. One problem in which the Duffing equation arises is the undamped motion of a mass subject to a nonlinear spring force. We extend the earlier analysis of a related problem in Example 4.23. To better formulate the perturbation method, we must consider first the physical dimensions of parameters and variables, and second how to scale all parameters and variables so that one is dealing with a problem where units with physical dimensions have been removed.

Consider, then, a body of mass m moving in the x direction (see Figure 5.14). All physical variables and parameters have units, which we take to be SI. Here m has units of kg, and the distance x has units of m. Initially the body is given a small positive displacement $x(0) = x_0$. The body has zero initial velocity $dx/dt(0) = 0$. Here time t has units of s. The body is subjected to a nonlinear spring force F_s oriented such that it will pull the body towards $x = 0$:

$$F_s = (k_0 + k_1 x^2)x. \tag{5.247}$$

Here we recognize k_0 and k_1 are dimensional constants with units N/m and N/m^3, respectively. Newton's second law gives us

$$m\frac{d^2x}{dt^2} + (k_0 + k_1 x^2)x = 0, \qquad x(0) = x_0, \qquad \frac{dx}{dt}(0) = 0. \tag{5.248}$$

Choose an as yet arbitrary length scale L and an as yet arbitrary time scale T with which to scale the problem, and take dimensionless variables $\tilde{x}$ and $\tilde{t}$ to be

$$\tilde{x} = \frac{x}{L}, \qquad \tilde{t} = \frac{t}{T}. \tag{5.249}$$

Eliminate the dimensional variables in favor of the dimensionless in Eq. (5.248) to find

$$\frac{mL}{T^2}\frac{d^2\tilde{x}}{d\tilde{t}^2} + k_0 L\tilde{x} + k_1 L^3 \tilde{x}^3 = 0, \qquad L\tilde{x}(0) = x_0, \qquad \frac{L}{T}\frac{d\tilde{x}}{d\tilde{t}}(0) = 0. \tag{5.250}$$

Rearrange to make all terms dimensionless:

$$\frac{d^2\tilde{x}}{d\tilde{t}^2} + \frac{k_0 T^2}{m}\tilde{x} + \frac{k_1 L^2 T^2}{m}\tilde{x}^3 = 0, \qquad \tilde{x}(0) = \frac{x_0}{L}, \qquad \frac{d\tilde{x}}{d\tilde{t}}(0) = 0. \tag{5.251}$$

Now we want to examine the effect of small nonlinearities. Choose the length and time scales such that the leading order motion has an amplitude which is $O(1)$ and a frequency

---

[8] Georg Duffing, 1861–1944, German engineer.

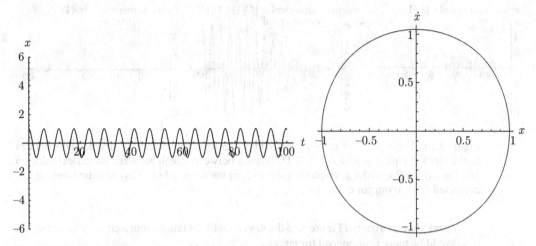

Figure 5.15. Numerical solution $x(t)$ and phase plane trajectory, $\dot{x}$ versus x for the Duffing equation, $\ddot{x} + x + \epsilon x^3 = 0$, $x(0) = 1$, $\dot{x}(0) = 0$, $\epsilon = 1/5$.

which is $O(1)$. One simple choice to achieve this end is to take

$$T \equiv \sqrt{\frac{m}{k_0}}, \qquad L \equiv x_0. \tag{5.252}$$

So

$$\frac{d^2\tilde{x}}{d\tilde{t}^2} + \tilde{x} + \frac{k_1 x_0^2 \frac{m}{k_0}}{m} \tilde{x}^3 = 0, \qquad \tilde{x}(0) = 1, \qquad \frac{d\tilde{x}}{d\tilde{t}}(0) = 0. \tag{5.253}$$

Choosing now to define a new dimensionless parameter ϵ in terms of the known dimensional parameters of the problem,

$$\epsilon \equiv \frac{k_1 x_0^2}{k_0}, \tag{5.254}$$

we get

$$\frac{d^2\tilde{x}}{d\tilde{t}^2} + \tilde{x} + \epsilon \tilde{x}^3 = 0, \qquad \tilde{x}(0) = 1, \qquad \frac{d\tilde{x}}{d\tilde{t}}(0) = 0. \tag{5.255}$$

Let us consider an asymptotic limit in which ϵ is small. Our solution will thus require

$$\epsilon \ll 1. \tag{5.256}$$

In terms of physical variables, this corresponds to

$$k_1 x_0^2 \ll k_0. \tag{5.257}$$

This implies the linear spring constant dominates the most important nonlinear spring effects. However, we have not neglected all nonlinear effects because we consider small, *finite* ϵ.

Now, let us drop the superscripts and focus on the mathematics. An accurate numerical approximation to the exact solution $x(t)$ for $\epsilon = 1/5$ and the so-called *phase plane* for this solution, giving a plot of $\dot{x}$ versus x, are shown in Figure 5.15. If $\epsilon = 0$, the problem is purely linear, and the solution is $x(t) = \cos t$, and thus $\dot{x} = -\sin t$. Thus, for $\epsilon = 0$, $x^2 + \dot{x}^2 = \cos^2 t + \sin^2 t = 1$. This is in fact a statement of mechanical energy conservation. The term x^2 is proportional to the potential energy of the linear spring, and $\dot{x}^2$ is proportional to the kinetic energy. The $\epsilon = 0$ phase plane solution is a unit circle. The

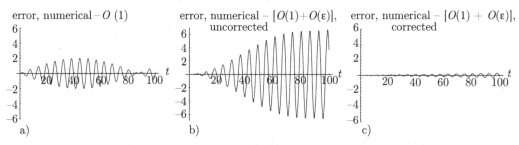

Figure 5.16. Error plots for various approximations from the method of strained coordinates to the Duffing equation with $\epsilon = 1/5$. Difference between a high accuracy numerical solution and the (a) leading order asymptotic solution, (b) uncorrected $O(\epsilon)$ asymptotic solution, (c) corrected $O(\epsilon)$ asymptotic solution.

phase plane portrait of Figure 5.15 displays a small deviation from a circle. This deviation would be more pronounced for larger ϵ.

Let us use an asymptotic method to try to capture the bulk of the nonlinear solution. Using the expansion

$$x(t) = x_0(t) + \epsilon x_1(t) + \epsilon^2 x_2(t) + \cdots, \tag{5.258}$$

and collecting terms, we find

$$O(\epsilon^0): \ddot{x}_0 + x_0 = 0, \quad x_0(0) = 1, \dot{x}_0(0) = 0 \Rightarrow x_0 = \cos t,$$
$$O(\epsilon^1): \ddot{x}_1 + x_1 = -x_0^3, x_1(0) = 0, \dot{x}_1(0) = 0 \Rightarrow x_1 = \tfrac{1}{32}(-\cos t + \cos 3t - 12t \sin t),$$
$$\vdots \tag{5.259}$$

The difference between the exact solution and the leading order solution, $x_{exact}(t) - x_0(t)$, is plotted in Figure 5.16a. The error is the same order of magnitude as the solution itself for moderate values of t. This is undesirable.

To $O(\epsilon)$, the solution is

$$x = \cos t + \frac{\epsilon}{32}\left(-\cos t + \cos 3t - \underbrace{12t \sin t}_{\text{secular term}}\right) + \cdots. \tag{5.260}$$

This series has a so-called *secular term*,[9] $-\epsilon \tfrac{3}{8} t \sin t$, that grows without bound. Thus, our solution is only valid for $t \ll \epsilon^{-1}$.

Now nature may or may not admit unbounded growth depending on the problem. Let us return to the original Eq. (5.246) to consider whether unbounded growth is admissible. Equation (5.246) can be integrated once via the following steps:

$$\dot{x}\left(\ddot{x} + x + \epsilon x^3\right) = 0, \tag{5.261}$$

$$\dot{x}\ddot{x} + \dot{x}x + \epsilon \dot{x}x^3 = 0, \tag{5.262}$$

$$\frac{d}{dt}\left(\frac{1}{2}\dot{x}^2 + \frac{1}{2}x^2 + \frac{\epsilon}{4}x^4\right) = 0, \tag{5.263}$$

$$\frac{1}{2}\dot{x}^2 + \frac{1}{2}x^2 + \frac{\epsilon}{4}x^4 = \left(\frac{1}{2}\dot{x}^2 + \frac{1}{2}x^2 + \frac{\epsilon}{4}x^4\right)\Big|_{t=0}, \tag{5.264}$$

$$= \frac{1}{4}(2 + \epsilon), \tag{5.265}$$

---

[9] Here the adjective secular is used in a different sense than that introduced in Section 4.2.1.

indicating that the solution is bounded. The physical interpretation is that the mechanical energy is again conserved. Here the spring potential energy, including nonlinear effects, is $x^2/2 + \epsilon x^4/4$. The kinetic energy is $\dot{x}^2/2$. The difference between the exact solution and the leading order solution, $x_{exact}(t) - (x_0(t) + \epsilon x_1(t))$, is plotted in Figure 5.16b. There is some improvement for early time, but the solution is actually worse for later time. This is because of the secularity.

To have a solution valid for all time, we strain the time coordinate

$$t = (1 + c_1\epsilon + c_2\epsilon^2 + \cdots)\tau, \tag{5.266}$$

where τ is the new time variable. The c_is should be chosen to avoid secular terms.

Differentiating

$$\dot{x} = \frac{dx}{d\tau}\frac{d\tau}{dt} = \frac{dx}{d\tau}\left(\frac{dt}{d\tau}\right)^{-1}, \tag{5.267}$$

$$= \frac{dx}{d\tau}(1 + c_1\epsilon + c_2\epsilon^2 + \cdots)^{-1}, \tag{5.268}$$

$$\ddot{x} = \frac{d^2x}{d\tau^2}(1 + c_1\epsilon + c_2\epsilon^2 + \cdots)^{-2}, \tag{5.269}$$

$$= \frac{d^2x}{d\tau^2}\left(1 - c_1\epsilon + (c_1^2 - c_2)\epsilon^2 + \cdots\right)^2, \tag{5.270}$$

$$= \frac{d^2x}{d\tau^2}(1 - 2c_1\epsilon + (3c_1^2 - 2c_2)\epsilon^2 + \cdots). \tag{5.271}$$

Furthermore, we write

$$x = x_0 + \epsilon x_1 + \epsilon^2 x_2 + \cdots. \tag{5.272}$$

Substituting into Eq. (5.246), we get

$$\underbrace{\left(\frac{d^2x_0}{d\tau^2} + \epsilon\frac{d^2x_1}{d\tau^2} + \epsilon^2\frac{d^2x_2}{d\tau^2} + \cdots\right)(1 - 2c_1\epsilon + (3c_1^2 - 2c_2)\epsilon^2 + \cdots)}_{\ddot{x}}$$

$$+ \underbrace{(x_0 + \epsilon x_1 + \epsilon^2 x_2 + \cdots)}_{x} + \epsilon\underbrace{(x_0 + \epsilon x_1 + \epsilon^2 x_2 + \cdots)^3}_{x^3} = 0. \tag{5.273}$$

Collecting terms, we get

$$O(\epsilon^0): \frac{d^2x_0}{d\tau^2} + x_0 = 0, \qquad\qquad x_0(0) = 1, \frac{dx_0}{d\tau}(0) = 0,$$
$$x_0(\tau) = \cos\tau,$$

$$O(\epsilon^1): \frac{d^2x_1}{d\tau^2} + x_1 = 2c_1\frac{d^2x_0}{d\tau^2} - x_0^3, \qquad\qquad x_1(0) = 0, \frac{dx_1}{d\tau}(0) = 0, \tag{5.274}$$
$$= -2c_1\cos\tau - \cos^3\tau,$$
$$= -(2c_1 + \tfrac{3}{4})\cos\tau - \tfrac{1}{4}\cos 3\tau.$$

If we choose $c_1 = -3/8$, we get the significant removal of the term which induces the secularity, yielding

$$x_1(\tau) = \frac{1}{32}(-\cos\tau + \cos 3\tau). \tag{5.275}$$

A higher order frequency component is now part of the solution at this order. One of the key steps here was to use standard trigonometric relations (see Eq. (A.84) to make the substitution

$$\cos^3\tau = \frac{3}{4}\cos\tau + \frac{1}{4}\cos 3\tau. \tag{5.276}$$

We are thus led to the approximate solution

$$x(\tau) = \cos \tau + \epsilon \frac{1}{32}(-\cos \tau + \cos 3\tau) + \cdots. \tag{5.277}$$

Because

$$t = \left(1 - \epsilon \frac{3}{8} + \cdots\right)\tau, \tag{5.278}$$

$$\tau = \left(1 + \epsilon \frac{3}{8} + \cdots\right)t, \tag{5.279}$$

we get the corrected solution approximation to be

$$x(t) = \cos\left(\underbrace{\left(1 + \epsilon \frac{3}{8} + \cdots\right)}_{\text{Frequency Modulation (FM)}} t\right)$$

$$+ \epsilon \frac{1}{32}\left(-\cos\left(\left(1 + \epsilon \frac{3}{8} + \cdots\right)t\right) + \cos\left(3\left(1 + \epsilon \frac{3}{8} + \cdots\right)t\right)\right) + \cdots. \tag{5.280}$$

Relative to the linear case, the nonlinear approximation exhibits a frequency modulation, colloquially known as "FM." The difference between the exact solution and the leading order solution, $x_{exact}(t) - (x_0(t) + \epsilon x_1(t))$, for the corrected solution to $O(\epsilon)$ is plotted in Figure 5.16c. The error is much smaller relative to the previous cases; there does appear to be a slight growth in the amplitude of the error with time.

EXAMPLE 5.20

Find the amplitude of the limit cycle oscillations of the van der Pol[10] equation

$$\ddot{x} - \epsilon(1 - x^2)\dot{x} + x = 0, \qquad x(0) = A, \qquad \dot{x}(0) = 0, \qquad 0 < \epsilon \ll 1. \tag{5.281}$$

Here A is the amplitude and is considered to be an adjustable parameter in this problem. If a limit cycle exists, it will be valid as $t \to \infty$; here we think of t as time. This could be thought of as a model for a mass-spring-damper system with a nonlinear damping coefficient of $-\epsilon(1 - x^2)$. For small $|x|$, the damping coefficient is negative. From our intuition from linear mass-spring-damper systems, we recognize that this will lead to amplitude growth, at least for sufficiently small $|x|$. However, when the amplitude grows to $|x| > 1/\sqrt{\epsilon}$, the damping coefficient again becomes positive, thus inducing decay of the amplitude. We might expect a *limit cycle* amplitude where there exists a balance between the tendency for amplitude to grow or decay.

Let us again strain the coordinate via

$$t = (1 + c_1\epsilon + c_2\epsilon^2 + \cdots)\tau, \tag{5.282}$$

so that Eq. (5.281) becomes

$$\frac{d^2x}{d\tau^2}(1 - 2c_1\epsilon + \cdots) - \epsilon(1 - x^2)\frac{dx}{d\tau}(1 - c_1\epsilon + \cdots) + x = 0. \tag{5.283}$$

We also use

$$x = x_0 + \epsilon x_1 + \epsilon^2 x_2 + \cdots. \tag{5.284}$$

---

[10] Balthasar van der Pol, 1889–1959, Dutch physicist.

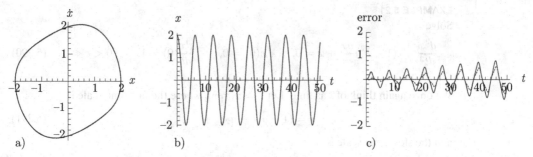

Figure 5.17. Results for the van der Pol equation, $\ddot{x} - \epsilon(1 - x^2)\dot{x} + x = 0$, $x(0) = 2$, $\dot{x}(0) = 0$, $\epsilon = 0.3$: (a) high precision numerical phase plane, (b) high precision numerical calculation of $x(t)$, (c) difference between the exact and the asymptotic leading order solution (solid), and difference between the exact and the corrected asymptotic solution to $O(\epsilon)$ (dashed) from the method of strained coordinates.

Thus, we get

$$x_0 = A \cos \tau, \tag{5.285}$$

to $O(\epsilon^0)$. To $O(\epsilon)$, the equation is

$$\frac{d^2 x_1}{d\tau^2} + x_1 = -2c_1 A \cos \tau - A \left(1 - \frac{A^2}{4}\right) \sin \tau + \frac{A^3}{4} \sin 3\tau. \tag{5.286}$$

Choosing $c_1 = 0$ and $A = 2$ to suppress secular terms, we get

$$x_1 = \frac{3}{4} \sin \tau - \frac{1}{4} \sin 3\tau. \tag{5.287}$$

The amplitude, to lowest order, is

$$A = 2, \tag{5.288}$$

so to $O(\epsilon)$ the solution is

$$x(t) = 2\cos\left(t + O(\epsilon^2)\right) + \epsilon \left(\frac{3}{4} \sin\left(t + O(\epsilon^2)\right) - \frac{1}{4} \sin\left(3\left(t + O(\epsilon^2)\right)\right)\right) + O(\epsilon^2). \tag{5.289}$$

The exact solution, x_{exact}, $\dot{x}_{exact}$, calculated by high precision numerics in the x, $\dot{x}$ phase plane, $x(t)$, and the difference between the exact solution and the asymptotic leading order solution, $x_{exact}(t) - x_0(t)$, and the difference between the exact solution and the asymptotic solution corrected to $O(\epsilon)$: $x_{exact}(t) - (x_0(t) + \epsilon x_1(t))$ is plotted in Figure 5.17.

Because of the special choice of initial conditions, the solution trajectory lies for all time on the limit cycle of the phase plane. The leading order solution is only marginally better than the corrected solution at this value of ϵ. For smaller values of ϵ, the relative errors between the two approximations would widen; that is, the asymptotic correction would become relatively speaking, more accurate.

5.4.4 Multiple Scales

The method of multiple scales is a strategy for isolating features of a solution that may evolve on widely disparate scales.

EXAMPLE 5.21

Solve

$$\frac{d^2x}{dt^2} - \epsilon(1-x^2)\frac{dx}{dt} + x = 0, \qquad x(0) = 0, \qquad \frac{dx}{dt}(0) = 1, \qquad 0 < \epsilon \ll 1. \quad (5.290)$$

We once again think of t as time. Let $x = x(\tau, \tilde{\tau})$, where the fast time scale is

$$\tau = (1 + a_1\epsilon + a_2\epsilon^2 + \cdots)t, \quad (5.291)$$

and the slow time scale is

$$\tilde{\tau} = \epsilon t. \quad (5.292)$$

Because

$$x = x(\tau, \tilde{\tau}), \quad (5.293)$$

we have

$$\frac{dx}{dt} = \frac{\partial x}{\partial \tau}\frac{d\tau}{dt} + \frac{\partial x}{\partial \tilde{\tau}}\frac{d\tilde{\tau}}{dt}. \quad (5.294)$$

The first derivative is

$$\frac{dx}{dt} = \frac{\partial x}{\partial \tau}(1 + a_1\epsilon + a_2\epsilon^2 + \cdots) + \frac{\partial x}{\partial \tilde{\tau}}\epsilon, \quad (5.295)$$

so

$$\frac{d}{dt} = (1 + a_1\epsilon + a_2\epsilon^2 + \cdots)\frac{\partial}{\partial \tau} + \epsilon\frac{\partial}{\partial \tilde{\tau}}. \quad (5.296)$$

Applying this operator to dx/dt, we get

$$\frac{d^2x}{dt^2} = (1 + a_1\epsilon + a_2\epsilon^2 + \cdots)^2\frac{\partial^2 x}{\partial \tau^2} + 2(1 + a_1\epsilon + a_2\epsilon^2 + \cdots)\epsilon\frac{\partial^2 x}{\partial \tau \partial \tilde{\tau}} + \epsilon^2\frac{\partial^2 x}{\partial \tilde{\tau}^2}. \quad (5.297)$$

Introduce

$$x = x_0 + \epsilon x_1 + \epsilon^2 x_2 + \cdots. \quad (5.298)$$

So to $O(\epsilon)$, Eq. (5.290) becomes

$$\underbrace{(1 + 2a_1\epsilon + \cdots)\frac{\partial^2 (x_0 + \epsilon x_1 + \cdots)}{\partial \tau^2} + 2\epsilon\frac{\partial^2 (x_0 + \cdots)}{\partial \tau \partial \tilde{\tau}} + \cdots}_{\ddot{x}}$$

$$\underbrace{-\epsilon(1 - x_0^2 - \cdots)\frac{\partial(x_0 + \cdots)}{\partial \tau} + \cdots}_{(1-x^2)\dot{x}} + \underbrace{(x_0 + \epsilon x_1 + \cdots)}_{x} = 0. \quad (5.299)$$

Collecting terms of $O(\epsilon^0)$, we have

$$\frac{\partial^2 x_0}{\partial \tau^2} + x_0 = 0, \qquad x_0(0,0) = 0, \quad \frac{\partial x_0}{\partial \tau}(0,0) = 1. \quad (5.300)$$

The solution is

$$x_0 = A(\tilde{\tau})\cos\tau + B(\tilde{\tau})\sin\tau, \qquad A(0) = 0, \ B(0) = 1. \quad (5.301)$$

The terms of $O(\epsilon^1)$ give

$$\frac{\partial^2 x_1}{\partial \tau^2} + x_1 = -2a_1\frac{\partial^2 x_0}{\partial \tau^2} - 2\frac{\partial^2 x_0}{\partial \tau \partial \tilde{\tau}} + (1 - x_0^2)\frac{\partial x_0}{\partial \tau}, \quad (5.302)$$

$$= \left(2a_1B + 2A' - A + \frac{A}{4}(A^2 + B^2)\right)\sin\tau$$

$$+\left(2a_1 A - 2B' + B - \frac{B}{4}(A^2 + B^2)\right)\cos\tau$$

$$+\frac{A}{4}(A^2 - 3B^2)\sin 3\tau - \frac{B}{4}(3A^2 - B^2)\cos 3\tau, \tag{5.303}$$

with

$$x_1(0,0) = 0, \tag{5.304}$$

$$\frac{\partial x_1}{\partial\tau}(0,0) = -a_1\frac{\partial x_0}{\partial\tau}(0,0) - \frac{\partial x_0}{\partial\tilde{\tau}}(0,0), \tag{5.305}$$

$$= -a_1 - \frac{\partial x_0}{\partial\tilde{\tau}}(0,0). \tag{5.306}$$

Because ϵt is already represented in $\tilde{\tau}$, choose $a_1 = 0$. Then

$$2A' - A + \frac{A}{4}(A^2 + B^2) = 0, \tag{5.307}$$

$$2B' - B + \frac{B}{4}(A^2 + B^2) = 0. \tag{5.308}$$

Because $A(0) = 0$, try $A(\tilde{\tau}) = 0$. Then

$$2B' - B + \frac{B^3}{4} = 0. \tag{5.309}$$

Multiplying by B, we find

$$2BB' - B^2 + \frac{B^4}{4} = 0, \tag{5.310}$$

$$(B^2)' - B^2 + \frac{B^4}{4} = 0. \tag{5.311}$$

Taking $F \equiv B^2$, we see that

$$F' - F + \frac{F^2}{4} = 0. \tag{5.312}$$

This is a first-order ordinary differential equation in F, which can be easily solved. Separating variables, integrating, and transforming from F back to B, we get

$$\frac{B^2}{1 - \frac{B^2}{4}} = Ce^{\tilde{\tau}}. \tag{5.313}$$

Because $B(0) = 1$, we get $C = 4/3$. From this,

$$B = \frac{2}{\sqrt{1 + 3e^{-\tilde{\tau}}}}, \tag{5.314}$$

so that

$$x(\tau, \tilde{\tau}) = \frac{2}{\sqrt{1 + 3e^{-\tilde{\tau}}}}\sin\tau + O(\epsilon) \tag{5.315}$$

$$x(t) = \underbrace{\frac{2}{\sqrt{1 + 3e^{-\epsilon t}}}}_{\text{Amplitude Modulation (AM)}}\sin\left((1 + O(\epsilon^2))t\right) + O(\epsilon). \tag{5.316}$$

The high precision numerical approximation for the solution trajectory in the $(x, \dot{x})$ phase plane, the high precision numerical solution $x_{exact}(t)$, and the difference between the exact solution and the asymptotic leading order solution, $x_{exact}(t) - x_0(t)$, and the difference between the exact solution and the asymptotic solution corrected to $O(\epsilon)$: $x_{exact}(t) - (x_0(t) + \epsilon x_1(t))$ are plotted in Figure 5.18.

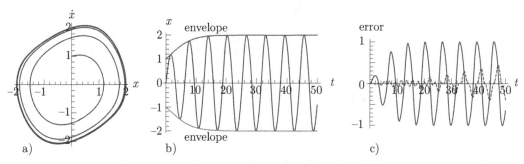

Figure 5.18. Results for the van der Pol equation, $\ddot{x} - \epsilon(1 - x^2)\dot{x} + x = 0$, $x(0) = 0, \dot{x}(0) = 1, \epsilon = 0.3$: (a) high precision numerical phase plane, (b) high precision numerical calculation of $x(t)$, along with the envelope $2/\sqrt{1 + 3e^{-\epsilon t}}$, (c) difference between the exact and the asymptotic leading order solution (solid), and difference between the exact and the corrected asymptotic solution to $O(\epsilon)$ (dashed) from the method of multiple scales.

The amplitude, which is initially 1, grows to a value of 2, the same value obtained in the previous example. This is obviously a form of amplitude modulation, in colloquial terms "AM." This is evident in the phase plane, where the initial condition does not lie on the long-time limit cycle. Here we have additionally obtained the time scale for the growth of the amplitude change. Also, the leading order approximation is poor for $t > 1/\epsilon$, while the corrected approximation is relatively good. Lastly, for $\epsilon = 0.3$, the segregation in time scales is not dramatic. The "fast" time scale is that of the oscillation and is $O(1)$. The slow time scale is $O(1/\epsilon)$, which here is around 3. For smaller ϵ, the effect would be more dramatic.

5.4.5 Boundary Layers

The method of boundary layers, also known as matched asymptotic expansions, can be used in some cases. It is most appropriate for cases in which a small parameter multiplies the highest order derivative. In such cases, a regular perturbation scheme will fail because we lose a boundary condition at leading order. Because the highest order term vanishes in such problems if the perturbation parameter is zero, the theory is classified as a *singular perturbation method*.

EXAMPLE 5.22

Solve

$$\epsilon y'' + y' + y = 0, \qquad y(0) = 0, \qquad y(1) = 1. \tag{5.317}$$

An exact solution exists, namely,

$$y(x) = \exp\left(\frac{1-x}{2\epsilon}\right) \frac{\sinh\left(\dfrac{x\sqrt{1-4\epsilon}}{2\epsilon}\right)}{\sinh\left(\dfrac{\sqrt{1-4\epsilon}}{2\epsilon}\right)}. \tag{5.318}$$

We could in principle simply expand this in a Taylor series in ϵ. However, for more difficult problems, exact solutions are not available. So here we will just use the exact solution to verify the validity of the method.

We begin with a regular perturbation expansion:

$$y(x) = y_0 + \epsilon y_1(x) + \epsilon^2 y_2(x) + \cdots. \tag{5.319}$$

Substituting and collecting terms, we get

$$O(\epsilon^0) : y_0' + y_0 = 0, \qquad y_0(0) = 0, \qquad y_0(1) = 1, \tag{5.320}$$

the solution to which is

$$y_0 = ae^{-x}. \tag{5.321}$$

It is not possible for the solution to satisfy the two boundary conditions simultaneously because we only have one free variable, a. So, we divide the region of interest $x \in [0, 1]$ into two parts, a thin *inner region* or *boundary layer* around $x = 0$, and a thick *outer region* elsewhere.

Equation (5.321) gives the solution in the outer region. To satisfy the boundary condition $y_0(1) = 1$, we find that $a = e$, so that

$$y = e^{1-x} + \cdots. \tag{5.322}$$

In the inner region, we choose a new independent variable X defined as $X = x/\epsilon$, so that the equation becomes

$$\frac{d^2 y}{dX^2} + \frac{dy}{dX} + \epsilon y = 0. \tag{5.323}$$

In essence we have trained a mathematical "microscope" onto the thin layer near $x = 0$ in such a way that its effect is amplified following the stretching coordinate transformation. Using a perturbation expansion, the lowest order equation is

$$\frac{d^2 y_0}{dX^2} + \frac{dy_0}{dX} = 0, \tag{5.324}$$

with a solution

$$y_0 = A + Be^{-X}. \tag{5.325}$$

Applying the boundary condition $y_0(0) = 0$, we get

$$y_0 = A(1 - e^{-X}). \tag{5.326}$$

Matching of the inner and outer solutions can be achieved by Prandtl's[11] method, which demands that

$$y_{inner}(X \to \infty) = y_{outer}(x \to 0), \tag{5.327}$$

which gives $A = e$. The solution in the inner region is

$$y(x) = e(1 - e^{-x/\epsilon}) + \cdots, \tag{5.328}$$

$$\lim_{x \to \infty} y = e, \tag{5.329}$$

and in the outer region is

$$y(x) = e^{1-x} + \cdots, \tag{5.330}$$

$$\lim_{x \to 0} y = e. \tag{5.331}$$

A composite solution can also be written by adding the two solutions. However, one must realize that this induces a double counting in the region where the inner layer solution

---

[11] Ludwig Prandtl, 1875–1953, German engineer.

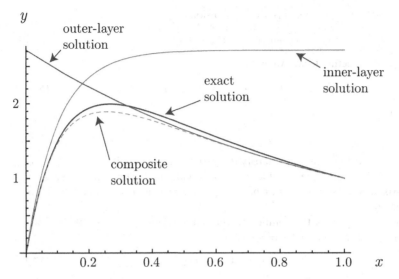

Figure 5.19. Exact, inner layer solution, outer layer solution, and composite solution for the boundary layer problem, $\epsilon y'' + y' + y = 0$, $y(0) = 0$, $y(1) = 1$, $\epsilon = 1/10$.

matches onto the outer layer solution. Thus, we need to subtract one term to account for this overlap. This is known as the *common part*. Thus, the correct composite solution is the summation of the inner and outer parts, with the common part subtracted:

$$y(x) = \underbrace{\left(e(1 - e^{-x/\epsilon}) + \cdots\right)}_{\text{inner}} + \underbrace{\left(e^{1-x} + \cdots\right)}_{\text{outer}} - \underbrace{e}_{\text{common part}}, \tag{5.332}$$

$$y = e(e^{-x} - e^{-x/\epsilon}) + \cdots. \tag{5.333}$$

The exact solution, the inner layer solution, the outer layer solution, and the composite solution are plotted in Figure 5.19.

EXAMPLE 5.23

Obtain the solution of the previous problem,

$$\epsilon y'' + y' + y = 0, \qquad y(0) = 0, \qquad y(1) = 1, \tag{5.334}$$

to the next order.

Keeping terms of the next order in ϵ, we have

$$y = e^{1-x} + \epsilon((1 - x)e^{1-x}) + \cdots, \tag{5.335}$$

for the outer solution and

$$y = A(1 - e^{-X}) + \epsilon\left(B - AX - (B + AX)e^{-X}\right) + \cdots, \tag{5.336}$$

for the inner solution.

Higher order matching via Van Dyke's[12] method is obtained by expanding the outer solution in terms of the inner variable, the inner solution in terms of the outer variable,

---

[12] Milton Denman Van Dyke, 1922–2010, American engineer and applied mathematician.

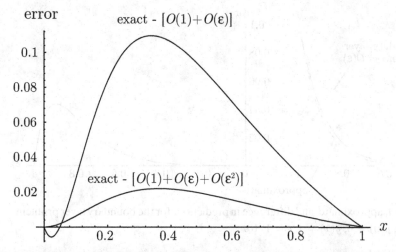

Figure 5.20. Difference between the exact and asymptotic solutions for two different orders of approximation for the boundary layer problem, $\epsilon y'' + y' + y = 0$, $y(0) = 0$, $y(1) = 1$, $\epsilon = 1/10$.

and comparing. Thus, the outer solution is, as $\epsilon \to 0$,

$$y = e^{1-\epsilon X} + \epsilon \left((1 - \epsilon X)e^{1-\epsilon X} \right) + \cdots, \tag{5.337}$$

$$= e(1 - \epsilon X) + \epsilon e(1 - \epsilon X)^2. \tag{5.338}$$

Ignoring terms that are $> O(\epsilon^2)$, we get

$$y = e(1 - \epsilon X) + \epsilon e, \tag{5.339}$$

$$= e + \epsilon e(1 - X), \tag{5.340}$$

$$= e + \epsilon e \left(1 - \frac{x}{\epsilon} \right), \tag{5.341}$$

$$= e + \epsilon e - e x. \tag{5.342}$$

Similarly, the inner solution as $\epsilon \to 0$ is

$$y = A(1 - e^{-x/\epsilon}) + \epsilon \left(B - A\frac{x}{\epsilon} - \left(B + A\frac{x}{\epsilon} \right) e^{-x/\epsilon} \right) + \cdots, \tag{5.343}$$

$$= A + B\epsilon - Ax. \tag{5.344}$$

Comparing, we get $A = B = e$, so that, in the inner region,

$$y(x) = e(1 - e^{-x/\epsilon}) + e \left(\epsilon - x - (\epsilon + x)e^{-x/\epsilon} \right) + \cdots, \tag{5.345}$$

and in the outer region,

$$y(x) = e^{1-x} + \epsilon(1 - x)e^{1-x} \cdots. \tag{5.346}$$

The composite solution, inner plus outer minus common part, reduces to

$$y = e^{1-x} - (1 + x)e^{1-x/\epsilon} + \epsilon \left((1 - x)e^{1-x} - e^{1-x/\epsilon} \right) + \cdots. \tag{5.347}$$

The difference between the exact solution and the approximation from the previous example, and the difference between the exact solution and approximation from this example, are plotted in Figure 5.20.

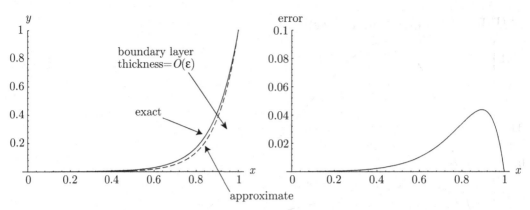

Figure 5.21. Exact, approximate, and difference in predictions for the boundary layer problem $\epsilon y'' - y' + y = 0$, $y(0) = 0$, $y(1) = 1$, $\epsilon = 1/10$.

EXAMPLE 5.24

For the previous problem, investigate the possibility of having the boundary layer at $x = 1$.

The outer solution now satisfies the condition $y(0) = 0$, giving $y = 0$. Let

$$X = \frac{x - 1}{\epsilon}. \tag{5.348}$$

The lowest order inner solution satisfying $y(X = 0) = 1$ is

$$y = A + (1 - A)e^{-X}. \tag{5.349}$$

However, as $X \to -\infty$, this becomes unbounded and cannot be matched with the outer solution. Thus, a boundary layer at $x = 1$ is not possible.

EXAMPLE 5.25

Solve

$$\epsilon y'' - y' + y = 0, \qquad y(0) = 0, \qquad y(1) = 1. \tag{5.350}$$

The boundary layer is at $x = 1$. The outer solution is $y = 0$. Taking

$$X = \frac{x - 1}{\epsilon}, \tag{5.351}$$

the inner solution is

$$y = A + (1 - A)e^X + \cdots. \tag{5.352}$$

Matching, we get

$$A = 0, \tag{5.353}$$

so that we have a composite solution

$$y(x) = e^{(x-1)/\epsilon} + \cdots. \tag{5.354}$$

The exact solution, the approximate solution to $O(\epsilon)$, and the difference between the exact solution and the approximation, all for $\epsilon = 1/10$, are plotted in Figure 5.21. Relative to the second previous example, notice that the sign of the coefficient on y' determines the location of the boundary layer.

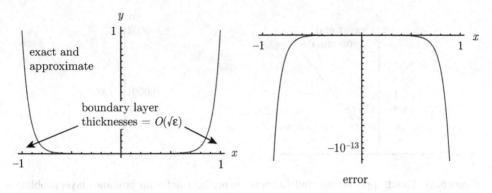

Figure 5.22. Exact, approximate, and difference in predictions for the boundary layer problem $\epsilon y'' - y = 0$, $y(-1) = y(1) = 1$, $\epsilon = 1/200$.

It is possible for a problem to have two boundary layers, as demonstrated in the following example.

EXAMPLE 5.26

Approximate the solution to

$$\epsilon y'' - y = 0, \qquad y(-1) = 1, \quad y(1) = 1. \tag{5.355}$$

The exact solution is available and is

$$y = \frac{\cosh\left(\dfrac{x}{\sqrt{\epsilon}}\right)}{\cosh\dfrac{1}{\sqrt{\epsilon}}}. \tag{5.356}$$

One can also apply perturbation methods. The outer solution is by inspection $y = 0$ and satisfies neither boundary condition. There are boundary layers at $x = -1$ and at $x = 1$.

Near $x = -1$, we take $X_{-1} = (x + 1)/\sqrt{\epsilon}$ and find solution

$$y = e^{-X_{-1}}. \tag{5.357}$$

Near $x = 1$, we take $X_1 = (x - 1)/\sqrt{\epsilon}$ and find solution

$$y = e^{X_1}. \tag{5.358}$$

Both of these solutions have a common part of zero with the outer solution. The composite asymptotic solution is thus

$$y = \exp\left(-\frac{x + 1}{\sqrt{\epsilon}}\right) + \exp\left(\frac{x - 1}{\sqrt{\epsilon}}\right). \tag{5.359}$$

The solutions and error are shown in Figure 5.22.

5.4.6 Interior Layers

Not all layers are at boundaries. It is also possible to have thin layers in the interior of the domain in which the solution adjusts to match outer solutions.

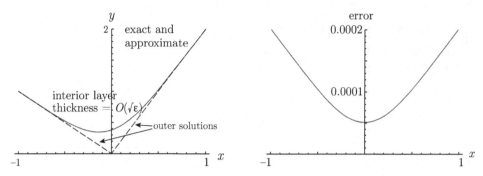

Figure 5.23. Exact, approximate, and difference in predictions for the boundary layer problem $\epsilon y'' a + xy - y = 0, y(-1) = 1, \ y(1) = 2, \epsilon = 1/10$.

EXAMPLE 5.27

Examine an example discussed in detail by Kevorkian and Cole (1981):

$$\epsilon y'' + xy' - y = 0, \qquad y(-1) = 1, \quad y(1) = 2.$$

When $\epsilon = 0$, we are led to the equation

$$y' = \frac{y}{x}, \tag{5.360}$$

which, by separation of variables, has solution

$$y = Cx. \tag{5.361}$$

Matching the boundary condition at $x = -1$ would lead us to take $y = -x$; matching the boundary condition at $x = 1$ would lead us to take $y = 2x$. Obviously neither solution is uniformly valid for $x \in [-1, 1]$, but both are valid at and away from either respective boundary. It so happens that near $x = 0$, there is an interior layer where one outer solution undergoes a transition into the other. The nature of the interior layer is revealed by transforming to $X = x/\sqrt{\epsilon}$ and then taking the limit as $\epsilon \to 0$. This leads us to

$$\frac{d^2 y}{dX^2} + X \frac{dy}{dX} - y = 0. \tag{5.362}$$

So near $x = 0$, we have rescaled to bring back the second derivative term so that it has equal weight with the other terms. One might also say this problem is almost as difficult as the original! In any case, the interior layer solution can be shown to be

$$y \sim \sqrt{\epsilon} \left(\frac{3 \left(\frac{\sqrt{\frac{\pi}{2}} x \left(\text{erf} \left(\frac{x}{\sqrt{2}\sqrt{\epsilon}} \right) + 1 \right)}{\sqrt{\epsilon}} + e^{-\frac{x^2}{2\epsilon}} \right)}{\sqrt{2\pi}} - \frac{x}{\sqrt{\epsilon}} \right). \tag{5.363}$$

Here "erf" is the error function, defined in Section A.7.2. The exact solution can be shown to be

$$y = \frac{e^{-\frac{x^2}{2\epsilon}} \left(\sqrt{2\pi} x \, \text{erf} \left(\frac{1}{\sqrt{2}\sqrt{\epsilon}} \right) e^{\frac{x^2}{2\epsilon} + \frac{1}{2\epsilon}} + 3\sqrt{2\pi} x e^{\frac{x^2}{2\epsilon} + \frac{1}{2\epsilon}} \, \text{erf} \left(\frac{x}{\sqrt{2}\sqrt{\epsilon}} \right) + 2x\sqrt{\epsilon} e^{\frac{x^2}{2\epsilon}} + 6 e^{\frac{1}{2\epsilon}} \sqrt{\epsilon} \right)}{2 \left(\sqrt{2\pi} e^{\frac{1}{2\epsilon}} \, \text{erf} \left(\frac{1}{\sqrt{2}\sqrt{\epsilon}} \right) + 2\sqrt{\epsilon} \right)}. \tag{5.364}$$

The solutions and error are shown in Figure 5.23.

5.4.7 WKBJ Method

Any equation of the form

$$\frac{d^2v}{dx^2} + P(x)\frac{dv}{dx} + Q(x)v = 0, \tag{5.365}$$

can be written as

$$\frac{d^2y}{dx^2} + R(x)y = 0, \tag{5.366}$$

where

$$v(x) = y(x)\exp\left(-\frac{1}{2}\int_0^x P(s)\,ds\right), \tag{5.367}$$

$$R(x) = Q(x) - \frac{1}{2}\frac{dP}{dx} - \frac{1}{4}\left(P(x)\right)^2. \tag{5.368}$$

So it is sufficient to study equations of the form of Eq. (5.366). The Wentzel,[13] Kramers,[14] Brillouin,[15] Jeffreys[16] (WKBJ) method is used for equations of the kind

$$\epsilon^2\frac{d^2y}{dx^2} = f(x)y, \tag{5.369}$$

where ϵ is a small parameter. This also includes an equation of the type

$$\epsilon^2\frac{d^2y}{dx^2} = (\lambda^2 p(x) + q(x))y, \tag{5.370}$$

where λ is a large parameter. Alternatively, by taking $x = \epsilon t$, Eq. (5.369) becomes

$$\frac{d^2y}{dt^2} = f(\epsilon t)y. \tag{5.371}$$

We can also write Eq. (5.369) as

$$\frac{d^2y}{dx^2} = g(x)y, \tag{5.372}$$

where $g(x)$ is slowly varying in the sense that $g'/g^{3/2} \sim O(\epsilon)$.

We seek solutions to Eq. (5.369) of the form

$$y(x) = \exp\left(\frac{1}{\epsilon}\int_{x_0}^x (S_0(s) + \epsilon S_1(s) + \epsilon^2 S_2(s) + \cdots)\,ds\right). \tag{5.373}$$

The derivatives are

$$\frac{dy}{dx} = \frac{1}{\epsilon}\left(S_0(x) + \epsilon S_1(x) + \epsilon^2 S_2(x) + \cdots\right)y(x), \tag{5.374}$$

$$\frac{d^2y}{dx^2} = \frac{1}{\epsilon^2}\left(S_0(x) + \epsilon S_1(x) + \epsilon^2 S_2(x) + \cdots\right)^2 y(x),$$

$$+ \frac{1}{\epsilon}\left(\frac{dS_0}{dx} + \epsilon\frac{dS_1}{dx} + \epsilon^2\frac{dS_2}{dx} + \cdots\right)y(x). \tag{5.375}$$

---

[13] Gregor Wentzel, 1898–1978, German physicist.
[14] Hendrik Anthony Kramers, 1894–1952, Dutch physicist.
[15] Léon Brillouin, 1889–1969, French physicist.
[16] Harold Jeffreys, 1891–1989, English mathematician.

Substituting into Eq. (5.369), we get

$$\underbrace{\left((S_0(x))^2 + 2\epsilon S_0(x)S_1(x) + \cdots\right) y(x) + \epsilon \left(\frac{dS_0}{dx} + \cdots\right) y(x)}_{=\epsilon^2 d^2 y/dx^2} = f(x)y(x). \quad (5.376)$$

Collecting terms, at $O(\epsilon^0)$, we have

$$S_0^2(x) = f(x), \quad (5.377)$$

from which

$$S_0(x) = \pm\sqrt{f(x)}. \quad (5.378)$$

To $O(\epsilon^1)$ we have

$$2S_0(x)S_1(x) + \frac{dS_0}{dx} = 0, \quad (5.379)$$

from which

$$S_1(x) = -\frac{\frac{dS_0}{dx}}{2S_0(x)}, \quad (5.380)$$

$$= -\frac{\pm\frac{1}{2\sqrt{f(x)}}\frac{df}{dx}}{2\left(\pm\sqrt{f(x)}\right)}, \quad (5.381)$$

$$= -\frac{\frac{df}{dx}}{4f(x)}. \quad (5.382)$$

Thus, we get the general solution

$$y(x) = C_1 \exp\left(\frac{1}{\epsilon}\int_{x_0}^{x}(S_0(s) + \epsilon S_1(s) + \cdots)\, ds\right)$$
$$+ C_2 \exp\left(\frac{1}{\epsilon}\int_{x_0}^{x}(S_0(s) + \epsilon S_1(s) + \cdots)\, ds\right), \quad (5.383)$$

$$= C_1 \exp\left(\frac{1}{\epsilon}\int_{x_0}^{x}(\sqrt{f(s)} - \epsilon\frac{\frac{df}{ds}}{4f(s)} + \cdots)\, ds\right)$$
$$+ C_2 \exp\left(\frac{1}{\epsilon}\int_{x_0}^{x}(-\sqrt{f(s)} - \epsilon\frac{\frac{df}{ds}}{4f(s)} + \cdots)\, ds\right), \quad (5.384)$$

$$= C_1 \exp\left(-\int_{f(x_0)}^{f(x)}\frac{df}{4f}\right)\exp\left(\frac{1}{\epsilon}\int_{x_0}^{x}(\sqrt{f(s)} + \cdots)\, ds\right)$$
$$+ C_2 \exp\left(-\int_{f(x_0)}^{f(x)}\frac{df}{4f}\right)\exp\left(-\frac{1}{\epsilon}\int_{x_0}^{x}(\sqrt{f(s)} + \cdots)\, ds\right), \quad (5.385)$$

$$= \frac{\hat{C}_1}{(f(x))^{1/4}}\exp\left(\frac{1}{\epsilon}\int_{x_0}^{x}\sqrt{f(s)}\, ds\right) + \frac{\hat{C}_2}{(f(x))^{1/4}}\exp\left(-\frac{1}{\epsilon}\int_{x_0}^{x}\sqrt{f(s)}\, ds\right) + \cdots.$$
$$(5.386)$$

This solution is not valid near $x = a$ for which $f(a) = 0$. These are called *turning points*. At such points, the solution changes from an oscillatory to an exponential character.

EXAMPLE 5.28

Find an approximate solution of the Airy[17] equation,

$$\epsilon^2 y'' + xy = 0, \qquad x > 0, \tag{5.387}$$

using WKBJ.

Before delving into the details of a WKBJ approximation, one may look at the equation for physical analogs. If we were to replace x by t and think of t as time, we could imagine the equation modeling a mass-spring system in which small ϵ indicates a small mass and the spring coefficient stiffens as time increases. We might thus expect some oscillatory solution whose frequency changes as the solution evolves.

Returning to the mathematics, in this case,

$$f(x) = -x. \tag{5.388}$$

Thus, $x = 0$ is a turning point. We find that

$$S_0(x) = \pm i\sqrt{x}, \tag{5.389}$$

and

$$S_1(x) = -\frac{S_0'}{2S_0} = -\frac{1}{4x}. \tag{5.390}$$

The solutions are of the form

$$y = \exp\left(\pm \frac{i}{\epsilon} \int \sqrt{x}\, dx - \int \frac{dx}{4x} \right) + \cdots, \tag{5.391}$$

$$= \frac{1}{x^{1/4}} \exp\left(\pm \frac{2x^{3/2} i}{3\epsilon} \right) + \cdots. \tag{5.392}$$

The general approximate solution is

$$y = \frac{C_1}{x^{1/4}} \sin\left(\frac{2x^{3/2}}{3\epsilon} \right) + \frac{C_2}{x^{1/4}} \cos\left(\frac{2x^{3/2}}{3\epsilon} \right) + \cdots. \tag{5.393}$$

The approximate solution has an obvious oscillatory behavior; both the amplitude and frequency are functions of x. The exact solution can be shown to be

$$y = C_1 \text{Ai}\left(-\epsilon^{-2/3} x \right) + C_2 \text{Bi}\left(-\epsilon^{-2/3} x \right). \tag{5.394}$$

Here Ai and Bi are Airy functions of the first and second kind, respectively (see Section A.7.5). For $x > 0$, these solutions are oscillatory.

EXAMPLE 5.29

Find a solution of $x^3 y'' = y$ for small, positive x.

Let $\epsilon^2 X = x$ so that X is of $O(1)$ if x is small. Then the equation becomes

$$\epsilon^2 \frac{d^2 y}{dX^2} = X^{-3} y. \tag{5.395}$$

---

[17] George Biddell Airy, 1801–1892, English applied mathematician.

The WKBJ method is applicable. We have $f = X^{-3}$. The general solution is

$$y = C_1' X^{3/4} \exp\left(-\frac{2}{\epsilon\sqrt{X}}\right) + C_2' X^{3/4} \exp\left(\frac{2}{\epsilon\sqrt{X}}\right) + \cdots. \tag{5.396}$$

In terms of the original variables,

$$y = C_1 x^{3/4} \exp\left(-\frac{2}{\sqrt{x}}\right) + C_2 x^{3/4} \exp\left(\frac{2}{\sqrt{x}}\right) + \cdots. \tag{5.397}$$

The exact solution can be shown to be

$$y = \sqrt{x}\left(C_1 I_1\left(\frac{2}{\sqrt{x}}\right) + C_2 K_1\left(\frac{2}{\sqrt{x}}\right)\right). \tag{5.398}$$

Here I_1 is a modified Bessel function of the first kind of $O(1)$, and K_1 is a modified Bessel function of the second kind of $O(1)$ (see Section 4.4.8).

5.4.8 Solutions of the Type $e^{S(x)}$

We can sometimes assume a solution of the form $e^{S(x)}$ and develop a good approximation, as illustrated in the following example.

EXAMPLE 5.30
Solve

$$x^3 y'' = y, \tag{5.399}$$

for small, positive x.

Let $y = e^{S(x)}$ so that $y' = S'e^S$, $y'' = (S')^2 e^S + S''e^S$, from which

$$S'' + (S')^2 = x^{-3}. \tag{5.400}$$

Assume that $S'' \ll (S')^2$, which will be checked soon. Thus, $S' = \pm x^{-3/2}$, and $S = \pm 2x^{-1/2}$. Checking, we get $S''/(S')^2 = x^{1/2} \to 0$ as $x \to 0$, confirming the assumption. Now we add a correction term so that $S(x) = 2x^{-1/2} + C(x)$, where we have taken the positive sign. Assume that $C \ll 2x^{-1/2}$. Substituting in the equation, we have

$$\frac{3}{2}x^{-5/2} + C'' - 2x^{-3/2}C' + (C')^2 = 0. \tag{5.401}$$

Because $C \ll 2x^{-1/2}$, we have $C' \ll x^{-3/2}$ and $C'' \ll (3/2)x^{-5/2}$. Thus

$$\frac{3}{2}x^{-5/2} - 2x^{-3/2}C' = 0, \tag{5.402}$$

from which $C' = (3/4)x^{-1}$ and $C = (3/4)\ln x$. We can now check the assumption on C. We have $S(x) = 2x^{-1/2} + (3/4)\ln x$, so that

$$y = x^{3/4} \exp\left(-\frac{2}{\sqrt{x}}\right) + \cdots. \tag{5.403}$$

Another solution is obtained by taking $S(x) = -2x^{-1/2} + C(x)$. This procedure is similar to that of the WKBJ method, and the solution is identical. The exact solution is, of course, the same as a portion of the previous example. Had we included both $\pm$ terms, we would have achieved the full solution of the previous example.

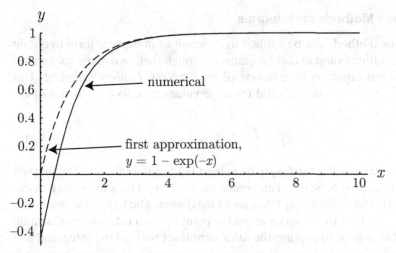

Figure 5.24. Numerical and first approximate solutions for repeated substitution problem, $y' = \exp(-xy)$, $y(\infty) = 1$.

5.4.9 Repeated Substitution

This so-called technique of repeated substitution sometimes works if the range of the independent variable is such that some term is small.

EXAMPLE 5.31

Solve

$$y' = e^{-xy}, \qquad y(\infty) \to c, \quad c > 0, \qquad (5.404)$$

for $y > 0$ and large x.

As $x \to \infty$, $y' \to 0$, so that $y \to c$. Substituting $y = c$ into Eq. (5.404), we get

$$y' = e^{-cx}, \qquad (5.405)$$

which can be integrated to get, after application of the boundary condition,

$$y = c - \frac{1}{c}e^{-cx}. \qquad (5.406)$$

Substituting Eq. (5.406) into the original Eq. (5.404), we find

$$y' = \exp\left(-x\left(c - \frac{1}{c}e^{-cx}\right)\right), \qquad (5.407)$$

$$= e^{-cx}\left(1 + \frac{x}{c}e^{-cx} + \cdots\right), \qquad (5.408)$$

which can be integrated to give

$$y = c - \frac{1}{c}e^{-cx} - \frac{1}{c^2}\left(x + \frac{1}{2c}\right)e^{-2cx} + \cdots. \qquad (5.409)$$

The series converges for large x. An accurate numerical solution along with the first approximation is plotted in Figure 5.24.

5.5 Asymptotic Methods for Integrals

Often asymptotic methods can be applied to expressions in integral form to obtain simpler representations valid in certain limits. Although there are many such methods, we confine our attention here to one of the simplest: *Laplace's method*. This method provides approximations, valid for large values of x, to expressions of the form

$$I(x) = \int_a^b e^{xf(s)} \, ds. \tag{5.410}$$

Here $a, b \in \mathbb{R}^1$, and we will need $f(s)$ to be at least twice differentiable. The method can work only when $f(s)$ possesses a maximum for $s \in [a, b]$. This allows one to focus attention on the region where $f(s)$ takes on its maximum. The key to the method is to approximate $f(s)$ by a Taylor series about the point where it takes on its maximum value. This will enable us to capture the most important parts of the integrand and ignore the rest.

EXAMPLE 5.32

Use Laplace's method to find an approximation, valid for $x \to \infty$, to

$$I(x) = \int_{-\infty}^{\infty} e^{-x \cosh s} \, ds. \tag{5.411}$$

We first see that $f(s) = -\cosh s$, $a = -\infty$, and $b = \infty$. We also note for $x > 0$ that the integrand takes on exponentially small values as $s \to \pm\infty$; it is thus not likely to contribute greatly to the value of $I(x)$ in such regions. This property will be critical for the success of the method. We next see that $f'(s) = -\sinh s$, which is zero when $s = 0$. We also see that $f''(s) = -\cosh s$ and thus $f''(0) = -1$; therefore, $f(s)$ is maximized at $s = 0$. So we take the Taylor series for $f(s)$ in the neighborhood of $s = 0$, that being

$$f(s) = -\cosh s = -1 - \frac{s^2}{2} + \cdots. \tag{5.412}$$

We focus here on only the first two terms. So

$$I(x) \sim \int_{-\infty}^{\infty} e^{\left(-x - xs^2/2 + \cdots\right)} \, ds, \tag{5.413}$$

$$\sim \int_{-\infty}^{\infty} e^{-x} e^{-xs^2/2} \, ds, \tag{5.414}$$

$$\sim e^{-x} \int_{-\infty}^{\infty} e^{-xs^2/2} \, ds. \tag{5.415}$$

This integrand is symmetric in s about $s = 0$ so that we only need to consider a semi-infinite domain and multiply the result by 2:

$$I(x) = 2e^{-x} \int_0^{\infty} e^{-xs^2/2} \, ds. \tag{5.416}$$

Now change variables so that

$$\xi^2 = \frac{1}{2}xs^2, \tag{5.417}$$

$$2\xi \, d\xi = xs \, ds, \tag{5.418}$$

$$2\sqrt{\frac{x}{2}} s \, d\xi = xs \, ds, \tag{5.419}$$

$$\sqrt{\frac{2}{x}} \, d\xi = ds. \tag{5.420}$$

Thus, our integral, Eq. (5.416), transforms to

$$I(x) \sim 2e^{-x} \int_0^\infty e^{-\xi^2} \sqrt{\frac{2}{x}} \, d\xi, \tag{5.421}$$

$$\sim 2\sqrt{\frac{2}{x}} e^{-x} \int_0^\infty e^{-\xi^2} \, d\xi, \tag{5.422}$$

$$\sim \sqrt{\frac{2}{x}} e^{-x} \int_{-\infty}^\infty e^{-\xi^2} \, d\xi. \tag{5.423}$$

The term $\int_{-\infty}^\infty e^{-\xi^2} \, d\xi$ has a numerical value; let us examine how to obtain it. Consider then the following auxiliary double integral, I_2, defined over the entire (ξ, η) plane, considered to be Cartesian:

$$I_2 = \int_{-\infty}^\infty \int_{-\infty}^\infty e^{-\xi^2 - \eta^2} \, d\xi \, d\eta, \tag{5.424}$$

$$= \int_{-\infty}^\infty e^{-\eta^2} \int_{-\infty}^\infty e^{-\xi^2} \, d\xi \, d\eta, \tag{5.425}$$

$$= \left(\int_{-\infty}^\infty e^{-\xi^2} \, d\xi \right) \left(\int_{-\infty}^\infty e^{-\eta^2} \, d\eta \right), \tag{5.426}$$

$$= \left(\int_{-\infty}^\infty e^{-\xi^2} \, d\xi \right)^2, \tag{5.427}$$

$$\sqrt{I_2} = \int_{-\infty}^\infty e^{-\xi^2} \, d\xi. \tag{5.428}$$

Now let us transform to polar coordinates with $\xi = r\cos\theta, \eta = r\sin\theta$. It is easily shown that $d\xi \, d\eta = r \, dr \, d\theta$ and $x^2 + y^2 = r^2$. Making these substitutions into Eq. (5.424) as well as transforming its limits of integration yields

$$I_2 = \int_0^{2\pi} \int_0^\infty e^{-r^2} r \, dr \, d\theta. \tag{5.429}$$

The inner integral can be evaluated to obtain

$$I_2 = \int_0^{2\pi} \left(-\frac{1}{2} e^{-r^2} \right) \Big|_0^\infty d\theta, \tag{5.430}$$

$$= \int_0^{2\pi} \left(\frac{1}{2} \right) d\theta, \tag{5.431}$$

$$= \pi. \tag{5.432}$$

Comparing this to Eq. (5.428), we determine that

$$\int_{-\infty}^\infty e^{-\xi^2} \, d\xi = \sqrt{\pi}. \tag{5.433}$$

Using this then in our actual problem, Eq. (5.423) reduces to

$$I(x) \sim \sqrt{\frac{2\pi}{x}} e^{-x}. \tag{5.434}$$

Had we retained more terms in the Taylor series of $\cosh s$, we would have obtained the corrected formula

$$I(x) = \sqrt{\frac{2\pi}{x}} \left(1 - \frac{1}{8x} + \frac{9}{128x^2} + \cdots \right). \tag{5.435}$$

These forms are superior to the original in that no explicit integration operation is required. For $x = 20$, high-accuracy numerical integration reveals that

$I = 1.14825 \times 10^{-9}$. Our one-term approximation of Eq. (5.434) is accurate, yielding $I \sim 1.15527 \times 10^{-9}$. The three-term approximation of Eq. (5.435) is more accurate, yielding $I \sim 1.14826 \times 10^{-9}$.

EXAMPLE 5.33

Use Laplace's method to find an approximation to the factorial function $y(x) = x!$ for large positive x.

One recalls that the factorial function is traditionally defined for nonnegative integers $x \in \mathbb{Z}^*$ such that $x! = x(x-1)(x-2)\ldots(3)(2)(1)$. It can be extended to the reals by use of the gamma function (see Section A.7.1). For $x \in \mathbb{R}^1$, we have

$$y(x) = x! = \Gamma(x+1) = \int_0^\infty e^{-t} t^x \, dt. \tag{5.436}$$

This can be rewritten as

$$y(x) = x! = \int_0^\infty e^{-t+x \ln t} \, dt. \tag{5.437}$$

Now, let us remove t in favor of s via the transformation $t = xs$. This gives $dt = x \, ds$, and our equation transforms to

$$y(x) = x! = \int_0^\infty e^{-xs+x \ln(xs)} x \, ds, \tag{5.438}$$

$$= \int_0^\infty e^{-xs+x \ln x+x \ln s} x \, ds, \tag{5.439}$$

$$= xe^{x \ln x} \int_0^\infty e^{-xs+x \ln s} \, ds, \tag{5.440}$$

$$= x^{x+1} \int_0^\infty e^{x(-s+\ln s)} \, ds. \tag{5.441}$$

So here we have an integral prepared for Laplace's method with $f(s) = -s + \ln s$. Differentiating, we find $f'(s) = -1 + 1/s$. Thus $f'(s) = 0$ when $s = 1$. And $f''(s) = -1/s^2$, so at the critical point, we have $f''(1) = -1$, so there is a maximum in $f(s)$ at $s = 1$. Let us then perform a Taylor series expansion of $f(s)$ in the neighborhood of $s = 1$. Doing so, we find

$$f(s) = -1 - \frac{1}{2}(s-1)^2 + \frac{1}{3}(s-1)^3 + \cdots. \tag{5.442}$$

Let us only retain the first two terms, recognizing that if we were to keep more, we might achieve higher accuracy approximations. So we will take

$$y(x) = x! \sim x^{x+1} \int_0^\infty e^{x(-1-(s-1)^2/2)} \, ds, \tag{5.443}$$

$$\sim x^{x+1} e^{-x} \int_0^\infty e^{-x(s-1)^2/2} \, ds. \tag{5.444}$$

Let us break this integral into two parts, one on either side of the maximum at $s = 1$:

$$y(x) = x! \sim x^{x+1} e^{-x} \left(\int_0^1 e^{-x(s-1)^2/2} \, ds + \int_1^\infty e^{-x(s-1)^2/2} \, ds \right). \tag{5.445}$$

Now take a transformation similar to that of a previous example:

$$\xi^2 = \frac{x}{2}(s-1)^2. \tag{5.446}$$

This will render the argument of the exponential to be much simpler and ultimately allow us to use previously obtained results. Thus,

$$2\xi \, d\xi = x(s-1) \, ds, \tag{5.447}$$

$$2\sqrt{\frac{x}{2}}(s-1) \, d\xi = x(s-1) \, ds, \tag{5.448}$$

$$\sqrt{\frac{2}{x}} \, d\xi = ds. \tag{5.449}$$

Then Eq. (5.445) transforms to

$$y(x) = x! \sim x^{x+1}e^{-x}\left(\int_{-\sqrt{x/2}}^{0} e^{-\xi^2}\sqrt{\frac{2}{x}} \, d\xi + \int_{0}^{\infty} e^{-\xi^2}\sqrt{\frac{2}{x}} \, d\xi\right), \tag{5.450}$$

$$\sim \sqrt{2}x^{x+1/2}e^{-x}\left(\int_{-\sqrt{x/2}}^{0} e^{-\xi^2} \, d\xi + \int_{0}^{\infty} e^{-\xi^2} \, d\xi\right). \tag{5.451}$$

For large x, we can approximate the lower limit of the first integral by $-\infty$, thus yielding

$$y(x) = x! \sim \sqrt{2}x^{x+1/2}e^{-x}\underbrace{\int_{-\infty}^{\infty} e^{-\xi^2} \, d\xi}_{=\sqrt{\pi}}. \tag{5.452}$$

The integral has the value $\sqrt{\pi}$ via the analysis that gave us Eq. (5.433). So

$$y(x) = x! \sim \sqrt{2\pi x}x^{x}e^{-x}. \tag{5.453}$$

This formula is often known as a one-term Stirling's[18] approximation (see Section A.7.1).

Let us see how well our estimate performs for x sufficiently moderate that we might expect a poor approximation. For $x = 3$, we have $3! = 3(2)(1) = 6$. Our approximate formula gives $3! \sim 5.83621$. This is surprisingly accurate. Let us try an x that is large. When $x = 1000$, we find $1000! = 4.023872600770938 \times 10^{2567}$. Our approximate formula gives $1000! \sim 4.023537292036775 \times 10^{2567}$. It can be shown that an even better approximation, especially for small x, is given by the closely related

$$x! \sim \sqrt{2\pi x + \frac{\pi}{3}}\, x^{x}e^{-x}. \tag{5.454}$$

EXERCISES

1. Verify that

$$\lim_{n\to\infty} \int_{\pi}^{2\pi} \frac{\sin nx}{nx} \, dx = 0.$$

2. Evaluate

$$\lim_{x\to 0} \frac{4\cosh x + \sinh(\arctan(\ln(\cos(2x)))) - 4}{e^{-x} + \arcsin x - \sqrt{1+x^2}}.$$

3. Write three terms of a Taylor series expansion for the function $f(x) = \exp(\tan x)$ about the point $x = \pi/4$. For what range of x is the series convergent?

---

[18] James Stirling, 1692–1770, Scottish mathematician.

4. Find the Taylor series expansions of $\sin x$ and $\cos x$, and show that the latter is the derivative of the former.

5. Find the first three derivatives of $f(x) = \exp(-1/x)$ as $x \to 0+$ and at $x = 1$.

6. Show that all derivatives of $\exp(-1/x^2)$ at $x = 0$ vanish.

7. Write a Taylor series expansion for the function $f(x, y) = x^2 \cos y$ about the point $(x, y) = (2, \pi)$. Include up to the x^2, y^2, and xy terms.

8. Approximate $\arctan x$ by a Padé approximant with three terms in the numerator and three in the denominator.

9. Find the solution (first as a power series and then in exact form) of

$$xy'' - y' + 4x^3 y = 0.$$

10. Find the power series solution of
   (a) $x^2 y'' + xy' + (x^2 - 1)y = 0$,
   (b) $y'' + xy = 0$,
   (c) $y'' + k^2 y = 0$,
   (d) $y'' + e^x y = 1$,
   (e) $x^2 y'' + x \left(\dfrac{1}{2} + 2x \right) y' + \left(x - \dfrac{1}{2} \right) y = 0$,
   (f) $y'' = \sqrt{x} y$,
   (g) $y''' - \sqrt{x} y = 0$,
   around $x = 0$.

11. Find the first two nonzero terms in the power series solution around $x = 1$ of

$$xy'' - y = 0,$$

with $y(1) = 0$, $y'(1) = 1$.

12. Solve as a power series in x for $x > 0$ about the point $x = 0$:
   (a) $x^2 y'' - 2xy' + (x + 1)y = 0$, $y(1) = 1$, $y(4) = 0$.
   (b) $xy'' + y' + 2x^2 y = 0$, $|y(0)| < \infty$, $y(1) = 1$.
   In each case, find the exact solution with a symbolic computer mathematics program, and compare graphically the first four terms of your series solution with the exact solution.

13. Obtain a power series solution for $y' + ky = 0$ about $x = 0$, where k is an arbitrary, nonzero constant. Compare to a Taylor series expansion of the exact solution.

14. Find an approximate general solution of

$$(x \sin x)\, y'' + (2x \cos x + x^2 \sin x)\, y' + (x \sin x + \sin x + x^2 \cos x)\, y = 0,$$

valid near $x = 0$.

15. Find the complementary functions of

$$y''' - xy = 0,$$

in terms of expansions near $x = 0$. Retain only two terms for each function.

16. Approximate $y(x)$ that satisfies

$$\epsilon y'' + y' + xy = 0, \qquad y(0) = 0,\ y(1) = 1,$$

for small ϵ.

17. Find two-term expansions for each of the roots of

$$(x - 1)(x + 3)(x - 3\lambda) + 1 = 0,$$

where λ is large.

18. Find two terms of an approximate solution of

$$y'' + \frac{\lambda}{\lambda + x}y = 0,$$

with $y(0) = 0, y(1) = 1$, where λ is a large parameter. For $\lambda = 20$, plot $y(x)$ for the two-term expansion. Also compute the exact solution by numerical methods. Plot the difference between the asymptotic and numerical solutions versus x.

19. Find the leading order solution for

$$(x - \epsilon y)\frac{dy}{dx} + xy = e^{-x},$$

where $y(1) = 1$ and $x \in [0, 1]$, $\epsilon \ll 1$. For $\epsilon = 0.2$, plot the asymptotic solution, the exact solution, and the difference versus x.

20. Find an approximate solution for

$$y'' - ye^{y/10} = 0,$$

with $y(0) = 1, y(1) = e$.

21. Find an approximate solution for

$$y'' - ye^{y/12} = 0, \qquad y(0) = 0.1, \ y'(0) = 1.2.$$

Compare with the numerical solution for $0 \le x \le 1$.

22. Find the lowest order boundary layer solution to

$$\epsilon y'' + (1 + \epsilon)y' + y = 0,$$

with $y(0) = 0, y(1) = 1$, where $\epsilon \ll 1$.

23. Find the lowest order solution for

$$\epsilon^2 y'' + \epsilon y^2 - y + 1 = 0,$$

with $y(0) = 1, y(1) = 3$, where ϵ is small. For $\epsilon = 0.2$, plot the asymptotic and exact solutions.

24. Show that for small ϵ, the solution of

$$\frac{dy}{dt} - y = \epsilon e^t,$$

with $y(0) = 1$, can be approximated as an exponential on a slightly different time scale, where t is taken to be time.

25. Obtain approximate general solutions of
    (a) $xy'' + y = 0$, through $O(x^2)$,
    (b) $xy'' + y' + xy = 0$, through $O(x^6)$,
    near $x = 0$.

26. Find all solutions through $O(\epsilon^2)$, where ϵ is a small parameter, and compare with the exact result for $\epsilon = 0.01$:
    (a) $4x^4 + 4(\epsilon + 1)x^3 + 3(2\epsilon - 5)x^2 + (2\epsilon - 16)x - 4 = 0$,
    (b) $2\epsilon x^4 + 2(2\epsilon + 1)x^3 + (7 - 2\epsilon)x^2 - 5x - 4 = 0$.

27. Find three terms of a solution of

$$x + \epsilon \cos(x + 2\epsilon) = \frac{\pi}{2},$$

where ϵ is a small parameter. For $\epsilon = 0.2$, compare the best asymptotic solution with the exact solution.

28. Find three terms of the solution of

$$\dot{x} + 2x + \epsilon x^2 = 0, \qquad x(0) = \cosh \epsilon,$$

where ϵ is a small parameter. Compare graphically with the exact solution for $\epsilon = 0.3$ and $0 \le t \le 2$.

29. Write an approximation for

$$\int_0^{\pi/2} \sqrt{1 + \epsilon \cos^2 x}\, dx,$$

if $\epsilon = 0.1$, so that the absolute error is less than 2×10^{-4}.

30. Solve

$$y'' + y = e^{\epsilon \sin x}, \qquad y(0) = y(1) = 0,$$

through $O(\epsilon)$, where ϵ is a small parameter. For $\epsilon = 0.25$, graphically compare the asymptotic solution with a numerically obtained solution.

31. Find approximations to all solutions of $\epsilon x^4 + x - 2 = 0$ if ϵ is small and positive. If $\epsilon = 0.001$, compare the exact solutions obtained numerically with the asymptotic solutions.

32. Obtain the first two terms of an approximate solution to

$$x'' + 3(1 + \epsilon)x' + 2x = 0, \qquad x(0) = 2(1 + \epsilon),\ x'(0) = -3(1 + 2\epsilon)$$

for small ϵ. Compare the approximate and exact solutions graphically in the range $0 \le x \le 1$ for (a) $\epsilon = 0.1$, (b) $\epsilon = 0.25$, and (c) $\epsilon = 0.5$.

33. Find an approximate solution to

$$x'' + (1 + \epsilon)x = 0, \qquad x(0) = A,\ x'(0) = B,$$

for small, positive ϵ. Compare with the exact solution. Plot both the exact solution and the approximate solution on the same graph for $A = 1, B = 0, \epsilon = 0.3$.

34. Find an approximate solution to

$$\epsilon^2 y'' - y = -1, \qquad y(0) = 0,\ y(1) = 0,$$

for small ϵ. Compare graphically with the exact solution for $\epsilon = 0.1$.

35. Solve to leading order

$$\epsilon y'' + yy' - y = 0, \qquad y(0) = 0,\ y(1) = 3.$$

Compare graphically to the exact solution for $\epsilon = 0.2$.

36. If $\ddot{x} + x + \epsilon x^3 = 0$ with $x(0) = A, \dot{x}(0) = 0$, where ϵ is small and t is time, a regular expansion gives $x(t) \sim A\cos t + \epsilon(A^3/32)(-\cos t + \cos 3t - 12t \sin t)$. Explain why this is not valid for all time, and obtain a better solution by inserting

$t = (1 + a_1\epsilon + \cdots)\tau$ into this solution, expanding in terms of ϵ, and choosing $a_1, a_2, \ldots$ properly (Pritulo's[19] method).

37. Use perturbations to find an approximate solution to

$$y'' + \lambda y' = \lambda, \qquad y(0) = 0, \ y(1) = 0,$$

where $\lambda \gg 1$.

38. Find, correct to $O(\epsilon)$, the solution of

$$\ddot{x} + (1 + \epsilon \cos 2t)\, x = 0, \qquad x(0) = 1, \ \dot{x}(0) = 0,$$

bounded for all t, where $\epsilon \ll 1$.

39. Find the function f to $O(\epsilon)$ that satisfies the integral equation

$$x = \int_0^{x + \epsilon \sin x} f(\xi)\, d\xi.$$

40. Find three terms of a perturbation solution of

$$y'' + \epsilon y^2 = 0,$$

with $y(0) = 0$, $y(1) = 1$ for $\epsilon \ll 1$. For $\epsilon = 0.25$, compare the $O(1), O(\epsilon)$, and $O(\epsilon^2)$ solutions to a numerically obtained solution in $x \in [0, 1]$.

41. Obtain two terms of an approximate solution for $\epsilon e^x = \cos x$ when ϵ is small. Graphically compare to the actual values obtained numerically, when $\epsilon = 0.2, 0.1, 0.01$.

42. Obtain three terms of a perturbation solution for the roots of the equation $(1 - \epsilon)x^2 - 2x + 1 = 0$.

43. The solution of the matrix equation $\mathbf{A} \cdot \mathbf{x} = \mathbf{b}$ can be written as $\mathbf{x} = \mathbf{A}^{-1} \cdot \mathbf{b}$. Find the nth term of the perturbation solution of $(\mathbf{A} + \epsilon\mathbf{B}) \cdot \mathbf{x} = \mathbf{b}$, where ϵ is a small parameter. Obtain the first three terms of the solution for

$$\mathbf{A} = \begin{pmatrix} 1 & 2 & 1 \\ 2 & 2 & 1 \\ 1 & 2 & 3 \end{pmatrix}, \ \mathbf{B} = \begin{pmatrix} 1/10 & 1/2 & 1/10 \\ 0 & 1/5 & 0 \\ 1/2 & 1/10 & 1/2 \end{pmatrix}, \ \mathbf{b} = \begin{pmatrix} 1/2 \\ 1/5 \\ 1/10 \end{pmatrix}.$$

44. Obtain leading and first-order approximations for u and v, governed by the following set of coupled differential equations, for small ϵ:

$$\frac{d^2u}{dx^2} + \epsilon v \frac{du}{dx} = 1, \ u(0) = 0, \ u(1) = \frac{1}{2} + \frac{1}{120}\epsilon$$

$$\frac{d^2v}{dx^2} + \epsilon u \frac{dv}{dx} = x, \ v(0) = 0, \ v(1) = \frac{1}{6} + \frac{1}{80}\epsilon.$$

Compare asymptotic and numerically obtained results for $\epsilon = 0.2$.

45. Obtain two terms of a perturbation solution to $\epsilon f_{xx} + f_x = -e^{-x}$ with boundary conditions $f(0) = 0$, $f(1) = 1$. Graph the solution for $\epsilon = 0.2, 0.1, 0.05, 0.025$ on $0 \le x \le 1$.

---

[19] Mikhail Fiodorovich Pritulo, 1929–2003, Russian aerodynamicist.

46. Using a two-variable expansion, find the lowest order solution of

(a) $\ddot{x} + \epsilon\dot{x} + x = 0,$ $\quad x(0) = 0,\ \dot{x}(0) = 1,$

(b) $\ddot{x} + \epsilon\dot{x}^3 + x = 0,$ $\quad x(0) = 0,\ \dot{x}(0) = 1,$

where $\epsilon \ll 1$. Compare asymptotic and numerically obtained results for $\epsilon = 0.01$.

47. Obtain a three-term solution of

$$\epsilon x'' - x' = 1, \qquad x(0) = 0,\ x(1) = 2,$$

where $\epsilon \ll 1$.

48. Find an approximate solution to

$$\epsilon^2 y'' - y = -1, \qquad y(0) = 0,\ y(1) = 0$$

for small ϵ. Compare graphically with the exact solution for $\epsilon = 0.1$.

49. Solve $y'' - \sqrt{x}y = 0$, $x > 0$ in each one of the following ways. (a) Substitute $x = \epsilon^{-4/5}X$, and then use the WKBJ method. (b) Substitute $x = \epsilon^{2/5}X$, and then use regular perturbation. (c) Find an approximate solution of the kind $y = e^{S(x)}$. In each case, assume ϵ is small when it appears.

50. For small ϵ, solve using the WKBJ method

$$\epsilon^2 y'' = (1 + x^2)^2 y, \qquad y(0) = 0,\ y(1) = 1.$$

51. Find two terms of the perturbation solution of

$$(1 + \epsilon y)y'' + \epsilon(y')^2 - N^2 y = 0,$$

with $y'(0) = 0, y(1) = 1$ for small ϵ. N is a constant. Plot the asymptotic and numerical solution for $\epsilon = 0.12, N = 10$.

52. Solve

$$\epsilon y'' + y' = \frac{1}{2},$$

with $y(0) = 0, y(1) = 1$ for small ϵ. Plot asymptotic and numerical solutions for $\epsilon = 0.12$.

53. Find if the van der Pol equation,

$$\ddot{y} - \epsilon(1 - y^2)\dot{y} + k^2 y = 0,$$

has an approximate limit cycle of the form $y = A\cos\omega t$.

54. For small ϵ, solve to lowest order using the method of multiple scales

$$\ddot{x} + \epsilon\dot{x} + x = 0, \qquad x(0) = 0,\ \dot{x}(0) = 1.$$

Compare exact and asymptotic results for $\epsilon = 0.3$.

55. For small ϵ, solve using the WKBJ method

$$\epsilon^2 y'' = (1 + x^2)^2 y, \qquad y(0) = 0,\ y(1) = 1.$$

Plot asymptotic and numerical solutions for $\epsilon = 0.11$.

56. Find the lowest order approximate solution to

$$\epsilon^2 y'' + \epsilon y^2 - y + 1 = 0, \qquad y(0) = 1,\ y(1) = 2,$$

where ϵ is small. Plot asymptotic and numerical solutions for $\epsilon = 0.23$.

57. Find two terms of an approximate solution of

$$y'' + \frac{\lambda}{\lambda + x} y = 0,$$

with $y(0) = 0, y(1) = 1$, where λ is a large parameter.

58. Find all solutions of $e^{\epsilon x} = x^2$ through $O(\epsilon^2)$, where ϵ is a small parameter.

59. Solve

$$(1 + \epsilon)y'' + \epsilon y^2 = 1,$$

with $y(0) = 0, y(1) = 1$ through $O(\epsilon^2)$, where ϵ is a small parameter.

60. Solve to lowest order

$$\epsilon y'' + y' + \epsilon y^2 = 1,$$

with $y(0) = -1, y(1) = 1$, where ϵ is a small parameter. For $\epsilon = 0.2$, plot asymptotic and numerical solutions to the full equation.

61. Find the solution of the transcendental equation

$$\sin x = \epsilon \cos 2x,$$

near $x = \pi$ for small positive ϵ.

62. Solve

$$\epsilon y'' - y' = 1,$$

with $y(0) = 0, y(1) = 2$ for small ϵ. Plot asymptotic and numerical solutions for $\epsilon = 0.04$.

63. Solve $y' = e^{-2xy}$ for large x where y is positive. Plot $y(x)$.

64. Use Laplace's method to find an approximation, valid for $x \to \infty$, to the integrals
   (a) $I(x) = \int_0^\pi e^{x \cos s} \, ds$,
   (b) $I(x) = \int_0^\pi e^{-x \cos s} \sin s \, ds$.

65. The motion of a pendulum is governed by the equation

$$\frac{d^2x}{dt^2} + \sin x = 0,$$

with $x(0) = \epsilon, dx/dt|_{t=0} = 0$. Using strained coordinates, find the approximate solution of $x(t)$ for small ϵ through $O(\epsilon^2)$. Plot your results for both your asymptotic analysis and by a numerical integration of the full equation.

66. A pendulum is used to measure the earth's gravity. The frequency of oscillation is measured, and the gravity is calculated assuming a small amplitude of motion and knowing the length of the pendulum. What must the maximum initial angular displacement of the pendulum be if the error in gravity is to be less than 1%? Neglect air resistance.

67. A projectile of mass m is launched at an angle α with respect to the horizontal and with an initial velocity V. Find the time it takes to reach its maximum height. Assume that the air resistance is small and can be written as k times the square of the velocity of the projectile. Choosing $m = 1$ kg, $k = 0.1$ kg/m, $\alpha = 45°$, and $V = 1$ m/s for the parameters, compare with the numerical result.

68. The temperature $T(x)$ in a composite fin (i.e., a fin made up of two different materials) with internal heating and cooling is governed by

$$\epsilon T'' - T = -1, \qquad 0 \le x \le 1,$$

$$T'' - T = 1, \qquad 1 < x \le 2,$$

with boundary conditions $T(0) = 0$ and $T(2) = 1$; ϵ represents the ratio of thermal conductivities of the two materials. The temperature and heat flux are continuous at $x = 1$, so that $T(1^-) = T(1^+)$ and $\epsilon T'(1^-) = T'(1^+)$. Find the exact solution for $T(x)$, and compare graphically with the $O(\epsilon)$ approximate solution for $\epsilon = 1, 0.1, 0.01, 0.001$.

69. A bead can slide along a circular hoop in a vertical plane. The bead is initially at the lowest position, $\theta = 0$, and given an initial velocity of $2\sqrt{gR}$, where g is the acceleration due to gravity and R is the radius of the hoop. If the friction coefficient is μ, find the maximum angle θ_{max} reached by the bead. Compare perturbation and numerical results. Present results on a θ_{max} versus μ plot, for $0 \le \mu \le 0.3$.

70. The initial velocity downward of a body of mass m immersed in a very viscous fluid is V. Find the velocity of the body as a function of time. Assume that the viscous force is proportional to the velocity. Assume that the inertia of the body is small, but not negligible, relative to viscous and gravity forces. Compare perturbation and exact solutions graphically.

6 Linear Analysis

This chapter introduces some more formal notions of what is known as *linear analysis*. It is in fact an abstraction of not only much of the analysis of previous chapters but also of Euclidean and non-Euclidean geometry. Importantly, we define in a more general sense the notion of a vector. In addition to traditional vectors that exist within a space of finite dimension, we see how what is known as function space can be thought of as vector space of infinite dimension. Additionally, we generalize the concepts of (1) projection of vectors onto spaces of lower dimension, (2) eigenvalues and eigenvectors, and (3) linear operators. Two important engineering analysis tools are discussed: the method of weighted residuals and uncertainty quantification via polynomial chaos. Others exist or could be easily inferred in any engineering system modeled by linear equations. Throughout the chapter, we introduce and use some of the more formal notation of modern mathematics, which, for many students, will be the key to enabling further studies in more mathematical realms.

For this and the next two chapters, we largely confine our discussion to linear systems. However, the ideas developed are of critical importance when we return to nonlinear systems in Chapter 9, which, once again, relies on local linearization to illuminate globally nonlinear behavior.

6.1 Sets

Consider two sets $\mathbb{A}$ and $\mathbb{B}$. We use the following notation:

$x \in \mathbb{A}$, x is an element of $\mathbb{A}$,

$x \notin \mathbb{A}$, x is not an element of $\mathbb{A}$,

$\mathbb{A} = \mathbb{B}$, $\mathbb{A}$ and $\mathbb{B}$ have the same elements,

$\mathbb{A} \subset \mathbb{B}$, the elements of $\mathbb{A}$ also belong to $\mathbb{B}$,

$\mathbb{A} \cup \mathbb{B}$, set of elements that belong to $\mathbb{A}$ or $\mathbb{B}$, that is, the *union* of $\mathbb{A}$ and $\mathbb{B}$,

$\mathbb{A} \cap \mathbb{B}$, set of elements that belong to $\mathbb{A}$ and $\mathbb{B}$, that is, the *intersection* of $\mathbb{A}$ and $\mathbb{B}$,

$\mathbb{A} - \mathbb{B}$, set of elements that belong to $\mathbb{A}$ but not to $\mathbb{B}$.

If $\mathbb{A} \subset \mathbb{B}$, then $\mathbb{B} - \mathbb{A}$ is the *complement* of $\mathbb{A}$ in $\mathbb{B}$. Some sets that are commonly used are

- $\mathbb{Z}$: set of all integers,
- $\mathbb{Z}^*$: set of all nonnegative integers,

- $\mathbb{N}$: set of all positive integers; this is occasionally defined to be the set of all nonnegative integers,
- $\mathbb{Q}$: set of all rational numbers,
- $\mathbb{R}$: set of all real numbers,
- $\mathbb{R}_+$: set of all nonnegative real numbers,
- $\mathbb{C}$: set of all complex numbers,

Some common definitions include the following:

- An *interval* is a portion of the real line.
- An *open* interval (a, b) does not include the endpoints, so that if $x \in (a, b)$, then $a < x < b$. In set notation, this is $\{x \in \mathbb{R} : a < x < b\}$ if x is real.
- A *closed* interval $[a, b]$ includes the endpoints. If $x \in [a, b]$, then $a \le x \le b$. In set notation, this is $\{x \in \mathbb{R} : a \le x \le b\}$ if x is real.
- A member x of a set $\mathbb{S}$ is on its *boundary* if it can be approached from points within $\mathbb{S}$ and from points outside of $\mathbb{S}$. The boundary of $\mathbb{S}$ is sometimes described as $\partial \mathbb{S}$.
- A member x of a set $\mathbb{S}$ is within the *interior* if it is not on the boundary.
- A set is *open* if it does not contain any of its boundary points.
- The complement of any open subset of $[a, b]$ is a *closed* set.
- A set $\mathbb{A} \subset \mathbb{R}$ is bounded from above if there exists a real number, called the *upper bound*, such that every $x \in \mathbb{A}$ is less than or equal to that number.
- The *least upper bound* or *supremum* is the minimum of all upper bounds.
- In a similar fashion, a set $\mathbb{A} \subset \mathbb{R}$ can be bounded from below, in which case it will have a *greatest lower bound* or *infimum*.
- A set $\mathbb{A} \subset \mathbb{R}$ is *bounded* if it has both upper and lower bounds.
- A set that has no elements is the empty set $\{\}$, also known as the null set $\emptyset$. The set with 0 as the only element, $\{0\}$, is not empty.
- A set that is either finite or for which each element can be associated with a member of $\mathbb{N}$ is said to be *countable*. Otherwise, the set is *uncountable*.
- For $x \in \mathbb{A}$, the *neighborhood* of x is a subset $\mathbb{V}$ of $\mathbb{A}$ which includes an open set $\mathbb{U}: x \in \mathbb{U} \subset \mathbb{V}$. Neighborhoods are often endowed with a distance restriction relating how far a nearby element may be from x. However, we have not yet defined a measure for distance.
- An ordered pair is $P = (x, y)$, where $x \in \mathbb{A}$ and $y \in \mathbb{B}$. Then $P \in \mathbb{A} \times \mathbb{B}$, where the symbol $\times$ is defined as a Cartesian product. If $x \in \mathbb{A}$ and $y \in \mathbb{A}$ also, then we write $P = (x, y) \in \mathbb{A}^2$.
- A real *function* of a single variable can be written as $f : \mathbb{X} \to \mathbb{Y}$ or $y = f(x)$, where f maps $x \in \mathbb{X} \subset \mathbb{R}$ to $y \in \mathbb{Y} \subset \mathbb{R}$. For each x, there is only one y, though there may be more than one x that maps to a given y. The set $\mathbb{X}$ is called the *domain* of f, y the *image* of x, and the *range* the set of all images.

6.2 Integration

We consider here some more formal ways of treating integration. Most importantly, we consider two approaches to integration, Riemann[1] and Lebesgue,[2] the second

---

[1] Georg Friedrich Bernhard Riemann, 1826–1866, Hanover-born German mathematician.
[2] Henri Léon Lebesgue, 1875–1941, French mathematician.

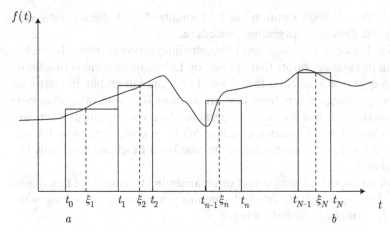

Figure 6.1. Riemann integration process.

of which admits a broader class of functions. This is necessary to allow our Fourier series representations of functions to be what is known as *complete*.

We first consider *Riemann integration*. Consider a function $f(t)$ defined in the interval $[a, b]$. Choose $t_1, t_2, \ldots, t_{N-1}$ such that

$$a = t_0 < t_1 < t_2 < \cdots < t_{N-1} < t_N = b. \tag{6.1}$$

Let $\xi_n \in [t_{n-1}, t_n]$ and

$$I_N = f(\xi_1)(t_1 - t_0) + f(\xi_2)(t_2 - t_1) + \cdots + f(\xi_N)(t_N - t_{N-1}). \tag{6.2}$$

Also, let $\max_n |t_n - t_{n-1}| \to 0$ as $N \to \infty$. Then $I_N \to I$, where

$$I = \int_a^b f(t) \, dt. \tag{6.3}$$

If I exists and is *independent of the manner of subdivision*, then $f(t)$ is Riemann integrable in $[a, b]$. The Riemann integration process is sketched in Figure 6.1.

EXAMPLE 6.1

Determine if the function $f(t)$ is Riemann integrable in $[0, 1]$, where

$$f(t) = \begin{cases} 0, & \text{if } t \text{ is rational}, \\ 1, & \text{if } t \text{ is irrational}. \end{cases} \tag{6.4}$$

On choosing ξ_n rational, $I = 0$, but if ξ_n is irrational, then $I = 1$. So $f(t)$ is not Riemann integrable.

The previous example shows that there are functions that cannot be integrated in the Riemann sense. However, the nature of the function of the previous example might lead one to wonder if there is a broader definition of integration that would be useful. Consider that it can be shown that there are infinitely many more irrational numbers on the real axis than rational numbers. One might even imagine that the set of numbers for which $f(t) = 1$ would be infinitely denser than the set of numbers for which $f(t) = 0$. And so one might be tempted to ignore the relatively much smaller set of rational numbers and say the function is well approximated by $f(t) = 1$, in which case $\int_0^1 f(t)\, dt = 1$. Such a notion can be formalized, with details not given here, by the so-called *Lebesgue integration* procedure.

The distinction between Riemann and Lebesgue integration is subtle. It can be shown that certain integral operators that operate on Lebesgue integrable functions are guaranteed to generate a function that is also Lebesgue integrable. In contrast, certain operators operating on functions that are at most Riemann integrable can generate functions that are not Riemann integrable. As a consequence, when we consider Fourier expansions of functions, such as has been done in Section 4.5, we need to impose the less restrictive condition that the basis functions need only be Lebesgue integrable.

We close with an important notion that often arises in integration of functions. If the integrand $f(x)$ of a definite integral contains a singularity at $x = x_0$ with $x_0 \in (a, b)$, then the *Cauchy principal value* is

$$\fint_a^b f(x)\, dx = \lim_{\epsilon \to 0} \left(\int_a^{x_0 - \epsilon} f(x)\, dx + \int_{x_0 + \epsilon}^b f(x)\, dx \right). \tag{6.5}$$

Notice that a Cauchy principal value may exist for cases in which a Riemann integral does not exist.

EXAMPLE 6.2

Find the Cauchy principal value of

$$\int_0^3 \frac{dx}{x - 1}. \tag{6.6}$$

We have a singularity at $x = x_0 = 1$. The Cauchy principal value is

$$\fint_0^3 \frac{dx}{x - 1} = \lim_{\epsilon \to 0} \left(\int_0^{1 - \epsilon} \frac{dx}{x - 1} + \int_{1 + \epsilon}^3 \frac{dx}{x - 1} \right), \tag{6.7}$$

$$= \lim_{\epsilon \to 0} \left(\ln(x - 1)\big|_0^{1 - \epsilon} + \ln(x - 1)\big|_{1 + \epsilon}^3 \right), \tag{6.8}$$

$$= \lim_{\epsilon \to 0} \left(\ln \left(\frac{-\epsilon}{-1} \right) + \ln \left(\frac{2}{\epsilon} \right) \right), \tag{6.9}$$

$$= \lim_{\epsilon \to 0} (\ln \epsilon + \ln 2 - \ln \epsilon), \tag{6.10}$$

$$= \ln 2. \tag{6.11}$$

Though $\lim_{\epsilon \to 0}(\int_0^{1 - \alpha\epsilon} dx/(x - 1) + \int_{1 + \epsilon}^3 dx/(x - 1))$ seems as though it should be equivalent, in fact it yields $\ln(2\alpha)$. Thus, the value of α affects the final value of the integral, even though it is modulating the small number ϵ. Thus, we are in fact considering the principal value of the integral to be that value realized when $\alpha = 1$. We might better think of our integration process as first splitting the domain via $\fint$, and then, of the many ways we can choose to let $\epsilon \to 0$, we select one, the Cauchy principal value, for which ϵ approaches zero at the same rate on either side of the point of consideration.

The Cauchy principal value can be considered in a broader context in which singularities at finite points are not important. Consider the function

$$f(x) = \frac{2x}{x^2 + 1}. \tag{6.12}$$

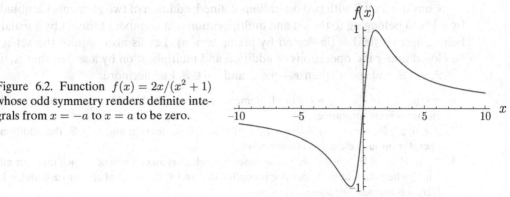

Figure 6.2. Function $f(x) = 2x/(x^2 + 1)$ whose odd symmetry renders definite integrals from $x = -a$ to $x = a$ to be zero.

This is plotted in Figure 6.2. The function is nonsingular, so in forming its integral, there is no need to split the domain via f. By inspection of the graph, any definite integral with symmetry about the origin, for example, from $x = -a$ to $x = a$, will be zero, easily verified by calculation:

$$\int_{-a}^{a} \frac{2x}{x^2 + 1}\, dx = \ln(1 + x^2)\big|_{-a}^{a} = \ln\left(\frac{1 + a^2}{1 + a^2}\right) = 0. \tag{6.13}$$

However, for any nonzero limits that are asymmetric about the origin, the integral will be nonzero.

Now consider a limiting process for evaluating the integral $I = \int_{-\infty}^{\infty} f(x)\, dx$:

$$I = \int_{-\infty}^{\infty} \frac{2x}{x^2 + 1}\, dx = \lim_{a \to \infty} \int_{-\alpha a}^{a} \frac{2x}{x^2 + 1}\, dx, \tag{6.14}$$

where $\alpha > 0$. Evaluating, we get

$$I = \lim_{a \to \infty} \ln\left(\frac{1 + a^2}{1 + \alpha a^2}\right) = \ln\left(\frac{a^2}{\alpha a^2}\right) = -\ln \alpha. \tag{6.15}$$

Unfortunately, the value of I depends on α, even though both upper and lower bounds of the integral are approaching the proper limits. If $\alpha = 1$, we get $I = 0$. Choosing $\alpha = 1$ implies that both limits of integration go to $\pm\infty$ at the same rate. So of the many values of α that yield corresponding values for I, one choice, the Cauchy principal value, is the natural choice, that being $\alpha = 1$, for which one can say

$$\fint_{-\infty}^{\infty} \frac{2x}{x^2 + 1}\, dx = 0. \tag{6.16}$$

6.3 Vector Spaces

Let us here build the necessary infrastructure so that we may formally define what are known as *vector spaces*. First, a *field* $\mathbb{F}$ is typically a set of numbers that contains the sum, difference, product, and quotient (excluding division by zero) of any two numbers in the field.[3] Examples are the sets of rational numbers $\mathbb{Q}$, real numbers $\mathbb{R}$, or complex numbers $\mathbb{C}$. We will usually use only $\mathbb{R}$ or $\mathbb{C}$. The integers $\mathbb{Z}$ are not a field as the quotient of two integers is not necessarily an integer.

---

[3] More formally, a field is what is known as a commutative ring with some special properties, not discussed here. What are known as function fields can also be defined.

Consider a set $\mathbb{S}$ with two operations defined: addition of two elements (denoted by $+$) both belonging to the set and multiplication of a member of the set by a scalar belonging to a field $\mathbb{F}$ (indicated by juxtaposition). Let us also require the set to be closed under the operations of addition and multiplication by a scalar, that is, if $x \in \mathbb{S}$, $y \in \mathbb{S}$, and $\alpha \in \mathbb{F}$, then $x + y \in \mathbb{S}$ and $\alpha x \in \mathbb{S}$. Furthermore:

1. $\forall\, x, y \in \mathbb{S}:\ x + y = y + x$. For all elements x and y in $\mathbb{S}$, the addition operator on such elements is commutative.
2. $\forall\, x, y, z \in \mathbb{S}:\ (x + y) + z = x + (y + z)$. For all elements x and y in $\mathbb{S}$, the addition operator on such elements is associative.
3. $\exists\, 0 \in \mathbb{S}\ |\ \forall\, x \in \mathbb{S},\ x + 0 = x$. There exists[4] a 0, which is an element of $\mathbb{S}$, such that[5] for all x in $\mathbb{S}$, when the addition operator is applied to 0 and x, the original element x is yielded. Here 0 is the *additive identity element*.
4. $\forall\, x \in \mathbb{S},\ \exists\, -x \in \mathbb{S}\ |\ x + (-x) = 0$. For all x in $\mathbb{S}$, there exists an element $-x$, also in $\mathbb{S}$, such that when added to x, yields the 0 element.
5. $\exists\, 1 \in \mathbb{F}\ |\ \forall\, x \in \mathbb{S},\ 1x = x$. There exists an element 1 in $\mathbb{F}$ such that for all x in $\mathbb{S}$, 1 multiplying the element x yields the element x. Here 1 is the *multiplicative identity element*.
6. $\forall\, a, b \in \mathbb{F}, \forall\, x \in \mathbb{S},\ (a + b)x = ax + bx$. For all a and b that are in $\mathbb{F}$ and for all x that are in $\mathbb{S}$, the addition operator distributes onto multiplication.
7. $\forall\, a \in \mathbb{F}, \forall\, x, y \in \mathbb{S},\ a(x + y) = ax + ay$.
8. $\forall\, a, b \in \mathbb{F}, \forall\, x \in \mathbb{S},\ a(bx) = (ab)x$.

Such a set is called a *linear space* or *vector space* over the field $\mathbb{F}$, and its elements are called *vectors*. We will see that our definition is inclusive enough to include elements that are traditionally thought of as vectors (in the sense of a directed line segment) and some that are outside of this tradition. For this chapter, as is common in the pure mathematics literature, typical vectors, such as x and y, are no longer indicated in bold. However, they are in general *not* scalars, though in special cases, they can be.

The element $0 \in \mathbb{S}$ is called the *null vector*. Examples of vector spaces $\mathbb{S}$ over the field of real numbers (i.e., $\mathbb{F} : \mathbb{R}$) are as follows:

- $\mathbb{S} : \mathbb{R}^1$. Set of real numbers, $x = x_1$, with addition and scalar multiplication defined as usual; also known as $\mathbb{S} : \mathbb{R}$.
- $\mathbb{S} : \mathbb{R}^2$. Set of ordered pairs of real numbers, $x = (x_1, x_2)^T$, with addition and scalar multiplication defined as

$$x + y = \begin{pmatrix} x_1 + y_1 \\ x_2 + y_2 \end{pmatrix} = (x_1 + y_1, x_2 + y_2)^T, \qquad (6.17)$$

$$\alpha x = \begin{pmatrix} \alpha x_1 \\ \alpha x_2 \end{pmatrix} = (\alpha x_1, \alpha x_2)^T, \qquad (6.18)$$

where

$$x = \begin{pmatrix} x_1 \\ x_2 \end{pmatrix} = (x_1, x_2)^T \in \mathbb{R}^2, \qquad y = \begin{pmatrix} y_1 \\ y_2 \end{pmatrix} = (y_1, y_2)^T \in \mathbb{R}^2, \qquad \alpha \in \mathbb{R}^1.$$
$$(6.19)$$

Note $\mathbb{R}^2 = \mathbb{R}^1 \times \mathbb{R}^1$, where the symbol $\times$ represents a Cartesian product.

---

[4] The notation $\exists$ connotes "there exists."
[5] The notation $|$ connotes "such that."

- $\mathbb{S} : \mathbb{R}^N$. Set of N real numbers, $x = (x_1, \ldots, x_N)^T$, with addition and scalar multiplication defined similarly to that just defined in $\mathbb{R}^2$.
- $\mathbb{S} : \mathbb{R}^\infty$. Set of an infinite number of real numbers, $x = (x_1, x_2, \ldots)^T$, with addition and scalar multiplication defined similarly to those defined for $\mathbb{R}^N$. One can interpret many well-behaved functions, for example, $x = 3t^2 + t$, $t \in \mathbb{R}^1$, to generate vectors $x \in \mathbb{R}^\infty$.
- $\mathbb{S} : \mathbb{C}$. Set of all complex numbers $z = z_1$, with $z_1 = a_1 + ib_1; a_1, b_1 \in \mathbb{R}^1$.
- $\mathbb{S} : \mathbb{C}^2$. Set of all ordered pairs of complex numbers $z = (z_1, z_2)^T$, with $z_1 = a_1 + ib_1, z_2 = a_2 + ib_2; a_1, a_2, b_1, b_2 \in \mathbb{R}^1$.
- $\mathbb{S} : \mathbb{C}^N$. Set of N complex numbers, $z = (z_1, \ldots, z_N)^T$.
- $\mathbb{S} : \mathbb{C}^\infty$. Set of an infinite number of complex numbers, $z = (z_1, z_2, \ldots)^T$. Scalar complex functions give rise to sets in $\mathbb{C}^\infty$.
- $\mathbb{S} : \mathbb{M}$. Set of all $M \times N$ matrices with addition and multiplication by a scalar defined as usual, and $M \in \mathbb{N}, N \in \mathbb{N}$.
- $\mathbb{S} : C[a, b]$. Set of real-valued continuous functions, $x(t)$ for $t \in [a, b] \in \mathbb{R}^1$, with addition and scalar multiplication defined as usual.
- $\mathbb{S} : C^N[a, b]$. Set of real-valued functions $x(t)$ for $t \in [a, b]$ with continuous N^{th} derivative, with addition and scalar multiplication defined as usual; $N \in \mathbb{N}$.
- $\mathbb{S} : \mathbb{L}_2[a, b]$. Set of real-valued functions $x(t)$ such that $x(t)^2$ is Lebesgue integrable in $t \in [a, b] \in \mathbb{R}^1, a < b$, with addition and multiplication by a scalar defined as usual. The integral must be finite.
- $\mathbb{S} : \mathbb{L}_p[a, b]$. Set of real-valued functions $x(t)$ such that $|x(t)|^p$, $p \in [1, \infty)$, is Lebesgue integrable for $t \in [a, b] \in \mathbb{R}^1, a < b$, with addition and multiplication by a scalar defined as usual. The integral must be finite.
- $\mathbb{S} : \overline{\mathbb{L}}_p[a, b]$. Set of complex-valued functions $x(t)$ such that $|x(t)|^p$, $p \in [1, \infty) \in \mathbb{R}^1$, is Lebesgue integrable for $t \in [a, b] \in \mathbb{R}^1, a < b$, with addition and multiplication by a scalar defined as usual.
- $\mathbb{S} : \mathbb{W}_2^1(G)$. Set of real-valued functions $u(x)$ such that $u(x)^2$ and $\sum_{n=1}^N (\partial u / \partial x_n)^2$ are Lebesgue integrable in G, where $x \in G \in \mathbb{R}^N, N \in \mathbb{N}$. This is an example of a Sobolov[6] space, which is useful in variational calculus and the finite element method. Sobolov space $W_2^1(G)$ is to Lebesgue space $\mathbb{L}_2[a, b]$ as the real space $\mathbb{R}^1$ is to the rational space $\mathbb{Q}^1$. That is, Sobolov space allows a broader class of functions to be solutions to physical problems.
- $\mathbb{S} : \mathbb{P}^N$. Set of all polynomials of degree $\leq N$ with addition and multiplication by a scalar defined as usual; $N \in \mathbb{N}$.

Some examples of sets that are *not* vector spaces are $\mathbb{Z}$ and $\mathbb{N}$ over the field $\mathbb{R}$ for the same reason that they do not form a field, namely, that they are not closed over the multiplication operation.

We next list a set of useful definitions and concepts.

- $\mathbb{S}'$ is a *subspace* of $\mathbb{S}$ if $\mathbb{S}' \subset \mathbb{S}$, and $\mathbb{S}'$ is itself a vector space. For example $\mathbb{R}^2$ is a subspace of $\mathbb{R}^3$.
- If $\mathbb{S}_1$ and $\mathbb{S}_2$ are subspaces of $\mathbb{S}$, then $\mathbb{S}_1 \cap \mathbb{S}_2$ is also a subspace. The set $\mathbb{S}_1 + \mathbb{S}_2$ of all $x_1 + x_2$ with $x_1 \in \mathbb{S}_1$ and $x_2 \in \mathbb{S}_2$ is also a subspace of $\mathbb{S}$.
- If $\mathbb{S}_1 + \mathbb{S}_2 = \mathbb{S}$, and $\mathbb{S}_1 \cap \mathbb{S}_2 = \{0\}$, then $\mathbb{S}$ is the *direct sum* of $\mathbb{S}_1$ and $\mathbb{S}_2$, written as $\mathbb{S} = \mathbb{S}_1 \oplus \mathbb{S}_2$.

---

[6] Sergei Lvovich Sobolev, 1908–1989, Soviet physicist and mathematician.

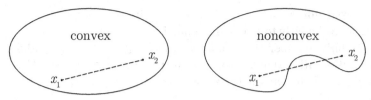

Figure 6.3. Sketches of convex and nonconvex spaces.

- If $x_1, x_2, \ldots, x_N$ are elements of a vector space $\mathbb{S}$, and $\alpha_1, \alpha_2, \ldots, \alpha_N$ belong to the field $\mathbb{F}$, then $x = \alpha_1 x_1 + \alpha_2 x_2 + \cdots + \alpha_N x_N \in \mathbb{S}$ is a *linear combination* of the vectors $x_1, x_2, \ldots, x_N$.
- Vectors $x_1, x_2, \ldots, x_N$ for which it is possible to have $\alpha_1 x_1 + \alpha_2 x_2 + \cdots + \alpha_N x_N = 0$ where the scalars α_n are not all zero are said to be *linearly dependent*. Otherwise, they are *linearly independent*.
- For $M \leq N$, the set of all linear combinations of M vectors $\{x_1, x_2, \ldots, x_M\}$ of a vector space constitute a subspace of an N-dimensional vector space.
- A set of N linearly independent vectors in an N-dimensional vector space is said to *span* the space.
- If the vector space $\mathbb{S}$ contains a set of N linearly independent vectors, and any set with $(N + 1)$ elements is linearly dependent, then the space is said to be *finite-dimensional*, and N is the *dimension* of the space. If N does not exist, the space is *infinite dimensional*.
- A *basis* of a finite-dimensional space of dimension N is a set of N linearly independent vectors $\{u_1, u_2, \ldots, u_N\}$, where all the basis vectors $u_1, u_2, \ldots, u_N$ are elements of the space. All elements of the vector space can be represented as linear combinations of the basis vectors.
- A set of vectors in a linear space $\mathbb{S}$ is *convex* iff $\forall \, x_1, x_2 \in \mathbb{S}$ and $\alpha \in [0, 1] \in \mathbb{R}^1$ implies $\alpha x_1 + (1 - \alpha) x_2 \in \mathbb{S}$. For example, if we consider $\mathbb{S}$ to be a subspace of $\mathbb{R}^2$, $\mathbb{S}$ is convex if for any two points in $\mathbb{S}$, all points on the line segment between them also lie in $\mathbb{S}$. Spaces with lobes are not convex. In simpler terms, *sets are convex if any line segment connecting two points of the set lie within the set*. A sketch depicting convex and nonconvex spaces is shown in Figure 6.3.
- Functions $f(x)$ of single variables are convex iff the space on which they operate are convex and if $f(\alpha x_1 + (1 - \alpha) x_2) \leq \alpha f(x_1) + (1 - \alpha) f(x_2) \, \forall \, x_1, x_2 \in \mathbb{S}, \alpha \in [0, 1] \in \mathbb{R}^1$. In simpler terms, *functions are convex if any line segment connecting two points on the function lies on or above the function*. Even simpler, if the set of points above the function forms a convex set, then the function is convex. For functions in higher dimensions, one must consider the Hessian matrix. A function of many variables $f(x_1, x_2, \ldots)$ is convex iff its Hessian is positive semi-definite, and if the Hessian is positive definite (see Sections 6.7.1 or 7.2.5), then f is strictly convex. A convex surface will have positive Gaussian curvature (see Section 2.7.3).

EXAMPLE 6.3

Determine the convexity of (a) $f(x) = 1 + x^2$ and (b) $f(x) = 1 - 4x^2 + 5x^4$ for $x \in [-1, 1]$.

Loosely speaking, our function f is convex if the line segment joining any two points on f is never below f. For case a, that is true, so $f(x) = 1 + x^2$ is convex. Let us check this

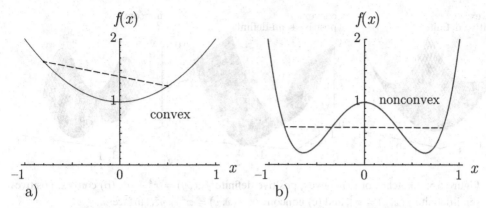

Figure 6.4. Plot demonstrating (a) the convex nature of $f(x) = 1 + x^2$ and the (b) nonconvex nature of $f(x) = 1 - 4x^2 + 5x^4$ for $x \in [-1, 1]$.

using the formal definition, taking $x_1 = -4/5$ and $x_2 = 1/2$. Points on the line segment connecting x_1 and x_2 are given by

$$\alpha f\left(-\frac{4}{5}\right) + (1 - \alpha)f\left(\frac{1}{2}\right) = \frac{5}{4} + \frac{39\alpha}{100}. \tag{6.20}$$

For convexity, this term must be greater than the value of the function on the same domain:

$$f(\alpha x_1 + (1 - \alpha)x_2) = 1 + \left(\frac{1 - \alpha}{2} - \frac{4\alpha}{5}\right)^2. \tag{6.21}$$

Subtracting the two, we find

$$\alpha f\left(-\frac{4}{5}\right) + (1 - \alpha)f\left(\frac{1}{2}\right) - f(\alpha x_1 + (1 - \alpha)x_2) = \frac{169}{100}\alpha(1 - \alpha). \tag{6.22}$$

Because $\alpha \in [0, 1]$, the difference is always positive, and our function is convex. This is illustrated in Figure 6.4a.

Leaving out details, for case b, we can see from Figure 6.4b that it is possible for points on a line segment to be below f, thus rendering the curve nonconvex.

EXAMPLE 6.4

Determine the convexity of

$$f(x, y) = x^2 + ay^2. \tag{6.23}$$

The Hessian matrix is

$$\mathbf{H} = \begin{pmatrix} \frac{\partial^2 f}{\partial x^2} & \frac{\partial^2 f}{\partial x \partial y} \\ \frac{\partial^2 f}{\partial x \partial y} & \frac{\partial^2 f}{\partial y^2} \end{pmatrix} = \begin{pmatrix} 2 & 0 \\ 0 & 2a \end{pmatrix}. \tag{6.24}$$

The eigenvalues of $\mathbf{H}$ are $2, 2a$. For convexity of f, we need positive semi-definite eigenvalues, so we need $a \geq 0$. We give plots of the convex positive definite surface when $a = 1$, the convex positive semi-definite surface when $a = 0$, and the nonconvex surface when $a = -1$ in Figure 6.5. One can easily visualize a line segment connecting two points

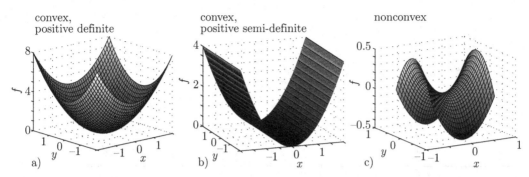

Figure 6.5. Sketches of (a) convex, positive definite $f(x,y) = x^2 + y^2$, (b) convex, positive semi-definite $f(x,y) = x^2$, and (c) nonconvex, $f(x,y) = x^2 - y^2$, surfaces.

on the convex surface always being above the surface; this would not be the case for the nonconvex surface.

6.3.1 Normed

A normed vector space is a vector space equipped with a so-called *norm*. The norm of a vector $x \in \mathbb{S}$, denoted $||x||$, is a real number that satisfies the following properties:

1. $||x|| \geq 0$,
2. $||x|| = 0$ iff $x = 0$,
3. $||\alpha x|| = |\alpha| \, ||x||$, $\quad \alpha \in \mathbb{C}^1$,
4. $||x + y|| \leq ||x|| + ||y||$, (triangle or Minkowski[7] inequality).

We can combine the first two properties to say that the norm has a positive definite character. The norm is a natural generalization of the length of a vector. All properties of a norm can be cast in terms of ordinary finite-dimensional Euclidean vectors, and thus have geometrical interpretations. The first property says length is greater than or equal to zero. The second says the only vector with zero length is the zero vector. The third says the length of a scalar multiple of a vector is equal to the product of the magnitude of the scalar and the length of the original vector. The Minkowski inequality is easily understood in terms of vector addition. If we add vectorially two vectors x and y, we will get a third vector whose length is less than or equal to the sum of the lengths of the original two vectors. We will get equality only if x and y point in the same direction. The interesting generalization is that these properties hold for the norms of *functions* as well as ordinary geometric vectors. An operator which satisfies properties 1, 3, and 4 of a norm, but not necessarily property 2 is a *semi-norm*. A semi-norm may have $||x|| = 0$ when $x \neq 0$. All norms are also semi-norms.

Examples of norms are as follows:

1. $x \in \mathbb{R}^1$, $||x|| = |x|$. This space is also written as $\ell_1(\mathbb{R}^1)$ or in abbreviated form ℓ_1^1. The subscript on ℓ in either case denotes the type of norm; the superscript in the second form denotes the dimension of the space. Another way to denote this norm is $||x||_1$.
2. $x \in \mathbb{R}^2$, $x = (x_1, x_2)^T$, the Euclidean norm $||x|| = ||x||_2 = +\sqrt{x_1^2 + x_2^2} = +\sqrt{x^T x}$. We can call this normed space $\mathbb{E}^2$, or $\ell_2(\mathbb{R}^2)$, or ℓ_2^2.

---

[7] Hermann Minkowski, 1864–1909, Russian/Lithuanian-born German-based mathematician and physicist.

3. $x \in \mathbb{R}^N$, $x = (x_1, x_2, \ldots, x_N)^T$, $||x|| = ||x||_2 = +\sqrt{x_1^2 + x_2^2 + \cdots + x_N^2} = +\sqrt{x^T x}$. We can call this norm the 2-norm, Euclidean norm, and the normed space Euclidean $\mathbb{E}^N$, or $\ell_2(\mathbb{R}^N)$ or ℓ_2^N.

4. $x \in \mathbb{R}^N$, $x = (x_1, x_2, \ldots, x_N)^T$, $||x|| = ||x||_1 = |x_1| + |x_2| + \cdots + |x_N|$, the 1-norm, taxi-cab norm, or Manhattan norm. This is also $\ell_1(\mathbb{R}^N)$ or ℓ_1^N.

5. $x \in \mathbb{R}^N$, $x = (x_1, x_2, \ldots, x_N)^T$, $||x|| = ||x||_p = (|x_1|^p + |x_2|^p + \cdots + |x_N|^p)^{1/p}$, where $1 \le p < \infty$. This space is called or $\ell_p(\mathbb{R}^N)$ or ℓ_p^N.

6. $x \in \mathbb{R}^N$, $x = (x_1, x_2, \ldots, x_N)^T$, $||x|| = ||x||_\infty = \max_{1 \le n \le N} |x_n|$. This space is called $\ell_\infty(\mathbb{R}^N)$ or ℓ_∞^N. This is often called the infinity norm, the maximum norm, the Chebyshev norm, the chessboard norm, or the supremum norm.

7. $x \in \mathbb{C}^N$, $x = (x_1, x_2, \ldots, x_N)^T$, $||x|| = ||x||_2 = +\sqrt{|x_1|^2 + |x_2|^2 + \cdots + |x_N|^2} = +\sqrt{\overline{x}^T x}$. This space is described as $\ell_2(\mathbb{C}^N)$. The operation of conjugation followed by transpose arises sufficiently often that we give it a special name, the *Hermitian transpose*. It is denoted by the superscript H: $\overline{x}^T \equiv x^H$.

8. $x \in C[a, b]$, $||x|| = \max_{a \le t \le b} |x(t)|$; $t \in [a, b] \in \mathbb{R}^1$.

9. $x \in C^1[a, b]$, $||x|| = \max_{a \le t \le b} |x(t)| + \max_{a \le t \le b} |x'(t)|$; $t \in [a, b] \in \mathbb{R}^1$.

10. $x \in \mathbb{L}_2[a, b]$, $||x|| = ||x||_2 = +\sqrt{\int_a^b x(t)^2 \, dt}$; $t \in [a, b] \in \mathbb{R}^1$.

11. $x \in \mathbb{L}_p[a, b]$, $||x|| = ||x||_p = +(\int_a^b |x(t)|^p \, dt)^{1/p}$; $t \in [a, b] \in \mathbb{R}^1$.

12. $x \in \overline{\mathbb{L}}_2[a, b]$, $||x|| = ||x||_2 = +\sqrt{\int_a^b |x(t)|^2 \, dt} = +\sqrt{\int_a^b \overline{x(t)} x(t) \, dt}$; $t \in [a, b] \in \mathbb{R}^1$.

13. $x \in \overline{\mathbb{L}}_p[a, b]$, $||x|| = ||x||_p = +(\int_a^b |x(t)|^p \, dt)^{1/p} = +\left(\int_a^b (\overline{x(t)} x(t))^{p/2} \, dt\right)^{1/p}$; $t \in [a, b] \in \mathbb{R}^1$.

14. $u \in \mathbb{W}_2^1(G)$, $||u|| = ||u||_{1,2} = +\sqrt{\int_G \left(u(x)u(x) + \sum_{n=1}^N (\partial u/\partial x_n)(\partial u/\partial x_n)\right) \, dx}$; $x \in G \in \mathbb{R}^N$, $u \in \mathbb{L}_2(G)$, $\partial u/\partial x_n \in \mathbb{L}_2(G)$. This is an example of a Sobolov space, which is useful in variational calculus and the finite element method.

Some additional notes on properties of norms include the following:

- The *metric* or *distance* between x and y is defined by $d(x, y) = ||x - y||$. This a natural metric induced by the norm. Thus, $||x||$ is the distance between x and the null vector.
- The *open ball* of radius ϵ centered at x_0 which lies within the normed vector space $\mathbb{S}$ is $\{x \in \mathbb{S} \mid d(x, x_0) < \epsilon\}$.
- The *closed ball* of radius ϵ centered at x_0 which lies within the normed vector space $\mathbb{S}$ is $\{x \in \mathbb{S} \mid d(x, x_0) \le \epsilon\}$.
- The diameter of a set of vectors is the supremum (i.e. the least upper bound) of the distance between any two vectors of the set.
- An ϵ-neighborhood of $x_0 \in \mathbb{S}$ may be defined such that it is composed of $\{x \in \mathbb{S} : ||x - x_0|| < \epsilon\}$.
- Let $\mathbb{S}_1$ and $\mathbb{S}_2$ be subsets of a normed vector space $\mathbb{S}$ such that $\mathbb{S}_1 \subset \mathbb{S}_2$. Then $\mathbb{S}_1$ is *dense* in $\mathbb{S}_2$ if for every $x^{(2)} \in \mathbb{S}_2$ and every $\epsilon > 0$, there is an $x^{(1)} \in \mathbb{S}_1$ for which $||x^{(2)} - x^{(1)}|| < \epsilon$.
- A *sequence* $x^{(1)}, x^{(2)}, \ldots \in \mathbb{S}$, where $\mathbb{S}$ is a normed vector space, is a *Cauchy sequence* if for every $\epsilon > 0$ there exists a number N_ϵ such that $||x^{(m)} - x^{(n)}|| < \epsilon$ for every m and n greater than N_ϵ. Informally, the individual terms of the sequence become closer and closer to each other as the sequence evolves.
- The sequence $x^{(1)}, x^{(2)}, \ldots \in \mathbb{S}$, where $\mathbb{S}$ is a normed vector space, converges if there exists an $x \in \mathbb{S}$ such that $\lim_{n \to \infty} ||x^{(n)} - x|| = 0$. Then x is the *limit point* of the sequence, and we write $\lim_{n \to \infty} x^{(n)} = x$ or $x^{(n)} \to x$.
- Every convergent sequence is a Cauchy sequence, but the converse is not true.
- A normed vector space $\mathbb{S}$ is *complete* if every Cauchy sequence in $\mathbb{S}$ is convergent, i.e. if $\mathbb{S}$ contains all the limit points.

- A complete normed vector space is also called a *Banach*[8] space.
- It can be shown that every finite-dimensional normed vector space is complete.
- Norms $||\cdot||_n$ and $||\cdot||_m$ in $\mathbb{S}$ are *equivalent* if there exist $a, b > 0$ such that, for any $x \in \mathbb{S}$,

$$a||x||_m \leq ||x||_n \leq b||x||_m. \tag{6.25}$$

- In a finite-dimensional vector space, any norm is equivalent to any other norm. So, the convergence of a sequence in such a space does not depend on the choice of norm.

We recall (see Section A.8) that if $z \in \mathbb{C}^1$, we can represent z as $z = a + ib$, where $a \in \mathbb{R}^1, b \in \mathbb{R}^1$; furthermore, the complex conjugate of z is represented as $\overline{z} = a - ib$. It can be shown for $z_1 \in \mathbb{C}^1, z_2 \in \mathbb{C}^1$ that

- $\overline{(z_1 + z_2)} = \overline{z_1} + \overline{z_2}$,
- $\overline{(z_1 - z_2)} = \overline{z_1} - \overline{z_2}$,
- $\overline{z_1 z_2} = \overline{z_1}\,\overline{z_2}$,
- $\overline{\left(\dfrac{z_1}{z_2}\right)} = \overline{z_1}/\overline{z_2}$.

We also recall that the modulus of z, $|z|$, has the following properties:

$$|z|^2 = z\overline{z}, \tag{6.26}$$

$$= (a + ib)(a - ib), \tag{6.27}$$

$$= a^2 + iab - iab - i^2 b^2, \tag{6.28}$$

$$= a^2 + b^2 \geq 0. \tag{6.29}$$

EXAMPLE 6.5

Consider $x \in \mathbb{R}^3$ and take

$$x = \begin{pmatrix} 1 \\ -4 \\ 2 \end{pmatrix}. \tag{6.30}$$

Find the norm if $x \in \ell_1^3$ (absolute value norm), $x \in \ell_2^3$ (Euclidean norm), $x = \ell_3^3$ (another norm), and $x \in \ell_\infty^3$ (maximum norm).

By the definition of the absolute value norm for $x \in \ell_1^3$,

$$||x|| = ||x||_1 = |x_1| + |x_2| + |x_3|, \tag{6.31}$$

we get

$$||x||_1 = |1| + |-4| + |2| = 1 + 4 + 2 = 7. \tag{6.32}$$

Now consider the Euclidean norm for $x \in \ell_2^3$:

$$||x|| = ||x||_2 = +\sqrt{x_1^2 + x_2^2 + x_3^2}. \tag{6.33}$$

Applying to our x, we get

$$||x||_2 = +\sqrt{1^2 + (-4)^2 + 2^2} = \sqrt{1 + 16 + 4} = +\sqrt{21} \approx 4.583. \tag{6.34}$$

Because the norm is Euclidean, this is the ordinary length of the vector.

---

[8] Stefan Banach, 1892–1945, Polish mathematician.

For the norm, $x \in \ell_3^3$, we have

$$\|x\| = \|x\|_3 = + \left(|x_1|^3 + |x_2|^3 + |x_3|^3 \right)^{1/3}, \tag{6.35}$$

so

$$\|x\|_3 = + \left(|1|^3 + |-4|^3 + |2|^3 \right)^{1/3} = (1 + 64 + 8)^{1/3} \approx 4.179. \tag{6.36}$$

For the maximum norm, $x \in \ell_\infty^3$, we have

$$\|x\| = \|x\|_\infty = \lim_{p \to \infty} + \left(|x_1|^p + |x_2|^p + |x_3|^p \right)^{1/p}, \tag{6.37}$$

so

$$\|x\|_\infty = \lim_{p \to \infty} + \left(|1|^p + |-4|^p + |2|^p \right)^{1/p} = 4. \tag{6.38}$$

This selects the magnitude of the component of x whose magnitude is maximum. As p increases, the norm of the vector decreases.

EXAMPLE 6.6

For $x \in \ell_p(\mathbb{R}^2)$, find the loci of points for which $\|x\|_p = 1$ for $p = 1, 2, 3, \infty$.

We are asking for a set of points which are equidistant from the origin in a general Banach space. The notion of distance is not confined to Euclidean spaces. It simply has different manifestations in different spaces. We require, by the definition of the norm, for

$$\left(|x_1|^p + |x_2|^p \right)^{1/p} = 1, \tag{6.39}$$

$$|x_1|^p + |x_2|^p = 1. \tag{6.40}$$

For example if we have a Euclidean norm, we get

$$x_1^2 + x_2^2 = 1, \tag{6.41}$$

which is an ordinary circle of radius unity. The other plots are easy to obtain by direct substitution and are plotted in Figure 6.6. Each of these may be thought of as p-unit circles in spaces equipped with p norms.

The concept of distance in a non-Euclidean Banach space need not seem alien. One might in fact be justified in thinking that the distance which is relevant to a traveler is that which is associated with the available routes of travel. For instance, if one were in a taxicab traversing a city with a Cartesian grid of streets, such as found in large parts of Manhattan, the number of city blocks traveled could be taken as a measure of distance; equidistant points on such a constricted grid are well described by the L_1 norm, also

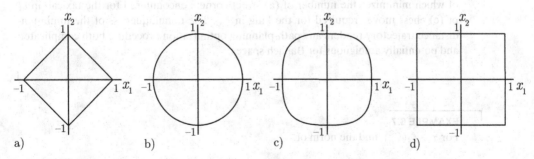

Figure 6.6. Plot of loci of points in $\ell_p(\mathbb{R}^2)$ for which $\|x\|_p = 1$ for (a) $p = 1$, (b) $p = 2$, (c) $p = 3$, and (d) $p \to \infty$.

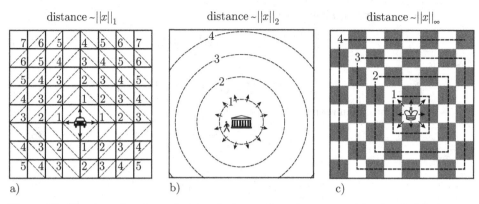

Figure 6.7. Sketches of motions of a (a) taxicab on a Cartesian street grid such as in Manhattan, (b) pedestrian on an open plain, and (c) king on a chessboard, all displaying equidistant contours under the (a) taxicab, Manhattan, ℓ_1, (b) Euclidean ℓ_2, and (c) chessboard, ℓ_∞ distance norms for Banach spaces.

known as the taxicab or Manhattan norm. For efficient travel in such a city, it is often the number of nodal points encountered at street corners that is of more relevance than the actual Euclidean distance. A pedestrian choosing a path from a central point on an open plain would likely be most interested in the Euclidean distance, and could select any direction of motion. If one were a king on a chessboard, the distance between any two points on the board can be thought of as the number of moves necessary to reach that point. Because the king, unlike the taxicab in much of Manhattan, can move diagonally, equidistant points for the king are better represented as an infinity norm or "chessboard norm." The contour lines of equal distance for a taxicab moving on a Cartesian street grid, a pedestrian moving from a central point, and a king moving on a chessboard are sketched in Figure 6.7.

Other constraints imposed by rules or topology are likely to induce more complicated distance norms. The non-Cartesian nature of streets found in many cities would induce an unusual norm, as would equidistant points for motion of a knight on a chessboard. Such unusual norms have application whenever any generalized distance is to be minimized, such as for problems in logistics, network theory, or traffic flow.

Let us further compare a feature which distinguishes Euclidean space from these two important non-Euclidean spaces. Certainly, one's intuition, supported formally by the calculus of variations, Eq. (1.188), shows that in the Euclidean ℓ_2, the unique trajectory of minimum distance between two points, however widely separated, is given by the straight line connecting those points. By inspection of Figure 6.7a,c, however, for both ℓ_1 and ℓ_∞ there exist multiple trajectories connecting many widely separate points each of which minimizes the number of (a) street corners encountered for the taxicab in ℓ_1 or (c) chess moves required for the king in ℓ_∞. The nonuniqueness of the minimum-distance trajectory renders any path-planning optimization procedure both complicated and potentially ambiguous for Banach spaces.

EXAMPLE 6.7

For $x \in \ell_2(\mathbb{C}^2)$, find the norm of

$$x = \begin{pmatrix} i \\ 1 \end{pmatrix} = \begin{pmatrix} 0 + 1i \\ 1 + 0i \end{pmatrix}. \tag{6.42}$$

The definition of the space defines the norm as a 2-norm ("Euclidean"):

$$||x|| = ||x||_2 = +\sqrt{\bar{x}^T x} = +\sqrt{x^H x} = +\sqrt{\bar{x}_1 x_1 + \bar{x}_2 x_2} = \sqrt{|x_1|^2 + |x_2|^2},\qquad(6.43)$$

so

$$||x||_2 = +\sqrt{(\overline{0 + 1i} \quad \overline{1 + 0i})\begin{pmatrix} 0 + 1i \\ 1 + 0i \end{pmatrix}},\qquad(6.44)$$

$$||x||_2 = +\sqrt{\overline{(0 + 1i)}(0 + 1i) + \overline{(1 + 0i)}(1 + 0i)}$$

$$= +\sqrt{(0 - 1i)(0 + 1i) + (1 - 0i)(1 + 0i)},\qquad(6.45)$$

$$||x||_2 = +\sqrt{-i^2 + 1} = +\sqrt{2}.\qquad(6.46)$$

If we had been negligent in the use of the conjugate and tried to define the norm as $+\sqrt{x^T x}$, we would have obtained the erroneous

$$+\sqrt{x^T x} = +\sqrt{(i \quad 1)\begin{pmatrix} i \\ 1 \end{pmatrix}} = +\sqrt{i^2 + 1} = +\sqrt{-1 + 1} = 0.\qquad(6.47)$$

This violates the property of the norm that $||x|| > 0$ if $x \neq 0$!

EXAMPLE 6.8

Consider $x \in \mathbb{L}_2[0, 1]$ where $x(t) = 2t$; $t \in [0, 1] \in \mathbb{R}^1$. Find $||x||$.

By the definition of the norm for this space, we have

$$||x|| = ||x||_2 = +\sqrt{\int_0^1 x^2(t)\, dt},\qquad(6.48)$$

$$||x||_2^2 = \int_0^1 x(t)x(t)\, dt = \int_0^1 (2t)(2t)\, dt = 4\int_0^1 t^2\, dt = 4\left(\frac{t^3}{3}\right)\Big|_0^1,\qquad(6.49)$$

$$= 4\left(\frac{1^3}{3} - \frac{0^3}{3}\right) = \frac{4}{3},\qquad(6.50)$$

$$||x||_2 = \frac{2\sqrt{3}}{3} \approx 1.1547.\qquad(6.51)$$

EXAMPLE 6.9

Consider $x \in \overline{\mathbb{L}}_3[-2, 3]$, where $x(t) = 1 + 2it$; $t \in [-2, 3] \in \mathbb{R}^1$. Find $||x||$.

By the definition of the norm, we have

$$||x|| = ||x||_3 = +\left(\int_{-2}^3 |1 + 2it|^3\, dt\right)^{1/3},\qquad(6.52)$$

$$||x||_3 = +\left(\int_{-2}^3 \left(\overline{(1 + 2it)}(1 + 2it)\right)^{3/2} dt\right)^{1/3},\qquad(6.53)$$

$$||x||_3^3 = \int_{-2}^3 \left(\overline{(1 + 2it)}(1 + 2it)\right)^{3/2} dt,\qquad(6.54)$$

$$= \int_{-2}^3 ((1 - 2it)(1 + 2it))^{3/2}\, dt,\qquad(6.55)$$

$$= \int_{-2}^{3} \left(1 + 4t^2\right)^{3/2} \, dt, \tag{6.56}$$

$$= \left. \left(\sqrt{1 + 4t^2} \left(\frac{5t}{8} + t^3\right) + \frac{3}{16} \sinh^{-1}(2t)\right) \right|_{-2}^{3}, \tag{6.57}$$

$$= \frac{37\sqrt{17}}{4} + \frac{3\sinh^{-1}(4)}{16} + \frac{3}{16} \left(154\sqrt{17} + \sinh^{-1}(6)\right) \approx 214.638, \tag{6.58}$$

$$||x||_3 \approx 5.98737. \tag{6.59}$$

EXAMPLE 6.10

Consider $x \in \overline{\mathbb{L}}_p[a, b]$, where $x(t) = c; \; t \in [a, b] \in \mathbb{R}^1, c \in \mathbb{C}^1$. Find $||x||$.

Let us take the complex constant $c = \alpha + i\beta, \alpha \in \mathbb{R}^1, \beta \in \mathbb{R}^1$. Then

$$|c| = \left(\alpha^2 + \beta^2\right)^{1/2}. \tag{6.60}$$

Now

$$||x|| = ||x||_p = \left(\int_a^b |x(t)|^p \, dt\right)^{1/p}, \tag{6.61}$$

$$||x||_p = \left(\int_a^b \left(\alpha^2 + \beta^2\right)^{p/2} \, dt\right)^{1/p}, \tag{6.62}$$

$$= \left(\left(\alpha^2 + \beta^2\right)^{p/2} \int_a^b dt\right)^{1/p}, \tag{6.63}$$

$$= \left(\left(\alpha^2 + \beta^2\right)^{p/2} (b - a)\right)^{1/p}, \tag{6.64}$$

$$= \left(\alpha^2 + \beta^2\right)^{1/2} (b - a)^{1/p}, \tag{6.65}$$

$$= |c|(b - a)^{1/p}. \tag{6.66}$$

The norm is proportional to the magnitude of the complex constant c. For finite p, it also increases with the extent of the domain $b - a$. For infinite p, it is independent of the length of the domain, and simply selects the value $|c|$. This is consistent with the norm in $\overline{\mathbb{L}}_\infty$ selecting the maximum value of the function.

EXAMPLE 6.11

Consider $x \in \mathbb{L}_p[0, b]$, where $x(t) = 2t^2; \; t \in [0, b] \in \mathbb{R}^1$. Find $||x||$.

Now

$$||x|| = ||x||_p = \left(\int_0^b |x(t)|^p \, dt\right)^{1/p}, \tag{6.67}$$

$$||x||_p = \left(\int_0^b |2t^2|^p \, dt\right)^{1/p}, \tag{6.68}$$

$$= \left(\int_0^b 2^p t^{2p} \, dt\right)^{1/p}, \tag{6.69}$$

$$= \left(\left(\frac{2^p t^{2p+1}}{2p+1} \right) \Big|_0^b \right)^{1/p}, \tag{6.70}$$

$$= \left(\frac{2^p b^{2p+1}}{2p+1} \right)^{1/p}, \tag{6.71}$$

$$= \frac{2 b^{\frac{2p+1}{p}}}{(2p+1)^{1/p}}. \tag{6.72}$$

As $p \to \infty$, $(2p+1)^{1/p} \to 1$, and $(2p+1)/p \to 2$, so

$$\lim_{p \to \infty} ||x|| = 2b^2. \tag{6.73}$$

This is the maximum value of $x(t) = 2t^2$ for $t \in [0, b]$, as expected.

EXAMPLE 6.12

Consider $u \in \mathbb{W}_2^1(G)$ with $u(x) = 2x^4$; $x \in [0, 3] \in \mathbb{R}^1$. Find $||u||$.

Here we require $u \in \mathbb{L}_2[0, 3]$ and $\partial u / \partial x \in \mathbb{L}_2[0, 3]$, which for our choice of u is satisfied. The formula for the norm in $\mathbb{W}_2^1[0, 3]$ is

$$||u|| = ||u||_{1,2} = +\sqrt{\int_0^3 \left(u(x)u(x) + \frac{du}{dx}\frac{du}{dx} \right) dx}, \tag{6.74}$$

$$||u||_{1,2} = +\sqrt{\int_0^3 ((2x^4)(2x^4) + (8x^3)(8x^3)) \ dx}, \tag{6.75}$$

$$= +\sqrt{\int_0^3 (4x^8 + 64x^6) \ dx}, \tag{6.76}$$

$$= +\sqrt{\left(\frac{4x^9}{9} + \frac{64x^7}{7} \right) \Big|_0^3} = 54\sqrt{\frac{69}{7}} \approx 169.539. \tag{6.77}$$

EXAMPLE 6.13

Consider the sequence of vectors $\{x_{(1)}, x_{(2)}, \ldots\} \in \mathbb{Q}^1$, where $\mathbb{Q}^1$ is the space of rational numbers over the field of rational numbers and

$$x_{(1)} = 1, \tag{6.78}$$

$$x_{(2)} = \frac{1}{1+1} = \frac{1}{2}, \tag{6.79}$$

$$x_{(3)} = \frac{1}{1 + \frac{1}{2}} = \frac{2}{3}, \tag{6.80}$$

$$x_{(4)} = \frac{1}{1 + \frac{2}{3}} = \frac{3}{5}, \tag{6.81}$$

$$\vdots$$

$$x_{(n)} = \frac{1}{1 + x_{(n-1)}}, \tag{6.82}$$

$$\vdots$$

for $n \geq 2$. Does this sequence have a limit point in $\mathbb{Q}^1$? Is this a Cauchy sequence?

The series has converged when the nth term is equal to the $(n-1)^{th}$ term:

$$x_{(n-1)} = \frac{1}{1 + x_{(n-1)}}. \tag{6.83}$$

Rearranging, it is found that

$$x_{(n-1)}^2 + x_{(n-1)} - 1 = 0. \tag{6.84}$$

Solving, one finds that

$$x_{(n-1)} = \frac{-1 \pm \sqrt{5}}{2}. \tag{6.85}$$

We find from numerical experimentation that it is the "+" root to which x converges:

$$\lim_{n \to \infty} x_{(n-1)} = \frac{\sqrt{5} - 1}{2}. \tag{6.86}$$

As $n \to \infty$,

$$x_{(n)} \to \frac{\sqrt{5} - 1}{2}. \tag{6.87}$$

This is an irrational number. Because the limit point for this sequence is *not* in $\mathbb{Q}^1$, the sequence is not convergent. Had the set been defined in $\mathbb{R}^1$, it would have been convergent.

However, the sequence *is a Cauchy sequence*. Consider, say, $\epsilon = 0.01$. Following the definition in Section 6.3.1, we then find by numerical experimentation that $N_\epsilon = 4$. Choosing, for example, $m = 5 > N_\epsilon$ and $n = 21 > N_\epsilon$, we get

$$x_{(5)} = \frac{5}{8}, \tag{6.88}$$

$$x_{(21)} = \frac{10946}{17711}, \tag{6.89}$$

$$||x_{(5)} - x_{(21)}||_2 = \left\| \frac{987}{141688} \right\|_2 = 0.00696 < 0.01. \tag{6.90}$$

This could be generalized for arbitrary ϵ, so the sequence can be shown to be a Cauchy sequence.

EXAMPLE 6.14

Does the infinite sequence of functions

$$v = \{v_1(t), v_2(t), \ldots, v_n(t), \ldots\} = \{t^2, t^3, \ldots, t^{n+1}, \ldots\}, \qquad n = 1, 2, \ldots \tag{6.91}$$

converge in $\mathbb{L}_2[0, 1]$? Does the sequence converge in $C[0, 1]$?

First, check if the sequence is a Cauchy sequence:

$$\lim_{n,m \to \infty} ||v_n(t) - v_m(t)||_2 = \sqrt{\int_0^1 (t^{n+1} - t^{m+1})^2 \, dt}$$

$$= \sqrt{\frac{1}{2n+3} - \frac{2}{m+n+3} + \frac{1}{2m+3}} = 0. \tag{6.92}$$

As this norm approaches zero, it will be possible for any $\epsilon > 0$ to find an integer N_ϵ such that $||v_n(t) - v_m(t)||_2 < \epsilon$. So the sequence is a Cauchy sequence. We also have

$$\lim_{n \to \infty} v_n(t) = \begin{cases} 0, & t \in [0, 1), \\ 1, & t = 1. \end{cases} \tag{6.93}$$

The function given in Eq. (6.93), the "limit point" to which the sequence converges, is in $\mathbb{L}_2[0, 1]$, which is sufficient condition for convergence of the sequence of functions in $\mathbb{L}_2[0, 1]$. However the "limit point" is not a continuous function, so despite the fact that the sequence is a Cauchy sequence and elements of the sequence are in $C[0, 1]$, the sequence does not converge in $C[0, 1]$.

EXAMPLE 6.15

Analyze the sequence of functions

$$v = \{v_1, v_2, \ldots, v_n, \ldots\} = \left\{\sqrt{2}\sin(\pi t), \sqrt{2}\sin(2\pi t), \ldots, \sqrt{2}\sin(n\pi t), \ldots\right\} \quad (6.94)$$

in $\mathbb{L}_2[0, 1]$.

This is simply a set of sine functions, which can be shown to form a basis; such a proof will not be given here. Each element of the set is orthonormal to other elements because

$$||v_n(t)||_2 = \left(\int_0^1 \left(\sqrt{2}\sin(n\pi t)\right)^2 dt\right)^{1/2} = 1, \quad (6.95)$$

and $\int_0^1 v_n(t)v_m(t)\, dt = 0$ for $n \neq m$. As $n \to \infty$, the norm of the basis function remains bounded and is, in fact, unity.

Consider the norm of the difference of the mth and n^{th} functions:

$$||v_n(t) - v_m(t)||_2 = \left(\int_0^1 \left(\sqrt{2}\sin(n\pi t) - \sqrt{2}\sin(m\pi t)\right)^2 dt\right)^{\frac{1}{2}} = \begin{cases} 0, & m = n, \\ \sqrt{2}, & m \neq n. \end{cases} \quad (6.96)$$

Because we can find a value of $\epsilon > 0$ that violates the conditions for a Cauchy sequence, this series of functions is not a Cauchy sequence. The functions do not get close to each other for large m and n.

6.3.2 Inner Product

The inner product $\langle x, y \rangle$ is, in general, a complex scalar, $\langle x, y \rangle \in \mathbb{C}^1$, associated with two elements x and y of a normed vector space satisfying the following rules. For $x, y, z \in \mathbb{S}$ and $\alpha, \beta \in \mathbb{C}$,

1. $\langle x, x \rangle \geq 0$,
2. $\langle x, x \rangle = 0$ iff $x = 0$,
3. $\langle x, \alpha y + \beta z \rangle = \alpha\langle x, y \rangle + \beta\langle x, z \rangle, \quad \alpha \in \mathbb{C}^1, \beta \in \mathbb{C}^1$,
4. $\langle x, y \rangle = \overline{\langle y, x \rangle}$, where $\overline{\langle \cdot \rangle}$ indicates the complex conjugate of the inner product.

The first two properties tell us that $\langle x, x \rangle$ is positive definite. We have seen specific cases of the inner product before in the scalar product or dot product of Section 2.2.2 and the inner product of functions in Section 4.4.1. Inner product spaces are subspaces of linear vector spaces and are sometimes called *pre-Hilbert*[9] *spaces*. A pre-Hilbert space is not necessarily complete, so it may or may not form a Banach space.

[9] David Hilbert, 1862–1943, German mathematician.

EXAMPLE 6.16
Show

$$\langle \alpha x, y \rangle = \overline{\alpha} \langle x, y \rangle. \tag{6.97}$$

Using the properties of the inner product and the complex conjugate, we have

$$\langle \alpha x, y \rangle = \overline{\langle y, \alpha x \rangle}, \tag{6.98}$$

$$= \overline{\alpha \langle y, x \rangle}, \tag{6.99}$$

$$= \overline{\alpha} \, \overline{\langle y, x \rangle}, \tag{6.100}$$

$$= \overline{\alpha} \, \langle x, y \rangle. \tag{6.101}$$

In a real vector space, we have

$$\langle x, \alpha y \rangle = \langle \alpha x, y \rangle = \alpha \langle x, y \rangle, \tag{6.102}$$

and also that

$$\langle x, y \rangle = \langle y, x \rangle, \tag{6.103}$$

because every scalar is equal to its complex conjugate. Some authors use $\langle \alpha y + \beta z, x \rangle = \alpha \langle y, x \rangle + \beta \langle z, x \rangle$ instead of Property 3, which we have chosen. This is acceptable if the rule is consistently applied.

Hilbert Space

A Banach space (i.e., a complete normed vector space) on which an inner product is defined is also called a *Hilbert* space. Whereas Banach spaces allow for the definition of several types of norms, Hilbert spaces are more restrictive: we *must* define the norm such that

$$\|x\| = \|x\|_2 = +\sqrt{\langle x, x \rangle}. \tag{6.104}$$

As a counterexample, if $x \in \mathbb{R}^2$, and we take $\|x\| = \|x\|_3 = (|x_1|^3 + |x_2|^3)^{1/3}$ (thus $x \in \ell_3^2$ which is a Banach space), we *cannot find* a definition of the inner product which satisfies all its properties. Thus, the space ℓ_3^2 cannot be a Hilbert space! Unless specified otherwise, the unsubscripted norm $\| \cdot \|$ can be taken to represent the Hilbert space norm $\| \cdot \|_2$ in much of the literature.

The *Cauchy-Schwarz*[10] *inequality* is embodied in the following:

Theorem: For x and y which are elements of a Hilbert space,

$$\|x\|_2 \, \|y\|_2 \geq |\langle x, y \rangle|. \tag{6.105}$$

Let us prove the theorem. If $y = 0$, both sides are zero, and the equality holds. So let us take $y \neq 0$. This gives, where α is any scalar,

$$\|x - \alpha y\|_2^2 = \langle x - \alpha y, x - \alpha y \rangle, \tag{6.106}$$

$$= \langle x, x \rangle - \langle x, \alpha y \rangle - \langle \alpha y, x \rangle + \langle \alpha y, \alpha y \rangle, \tag{6.107}$$

$$= \langle x, x \rangle - \alpha \langle x, y \rangle - \overline{\alpha} \, \langle y, x \rangle + \alpha \overline{\alpha} \, \langle y, y \rangle. \tag{6.108}$$

[10] Karl Hermann Amandus Schwarz, 1843–1921, Silesia-born German mathematician.

Now choose

$$\alpha = \frac{\langle y, x \rangle}{\langle y, y \rangle} = \frac{\overline{\langle x, y \rangle}}{\langle y, y \rangle}, \tag{6.109}$$

so that

$$\|x - \alpha y\|_2^2 = \langle x, x \rangle - \frac{\overline{\langle x, y \rangle}}{\langle y, y \rangle} \langle x, y \rangle$$

$$\underbrace{- \frac{\langle x, y \rangle}{\langle y, y \rangle} \langle y, x \rangle + \frac{\langle y, x \rangle \langle x, y \rangle}{\langle y, y \rangle^2} \langle y, y \rangle}_{=0}, \tag{6.110}$$

$$= \|x\|_2^2 - \frac{|\langle x, y \rangle|^2}{\|y\|_2^2}, \tag{6.111}$$

$$\|x - \alpha y\|_2^2 \|y\|_2^2 = \|x\|_2^2 \|y\|_2^2 - |\langle x, y \rangle|^2. \tag{6.112}$$

Because $\|x - \alpha y\|_2^2 \|y\|_2^2 \geq 0$,

$$\|x\|_2^2 \|y\|_2^2 - |\langle x, y \rangle|^2 \geq 0, \tag{6.113}$$

$$\|x\|_2^2 \|y\|_2^2 \geq |\langle x, y \rangle|^2, \tag{6.114}$$

$$\|x\|_2 \|y\|_2 \geq |\langle x, y \rangle|, \qquad \text{QED}. \tag{6.115}$$

The theorem is proved.[11]

This effectively defines the angle between two vectors. Because of the Cauchy-Schwarz inequality, we have

$$\frac{\|x\|_2 \|y\|_2}{|\langle x, y \rangle|} \geq 1, \tag{6.116}$$

$$\frac{|\langle x, y \rangle|}{\|x\|_2 \|y\|_2} \leq 1. \tag{6.117}$$

Defining α to be the angle between the vectors x and y, we recover the familiar result from vector analysis

$$\cos \alpha = \frac{\langle x, y \rangle}{\|x\|_2 \|y\|_2}. \tag{6.118}$$

This reduces to the ordinary relationship we find in Euclidean geometry when $x, y \in \mathbb{R}^3$. The Cauchy-Schwarz inequality is actually a special case of the so-called Hölder[12] inequality:

$$\|x\|_p \|y\|_q \geq |\langle x, y \rangle|, \qquad \frac{1}{p} + \frac{1}{q} = 1. \tag{6.119}$$

The Hölder inequality reduces to the Cauchy-Schwarz inequality when $p = q = 2$.

Examples of Hilbert spaces include the following:

- Finite-dimensional vector spaces

---

[11] The common abbreviation QED at the end of the proof stands for the Latin *quod erat demonstrandum*, "that which was to be demonstrated."

[12] Otto Hölder, 1859–1937, German mathematician.

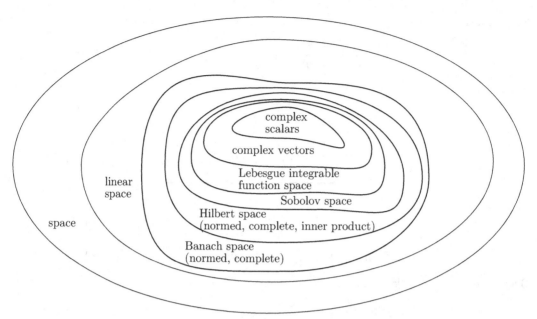

Figure 6.8. Venn diagram showing relationship between various classes of spaces.

    ○ $x \in \mathbb{R}^3, y \in \mathbb{R}^3$ with $\langle x, y \rangle = x^T y = x_1 y_1 + x_2 y_2 + x_3 y_3$,   where $x = (x_1, x_2, x_3)^T$, and $y = (y_1, y_2, y_3)^T$. This is the ordinary dot product for three-dimensional Cartesian vectors. With this definition of the inner product $\langle x, x \rangle = ||x||^2 = x_1^2 + x_2^2 + x_3^2$, so the space is the Euclidean space, $\mathbb{E}^3$. The space is also $\ell_2(\mathbb{R}^3)$ or ℓ_2^3.

    ○ $x \in \mathbb{R}^N, y \in \mathbb{R}^N$ with $\langle x, y \rangle = x^T y = x_1 y_1 + x_2 y_2 + \cdots + x_N y_N$, where $x = (x_1, x_2, \ldots, x_N)^T$, and $y = (y_1, y_2, \ldots, y_N)^T$. This is the ordinary dot product for N-dimensional Cartesian vectors; the space is the Euclidean space, $\mathbb{E}^N$, or $\ell_2(\mathbb{R}^N)$, or ℓ_2^N.

    ○ $x \in \mathbb{C}^N, y \in \mathbb{C}^N$ with $\langle x, y \rangle = \overline{x}^T y = x^H y = \overline{x}_1 y_1 + \overline{x}_2 y_2 + \cdots + \overline{x}_N y_N$, where $x = (x_1, x_2, \ldots, x_N)^T$, and $y = (y_1, y_2, \ldots, y_N)^T$. This space is also $\ell_2(\mathbb{C}^N)$. Note that

      * $\langle x, x \rangle = \overline{x}_1 x_1 + \overline{x}_2 x_2 + \cdots + \overline{x}_N x_N = |x_1|^2 + |x_2|^2 + \cdots + |x_N|^2 = ||x||_2^2$.

      * $\langle x, y \rangle = \overline{x}_1 y_1 + \overline{x}_2 y_2 + \cdots + \overline{x}_N y_N$.

      * It is easily shown that this definition guarantees $||x||_2 \geq 0$ and $\langle x, y \rangle = \overline{\langle y, x \rangle}$.

- Lebesgue spaces
    ○ $x \in \mathbb{L}_2[a, b], y \in \mathbb{L}_2[a, b], t \in [a, b] \in \mathbb{R}^1$ with $\langle x, y \rangle = \int_a^b x(t) y(t) \, dt$,
    ○ $x \in \overline{\mathbb{L}}_2[a, b], y \in \overline{\mathbb{L}}_2[a, b], t \in [a, b] \in \mathbb{R}^1$ with $\langle x, y \rangle = \int_a^b \overline{x(t)} y(t) \, dt$.

- Sobolov spaces
    ○ $u \in \mathbb{W}_2^1(G), v \in \mathbb{W}_2^1(G), x \in G \in \mathbb{R}^N, N \in \mathbb{N}, u \in \mathbb{L}_2(G), \partial u / \partial x_n \in \mathbb{L}_2(G), v \in \mathbb{L}_2(G), \partial v / \partial x_n \in \mathbb{L}_2(G)$ with

$$\langle u, v \rangle = \int_G \left(u(x)v(x) + \sum_{n=1}^{N} \frac{\partial u}{\partial x_n} \frac{\partial v}{\partial x_n} \right) dx \qquad (6.120)$$

A Venn[13] diagram of some of the common spaces is shown in Figure 6.8.

---

[13] John Venn, 1834–1923, English mathematician.

Noncommutation of the Inner Product

By the fourth property of inner products, we see that the inner product operation is not commutative in general. Specifically, when the vectors are complex, $\langle x, y \rangle \neq \langle y, x \rangle$. When the vectors x and y are real, the inner product is real, and the inner product commutes, that is, $\forall\, x \in \mathbb{R}^N, y \in \mathbb{R}^N, \langle x, y \rangle = \langle y, x \rangle$. At first glance one may wonder why one would define a noncommutative operation. It is done to preserve the positive definite character of the norm. If, for example, we had instead defined the inner product to commute for complex vectors, we might have taken $\langle x, y \rangle = x^T y$. Then if we had taken $x = (i, 1)^T$ and $y = (1, 1)^T$, we would have $\langle x, y \rangle = \langle y, x \rangle = 1 + i$. However, we would also have $\langle x, x \rangle = ||x||_2^2 = (i, 1)(i, 1)^T = 0$. Obviously, this would violate the property of the norm because we must have $||x||_2^2 > 0$ for $x \neq 0$.

Interestingly, one can interpret the Heisenberg[14] uncertainty principle to be entirely consistent with our definition of an inner product which does not commute in a complex space. In quantum mechanics, the superposition of physical states of a system is defined by a complex-valued vector field. Position is determined by application of a position operator, and momentum is determined by application of a momentum operator. If one wants to know both position and momentum, both operators are applied. However, they do not commute, and application of them in different orders leads to a result which varies by a factor related to Planck's[15] constant.

Matrix multiplication is another example of an inner product that does not commute, in general. Such topics are considered in the more general group theory. Operators that commute are known as Abelian, and those that do not are known as non-Abelian.

EXAMPLE 6.17

For x and y belonging to a Hilbert space, prove the parallelogram equality:

$$||x + y||_2^2 + ||x - y||_2^2 = 2||x||_2^2 + 2||y||_2^2. \tag{6.121}$$

The left side is

$$\langle x + y, x + y \rangle + \langle x - y, x - y \rangle = (\langle x, x \rangle + \langle x, y \rangle + \langle y, x \rangle + \langle y, y \rangle)$$
$$+ (\langle x, x \rangle - \langle x, y \rangle - \langle y, x \rangle + \langle y, y \rangle), \tag{6.122}$$
$$= 2\langle x, x \rangle + 2\langle y, y \rangle, \tag{6.123}$$
$$= 2||x||_2^2 + 2||y||_2^2, \quad \text{QED.} \tag{6.124}$$

EXAMPLE 6.18

For $x, y \in \ell_2(\mathbb{R}^2)$, find $\langle x, y \rangle$ if

$$x = \begin{pmatrix} 1 \\ 3 \end{pmatrix}, \quad y = \begin{pmatrix} 2 \\ -2 \end{pmatrix}. \tag{6.125}$$

The solution is

$$\langle x, y \rangle = x^T y = (1 \quad 3) \begin{pmatrix} 2 \\ -2 \end{pmatrix} = (1)(2) + (3)(-2) = -4. \tag{6.126}$$

---

[14] Werner Karl Heisenberg, 1901–1976, German physicist.
[15] Max Karl Ernst Ludwig Planck, 1858–1947, German physicist.

The inner product yields a real scalar, but in contrast to the norm, it can be negative. Also that the Cauchy-Schwarz inequality holds as $||x||_2||y||_2 = \sqrt{10}\sqrt{8} \approx 8.944 > |-4|$. Also, the Minkowski inequality holds as $||x + y||_2 = ||(3,1)^T||_2 = +\sqrt{10} < ||x||_2 + ||y||_2 = \sqrt{10} + \sqrt{8}$.

EXAMPLE 6.19

For $x, y \in \ell_2(\mathbb{C}^2)$, find $\langle x, y \rangle$ if

$$x = \begin{pmatrix} -1 + i \\ 3 - 2i \end{pmatrix}, \qquad y = \begin{pmatrix} 1 - 2i \\ -2 \end{pmatrix}. \tag{6.127}$$

The solution is

$$\langle x, y \rangle = \overline{x}^T y = x^H y = (-1 - i \quad 3 + 2i) \begin{pmatrix} 1 - 2i \\ -2 \end{pmatrix}$$

$$= (-1 - i)(1 - 2i) + (3 + 2i)(-2) = -9 - 3i. \tag{6.128}$$

Here the inner product is a complex scalar which has negative components. It is easily shown that $||x||_2 = 3.870$ and $||y||_2 = 3$ and $||x + y||_2 = 2.4495$. Also, $|\langle x, y \rangle| = 9.4868$. The Cauchy-Schwarz inequality holds as $(3.870)(3) = 11.61 > 9.4868$. The Minkowski inequality holds as $2.4495 < 3.870 + 3 = 6.870$.

EXAMPLE 6.20

For $x, y \in \mathbb{L}_2[0, 1]$, find $\langle x, y \rangle$ if

$$x(t) = 3t + 4, \qquad y(t) = -t - 1. \tag{6.129}$$

The solution is

$$\langle x, y \rangle = \int_0^1 (3t + 4)(-t - 1) \, dt = \left(-4t - \frac{7t^2}{2} - t^3 \right) \Big|_0^1 = -\frac{17}{2} = -8.5. \tag{6.130}$$

Once more, the inner product is a negative scalar. It is easily shown that $||x||_2 = 5.56776$ and $||y||_2 = 1.52753$ and $||x + y||_2 = 4.04145$. Also, $|\langle x, y \rangle| = 8.5$. It is easily seen that the Cauchy-Schwarz inequality holds as $(5.56776)(1.52753) = 8.505 > 8.5$. The Minkowski inequality holds as $4.04145 < 5.56776 + 1.52753 = 7.095$.

EXAMPLE 6.21

For $x, y \in \overline{\mathbb{L}}_2[0, 1]$, find $\langle x, y \rangle$ if

$$x(t) = it, \qquad y(t) = t + i. \tag{6.131}$$

We recall that

$$\langle x, y \rangle = \int_0^1 \overline{x(t)} y(t) \, dt. \tag{6.132}$$

The solution is

$$\langle x, y \rangle = \int_0^1 (-it)(t + i) \, dt = \left(\frac{t^2}{2} - \frac{it^3}{3} \right) \Big|_0^1 = \frac{1}{2} - \frac{i}{3}. \tag{6.133}$$

The inner product is a complex scalar. It is easily shown that $||x||_2 = 0.5776$ and $||y||_2 = 1.1547$ and $||x + y||_2 = 1.6330$. Also, $|\langle x, y \rangle| = 0.601$. The Cauchy-Schwarz inequality holds as $(0.57735)(1.1547) = 0.6667 > 0.601$. The Minkowski inequality holds as $1.63299 < 0.57735 + 1.1547 = 1.7321$.

EXAMPLE 6.22

For $u, v \in W_2^1(G))$, find $\langle u, v \rangle$ if

$$u(x) = x_1 + x_2, \qquad v(x) = -x_1 x_2, \tag{6.134}$$

and G is the square region in the (x_1, x_2) plane with $x_1 \in [0, 1], x_2 \in [0, 1]$.

We recall that

$$\langle u, v \rangle = \int_G \left(u(x)v(x) + \frac{\partial u}{\partial x_1}\frac{\partial v}{\partial x_1} + \frac{\partial u}{\partial x_2}\frac{\partial v}{\partial x_2} \right) dx, \tag{6.135}$$

$$\langle u, v \rangle = \int_0^1 \int_0^1 \left((x_1 + x_2)(-x_1 x_2) + (1)(-x_2) + (1)(-x_1) \right) dx_1\, dx_2$$

$$= -\frac{4}{3} = -1.33333. \tag{6.136}$$

The inner product here is negative real scalar. It is easily shown that $\|u\|_{1,2} = 1.77951$ and $\|v\|_{1,2} = 0.881917$ and $\|u + v\|_{1,2} = 1.13039$. Also, $|\langle u, v \rangle| = 1.33333$. The Cauchy-Schwarz inequality holds as $(1.77951)(0.881917) = 1.56938 > 1.33333$. The Minkowski inequality holds as $1.13039 < 1.77951 + 0.881917 = 2.66143$.

Orthogonality

One of the primary advantages of working in Hilbert spaces is that the inner product allows one to utilize the useful concept of orthogonality:

- Vectors x and y are said to be *orthogonal* to each other if

$$\langle x, y \rangle = 0. \tag{6.137}$$

- For an orthogonal set of vectors $\{v_1, v_2, \ldots\}$, the elements of the set are all orthogonal to each other, so that $\langle v_n, v_m \rangle = 0$ if $n \neq m$.
- If a set $\{\varphi_1, \varphi_2, \ldots\}$ exists such that $\langle \varphi_n, \varphi_m \rangle = \delta_{nm}$, then the elements of the set are *orthonormal*.
- A basis $\{v_1, v_2, \ldots, v_N\}$ of a finite-dimensional space that is also orthogonal is an orthogonal basis. On dividing each vector by its norm, we get

$$\varphi_n = \frac{v_n}{\sqrt{\langle v_n, v_n \rangle}}, \tag{6.138}$$

to give us an orthonormal basis $\{\varphi_1, \varphi_2, \ldots, \varphi_N\}$.

EXAMPLE 6.23

If elements x and y of an inner product space are orthogonal to each other, prove Pythagoras' theorem:

$$\|x\|_2^2 + \|y\|_2^2 = \|x + y\|_2^2. \tag{6.139}$$

The right side is

$$\langle x + y, x + y \rangle = \langle x, x \rangle + \underbrace{\langle x, y \rangle}_{=0} + \underbrace{\langle y, x \rangle}_{=0} + \langle y, y \rangle, \tag{6.140}$$

$$= \langle x, x \rangle + \langle y, y \rangle, \tag{6.141}$$

$$= \|x\|_2^2 + \|y\|_2^2, \qquad \textbf{QED}. \tag{6.142}$$

EXAMPLE 6.24

Show that an orthogonal set of vectors in an inner product space is linearly independent.

Let $\{v_1, v_2, \ldots, v_n, \ldots, v_N\}$ be an orthogonal set of vectors. Then consider

$$\alpha_1 v_1 + \alpha_2 v_2 + \cdots + \alpha_n v_n + \cdots + \alpha_N v_N = 0. \tag{6.143}$$

Taking the inner product with v_n, we get

$$\langle v_n, (\alpha_1 v_1 + \alpha_2 v_2 + \cdots + \alpha_n v_n + \cdots + \alpha_N v_N)\rangle = \langle v_n, 0\rangle, \tag{6.144}$$

$$\alpha_1 \underbrace{\langle v_n, v_1\rangle}_{0} + \alpha_2 \underbrace{\langle v_n, v_2\rangle}_{0} + \cdots + \alpha_n \underbrace{\langle v_n, v_n\rangle}_{\neq 0} + \cdots + \alpha_N \underbrace{\langle v_n, v_N\rangle}_{0} = 0, \tag{6.145}$$

$$\alpha_n \langle v_n, v_n\rangle = 0, \tag{6.146}$$

because all the other inner products, $\langle v_n, v_m\rangle$, $m \neq n$, are zero. Because $v_n \neq 0$, we have $\langle v_n, v_n\rangle \neq 0$. Thus, $\alpha_n = 0$, indicating that the set $\{v_1, v_2, \ldots, v_n, \ldots, v_N\}$ is linearly independent.

6.4 Gram-Schmidt Procedure

In a given inner product space, the *Gram-Schmidt*[16] procedure can be used to identify an orthonormal set of vectors by building them from a given set of linearly independent vectors. The following example gives one approach to this procedure.

EXAMPLE 6.25

Find an orthonormal set of vectors $\{\varphi_1, \varphi_2, \ldots\}$ in $\mathbb{L}_2[-1, 1]$ using linear combinations of the linearly independent set of vectors $\{1, t, t^2, t^3, \ldots\}$ where $t \in [-1, 1]$.

Choose

$$v_1(t) = 1. \tag{6.147}$$

Now choose the second vector, linearly independent of v_1, as a linear combination of the first two vectors:

$$v_2(t) = a + bt. \tag{6.148}$$

This should be orthogonal to v_1, so that

$$\int_{-1}^{1} v_1(t) v_2(t)\, dt = 0, \tag{6.149}$$

$$\int_{-1}^{1} \underbrace{(1)}_{=v_1(t)}\ \underbrace{(a + bt)}_{=v_2(t)}\, dt = 0, \tag{6.150}$$

$$\left(at + \frac{bt^2}{2}\right)\Bigg|_{-1}^{1} = 0, \tag{6.151}$$

$$a(1 - (-1)) + \frac{b}{2}(1^2 - (-1)^2) = 0, \tag{6.152}$$

---

[16] Jørgen Pedersen Gram, 1850–1916, Danish actuary and mathematician, and Erhard Schmidt, 1876–1959, Livonian-born German mathematician.

from which

$$a = 0. \tag{6.153}$$

Taking $b = 1$ arbitrarily, because orthogonality does not depend on the magnitude of $v_2(t)$, we have

$$v_2 = t. \tag{6.154}$$

Choose the third vector to be linearly independent of $v_1(t)$ and $v_2(t)$. We can achieve this by taking

$$v_3(t) = a + bt + ct^2. \tag{6.155}$$

For $v_3(t)$ to be orthogonal to $v_1(t)$ and $v_2(t)$, we get the conditions

$$\int_{-1}^{1} \underbrace{(1)}_{=v_1(t)} \underbrace{(a + bt + ct^2)}_{=v_3(t)} \, dt = 0, \tag{6.156}$$

$$\int_{-1}^{1} \underbrace{t}_{=v_2(t)} \underbrace{(a + bt + ct^2)}_{=v_3(t)} \, dt = 0. \tag{6.157}$$

The first of these gives $c = -3a$. Taking $a = 1$ arbitrarily, we have $c = -3$. The second relation gives $b = 0$. Thus

$$v_3 = 1 - 3t^2. \tag{6.158}$$

In this manner we can find as many orthogonal vectors as we want. We can make them orthonormal by dividing each by its norm, so that we have

$$\varphi_1 = \frac{1}{\sqrt{2}}, \tag{6.159}$$

$$\varphi_2 = \sqrt{\frac{3}{2}} \, t, \tag{6.160}$$

$$\varphi_3 = \sqrt{\frac{5}{8}} \, (1 - 3t^2), \tag{6.161}$$

$$\vdots$$

Scalar multiples of these functions, with the functions set to unity at $t = 1$, are the Legendre polynomials: $P_0(t) = 1$, $P_1(t) = t$, $P_2(t) = (1/2)(3t^2 - 1)$. As studied earlier in Section 4.4, some other common orthonormal sets can be formed on the foundation of several eigenfunctions to Sturm-Liouville differential equations.

We can systematize the Gram-Schmidt procedure as follows. It is not hard to show that the projection of a vector u onto a vector v is enabled by the operator $\mathbf{L}_{GS}$, where $\mathbf{L}_{GS}$ is defined such that

$$\mathbf{L}_{GS}(u, v) = \frac{\langle v, u \rangle}{\langle v, v \rangle} v = \frac{\langle v, u \rangle}{\sqrt{\langle v, v \rangle}} \frac{v}{\sqrt{\langle v, v \rangle}} = \frac{\langle v, u \rangle}{||v||_2} \frac{v}{||v||_2}. \tag{6.162}$$

This can be interpreted as follows. The scalar $\langle v, u \rangle / ||v||_2$ is the magnitude of the component of u which points in the same direction as v. This magnitude is multiplied by the unit vector $v/||v||_2$ which points in the direction of v. Then given any set of linearly independent vectors $\{u_1, u_2, \ldots\}$, we can obtain an orthonormal set

$\{\varphi_1, \varphi_2, \ldots\}$ as follows:

$$v_1 = u_1, \tag{6.163}$$

$$v_2 = u_2 - \mathbf{L}_{GS}(u_2, v_1), \tag{6.164}$$

$$v_3 = u_3 - \mathbf{L}_{GS}(u_3, v_1) - \mathbf{L}_{GS}(u_3, v_2), \tag{6.165}$$

$$v_4 = u_4 - \mathbf{L}_{GS}(u_4, v_1) - \mathbf{L}_{GS}(u_4, v_2) - \mathbf{L}_{GS}(u_4, v_3), \tag{6.166}$$

$$\vdots$$

With this set of orthogonal v_i, we can form the orthonormal φ_i via

$$\varphi_1 = \frac{v_1}{\|v_1\|_2}, \tag{6.167}$$

$$\varphi_2 = \frac{v_2}{\|v_2\|_2}, \tag{6.168}$$

$$\varphi_3 = \frac{v_3}{\|v_3\|_2}, \tag{6.169}$$

$$\varphi_4 = \frac{v_4}{\|v_4\|_2}, \tag{6.170}$$

$$\vdots$$

EXAMPLE 6.26

Given the linearly independent vectors $u_1 = (1, 0, 0)^T$, $u_2 = (1, 1, 1)^T$, and $u_3 = (1, 2, 3)^T$, use the Gram-Schmidt procedure to find an associated orthonormal set φ_1, φ_2, and φ_3.

The first vector v_1 is given by

$$v_1 = u_1 = \begin{pmatrix} 1 \\ 0 \\ 0 \end{pmatrix}. \tag{6.171}$$

Knowing v_1, we can find v_2:

$$v_2 = u_2 - \frac{\langle v_1, u_2 \rangle}{\langle v_1, v_1 \rangle} v_1, \tag{6.172}$$

$$= \begin{pmatrix} 1 \\ 1 \\ 1 \end{pmatrix} - \frac{(1)(1) + (0)(1) + (0)(1)}{(1)(1) + (0)(0) + (0)(0)} \begin{pmatrix} 1 \\ 0 \\ 0 \end{pmatrix}, \tag{6.173}$$

$$= \begin{pmatrix} 0 \\ 1 \\ 1 \end{pmatrix}. \tag{6.174}$$

Knowing v_1 and v_2, we can find v_3:

$$v_3 = u_3 - \frac{\langle v_1, u_3 \rangle}{\langle v_1, v_1 \rangle} v_1 - \frac{\langle v_2, u_3 \rangle}{\langle v_2, v_2 \rangle} v_2, \tag{6.175}$$

$$= \begin{pmatrix} 1 \\ 2 \\ 3 \end{pmatrix} - \frac{(1)(1) + (0)(2) + (0)(3)}{(1)(1) + (0)(0) + (0)(0)} \begin{pmatrix} 1 \\ 0 \\ 0 \end{pmatrix} - \frac{(0)(1) + (1)(2) + (1)(3)}{(0)(0) + (1)(1) + (1)(1)} \begin{pmatrix} 0 \\ 1 \\ 1 \end{pmatrix}, \tag{6.176}$$

$$= \begin{pmatrix} 0 \\ -\frac{1}{2} \\ \frac{1}{2} \end{pmatrix}. \tag{6.177}$$

Then, normalizing, we find

$$\varphi_1 = \frac{v_1}{||v_1||_2} = \begin{pmatrix} 1 \\ 0 \\ 0 \end{pmatrix}, \qquad \varphi_2 = \frac{v_2}{||v_2||_2} = \begin{pmatrix} 0 \\ \frac{1}{\sqrt{2}} \\ \frac{1}{\sqrt{2}} \end{pmatrix}, \qquad \varphi_3 = \frac{v_3}{||v_3||_2} = \begin{pmatrix} 0 \\ -\frac{1}{\sqrt{2}} \\ \frac{1}{\sqrt{2}} \end{pmatrix}.$$

$$(6.178)$$

6.5 Projection of Vectors onto New Bases

Here we consider how to project N-dimensional vectors x, first onto general nonorthogonal bases of dimension $M \leq N$, and then specialize for orthogonal bases of dimension $M \leq N$. This important discussion will encapsulate and abstract the concept of geometrical projection, which we have done with less generality in previous portions of the text, such as Section 4.4. We will continue this theme into later chapters as well.

For ordinary vectors in Euclidean space, N and M will be integers. When $M < N$, we will usually lose information in projecting the N-dimensional x onto a lower M-dimensional basis. When $M = N$, we will lose no information, and the projection can be better characterized as a new *representation*. While much of our discussion is most easily digested when M and N take on finite values, the analysis will be easily extended to infinite dimension, which is appropriate for a space of vectors which are functions.

6.5.1 Nonorthogonal

We are given M linearly independent nonorthogonal basis vectors $\{u_1, u_2, \ldots, u_M\}$ on which to project the N-dimensional x, with $M \leq N$. Each of the M basis vectors, u_m, is taken for convenience to be a vector of length N; we must realize that both x and u_m could be functions as well, in which case saying they have length N would be meaningless.

The general task here is to find expressions for the coefficients α_m, $m = 1, 2, \ldots M$, to best represent x in the linear combination

$$\alpha_1 u_1 + \alpha_2 u_2 + \cdots + \alpha_M u_M = \sum_{m=1}^{M} \alpha_m u_m \approx x. \qquad (6.179)$$

We use the notation for an approximation, $\approx$, because for $M < N$, x most likely will not be exactly equal to the linear combination of basis vectors. Because $u \in \mathbb{C}^N$, we can define $\mathbf{U}$ as the $N \times M$ matrix whose M columns are populated by the M basis vectors of length N, $u_1, u_2, \ldots, u_M$:

$$\mathbf{U} = \begin{pmatrix} \vdots & \vdots & & \vdots \\ u_1 & u_2 & \cdots & u_M \\ \vdots & \vdots & & \vdots \end{pmatrix}. \qquad (6.180)$$

We can thus rewrite Eq. (6.179) as

$$\mathbf{U} \cdot \boldsymbol{\alpha} \approx \mathbf{x}, \qquad (6.181)$$

where $\boldsymbol{\alpha}$ is a column vector of length M containing α_m, $m = 1, \ldots, M$. If $M = N$, the approximation would become an equality; thus, we could invert Eq. (6.181) and find simply that $\alpha = \mathbf{U}^{-1} \cdot \mathbf{x}$. When $M = N$, we have an analog from $\mathbf{U} \cdot \alpha = \mathbf{x}$ to the linear homogeneous coordinate transformation of Eq. (1.281), $\mathbf{J} \cdot \mathbf{x} = \boldsymbol{\xi}$, with $\mathbf{U}$ playing the role of the constant Jacobian $\mathbf{J}$, the Fourier coefficients α playing the role of the transformed coordinates $\mathbf{x}$, and $\mathbf{x}$ playing the role of the Cartesian coordinates $\boldsymbol{\xi}$. Carrying forward the analogy, one could imagine the matrix product $\mathbf{U}^H \cdot \mathbf{U}$ playing the role of the metric tensor $\mathbf{G}$. The matrix $\mathbf{G} = \mathbf{U}^H \cdot \mathbf{U}$ is sometimes known as the *Gramian matrix*.

If $M < N$, $\mathbf{U}^{-1}$ does not exist, and we cannot use this approach to find α. We need another strategy. To get the values of α_m in the most general of cases, we begin by taking inner products of Eq. (6.179) with u_1 to get

$$\langle u_1, \alpha_1 u_1 \rangle + \langle u_1, \alpha_2 u_2 \rangle + \cdots + \langle u_1, \alpha_M u_M \rangle = \langle u_1, x \rangle. \tag{6.182}$$

Using the properties of an inner product and performing the procedure for all $u_m, m = 1, \ldots, M$, we get

$$\alpha_1 \langle u_1, u_1 \rangle + \alpha_2 \langle u_1, u_2 \rangle + \cdots + \alpha_M \langle u_1, u_M \rangle = \langle u_1, x \rangle, \tag{6.183}$$

$$\alpha_1 \langle u_2, u_1 \rangle + \alpha_2 \langle u_2, u_2 \rangle + \cdots + \alpha_M \langle u_2, u_M \rangle = \langle u_2, x \rangle, \tag{6.184}$$

$$\vdots$$

$$\alpha_1 \langle u_M, u_1 \rangle + \alpha_2 \langle u_M, u_2 \rangle + \cdots + \alpha_M \langle u_M, u_M \rangle = \langle u_M, x \rangle. \tag{6.185}$$

Knowing x and $u_1, u_2, \ldots, u_M$, all the inner products can be determined, and Eqs. (6.183–6.185) can be posed as the linear algebraic system:

$$\underbrace{\begin{pmatrix} \langle u_1, u_1 \rangle & \langle u_1, u_2 \rangle & \cdots & \langle u_1, u_M \rangle \\ \langle u_2, u_1 \rangle & \langle u_2, u_2 \rangle & \cdots & \langle u_2, u_M \rangle \\ \vdots & \vdots & \cdots & \vdots \\ \langle u_M, u_1 \rangle & \langle u_M, u_2 \rangle & \cdots & \langle u_M, u_M \rangle \end{pmatrix}}_{\mathbf{U}^H \cdot \mathbf{U}} \underbrace{\begin{pmatrix} \alpha_1 \\ \alpha_2 \\ \vdots \\ \alpha_M \end{pmatrix}}_{\alpha} = \underbrace{\begin{pmatrix} \langle u_1, x \rangle \\ \langle u_2, x \rangle \\ \vdots \\ \langle u_M, x \rangle \end{pmatrix}}_{\mathbf{U}^H \cdot \mathbf{x}}. \tag{6.186}$$

Equation (6.186) can also be written compactly as

$$\langle u_i, u_m \rangle \alpha_m = \langle u_i, x \rangle. \tag{6.187}$$

In either case, any standard methods such as Cramer's rule (see Section A.2) or Gaussian elimination (see Section A.3) can be used to determine the unknown coefficients, α_m.

We can understand this in another way by considering an approach using Gibbs notation, valid when each of the M basis vectors $u_m \in \mathbb{C}^N$. The Gibbs notation does not suffice for other classes of basis vectors, e.g. when the vectors are functions, $u_m \in \mathbb{L}_2$. Operate on Eq. (6.181) with $\mathbf{U}^H$ to get

$$\left(\mathbf{U}^H \cdot \mathbf{U} \right) \cdot \alpha = \mathbf{U}^H \cdot \mathbf{x}. \tag{6.188}$$

This is the Gibbs notation equivalent of Eq. (6.186). We cannot expect $\mathbf{U}^{-1}$ to always exist; however, as long as the $M \leq N$ basis vectors are linearly independent, we can expect the $M \times M$ matrix $\left(\mathbf{U}^H \cdot \mathbf{U} \right)^{-1}$ to exist. We can then solve for the coefficients α via

$$\alpha = \left(\mathbf{U}^H \cdot \mathbf{U} \right)^{-1} \cdot \mathbf{U}^H \cdot \mathbf{x}, \qquad M \leq N. \tag{6.189}$$

In this case, one is projecting $\mathbf{x}$ onto a basis of equal or lower dimension than itself, and we recover the $M \times 1$ vector $\boldsymbol{\alpha}$. If one then operates on both sides of Eq. (6.189) with the $N \times M$ operator $\mathbf{U}$, one gets

$$\mathbf{U} \cdot \boldsymbol{\alpha} = \underbrace{\mathbf{U} \cdot \left(\mathbf{U}^H \cdot \mathbf{U}\right)^{-1} \cdot \mathbf{U}^H}_{\mathbf{P}} \cdot \mathbf{x} = \mathbf{x}_p. \tag{6.190}$$

Here we have defined the $N \times N$ *projection matrix* $\mathbf{P}$ as

$$\mathbf{P} = \mathbf{U} \cdot \left(\mathbf{U}^H \cdot \mathbf{U}\right)^{-1} \cdot \mathbf{U}^H. \tag{6.191}$$

We have also defined $\mathbf{x}_p = \mathbf{P} \cdot \mathbf{x}$ as the projection of $\mathbf{x}$ onto the basis $\mathbf{U}$. These topics will be considered later in a strictly linear algebraic context in Section 7.10. When there are $M = N$ linearly independent basis vectors, Eq. (6.191) can be reduced to show $\mathbf{P} = \mathbf{I}$. In this case $\mathbf{U}^{-1}$ exists, and we get

$$\mathbf{P} = \underbrace{\mathbf{U} \cdot \mathbf{U}^{-1}}_{\mathbf{I}} \cdot \underbrace{\mathbf{U}^{H-1} \cdot \mathbf{U}^H}_{\mathbf{I}} = \mathbf{I}. \tag{6.192}$$

So with $M = N$ linearly independent basis vectors, we have $\mathbf{U} \cdot \boldsymbol{\alpha} = \mathbf{x}$, and recover the much simpler

$$\boldsymbol{\alpha} = \mathbf{U}^{-1} \cdot \mathbf{x}, \qquad M = N. \tag{6.193}$$

EXAMPLE 6.27

Project the vector $\mathbf{x} = \begin{pmatrix} 6 \\ -3 \end{pmatrix}$ onto the nonorthogonal basis composed of $u_1 = \begin{pmatrix} 2 \\ 1 \end{pmatrix}$, $u_2 = \begin{pmatrix} 1 \\ -1 \end{pmatrix}$.

Here we have the length of $\mathbf{x}$ as $N = 2$, and we have $M = N = 2$ linearly independent basis vectors. When the basis vectors are combined into a set of column vectors, they form the matrix

$$\mathbf{U} = \begin{pmatrix} 2 & 1 \\ 1 & -1 \end{pmatrix}. \tag{6.194}$$

Because we have a sufficient number of basis vectors to span the space, to get $\boldsymbol{\alpha}$, we can simply apply Eq. (6.193) to get

$$\boldsymbol{\alpha} = \mathbf{U}^{-1} \cdot \mathbf{x}, \tag{6.195}$$

$$= \begin{pmatrix} 2 & 1 \\ 1 & -1 \end{pmatrix}^{-1} \begin{pmatrix} 6 \\ -3 \end{pmatrix}, \tag{6.196}$$

$$= \begin{pmatrix} \frac{1}{3} & \frac{1}{3} \\ \frac{1}{3} & -\frac{2}{3} \end{pmatrix} \begin{pmatrix} 6 \\ -3 \end{pmatrix}, \tag{6.197}$$

$$= \begin{pmatrix} 1 \\ 4 \end{pmatrix}. \tag{6.198}$$

Thus

$$\mathbf{x} = \alpha_1 u_1 + \alpha_2 u_2 = 1 \begin{pmatrix} 2 \\ 1 \end{pmatrix} + 4 \begin{pmatrix} 1 \\ -1 \end{pmatrix} = \begin{pmatrix} 6 \\ -3 \end{pmatrix}. \tag{6.199}$$

The projection matrix $\mathbf{P} = \mathbf{I}$, and $\mathbf{x}_p = \mathbf{x}$. Thus, the projection is actually a representation, with no lost information.

EXAMPLE 6.28

Project the vector $\mathbf{x} = \left(\begin{smallmatrix} 6 \\ -3 \end{smallmatrix}\right)$ on the basis composed of $u_1 = \left(\begin{smallmatrix} 2 \\ 1 \end{smallmatrix}\right)$.

Here we have a vector $\mathbf{x}$ with $N = 2$ and an $M = 1$ linearly independent basis vector which, when cast into columns, forms

$$\mathbf{U} = \begin{pmatrix} 2 \\ 1 \end{pmatrix}. \tag{6.200}$$

This vector does not span the space, so to get the projection, we must use the more general Eq. (6.189), which reduces to

$$\boldsymbol{\alpha} = \left(\underbrace{(2 \; 1)}_{\mathbf{U}^H} \underbrace{\begin{pmatrix} 2 \\ 1 \end{pmatrix}}_{\mathbf{U}} \right)^{-1} \underbrace{(2 \; 1)}_{\mathbf{U}^H} \underbrace{\begin{pmatrix} 6 \\ -3 \end{pmatrix}}_{\mathbf{x}} = (5)^{-1}(9) = \left(\tfrac{9}{5} \right). \tag{6.201}$$

So the projection is

$$\mathbf{x}_p = \alpha_1 u_1 = \left(\tfrac{9}{5} \right) \begin{pmatrix} 2 \\ 1 \end{pmatrix} = \begin{pmatrix} \tfrac{18}{5} \\ \tfrac{9}{5} \end{pmatrix}. \tag{6.202}$$

The projection is not obtained by simply setting $\alpha_2 = 0$ from the previous example. This is because the component of $\mathbf{x}$ aligned with u_2 itself has a projection onto u_1. Had u_1 been orthogonal to u_2, one could have obtained the projection onto u_1 by setting $\alpha_2 = 0$.

The projection matrix is

$$\mathbf{P} = \underbrace{\begin{pmatrix} 2 \\ 1 \end{pmatrix}}_{\mathbf{U}} \left(\underbrace{(2 \; 1)}_{\mathbf{U}^H} \underbrace{\begin{pmatrix} 2 \\ 1 \end{pmatrix}}_{\mathbf{U}} \right)^{-1} \underbrace{(2 \; 1)}_{\mathbf{U}^H} = \begin{pmatrix} \tfrac{4}{5} & \tfrac{2}{5} \\ \tfrac{2}{5} & \tfrac{1}{5} \end{pmatrix}. \tag{6.203}$$

It is easily verified that $\mathbf{x}_p = \mathbf{P} \cdot \mathbf{x}$.

EXAMPLE 6.29

Project the function $x(t) = t^3$, $t \in [0, 1]$ onto the space spanned by the nonorthogonal basis functions $u_1 = t$, $u_2 = \sin(4t)$.

This is an unusual projection. The $M = 2$ basis functions are not orthogonal. In fact they bear no clear relation to each other. The success in finding accurate approximations to the original function depends on how well the chosen basis functions approximate the original function. The appropriateness of the basis functions notwithstanding, it is not difficult to calculate the projection. Equation (6.186) reduces to

$$\begin{pmatrix} \int_0^1 (t)(t) \, dt & \int_0^1 (t) \sin 4t \, dt \\ \int_0^1 (\sin 4t)(t) \, dt & \int_0^1 \sin^2 4t \, dt \end{pmatrix} \begin{pmatrix} \alpha_1 \\ \alpha_2 \end{pmatrix} = \begin{pmatrix} \int_0^1 (t)(t^3) \, dt \\ \int_0^1 (\sin 4t)(t^3) \, dt \end{pmatrix}. \tag{6.204}$$

Evaluating the integrals gives

$$\begin{pmatrix} 0.333333 & 0.116111 \\ 0.116111 & 0.438165 \end{pmatrix} \begin{pmatrix} \alpha_1 \\ \alpha_2 \end{pmatrix} = \begin{pmatrix} 0.2 \\ -0.0220311 \end{pmatrix}. \tag{6.205}$$

Inverting and solving gives

$$\begin{pmatrix} \alpha_1 \\ \alpha_2 \end{pmatrix} = \begin{pmatrix} 0.680311 \\ -0.230558 \end{pmatrix}. \tag{6.206}$$

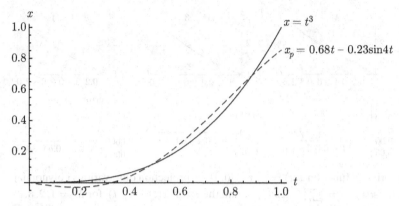

Figure 6.9. Projection of $x(t) = t^3$ onto a two-term nonorthogonal basis composed of functions $u_1 = t$, $u_2 = \sin 4t$.

So our projection of $x(t) = t^3$ onto the basis functions yields the approximation $x_p(t)$:

$$x(t) = t^3 \approx x_p(t) = \alpha_1 u_1 + \alpha_2 u_2 = 0.680311t - 0.230558 \sin 4t. \qquad (6.207)$$

Figure 6.9 shows the original function and its two-term approximation. It seems the approximation is not bad; however, there is no clear path to improvement by adding more basis functions. So one might imagine in a specialized problem that the ability to project onto an unusual basis could be useful. But in general this approach is rarely taken.

EXAMPLE 6.30

Project the function $x = e^t$, $t \in [0,1]$ onto the space spanned by the functions $u_m = t^{m-1}, m = 1, \ldots, M$, for $M = 4$.

Similar to the previous example, the basis functions are nonorthogonal. Unlike the previous example, there is a clear way to improve the approximation by increasing M. For $M = 4$, Eq. (6.186) reduces to

$$\begin{pmatrix} \int_0^1 (1)(1)\, dt & \int_0^1 (1)(t)\, dt & \int_0^1 (1)(t^2)\, dt & \int_0^1 (1)(t^3)\, dt \\ \int_0^1 (t)(1)\, dt & \int_0^1 (t)(t)\, dt & \int_0^1 (t)(t^2)\, dt & \int_0^1 (t)(t^3)\, dt \\ \int_0^1 (t^2)(1)\, dt & \int_0^1 (t^2)(t)\, dt & \int_0^1 (t^2)(t^2)\, dt & \int_0^1 (t^2)(t^3)\, dt \\ \int_0^1 (t^3)(1)\, dt & \int_0^1 (t^3)(t)\, dt & \int_0^1 (t^3)(t^2)\, dt & \int_0^1 (t^3)(t^3)\, dt \end{pmatrix} \begin{pmatrix} \alpha_1 \\ \alpha_2 \\ \alpha_3 \\ \alpha_4 \end{pmatrix} = \begin{pmatrix} \int_0^1 (1)(e^t)\, dt \\ \int_0^1 (t)(e^t)\, dt \\ \int_0^1 (t^2)(e^t)\, dt \\ \int_0^1 (t^3)(e^t)\, dt \end{pmatrix}.$$
$$(6.208)$$

Evaluating the integrals, this becomes

$$\begin{pmatrix} 1 & \frac{1}{2} & \frac{1}{3} & \frac{1}{4} \\ \frac{1}{2} & \frac{1}{3} & \frac{1}{4} & \frac{1}{5} \\ \frac{1}{3} & \frac{1}{4} & \frac{1}{5} & \frac{1}{6} \\ \frac{1}{4} & \frac{1}{5} & \frac{1}{6} & \frac{1}{7} \end{pmatrix} \begin{pmatrix} \alpha_1 \\ \alpha_2 \\ \alpha_3 \\ \alpha_4 \end{pmatrix} = \begin{pmatrix} -1+e \\ 1 \\ -2+e \\ 6-2e \end{pmatrix}. \qquad (6.209)$$

Solving for α_m and composing, the approximation gives

$$x_p(t) = 0.999060 + 1.01830t + 0.421246t^2 + 0.278625t^3. \qquad (6.210)$$

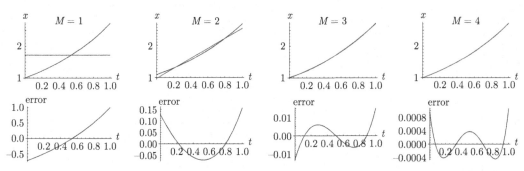

Figure 6.10. The original function $x(t) = e^t$, $t \in [0, 1]$, its projection onto various polynomial basis functions $x(t) \approx x_p(t) = \sum_{m=1}^{M} \alpha_m t^{m-1}$, and the error, $x(t) - x_p(t)$, for $M = 1, 2, 3, 4$.

We can compare this to $x_T(t)$, the four-term Taylor series approximation of e^t about $t = 0$:

$$x_T(t) = 1 + t + \frac{t^2}{2} + \frac{t^3}{6} \approx e^t, \tag{6.211}$$

$$= 1.00000 + 1.00000t - 0.500000t^2 + 0.166667t^3. \tag{6.212}$$

Obviously, the Taylor series approximation is close to the $M = 4$ projection. The Taylor approximation, $x_T(t)$, gains accuracy as $t \to 0$, while our $x_p(t)$ is better suited to the entire domain $t \in [0, 1]$. We can expect as $M \to \infty$ for the value of each α_m to approach those given by the independent Taylor series approximation. Figure 6.10 shows the original function against its $M = 1, 2, 3, 4$-term approximations, as well as the error. Clearly the approximation improves as M increases; for $M = 4$, the graphs of the original function and its approximation are indistinguishable at this scale.

Also, we note that the so-called root-mean-square (rms) error, E_2, is lower for our approximation relative to the Taylor series approximation about $t = 0$. We define rms errors, E_2^p, E_2^T, in terms of a norm, for both our projection and the Taylor approximation, respectively, and find for four-term approximations

$$E_2^p = ||x(t) - x_p(t)||_2 = \sqrt{\int_0^1 (e^t - x_p(t))^2 \, dt} = 0.000331, \tag{6.213}$$

$$E_2^T = ||x(t) - x_T(t)||_2 = \sqrt{\int_0^1 (e^t - x_T(t))^2 \, dt} = 0.016827. \tag{6.214}$$

Our $M = 4$ approximation is better, when averaged over the entire domain, than the $M = 4$ Taylor series approximation. For larger M, the relative differences become more dramatic. For example, for $M = 10$, we find $E_2^p = 5.39 \times 10^{-13}$ and $E_2^T = 6.58 \times 10^{-8}$.

These effects are summarized in Figure 6.11, which plots the rms error as a function of M on (a) a log-log scale and (b) a semi-log scale. A weighted least squares curve fit, with a weighting factor proportional to M^8, shows that near $M = 15$, the error is well described by

$$||x(t) - x_p(t)||_2 \approx 1.28 \times 10^{44} \, M^{-55.2}. \tag{6.215}$$

While we are concerned that the value of 1.28×10^{44} is large indeed, it is more than compensated by the $M^{-55.2}$ term. In fact, if we study Figure 6.11a carefully, we see in this log-log plot, the slope of the curve continues to steepen as M increases. That is to say, the convergence rate itself is a function of the number of terms in the series. Instead of modeling the error by a power law model, it is in fact better to model it as an exponential curve. Doing so, we find the error is well approximated by

$$||x(t) - x_p(t)||_2 \approx 3.19 \times 10^5 e^{-4.04M}. \tag{6.216}$$

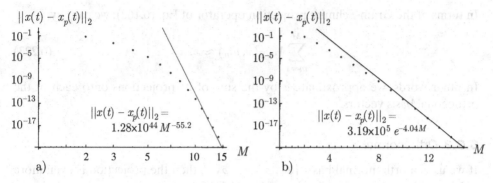

Figure 6.11. Convergence of $x(t) = e^t$ onto space spanned by $u_m(t) = t^{m-1}, m = 1, \ldots, M$: (a) log-log plot, (b) semi-log plot.

Taking the log of both sides, we find

$$\ln \|x(t) - x_p(t)\|_2 \approx \ln(3.19 \times 10^5) - 4.04M. \tag{6.217}$$

Thus, our estimate of error should be linear with M in a semi-log plot, such as shown in Figure 6.11b. When this is the case, we say we have achieved *exponential convergence*; this is also known as *spectral convergence*. We achieved a remarkably high convergence rate in spite of the fact that

- our basis functions were not explicitly constructed to match the original function at any interior or boundary point,
- our basis functions were not normalized,
- our basis functions were not even orthogonal.

<hr style="width:30%" />

6.5.2 Orthogonal

The process is simpler if the basis vectors are orthogonal. If orthogonal,

$$\langle u_i, u_m \rangle = 0, \qquad i \neq m, \tag{6.218}$$

and substituting this into Eq. (6.186), we get

$$\begin{pmatrix} \langle u_1, u_1 \rangle & 0 & \cdots & 0 \\ 0 & \langle u_2, u_2 \rangle & \cdots & 0 \\ \vdots & \vdots & \cdots & \vdots \\ 0 & 0 & \cdots & \langle u_M, u_M \rangle \end{pmatrix} \begin{pmatrix} \alpha_1 \\ \alpha_2 \\ \vdots \\ \alpha_M \end{pmatrix} = \begin{pmatrix} \langle u_1, x \rangle \\ \langle u_2, x \rangle \\ \vdots \\ \langle u_M, x \rangle \end{pmatrix}. \tag{6.219}$$

Equation (6.219) can be solved directly for the coefficients, yielding

$$\alpha_m = \frac{\langle u_m, x \rangle}{\langle u_m, u_m \rangle}. \tag{6.220}$$

So, if the basis vectors are orthogonal, we can write Eq. (6.179) as

$$\frac{\langle u_1, x \rangle}{\langle u_1, u_1 \rangle} u_1 + \frac{\langle u_2, x \rangle}{\langle u_2, u_2 \rangle} u_2 + \cdots + \frac{\langle u_M, x \rangle}{\langle u_M, u_M \rangle} u_M \approx x, \tag{6.221}$$

$$\sum_{m=1}^{M} \frac{\langle u_m, x \rangle}{\langle u_m, u_m \rangle} u_m = \sum_{m=1}^{M} \alpha_m u_m \approx x. \tag{6.222}$$

In terms of the Gram-Schmidt projection operator of Eq. (6.162), we can say

$$\sum_{m=1}^{M} \mathbf{L}_{GS}(x, u_m) \approx x. \tag{6.223}$$

In other words, we approximate x by the sum of its projections onto each of the orthogonal basis vectors.

6.5.3 Orthonormal

If we use an orthonormal basis $\{\varphi_1, \varphi_2, \ldots, \varphi_M\}$, then the projection is even more efficient. We get the generalization of Eq. (4.511):

$$\alpha_m = \langle \varphi_m, x \rangle, \tag{6.224}$$

which yields

$$\sum_{m=1}^{M} \underbrace{\langle \varphi_m, x \rangle}_{\alpha_m} \varphi_m \approx x. \tag{6.225}$$

In all cases, if $M = N$, we can replace the "$\approx$" by an "$=$," and the approximation becomes in fact a representation.

Similar expansions apply to vectors in infinite-dimensional spaces, except that one must be careful that the orthonormal set is *complete*. Only then is there any guarantee that any vector can be represented as linear combinations of this orthonormal set. If $\{\varphi_1, \varphi_2, \ldots\}$ is a complete orthonormal set of vectors in some domain Ω, then any vector x can be represented as

$$x = \sum_{n=1}^{\infty} \alpha_n \varphi_n, \tag{6.226}$$

where

$$\alpha_n = \langle \varphi_n, x \rangle. \tag{6.227}$$

This is a *Fourier series* representation, as previously studied in Section 4.4, and the values of α_n are the Fourier coefficients. It is a representation and not just a projection because the summation runs to infinity.

EXAMPLE 6.31

Expand the top hat function $x(t) = H(t - 1/4) - H(t - 3/4)$ in a Fourier-sine series in the domain $t \in [0, 1]$.

Here, the function $x(t)$ is discontinuous at $t = 1/4$ and $t = 3/4$. While $x(t)$ is not a member of $C[0, 1]$, it is a member of $\mathbb{L}_2[0, 1]$. Here we will see that the Fourier-sine series projection, composed of functions which are continuous in $[0, 1]$, converges to the discontinuous function $x(t)$, except at the discontinuity, where a finite error persists.

Building on previous work, we know from Eq. (4.347) that the functions

$$\varphi_n(t) = \sqrt{2} \sin(n\pi t), \qquad n = 1, \ldots, \infty, \tag{6.228}$$

form an orthonormal set for $t \in [0, 1]$. We then find for the Fourier coefficients

$$\alpha_n = \sqrt{2} \int_0^1 \left(H\left(t - \frac{1}{4}\right) - H\left(t - \frac{3}{4}\right) \right) \sin(n\pi t) \, dt = \sqrt{2} \int_{1/4}^{3/4} \sin(n\pi t) \, dt. \tag{6.229}$$

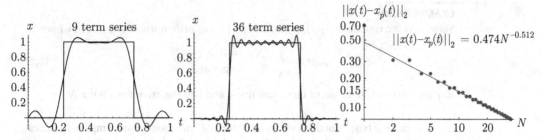

Figure 6.12. Expansion of top hat function $x(t) = H(t - 1/4) - H(t - 3/4)$ in terms of sinusoidal basis functions for two levels of approximation, $N = 9$, $N = 36$, along with a plot of how the rms error converges as the number of terms increases.

Performing the integration for the first nine terms, we find

$$\alpha_n = \frac{2}{\pi} \left(1, 0, -\frac{1}{3}, 0, -\frac{1}{5}, 0, \frac{1}{7}, 0, \frac{1}{9}, \dots \right). \tag{6.230}$$

Note that α_n is zero if n is even. Forming an approximation from these nine terms, we find

$$H \left(t - \frac{1}{4} \right) - H \left(t - \frac{3}{4} \right)$$

$$= \frac{2\sqrt{2}}{\pi} \left(\sin(\pi t) - \frac{\sin(3\pi t)}{3} - \frac{\sin(5\pi t)}{5} + \frac{\sin(7\pi t)}{7} + \frac{\sin(9\pi t)}{9} + \cdots \right). \tag{6.231}$$

Generalizing, we can show

$$H \left(t - \frac{1}{4} \right) - H \left(t - \frac{3}{4} \right) = \frac{2\sqrt{2}}{\pi} \sum_{k=1}^{\infty} (-1)^{k-1} \left(\frac{\sin((4k - 3)\pi t)}{4k - 3} - \frac{\sin((4k - 1)\pi t)}{4k - 1} \right). \tag{6.232}$$

The discontinuous function $x(t)$, two continuous approximations to it, and a plot revealing how the rms error decreases as the number of terms in the approximation increase are shown in Figure 6.12. The stair-stepping effect is because every other α_n is zero. As more terms are added, the approximation gets better at most points in the domain. But there is always a persistently large error at the discontinuities $t = 1/4$, $t = 3/4$. We say this function is convergent in $\mathbb{L}_2[0, 1]$, but is not convergent in $\mathbb{L}_\infty[0, 1]$. This simply says that the rms error norm converges, while the maximum error norm does not. This is an example of the well-known *Gibbs phenomenon*. Convergence in $\mathbb{L}_2[0, 1]$ is shown in Figure 6.12. The achieved convergence rate is $\|x(t) - x_p(t)\|_2 \approx 0.474088 N^{-0.512}$. This suggests that

$$\lim_{N \to \infty} \|x(t) - x_p(t)\|_2 \approx \frac{1}{\sqrt{N}}, \tag{6.233}$$

where N is the number of terms retained in the projection.

The previous example showed one could use continuous functions to approximate a discontinuous function. The converse is also true: discontinuous functions can be used to approximate continuous functions.

EXAMPLE 6.32

Show that the functions $\varphi_1(t), \varphi_2(t), \ldots, \varphi_N(t)$ are orthonormal in $\mathbb{L}_2(0,1]$, where

$$\varphi_n(t) = \begin{cases} \sqrt{N}, & \frac{n-1}{N} < t \le \frac{n}{N}, \\ 0, & \text{otherwise.} \end{cases} \tag{6.234}$$

Expand $x(t) = t^2$ in terms of these functions, and find the error for a finite N.

We note that the basis functions are a set of "top hat" functions whose amplitude increases and width decreases as N increases. For fixed N, the basis functions are a series of top hats that fills the domain $[0,1]$. The area enclosed by a single basis function is $1/\sqrt{N}$. If $n \ne m$, the inner product

$$\langle \varphi_n, \varphi_m \rangle = \int_0^1 \varphi_n(t) \varphi_m(t)\, dt = 0, \tag{6.235}$$

because the integrand is zero everywhere. If $n = m$, the inner product is

$$\int_0^1 \varphi_n(t) \varphi_n(t)\, dt = \int_0^{\frac{n-1}{N}} (0)(0)\, dt + \int_{\frac{n-1}{N}}^{\frac{n}{N}} \sqrt{N}\sqrt{N}\, dt + \int_{\frac{n}{N}}^1 (0)(0)\, dt, \tag{6.236}$$

$$= N\left(\frac{n}{N} - \frac{n-1}{N}\right), \tag{6.237}$$

$$= 1. \tag{6.238}$$

So, $\{\varphi_1, \varphi_2, \ldots, \varphi_N\}$ is an orthonormal set. We can expand the function $f(t) = t^2$ in the form

$$t^2 = \sum_{n=1}^N \alpha_n \varphi_n. \tag{6.239}$$

Taking the inner product of both sides with $\varphi_m(t)$, we get

$$\int_0^1 \varphi_m(t) t^2\, dt = \int_0^1 \varphi_m(t) \sum_{n=1}^N \alpha_n \varphi_n(t)\, dt, \tag{6.240}$$

$$= \sum_{n=1}^N \alpha_n \underbrace{\int_0^1 \varphi_m(t)\varphi_n(t)\, dt}_{= \delta_{nm}}, \tag{6.241}$$

$$= \sum_{n=1}^N \alpha_n \delta_{nm}, \tag{6.242}$$

$$= \alpha_m, \tag{6.243}$$

$$\int_0^1 \varphi_n(t) t^2\, dt = \alpha_n. \tag{6.244}$$

Thus,

$$\alpha_n = 0 + \int_{\frac{n-1}{N}}^{\frac{n}{N}} t^2 \sqrt{N}\, dt + 0, \tag{6.245}$$

$$= \frac{1}{3N^{5/2}}\left(3n^2 - 3n + 1\right). \tag{6.246}$$

The functions t^2 and the partial sums $f_N(t) = \sum_{n=1}^N \alpha_n \varphi_n(t)$ for $N = 5$ and $N = 10$ are shown in Figure 6.13. Detailed analysis not shown here reveals the $\mathbb{L}_2$ error for the partial

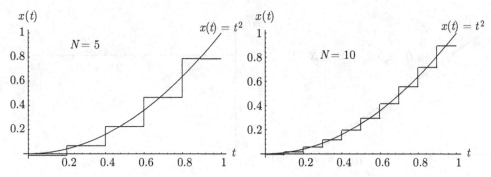

Figure 6.13. Expansion of $x(t) = t^2$ in terms of "top hat" basis functions for two levels of approximation, $N = 5, N = 10$.

sums can be calculated as Δ_N, where

$$\Delta_N^2 = \|f(t) - f_N(t)\|_2^2, \tag{6.247}$$

$$= \int_0^1 \left(t^2 - \sum_{n=1}^N \alpha_n \varphi_n(t) \right)^2 \, dt, \tag{6.248}$$

$$= \frac{1}{9N^2} \left(1 - \frac{1}{5N^2} \right), \tag{6.249}$$

$$\Delta_N = \frac{1}{3N} \sqrt{1 - \frac{1}{5N^2}}, \tag{6.250}$$

which vanishes as $N \to \infty$ at a rate of convergence proportional to $1/N$.

EXAMPLE 6.33

Demonstrate the Fourier-sine series for $x(t) = 2t$ converges at a rate proportional to $1/\sqrt{N}$, where N is the number of terms used to approximate $x(t)$, in $\mathbb{L}_2[0, 1]$.

Consider the sequence of orthonormal functions

$$\varphi_n(t) = \left\{ \sqrt{2} \sin(\pi t), \sqrt{2} \sin(2\pi t), \dots, \sqrt{2} \sin(n\pi t), \dots \right\}. \tag{6.251}$$

It is easy to show linear independence for these functions. They are orthonormal in the Hilbert space $\mathbb{L}_2[0, 1]$, for example,

$$\langle \varphi_2, \varphi_3 \rangle = \int_0^1 \left(\sqrt{2} \sin(2\pi t) \right) \left(\sqrt{2} \sin(3\pi t) \right) \, dt = 0, \tag{6.252}$$

$$\langle \varphi_3, \varphi_3 \rangle = \int_0^1 \left(\sqrt{2} \sin(3\pi t) \right) \left(\sqrt{2} \sin(3\pi t) \right) \, dt = 1. \tag{6.253}$$

While the basis functions evaluate to 0 at both $t = 0$ and $t = 1$, the function itself only has value 0 at $t = 0$. We must tolerate a large error at $t = 1$, but hope that this error is confined to an ever collapsing neighborhood around $t = 1$ as more terms are included in the approximation.

The Fourier coefficients are

$$\alpha_n = \langle 2t, \varphi_n(t) \rangle = \int_0^1 (2t) \sqrt{2} \sin(n\pi t) \, dt = \frac{2\sqrt{2}(-1)^{n+1}}{n\pi}. \tag{6.254}$$

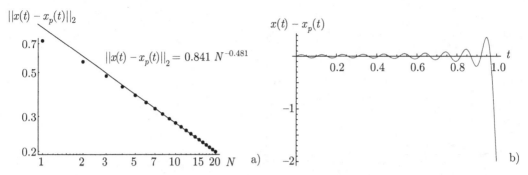

Figure 6.14. (a) Behavior of the error norm of the Fourier-sine series approximation to $x(t) = 2t$ on $t \in [0, 1]$ with the number N of terms included in the series; (b) error distribution for $N = 20$ approximation.

The approximation then is

$$x_p(t) = \sum_{n=1}^{N} \frac{4(-1)^{n+1}}{n\pi} \sin(n\pi t). \qquad (6.255)$$

The norm of the error is then

$$\|x(t) - x_p(t)\|_2 = \sqrt{\int_0^1 \left(2t - \left(\sum_{n=1}^{N} \frac{4(-1)^{n+1}}{n\pi} \sin(n\pi t)\right)\right)^2 dt}. \qquad (6.256)$$

This is difficult to evaluate analytically. It is straightforward to examine this with symbolic computer mathematics.

A plot of the norm of the error as a function of the number of terms in the approximation, N, is given in the log-log plot of Figure 6.14a. A weighted least squares curve fit, with a weighting factor proportional to N^2 so that priority is given to data as $N \to \infty$, shows that the function

$$\|x(t) - x_p(t)\|_2 \approx 0.841 \, N^{-0.481}, \qquad (6.257)$$

approximates the convergence performance well. In the log-log plot the exponent on N is the slope, because

$$\ln\left(\|x(t) - x_p(t)\|_2\right) \approx \ln(0.841) - 0.481 \ln N. \qquad (6.258)$$

It appears from the graph that the slope may be approaching a limit, in which it is likely that

$$\|x(t) - x_p(t)\|_2 \approx \frac{1}{\sqrt{N}}. \qquad (6.259)$$

This indicates convergence of this series. The series converges even though the norm of the nth basis function does not approach zero as $n \to \infty$:

$$\lim_{n \to \infty} \|\varphi_n\|_2 = 1, \qquad (6.260)$$

because the basis functions are orthonormal. Also, note that the behavior of the norm of the final term in the series,

$$\|\alpha_N \varphi_N(t)\|_2 = \sqrt{\int_0^1 \left(\frac{2\sqrt{2}(-1)^{N+1}}{N\pi} \sqrt{2} \sin(N\pi t)\right)^2 dt} = \frac{2\sqrt{2}}{N\pi}, \qquad (6.261)$$

does not tell us how the series actually converges. Figure 6.14b gives the error distribution for the $N = 20$ approximation. Obviously, there is nonuniform error distribution with an $O(1)$ error at the $x = 1$ boundary.

EXAMPLE 6.34

Show the Fourier-sine series for $x(t) = t - t^2$ converges at a rate proportional to $1/N^{5/2}$, where N is the number of terms used to approximate $x(t)$, in $\mathbb{L}_2[0, 1]$.

Again, consider the sequence of orthonormal functions

$$\varphi_n(t) = \left\{ \sqrt{2}\sin(\pi t), \sqrt{2}\sin(2\pi t), \ldots, \sqrt{2}\sin(n\pi t), \ldots \right\}, \qquad (6.262)$$

which are, as before, linearly independent and, moreover, orthonormal. In this case, as opposed to the previous example, both the basis functions and the function to be approximated vanish identically at both $t = 0$ and $t = 1$. Consequently, there will be *no error* in the approximation at either endpoint.

The Fourier coefficients are found to be

$$\alpha_n = \frac{2\sqrt{2}\left(1 + (-1)^{n+1}\right)}{n^3\pi^3}. \qquad (6.263)$$

Note that $\alpha_n = 0$ for even values of n. Taking this into account and retaining only the necessary basis functions, we can write the Fourier-sine series as

$$x(t) = t(1 - t) \approx x_p(t) = \sum_{m=1}^{N} \frac{4\sqrt{2}}{(2m - 1)^3\pi^3} \sin((2m - 1)\pi t). \qquad (6.264)$$

The norm of the error is then

$$\|x(t) - x_p(t)\|_2 = \sqrt{\int_0^1 \left(t(1 - t) - \left(\sum_{m=1}^{N} \frac{4\sqrt{2}}{(2m - 1)^3\pi^3} \sin((2m - 1)\pi t) \right) \right)^2 dt}. \qquad (6.265)$$

Again, this is difficult to address analytically, but symbolic computer mathematics allows computation of the error norm as a function of N.

A plot of the norm of the error as a function of the number of terms in the approximation, N, is given in the log-log plot of Figure 6.15. A weighted least squares curve fit, with a weighting factor proportional to N^2 so that priority is given to data as $N \to \infty$,

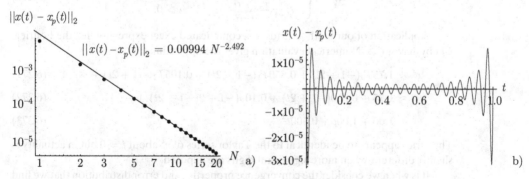

Figure 6.15. (a) Behavior of the error norm of the Fourier-sine series approximation to $x(t) = t(1 - t)$ on $t \in [0, 1]$ with the number N of nonzero terms included in the series; (b) error distribution for $N = 20$ approximation.

shows that the function

$$\|x(t) - x_p(t)\|_2 \approx 0.00994 \, N^{-2.492}, \tag{6.266}$$

approximates the convergence performance well. Thus, we might suspect that

$$\lim_{n \to \infty} \|x(t) - x_p(t)\|_2 \approx \frac{1}{N^{5/2}}. \tag{6.267}$$

The convergence is much more rapid than in the previous example! This can be critically important in numerical calculations and demonstrates that a judicious selection of basis functions can have fruitful consequences.

Figure 6.15b gives the error distribution for the $N = 20$ approximation. There are 40 zero crossings in the error distribution. This is because every other term in the expansion was actually zero, and here we only included the nonzero terms in counting N. Here the error at either boundary is zero. However, there is a nonuniform error distribution of $O(10^{-5})$, with the errors near the boundaries at $x = 0$ and $x = 1$ being slightly larger than those in the interior.

EXAMPLE 6.35

Show that the Fourier-Chebyshev approximation for $x(t) = e^t$ converges exponentially with N, where N is the number of terms used to approximate $x(t)$ in $\mathbb{L}_2[0, 1]$ with an error distribution whose amplitude is uniform throughout the domain.

Here, we employ as basis functions the Chebyshev polynomials, which are orthogonal on the domain $[-1, 1]$ with an appropriate weighting function. The mapping $\tilde{t} = -1 + 2t$ takes our domain of $t \in [0, 1]$ onto the domain $\tilde{t} = [-1, 1]$. Thus we seek α_n such that

$$e^t \approx \sum_{n=0}^{N} \alpha_n T_n(\tilde{t}) = \sum_{n=0}^{N} \alpha_n T_n(-1 + 2t). \tag{6.268}$$

Because the basis functions are orthogonal, under suitable weighting, we can apply Eq. (6.220) to determine the values of α_n:

$$\alpha_n = \frac{\int_{-1}^{1} \frac{T_n(\tilde{t}) f(\tilde{t})}{\sqrt{1 - \tilde{t}^2}} \, d\tilde{t}}{\int_{-1}^{1} \frac{T_n(\tilde{t}) T_n(\tilde{t})}{\sqrt{1 - \tilde{t}^2}} \, d\tilde{t}}. \tag{6.269}$$

Applying the results given in Eq. (4.405), we find then that

$$\alpha_n = \begin{cases} \frac{1}{\pi} \int_{-1}^{1} \frac{T_n(\tilde{t}) f(\tilde{t})}{\sqrt{1 - \tilde{t}^2}} \, d\tilde{t}, & n = 0, \\[2mm] \frac{2}{\pi} \int_{-1}^{1} \frac{T_n(\tilde{t}) f(\tilde{t})}{\sqrt{1 - \tilde{t}^2}} \, d\tilde{t}, & n > 0. \end{cases} \tag{6.270}$$

Application of our method leads to a complicated exact expression for the Fourier-Chebyshev series. Numerical evaluation gives the approximation

$$e^t \approx 1.75 T_0(-1 + 2t) + 0.850 T_1(-1 + 2t) + 0.105 T_2(-1 + 2t) + \cdots, \tag{6.271}$$

$$\approx 1.75 + 0.850(-1 + 2t) + 0.105(-1 + 2(-1 + 2t)^2) + \cdots, \tag{6.272}$$

$$\approx 1.00 + 1.00t + 0.500t^2 + \cdots. \tag{6.273}$$

The series appears to be identical to the Taylor series of e^t about $t = 0$ but in actuality is slightly different when more significant digits are exposed.

It is when we consider the convergence properties and error distribution that we find remarkable properties of the Fourier-Chebyshev expansion. Figure 6.16a gives a log-log plot of the error in the approximation as a function of N. First, simply note that with only $N = 20$ terms, we have driven the error to near the remarkably small 10^{-30}.

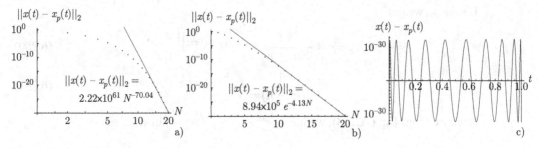

Figure 6.16. (a) Behavior of the error norm of the Fourier-Chebyshev series approximation to $x(t) = e^t$ on $t \in [0, 1]$ with the number N of terms included in the series, log-log scale; (b) same as (a), except semi-log scale; (c) error distribution for $N = 20$ approximation.

A weighted least squares curve fit, with a weighting factor proportional to N^2, shows that near $N = 20$, the error is well described by

$$\|x(t) - x_p(t)\|_2 \approx 2.22 \times 10^{61} \, N^{-70.04}. \tag{6.274}$$

It appears, however, that the data are better modeled by an exponential convergence, demonstrated in the semi-log plot of Figure 6.16b, where we find the error is well approximated by

$$\|x(t) - x_p(t)\|_2 \approx 8.94 \times 10^5 e^{-4.13N}. \tag{6.275}$$

Lastly, we note in Figure 6.16c that the error distribution has a uniform amplitude, in contrast with the two previous examples, where the error was concentrated at or near the boundaries. We summarize the properties of the Fourier-Chebyshev approximation as follows. For smooth functions which need not have periodicity and do not necessarily vanish at domain boundaries, one can nevertheless obtain an approximation that

- has an error that decreases *exponentially* as the number of terms in the approximation increases,
- relative to other methods, has a uniform distribution of the error.

Let us extend our discussion to complex Fourier series. For example, we can consider a set of basis functions to be

$$\varphi_m(t) = e^{2m\pi i t}, \qquad m = 0, \pm 1, \pm 2, \ldots, \pm M. \tag{6.276}$$

In contrast to earlier analysis, we find it convenient to consider positive and negative integers for the expansion. A simple transformation could recast these as nonnegative integers if desired, e.g. $\hat{m} = m + M$. It is easy to see that these basis functions form an orthonormal set on $t \in [-1/2, 1/2]$:

$$\langle \varphi_n, \varphi_m \rangle = \int_{-1/2}^{1/2} \overline{e^{2n\pi i t}} e^{2m\pi i t} \, dt, \tag{6.277}$$

$$= \int_{-1/2}^{1/2} e^{2(m-n)\pi i t} \, dt. \tag{6.278}$$

If $m = n$, we get

$$\langle \varphi_n, \varphi_n \rangle = \int_{-1/2}^{1/2} e^0 \, dt = 1. \tag{6.279}$$

If $m \neq n$, we get

$$\langle \varphi_n, \varphi_m \rangle = \frac{1}{2(m-n)\pi i} \, e^{2(m-n)\pi i t} \Big|_{-1/2}^{1/2}, \tag{6.280}$$

$$= \frac{1}{2(m-n)\pi i} \left(e^{(m-n)\pi i} - e^{(n-m)\pi i} \right). \tag{6.281}$$

Because from Eq. (A.138), we have $(e^{is} - e^{-is})/(2i) = \sin s$, we get

$$\langle \varphi_n, \varphi_m \rangle = \frac{1}{(m-n)\pi} \sin \left((m-n)\pi \right), \tag{6.282}$$

$$= 0. \tag{6.283}$$

As long as m and n are integers, the integral evaluates to zero for $m \neq n$ because of the nature of the sine function. Thus, $\langle \varphi_n, \varphi_m \rangle = \delta_{nm}$.

Now consider the Fourier series approximation of a function:

$$x(t) \approx \sum_{m=-M}^{M} \alpha_m \varphi_m(t), \tag{6.284}$$

$$\langle \varphi_n(t), x(t) \rangle = \langle \varphi_n(t), \sum_{m=-M}^{M} \alpha_m \varphi_m(t) \rangle, \tag{6.285}$$

$$\int_{-1/2}^{1/2} \overline{\varphi_n(t)} x(t) \, dt = \int_{-1/2}^{1/2} \overline{\varphi_n(t)} \sum_{m=-M}^{M} \alpha_m \varphi_m(t) \, dt, \tag{6.286}$$

$$= \sum_{m=-M}^{M} \alpha_m \int_{-1/2}^{1/2} \overline{\varphi_n(t)} \varphi_m(t) \, dt, \tag{6.287}$$

$$= \sum_{m=-M}^{M} \alpha_m \delta_{nm}, \tag{6.288}$$

$$= \alpha_n. \tag{6.289}$$

Thus, with $\varphi_m(t) = e^{2m\pi i t}$, we have

$$\alpha_m = \langle \varphi_m, x \rangle = \int_{-1/2}^{1/2} e^{-2m\pi i t} x(t) \, dt, \qquad m = 0, \pm 1, \pm 2, \ldots, \pm M. \tag{6.290}$$

This expansion is valid for either real or complex $x(t)$. Changing t to the new dummy variable s, and using de Moivre's formula, Eq. (A.136), we can write the complex exponential in terms of trigonometric functions, giving

$$\alpha_m = \int_{-1/2}^{1/2} \left(\cos 2m\pi s - i \sin 2m\pi s \right) x(s) \, ds, \tag{6.291}$$

$$= \underbrace{\int_{-1/2}^{1/2} x(s) \cos(2m\pi s) \, ds}_{a_m} - i \underbrace{\int_{-1/2}^{1/2} x(s) \sin(2m\pi s) \, ds}_{b_m}, \tag{6.292}$$

$$= a_m - i b_m, \qquad m = 0, \pm 1, \pm 2, \ldots, \pm M. \tag{6.293}$$

The approximation then is

$$x(t) \approx \sum_{m=-M}^{M} \underbrace{(a_m - ib_m)}_{\alpha_m} \underbrace{(\cos(2m\pi t) + i\sin(2m\pi t))}_{e^{2m\pi it}}, \tag{6.294}$$

$$\approx \sum_{m=-M}^{M} a_m \cos(2m\pi t) + b_m \sin(2m\pi t) + i\left(\sin(2m\pi t)a_m - \cos(2m\pi t)b_m\right). \tag{6.295}$$

Using the definitions of a_m and b_m along with trigonometric identities gives

$$x(t) \approx \sum_{m=-M}^{M} \int_{-1/2}^{1/2} x(s)\cos(2\pi m(s-t))\,ds - i\int_{-1/2}^{1/2} x(s)\sin(2\pi m(s-t))\,ds. \tag{6.296}$$

Because $\sin(2\pi m(s-t)) = -\sin(2\pi(-m)(s-t))$ and because for every negative value of m, there is a balancing positive value in the summation, the terms involving the sine function sum to zero giving

$$x(t) \approx \sum_{m=-M}^{M} \int_{-1/2}^{1/2} x(s)\cos(2\pi m(s-t))\,ds. \tag{6.297}$$

Using trigonometric identities to expand, we find

$$x(t) \approx \sum_{m=-M}^{M} \int_{-1/2}^{1/2} x(s)\left(\cos(2\pi ms)\cos(2\pi mt) + \sin(2\pi ms)\sin(2\pi mt)\right)\,ds, \tag{6.298}$$

$$\approx \sum_{m=-M}^{M} \cos(2\pi mt) \underbrace{\int_{-1/2}^{1/2} x(s)\cos(2\pi ms)\,ds}_{a_m}$$

$$+ \sum_{m=-M}^{M} \sin(2\pi mt) \underbrace{\int_{-1/2}^{1/2} x(s)\sin(2\pi ms)\,ds}_{b_m}. \tag{6.299}$$

So we have our final form

$$x(t) \approx \sum_{m=-M}^{M} a_m \cos(2m\pi t) + \sum_{m=-M}^{M} b_m \sin(2m\pi t), \tag{6.300}$$

with the Fourier coefficients given by

$$a_m = \int_{-1/2}^{1/2} x(t)\cos(2m\pi t)\,dt, \qquad m = 0, \pm 1, \pm 2, \ldots \pm M, \tag{6.301}$$

$$b_m = \int_{-1/2}^{1/2} x(t)\sin(2m\pi t)\,dt, \qquad m = 0, \pm 1, \pm 2, \ldots \pm M. \tag{6.302}$$

Because $b_0 = 0$, it is common to omit this term from the series. If $x(t)$ is an odd function for $t \in [-1/2, /1/2]$, we would find $a_m = 0, \forall\, m$. If $x(t)$ is an even function

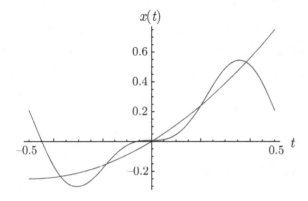

Figure 6.17. Plot of $x(t) = t + t^2$ and its $M = 2$ general Fourier series approximation.

for $t \in [-1/2, 1/2]$, we would find $b_m = 0, \forall\, m$. For functions that are neither even nor odd, we would find $a_m \neq 0$ and $b_m \neq 0, \forall\, m$.

EXAMPLE 6.36

Use the approach for general complex Fourier series to expand the real function $x(t) = t + t^2$ for $t \in [-1/2, 1/2]$.

Note that $x(t)$ is neither an even or odd function in the domain, and it takes on finite values at the endpoints. Let us choose $M = 2$ so that we seek α_m to approximate

$$x(t) = t + t^2 \approx \sum_{m=-2}^{2} \alpha_m e^{2m\pi i t}. \tag{6.303}$$

We get the α_m by use of Eq. (6.290) for our $x(t)$, which gives

$$\alpha_m = \int_{-1/2}^{1/2} e^{-2m\pi i t}(t + t^2)\, dt. \tag{6.304}$$

We find

$$\alpha_{\pm 2} = \frac{1 \pm 2i\pi}{8\pi^2}, \qquad \alpha_{\pm 1} = -\frac{1 \pm i\pi}{2\pi^2}, \qquad \alpha_0 = \frac{1}{12}. \tag{6.305}$$

Our Fourier approximation is

$$x(t) = t + t^2 \approx \frac{(1 - 2i\pi)e^{-4i\pi t}}{8\pi^2} - \frac{(1 - i\pi)e^{-2i\pi t}}{2\pi^2} + \frac{1}{12} - \frac{(1 + i\pi)e^{2i\pi t}}{2\pi^2}$$
$$+ \frac{(1 + 2i\pi)e^{4i\pi t}}{8\pi^2}. \tag{6.306}$$

Application of de Moivre's formula shows the approximation is purely real and is

$$x(t) = t + t^2 \approx \frac{1}{12} + \frac{\sin(2\pi t)}{\pi} - \frac{\sin(4\pi t)}{2\pi} - \frac{\cos(2\pi t)}{\pi^2} + \frac{\cos(4\pi t)}{4\pi^2}. \tag{6.307}$$

We show a plot of the original function and its approximation in Figure 6.17.

6.5.4 Reciprocal

Here we expand on a topic introduced in Section 1.6.5. Let $\{u_1, \ldots, u_N\}$ be a basis of a finite-dimensional inner product space. Also, let $\{u_1^R, \ldots, u_N^R\}$ be elements of the same space such that

$$\langle u_n, u_m^R \rangle = \delta_{nm}. \tag{6.308}$$

Then $\{u_1^R, \ldots, u_N^R\}$ is called the reciprocal (or dual) basis of $\{u_1, \ldots, u_N\}$. Of course, an orthonormal basis is its own reciprocal. Because $\{u_1, \ldots, u_N\}$ is a basis, we can write any vector x as

$$x = \sum_{m=1}^{N} \alpha_m u_m. \tag{6.309}$$

Taking the inner product of both sides with u_n^R, we get

$$\langle u_n^R, x \rangle = \langle u_n^R, \sum_{m=1}^{N} \alpha_m u_m \rangle, \tag{6.310}$$

$$= \sum_{m=1}^{N} \langle u_n^R, \alpha_m u_m \rangle, \tag{6.311}$$

$$= \sum_{m=1}^{N} \alpha_m \langle u_n^R, u_m \rangle, \tag{6.312}$$

$$= \sum_{m=1}^{N} \alpha_m \delta_{nm}, \tag{6.313}$$

$$= \alpha_n, \tag{6.314}$$

so that

$$x = \sum_{n=1}^{N} \underbrace{\langle u_n^R, x \rangle}_{=\alpha_n} u_n. \tag{6.315}$$

The transformation of the representation of a vector x from a basis to a dual basis is a type of alias transformation.

EXAMPLE 6.37

A vector $\mathbf{v}$ resides in $\mathbb{R}^2$. Its representation in Cartesian coordinates is $\mathbf{v} = \xi = \begin{pmatrix} 3 \\ 5 \end{pmatrix}$. The vectors $u_1 = \begin{pmatrix} 2 \\ 0 \end{pmatrix}$ and $u_2 = \begin{pmatrix} 1 \\ 3 \end{pmatrix}$ span the space $\mathbb{R}^2$ and thus can be used as a basis on which to represent $\mathbf{v}$. Find the reciprocal basis u_1^R, u_2^R, and use Eq. (6.315) to represent $\mathbf{v}$ in terms of both the basis u_1, u_2 and then the reciprocal basis u_1^R, u_2^R.

We adopt the dot product as our inner product. Let us get α_1, α_2. To do this, we first need the reciprocal basis vectors, which are defined by the inner product:

$$\langle u_n, u_m^R \rangle = \delta_{nm}. \tag{6.316}$$

We take

$$u_1^R = \begin{pmatrix} a_{11} \\ a_{21} \end{pmatrix}, \qquad u_2^R = \begin{pmatrix} a_{12} \\ a_{22} \end{pmatrix}. \tag{6.317}$$

Expanding Eq. (6.316), we get,

$$\langle u_1, u_1^R \rangle = u_1^T u_1^R = (2, 0) \begin{pmatrix} a_{11} \\ a_{21} \end{pmatrix} = (2)a_{11} + (0)a_{21} = 1, \tag{6.318}$$

$$\langle u_1, u_2^R \rangle = u_1^T u_2^R = (2, 0) \begin{pmatrix} a_{12} \\ a_{22} \end{pmatrix} = (2)a_{12} + (0)a_{22} = 0, \tag{6.319}$$

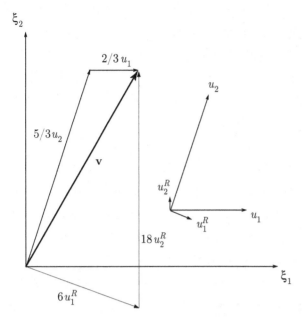

Figure 6.18. Representation of a vector **v** on a nonorthogonal contravariant basis u_1, u_2 and its reciprocal covariant basis u_1^R, u_2^R.

$$\langle u_2, u_1^R \rangle = u_2^T u_1^R = (1,3) \begin{pmatrix} a_{11} \\ a_{21} \end{pmatrix} = (1)a_{11} + (3)a_{21} = 0, \tag{6.320}$$

$$\langle u_2, u_2^R \rangle = u_2^T u_2^R = (1,3) \begin{pmatrix} a_{12} \\ a_{22} \end{pmatrix} = (1)a_{12} + (3)a_{22} = 1. \tag{6.321}$$

Solving, we get

$$a_{11} = \frac{1}{2}, \qquad a_{21} = -\frac{1}{6}, \qquad a_{12} = 0, \qquad a_{22} = \frac{1}{3}, \tag{6.322}$$

so substituting into Eq. (6.317), we get expressions for the reciprocal basis vectors:

$$u_1^R = \begin{pmatrix} \frac{1}{2} \\ -\frac{1}{6} \end{pmatrix}, \qquad u_2^R = \begin{pmatrix} 0 \\ \frac{1}{3} \end{pmatrix}. \tag{6.323}$$

What we have really done here is invert the matrix whose columns are u_1 and u_2:

$$\begin{pmatrix} \vdots & \vdots \\ u_1 & u_2 \\ \vdots & \vdots \end{pmatrix}^{-1} = \begin{pmatrix} 2 & 1 \\ 0 & 3 \end{pmatrix}^{-1} = \begin{pmatrix} \frac{1}{2} & -\frac{1}{6} \\ 0 & \frac{1}{3} \end{pmatrix} = \begin{pmatrix} \cdots & u_1^{RT} & \cdots \\ \cdots & u_2^{RT} & \cdots \end{pmatrix}. \tag{6.324}$$

We can now get the coefficients α_i:

$$\alpha_1 = \langle u_1^R, \xi \rangle = (\tfrac{1}{2} \quad -\tfrac{1}{6}) \begin{pmatrix} 3 \\ 5 \end{pmatrix} = \frac{3}{2} - \frac{5}{6} = \frac{2}{3}, \tag{6.325}$$

$$\alpha_2 = \langle u_2^R, \xi \rangle = (0 \quad \tfrac{1}{3}) \begin{pmatrix} 3 \\ 5 \end{pmatrix} = 0 + \frac{5}{3} = \frac{5}{3}. \tag{6.326}$$

So on the new basis, **v** can be represented as

$$\mathbf{v} = \frac{2}{3} u_1 + \frac{5}{3} u_2. \tag{6.327}$$

The representation is shown geometrically in Figure 6.18. Note that u_1^R is orthogonal to u_2 and that u_2^R is orthogonal to u_1. Furthermore, because $||u_1||_2 > 1, ||u_2||_2 > 1$, we get $||u_1^R||_2 < 1$ and $||u_2^R||_2 < 1$ to have $\langle u_i, u_j^R \rangle = \delta_{ij}$.

In a similar manner, it is easily shown that $\mathbf{v}$ can be represented in terms of the reciprocal basis as

$$\mathbf{v} = \sum_{n=1}^{N} \beta_n u_n^R = \beta_1 u_1^R + \beta_2 u_2^R, \qquad (6.328)$$

where

$$\beta_n = \langle u_n, \xi \rangle. \qquad (6.329)$$

For this problem, this yields

$$\mathbf{v} = 6u_1^R + 18u_2^R. \qquad (6.330)$$

Thus, we see for the nonorthogonal basis that two natural representations of the same vector exist. One of these is actually a covariant representation; the other is contravariant.

Let us show this is consistent with the earlier described notions using "upstairs-downstairs" notation of Section 1.6. Our nonorthogonal coordinate system is a transformation of the form

$$\xi^i = \frac{\partial \xi^i}{\partial x^j} x^j, \qquad (6.331)$$

where ξ^i is the Cartesian representation and x^j is the contravariant representation in the transformed system. In Gibbs form, this is

$$\xi = \mathbf{J} \cdot \mathbf{x}. \qquad (6.332)$$

Inverting, we also have

$$\mathbf{x} = \mathbf{J}^{-1} \cdot \xi. \qquad (6.333)$$

For this problem, we have

$$\frac{\partial \xi^i}{\partial x^j} = \mathbf{J} = \begin{pmatrix} 2 & 1 \\ 0 & 3 \end{pmatrix} = \begin{pmatrix} \vdots & \vdots \\ u_1 & u_2 \\ \vdots & \vdots \end{pmatrix}, \qquad (6.334)$$

so that

$$\begin{pmatrix} \xi^1 \\ \xi^2 \end{pmatrix} = \begin{pmatrix} 2 & 1 \\ 0 & 3 \end{pmatrix} \cdot \begin{pmatrix} x^1 \\ x^2 \end{pmatrix}. \qquad (6.335)$$

The unit vector in the transformed space,

$$\begin{pmatrix} x^1 \\ x^2 \end{pmatrix} = \begin{pmatrix} 1 \\ 0 \end{pmatrix}, \qquad (6.336)$$

has representation in Cartesian space of $(2,0)^T$, and the other unit vector in the transformed space

$$\begin{pmatrix} x^1 \\ x^2 \end{pmatrix} = \begin{pmatrix} 0 \\ 1 \end{pmatrix}, \qquad (6.337)$$

has representation in Cartesian space of $(1,3)^T$.

Now the metric tensor is

$$g_{ij} = \mathbf{G} = \mathbf{J}^T \cdot \mathbf{J} = \begin{pmatrix} 2 & 0 \\ 1 & 3 \end{pmatrix} \begin{pmatrix} 2 & 1 \\ 0 & 3 \end{pmatrix} = \begin{pmatrix} 4 & 2 \\ 2 & 10 \end{pmatrix}. \qquad (6.338)$$

The Cartesian vector $\boldsymbol{\xi} = (3,5)^T$ has a contravariant representation in the transformed space of

$$\mathbf{x} = \mathbf{J}^{-1} \cdot \boldsymbol{\xi} = \begin{pmatrix} 2 & 1 \\ 0 & 3 \end{pmatrix}^{-1} \begin{pmatrix} 3 \\ 5 \end{pmatrix} = \begin{pmatrix} \frac{1}{2} & -\frac{1}{6} \\ 0 & \frac{1}{3} \end{pmatrix} \begin{pmatrix} 3 \\ 5 \end{pmatrix} = \begin{pmatrix} \frac{2}{3} \\ \frac{5}{3} \end{pmatrix} = x^j. \tag{6.339}$$

This is consistent with our earlier finding.

This vector has a covariant representation as well by the formula

$$x_i = g_{ij} x^j = \begin{pmatrix} 4 & 2 \\ 2 & 10 \end{pmatrix} \begin{pmatrix} \frac{2}{3} \\ \frac{5}{3} \end{pmatrix} = \begin{pmatrix} 6 \\ 18 \end{pmatrix}. \tag{6.340}$$

Once again, this is consistent with our earlier finding.

Note further that

$$\mathbf{J}^{-1} = \begin{pmatrix} \frac{1}{2} & -\frac{1}{6} \\ 0 & \frac{1}{3} \end{pmatrix} = \begin{pmatrix} \cdots u_1^{RT} \cdots \\ \cdots u_2^{RT} \cdots \end{pmatrix}. \tag{6.341}$$

The *rows* of this matrix describe the reciprocal basis vectors and are also consistent with our earlier finding. So if we think of the columns of any matrix as forming a basis, the rows of the inverse of that matrix form the reciprocal basis:

$$\underbrace{\begin{pmatrix} \cdots u_1^{RT} \cdots \\ \cdots\cdots\cdots \\ \cdots u_N^{RT} \cdots \end{pmatrix}}_{\mathbf{J}^{-1}} \underbrace{\begin{pmatrix} \vdots & \vdots & \vdots \\ u_1 & \vdots & u_N \\ \vdots & \vdots & \vdots \end{pmatrix}}_{\mathbf{J}} = \mathbf{I}. \tag{6.342}$$

Lastly, note that $\det \mathbf{J} = 6$, so the transformation is orientation-preserving but not volume-preserving. A unit volume element in $\boldsymbol{\xi}$-space is larger than one in $\mathbf{x}$-space. Moreover, the mapping $\boldsymbol{\xi} = \mathbf{J} \cdot \mathbf{x}$ can be shown to involve both stretching and rotation.

EXAMPLE 6.38

For the previous example problem, consider the tensor $\mathbf{A}$, whose representation in the Cartesian space is

$$\mathbf{A} = \begin{pmatrix} 3 & 4 \\ 1 & 2 \end{pmatrix}. \tag{6.343}$$

Demonstrate the invariance of the scalar $\boldsymbol{\xi}^T \cdot \mathbf{A} \cdot \boldsymbol{\xi}$ in the non-Cartesian space.

First, in the Cartesian space, we have

$$\boldsymbol{\xi}^T \cdot \mathbf{A} \cdot \boldsymbol{\xi} = (3 \quad 5) \begin{pmatrix} 3 & 4 \\ 1 & 2 \end{pmatrix} \begin{pmatrix} 3 \\ 5 \end{pmatrix} = 152. \tag{6.344}$$

Now $\mathbf{A}$ has a different representation, $\mathbf{A}'$, in the transformed coordinate system via the definition of a tensor, Eq. (1.354), which, for this linear alias transformation, reduces to:[17]

$$\mathbf{A}' = \mathbf{J}^{-1} \cdot \mathbf{A} \cdot \mathbf{J}. \tag{6.345}$$

---

[17] If $\mathbf{J}$ had been a rotation matrix $\mathbf{Q}$, for which $\mathbf{Q}^T = \mathbf{Q}^{-1}$ and $\det \mathbf{Q} = 1$, then $\mathbf{A}' = \mathbf{Q}^T \cdot \mathbf{A} \cdot \mathbf{Q}$ from Eq. (2.84). Here our linear transformation has both stretching and rotation associated with it. Also, so as not to confuse with the complex conjugate, we use the $'$ notation for the representation of $\mathbf{A}$ in the transformed coordinate system.

So

$$\mathbf{A}' = \underbrace{\begin{pmatrix} \frac{1}{2} & -\frac{1}{6} \\ 0 & \frac{1}{3} \end{pmatrix}}_{\mathbf{J}^{-1}} \underbrace{\begin{pmatrix} 3 & 4 \\ 1 & 2 \end{pmatrix}}_{\mathbf{A}} \underbrace{\begin{pmatrix} 2 & 1 \\ 0 & 3 \end{pmatrix}}_{\mathbf{J}}, \tag{6.346}$$

$$= \begin{pmatrix} \frac{8}{3} & \frac{19}{3} \\ \frac{2}{3} & \frac{7}{3} \end{pmatrix}. \tag{6.347}$$

We also see by inversion that

$$\mathbf{A} = \mathbf{J} \cdot \mathbf{A}' \cdot \mathbf{J}^{-1}. \tag{6.348}$$

Because $\boldsymbol{\xi} = \mathbf{J} \cdot \mathbf{x}$, our tensor invariant becomes in the transformed space

$$\boldsymbol{\xi}^T \cdot \mathbf{A} \cdot \boldsymbol{\xi} = (\mathbf{J} \cdot \mathbf{x})^T \cdot (\mathbf{J} \cdot \mathbf{A}' \cdot \mathbf{J}^{-1}) \cdot (\mathbf{J} \cdot \mathbf{x}), \tag{6.349}$$

$$= \mathbf{x}^T \cdot \underbrace{\mathbf{J}^T \cdot \mathbf{J}}_{\mathbf{G}} \cdot \mathbf{A}' \cdot \mathbf{x}, \tag{6.350}$$

$$= \underbrace{\mathbf{x}^T \cdot \mathbf{G}}_{\text{covariant } \mathbf{x}} \cdot \mathbf{A}' \cdot \mathbf{x}, \tag{6.351}$$

$$= (\begin{matrix} \frac{2}{3} & \frac{5}{3} \end{matrix}) \begin{pmatrix} 4 & 2 \\ 2 & 10 \end{pmatrix} \begin{pmatrix} \frac{8}{3} & \frac{19}{3} \\ \frac{2}{3} & \frac{7}{3} \end{pmatrix} \begin{pmatrix} \frac{2}{3} \\ \frac{5}{3} \end{pmatrix}, \tag{6.352}$$

$$= \underbrace{(\begin{matrix} 6 & 18 \end{matrix})}_{\text{covariant } \mathbf{x}} \underbrace{\begin{pmatrix} \frac{8}{3} & \frac{19}{3} \\ \frac{2}{3} & \frac{7}{3} \end{pmatrix}}_{\mathbf{A}'} \underbrace{\begin{pmatrix} \frac{2}{3} \\ \frac{5}{3} \end{pmatrix}}_{\text{contravariant } \mathbf{x}}, \tag{6.353}$$

$$= 152. \tag{6.354}$$

Note that $\mathbf{x}^T \cdot \mathbf{G}$ gives the covariant representation of $\mathbf{x}$.

EXAMPLE 6.39

Given a space spanned by the functions $u_1 = 1$, $u_2 = t$, $u_3 = t^2$, for $t \in [0, 1]$, find a reciprocal basis u_1^R, u_2^R, u_3^R within this space.

We insist that

$$\langle u_n, u_m^R \rangle = \int_0^1 u_n(t) u_m^R(t) \, dt = \delta_{nm}. \tag{6.355}$$

If we assume that

$$u_1^R = a_1 + a_2 t + a_3 t^2, \tag{6.356}$$

$$u_2^R = b_1 + b_2 t + b_3 t^2, \tag{6.357}$$

$$u_3^R = c_1 + c_2 t + c_3 t^2, \tag{6.358}$$

and substitute directly into Eq. (6.355), it is easy to find that

$$u_1^R = 9 - 36t + 30t^2, \tag{6.359}$$

$$u_2^R = -36 + 192t - 180t^2, \tag{6.360}$$

$$u_3^R = 30 - 180t + 180t^2. \tag{6.361}$$

6.6 Parseval's Equation, Convergence, and Completeness

We consider Parseval's[18] equation and associated issues here. For a basis to be complete, we require that the norm of the difference of the series representation of all functions and the functions themselves converge to zero in $\mathbb{L}_2$ as the number of terms in the series approaches infinity. For an orthonormal basis $\varphi_n(t)$, this is

$$\lim_{N \to \infty} \left\| x(t) - \sum_{n=1}^{N} \alpha_n \varphi_n(t) \right\|_2 = 0. \tag{6.362}$$

Now, for the orthonormal basis, we can show that this reduces to a particularly simple form. Consider, for instance, the error for a one-term Fourier expansion

$$\|x - \alpha\varphi\|_2^2 = \langle x - \alpha\varphi, x - \alpha\varphi \rangle, \tag{6.363}$$

$$= \langle x, x \rangle - \langle x, \alpha\varphi \rangle - \langle \alpha\varphi, x \rangle + \langle \alpha\varphi, \alpha\varphi \rangle, \tag{6.364}$$

$$= \|x\|_2^2 - \alpha\langle x, \varphi \rangle - \overline{\alpha}\langle \varphi, x \rangle + \overline{\alpha}\alpha\langle \varphi, \varphi \rangle, \tag{6.365}$$

$$= \|x\|_2^2 - \alpha\overline{\langle \varphi, x \rangle} - \overline{\alpha}\langle \varphi, x \rangle + \overline{\alpha}\alpha\langle \varphi, \varphi \rangle, \tag{6.366}$$

$$= \|x\|_2^2 - \alpha\overline{\alpha} - \overline{\alpha}\alpha + \overline{\alpha}\alpha(1), \tag{6.367}$$

$$= \|x\|_2^2 - \alpha\overline{\alpha}, \tag{6.368}$$

$$= \|x\|_2^2 - |\alpha|^2. \tag{6.369}$$

Here we have used the definition of the Fourier coefficient $\langle \varphi, x \rangle = \alpha$ and orthonormality $\langle \varphi, \varphi \rangle = 1$. This is easily extended to multiterm expansions to give

$$\left\| x(t) - \sum_{n=1}^{N} \alpha_n \varphi_n(t) \right\|_2^2 = \|x(t)\|_2^2 - \sum_{n=1}^{N} |\alpha_n|^2. \tag{6.370}$$

So convergence, and thus completeness of the basis, is equivalent to requiring that what we call Parseval's equation,

$$\|x(t)\|_2^2 = \lim_{N \to \infty} \sum_{n=1}^{N} |\alpha_n|^2, \tag{6.371}$$

hold for all functions $x(t)$. This requirement is stronger than just requiring that the last Fourier coefficient vanish for large N; also note that it does not address the important question of the rate of convergence, which can be different for different functions $x(t)$, for the same basis.

6.7 Operators

We next consider what is broadly known as operators.[19] We begin with general definitions and concepts for all operators, then specialize to what is known as linear operators.

---

[18] Marc-Antoine Parseval des Chênes, 1755–1835, French mathematician.

[19] We generally will use notation such $\mathbf{L}x$ to denote the operator $\mathbf{L}$ operating on the operand x. Occasionally, it will be written as $\mathbf{L}(x)$. Sometimes one will find similar notations, such as $f : \mathbb{X} \to \mathbb{Y}$ to denote an operator f that maps elements from the domain $\mathbb{X}$ to the range $\mathbb{Y}$.

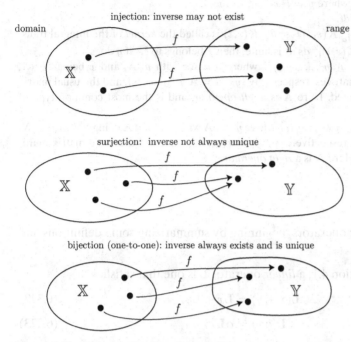

Figure 6.19. Diagram showing classes of operators.

- For two sets $\mathbb{X}$ and $\mathbb{Y}$, an *operator* (or *mapping*, or *transformation*) f is a rule that associates every $x \in \mathbb{X}$ with an *image* $y \in \mathbb{Y}$. We can write $f : \mathbb{X} \to \mathbb{Y}$, $\mathbb{X} \xrightarrow{f} \mathbb{Y}$ or $x \mapsto y$. $\mathbb{X}$ is the *domain* of the operator, and $\mathbb{Y}$ is the *range*.
- If every element of $\mathbb{Y}$ is not necessarily an image, then $\mathbb{X}$ is mapped *into* $\mathbb{Y}$; this map is called an *injection*.
- If, on the other hand, every element of $\mathbb{Y}$ is an image of some element of $\mathbb{X}$, then $\mathbb{X}$ is mapped *onto* $\mathbb{Y}$ and the map is a *surjection*.
- If, $\forall\, x \in \mathbb{X}$ there is a unique $y \in \mathbb{Y}$, and for every $y \in \mathbb{Y}$ there is a unique $x \in \mathbb{X}$, the operator is *one-to-one* or *invertible*; it is a *bijection*.
- f and g are inverses of each other if $\mathbb{X} \xrightarrow{f} \mathbb{Y}$ and $\mathbb{Y} \xrightarrow{g} \mathbb{X}$.
- The operator $f : \mathbb{X} \to \mathbb{Y}$ is a *homeomorphism* if f is a bijection, continuous, and has a continuous inverse. Geometrically, homeomorphisms imply that the two spaces $\mathbb{X}$ and $\mathbb{Y}$ are topologically equivalent. For example, a square is topologically equivalent to an ellipse, but a sphere is not topologically equivalent to a torus.
- $f : \mathbb{X} \to \mathbb{Y}$ is continuous at $x_0 \in \mathbb{X}$ if, for every $\epsilon > 0$, there is a $\delta > 0$, such that $\|f(x) - f(x_0)\| < \epsilon\ \forall\, x$ satisfying $\|x - x_0\| < \delta$.
- If for every bounded sequence x_n in a Hilbert space the sequence $f(x_n)$ contains a convergent subsequence, then f is said to be *compact*.
- $f : \mathbb{X} \to \mathbb{Y}$ is *positive definite* if $\forall\, x \neq 0$, $f(x) > 0$, $f(0) = 0$.
- $Q : \mathbb{X} \to \mathbb{Y}$ is *unitary* iff Q is linear, surjective and $\langle Qx_1, Qx_2 \rangle = \langle x_1, x_2 \rangle$, $\forall\, x_1, x_2 \in \mathbb{X}$ (see also Section 7.7).

A diagram showing various classes of operators is given in Figure 6.19. Examples of continuous operators are as follows:

1. $(x_1, x_2, \ldots, x_N) \mapsto y$, where $y = f(x_1, x_2, \ldots, x_N)$.
2. $f \mapsto g$, where $g = df/dt$.
3. $f \mapsto g$, where $g(t) = \int_a^b K(s,t)f(s)\,ds$. $K(s,t)$ is called the *kernel* of the integral transformation. If $\int_a^b \int_a^b |K(s,t)|^2\,ds\,dt$ is finite, then f belongs to $\mathbb{L}_2$ if g does.
4. $(x_1, x_2, \ldots, x_M)^T \mapsto (y_1, y_2, \ldots, y_N)^T$, where $y = \mathbf{A}x$ with y, $\mathbf{A}$, and x being $N \times 1$, $N \times M$, and $M \times 1$ matrices, respectively ($y_{N \times 1} = \mathbf{A}_{N \times M} x_{M \times 1}$), and the usual matrix multiplication is assumed. Here $\mathbf{A}$ is a *left operator*, and is the most common type of matrix operator.
5. $(x_1, x_2, \ldots, x_N) \mapsto (y_1, y_2, \ldots, y_M)$, where $y = x\mathbf{A}$ with y, x, and $\mathbf{A}$ being $1 \times M$, $1 \times N$, and $N \times M$ matrices, respectively ($y_{1 \times M} = x_{1 \times N} \mathbf{A}_{N \times M}$), and the usual matrix multiplication is assumed. Here $\mathbf{A}$ is a *right operator*.

6.7.1 Linear

We consider here linear operators, beginning by summarizing some definitions and properties.

- As discussed in Section 4.1, a *linear* operator $\mathbf{L}$ is one that satisfies

$$\mathbf{L}(x + y) = \mathbf{L}x + \mathbf{L}y, \qquad (6.372)$$

$$\mathbf{L}(\alpha x) = \alpha \mathbf{L}x. \qquad (6.373)$$

- An operator $\mathbf{L}$ is *bounded* if $\forall\, x \in \mathbb{X}\ \exists$ a constant c such that

$$||\mathbf{L}x|| \le c||x||. \qquad (6.374)$$

A derivative is an example of an *unbounded* linear operator. For example, if $x(t)$ is a unit step function, $x(t) = 0, t \in (-\infty, 0)$, $x(t) = 1, t \in [0, \infty)$, which is bounded between 0 and 1, the derivative of x is unbounded at $t = 0$.
- A special operator is the *identity operator* $\mathbf{I}$, which is defined by $\mathbf{I}x = x$.
- The *null space* or *kernel* of an operator $\mathbf{L}$ is the set of all x such that $\mathbf{L}x = 0$. The null space is a vector space.
- The norm of an operator $\mathbf{L}$ can be defined as

$$||\mathbf{L}|| = \sup_{x \ne 0} \frac{||\mathbf{L}x||}{||x||}. \qquad (6.375)$$

- An operator $\mathbf{L}$ is
  ○ *positive definite* if $\langle \mathbf{L}x, x \rangle > 0$,
  ○ *positive semi-definite* if $\langle \mathbf{L}x, x \rangle \ge 0$,
  ○ *negative definite* if $\langle \mathbf{L}x, x \rangle < 0$,
  ○ *negative semi-definite* if $\langle \mathbf{L}x, x \rangle \le 0$,
  $\forall\, x \ne 0$. Operators that are positive semi-definite are also known simply as *positive*.
- For a matrix $\mathbf{A}\colon \mathbb{C}^M \to \mathbb{C}^N$, the *spectral norm* $||\mathbf{A}||_2$ is defined as

$$||\mathbf{A}||_2 = \sup_{x \ne 0} \frac{||\mathbf{A}x||_2}{||x||_2}. \qquad (6.376)$$

This can be shown to reduce to

$$||\mathbf{A}||_2 = \sqrt{\kappa_{max}}, \qquad (6.377)$$

where κ_{max} is the largest eigenvalue of the matrix $\overline{\mathbf{A}}^T \cdot \mathbf{A}$. It will soon be shown in Section 6.8 that because of the nature of $\overline{\mathbf{A}}^T \cdot \mathbf{A}$, all of its eigenvalues are guaranteed real. Moreover, it can be shown that they are also all greater than or equal to zero. Hence, the definition will satisfy all properties of the norm. This holds only for Hilbert spaces and not for arbitrary Banach spaces. There are also other valid definitions of norms for matrix operators. For example, the *Schatten p-norm* of a matrix **A** is

$$||\mathbf{A}||_p = \sup_{x \neq 0} \frac{||\mathbf{A}x||_p}{||x||_p}. \tag{6.378}$$

EXAMPLE 6.40

Find the Schatten 4-norm of the matrix operator

$$\mathbf{A} = \begin{pmatrix} 1 & 1 \\ 1 & 2 \end{pmatrix}, \tag{6.379}$$

and give a geometric interpretation.

We have det **A** $= 1$, so the mapping is in some sense area- and orientation-preserving; however, the notion of area in a space with a 4-norm has not been well established. It is easy to show that Eq. (6.378) is equivalent to

$$||\mathbf{A}||_p = \sup ||\mathbf{A}x||_p, \qquad ||x||_p = 1. \tag{6.380}$$

Now, for the 4-norm, we have

$$||x||_4 = (x_1^4 + x_2^4)^{1/4}. \tag{6.381}$$

Because we wish to consider $||x||_4 = 1$, we can thus restrict attention to the action of **A** on points for which

$$x_1^4 + x_2^4 = 1. \tag{6.382}$$

Now we have

$$\mathbf{A}x = \begin{pmatrix} 1 & 1 \\ 1 & 2 \end{pmatrix} \begin{pmatrix} x_1 \\ x_2 \end{pmatrix} = \begin{pmatrix} x_1 + x_2 \\ x_1 + 2x_2 \end{pmatrix}. \tag{6.383}$$

And we have then

$$||\mathbf{A}x||_4 = \left((x_1 + x_2)^4 + (x_1 + 2x_2)^4 \right)^{1/4}. \tag{6.384}$$

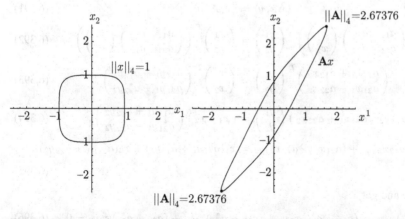

Figure 6.20. Locus of points with $||x||_4 = 1$ and their alibi mapping under an area- and orientation-preserving linear map.

Maximizing $||\mathbf{A}x||_4$ subject to the constraint of $||x||_4 = 1$ is a problem for optimization via Lagrange multipliers (see Section 1.5.3). Doing so, we find

$$||\mathbf{A}||_4 = 2.67376. \tag{6.385}$$

We show in Figure 6.20 the locus of points for which $||x||_4 = 1$ and how they map under the action of $\mathbf{A}$. At the points in the mapped space marked with a dot, $(x_1, x_2)^T = (1.66, 2.57)^T$, $(-1.66, -2.56)^T$, the quantity $||\mathbf{A}x||_4$ takes on its maximum value of 2.67376. Had we been in a Euclidean space, the circle defined by $||x||_2 = 1$ would have mapped into an ellipse, and $||\mathbf{A}||_2 = 2.61803$ would be that given by Eq. (6.377).

6.7.2 Adjoint

The operator $\mathbf{L}^*$ is the *adjoint* of the operator $\mathbf{L}$ if

$$\langle \mathbf{L}x, y \rangle = \langle x, \mathbf{L}^*y \rangle. \tag{6.386}$$

If $\mathbf{L}^* = \mathbf{L}$, the operator is self-adjoint.

EXAMPLE 6.41

Find the adjoint of the real matrix $\mathbf{A} : \mathbb{R}^2 \to \mathbb{R}^2$, where

$$\mathbf{A} = \begin{pmatrix} a_{11} & a_{12} \\ a_{21} & a_{22} \end{pmatrix}. \tag{6.387}$$

Assume $a_{11}, a_{12}, a_{21}, a_{22}$ are known constants.

Let the adjoint of $\mathbf{A}$ be

$$\mathbf{A}^* = \begin{pmatrix} a_{11}^* & a_{12}^* \\ a_{21}^* & a_{22}^* \end{pmatrix}. \tag{6.388}$$

Here the starred quantities are to be determined. We also have for x and y:

$$x = \begin{pmatrix} x_1 \\ x_2 \end{pmatrix}, \tag{6.389}$$

$$y = \begin{pmatrix} y_1 \\ y_2 \end{pmatrix}. \tag{6.390}$$

We take Eq. (6.386) and expand:

$$\langle \mathbf{A}x, y \rangle = \langle x, \mathbf{A}^*y \rangle, \tag{6.391}$$

$$\left(\begin{pmatrix} a_{11} & a_{12} \\ a_{21} & a_{22} \end{pmatrix} \begin{pmatrix} x_1 \\ x_2 \end{pmatrix} \right)^T \begin{pmatrix} y_1 \\ y_2 \end{pmatrix} = \begin{pmatrix} x_1 \\ x_2 \end{pmatrix}^T \left(\begin{pmatrix} a_{11}^* & a_{12}^* \\ a_{21}^* & a_{22}^* \end{pmatrix} \begin{pmatrix} y_1 \\ y_2 \end{pmatrix} \right), \tag{6.392}$$

$$\begin{pmatrix} a_{11}x_1 + a_{12}x_2 \\ a_{21}x_1 + a_{22}x_2 \end{pmatrix}^T \begin{pmatrix} y_1 \\ y_2 \end{pmatrix} = \begin{pmatrix} x_1 \\ x_2 \end{pmatrix}^T \begin{pmatrix} a_{11}^*y_1 + a_{12}^*y_2 \\ a_{21}^*y_1 + a_{22}^*y_2 \end{pmatrix}, \tag{6.393}$$

$$\begin{pmatrix} a_{11}x_1 + a_{12}x_2 & a_{21}x_1 + a_{22}x_2 \end{pmatrix} \begin{pmatrix} y_1 \\ y_2 \end{pmatrix} = \begin{pmatrix} x_1 & x_2 \end{pmatrix} \begin{pmatrix} a_{11}^*y_1 + a_{12}^*y_2 \\ a_{21}^*y_1 + a_{22}^*y_2 \end{pmatrix}, \tag{6.394}$$

$$(a_{11}x_1 + a_{12}x_2)y_1 + (a_{21}x_1 + a_{22}x_2)y_2 = x_1(a_{11}^*y_1 + a_{12}^*y_2) + x_2(a_{21}^*y_1 + a_{22}^*y_2). \tag{6.395}$$

Rearrange and get

$$(a_{11} - a_{11}^*)x_1y_1 + (a_{21} - a_{12}^*)x_1y_2 + (a_{12} - a_{21}^*)x_2y_1 + (a_{22} - a_{22}^*)x_2y_2 = 0. \tag{6.396}$$

Because this must hold for any x_1, x_2, y_1, y_2, we have

$$a_{11}^* = a_{11}, \tag{6.397}$$

$$a_{12}^* = a_{21}, \tag{6.398}$$

$$a_{21}^* = a_{12}, \tag{6.399}$$

$$a_{22}^* = a_{22}. \tag{6.400}$$

Thus,

$$\mathbf{A}^* = \begin{pmatrix} a_{11} & a_{21}, \\ a_{12} & a_{22} \end{pmatrix}, \tag{6.401}$$

$$= \mathbf{A}^T. \tag{6.402}$$

Thus, a symmetric matrix is self-adjoint. This result is easily extended to complex matrices $\mathbf{A} : \mathbb{C}^N \to \mathbb{C}^M$:

$$\mathbf{A}^* = \overline{\mathbf{A}}^T = \mathbf{A}^H. \tag{6.403}$$

EXAMPLE 6.42

Find the adjoint of the differential operator $\mathbf{L} : \mathbb{X} \to \mathbb{X}$, where

$$\mathbf{L} = \frac{d^2}{ds^2} + \frac{d}{ds}, \tag{6.404}$$

and $\mathbb{X}$ is the subspace of $\mathbb{L}_2[0, 1]$ with $x(0) = x(1) = 0$ if $x \in \mathbb{X}$.

Using integration by parts on the inner product,

$$\langle \mathbf{L}x, y \rangle = \int_0^1 (x''(s) + x'(s))\, y(s)\, ds, \tag{6.405}$$

$$= \int_0^1 x''(s)y(s)\, ds + \int_0^1 x'(s)y(s)\, ds, \tag{6.406}$$

$$= \left(x'(1)y(1) - x'(0)y(0) - \int_0^1 x'(s)y'(s)\, ds \right)$$

$$+ \left(\underbrace{x(1)}_{=0}\, y(1) - \underbrace{x(0)}_{=0}\, y(0) - \int_0^1 x(s)y'(s)\, ds \right), \tag{6.407}$$

$$= x'(1)y(1) - x'(0)y(0) - \int_0^1 x'(s)y'(s)\, ds - \int_0^1 x(s)y'(s)\, ds, \tag{6.408}$$

$$= x'(1)y(1) - x'(0)y(0) - \left(\underbrace{x(1)}_{=0}\, y'(1) - \underbrace{x(0)}_{=0}\, y'(0) - \int_0^1 x(s)y''(s)\, ds \right)$$

$$- \int_0^1 x(s)y'(s)\, ds, \tag{6.409}$$

$$= x'(1)y(1) - x'(0)y(0) + \int_0^1 x(s)y''(s)\, ds - \int_0^1 x(s)y'(s)\, ds, \tag{6.410}$$

$$= x'(1)y(1) - x'(0)y(0) + \int_0^1 x(s)\, (y''(s) - y'(s))\, ds. \tag{6.411}$$

This maintains the form of an inner product in $\mathbb{L}_2[0,1]$ if we require $y(0) = y(1) = 0$; doing this, we get

$$\langle \mathbf{L}x, y \rangle = \int_0^1 x(s) \left(y''(s) - y'(s) \right) \, ds = \langle x, \mathbf{L}^* y \rangle. \tag{6.412}$$

Functions that do not vanish at the boundaries do not form a vector space. We see by inspection that the adjoint operator is

$$\mathbf{L}^* = \frac{d^2}{ds^2} - \frac{d}{ds}. \tag{6.413}$$

Because the adjoint operator is not equal to the operator itself, the operator is not self-adjoint.

EXAMPLE 6.43

Find the adjoint of the differential operator $\mathbf{L} : \mathbb{X} \to \mathbb{X}$, where $\mathbf{L} = d^2/ds^2$, and $\mathbb{X}$ is the subspace of $\mathbb{L}_2[0,1]$ with $x(0) = x(1) = 0$ if $x \in \mathbb{X}$.

Using integration by parts on the inner product:

$$\langle \mathbf{L}x, y \rangle = \int_0^1 x''(s)y(s) \, ds, \tag{6.414}$$

$$= x'(1)y(1) - x'(0)y(0) - \int_0^1 x'(s)y'(s) \, ds, \tag{6.415}$$

$$= x'(1)y(1) - x'(0)y(0) - \left(\underbrace{x(1)}_{=0} y'(1) - \underbrace{x(0)}_{=0} y'(0) - \int_0^1 x(s)y''(s) \, ds \right), \tag{6.416}$$

$$= x'(1)y(1) - x'(0)y(0) + \int_0^1 x(s)y''(s) \, ds. \tag{6.417}$$

If we require $y(0) = y(1) = 0$, then

$$\langle \mathbf{L}x, y \rangle = \int_0^1 x(s)y''(s) \, dt = \langle x, \mathbf{L}^* y \rangle. \tag{6.418}$$

In this case, we see that $\mathbf{L} = \mathbf{L}^*$, so the operator is self-adjoint.

EXAMPLE 6.44

Find the adjoint of the linear integral operator $\mathbf{L} : \mathbb{L}_2[a,b] \to \mathbb{L}_2[a,b]$, where

$$\mathbf{L}x = \int_a^b K(s,t)x(s) \, ds. \tag{6.419}$$

The inner product is

$$\langle \mathbf{L}x, y \rangle = \int_a^b \left(\int_a^b K(s,t)x(s) \, ds \right) y(t) \, dt, \tag{6.420}$$

$$= \int_a^b \int_a^b K(s,t)x(s)y(t) \, ds \, dt, \tag{6.421}$$

$$= \int_a^b \int_a^b x(s)K(s,t)y(t) \, dt \, ds, \tag{6.422}$$

$$= \int_a^b x(s) \left(\int_a^b K(s,t) y(t) \, dt \right) \, ds, \tag{6.423}$$

$$= \langle x, \mathbf{L}^* y \rangle, \tag{6.424}$$

where

$$\mathbf{L}^* y = \int_a^b K(s,t) y(t) \, dt, \tag{6.425}$$

or equivalently,

$$\mathbf{L}^* y = \int_a^b K(t,s) y(s) \, ds. \tag{6.426}$$

In the definition of $\mathbf{L}x$, the second argument of K is a free variable, whereas in the consequent definition of $\mathbf{L}^* y$, the first argument of K is a free argument. So in general, the operator and its adjoint are different. However, if $K(s,t) = K(t,s)$, the operator is self-adjoint. That is, a symmetric kernel yields a self-adjoint operator. The theory of integral equations with symmetric kernels is sometimes known as *Hilbert-Schmidt theory*. This will be considered briefly in the upcoming Chapter 8.

Properties of adjoint operators include the following:

$$||\mathbf{L}^*|| = ||\mathbf{L}||, \tag{6.427}$$

$$(\mathbf{L}_1 + \mathbf{L}_2)^* = \mathbf{L}_1^* + \mathbf{L}_2^*, \tag{6.428}$$

$$(\alpha \mathbf{L})^* = \overline{\alpha} \mathbf{L}^*, \tag{6.429}$$

$$(\mathbf{L}_1 \mathbf{L}_2)^* = \mathbf{L}_2^* \mathbf{L}_1^*, \tag{6.430}$$

$$(\mathbf{L}^*)^* = \mathbf{L}. \tag{6.431}$$

And if $\mathbf{L}^{-1}$ exists, we have

$$(\mathbf{L}^{-1})^* = (\mathbf{L}^*)^{-1}. \tag{6.432}$$

6.7.3 Inverse

Consider next in more detail the inverse of a linear operator. Let

$$\mathbf{L}x = y. \tag{6.433}$$

If an *inverse* of $\mathbf{L}$ exists, which we will call $\mathbf{L}^{-1}$, then

$$x = \mathbf{L}^{-1} y. \tag{6.434}$$

Using Eq. (6.434) to eliminate x in favor of y in Eq. (6.433), we get

$$\mathbf{L} \underbrace{\mathbf{L}^{-1} y}_{=x} = y, \tag{6.435}$$

so that

$$\mathbf{L}\mathbf{L}^{-1} = \mathbf{I}. \tag{6.436}$$

A property of the inverse operator is

$$(\mathbf{L}_a \mathbf{L}_b)^{-1} = \mathbf{L}_b^{-1} \mathbf{L}_a^{-1}. \tag{6.437}$$

Let us show this. Say

$$y = \mathbf{L}_a \mathbf{L}_b x. \tag{6.438}$$

Then

$$\mathbf{L}_a^{-1} y = \mathbf{L}_b x, \tag{6.439}$$

$$\mathbf{L}_b^{-1} \mathbf{L}_a^{-1} y = x. \tag{6.440}$$

Consequently, we see that

$$(\mathbf{L}_a \mathbf{L}_b)^{-1} = \mathbf{L}_b^{-1} \mathbf{L}_a^{-1}. \tag{6.441}$$

EXAMPLE 6.45

Let **L** be the operator defined by

$$\mathbf{L}x = \left(\frac{d^2}{dt^2} + k^2 \right) x(t) = f(t), \tag{6.442}$$

where x belongs to the subspace of $\mathbb{L}_2[0, \pi]$ with $x(0) = a$ and $x(\pi) = b$. Show that the inverse operator $\mathbf{L}^{-1}$ is given by

$$x(t) = \mathbf{L}^{-1} f(t) = b \frac{\partial g}{\partial \tau}(\pi, t) - a \frac{\partial g}{\partial \tau}(0, t) + \int_0^\pi g(\tau, t) f(\tau) \, d\tau, \tag{6.443}$$

where $g(\tau, t)$ is the Green's function (see Section 4.3.3).

From the definition of **L** and $\mathbf{L}^{-1}$ in Eqs. (6.442,6.443), we get

$$\mathbf{L}^{-1}(\mathbf{L}x) = b \frac{\partial g}{\partial \tau}(\pi, t) - a \frac{\partial g}{\partial \tau}(0, t) + \int_0^\pi g(\tau, t) \underbrace{\left(\frac{d^2 x(\tau)}{d\tau^2} + k^2 x(\tau) \right)}_{=f(\tau)} \, d\tau. \tag{6.444}$$

Using integration by parts and the property that $g(0, t) = g(\pi, t) = 0$, the integral on the right side of Eq. (6.444) can be simplified as

$$\int_0^\pi g(\tau, t) \underbrace{\left(\frac{d^2 x(\tau)}{d\tau^2} + k^2 x(\tau) \right)}_{=f(\tau)} \, d\tau = - \underbrace{x(\pi)}_{=b} \frac{\partial g}{\partial \tau}(\pi, t) + \underbrace{x(0)}_{=a} \frac{\partial g}{\partial \tau}(0, t)$$

$$+ \int_0^\pi x(\tau) \underbrace{\left(\frac{\partial^2 g}{\partial \tau^2} + k^2 g \right)}_{=\delta(t-\tau)} \, d\tau. \tag{6.445}$$

Because $x(0) = a$, $x(\pi) = b$, and

$$\frac{\partial^2 g}{\partial \tau^2} + k^2 g = \delta(t - \tau), \tag{6.446}$$

we have

$$\mathbf{L}^{-1}(\mathbf{L}x) = \int_0^\pi x(\tau) \delta(t - \tau) \, d\tau, \tag{6.447}$$

$$= x(t). \tag{6.448}$$

Thus, $\mathbf{L}^{-1}\mathbf{L} = \mathbf{I}$, proving the proposition.

It is easily shown for this problem that the Green's function is

$$g(\tau, t) = -\frac{\sin(k(\pi - \tau))\sin(kt)}{k\sin(k\pi)}, \qquad t \leq \tau, \tag{6.449}$$

$$= -\frac{\sin(k\tau)\sin(k(\pi - t))}{k\sin(k\pi)}, \qquad \tau \leq t, \tag{6.450}$$

so that we can write $x(t)$ explicitly in terms of the forcing function $f(t)$, including the inhomogeneous boundary conditions, as follows:

$$x(t) = \frac{b\sin(kt)}{\sin(k\pi)} + \frac{a\sin(k(\pi - t))}{\sin(k\pi)} \tag{6.451}$$

$$-\frac{\sin(k(\pi - t))}{k\sin(k\pi)} \int_0^t f(\tau)\sin(k\tau)\, d\tau - \frac{\sin(kt)}{k\sin(k\pi)} \int_t^\pi f(\tau)\sin(k(\pi - \tau))\, d\tau. \tag{6.452}$$

6.8 Eigenvalues and Eigenvectors

Let us consider here in a more formal fashion topics that have been previously introduced in Sections 2.1.5 and 4.4. If $\mathbf{L}$ is a linear operator, its eigenvalue problem consists of finding a nontrivial solution of the equation

$$\mathbf{L}e = \lambda e, \tag{6.453}$$

where e is called an *eigenvector* and λ an *eigenvalue*. We begin by presenting several relevant theorems valid for general linear operators. We first have the following:

Theorem: The eigenvalues of a linear operator and its adjoint are complex conjugates of each other.

Let us prove this. Let λ and λ^* be the eigenvalues of $\mathbf{L}$ and $\mathbf{L}^*$, respectively, and let e and e^* be the corresponding eigenvectors. Consider, then,

$$\langle \mathbf{L}e, e^* \rangle = \langle e, \mathbf{L}^* e^* \rangle, \tag{6.454}$$

$$\langle \lambda e, e^* \rangle = \langle e, \lambda^* e^* \rangle, \tag{6.455}$$

$$\overline{\lambda}\langle e, e^* \rangle = \lambda^* \langle e, e^* \rangle. \tag{6.456}$$

Because by definition eigenvectors are nontrivial, $\langle e, e^* \rangle \neq 0$, we deduce that

$$\overline{\lambda} = \lambda^*, \qquad \text{QED}. \tag{6.457}$$

We next consider the following:

Theorem: The eigenvalues of a self-adjoint linear operator are real.

The proof is as follows. Because the operator is self-adjoint, we have

$$\langle \mathbf{L}e, e \rangle = \langle e, \mathbf{L}e \rangle, \tag{6.458}$$

$$\langle \lambda e, e \rangle = \langle e, \lambda e \rangle, \tag{6.459}$$

$$\overline{\lambda}\langle e, e \rangle = \lambda \langle e, e \rangle, \tag{6.460}$$

$$\overline{\lambda} = \lambda, \tag{6.461}$$

$$\lambda_R - i\lambda_I = \lambda_R + i\lambda_I; \qquad \lambda_R, \lambda_I \in \mathbb{R}^2, \tag{6.462}$$

$$\lambda_R = \lambda_R, \tag{6.463}$$

$$-\lambda_I = \lambda_I, \tag{6.464}$$

$$\lambda_I = 0, \qquad \text{QED.} \tag{6.465}$$

Once again, because eigenvectors are nontrivial, $\langle e, e \rangle \neq 0$; thus, the division can be performed. The only way a complex number can equal its conjugate is if its imaginary part is zero; consequently, the eigenvalue must be strictly real.

Next, examine the following:

Theorem: The eigenvectors of a self-adjoint linear operator corresponding to distinct eigenvalues are orthogonal.

The proof is as follows. Let λ_i and λ_j be two distinct, $\lambda_i \neq \lambda_j$, real, $\lambda_i, \lambda_j \in \mathbb{R}^1$, eigenvalues of the self-adjoint operator $\mathbf{L}$, and let e_i and e_j be the corresponding eigenvectors. Then,

$$\langle \mathbf{L}e_i, e_j \rangle = \langle e_i, \mathbf{L}e_j \rangle, \tag{6.466}$$

$$\langle \lambda_i e_i, e_j \rangle = \langle e_i, \lambda_j e_j \rangle, \tag{6.467}$$

$$\lambda_i \langle e_i, e_j \rangle = \lambda_j \langle e_i, e_j \rangle, \tag{6.468}$$

$$\langle e_i, e_j \rangle \, (\lambda_i - \lambda_j) = 0, \tag{6.469}$$

$$\langle e_i, e_j \rangle = 0, \qquad \text{QED.} \tag{6.470}$$

because $\lambda_i \neq \lambda_j$.

Lastly, consider the following:

Theorem: The eigenvectors of any self-adjoint operator on vectors of a finite-dimensional vector space constitute a basis for the space.

Proof is left to the reader.

The following conditions are sufficient for the eigenvectors in an infinite-dimensional Hilbert space to be form a complete basis:

- the operator must be self-adjoint,
- the operator is defined on a finite domain,
- the operator has no singularities in its domain.

If the operator is not self-adjoint, Friedman (1956, Chapter 4), discusses how the eigenfunctions of the adjoint operator can be used to obtain the coefficients α_k on the eigenfunctions of the operator.

EXAMPLE 6.46

For $x \in \mathbb{R}^2, \mathbf{A} : \mathbb{R}^2 \to \mathbb{R}^2$, find the eigenvalues and eigenvectors of

$$\mathbf{A} = \begin{pmatrix} 2 & 1 \\ 1 & 2 \end{pmatrix}. \tag{6.471}$$

The eigenvalue problem is

$$\mathbf{A}x = \lambda x, \tag{6.472}$$

which can be written as

$$\mathbf{A}x = \lambda \mathbf{I}x, \tag{6.473}$$

$$(\mathbf{A} - \lambda \mathbf{I})x = 0, \tag{6.474}$$

where the identity matrix is

$$\mathbf{I} = \begin{pmatrix} 1 & 0 \\ 0 & 1 \end{pmatrix}. \tag{6.475}$$

If we write

$$x = \begin{pmatrix} x_1 \\ x_2 \end{pmatrix}, \tag{6.476}$$

then

$$\begin{pmatrix} 2 - \lambda & 1 \\ 1 & 2 - \lambda \end{pmatrix} \begin{pmatrix} x_1 \\ x_2 \end{pmatrix} = \begin{pmatrix} 0 \\ 0 \end{pmatrix}. \tag{6.477}$$

By Cramer's rule, we could say

$$x_1 = \frac{\det \begin{pmatrix} 0 & 1 \\ 0 & 2 - \lambda \end{pmatrix}}{\det \begin{pmatrix} 2 - \lambda & 1 \\ 1 & 2 - \lambda \end{pmatrix}} = \frac{0}{\det \begin{pmatrix} 2 - \lambda & 1 \\ 1 & 2 - \lambda \end{pmatrix}}, \tag{6.478}$$

$$x_2 = \frac{\det \begin{pmatrix} 2 - \lambda & 0 \\ 1 & 0 \end{pmatrix}}{\det \begin{pmatrix} 2 - \lambda & 1 \\ 1 & 2 - \lambda \end{pmatrix}} = \frac{0}{\det \begin{pmatrix} 2 - \lambda & 1 \\ 1 & 2 - \lambda \end{pmatrix}}. \tag{6.479}$$

An obvious but uninteresting solution is the trivial solution $x_1 = 0, x_2 = 0$. Nontrivial solutions of x_1 and x_2 can be obtained only if

$$\begin{vmatrix} 2 - \lambda & 1 \\ 1 & 2 - \lambda \end{vmatrix} = 0, \tag{6.480}$$

which gives the characteristic equation

$$(2 - \lambda)^2 - 1 = 0. \tag{6.481}$$

Solutions are $\lambda_1 = 1$ and $\lambda_2 = 3$.

The eigenvector corresponding to each eigenvalue is found in the following manner. The eigenvalue is substituted in Eq. (6.477). A dependent set of equations in x_1 and x_2 is obtained. The eigenvector solution is thus not unique.

For $\lambda = 1$, Eq. (6.477) gives

$$\begin{pmatrix} 2 - 1 & 1 \\ 1 & 2 - 1 \end{pmatrix} \begin{pmatrix} x_1 \\ x_2 \end{pmatrix} = \begin{pmatrix} 1 & 1 \\ 1 & 1 \end{pmatrix} \begin{pmatrix} x_1 \\ x_2 \end{pmatrix} = \begin{pmatrix} 0 \\ 0 \end{pmatrix}, \tag{6.482}$$

which are the two identical equations,

$$x_1 + x_2 = 0. \tag{6.483}$$

If we choose $x_1 = \gamma$, then $x_2 = -\gamma$. So the eigenvector corresponding to $\lambda = 1$ is

$$e_1 = \gamma \begin{pmatrix} 1 \\ -1 \end{pmatrix}. \tag{6.484}$$

Because the magnitude of an eigenvector is arbitrary, we can take $\gamma = 1$ and thus have

$$e_1 = \begin{pmatrix} 1 \\ -1 \end{pmatrix}. \tag{6.485}$$

For $\lambda = 3$, the equations are

$$\begin{pmatrix} 2-3 & 1 \\ 1 & 2-3 \end{pmatrix} \begin{pmatrix} x_1 \\ x_2 \end{pmatrix} = \begin{pmatrix} -1 & 1 \\ 1 & -1 \end{pmatrix} \begin{pmatrix} x_1 \\ x_2 \end{pmatrix} = \begin{pmatrix} 0 \\ 0 \end{pmatrix}, \tag{6.486}$$

which yield the two identical equations,

$$- x_1 + x_2 = 0. \tag{6.487}$$

This yields an eigenvector of

$$e_2 = \beta \begin{pmatrix} 1 \\ 1 \end{pmatrix}. \tag{6.488}$$

We take $\beta = 1$, so that

$$e_2 = \begin{pmatrix} 1 \\ 1 \end{pmatrix}. \tag{6.489}$$

Comments:

- Because the real matrix is symmetric (thus, self-adjoint), the eigenvalues are real, and the eigenvectors are orthogonal.
- We have actually solved for the *right eigenvectors*. This is the usual set of eigenvectors. The *left eigenvectors* can be found from $x^H \mathbf{A} = x^H \mathbf{I} \lambda$. Because here $\mathbf{A}$ is equal to its conjugate transpose, $x^H \mathbf{A} = \mathbf{A}x$, the left eigenvectors are the same as the right eigenvectors. More generally, we can say the left eigenvectors of an operator are the right eigenvectors of the adjoint of that operator, $\mathbf{A}^H$.
- Multiplication of an eigenvector by any scalar is also an eigenvector.
- The normalized eigenvectors are

$$e_1 = \begin{pmatrix} \frac{1}{\sqrt{2}} \\ -\frac{1}{\sqrt{2}} \end{pmatrix}, \qquad e_2 = \begin{pmatrix} \frac{1}{\sqrt{2}} \\ \frac{1}{\sqrt{2}} \end{pmatrix}. \tag{6.490}$$

- A natural way to express a vector is on orthonormal basis as given here

$$x = \alpha_1 \begin{pmatrix} \frac{1}{\sqrt{2}} \\ -\frac{1}{\sqrt{2}} \end{pmatrix} + \alpha_2 \begin{pmatrix} \frac{1}{\sqrt{2}} \\ \frac{1}{\sqrt{2}} \end{pmatrix} = \underbrace{\begin{pmatrix} \frac{1}{\sqrt{2}} & \frac{1}{\sqrt{2}} \\ -\frac{1}{\sqrt{2}} & \frac{1}{\sqrt{2}} \end{pmatrix}}_{=\mathbf{Q}} \begin{pmatrix} \alpha_1 \\ \alpha_2 \end{pmatrix}. \tag{6.491}$$

- The set of orthonormalized eigenvectors forms an orthogonal matrix $\mathbf{Q}$; see Section 2.1.2 or the upcoming Section 7.7. Note that $\mathbf{Q}$ has determinant of unity, so it is a rotation. As suggested by Eq. (2.54), the angle of rotation here is $\alpha = \sin^{-1}(-1/\sqrt{2}) = -\pi/4$.

EXAMPLE 6.47

For $x \in \mathbb{C}^2$, $\mathbf{A} : \mathbb{C}^2 \rightarrow \mathbb{C}^2$, find the eigenvalues and eigenvectors of

$$\mathbf{A} = \begin{pmatrix} 0 & -2 \\ 2 & 0 \end{pmatrix}. \tag{6.492}$$

This matrix is antisymmetric. We find the eigensystem by solving

$$(\mathbf{A} - \lambda \mathbf{I}) e = 0. \tag{6.493}$$

The characteristic equation that results is

$$\lambda^2 + 4 = 0, \tag{6.494}$$

which has two imaginary roots that are complex conjugates: $\lambda_1 = 2i$, $\lambda_2 = -2i$. The corresponding eigenvectors are

$$e_1 = \alpha \begin{pmatrix} i \\ 1 \end{pmatrix}, \qquad e_2 = \beta \begin{pmatrix} -i \\ 1 \end{pmatrix}, \tag{6.495}$$

where α and β are arbitrary scalars. Let us take $\alpha = -i$, $\beta = 1$, so

$$e_1 = \begin{pmatrix} 1 \\ -i \end{pmatrix}, \qquad e_2 = \begin{pmatrix} -i \\ 1 \end{pmatrix}. \tag{6.496}$$

Note that

$$\langle e_1, e_2 \rangle = e_1{}^H e_2 = \begin{pmatrix} 1 & i \end{pmatrix} \begin{pmatrix} -i \\ 1 \end{pmatrix} = (-i) + i = 0, \tag{6.497}$$

so this is an orthogonal set of vectors, even though the generating matrix was not self-adjoint. We can render it orthonormal by scaling by the magnitude of each eigenvector. The orthonormal eigenvector set is

$$e_1 = \begin{pmatrix} \frac{1}{\sqrt{2}} \\ \frac{-i}{\sqrt{2}} \end{pmatrix}, \qquad e_2 = \begin{pmatrix} -\frac{i}{\sqrt{2}} \\ \frac{1}{\sqrt{2}} \end{pmatrix}. \tag{6.498}$$

These two orthonormalized vectors can form a matrix $\mathbf{Q}$:

$$\mathbf{Q} = \begin{pmatrix} \frac{1}{\sqrt{2}} & -\frac{i}{\sqrt{2}} \\ -\frac{i}{\sqrt{2}} & \frac{1}{\sqrt{2}} \end{pmatrix}. \tag{6.499}$$

It is easy to check that $\|\mathbf{Q}\|_2 = 1$ and $\det \mathbf{Q} = 1$, so it is a rotation. However, for the complex basis vectors, it is difficult to define an angle of rotation in the traditional sense. Our special choices of α and β were actually made to ensure $\det \mathbf{Q} = 1$.

EXAMPLE 6.48

For $x \in \mathbb{C}^2$, $\mathbf{A} : \mathbb{C}^2 \to \mathbb{C}^2$, find the eigenvalues and eigenvectors of

$$\mathbf{A} = \begin{pmatrix} 1 & -1 \\ 0 & 1 \end{pmatrix}. \tag{6.500}$$

This matrix is asymmetric. We find the eigensystem by solving

$$(\mathbf{A} - \lambda \mathbf{I}) e = 0. \tag{6.501}$$

The characteristic equation that results is

$$(1 - \lambda)^2 = 0, \tag{6.502}$$

which has repeated roots $\lambda = 1$, $\lambda = 1$. For this eigenvalue, there is only one ordinary eigenvector

$$e = \alpha \begin{pmatrix} 1 \\ 0 \end{pmatrix}. \tag{6.503}$$

We take arbitrarily $\alpha = 1$ so that

$$e = \begin{pmatrix} 1 \\ 0 \end{pmatrix}. \tag{6.504}$$

We can however find a *generalized eigenvector* g such that

$$(\mathbf{A} - \lambda \mathbf{I}) g = e. \tag{6.505}$$

Note then that

$$(\mathbf{A} - \lambda\mathbf{I})(\mathbf{A} - \lambda\mathbf{I})g = (\mathbf{A} - \lambda\mathbf{I})e, \tag{6.506}$$

$$(\mathbf{A} - \lambda\mathbf{I})^2 g = 0. \tag{6.507}$$

Now

$$(\mathbf{A} - \lambda\mathbf{I}) = \begin{pmatrix} 0 & -1 \\ 0 & 0 \end{pmatrix}. \tag{6.508}$$

So with $g = (\beta, \gamma)^T$, take from Eq. (6.505)

$$\underbrace{\begin{pmatrix} 0 & -1 \\ 0 & 0 \end{pmatrix}}_{\mathbf{A}-\lambda\mathbf{I}} \underbrace{\begin{pmatrix} \beta \\ \gamma \end{pmatrix}}_{g} = \underbrace{\begin{pmatrix} 1 \\ 0 \end{pmatrix}}_{e}. \tag{6.509}$$

We get a solution if $\beta \in \mathbb{R}^1, \gamma = -1$. That is,

$$g = \begin{pmatrix} \beta \\ -1 \end{pmatrix}. \tag{6.510}$$

Take $\beta = 0$ to give an orthogonal generalized eigenvector. So

$$g = \begin{pmatrix} 0 \\ -1 \end{pmatrix}. \tag{6.511}$$

The ordinary eigenvector and the generalized eigenvector combine to form a basis, in this case, an orthonormal basis.

More properly, we should distinguish the generalized eigenvector we have found as a *generalized eigenvector in the first sense*. There is another common, unrelated generalization in usage, which we study later in Section 7.4.2.

EXAMPLE 6.49

For $x \in \mathbb{C}^2, \mathbf{A} : \mathbb{C}^2 \to \mathbb{C}^2$, find the eigenvalues, right eigenvectors, and left eigenvectors if

$$\mathbf{A} = \begin{pmatrix} 1 & 2 \\ -3 & 1 \end{pmatrix}. \tag{6.512}$$

The matrix is asymmetric. The right eigenvector problem is the usual

$$\mathbf{A}e_R = \lambda\mathbf{I}e_R. \tag{6.513}$$

The characteristic equation is

$$(1 - \lambda)^2 + 6 = 0, \tag{6.514}$$

which has complex roots. The eigensystem is

$$\lambda_1 = 1 - \sqrt{6}i, \ e_{1R} = \begin{pmatrix} \sqrt{\frac{2}{3}}i \\ 1 \end{pmatrix}, \qquad \lambda_2 = 1 + \sqrt{6}i, \ e_{2R} = \begin{pmatrix} -\sqrt{\frac{2}{3}}i \\ 1 \end{pmatrix}. \tag{6.515}$$

As the operator is not self-adjoint, we are not guaranteed real eigenvalues. The right eigenvectors are not orthogonal as $e_{1R}{}^H e_{2R} = 1/3$.

For the left eigenvectors, we have

$$e_L^H \mathbf{A} = e_L^H \mathbf{I}\lambda. \tag{6.516}$$

We can put this in a slightly more standard form by taking the conjugate transpose of both sides:

$$(e_L^H \mathbf{A})^H = (e_L^H \mathbf{I}\lambda)^H, \tag{6.517}$$

$$\mathbf{A}^H e_L = (\mathbf{I}\lambda)^H e_L, \tag{6.518}$$

$$\mathbf{A}^* e_L = \mathbf{I}\bar{\lambda} e_L, \tag{6.519}$$

$$\mathbf{A}^* e_L = \mathbf{I}\lambda^* e_L. \tag{6.520}$$

So the left eigenvectors of $\mathbf{A}$ are the right eigenvectors of the adjoint of $\mathbf{A}$. Now we have

$$\mathbf{A}^H = \begin{pmatrix} 1 & -3 \\ 2 & 1 \end{pmatrix}. \tag{6.521}$$

The resulting eigensystem is

$$\lambda_1^* = 1 + \sqrt{6}i, \quad e_{1L} = \begin{pmatrix} \sqrt{\frac{3}{2}}i \\ 1 \end{pmatrix}, \qquad \lambda_2^* = 1 - \sqrt{6}i, \quad e_{2L} = \begin{pmatrix} -\sqrt{\frac{3}{2}}i \\ 1 \end{pmatrix}. \tag{6.522}$$

In addition to being complex conjugates of themselves, which does not hold for general complex matrices, the eigenvalues of the adjoint are complex conjugates of those of the original matrix, $\lambda^* = \bar{\lambda}$. The left eigenvectors are not orthogonal as $e_{1L}{}^H e_{2L} = -1/2$. It is easily shown by taking the conjugate transpose of the adjoint eigenvalue problem, however, that

$$e_L^H \mathbf{A} = e_L^H \lambda, \tag{6.523}$$

as desired. The eigenvalues for both the left and right eigensystems are the same.

EXAMPLE 6.50
Consider a small change from the previous example. For $x \in \mathbb{C}^2$, $\mathbf{A} : \mathbb{C}^2 \to \mathbb{C}^2$, find the eigenvalues, right eigenvectors, and left eigenvectors if

$$\mathbf{A} = \begin{pmatrix} 1 & 2 \\ -3 & 1+i \end{pmatrix}. \tag{6.524}$$

The right eigenvector problem is the usual

$$\mathbf{A} e_R = \lambda \mathbf{I} e_R. \tag{6.525}$$

The characteristic equation is

$$\lambda^2 - (2+i)\lambda + (7+i) = 0, \tag{6.526}$$

which has complex roots. The eigensystem is

$$\lambda_1 = 1 - 2i, \quad e_{1R} = \begin{pmatrix} i \\ 1 \end{pmatrix}, \qquad \lambda_2 = 1 + 3i, \quad e_{2R} = \begin{pmatrix} -2i \\ 3 \end{pmatrix}. \tag{6.527}$$

As the operator is not self-adjoint, we are not guaranteed real eigenvalues. The right eigenvectors are not orthogonal as $e_{1R}{}^H e_{2R} = 1 \neq 0$.

For the left eigenvectors, we solve the corresponding right eigensystem for the adjoint of $\mathbf{A}$, which is $\mathbf{A}^* = \mathbf{A}^H$:

$$\mathbf{A}^H = \begin{pmatrix} 1 & -3 \\ 2 & 1-i \end{pmatrix}. \tag{6.528}$$

The eigenvalue problem is $\mathbf{A}^H e_L = \lambda^* e_L$. The eigensystem is

$$\lambda_1^* = 1 + 2i, \quad e_{1L} = \begin{pmatrix} 3i \\ 2 \end{pmatrix}, \tag{6.529}$$

$$\lambda_2^* = 1 - 3i, \quad e_{2L} = \begin{pmatrix} -i \\ 1 \end{pmatrix}. \tag{6.530}$$

Here the eigenvalues λ_1^*, λ_2^* have no relation to each other, but they are complex conjugates of the eigenvalues, λ_1, λ_2, of the right eigenvalue problem of the original matrix. The left eigenvectors are not orthogonal as $e_{1L}{}^H e_{2L} = -1$. It is easily shown, however, that

$$e_L^H \mathbf{A} = e_L^H \lambda \mathbf{I}, \tag{6.531}$$

as desired.

EXAMPLE 6.51

For $x \in \mathbb{R}^3, \mathbf{A} : \mathbb{R}^3 \to \mathbb{R}^3$, find the eigenvalues and eigenvectors of

$$\mathbf{A} = \begin{pmatrix} 2 & 0 & 0 \\ 0 & 1 & 1 \\ 0 & 1 & 1 \end{pmatrix}. \tag{6.532}$$

From

$$\begin{vmatrix} 2-\lambda & 0 & 0 \\ 0 & 1-\lambda & 1 \\ 0 & 1 & 1-\lambda \end{vmatrix} = 0, \tag{6.533}$$

the characteristic equation is

$$(2 - \lambda)\left((1 - \lambda)^2 - 1\right) = 0. \tag{6.534}$$

The solutions are $\lambda = 0, 2, 2$. The second eigenvalue is of multiplicity 2. Next, we find the eigenvectors

$$e = \begin{pmatrix} x_1 \\ x_2 \\ x_3 \end{pmatrix}. \tag{6.535}$$

For $\lambda = 0$, the equations for the components of the eigenvectors are

$$\begin{pmatrix} 2 & 0 & 0 \\ 0 & 1 & 1 \\ 0 & 1 & 1 \end{pmatrix} \begin{pmatrix} x_1 \\ x_2 \\ x_3 \end{pmatrix} = \begin{pmatrix} 0 \\ 0 \\ 0 \end{pmatrix}, \tag{6.536}$$

$$2x_1 = 0, \tag{6.537}$$

$$x_2 + x_3 = 0, \tag{6.538}$$

from which

$$e_1 = \alpha \begin{pmatrix} 0 \\ 1 \\ -1 \end{pmatrix}. \tag{6.539}$$

For $\lambda = 2$, we have

$$\begin{pmatrix} 0 & 0 & 0 \\ 0 & -1 & 1 \\ 0 & 1 & -1 \end{pmatrix} \begin{pmatrix} x_1 \\ x_2 \\ x_3 \end{pmatrix} = \begin{pmatrix} 0 \\ 0 \\ 0 \end{pmatrix}. \tag{6.540}$$

This yields only

$$-x_2 + x_3 = 0. \tag{6.541}$$

We then see that the eigenvector

$$e = \begin{pmatrix} \beta \\ \gamma \\ \gamma \end{pmatrix} \tag{6.542}$$

satisfies Eq. (6.541). Here we have two free parameters, β and γ; we can thus extract two independent eigenvectors from this. For e_2, we arbitrarily take $\beta = 0$ and $\gamma = 1$ to get

$$e_2 = \begin{pmatrix} 0 \\ 1 \\ 1 \end{pmatrix}. \tag{6.543}$$

For e_3, we arbitrarily take $\beta = 1$ and $\gamma = 0$ to get

$$e_3 = \begin{pmatrix} 1 \\ 0 \\ 0 \end{pmatrix}. \tag{6.544}$$

In this case, e_1, e_2, e_3 are orthogonal, even though e_2 and e_3 correspond to the same eigenvalue.

EXAMPLE 6.52

For $y \in \mathbb{L}_2[0,1]$, find the eigenvalues and eigenvectors of $\mathbf{L} = -d^2/dt^2$, operating on functions that vanish at 0 and 1. Also, find $\|\mathbf{L}\|_2$.

This is a slight variant on a problem discussed earlier in Section 4.4.3. The eigenvalue problem is

$$\mathbf{L}y = -\frac{d^2y}{dt^2} = \lambda y, \qquad y(0) = y(1) = 0, \tag{6.545}$$

or

$$\frac{d^2y}{dt^2} + \lambda y = 0, \qquad y(0) = y(1) = 0. \tag{6.546}$$

The solution of this differential equation is

$$y(t) = a \sin \lambda^{1/2} t + b \cos \lambda^{1/2} t. \tag{6.547}$$

The boundary condition $y(0) = 0$ gives $b = 0$. The other condition $y(1) = 0$ gives $a \sin \lambda^{1/2} = 0$. A nontrivial solution can only be obtained if

$$\sin \lambda^{1/2} = 0. \tag{6.548}$$

There are an infinite but countable number of values of λ for which this can be satisfied. These are $\lambda_n = n^2 \pi^2$, $n = 1, 2, \ldots$. The eigenvectors (also called *eigenfunctions* in this case) $y_n(t)$, $n = 1, 2, \ldots$ are

$$y_n(t) = \sin n\pi t. \tag{6.549}$$

The differential operator is self-adjoint so that the eigenvalues are real, and the eigenfunctions are orthogonal.

Consider $||\mathbf{L}||_2$. Referring to the definition of Eq. (6.375), we see $||\mathbf{L}||_2 = \infty$, because by allowing y to be any eigenfunction, we have

$$\frac{||\mathbf{L}y||_2}{||y||_2} = \frac{||\lambda y||_2}{||y||_2}, \tag{6.550}$$

$$= \frac{|\lambda| \, ||y||_2}{||y||_2}, \tag{6.551}$$

$$= |\lambda|. \tag{6.552}$$

And because $\lambda = n^2\pi^2$, $n = 1, 2, \ldots, \infty$, the largest value that can be achieved by $||\mathbf{L}y||_2/||y||_2$ is infinite. Had the operator $\mathbf{L}$ not been self-adjoint, this analysis would require modification (see, e.g., Eq. (6.377) or the discussion on Section 7.4.1).

EXAMPLE 6.53

For $x \in \mathbb{L}_2[0, 1]$, and $\mathbf{L} = d^2/ds^2 + d/ds$ with $x(0) = x(1) = 0$, find the Fourier expansion of an arbitrary function $f(s)$ in terms of the eigenfunctions of $\mathbf{L}$. Then find the series representation of the "top hat" function

$$f(s) = H\left(s - \frac{1}{4}\right) - H\left(s - \frac{3}{4}\right). \tag{6.553}$$

This is related to the analysis first considered in Section 4.5, with important extensions due to the non-self-adjoint nature of $\mathbf{L}$. We thus seek expressions for α_n in

$$f(s) = \sum_{n=1}^{N} \alpha_n x_n(s). \tag{6.554}$$

Here $x_n(s)$ is an eigenfunction of $\mathbf{L}$. The eigenvalue problem is

$$\mathbf{L}x = \frac{d^2x}{ds^2} + \frac{dx}{ds} = \lambda x, \qquad x(0) = x(1) = 0. \tag{6.555}$$

It is easily shown that the eigenvalues of $\mathbf{L}$ are given by

$$\lambda_n = -\frac{1}{4} - n^2\pi^2, \qquad n = 1, 2, 3, \ldots, \tag{6.556}$$

where n is a positive integer, and the unnormalized eigenfunctions of $\mathbf{L}$ are

$$x_n(s) = e^{-s/2} \sin(n\pi s), \qquad n = 1, 2, 3, \ldots. \tag{6.557}$$

Although the eigenvalues happen to be real, the eigenfunctions are not orthogonal. We see this, for example, by forming $\langle x_1, x_2 \rangle$:

$$\langle x_1, x_2 \rangle = \int_0^1 \underbrace{e^{-s/2} \sin(\pi s)}_{=x_1(s)} \underbrace{e^{-s/2} \sin(2\pi s)}_{=x_2(s)} \, ds, \tag{6.558}$$

$$= \frac{4(1 + e)\pi^2}{e(1 + \pi^2)(1 + 9\pi^2)} \neq 0. \tag{6.559}$$

By using integration by parts, we find the adjoint operator to be

$$\mathbf{L}^* y = \frac{d^2 y}{ds^2} - \frac{dy}{ds} = \lambda^* y, \qquad y(0) = y(1) = 0. \tag{6.560}$$

We then find the eigenvalues of the adjoint operator to be the same as those of the operator (this is true because the eigenvalues are real; in general they are complex conjugates of one another):

$$\lambda_m^* = \overline{\lambda_m} = -\frac{1}{4} - m^2\pi^2, \qquad m = 1, 2, 3, \ldots, \tag{6.561}$$

where m is a positive integer.

The unnormalized eigenfunctions of the adjoint are

$$y_m(s) = e^{s/2} \sin(m\pi s), \qquad m = 1, 2, 3, \ldots. \tag{6.562}$$

Now, because by definition $\langle y_m, \mathbf{L}x_n \rangle = \langle \mathbf{L}^* y_m, x_n \rangle$, we have

$$\langle y_m, \mathbf{L}x_n \rangle - \langle \mathbf{L}^* y_m, x_n \rangle = 0, \tag{6.563}$$

$$\langle y_m, \lambda_n x_n \rangle - \langle \lambda_m^* y_m, x_n \rangle = 0, \tag{6.564}$$

$$\lambda_n \langle y_m, x_n \rangle - \overline{\lambda_m^*} \langle y_m, x_n \rangle = 0, \tag{6.565}$$

$$(\lambda_n - \lambda_m)\langle y_m, x_n \rangle = 0. \tag{6.566}$$

So, for $m = n$, we get $\langle y_n, x_n \rangle \neq 0$, and for $m \neq n$, we get $\langle y_m, x_n \rangle = 0$. Thus, we must have the so-called bi-orthogonality condition

$$\langle y_m, x_n \rangle = D_{mn}, \tag{6.567}$$

$$D_{mn} = 0 \qquad \text{if} \qquad m \neq n. \tag{6.568}$$

Here D_{mn} is a diagonal matrix that can be reduced to the identity matrix with proper normalization. If so normalized, y is a reciprocal basis to x, assuming both to be complete (see Section 6.5.4).

Now consider the following series of operations on the original form of the expansion we seek:

$$f(s) = \sum_{n=1}^{N} \alpha_n x_n(s), \tag{6.569}$$

$$\langle y_j(s), f(s) \rangle = \langle y_j(s), \sum_{n=1}^{N} \alpha_n x_n(s) \rangle, \tag{6.570}$$

$$\langle y_j(s), f(s) \rangle = \sum_{n=1}^{N} \alpha_n \langle y_j(s), x_n(s) \rangle, \tag{6.571}$$

$$\langle y_j(s), f(s) \rangle = \alpha_j \langle y_j(s), x_j(s) \rangle, \tag{6.572}$$

$$\alpha_j = \frac{\langle y_j(s), f(s) \rangle}{\langle y_j(s), x_j(s) \rangle}, \tag{6.573}$$

$$\alpha_n = \frac{\langle y_n(s), f(s) \rangle}{\langle y_n(s), x_n(s) \rangle}, \qquad n = 1, 2, 3, \ldots. \tag{6.574}$$

Now, in the case at hand, it is easily shown that

$$\langle y_n(s), x_n(s) \rangle = \frac{1}{2}, \qquad n = 1, 2, 3, \ldots, \tag{6.575}$$

so we have

$$\alpha_n = 2\langle y_n(s), f(s) \rangle. \tag{6.576}$$

The N-term approximate representation of $f(s)$ is thus given by

$$f(s) \approx \sum_{n=1}^{N} \underbrace{\left(2 \int_0^1 e^{t/2} \sin(n\pi t) f(t)\, dt \right)}_{=\alpha_n} \underbrace{e^{-s/2} \sin(n\pi s)}_{=x_n(s)}, \tag{6.577}$$

$$\approx 2 \int_0^1 e^{(t-s)/2} f(t) \sum_{n=1}^{N} \sin(n\pi t) \sin(n\pi s)\, dt, \tag{6.578}$$

$$\approx \int_0^1 e^{(t-s)/2} f(t) \sum_{n=1}^{N} \left(\cos(n\pi(s-t)) - \cos(n\pi(s+t)) \right) dt. \tag{6.579}$$

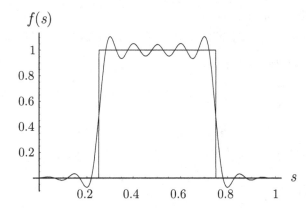

Figure 6.21. Twenty-term Fourier series approximation to a top hat function in terms of a nonorthogonal basis.

For the top hat function, a two-term expansion yields

$$f(s) \approx \underbrace{\frac{2\sqrt{2}e^{1/8}\left(-1 + 2\pi + e^{1/4}(1 + 2\pi)\right)}{1 + 4\pi^2}}_{=\alpha_1} \underbrace{e^{-s/2}\sin(\pi s)}_{=x_1(s)} - \underbrace{\frac{4(e^{1/8} + e^{3/8})}{1 + 16\pi^2}}_{=\alpha_2} \underbrace{e^{-s/2}\sin(2\pi s)}_{=x_2(s)}$$

$$+ \cdots. \tag{6.580}$$

A plot of a 20-term series expansion of the top hat function is shown in Figure 6.21.

In closing, we could have easily adjusted the constants on the eigenfunctions to obtain a true reciprocal basis. Taking

$$x_n = \sqrt{2}e^{-s/2}\sin(n\pi s), \tag{6.581}$$

$$y_m = \sqrt{2}e^{s/2}\sin(m\pi s), \tag{6.582}$$

gives $\langle y_m, x_n \rangle = \delta_{mn}$, as desired for a set of reciprocal basis functions. We see that getting the Fourier coefficients for eigenfunctions of a non-self-adjoint operator is facilitated by consideration of the adjoint operator. We also note that it is often a difficult exercise in problems with practical significance to actually find the adjoint operator and its eigenfunctions.

6.9 Rayleigh Quotient

Let us say that we wish to develop an estimate for the norm of a self-adjoint linear operator for vectors in a Hilbert space. The *Rayleigh*[20] *quotient* provides a simple method. Recall from Eq. (6.375) that $||\mathbf{L}||_2 = \sup_{x \neq 0} ||\mathbf{L}x||_2/||x||_2$. By definition of the norm of a vector, we can also say

$$||\mathbf{L}||_2^2 = \sup_{x \neq 0} \frac{\langle \mathbf{L}x, \mathbf{L}x \rangle}{\langle x, x \rangle}. \tag{6.583}$$

Now, because $\mathbf{L}$ is self-adjoint, it is equipped with a complete set of real eigenvalues λ_n and orthonormal eigenfunctions φ_n. And we can represent any arbitrary vector x in an eigenfunction expansion as

$$x = \sum_{n=1}^{\infty} \alpha_n \varphi_n. \tag{6.584}$$

---

[20] John William Strutt (Lord Rayleigh), 1842–1919, English physicist and mathematician.

Let us define the so-called *Rayleigh quotient* R as

$$R(x) = \frac{\langle \mathbf{L}x, x \rangle}{\langle x, x \rangle}. \tag{6.585}$$

Note that $R(x)$ is defined for any x, which may or may not be an eigenvector of $\mathbf{L}$.

Let us apply $\mathbf{L}$ to Eq. (6.584) and then operate further so as to consider $\|\mathbf{L}\|_2$:

$$\mathbf{L}x = \sum_{n=1}^{\infty} \alpha_n \mathbf{L}\varphi_n, \tag{6.586}$$

$$= \sum_{n=1}^{\infty} \alpha_n \lambda_n \varphi_n, \tag{6.587}$$

$$\langle \mathbf{L}x, x \rangle = \left\langle \sum_{n=1}^{\infty} \alpha_n \lambda_n \varphi_n, x \right\rangle, \tag{6.588}$$

$$= \sum_{n=1}^{\infty} \overline{\alpha_n} \lambda_n \langle \varphi_n, x \rangle, \tag{6.589}$$

$$= \sum_{n=1}^{\infty} \overline{\alpha_n} \lambda_n \left\langle \varphi_n, \sum_{m=1}^{\infty} \alpha_m \varphi_m \right\rangle, \tag{6.590}$$

$$= \sum_{n=1}^{\infty} \overline{\alpha_n} \lambda_n \sum_{m=1}^{\infty} \alpha_m \langle \varphi_n, \varphi_m \rangle, \tag{6.591}$$

$$= \sum_{n=1}^{\infty} \overline{\alpha_n} \lambda_n \sum_{m=1}^{\infty} \alpha_m \delta_{nm}, \tag{6.592}$$

$$= \sum_{n=1}^{\infty} \overline{\alpha_n} \alpha_n \lambda_n, \tag{6.593}$$

$$= \sum_{n=1}^{\infty} |\alpha_n|^2 \lambda_n. \tag{6.594}$$

In a similar set of operations in which we use Eq. (6.584) to form $\langle x, x \rangle$, we find

$$\langle x, x \rangle = \langle \sum_{n=1}^{\infty} \alpha_n \varphi_n, \sum_{m=1}^{\infty} \alpha_m \varphi_m \rangle, \tag{6.595}$$

$$= \sum_{n=1}^{\infty} \overline{\alpha_n} \sum_{m=1}^{\infty} \alpha_m \langle \varphi_n, \varphi_m \rangle, \tag{6.596}$$

$$= \sum_{n=1}^{\infty} |\alpha_n|^2. \tag{6.597}$$

So the Rayleigh quotient for $\mathbf{L}$ is

$$R = \frac{\sum_{n=1}^{\infty} |\alpha_n|^2 \lambda_n}{\sum_{n=1}^{\infty} |\alpha_n|^2}. \tag{6.598}$$

A straightforward exercise in optimization shows that R takes on extreme values for $\alpha_k \neq 0$ by requiring $\alpha_{n,n\neq k} = 0$. Thus, we see that R takes on extreme values when x is an eigenvector of $\mathbf{L}$, and those extreme values are the corresponding eigenvalues.

Now the eigenvalues of $\mathbf{L}$ are real and can be ordered such that $|\lambda_1| \geq |\lambda_2| \geq \ldots$. Thus, we can say

$$|R| = \left| \frac{\sum_{n=1}^{\infty} |\alpha_n|^2 \lambda_n}{\sum_{n=1}^{\infty} |\alpha_n|^2} \right| \leq \frac{\sum_{n=1}^{\infty} |\alpha_n|^2 |\lambda_1|}{\sum_{n=1}^{\infty} |\alpha_n|^2} = |\lambda_1|. \tag{6.599}$$

So we have established an upper bound of $|\lambda_1|$ for $|R|$. Thus,

$$\sup_{x \neq 0} \frac{\langle \mathbf{L}x, x \rangle}{\langle x, x \rangle} = |\lambda_1|. \tag{6.600}$$

Now, we can perform a similar exercise on Eq. (6.583) and learn that

$$||\mathbf{L}||_2 = |\lambda_1|. \tag{6.601}$$

It is important to note that this result only applies to self-adjoint operators. It can be shown that if one has a vector x which is within $O(\epsilon)$ of an eigenvector of $\mathbf{L}$, the Rayleigh quotient will give an estimate of λ_1 that has an error of $O(\epsilon^2)$. This renders the Rayleigh quotient to be of use when estimating an extreme eigenvalue. It can also be extended in an iterative scheme.

EXAMPLE 6.54

Estimate the eigenvalue with the largest magnitude of the self-adjoint matrix

$$\mathbf{A} = \begin{pmatrix} 1 & 2 \\ 2 & 1 \end{pmatrix}. \tag{6.602}$$

The matrix has eigenvalues and associated eigenvectors of $\lambda_1 = 3$, $e_1 = (1,1)^T$ and $\lambda_2 = -1$, $e_2 = (-1,1)^T$. Let us say we happen to have an estimate of e_1 as $\mathbf{x} = (1, 1.1)^T$. Then the Rayleigh quotient estimates λ_1:

$$\lambda_1 = ||\mathbf{A}||_2 \approx \frac{(\mathbf{A} \cdot \mathbf{x})^T \cdot \mathbf{x}}{\mathbf{x}^T \cdot \mathbf{x}} = \frac{(1 \quad 1.1) \begin{pmatrix} 1 & 2 \\ 2 & 1 \end{pmatrix} \begin{pmatrix} 1 \\ 1.1 \end{pmatrix}}{(1 \quad 1.1) \begin{pmatrix} 1 \\ 1.1 \end{pmatrix}} = \frac{6.61}{2.21} = 2.99095. \tag{6.603}$$

The error in the estimate of λ_1 is $3 - 2.99095 = 0.00905 \approx (0.1)^2$, in response to an error of 0.1 for the eigenvector.

If we do not have a good guess, we can iterate and will converge to e_1 and thus can get λ_1. The bad guess of $\mathbf{x} = (1, 10)^T$ gives a Rayleigh quotient of

$$\lambda_1 = ||\mathbf{A}||_2 \approx \frac{(\mathbf{A} \cdot \mathbf{x})^T \cdot \mathbf{x}}{\mathbf{x}^T \cdot \mathbf{x}} = \frac{(1 \quad 10) \begin{pmatrix} 1 & 2 \\ 2 & 1 \end{pmatrix} \begin{pmatrix} 1 \\ 10 \end{pmatrix}}{(1 \quad 10) \begin{pmatrix} 1 \\ 10 \end{pmatrix}} = 1.396. \tag{6.604}$$

If we take our next guess as $\mathbf{A} \cdot (1, 10)^T = (21, 12)^T$, we estimate

$$\lambda_1 = ||\mathbf{A}||_2 \approx \frac{(\mathbf{A} \cdot \mathbf{x})^T \cdot \mathbf{x}}{\mathbf{x}^T \cdot \mathbf{x}} = \frac{(21 \quad 12) \begin{pmatrix} 1 & 2 \\ 2 & 1 \end{pmatrix} \begin{pmatrix} 21 \\ 12 \end{pmatrix}}{(21 \quad 12) \begin{pmatrix} 21 \\ 12 \end{pmatrix}} = 2.723. \tag{6.605}$$

Ten iterations give us a good estimate of

$$\lambda_1 = 2.9999999930884. \tag{6.606}$$

In summary, if x_0 is an initial estimate of an eigenvector, however bad, we expect $\lim_{n\to\infty} \mathbf{A}^n \cdot x$ to converge to the eigenvector of $\mathbf{A}$ whose associated eigenvalue has the largest magnitude. This is will be presented more formally in the upcoming Eq. (7.127). An exception is if the seed value of x_0 is exactly an eigenvector associated with a non-extremal eigenvalue, in which case the Rayleigh quotient and all iterates will generate that eigenvalue. This will not hold for non-self-adjoint $\mathbf{A}$.

If $\mathbf{L}$ is a self-adjoint, one can prove the *Rayleigh-Ritz*[21] *theorem*, also known as the *min-max theorem*, which essentially holds that for any x,

$$\lambda_{min} \text{ (if any)} \le R(x) \le \lambda_{max} \text{ (if any)}. \tag{6.607}$$

We use the qualification "if any" as the minimum or maximum value of λ may not exist. If λ_{min} or λ_{max} exist, they are such that

$$\lambda_{max} = \max R(x) = \max_{x\ne 0} \frac{\langle \mathbf{L}x, x \rangle}{\langle x, x \rangle}, \tag{6.608}$$

$$\lambda_{min} = \min R(x) = \min_{x\ne 0} \frac{\langle \mathbf{L}x, x \rangle}{\langle x, x \rangle}. \tag{6.609}$$

EXAMPLE 6.55

Use the stratagem of extremization of the Rayleigh quotient to estimate an extreme eigenvalue of the operator $\mathbf{L} = -d^2/dt^2$ that operates on $x(t)$ with boundary conditions $x(0) = x(1) = 0$.

We actually know the eigenfunctions and eigenvalues to be $\sin(n\pi t)$ and $n^2\pi^2$ for $n = 1, 2, \ldots$. There is no maximum eigenvalue, whereas there is a minimum eigenvalue of π^2. Let us see if we can recover the minimum eigenvalue with the aid of the Rayleigh-Ritz theorem. To begin, we guess that x is a linear combination of trial functions, each of which satisfy the boundary conditions on x:

$$x(t) = \sum_{n=1}^{N} \alpha_n \phi_n(t). \tag{6.610}$$

We can choose a wide variety of $\phi_n(t)$. We will take

$$\phi_n(t) = t^n(1-t), \qquad n = 1, 2, \ldots, N. \tag{6.611}$$

Now compute the Rayleigh quotient:

$$R = \frac{\int_0^1 \left(-\frac{d^2}{dt^2}\left(\sum_{n=1}^N \alpha_n t^n(1-t)\right)\right)\left(\left(\sum_{n=1}^N \alpha_n t^n(1-t)\right)\right) dt}{\int_0^1 \left(\left(\sum_{n=1}^N \alpha_n t^n(1-t)\right)\right)\left(\left(\sum_{n=1}^N \alpha_n t^n(1-t)\right)\right) dt}. \tag{6.612}$$

When $N = 1$, we get, independent of α_1,

$$R = 10. \tag{6.613}$$

This is not far from the smallest eigenvalue, π^2. For $N = 2$, we get

$$R = \frac{14\left(5\alpha_1^2 + 5\alpha_2\alpha_1 + 2\alpha_2^2\right)}{7\alpha_1^2 + 7\alpha_2\alpha_1 + 2\alpha_2^2}. \tag{6.614}$$

---

[21] Walther Ritz, 1878–1909, Swiss physicist.

Standard optimization strategies show that R is minimized at $R = 10$ for $\alpha_1 = -1, \alpha_2 = 0$. And R is maximized at $R = 42$ for $\alpha_1 = -1$, $\alpha_2 = 2$. We did not improve our estimate for the smallest eigenvalue.

For $N = 3$, we find

$$R = \frac{12\left(35\alpha_1^2 + 7(5\alpha_2 + 3\alpha_3)\alpha_1 + 14\alpha_2^2 + 9\alpha_3^2 + 21\alpha_2\alpha_3\right)}{42\alpha_1^2 + 6(7\alpha_2 + 4\alpha_3)\alpha_1 + 12\alpha_2^2 + 5\alpha_3^2 + 15\alpha_2\alpha_3}. \tag{6.615}$$

Standard optimization reveals a minimum of $R = 9.86975$ for $\alpha_1 = -1$, $\alpha_2 = -1.331$, $\alpha_3 = 1.331$. This is extremely close to $\lambda = \pi^2 = 9.86960$. As expected, a maximum exists as well, but its value is not converging. Here the maximum is $R = 102.13$.

EXAMPLE 6.56

Use the stratagem of extremization of the Rayleigh quotient to obtain the extreme eigenvalues of the matrix operator:

$$\mathbf{A} = \begin{pmatrix} 1 & 2 & 3 \\ 2 & 4 & 1 \\ 3 & 1 & 5 \end{pmatrix}. \tag{6.616}$$

Now the exact eigenvalues of $\mathbf{A}$ are $\lambda = 7.5896, 3.38385, -0.973433$. Let us try a one-term approximation. Assigning $\mathbf{x} = \alpha_1(1, 0, 0)^T$, we find

$$R = 1, \tag{6.617}$$

which is not a particularly good approximation of any eigenvalue. That is because our initial guess was not close to any eigenvector. Trying a two-term approximation with $\mathbf{x} = \alpha_1(1, 1, 0)^T + \alpha_2(1, 2, 0)^T$, we find

$$R = \frac{(3\alpha_1 + 5\alpha_2)^2}{2\alpha_1^2 + 6\alpha_2\alpha_1 + 5\alpha_2^2}. \tag{6.618}$$

Extremizing gives $R = 5$ and $R = 0$, which is a better approximation of the largest and smallest eigenvalues.

The three-term approximation with $\mathbf{x} = \alpha_1(1, 1, 1)^T + \alpha_2(1, 2, 4)^T + \alpha_3(1, 3, 9)^T$ gives

$$R = \frac{22\alpha_1^2 + 8(14\alpha_2 + 27\alpha_3)\alpha_1 + 145\alpha_2^2 + 562\alpha_3^2 + 568\alpha_2\alpha_3}{3\alpha_1^2 + 2(7\alpha_2 + 13\alpha_3)\alpha_1 + 21\alpha_2^2 + 91\alpha_3^2 + 86\alpha_2\alpha_3}. \tag{6.619}$$

Extremizing yields the exact values for the extreme eigenvalues of $\lambda_{max} = 7.5896$, $\lambda_{min} = -0.973443$. The procedure yields no information on the intermediate eigenvalue. Any three-term guess for $\mathbf{x}$, following optimization, would yield the correct extreme eigenvalues. Also note that in contrast to the differential operator, the procedure applied to a matrix operator yielded both maximum and minimum estimates.

6.10 Linear Equations

The existence and uniqueness of the solution x of the equation

$$\mathbf{L}x = y, \tag{6.620}$$

for given linear operator $\mathbf{L}$ and y, are governed by the following:

Theorem: If the range of $\mathbf{L}$ is closed, $\mathbf{L}x = y$ has a solution if and only if y is orthogonal to every solution of the adjoint homogeneous equation $\mathbf{L}^*z = 0$.

Theorem: The solution of $\mathbf{L}x = y$ is nonunique if the solution of the homogeneous equation $\mathbf{L}x = 0$ is also nonunique, and conversely.

There are two basic ways in which the equation can be solved.

- Inverse: If an inverse of $\mathbf{L}$ exists, then

$$x = \mathbf{L}^{-1}y. \tag{6.621}$$

- Eigenvector expansion: Assume that x, y belong to a vector space $\mathbb{S}$ and the eigenvectors $\{e_1, e_2, \ldots\}$ of $\mathbf{L}$ span $\mathbb{S}$. Then we can write

$$y = \sum_n \alpha_n e_n; \tag{6.622}$$

$$x = \sum_n \beta_n e_n, \tag{6.623}$$

where the αs are known and the βs are unknown. We get

$$\mathbf{L}x = y, \tag{6.624}$$

$$\mathbf{L}\underbrace{\left(\sum_n \beta_n e_n\right)}_{x} = \underbrace{\sum_n \alpha_n e_n}_{y}, \tag{6.625}$$

$$\sum_n \mathbf{L}\beta_n e_n = \sum_n \alpha_n e_n, \tag{6.626}$$

$$\sum_n \beta_n \mathbf{L} e_n = \sum_n \alpha_n e_n, \tag{6.627}$$

$$\sum_n \beta_n \lambda_n e_n = \sum_n \alpha_n e_n, \tag{6.628}$$

$$\sum_n \underbrace{(\beta_n \lambda_n - \alpha_n)}_{=0} e_n = 0, \tag{6.629}$$

where the λs are the eigenvalues of $\mathbf{L}$. Because the e_n are linearly independent, we must demand for all n that

$$\beta_n \lambda_n = \alpha_n. \tag{6.630}$$

If all $\lambda_n \neq 0$, then $\beta_n = \alpha_n/\lambda_n$, and we have the unique solution

$$x = \sum_n \frac{\alpha_n}{\lambda_n} e_n. \tag{6.631}$$

If, however, one of the λs, λ_k say, is zero, we still have $\beta_n = \alpha_n/\lambda_n$ for $n \neq k$. For $n = k$, there are two possibilities:

○ If $\alpha_k \neq 0$, no solution is possible because equation (6.630) is not satisfied for $n = k$.

○ If $\alpha_k = 0$, we have the nonunique solution

$$x = \sum_{n \neq k} \frac{\alpha_n}{\lambda_n} e_n + \gamma e_k, \tag{6.632}$$

where γ is an arbitrary scalar. Equation (6.630) is satisfied $\forall n$.

This discussion is related to the Fredholm alternative, introduced in Section 4.6. In terms of the Fredholm alternative, we could say that given $\mathbf{L}$, nonzero λ, and y, there exists either

- a unique x such that $\mathbf{L}x - \lambda x = y$, or
- a nonunique x such that $\mathbf{L}x - \lambda x = 0$.

If one lets $\hat{\mathbf{L}} \equiv \mathbf{L} - \lambda$, the notation of the Fredholm alternative can be reconciled with that of this section.

EXAMPLE 6.57

Solve for x in $\mathbf{L}x = y$ if $\mathbf{L} = -d^2/dt^2$, with boundary conditions $x(0) = x(1) = 0$, and $y(t) = -2t$, via an eigenfunction expansion.

This problem has an exact solution obtained via straightforward integration:

$$\frac{d^2 x}{dt^2} = 2t, \qquad x(0) = x(1) = 0, \tag{6.633}$$

which integrates to yield

$$x_{exact}(t) = \frac{t}{3}(t^2 - 1). \tag{6.634}$$

However, let us use the series expansion technique. This can be more useful in other problems in which exact solutions do not exist. First, find the eigenvalues and eigenfunctions of the operator:

$$-\frac{d^2 x}{dt^2} = \lambda x, \qquad x(0) = x(1) = 0. \tag{6.635}$$

This has general solution

$$x(t) = A \sin\left(\sqrt{\lambda}t\right) + B \cos\left(\sqrt{\lambda}t\right). \tag{6.636}$$

To satisfy the boundary conditions, we require that $B = 0$ and $\lambda = n^2\pi^2$, so

$$x(t) = A \sin(n\pi t). \tag{6.637}$$

This suggests that we expand $y(t) = -2t$ in a Fourier-sine series. We deduce from Eq. (6.255) that the Fourier-sine series for $y(t) = -2t$ is

$$-2t = \sum_{n=1}^{\infty} \frac{4(-1)^n}{(n\pi)} \sin(n\pi t). \tag{6.638}$$

For $x(t)$, then we have

$$x(t) = \sum_{n=1}^{\infty} \frac{\alpha_n e_n}{\lambda_n} = \sum_{n=1}^{\infty} \frac{4(-1)^n}{(n\pi)\lambda_n} \sin(n\pi t). \tag{6.639}$$

Substituting in for $\lambda_n = n^2\pi^2$, we get

$$x(t) = \sum_{n=1}^{\infty} \frac{4(-1)^n}{(n\pi)^3} \sin(n\pi t). \tag{6.640}$$

Retaining only two terms in the expansion for $x(t)$,

$$x(t) \approx -\frac{4}{\pi^3} \sin(\pi t) + \frac{1}{2\pi^3} \sin(2\pi t), \tag{6.641}$$

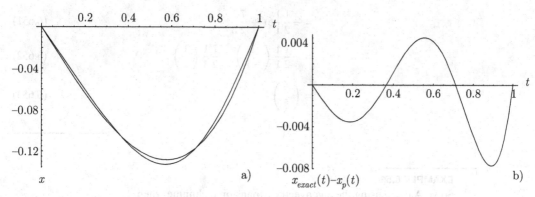

Figure 6.22. (a) Approximate and exact solution for $x(t)$; (b) error in solution $x_{exact}(t) - x_p(t)$.

gives a good approximation for the solution, which as shown in Figure 6.22, has a peak error amplitude of about 0.008.

EXAMPLE 6.58

Solve $\mathbf{A}x = y$ using the eigenvector expansion technique when

$$\mathbf{A} = \begin{pmatrix} 2 & 1 \\ 1 & 2 \end{pmatrix}, \qquad y = \begin{pmatrix} 3 \\ 4 \end{pmatrix}. \tag{6.642}$$

We already know from an earlier example (Example 6.46), that for $\mathbf{A}$,

$$\lambda_1 = 1, \qquad e_1 = \begin{pmatrix} 1 \\ -1 \end{pmatrix}, \tag{6.643}$$

$$\lambda_2 = 3, \qquad e_2 = \begin{pmatrix} 1 \\ 1 \end{pmatrix}. \tag{6.644}$$

We want to express y as

$$y = \alpha_1 e_1 + \alpha_2 e_2. \tag{6.645}$$

Because the eigenvectors are orthogonal, we have from Eq. (6.220)

$$\alpha_1 = \frac{\langle e_1, y \rangle}{\langle e_1, e_1 \rangle} = \frac{3-4}{1+1} = -\frac{1}{2}, \tag{6.646}$$

$$\alpha_2 = \frac{\langle e_2, y \rangle}{\langle e_2, e_2 \rangle} = \frac{3+4}{1+1} = \frac{7}{2}, \tag{6.647}$$

so

$$y = -\frac{1}{2}e_1 + \frac{7}{2}e_2. \tag{6.648}$$

Then

$$x = \frac{\alpha_1}{\lambda_1}e_1 + \frac{\alpha_2}{\lambda_2}e_2, \tag{6.649}$$

$$= -\frac{1}{2}\frac{1}{\lambda_1}e_1 + \frac{7}{2}\frac{1}{\lambda_2}e_2, \tag{6.650}$$

$$= -\frac{1}{2}\frac{1}{1}e_1 + \frac{7}{2}\frac{1}{3}e_2, \tag{6.651}$$

$$= -\frac{1}{2}\frac{1}{1}\begin{pmatrix} 1 \\ -1 \end{pmatrix} + \frac{7}{2}\frac{1}{3}\begin{pmatrix} 1 \\ 1 \end{pmatrix}, \tag{6.652}$$

$$= \begin{pmatrix} \frac{2}{3} \\ \frac{5}{3} \end{pmatrix}. \tag{6.653}$$

EXAMPLE 6.59

Solve $\mathbf{A}x = y$ using the eigenvector expansion technique when

$$\mathbf{A} = \begin{pmatrix} 2 & 1 \\ 4 & 2 \end{pmatrix}, \qquad y = \begin{pmatrix} 3 \\ 4 \end{pmatrix}, \qquad y = \begin{pmatrix} 3 \\ 6 \end{pmatrix}. \tag{6.654}$$

We first note that the two column space vectors,

$$\begin{pmatrix} 2 \\ 4 \end{pmatrix}, \qquad \begin{pmatrix} 1 \\ 2 \end{pmatrix}, \tag{6.655}$$

are linearly dependent. They span $\mathbb{R}^1$, but not $\mathbb{R}^2$.

It is easily shown that, for $\mathbf{A}$,

$$\lambda_1 = 4, \qquad e_1 = \begin{pmatrix} 1 \\ 2 \end{pmatrix}, \tag{6.656}$$

$$\lambda_2 = 0, \qquad e_2 = \begin{pmatrix} -1 \\ 2 \end{pmatrix}. \tag{6.657}$$

First consider $y = \begin{pmatrix} 3 \\ 4 \end{pmatrix}$. We want to express y as

$$y = \alpha_1 e_1 + \alpha_2 e_2. \tag{6.658}$$

For this asymmetric matrix, the eigenvectors are linearly independent, so they form a basis. However, they are not orthogonal, so we cannot directly compute α_1 and α_2. Matrix inversion can be done and shows that $\alpha_1 = 5/2$ and $\alpha_2 = -1/2$, so

$$y = \frac{5}{2}e_1 - \frac{1}{2}e_2. \tag{6.659}$$

Because the eigenvectors form a basis, y can be represented with an eigenvector expansion. However, no solution for x exists because $\lambda_2 = 0$ and $\alpha_2 \neq 0$, hence the coefficient $\beta_2 = \alpha_2/\lambda_2$ does not exist.

However, for $y = \begin{pmatrix} 3 \\ 6 \end{pmatrix}$, we can say that

$$y = 3e_1 + 0e_2. \tag{6.660}$$

We note that $(3, 6)^T$ is a scalar multiple of the so-called column space vector of $\mathbf{A}$, $(2, 4)^T$. Consequently,

$$x = \frac{\alpha_1}{\lambda_1}e_1 + \frac{\alpha_2}{\lambda_2}e_2, \tag{6.661}$$

$$= \frac{\alpha_1}{\lambda_1}e_1 + \frac{0}{0}e_2, \tag{6.662}$$

$$= \frac{3}{4}e_1 + \gamma e_2, \tag{6.663}$$

$$= \frac{3}{4} \begin{pmatrix} 1 \\ 2 \end{pmatrix} + \gamma \begin{pmatrix} -1 \\ 2 \end{pmatrix}, \tag{6.664}$$

$$= \begin{pmatrix} 3/4 - \gamma \\ 3/2 + 2\gamma \end{pmatrix}, \tag{6.665}$$

where γ is an arbitrary constant. The vector $e_2 = (-1, 2)^T$ lies in the null space of $\mathbf{A}$ because

$$\mathbf{A}e_2 = \begin{pmatrix} 2 & 1 \\ 4 & 2 \end{pmatrix} \begin{pmatrix} -1 \\ 2 \end{pmatrix} = \begin{pmatrix} 0 \\ 0 \end{pmatrix}. \tag{6.666}$$

Because e_2 lies in the null space, any scalar multiple of e_2, say, γe_2, also lies in the null space. We can conclude that for arbitrary y, the inverse does not exist. For vectors y that lie in the column space of $\mathbf{A}$, the inverse exists, but it is not unique; arbitrary vectors from the null space of $\mathbf{A}$ are admitted as part of the solution.

6.11 Method of Weighted Residuals

The method of weighted residuals is a general technique that provides a unifying framework for many methods to solve differential equations. Nearly all of the common numerical strategies used to generate approximate solutions to equations that arise in engineering, for example, the finite difference and finite element methods, can be considered special cases of this method. Consider then the differential equation

$$\mathbf{L}y(t) = f(t), \qquad t \in [a, b], \tag{6.667}$$

with homogeneous boundary conditions. Here, $\mathbf{L}$ is a differential operator that is not necessarily linear, $f(t)$ is known, and $y(t)$ is to be determined. We will work with functions and inner products in $\mathbb{L}_2[a, b]$ space. Let us define Eq. (6.667) along with appropriate boundary conditions as the *strong formulation* of the problem.

Within the context of the method of weighted residuals, we are actually seeking what are known as *weak solutions*, which are defined as follows. A function $y(t)$ is a weak solution of $\mathbf{L}y(t) - f(t) = 0$ for $t \in [a, b]$ if the *weak formulation*,

$$\langle \psi(t), \mathbf{L}y(t) - f(t) \rangle = 0, \tag{6.668}$$

holds for all admissible weighting functions $\psi(t)$. How admissible is defined depends on the nature of $\mathbf{L}$ and the boundary conditions. Ideally, the admissible function space is chosen so that the strong and weak formulations are equivalent. Typically, one requires that $\psi(t)$ have certain continuity properties. These become especially important within the finite element method, but for the problems we consider, such subtle issues will not be relevant, allowing us to specify rather ordinary weighting functions.

In integral form for real-valued functions, the inner product reduces to

$$\int_a^b \psi(t) \left(\mathbf{L}y(t) - f(t) \right) dt = 0, \qquad \forall \, \psi(t). \tag{6.669}$$

One also typically insists that $\psi(t)$ have what is known as *compact support* in that it is identically zero outside of some bounded region on the t axis. Because one can often use integration by parts to lower the order of the differential operator $\mathbf{L}$, the weak formulation allows the existence of solutions $y(t)$ with fewer continuity properties than one finds for so-called strong solutions of the original $\mathbf{L}y(t) - f(t) = 0$.[22]

Now, let us represent $y(t)$ by

$$y(t) = \sum_{n=1}^{\infty} \alpha_n \phi_n(t), \tag{6.670}$$

where $\phi_n(t), n = 1, \ldots, \infty$, are linearly independent functions (called *trial functions*) that satisfy the necessary differentiability requirements imposed by $\mathbf{L}$ and the homogeneous boundary conditions. They should form a complete basis within the function space to which y belongs. Forcing the trial functions to satisfy the boundary conditions allows us to concentrate solely on finding the α_n such that y satisfies the differential equation. The trial functions can be orthogonal or nonorthogonal.

Although much could be said about the solutions in terms of the infinite series, let us move forward to consider approximate solutions. We thus truncate the series at N terms so as to represent an approximate solution $y_p(t)$ as follows:

$$y(t) \approx y_p(t) = \sum_{n=1}^{N} \alpha_n \phi_n(t). \tag{6.671}$$

The constants $\alpha_n, n = 1, \ldots, N$, are to be determined. Substituting our approximation $y_p(t)$ into Eq. (6.667), we get a *residual*

$$r(t) = \mathbf{L}y_p(t) - f(t). \tag{6.672}$$

The residual $r(t)$ is not the error in the solution, $e(t)$, where

$$e(t) = y(t) - y_p(t). \tag{6.673}$$

Also, if we knew the error, $e(t)$, we would know the solution, and there would be no need for an approximate technique! The residual will almost always be nonzero for $t \in [a, b]$.

We can choose α_n such that the residual, computed in a weighted average over the domain, is zero. To achieve this, we select now a set of linearly independent *weighting functions* $\psi_m(t), m = 1, \ldots, N$, and make them orthogonal to the residual. Thus, one has an N-term discrete form of Eq. (6.668):

$$\langle \psi_m, r \rangle = 0, \qquad m = 1, \ldots, N, \tag{6.674}$$

which form N equations for the constants α_n.

There are several special ways in which the weighting functions can be selected:

- Galerkin[23] : $\psi_m(t) = \phi_m(t)$.
- Collocation: $\psi_m(t) = \delta(t - t_m)$, with $t_m \in [a, b]$; thus, $r(t_m) = 0$.
- Subdomain $\psi_m(t) = 1$ for $t \in [t_{m-1}, t_m)$ and zero everywhere else; these weighting functions are orthogonal to each other.

---

[22] One might wonder what value there is in considering the abstraction that is Eq. (6.669) in favor of the more straightforward Eq. (6.667). The best argument is that the modeling of many problems in nature, especially those with embedded discontinuities, requires the additional formalism of Eq. (6.669), which has less stringent continuity requirements. And even for problems that have no embedded discontinuities, the weak formalism provides a systematic means to compute approximations to y and often admits additional physical insights into the nature of the problem.

[23] Boris Gigorievich Galerkin, 1871–1945, Belarussian-born Russian-based engineer and mathematician.

- Least squares: Minimize $||r(t)||$; this gives

$$0 = \frac{\partial ||r||^2}{\partial \alpha_m} = \frac{\partial}{\partial \alpha_m} \int_a^b r^2 \, dt, \tag{6.675}$$

$$= 2 \int_a^b r \underbrace{\frac{\partial r}{\partial \alpha_m}}_{=\psi_m(t)} \, dt. \tag{6.676}$$

So this method corresponds to $\psi_n = \partial r / \partial \alpha_n$.
- Moments: $\psi_m(t) = t^{m-1}, m = 1, 2, \ldots$.

If the trial functions are orthogonal and the method is Galerkin, we will, following Fletcher (1991), who builds on the work of Finlayson (1972), define the method to be a *spectral method*. Other less restrictive definitions are in common usage in the present literature, and there is no single consensus on what precisely constitutes a spectral method.[24]

Another concern that arises with methods of this type is how many terms are necessary to properly model the desired frequency level. For example, take our equation to be $d^2u/dt^2 = 1 + u^2; u(0) = u(\pi) = 0$, and take $u = \sum_{n=1}^N a_n \sin(nt)$. If $N = 1$, we get $r(t) = -a_1 \sin t - 1 - a_1^2 \sin^2 t$. Expanding the square of the sin term, we see the error has higher order frequency content: $r(t) = -a_1 \sin t - 1 - a_1^2 (1/2 - 1/2 \cos(2t))$. The result is that if we want to get things right at a given level, we may have to reach outside that level. How far outside we have to reach will be problem dependent.

EXAMPLE 6.60

For $y \in \mathbb{L}_2[0, 1]$, find a one-term approximate solution of the linear equation

$$\frac{d^2y}{dt^2} + y = -t, \qquad y(0) = y(1) = 0. \tag{6.677}$$

It is easy to show that the exact solution is

$$y_{exact}(t) = -t + \csc(1) \sin(t). \tag{6.678}$$

Here we will see how well the method of weighted residuals can approximate this known solution. However, the real value of the method is for problems in which exact solutions are not known.

---

[24] An important school in spectral methods, exemplified in the work of Gottlieb and Orszag (1977), Canuto et al. (1988), and Fornberg (1998), uses a looser nomenclature, which is not always precisely defined. In these works, spectral methods are distinguished from finite difference methods and finite element methods in that spectral methods employ basis functions that have global rather than local support; that is, spectral methods' basis functions have nonzero values throughout the entire domain. While orthogonality of the basis functions within a Galerkin framework is often employed, it is not demanded that this be the distinguishing feature by those authors. Within this school, less emphasis is placed on the framework of the method of weighted residuals, and the spectral method is divided into subclasses known as Galerkin, tau, and collocation. The collocation method this school defines is identical to that defined here, and is also called by this school the "pseudospectral" method. In nearly all understandings of the word *spectral*, a convergence rate that is more rapid than those exhibited by finite difference or finite element methods exists. In fact, the accuracy of a spectral method should grow exponentially with the number of nodes for a spectral method, as opposed to that for a finite difference or finite element, whose accuracy grows only with the number of nodes raised to some power.

Let us consider a one-term approximation:

$$y \approx y_p(t) = \alpha\phi(t). \tag{6.679}$$

There are many choices of basis functions $\phi(t)$. Let us try finite-dimensional nontrivial polynomials that match the boundary conditions. If we choose $\phi(t) = a$, a constant, we must take $a = 0$ to satisfy the boundary conditions, so this does not work. If we choose $\phi(t) = a + bt$, we must take $a = 0, b = 0$ to satisfy both boundary conditions, so this also does not work. We can find a quadratic polynomial that is nontrivial and satisfies both boundary conditions:

$$\phi(t) = t(1 - t). \tag{6.680}$$

Then

$$y_p(t) = \alpha t(1 - t). \tag{6.681}$$

We have to determine α. Substituting into Eq. (6.672), the residual is found to be

$$r(t) = \mathbf{L}y_p - f(t) = \frac{d^2 y_p}{dt^2} + y_p - f(t), \tag{6.682}$$

$$= \underbrace{-2\alpha}_{d^2y_p/dt^2} + \underbrace{\alpha t(1-t)}_{y_p} - \underbrace{(-t)}_{f(t)} = t - \alpha(t^2 - t + 2). \tag{6.683}$$

Then, we choose α such that

$$\langle \psi, r \rangle = \int_0^1 \psi(t) \underbrace{\left(t - \alpha(t^2 - t + 2)\right)}_{=r(t)} dt = 0. \tag{6.684}$$

The form of the weighting function $\psi(t)$ is dictated by the particular method we choose:

1. Galerkin: $\psi(t) = \phi(t) = t(1 - t)$. The inner product gives $1/12 - (3/10)\alpha = 0$, so that for nontrivial solution, $\alpha = 5/18 = 0.277$.

$$y_p(t) = 0.277t(1 - t). \tag{6.685}$$

2. Collocation: Choose $\psi(t) = \delta(t - 1/2)$, which gives $-(7/2)\alpha + 1 = 0$, from which $\alpha = 2/7 = 0.286$:

$$y_p(t) = 0.286t(1 - t). \tag{6.686}$$

3. Subdomain: $\psi(t) = 1$, from which $-(11/6)\alpha + 1/2 = 0$, and $\alpha = 3/11 = 0.273$:

$$y_p(t) = 0.273t(1 - t). \tag{6.687}$$

4. Least squares: $\psi(t) = \partial r(t)/\partial\alpha = -t^2 + t - 2$. Thus, $-11/12 + 101/30\alpha = 0$, from which $\alpha = 55/202 = 0.273$:

$$y_p(t) = 0.273t(1 - t). \tag{6.688}$$

5. Moments: $\psi(t) = 1$ which, for this case, is the same as the subdomain method previously reported:

$$y_p(t) = 0.273t(1 - t). \tag{6.689}$$

The approximate solution determined by the Galerkin method is overlaid against the exact solution in Figure 6.23. Also shown is the error, $e(t) = y_{exact}(t) - y_p(t)$, in the

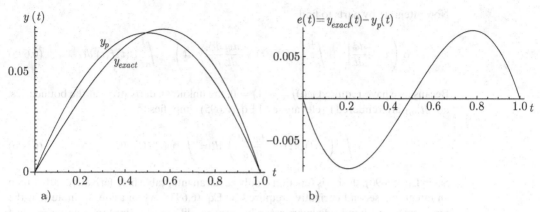

Figure 6.23. (a) One-term Galerkin estimate $y_p(t)$ and exact solution $y(t)$ to $d^2y/dt^2 + y = -t$, $y(0) = 0$, $y(1) = 0$; (b) error in solution $e(t) = y_{exact}(t) - y_p(t)$.

approximation. The approximation is surprisingly accurate. The error is available because in this case we have the exact solution.

Some simplification can arise through use of integration by parts. This has the result of admitting basis functions that have less stringent requirements on the continuity of their derivatives. It is also a commonly used strategy in the finite element technique.

EXAMPLE 6.61

Consider a slight variant of the previous example problem, and employ integration by parts,

$$\frac{d^2y}{dt^2} + y = f(t), \qquad y(0) = 0, \quad y(1) = 0, \tag{6.690}$$

to find a one-term approximate solution via a Galerkin method.

Again, take a one-term expansion:

$$y_p(t) = \alpha\phi(t). \tag{6.691}$$

At this point, we will only require $\phi(t)$ to satisfy the homogeneous boundary conditions and will specify it later. The residual in the approximation is

$$r(t) = \frac{d^2y_p}{dt^2} + y_p - f(t) = \alpha\frac{d^2\phi}{dt^2} + \alpha\phi - f(t). \tag{6.692}$$

Now set a weighted residual to zero. We will also require the weighting function $\psi(t)$ to vanish at the boundaries:

$$\langle\psi, r\rangle = \int_0^1 \psi(t) \underbrace{\left(\alpha\frac{d^2\phi}{dt^2} + c\phi(t) - f(t)\right)}_{=r(t)} dt = 0. \tag{6.693}$$

Rearranging, we get

$$\alpha\int_0^1 \left(\psi(t)\frac{d^2\phi}{dt^2} + \psi(t)\phi(t)\right) dt = \int_0^1 \psi(t)f(t)\, dt. \tag{6.694}$$

Now integrate by parts to find

$$\alpha \left(\psi(t)\frac{d\phi}{dt}\Big|_0^1 + \int_0^1 \left(\psi(t)\phi(t) - \frac{d\psi}{dt}\frac{d\phi}{dt} \right) dt \right) = \int_0^1 \psi(t)f(t)\, dt. \qquad (6.695)$$

Because we have required $\psi(0) = \psi(1) = 0$, the unknown derivatives at the boundaries, $d\phi/dt|_{0,1}$ are rendered irrelevant, and Eq. (6.695) simplifies to

$$\alpha \int_0^1 \left(\psi(t)\phi(t) - \frac{d\psi}{dt}\frac{d\phi}{dt} \right) dt = \int_0^1 \psi(t)f(t)\, dt. \qquad (6.696)$$

So, in Eq. (6.696), the basis function ϕ only needs an integrable first derivative rather than an integrable second derivative required by Eq. (6.693). As an aside, we note that the term on the left-hand side bears resemblance (but differs by a sign) to an inner product in the Sobolov space $\overline{\overline{W}}_2^1[0,1]$, (see Section 6.3), in which the Sobolov inner product $\langle .,. \rangle_s$ is $\langle \psi, \phi \rangle_s = \int_0^1 \left(\overline{\psi(t)}\phi(t) + \overline{(d\psi/dt)}(d\phi/dt) \right)\, dt$.

Taking now, as before, $\phi = t(1-t)$ and then choosing a Galerkin method so $\psi(t) = \phi(t) = t(1-t)$, and $f(t) = -t$, we get

$$\alpha \int_0^1 \left(t^2(1-t)^2 - (1-2t)^2 \right)\, dt = \int_0^1 t(1-t)(-t)\, dt, \qquad (6.697)$$

which gives

$$\alpha \left(-\frac{3}{10} \right) = -\frac{1}{12}, \qquad (6.698)$$

so

$$\alpha = \frac{5}{18}, \qquad (6.699)$$

as was found earlier. So

$$y_p = \frac{5}{18}t(1-t), \qquad (6.700)$$

with the Galerkin method.

EXAMPLE 6.62

For $y \in \mathbb{L}_2[0,1]$, find a two-term spectral approximation (which by our definition of "spectral" mandates a Galerkin formulation) to the solution of the linear equation

$$\frac{d^2y}{dt^2} + \sqrt{t}\, y = 1, \qquad y(0) = 0, \qquad y(1) = 0. \qquad (6.701)$$

We first note there is actually a complicated exact solution, revealed by symbolic computer mathematics to be

$$y_{exact}(t) = \frac{a(t)}{50\,(t^{5/4})^{2/5}\, J_{\frac{2}{5}}\left(\frac{4}{5}\right)}, \qquad (6.702)$$

where $a(t)$ is

$$a(t) = -2\,2^{2/5}5^{3/5}\left(t^{5/4}\right)^{2/5}\sqrt{t}\Gamma\left(-\frac{2}{5}\right)J_{-\frac{2}{5}}\left(\frac{4}{5}\right)J_{\frac{2}{5}}\left(\frac{4t^{5/4}}{5}\right){}_1F_2\left(\frac{4}{5};\frac{7}{5},\frac{9}{5};-\frac{4}{25}\right)$$

$$-25\,2^{3/5}5^{2/5}\left(t^{5/4}\right)^{2/5}\sqrt{t}\Gamma\left(\frac{7}{5}\right)J_{\frac{2}{5}}\left(\frac{4}{5}\right)J_{\frac{2}{5}}\left(\frac{4t^{5/4}}{5}\right){}_1F_2\left(\frac{2}{5};\frac{3}{5},\frac{7}{5};-\frac{4}{25}\right)$$

$$+25\,2^{3/5}5^{2/5}t^2\Gamma\left(\frac{7}{5}\right)J_{\frac{2}{5}}\left(\frac{4}{5}\right)J_{\frac{2}{5}}\left(\frac{4t^{5/4}}{5}\right){}_1F_2\left(\frac{2}{5};\frac{3}{5},\frac{7}{5};-\frac{4t^{5/2}}{25}\right)$$

$$+2\,2^{2/5}5^{3/5}\left(t^{5/4}\right)^{4/5}t^2\Gamma\left(-\frac{2}{5}\right)J_{\frac{2}{5}}\left(\frac{4}{5}\right)J_{-\frac{2}{5}}\left(\frac{4t^{5/4}}{5}\right){}_1F_2\left(\frac{4}{5};\frac{7}{5},\frac{9}{5};-\frac{4t^{5/2}}{25}\right).$$

$$(6.703)$$

This is unwieldy! Let us now try to get an approximate solution, $y_p = \sum_{n=1}^{N}\alpha_n\phi_n$, which may in fact yield more insight. We can try polynomial trial functions. At a minimum, these trial functions must satisfy the boundary conditions. Assumption of the first trial function to be a constant or linear gives rise to a trivial trial function when the boundary conditions are enforced. Thus, the first nontrivial trial function is a quadratic:

$$\phi_1(t) = a_0 + a_1 t + a_2 t^2. \tag{6.704}$$

We need $\phi_1(0) = 0$ and $\phi_1(1) = 0$. The first condition gives $a_0 = 0$; the second gives $a_1 = -a_2$, so we have $\phi_1 = a_1(t - t^2)$. Because the magnitude of a trial function is arbitrary, a_1 can be set to unity to give

$$\phi_1(t) = t(1 - t). \tag{6.705}$$

Alternatively, we could have chosen the magnitude in such a fashion to guarantee an orthonormal trial function with $\|\phi_1\|_2 = 1$, but that is a secondary concern for the purposes of this example.

We need a second linearly independent trial function for the two-term approximation. We try a third-order polynomial:

$$\phi_2(t) = b_0 + b_1 t + b_2 t^2 + b_3 t^3. \tag{6.706}$$

Enforcing the boundary conditions as before gives $b_0 = 0$ and $b_1 = -(b_2 + b_3)$, so

$$\phi_2(t) = -(b_2 + b_3)t + b_2 t^2 + b_3 t^3. \tag{6.707}$$

To achieve a spectral method (which in general is not necessary to achieve an approximate solution), we enforce orthogonality, $\langle\phi_1, \phi_2\rangle = 0$:

$$\int_0^1 \underbrace{t(1 - t)}_{=\phi_1(t)}\underbrace{\left(-(b_2 + b_3)t + b_2 t^2 + b_3 t^3\right)}_{=\phi_2(t)}\,dt = 0, \tag{6.708}$$

$$-\frac{b_2}{30} - \frac{b_3}{20} = 0, \tag{6.709}$$

$$b_2 = -\frac{3}{2}b_3. \tag{6.710}$$

Substituting and factoring gives

$$\phi_2(t) = \frac{b_3}{2}\,t(1 - t)(2t - 1). \tag{6.711}$$

Again, because ϕ_2 is a trial function, the lead constant is arbitrary; we take for convenience $b_3 = 2$ to give

$$\phi_2 = t(1 - t)(2t - 1). \tag{6.712}$$

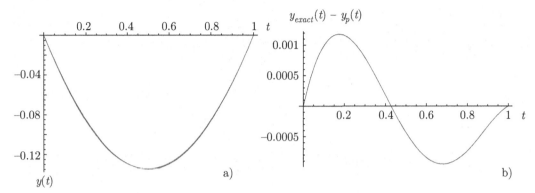

Figure 6.24. (a) Two-term spectral (Galerkin) estimate $y_p(t)$ and $y_{exact}(t)$; (b) error in approximation, $y_{exact}(t) - y_p(t)$.

Again, b_3 could alternatively have been chosen to yield an orthonormal trial function with $\|\phi_2\|_2 = 1$.

Now we want to choose α_1 and α_2 so that our approximate solution,

$$y_p(t) = \alpha_1 \phi_1(t) + \alpha_2 \phi_2(t), \tag{6.713}$$

has a zero weighted residual. With

$$\mathbf{L} = \left(\frac{d^2}{dt^2} + \sqrt{t} \right), \tag{6.714}$$

we have the residual as

$$r(t) = \mathbf{L}y_p(t) - f(t) = \mathbf{L}\left(\alpha_1 \phi_1(t) + \alpha_2 \phi_2(t)\right) - 1 = \alpha_1 \mathbf{L}\phi_1(t) + \alpha_2 \mathbf{L}\phi_2(t) - 1. \tag{6.715}$$

To drive the weighted residual to zero, take $\psi = \psi_1$ and $\psi = \psi_2$ so as to achieve

$$\langle \psi_1, r \rangle = \alpha_1 \langle \psi_1, \mathbf{L}\phi_1 \rangle + \alpha_2 \langle \psi_1, \mathbf{L}\phi_2 \rangle - \langle \psi_1, 1 \rangle = 0, \tag{6.716}$$

$$\langle \psi_2, r \rangle = \alpha_1 \langle \psi_2, \mathbf{L}\phi_1 \rangle + \alpha_2 \langle \psi_2, \mathbf{L}\phi_2 \rangle - \langle \psi_2, 1 \rangle = 0. \tag{6.717}$$

This is easily cast in matrix form as a linear system of equations for the unknowns α_1 and α_2:

$$\begin{pmatrix} \langle \psi_1, \mathbf{L}\phi_1 \rangle & \langle \psi_1, \mathbf{L}\phi_2 \rangle \\ \langle \psi_2, \mathbf{L}\phi_1 \rangle & \langle \psi_2, \mathbf{L}\phi_2 \rangle \end{pmatrix} \begin{pmatrix} \alpha_1 \\ \alpha_2 \end{pmatrix} = \begin{pmatrix} \langle \psi_1, 1 \rangle \\ \langle \psi_2, 1 \rangle \end{pmatrix}. \tag{6.718}$$

As we chose the Galerkin method, we set $\psi_1 = \phi_1$ and $\psi_2 = \phi_2$, so

$$\begin{pmatrix} \langle \phi_1, \mathbf{L}\phi_1 \rangle & \langle \phi_1, \mathbf{L}\phi_2 \rangle \\ \langle \phi_2, \mathbf{L}\phi_1 \rangle & \langle \phi_2, \mathbf{L}\phi_2 \rangle \end{pmatrix} \begin{pmatrix} \alpha_1 \\ \alpha_2 \end{pmatrix} = \begin{pmatrix} \langle \phi_1, 1 \rangle \\ \langle \phi_2, 1 \rangle \end{pmatrix}. \tag{6.719}$$

Each of the inner products represents a definite integral that is easily evaluated via symbolic computer mathematics. For example,

$$\langle \phi_1, \mathbf{L}\phi_1 \rangle = \int_0^1 \underbrace{t(1-t)}_{\phi_1} \underbrace{\left(-2 + (1-t)t^{3/2}\right)}_{\mathbf{L}\phi_1} dt = -\frac{215}{693}. \tag{6.720}$$

When each inner product is evaluated, the following system results:

$$\begin{pmatrix} -\frac{215}{693} & \frac{16}{9009} \\ \frac{16}{9009} & -\frac{197}{1001} \end{pmatrix} \begin{pmatrix} \alpha_1 \\ \alpha_2 \end{pmatrix} = \begin{pmatrix} \frac{1}{6} \\ 0 \end{pmatrix}. \tag{6.721}$$

Figure 6.25. Orthonormal trial functions $\varphi_1(t) = \sqrt{30}t(1-t)$, $\varphi_2(t) = \sqrt{210}t(1-t)(2t-1)$ employed in two-term spectral solution of $d^2y/dt^2 + \sqrt{t}y = 1, y(0) = 0, y(1) = 0$.

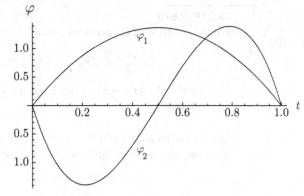

Inverting the system, it is found that

$$\alpha_1 = -\frac{760617}{1415794} = -0.537, \qquad \alpha_2 = -\frac{3432}{707897} = -0.00485. \qquad (6.722)$$

Thus, the estimate for the solution is

$$y_p(t) = -0.537 \, t(1-t) - 0.00485 \, t(1-t)(2t-1). \qquad (6.723)$$

The two-term approximate solution is overlaid against the exact solution in Figure 6.24. Also shown is the error in the approximation. The two-term solution is surprisingly accurate.

By normalizing the trial functions, we can find an orthonormal expansion. One finds that

$$\|\phi_1\|_2 = \sqrt{\int_0^1 \phi_1^2 \, dt}, \qquad (6.724)$$

$$= \sqrt{\int_0^1 t^2(1-t)^2 \, dt}, \qquad (6.725)$$

$$= \frac{1}{\sqrt{30}}, \qquad (6.726)$$

$$\|\phi_2\|_2 = \sqrt{\int_0^1 \phi_2^2 \, dt}, \qquad (6.727)$$

$$= \sqrt{\int_0^1 t^2(1-t)^2(2t-1)^2 \, dt}, \qquad (6.728)$$

$$= \frac{1}{\sqrt{210}}. \qquad (6.729)$$

The approximate solution can then be rewritten as an orthonormal expansion:

$$y_p(t) = -\frac{760617}{1415794\sqrt{30}}(\sqrt{30}t(1-t)) - \frac{3432}{707897\sqrt{210}}(\sqrt{210}t(1-t)(2t-1)), \quad (6.730)$$

$$= -0.981 \underbrace{(\sqrt{30}t(1-t))}_{\varphi_1} - 0.000335 \underbrace{(\sqrt{210}t(1-t)(2t-1))}_{\varphi_2}. \qquad (6.731)$$

Because the trial functions have been normalized, one can directly compare the coefficients' magnitude. It is seen that the bulk of the solution is captured by the first term. The orthonormalized trial functions φ_1 and φ_2 are plotted in Figure 6.25.

EXAMPLE 6.63

For the equation of the previous example,

$$\frac{d^2y}{dt^2} + \sqrt{t}\, y = 1, \qquad y(0) = 0, \qquad y(1) = 0, \tag{6.732}$$

examine the convergence rates for a collocation method as the number of trial functions becomes large.

Let us consider a set of trial functions that do not happen to be orthogonal but are linearly independent, one twice differentiable, and satisfy the boundary conditions. Take

$$\phi_n(t) = t^n(t - 1), \qquad n = 1, \ldots, N. \tag{6.733}$$

So we seek to find a vector $\boldsymbol{\alpha} = \alpha_n, n = 1, \ldots, N$, such that for a given number of collocation points N, the approximation

$$y_N(t) = \alpha_1 \phi_1(t) + \cdots \alpha_n \phi_n(t) + \cdots + \alpha_N \phi_N(t), \tag{6.734}$$

drives a weighted residual to zero. These trial functions have the advantage of being easy to program for an arbitrary N, as no Gram-Schmidt orthogonalization process is necessary. The details of the analysis are similar to those of the previous example, except we perform it many times, varying N in each calculation. For the collocation method, we take the weighting functions to be

$$\psi_n(t) = \delta(t - t_n), \qquad n = 1, \ldots, N. \tag{6.735}$$

We could choose $t_n = n/(N+1)$, $n = 1, \ldots, N$, so that the collocation points would be evenly distributed in $t \in [0, 1]$. However, we can get faster convergence and more uniform error distribution by concentrating the collocation points at either end of the domain. One way to achieve this is by taking

$$t_n = \frac{1}{2}\left(1 - \cos\left(\frac{\pi(n - 1/2)}{N}\right)\right), \qquad n = 1, \ldots, N. \tag{6.736}$$

These values of t_n are the zeros of the Chebyshev polynomial, $T_N(2t - 1)$, that is to say, $T_N(2t_n - 1) = 0$.

We then form the matrix

$$\mathbf{A} = \begin{pmatrix} \langle \psi_1, \mathbf{L}\phi_1 \rangle, & \langle \psi_1, \mathbf{L}\phi_2 \rangle & \ldots & \langle \psi_1, \mathbf{L}\phi_N \rangle \\ \langle \psi_2, \mathbf{L}\phi_1 \rangle, & \langle \psi_2, \mathbf{L}\phi_2 \rangle & \ldots & \langle \psi_2, \mathbf{L}\phi_N \rangle \\ \vdots & \vdots & \vdots & \vdots \\ \langle \psi_N, \mathbf{L}\phi_1 \rangle, & \langle \psi_N, \mathbf{L}\phi_2 \rangle & \ldots & \langle \psi_N, \mathbf{L}\phi_N \rangle \end{pmatrix}, \tag{6.737}$$

and the vector

$$\mathbf{b} = \begin{pmatrix} \langle \psi_1, 1 \rangle \\ \vdots \\ \langle \psi_N, 1 \rangle \end{pmatrix}, \tag{6.738}$$

and then solve for $\boldsymbol{\alpha}$ in

$$\mathbf{A} \cdot \boldsymbol{\alpha} = \mathbf{b}. \tag{6.739}$$

We perform this calculation for $N = 1, \ldots, N_{max}$ and calculate an error for each N by finding the norm of the difference between $y_{exact}(t)$ and our approximation $y_N(t)$:

$$e_N = \|y_{exact} - y_N\|_2 = \sqrt{\int_0^1 (y_{exact}(t) - y_N(t))^2\, dt}. \tag{6.740}$$

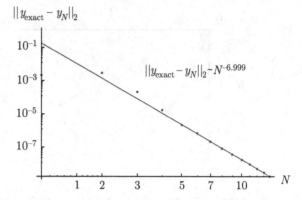

Figure 6.26. Error in solution, $\|y_{exact}(t) - y_N(t)\|_2$, as a function of number of collocation points N.

A plot of the error e_N as a function of N is given in Figure 6.26. We notice a rapid convergence rate of $O(N^{-6.999})$.

The linear differential equations of the previous examples led to linear algebra problems for the coefficients α_n. As demonstrated in the following example, non-linear differential equations subjected to the method of weighted residuals typically lead to nonlinear algebra problems for α_n, which may have nonunique solutions or for which solutions may not exist. These nonunique (or nonexistent) solutions often reflect the nonunique (or nonexistent) solutions of the underlying differential equation.

EXAMPLE 6.64

Use a one-term Galerkin method to identify solutions to the nonlinear problem

$$\frac{d^2y}{dt^2} + y^2 = 1, \qquad y(0) = y(1) = 0. \tag{6.741}$$

Let us assume a solution of the form $y_p(t) = \alpha t(1 - t)$ and take a one-term Galerkin approach with $\psi(t) = \phi(t) = t(1 - t)$. Then the residual is

$$r(t) = \frac{d^2y_p}{dt^2} + y_p^2 - 1 = -2\alpha + \alpha^2(1 - t)^2 t^2 - 1. \tag{6.742}$$

We require the weighted residual to be zero, yielding

$$\langle \psi, r \rangle = \int_0^1 \psi(t) r(t)\, dt = \int_0^1 t(1 - t)\left(-2\alpha + \alpha^2(1 - t)^2 t^2 - 1\right) dt = 0, \tag{6.743}$$

$$-\frac{1}{6} - \frac{\alpha}{3} + \frac{\alpha^2}{140} = 0. \tag{6.744}$$

This has two roots, corresponding to two distinct solutions, both of which satisfy the boundary conditions and approximate the solution of the nonlinear differential equation. The two approximate solutions are

$$y_p(t) = \frac{70 \pm \sqrt{5110}}{3} t(1 - t). \tag{6.745}$$

An exact solution is difficult to write in a simple fashion. However, via a numerical method, two highly accurate solutions can be attained and compared with our simple Galerkin approximations. These are shown in Figure 6.27. One could improve the

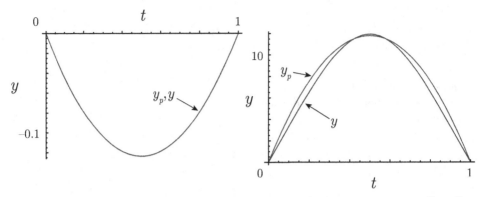

Figure 6.27. Nonunique solutions and their one-term Galerkin approximations to $d^2y/dt^2 + y^2 = 1$, $y(0) = y(1) = 0$.

accuracy by including more terms in the Galerkin expansion. This would lead to more complicated nonlinear algebraic equations to solve.

The method of weighted residuals can also be used for eigenvalue problems, which are necessarily linear but have an additional unknown parameter, λ, which is to be determined.

EXAMPLE 6.65

Consider the eigenvalue problem

$$\frac{d^2y}{dt^2} + \lambda t y = 0, \qquad y(0) = y(1) = 0, \tag{6.746}$$

and determine the eigenvalues through use of a method of weighted residuals.

We first note Eq. (6.746) is a type of Airy equation (see Section A.7.5). We are going to assume a one-term approximate solution of the form

$$y_p(t) = \alpha \phi(t), \tag{6.747}$$

where $\phi(t)$ satisfies the boundary conditions, so $\phi(0) = \phi(1) = 0$. This will induce a residual of

$$r(t) = \alpha \left(\frac{d^2\phi}{dt^2} + \lambda t \phi \right). \tag{6.748}$$

We then set the weighted residual to zero:

$$\alpha \int_0^1 \psi(t) \left(\frac{d^2\phi}{dt^2} + \lambda t \phi \right) dt = 0. \tag{6.749}$$

We can integrate by parts to obtain

$$\psi \frac{d\phi}{dt} \Big|_0^1 + \int_0^1 \left(-\frac{d\psi}{dt} \frac{d\phi}{dt} + \lambda t \psi \phi \right) dt = 0. \tag{6.750}$$

If we insist that ψ satisfy the same boundary conditions as ϕ, we get

$$\int_0^1 \left(-\frac{d\psi}{dt} \frac{d\phi}{dt} + \lambda t \psi \phi \right) dt = 0. \tag{6.751}$$

Table 6.1. *Eigenvalue Estimates as Number of Trial Functions Varies*

terms	λ_1	λ_2	λ_3	λ_4
1	20			
2	19.1884	102.145		
3	19.0093	85.9125	311.078	
$\vdots$	$\vdots$	$\vdots$	$\vdots$	
8	18.9563	81.8866	189.330	343.717

Let us next choose a Galerkin method so that $\psi = \phi$ and thus

$$\int_0^1 \left(-\left(\frac{d\phi}{dt}\right)^2 + \lambda t\phi^2 \right) dt = 0. \tag{6.752}$$

Now we choose a trial function that is at least once-differentiable and satisfies the homogeneous boundary condition

$$\phi(t) = t(1 - t). \tag{6.753}$$

The integral thus reduces to

$$\int_0^1 \left(-(1 - 2t)^2 + \lambda t^2 (1 - t)^2 \right) dt = 0, \tag{6.754}$$

$$-\frac{1}{3} + \frac{\lambda}{60} = 0, \tag{6.755}$$

$$\lambda = 20. \tag{6.756}$$

For more trial functions, which we take as $\phi_n(t) = t^n(1 - t)$, $n = 1, 2, \ldots$, we are led to a linear algebra problem of the type $\mathbf{A} \cdot \boldsymbol{\alpha} = \mathbf{0}$, where the matrix $\mathbf{A}$ is a function of λ. A nontrivial solution to this equation requires $\mathbf{A}$ to be singular, which implies that $\det \mathbf{A} = 0$. Solving the characteristic equation for λ, we can find as many eigenvalues as we choose. Doing so, we obtain the eigenvalue estimates given in Table 6.1 as the number of terms of the approximation is varied. Now the exact solution that satisfies the differential equation and the boundary condition at $t = 0$ is expressed in terms of the Airy functions (see Secion A.7.5) as

$$y(t) = \beta \left(\sqrt{3}\text{Ai}(-\lambda^{1/3}t) - \text{Bi}(-\lambda^{1/3}t) \right), \tag{6.757}$$

where β is an arbitrary constant. To match the boundary condition at $t = 1$, we require

$$y(1) = 0 = \sqrt{3}\text{Ai}(-\lambda^{1/3}) - \text{Bi}(-\lambda^{1/3}). \tag{6.758}$$

We can use a numerical trial and error procedure to obtain as many eigenvalues as we like. Doing so, we find the first four exact eigenvalues to be 18.95626559137320, 81.88658337813677, 189.2209332930337, and 340.9669590647526. The agreement is good with the predictions of our Galerkin method.

6.12 Ritz and Rayleigh-Ritz Methods

What is known by many as the *Ritz method* is one in which y is approximated by a linear combination of trial functions within the context of a variational principle (see

Example 1.16). The extremization problem is then directly solved to determine the coefficients for each trial function. It shares with the method of weighted residuals a similar usage for trial functions but is an otherwise independent method. When the method of weighted residuals is used with a Galerkin method on a problem possessing a variational principle, it yields identical results as the Ritz method. When the Ritz method is applied with trial functions that are piecewise polynomials with compact support, the method reduces to the finite element method.

If the Ritz method is applied to an eigenvalue problem, it is often known as the *Rayleigh-Ritz method*,[25] not to be conflated with the Rayleigh-Ritz theorem Eq. (6.607). Let us next consider the Rayleigh-Ritz method applied to what amounts to exactly the same problem considered in the previous example which used the method of weighted residuals.

EXAMPLE 6.66

Extremize I, where

$$I = \int_0^1 (-\dot{y}^2 + \lambda t y^2) \, dt, \qquad y(0) = y(1) = 0, \tag{6.759}$$

with the Rayleigh-Ritz method.

Our approach here will not define any residual, weighted or otherwise. In the language of the calculus of variations, the extremization of I is the variational principle. Moreover, using the nomenclature of the calculus of variations, we have for our problem

$$f(y, \dot{y}) = -\dot{y}^2 + \lambda t y^2. \tag{6.760}$$

The associated Euler-Lagrange equation (see Eq. (1.166)), for the extremization of I,

$$\frac{\partial f}{\partial y} - \frac{d}{dt}\left(\frac{\partial f}{\partial \dot{y}}\right) = 0, \tag{6.761}$$

$$2\lambda t y - \frac{d}{dt}(-2\dot{y}) = 0, \tag{6.762}$$

$$\lambda t y + \frac{d^2 y}{dt^2} = 0, \tag{6.763}$$

along with its boundary conditions, is identical to Eq. (6.746) from the previous example. Rearranging, it takes on a more transparent form of an eigenvalue problem:

$$-\frac{1}{t}\frac{d^2}{dt^2}y = \lambda y. \tag{6.764}$$

Thus, Eq. (6.759) constitutes an eigenvalue problem for an Airy equation, though it is not obvious from its formulation. Its reconstitution as Eq. (6.763), along with the boundary conditions, is the strong formulation. Also note that with $\mathbf{L} = -(1/t)d^2/dt^2$ and the Rayleigh quotient $R = \langle \mathbf{L}y, y \rangle / \langle y, y \rangle$, we seek y that extremizes R, which gives us $R = \lambda$. In that sense the Rayleigh-Ritz theorem of Section 6.9 has a link to the present Rayleigh-Ritz method; we take a different approach, however.

Let us take again $y \approx y_p = \alpha\phi(t)$, where $\phi(t)$ satisfies the boundary conditions on y. This yields

$$I = \alpha^2 \int_0^1 \left(-\left(\frac{d\phi}{dt}\right)^2 + \lambda t \phi^2\right) dt. \tag{6.765}$$

---

[25] The confusing taxonomy of terms is discussed in Finlayson, who also notes that Rayleigh used what some now call the Ritz method before the birth of Ritz.

Note that I is proportional to the left side of Eq. (6.752), which made no appeal to any extremization principle. We now select $\phi(t) = t(1 - t)$ and evaluate I to get

$$I = \alpha^2 \int_0^1 \left(-(1 - 2t)^2 + \lambda t^3 (1 - t)^2\right) dt, \tag{6.766}$$

$$= \alpha^2 \left(-\frac{1}{3} + \frac{\lambda}{60}\right). \tag{6.767}$$

For an extreme value of I, we need $dI/d\alpha = 0$. Doing so yields

$$\frac{dI}{d\alpha} = 2\alpha \left(-\frac{1}{3} + \frac{\lambda}{60}\right) = 0. \tag{6.768}$$

Ignoring the trivial solution associated with $\alpha = 0$ gives the eigenvalue solution

$$\lambda = 20, \tag{6.769}$$

which is precisely the estimate we found for the one-term weighted residual Galerkin approximation of the previous example, which did not appeal to any variational principle. For problems such as we have examined here, when the Galerkin method is used within a weighted residual scheme, it generates integrals that are also found in the variational principle. However, not all problems have an associated variational principle. Fortunately, for such cases, the method of weighted residuals remains a viable solution strategy.

6.13 Uncertainty Quantification Via Polynomial Chaos

The methods of this chapter can be applied to account for how potential uncertainties present in model parameters affect the solutions of differential equations. To study this, we introduce a stochastic nature into our parameters. There are many ways to deal with these so-called stochastic differential equations. One important method is known variously as "polynomial chaos," "Wiener[26]-Askey[27] chaos," and other names. The term *chaos* in this context was introduced by Wiener; it is in no way connected to the more modern interpretation of chaos from nonlinear dynamics, as considered in Section 9.11.3.

Polynomial chaos is relevant, for example, to a differential equation of the form

$$\frac{dy}{dt} = f(y; k), \qquad y(0) = y_0, \tag{6.770}$$

where k is a parameter. For an individual calculation, k is a fixed constant. But because k is taken to possess an intrinsic uncertainty, it is allowed to take on a slightly different value for the next calculation. We expect a solution of the form $y = y(t; k)$; that is, the effect of the parameter will be realized in the solution. One way to handle the uncertainty in k is to examine a large number of solutions, each for a different value of k. The values chosen for k are driven by its uncertainty distribution, assumed to be known. We thus see how uncertain k is manifested in the solution y. This is known as the Monte Carlo method; it is an effective strategy, although potentially expensive.

For many problems, we can more easily quantify the uncertainty of the output y by propagating the known uncertainty of k via polynomial chaos. The method has

---

[26] Norbert Wiener, 1894–1964, American mathematician.
[27] Richard Askey, 1933–, American mathematician.

the advantage of being a fully deterministic way to account for stochastic effects in parameters appearing in differential equations. There are many variants on this method; we focus only on one canonical linear example that illustrates key aspects of the technique for ordinary differential equations. The method can be extended to algebraic and partial differential equations, both for scalar equations and for systems.

EXAMPLE 6.67

Given that

$$\frac{dy}{dt} = -ky, \qquad y(0) = 1, \tag{6.771}$$

and that k has an associated uncertainty, such that

$$k = \mu + \sigma\xi, \tag{6.772}$$

where μ and σ are known constants, t is a time-like independent variable, y is a dependent variable, and $\xi \in (-\infty, \infty)$ is a random variable with a Gaussian distribution about a mean of zero with standard deviation of unity, find a two-term estimate of the behavior of $y(t)$ that accounts for the uncertainty in k.

For our $k = \mu + \sigma\xi$, the mean value of k can be easily shown to be be μ, and the standard deviation of k is σ. The solution to Eq. (6.771) is

$$y = e^{-kt} = e^{-(\mu+\sigma\xi)t}, \tag{6.773}$$

and will have different values, depending on the value k possesses for that calculation. If there is no uncertainty in k, that is, $\sigma = 0$, the solution to Eq. (6.771) is obviously

$$y = e^{-\mu t}. \tag{6.774}$$

Let us now try to account for the uncertainty in k in predicting the behavior of y when $\sigma \neq 0$. Let us imagine that k has an $N + 1$-term Fourier expansion of

$$k(\xi) = \sum_{n=0}^{N} \alpha_n \phi_n(\xi), \tag{6.775}$$

where $\phi_n(\xi)$ are a known set of basis functions. Now as the random input ξ is varied, k will vary. And we expect the output $y(t)$ to vary, so we can imagine that we really seek $y(t, k) = y(t, k(\xi))$. Dispensing with k in favor of ξ, we can actually seek $y(t, \xi)$. Let us assume that $y(t, \xi)$ has a similar Fourier expansion,

$$y(t, \xi) = \sum_{n=0}^{N} y_n(t) \phi_n(\xi), \tag{6.776}$$

where we have also employed a separation of variables technique, with $\phi_n(\xi)$, $n = 0, \ldots, N$, as a set of basis functions and $y_n(t)$ as the time-dependent amplitude of each basis function. Let us choose the basis functions to be orthogonal:

$$\langle \phi_n(\xi), \phi_m(\xi) \rangle = 0, \qquad n \neq m. \tag{6.777}$$

Because the domain of ξ is doubly infinite, a good choice for the basis functions is the Hermite polynomials; following standard practice, we choose the probabilists' form, $\phi_n(\xi) = He_n(\xi)$ (Section 4.4.6), recalling $He_0(\xi) = 1$, $He_1(\xi) = \xi$, $He_2(\xi) = -1 + \xi^2, \ldots$ Non-Gaussian distributions of parametric uncertainty can render other basis functions to be better choices.

When we equip the inner product with the weighting function,

$$w(\xi) = \frac{1}{\sqrt{2\pi}} e^{-\xi^2/2}, \tag{6.778}$$

we find our chosen basis functions are orthogonal:

$$\langle \phi_n(\xi), \phi_m(\xi) \rangle = \int_{-\infty}^{\infty} \phi_n(\xi)\phi_m(\xi)w(\xi)\,d\xi, \tag{6.779}$$

$$= \int_{-\infty}^{\infty} He_n(\xi)He_m(\xi)\frac{1}{\sqrt{2\pi}}e^{-\xi^2/2}\,d\xi, \tag{6.780}$$

$$= n!\delta_{nm}. \tag{6.781}$$

Let us first find the coefficients α_n in the Fourier-Hermite expansion of $k(\xi)$:

$$k(\xi) = \sum_{n=0}^{N} \alpha_n \phi_n(\xi), \tag{6.782}$$

$$\langle \phi_m(\xi), k(\xi) \rangle = \langle \phi_m(\xi), \sum_{n=0}^{\infty} \alpha_n \phi_n(\xi) \rangle, \tag{6.783}$$

$$= \sum_{n=0}^{\infty} \alpha_n \langle \phi_m(\xi), \phi_n(\xi) \rangle, \tag{6.784}$$

$$= \sum_{n=0}^{N} \alpha_n n!\delta_{mn}, \tag{6.785}$$

$$= m!\alpha_m, \tag{6.786}$$

$$\alpha_n = \frac{\langle \phi_n(\xi), k(\xi) \rangle}{n!}, \tag{6.787}$$

$$= \frac{\langle \phi_n(\xi), \mu + \sigma\xi \rangle}{n!}, \tag{6.788}$$

$$= \frac{1}{\sqrt{2\pi}n!} \int_{-\infty}^{\infty} He_n(\xi)\,(\mu + \sigma\xi)\,e^{-\xi^2/2}\,d\xi. \tag{6.789}$$

Because of the polynomial nature of $k = \mu + \sigma\xi$ and the orthogonality of the polynomial functions, there are only two nonzero terms in the expansion: $\alpha_0 = \mu$ and $\alpha_1 = \sigma$; thus,

$$k(\xi) = \mu + \sigma\xi = \alpha_0 He_0(\xi) + \alpha_1 He_1(\xi) = \mu He_0(\xi) + \sigma He_1(\xi). \tag{6.790}$$

So for this simple distribution of $k(\xi)$, the infinite Fourier series expansion of Eq. (6.776) is a finite two-term expansion. We actually could have seen this by inspection, but it was useful to go through the formal exercise.

Now, substitute the expansions of Eqs. (6.775, 6.776) into the governing Eq. (6.771):

$$\frac{d}{dt}\left(\underbrace{\sum_{n=0}^{N} y_n(t)\phi_n(\xi)}_{y}\right) = -\left(\underbrace{\sum_{n=0}^{N} \alpha_n \phi_n(\xi)}_{k}\right)\left(\underbrace{\sum_{m=0}^{N} y_m(t)\phi_m(\xi)}_{y}\right). \tag{6.791}$$

Equation (6.791) forms $N + 1$ ordinary differential equations, still with an explicit dependency on ξ. We need an initial condition for each of them. The initial condition $y(0) = 1$ can be recast as

$$y(0, \xi) = 1 = \sum_{n=0}^{N} y_n(0)\phi_n(\xi). \tag{6.792}$$

Now we could go through the same formal exercise as for k to determine the Fourier expansion of $y(0) = 1$. But because $\phi_0(\xi) = 1$, we see by inspection that the set of $N + 1$ initial conditions are

$$y_0(0) = 1, \quad y_1(0) = 0, \quad y_2(0) = 0, \quad \ldots, \quad y_N(0) = 0. \tag{6.793}$$

Let us now rearrange Eq. (6.791) to get

$$\sum_{n=0}^{N} \frac{dy_n}{dt} \phi_n(\xi) = -\sum_{n=0}^{N} \sum_{m=0}^{N} \alpha_n y_m(t) \phi_n(\xi) \phi_m(\xi). \tag{6.794}$$

We still need to remove the explicit dependency on the random variable ξ. To achieve this, we will take the inner product of Eq. (6.794) with a set of functions, so as to simplify the system into a cleaner system of ordinary differential equations. Let us choose to invoke a Galerkin procedure by taking the inner product of Eq. (6.794) with $\phi_l(\xi)$:

$$\langle \phi_l(\xi), \sum_{n=0}^{N} \frac{dy_n}{dt} \phi_n(\xi) \rangle = -\langle \phi_l(\xi), \sum_{n=0}^{N} \sum_{m=0}^{N} \alpha_n y_m(t) \phi_n(\xi) \phi_m(\xi) \rangle, \tag{6.795}$$

$$\sum_{n=0}^{N} \frac{dy_n}{dt} \langle \phi_l(\xi), \phi_n(\xi) \rangle = -\sum_{n=0}^{N} \sum_{m=0}^{N} \alpha_n y_m(t) \langle \phi_l(\xi), \phi_n(\xi) \phi_m(\xi) \rangle, \tag{6.796}$$

$$\frac{dy_l}{dt} \langle \phi_l(\xi), \phi_l(\xi) \rangle = -\sum_{n=0}^{N} \sum_{m=0}^{N} \alpha_n y_m(t) \langle \phi_l(\xi), \phi_n(\xi) \phi_m(\xi) \rangle, \tag{6.797}$$

$$\frac{dy_l}{dt} = -\frac{1}{\langle \phi_l(\xi), \phi_l(\xi) \rangle} \sum_{n=0}^{N} \sum_{m=0}^{N} \alpha_n y_m(t) \langle \phi_l(\xi), \phi_n(\xi) \phi_m(\xi) \rangle, \tag{6.798}$$

$$\frac{dy_l}{dt} = -\frac{1}{l!} \sum_{n=0}^{N} \sum_{m=0}^{N} \alpha_n y_m(t) \langle \phi_l(\xi), \phi_n(\xi) \phi_m(\xi) \rangle, \qquad l = 0, \ldots, N. \tag{6.799}$$

Equation (6.799) forms $N + 1$ ordinary differential equations, with $N + 1$ initial conditions provided by Eq. (6.793). All dependency on ξ is removed by explicit evaluation of the inner products for all l, n, and m. We could have arrived at an analogous system of ordinary differential equations had we chosen any of the other standard set of functions for the inner product. For example, Dirac delta functions would have led to a collocation method. The full expression of the unusual inner product which appears in Eq. (6.799) is

$$\langle \phi_l(\xi), \phi_n(\xi) \phi_m(\xi) \rangle = \int_{-\infty}^{\infty} \phi_l(\xi) \phi_n(\xi) \phi_m(\xi) \frac{1}{\sqrt{2\pi}} e^{-\xi^2/2} \, d\xi. \tag{6.800}$$

This relation can be reduced further, but it is not straightforward.

When $N = 1$, we have a two-term series, with $l = 0, 1$. Detailed evaluation of all inner products yields two ordinary differential equations:

$$\frac{dy_0}{dt} = -\mu y_0 - \sigma y_1, \qquad y_0(0) = 1, \tag{6.801}$$

$$\frac{dy_1}{dt} = -\sigma y_0 - \mu y_1, \qquad y_1(0) = 0. \tag{6.802}$$

When $\sigma = 0$, $y_0(t) = e^{-\mu t}$, $y_1(t) = 0$, and we recover our original nonstochastic result. For $\sigma \neq 0$, this linear system can be solved exactly using methods of the upcoming Section 9.5.2. Direct substitution reveals that the solution is in fact

$$y_0(t) = e^{-\mu t} \cosh(\sigma t), \tag{6.803}$$

$$y_1(t) = -e^{-\mu t} \sinh(\sigma t). \tag{6.804}$$

Thus, the two-term approximation is

$$y(t, \xi) \approx y_0(t) \phi_0(\xi) + y_1(t) \phi_1(\xi), \tag{6.805}$$

$$= e^{-\mu t} (\cosh(\sigma t) - \sinh(\sigma t) \xi). \tag{6.806}$$

The nonstochastic solution $e^{-\mu t}$ is obviously modulated by the uncertainty. Even when $\xi = 0$, there is a weak modulation by $\cosh(\sigma t) \approx 1 + \sigma^2 t^2/2 + \sigma^4 t^4/24 + \cdots$.

Standard probability theory lets us estimate the mean value of $y(t, \xi)$, which we call $\overline{y}(t)$, over a range of normally distributed values of ξ:

$$\overline{y}(t) = \int_{-\infty}^{\infty} y(t, \xi) \frac{1}{\sqrt{2\pi}} e^{-\xi^2/2} \, d\xi, \tag{6.807}$$

$$= \int_{-\infty}^{\infty} e^{-\mu t} \left(\cosh\left(\sigma t\right) - \sinh\left(\sigma t\right) \xi \right) \frac{1}{\sqrt{2\pi}} e^{-\xi^2/2} \, d\xi, \tag{6.808}$$

$$= e^{-\mu t} \cosh\left(\sigma t\right). \tag{6.809}$$

Thus, the mean value of y is $y_0(t) = e^{-\mu t} \cosh(\sigma t)$. The standard deviation, $\sigma_s(t)$, of the solution is found by a similar process

$$\sigma_s(t) = \sqrt{\int_{-\infty}^{\infty} (y(t, \xi) - \overline{y}(t))^2 \frac{1}{\sqrt{2\pi}} e^{-\xi^2/2} \, d\xi}, \tag{6.810}$$

$$= \sqrt{\int_{-\infty}^{\infty} (-e^{-\mu t} \sinh\left(\sigma t\right) \xi)^2 \frac{1}{\sqrt{2\pi}} e^{-\xi^2/2} \, d\xi}, \tag{6.811}$$

$$= e^{-\mu t} \sinh(\sigma t). \tag{6.812}$$

Note that $\sigma_s(t) = |y_1(t)|$. Also, note that σ_s is distinct from σ, the standard deviation of the parameter k.

All of this is easily verified by direct calculation. If we take $\mu = 1$ and $\sigma = 1/3$, we have $k = 1 + \xi/3$, recalling that ξ is a random number, with a Gaussian distribution with unity standard deviation about zero. Let us examine various predictions at $t = 1$. Ignoring all stochastic effects, we might naïvely predict that the expected value of y should be

$$y(t = 1) = e^{-\mu t} = e^{-(1)(1)} = 0.367879, \tag{6.813}$$

with no standard deviation. However, if we execute so-called Monte Carlo simulations where k is varied through its range, calculate y at $t = 1$ for each realization of k, and then take the mean value of all predictions, we find for 10^8 simulations that the mean value is

$$y_{Monte\ Carlo}(t = 1) = 0.3889. \tag{6.814}$$

This number will slightly change if a different set of random values of k are tested. Remarkably though, $y_{Monte\ Carlo}$ is well predicted by our polynomial chaos estimate of

$$y_0(t = 1) = e^{-\mu t} \cosh(\sigma t) = e^{-(1)(1)} \cosh\left(\frac{1}{3}\right) = 0.388507. \tag{6.815}$$

We could further improve the Monte Carlo estimate by taking more samples. We could further improve the polynomial chaos estimate by including more terms in the expansion. As the number of Monte Carlo estimates and the number terms in the polynomial chaos expansion approach infinity, the two estimates would converge. And they would converge to a number different than that of the naïve estimate. The exponential function warped the effect of the Gaussian-distributed k such that the realization of y at $t = 1$ was distorted to a greater value.

For the same 10^8 simulations, the Monte Carlo method predicts a standard deviation of y at $t = 1$ of

$$\sigma_{s,\ Monte\ Carlo} = 0.1333. \tag{6.816}$$

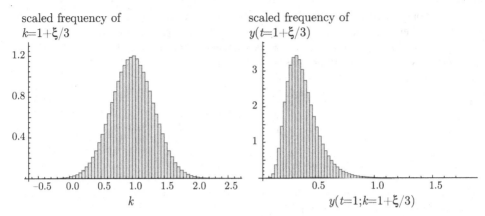

Figure 6.28. Histograms for distribution of k and $y(t = 1; k = 1 + \xi/3)$ for 10^6 Monte Carlo simulations for various values of k.

This number is well estimated by the magnitude, $|y_1(t = 1)|$:

$$|y_1(t = 1)| = e^{-\mu t} \sinh(\sigma t) = e^{-(1)(1)} \sinh\left(\frac{1}{3}\right) = 0.124910. \qquad (6.817)$$

Again, both estimates could be improved by more samples, and more terms in the expansion, respectively. For example, a three-term expansion yields a mean and standard deviation of 0.38891 and 0.132918, respectively.

Histograms of the scaled frequency of occurrence of k and $y(t = 1; k = 1 + \xi/3)$ within bins of specified width from the Monte Carlo method for 10^6 realizations are plotted in Figure 6.28. We show 50 bins within which the scaled number of occurrences of k and $y(t = 1)$ are realized. The scaling factor applied to the number of occurrences was selected so that the area under the curve is unity. This is achieved by scaling the number of occurrences within a bin by the product of the total number of occurrences and the bin width; this allows the scaled number of occurrences to be thought of as a probability density. As designed, k appears symmetric about its mean value of unity, with a standard deviation of $1/3$. Detailed analysis would reveal that k in fact has a Gaussian distribution. But $y(t = 1; k = 1 + \xi/3)$ does not have a Gaussian distribution about its mean, as it has been skewed by the dynamics of the differential equation for y. The time-evolution of estimates of y and error bounds are plotted in Figure 6.29.

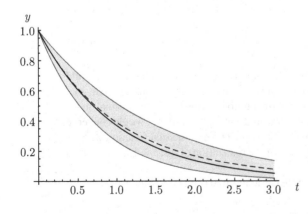

Figure 6.29. Estimates of $y(t)$ that satisfies $dy/dt = -ky$, $y(0) = 1$, for $k = 1 + \xi/3$, where ξ is a random variable, normally distributed about zero. The thick solid line gives the naïve estimate, e^{-t}. The thick dashed line gives $y_0(t)$, and the two thin outer lines bounding the gray region give $y_0(t) \pm y_1(t)$, that is, the mean value of y, plus or minus one standard deviation.

EXERCISES

1. Find the supremum and infimum of the set $\mathbb{S} = \{1/n\}$, where $n = 1, 2, \ldots$.

2. Show that the limits

$$\lim_{a \to 0+} \left(\int_{-1}^{-a} \frac{dx}{x} + \int_{a}^{1} \frac{dx}{x} \right)$$

and

$$\lim_{a \to 0+} \left(\int_{-1}^{-a} \frac{dx}{x} + \int_{2a}^{1} \frac{dx}{x} \right)$$

are different.

3. Determine the Cauchy principal value of

$$\int_{-2}^{1} \frac{dx}{x^3}.$$

4. Show that the set of solutions of the linear equations

$$x_1 + 3x_2 + x_3 - x_4 = 0$$

$$-2x_1 + 2x_2 - x_3 + x_4 = 0$$

form a vector space. Find its dimension and a set of basis vectors.

5. Show that the set of all matrices $\mathbb{A} : \mathbb{R}^N \to \mathbb{R}^N$ is a vector space under the usual rules of matrix manipulation.

6. Apply the Cauchy-Schwarz inequality to elements of the following vector spaces: (a) $\mathbb{R}^n$ and (b) $\mathbb{L}_2$.

7. Show that

$$\langle x, y \rangle = \int_{a}^{b} x(t) \overline{y}(t) w(t) \, dt,$$

with a known weighting function $w(t) > 0$, $t \in [a, b]$, is a suitable inner product.

8. Find the first three terms of an orthonormal series of functions with $t \in [0, 1]$, choosing them among linear combinations of $1, t, t^2, \ldots$. Use the inner product

$$\langle x, y \rangle = \int_{0}^{1} t x(t) y(t) \, dt.$$

9. Show that e^{-x^2} and its derivative are orthogonal in $\mathbb{L}_2(-\infty, \infty)$.

10. Are the basis vectors $u_1 = 1$, $u_2 = t$, $u_3 = t^2$, with inner product defined as

$$\langle u_i, u_j \rangle = \int_{0}^{1} u_i(t) u_j(t) \, dt,$$

orthogonal? If not, find the reciprocal basis.

11. Show that the n functions defined by

$$\phi_i(x) = \begin{cases} 1, & \frac{i-1}{n} < x \le \frac{i}{n} \\ 0, & \text{otherwise} \end{cases}$$

for $i = 1, 2, \ldots, n$, are orthogonal in $\mathbb{L}_2[0, 1]$. Expand $f(x) = x^2$ in terms of these orthogonal functions.

12. The following norms can be used in $\mathbb{R}^N$, where $x = (\xi_1, \ldots, \xi_N) \in \mathbb{R}^N$:

   (a) $\|x\|_\infty = \max_{1 \le n \le N} |\xi_n|$,
   (b) $\|x\|_1 = \sum_{n=1}^{N} |\xi_n|$,
   (c) $\|x\|_2 = (\sum_{n=1}^{N} |\xi_n|^2)^{1/2}$,
   (d) $\|x\|_p = (\sum_{n=1}^{N} |\xi_n|^p)^{1/p}$, $1 \le p < \infty$.
   Demonstrate with examples that these are all valid norms.

13. The function $f(x) = x + 1$ is a member of $\mathbb{L}_2[0, 1]$. Find its norm.

14. Find the norm of the operator $\mathbf{L}$, where $\mathbf{L}x = t^2 x(t)$ and $x \in \mathbb{L}_2[0, 1]$. Show that $\mathbf{L}$ is positive definite.

15. Find the distance between the functions x and x^3 under the $\mathbb{L}_2[0, 1]$ norm.

16. Find the inner product of the functions x and x^3 using the $\mathbb{L}_2[0,1]$ definition.

17. If $x_1, x_2, \ldots, x_N$ and $y_1, y_2, \ldots, y_N$ are real numbers, show that

$$\left(\sum_{n=1}^{N} x_n y_n \right)^2 \le \left(\sum_{n=1}^{N} x_n^2 \right) \left(\sum_{n=1}^{N} y_n^2 \right).$$

18. If $x, y \in \mathbb{X}$, an inner product space, and x is orthogonal to y, show that $\|x + \alpha y\| = \|x - \alpha y\|$, where α is a scalar.

19. Consider the sequence $\{\frac{1 + \frac{1}{N}}{2 + \frac{1}{N}}\}$ in $\mathbb{R}^N$. Show that this is a Cauchy sequence. Does it converge?

20. For an inner product space, show that

   (a) $\langle x, y + z \rangle = \langle x, y \rangle + \langle x, z \rangle$,
   (b) $\langle \alpha x, y \rangle = \overline{\alpha} \langle x, y \rangle$,
   (c) $\langle x, y \rangle = \langle y, x \rangle$ in a real vector space.

21. Let $\mathbb{Q}, \mathbb{C}$, and $\mathbb{R}$ be the sets of all rational, complex and real numbers, respectively. For the following, determine if $\mathbb{A}$ is a vector space over the field $\mathbb{F}$. For finite-dimensional vector spaces, find also a set of basis vectors.

   (a) $\mathbb{A}$ is the set of all polynomials that are all exactly of degree n, $\mathbb{F} = \mathbb{R}$.
   (b) $\mathbb{A}$ is the set of all functions with continuous second derivatives over the interval $[0, L]$ satisfying the differential equation $y'' + 2y' + y = 0$, $\mathbb{F} = \mathbb{R}$.
   (c) $\mathbb{A} = \mathbb{R}, \mathbb{F} = \mathbb{R}$.
   (d) $\mathbb{A} = \{(a_1, a_2, a_3) \mid a_1, a_2 \in \mathbb{Q}, 2a_1 + a_2 = 4a_3\}, \mathbb{F} = \mathbb{Q}$.
   (e) $\mathbb{A} = \mathbb{C}, \mathbb{F} = \mathbb{Q}$.
   (f) $\mathbb{A} = \{ae^x + be^{-2x} \mid a, b \in \mathbb{R}, x \in [0, 1]\}, \mathbb{F} = \mathbb{R}$.

22. Which of the following subsets of $\mathbb{R}^3$ constitute a subspace of $\mathbb{R}^3$ where $x = (x_1, x_2, x_3)^T \in \mathbb{R}^3$:

   (a) All x with $x_1 = x_2$ and $x_3 = 0$.
   (b) All x with $x_1 = x_2 + 1$.
   (c) All x with positive x_1, x_2, x_3.
   (d) All x with $x_1 - x_2 + x_3 = $ constant k.

23. Given a set $\mathbb{S}$ of linearly independent vectors in a vector space $\mathbb{V}$, show that any subset of $\mathbb{S}$ is also linearly independent.

24. Do the following vectors, $(3, 1, 4, -1)^T, (1, -4, 0, 4)^T, (-1, 2, 2, 1)^T, (-1, 9, 5, -6)^T$, form a basis in $\mathbb{R}^4$?

25. Given x_1, the iterative procedure $x_{n+1} = \mathbf{L}x_n$ generates $x_2, x_3, x_4, \ldots$, where $\mathbf{L}$ is a linear operator and all the xs belong to a complete normed space. Show that $\{x_n, n = 1, 2, \ldots\}$ is a Cauchy sequence if $\|\mathbf{L}\| < 1$. Does it converge? If so, find the limit.

26. If $\{e_n, n = 1, 2, \ldots\}$ is an orthonormal set in a Hilbert space $\mathbb{H}$, show that for every $x \in \mathbb{H}$, the vector $y = \sum_{n=1}^{N} \langle x, e_n \rangle e_n$ exists in $\mathbb{H}$, and that $x - y$ is orthogonal to every e_n.

27. Consider the inner product

$$(x, y) = \int_a^b w(t) x(t) y(t) \, dt,$$

where $w(t) > 0$. For $a \leq t \leq b$, show that the Sturm-Liouville operator

$$\mathbf{L} = \frac{1}{w(t)} \left(\frac{d}{dt} \left(p(t) \frac{d}{dt} \right) + r(t) \right),$$

with $\alpha x(a) + \beta x'(a) = 0$ and $\gamma x(b) + \delta x'(b) = 0$, is self-adjoint.

28. With $f_1(x) = 1 + i + x$ and $f_2(x) = 1 + ix + ix^2$, where f_1 and f_2 both belong to $\overline{\mathbb{L}}_2[0, 1]$, find the
    (a) norms of $f_1(x)$ and $f_2(x)$.
    (b) inner product of $f_1(x)$ and $f_2(x)$.
    (c) "distance" between $f_1(x)$ and $f_2(x)$.

29. For elements x, y, and z of an inner product space, prove the Apollonius[28] identity:

$$\|z - x\|_2^2 + \|z - y\|_2^2 = \frac{1}{2}\|x - y\|_2^2 + 2 \left\| z - \frac{1}{2}(x + y) \right\|_2^2.$$

30. If $x, y \in \mathbb{X}$, an inner product space, and x is orthogonal to y, show that $\|x + ay\|_2 = \|x - ay\|_2$, where a is a scalar.

31. Find the first three terms of an orthonormal series of functions in $t \in [0, 1]$, choosing them among linear combinations of $1, t, t^2, \ldots$. Use the inner product

$$\langle x, y \rangle = \int_0^1 tx(t)y(t) \, dt.$$

32. Let $C(0,1)$ be the space of all continuous functions in $(0,1)$ with the norm

$$\|f\|_2 = \sqrt{\int_0^1 |f(t)|^2 \, dt}.$$

Show that

$$f_n(t) = \begin{cases} 2^n t^{n+1}, & 0 \leq t < \frac{1}{2}, \\ 1 - 2^n(1-t)^{n+1}, & \frac{1}{2} \leq t \leq 1, \end{cases}$$

belongs to $C(0,1)$. Show also that $\{f_n, n = 1, \ldots\}$ is a Cauchy sequence and that $C(0,1)$ is not complete.

---

[28] Apollonius of Perga , ca. 262–190 B.C., Greek astronomer and geometer.

33. Show that the set of solutions of the linear equations

$$x_1 + 3x_2 + x_3 - x_4 = 0$$

$$-2x_1 + 2x_2 - x_3 + x_4 = 0$$

forms a vector space. Find its dimension and a set of basis vectors.

34. Let

$$\mathbf{A} = \begin{pmatrix} 1 & 1 & 1 \\ 0 & 1 & 1 \\ 0 & 0 & 1 \end{pmatrix}.$$

For $\mathbf{A} : \mathbb{R}^3 \to \mathbb{R}^3$, find $\|\mathbf{A}\|$ if the norm of $\mathbf{x} = (x_1, x_2, x_3)^T \in \mathbb{R}^3$ is given by

$$\|\mathbf{x}\|_\infty = \max(|x_1|, |x_2|, |x_3|).$$

35. For any complete orthonormal set $\{\varphi_i, \ i = 1, 2, \ldots\}$ in a Hilbert space $\mathbb{H}$, show that

$$u = \sum_i \langle u, \varphi_i \rangle \varphi_i,$$

$$\langle u, v \rangle = \sum_i \langle u, \varphi_i \rangle \langle v, \varphi_i \rangle,$$

$$\|u\|_2^2 = \sum_i |\langle u, \varphi_i \rangle|^2,$$

where u and v belong to $\mathbb{H}$.

36. Show that the set $\mathbb{P}^4[0,1]$ of all polynomials of degree 4 or less in the interval $0 < x < 1$ is a vector space. What is the dimension of this space?

37. Show that the functions $u_1(t), u_2(t), \ldots, u_N(t)$ are orthogonal in $\mathbb{L}_2(0,1]$, where

$$u_n(t) = \begin{cases} 1, & \frac{n-1}{N} < t \le \frac{n}{N} \\ 0, & \text{otherwise.} \end{cases}$$

Expand t^2 in terms of these functions.

38. Show that

$$\sqrt{\int_a^b (f(x) + g(x))^2 \, dx} \le \sqrt{\int_a^b (f(x))^2 \, dx} + \sqrt{\int_a^b (g(x))^2 \, dx},$$

where $f(x)$ and $y(x)$ belong to $\mathbb{L}_2[a, b]$.

39. If x and y belong to a suitably defined Hilbert space and are orthogonal, show that

$$\|x + y\|^2 = \|x\|^2 + \|y\|^2.$$

40. Using the Gram-Schmidt procedure, find the first three members of the orthonormal set belonging to $\mathbb{L}_2(-\infty, \infty)$ using the basis functions $\{\exp(-t^2/2), t\exp(-t^2/2), t^2\exp(-t^2/2), \ldots\}$. You may need the following definite integral:

$$\int_{-\infty}^{\infty} \exp(-t^2/2) \, dt = \sqrt{2\pi}.$$

41. Starting from the unit vector $(1/2, 1/2, 1/2, 1/2)^T$ in $\mathbb{R}^4$, use the Gram-Schmidt procedure to find other vectors to complete the orthonormal set.

42. For each set of basis vectors

    (a) $u_1 = (1, 0, 0, 0)^T$, $u_2 = (1, 1, 1, 1)^T$, $u_3 = (1, 2, 3, 4)^T$, $u_4 = (1, 2, 3, 5)^T$,
    (b) $u_1 = (1 + i, 0, 0, 0)^T$, $u_2 = (i, i, i, i)^T$, $u_3 = (1 - i, 2, 3, 4)^T$,
        $u_4 = (i, 2i, 3i, 5i)^T$,

    use the Gram-Schmidt procedure to find an associated orthonormal set of basis vectors.

43. Find an orthonormal set of vectors $\{\varphi_1, \varphi_2, \varphi_3, \ldots\}$ in $\mathbb{L}_2[0, 1]$ using linear combinations of the linearly independent set of vectors

    (a) $\{1, t^2, t^4, \ldots, t^{2n}\}$,
    (b) $\{1, it, -t^2, -it^3, t^4, \ldots, (it)^n\}$,
    where $t \in [0, 1]$. Consider $n = 3$ as well as arbitrary n. In each case does the set of orthonormal vectors form a basis?

44. Find an orthonormal set of vectors $\{\varphi_1, \varphi_2, \varphi_3\}$ in $\mathbb{L}_2[0, 1]$ using linear combinations of the linearly independent set of vectors $\{t, \sin t, e^t\}$, where $t \in [0, 1]$.

45. Let $\{u_1, \ldots, u_N\}$ be an orthonormal set in an inner product space $\mathbb{S}$. Approximate $x \in \mathbb{S}$ by $y = \beta_1 u_1 + \cdots + \beta_N u_N$, where the βs are to be selected. Show that $\|x - y\|$ is a minimum if we choose $\beta_i = \langle x, u_i \rangle$.

46. If $\{x_i\}$ is a sequence in an inner product space such that the series $\|x_1\| + \|x_2\| + \cdots$ converges, show that $\{s_N\}$ is a Cauchy sequence, where $s_N = x_1 + x_2 + \cdots + x_N$.

47. Using a dual basis, expand the vector $(1, 3, 2)^T$ in terms of the basis vectors $(1, 1, 1)^T$, $(1, 0, -1)^T$ and $(1, 0, 1)^T$ in $\mathbb{R}^3$. The inner product is defined as usual.

48. Show the vectors $u_1 = (-i, 0, 2, 1 + i)^T$, $u_2 = (1, 2, i, 3)^T$, $u_3 = (3 + i, 3 - i, 0, -2)^T$, $u_4 = (1, 0, 1, 3)^T$ form a basis in $\mathbb{C}^4$. Find the set of reciprocal basis vectors. For $x \in \mathbb{C}^4$ and $x = (i, 3 - i, -2, 2)^T$, express x as an expansion in the previously defined basis vectors; that is, find α_i such that $x = \alpha_i u_i$.

49. Consider the function $x(t) = \sin(4t)$ for $t \in [0, 1]$. Project $x(t)$ onto the space spanned by the functions $u_m(t)$ so as to find the coefficients α_m, where $x(t) \approx x_p(t) = \sum_{m=1}^{M} \alpha_m u_m(t)$ if the basis functions are

    (a) $M = 2$; $u_1(t) = t$, $u_2(t) = t^2$.
    (b) $M = 3$; $u_1(t) = 1$, $u_2(t) = t^2$, $u_3(t) = \tan t$.
    In each case, show $x(t)$ and its approximation on the same plot.

50. Project the vector $x = (1, 2, 3, 4)^T$ onto the space spanned by the vectors, u_1, u_2, so as to find the projection $x \approx x_p = \alpha_1 u_1 + \alpha_2 u_2$:

    (a) $u_1 = \begin{pmatrix} 1 \\ 0 \\ 0 \\ 0 \end{pmatrix}$, $u_2 = \begin{pmatrix} 1 \\ 1 \\ 1 \\ 1 \end{pmatrix}$,

    (b) $u_1 = \begin{pmatrix} i \\ 0 \\ 0 \\ 0 \end{pmatrix}$, $u_2 = \begin{pmatrix} i \\ 1 \\ i \\ 1 \end{pmatrix}$.

51. Show the vectors $u_1 = (-i, 3, 2 - i, 1 + i)^T$, $u_2 = (i + 1, 2, i, 3)^T$, $u_3 = (3 + i, 3 - i, 0, -2)^T$, $u_4 = (1, 0, 2, 3)^T$ form a basis in $\mathbb{C}^4$. Find the set of reciprocal basis vectors. For $x \in \mathbb{C}^4$ and $x = (i, 3 - i, -5, 2 + i)^T$,

(a) Express x as an expansion in the basis vectors; that is, find α_i such that $x = \sum_{i=1}^{4} \alpha_i u_i$

(b) Project onto the space spanned by u_1, u_2, and u_3; that is, find the best set of α_i such that $x \approx x_p = \sum_{i=1}^{3} \alpha_i u_i$.

52. Apply Parseval's equation to the coefficients obtained from a trigonometric Fourier analysis of the periodic square wave defined by

$$f(x) = \begin{cases} 0, & -1 \le x < 0, \\ 1, & 0 \le x \le 1. \end{cases}$$

53. The linear operator $\mathbf{A} : \mathbb{X} \to \mathbb{Y}$, where $\mathbb{X} = \mathbb{R}^2$, $\mathbb{Y} = \mathbb{R}^2$. The norms in $\mathbb{X}$ and $\mathbb{Y}$ are defined by

$$x = (\xi_1, \xi_2)^T \in \mathbb{X}, \|x\|_\infty = \max\left(|\xi_1|, |\xi_2|\right),$$
$$y = (\eta_1, \eta_2)^T \in \mathbb{Y}, \|y\|_1 = |\eta_1| + |\eta_2|.$$

Find $\|\mathbf{A}\|$ if

$$\mathbf{A} = \begin{pmatrix} 3 & -1 \\ 5 & -2 \end{pmatrix}.$$

54. Show that

$$y'' + y' + y = 0$$

is not formally self-adjoint. Multiply the equation by a suitable factor to make it formally self-adjoint.

55. Let $\mathbf{x} = \{x_1, x_2, \ldots, x_n\}$ and $\mathbf{y} = \{y_1, y_2, \ldots, y_n\}$. Show that the discrete form of the integral (use the trapezoidal rule, Section A.4, for integration),

$$y(t) = \int_a^b K(s,t) x(s)\, ds,$$

is of the form $\mathbf{y} = \mathbf{K} \cdot \mathbf{x}$. Identify $\mathbf{K}$, and find the condition under which it is self-adjoint.

56. A differential operator $\mathbf{L}$ is defined by $\mathbf{L}y = dy/dt + y$, where $y \in \mathbb{L}_2[0,1]$ and $y(0) = 0$. Find its adjoint.

57. Find the norm of the operator $\mathbf{L}$, where $\mathbf{L}x = t^2 x(t)$ and $x \in \mathbb{L}_2[0,1]$. Show that $\mathbf{L}$ is positive definite.

58. Let the linear operator $\mathbf{A} : \mathbb{C}^2 \to \mathbb{C}^2$ be represented by the matrix $\mathbf{A}$. Find $\|\mathbf{A}\|$ if all vectors in the domain and range are within a Hilbert space:

(a) $\mathbf{A} = \begin{pmatrix} 2 & -4 \\ 1 & 5 \end{pmatrix}.$

(b) $\mathbf{A} = \begin{pmatrix} 2+i & -4 \\ 1 & 5 \end{pmatrix}.$

59. Prove that $(\mathbf{L}_a \mathbf{L}_b)^* = \mathbf{L}_b^* \mathbf{L}_a^*$ when $\mathbf{L}_a$ and $\mathbf{L}_b$ are linear operators which operate on vectors in a Hilbert space.

60. Find the null space of the

(a) matrix operator

$$A = \begin{pmatrix} 1 & 1 & 1 \\ 2 & 2 & 1 \\ 2 & 2 & 1 \end{pmatrix},$$

(b) differential operator

$$L = \frac{d^2}{dt^2} + k^2.$$

61. A differential operator L is defined by $Ly = dy/dt + y$, where $y \in L_2[0, 1]$ and $y(0) = 0$. Find its adjoint.

62. Consider functions of two variables in a domain Ω with the inner product defined as

$$\langle u, v \rangle = \iint_\Omega u(x, y)v(x, y) \, dx \, dy.$$

Find the space of functions to which u and v belong such that the Laplacian operator is self-adjoint. Hint: Properly account for the effect of the boundary of Ω.

63. Test the positive definiteness of a diagonal matrix with positive real numbers on the diagonal.

64. If $Lx = a_0(t)(d^2x/dt^2) + a_1(t)(dx/dt) + a_2(t)x$, find the operator that is formally adjoint to it.

65. Show that if A is a linear operator such that

(a) $A : (\mathbb{R}^N, || \cdot ||_\infty) \to (\mathbb{R}^N, || \cdot ||_1)$, then $||A|| = \sum_{i,j=1}^{N} A_{ij}$.

(b) $A : (\mathbb{R}^N, || \cdot ||_\infty) \to (\mathbb{R}^N, || \cdot ||_\infty)$, then $||A|| = \max_{1 \leq i \leq N} \sum_{j=1}^{N} A_{ij}$.

66. If

$$Lu = a(x)\frac{d^2u}{dx^2} + b(x)\frac{du}{dx} + c(x)u,$$

show that

$$L^*u = \frac{d^2}{dx^2}(au) - \frac{d}{dx}(bu) + cu.$$

67. Find the eigenvalues and eigenfunctions of the operator

$$L = -\left(\frac{d^2}{dx^2} + 2\frac{d}{dx} + 1 \right),$$

which operates on functions $y \in L_2[0, 5]$, which vanish at $x = 0$ and $x = 5$.

68. Let $\mathbb{S}$ be a subspace of $L_2[0, 1]$ such that for every $y(t) \in \mathbb{S}$, $y(0) = y'(1) = 0$. Find the eigenvalues and eigenvectors of L where $L = d^2/dt^2$.

69. Let $\mathbb{S}$ be a subspace of $L_2[0, 1]$ such that for every $x \in \mathbb{S}$, $x(0) = 0$, and $\dot{x}(0) = 1$. Find the eigenvalues and eigenfunctions of $L = -d^2/dt^2$ operating on elements of $\mathbb{S}$.

70. Find the eigenvalues and eigenfunctions of the operator L where

$$Ly = (1 - t^2)\frac{d^2y}{dt^2} - t\frac{dy}{dt},$$

with $t \in [-1, 1]$ and $y(-1) = y(1) = 0$. Show that there exists a weighting function $r(x)$ such that the eigenfunctions are orthogonal in $[-1, 1]$ with respect to it.

71. Find the null space of the operator $\mathbf{L}$ defined by $\mathbf{L}x = (d^2/dt^2)x(t)$. Also find the eigenvalues and eigenfunctions (in terms of real functions) of $\mathbf{L}$ with $x(0) = 1, dx/dt|_{t=0} = 0$.

72. If

$$y(t) = \mathbf{L}x(t) = \int_0^t x(\tau)\, d\tau,$$

where $y(t)$ and $x(t)$ are real functions in some properly defined space, find the eigenvalues and eigenfunctions of the operator $\mathbf{L}$.

73. Approximate an eigenvalue of

$$\begin{pmatrix} 1 & 3 \\ 3 & 2 \end{pmatrix}$$

using the Rayleigh quotient method.

74. Solve for x in $\mathbf{L}x = y$ if $\mathbf{L} = -\left(d^2/dt^2 + 2d/dt + 1\right)$, with boundary conditions $x(0) = x(5) = 0$, and $y(t) = 2t$, via an eigenfunction expansion.

75. Using an eigenvector expansion, find the general solution of $\mathbf{A} \cdot \mathbf{x} = \mathbf{y}$ where

$$\mathbf{A} = \begin{pmatrix} 2 & 0 & 0 \\ 0 & 1 & 1 \\ 0 & 1 & 1 \end{pmatrix},$$

$$\mathbf{y} = \begin{pmatrix} 2 \\ 3 \\ 5 \end{pmatrix}.$$

76. Solve

$$\begin{pmatrix} 0 & 1 \\ -1 & -2 \end{pmatrix} \begin{pmatrix} x_1 \\ x_2 \end{pmatrix} = \begin{pmatrix} 2 \\ -5 \end{pmatrix}$$

using an eigenvector expansion method.

77. For each one of the matrix equations of the form $\mathbf{A} \cdot \mathbf{x} = \mathbf{y}$, where

(a).

$$\mathbf{A} = \begin{pmatrix} 1 & 1 \\ 2 & 2 \end{pmatrix}, \quad \mathbf{y} = \begin{pmatrix} 1 \\ 4 \end{pmatrix},$$

(b)

$$\mathbf{A} = \begin{pmatrix} 1 & 1 \\ 2 & 2 \end{pmatrix}, \quad \mathbf{y} = \begin{pmatrix} 1 \\ 2 \end{pmatrix},$$

(c)

$$\mathbf{A} = \begin{pmatrix} 1 & 2 \\ 2 & 1 \end{pmatrix}, \quad \mathbf{y} = \begin{pmatrix} 5 \\ 4 \end{pmatrix},$$

(d)

$$A = \begin{pmatrix} 1 & 2 & 0 \\ 0 & 1 & 1 \\ 1 & 2 & 0 \end{pmatrix}, \quad y = \begin{pmatrix} 1 \\ 2 \\ 1 \end{pmatrix},$$

(e)

$$A = \begin{pmatrix} 1 & 2 & 0 \\ 0 & 1 & 1 \\ 1 & 2 & 0 \end{pmatrix}, \quad y = \begin{pmatrix} 1 \\ 2 \\ -1 \end{pmatrix},$$

find the solutions of $A^* \cdot z = 0$, check if they are all orthogonal to y, and thus determine whether the original equation has at least one solution.

78. Solve

$$\frac{d^2 y}{dt^2} + 2\frac{dy}{dt} + y = 2e^{-t}$$

with $y(0) = 0$, $y(1) = 1$, using the method of moments and a single quadratic trial function.

79. Use two-term spectral, collocation, subdomain, least squares, and moments methods to solve the equation

$$y'''' + (1 + x)y = 1,$$

with $y(0) = y'(0) = y(1) = y''(1) = 0$. Compare graphically with the exact solution.

80. Consider

$$\frac{d^3 y}{dt^3} + 2t^3 y = 1 - t, \qquad y(0) = 0, \ y(2) = 0, \ \frac{dy}{dt}(0) = 0.$$

Choosing polynomials as the basis functions, use Galerkin and moments methods to obtain two-term estimates to $y(t)$. Plot your approximations and the exact solution on a single curve. Plot the residual in both methods for $t \in [0, 2]$.

81. Solve

$$\ddot{x} + 2x\dot{x} + t = 0,$$

with $x(0) = 0$, $x(4) = 0$, approximately using a two-term weighted residual method where the basis functions are of the type $\sin \lambda t$. Do both a spectral (as a consequence Galerkin) and pseudospectral (as a consequence collocation) method. Plot your approximations and the exact solution on a single curve. Plot the residual in both methods for $x \in [0, 4]$.

82. Find one-term collocation approximations for *all* solutions of

$$\frac{d^2 y}{dx^2} + y^4 = 1,$$

with $y(0) = 0$, $y(1) = 0$.

83. Find all approximate solutions of the boundary value problem

$$\frac{d^2 y}{dx^2} + y + 5y^2 = -x,$$

with $y(0) = y(1) = 0$, using a two-term collocation method. Compare graphically with a numerical solution.

84. Find a one-term approximation for the boundary value problem

$$y'' - y = -x^3,$$

with $y(0) = y(1) = 0$, using the collocation, Galerkin, least squares, and moments methods. Compare graphically with the exact solution.

85. Consider the eigenvalue problem

$$\frac{d^2y}{dt^2} + \lambda y = 0, \qquad y(0) = y(1) = 0,$$

and determine the eigenvalues through use of a method of weighted residuals with trial functions $\phi(t) = t(1-t)^n$, $n = 1, 2, \ldots, N$. Find estimates for all eigenvalues available for $N = 1$, $N = 2$, and $N = 3$. Show the same results are achieved with the Rayleigh-Ritz method involving extremization of

$$I = \int_0^1 (-\dot{y}^2 + \lambda y^2) \, dt, \qquad y(0) = y(1) = 0.$$

Compare your estimates to the exact solution for the eigenvalues.

86. Consider the eigenvalue problem

$$\frac{d^2y}{dt^2} + \lambda\sqrt{t}\,y = 0, \qquad y(0) = y(1) = 0,$$

and determine the eigenvalues through use of a method of weighted residuals with trial functions $\phi(t) = t(1-t)^n$, $n = 1, 2, \ldots, N$. Find estimates for all eigenvalues available for $N = 1$, $N = 2$, and $N = 3$. Show the same results are achieved with the Rayleigh-Ritz method involving extremization of

$$I = \int_0^1 (-\dot{y}^2 + \lambda\sqrt{t}\,y^2) \, dt, \qquad y(0) = y(1) = 0.$$

Find an exact solution for the eigenvalues via computer software. Compare your estimates to the exact solution for the eigenvalues.

87. Consider

$$\frac{d^2y}{dt^2} = -ky, \qquad y(0) = 1, \quad \frac{dy}{dt}(0) = 0.$$

With $\xi \in (-\infty, \infty)$ a random normally distributed variable with mean of zero and standard deviation of unity, consider $k = \mu + \sigma\xi$. Use the method of polynomial chaos to get a two-term estimate for y when $\mu = 1$, $\sigma = 1/10$. Compare the expected value of $y(t = 10)$ with that of a Monte Carlo simulation and that when $\sigma = 0$. Repeat the problem using a three-term estimate for y.

88. The following norms and names for $x(t)$ with $t \in [0, 1]$ are frequently used in numerical methods:

$$\|x\|_E = \left(\int_0^1 \left(\left(\frac{dx}{dt}\right)^2 + x^2 \right) dt \right)^{1/2}, \quad \text{energy norm,}$$

$$\|x\|_0 = \left(\int_0^1 x^2 \, dt \right)^{1/2}, \quad \text{root-mean-square norm,}$$

$$\|x\|_\infty = \max_{0 \le t \le 1} |x(t)|, \quad \text{maximum norm.}$$

Approximate the function $f(t) = 1 + e^t$, $t \in [0, 1]$ by its Taylor series around $t = 0.5$ up to n terms. Find the norm of the error using the three definitions, and plot them versus n.

89. The Euler-Bernoulli bending theory for an elastic beam gives

$$\frac{d^4 y}{dx^4} = P(x),$$

where x is a the coordinate along the length of the beam, $y(x)$ is the deflection of the beam, and P is proportional to the transverse force per unit length. If the beam is fixed at $x = 0$ and $x = 1$, the boundary conditions are $y(0) = y'(0) = y(1) = y'(1) = 0$. Find the lowest order polynomial that satisfies the boundary conditions. Using a one-term Galerkin method, find the approximate deflection of the beam in terms of $P(x)$. Find the maximum deflection if $P(x) = P_0$ is a constant.

90. Find the eigenfunctions of the linear operator $\mathbf{L}$ defined by $\mathbf{L}y = d^2 y/dt^2$, where $y(0) = y(1) = 0$. Expand the solution of the differential equation $\mathbf{L}y = t^2$ in terms of its first five orthonormal eigenfunctions. Use the Galerkin, collocation, subdomain, least squares, and moments methods to obtain the constants in the expansion (use numerical or symbolic computer mathematics methods to evaluate integrals, if necessary). Plot the solutions.

7 Linear Algebra

Linear algebra is part of the foundation of mathematics and has widespread usage in engineering. In this chapter, we specialize the linear analysis of Chapter 6 to finite-dimensional vector spaces in which the linear operator is a constant matrix. Many of the topics will be familiar, and some will likely be new. Considerable effort is spent defining terms and finding the best solution to systems of linear algebraic equations. As nearly all computational methods for solution of equations modeling physical systems rely on linear algebra, our expansive treatment is justified. Throughout the chapter, geometric interpretations are applied when appropriate. Some topics introduced in previous chapters are more fully explored, including matrices that effect rotation and reflection, projection matrices, eigenvalues and eigenvectors, and quadratic forms. New topics include a variety of matrix decompositions that are widely used in computational linear algebra. Of these the most important is the so-called singular value decomposition (SVD). We also give a matrix interpretation of two methods in wide use in engineering: (1) the least squares method and (2) the discrete Fourier transform. We close with a general strategy to find the best solution to linear algebra systems based on the SVD. In contrast to Chapter 6, we return in this chapter to Gibbs notation for vectors and matrices. Thus, matrices will be represented by uppercase bold-faced letters, such as $\mathbf{A}$, and vectors by lowercase bold-faced letters, such as $\mathbf{x}$.

7.1 Paradigm Problem

One of the most important problems in linear algebra lies in addressing the equation

$$\mathbf{A} \cdot \mathbf{x} = \mathbf{b}, \tag{7.1}$$

where $\mathbf{A}$ is a known constant matrix, $\mathbf{b}$ is a known column vector, and $\mathbf{x}$ is an unknown column vector. We note the analog to linear differential equations with the general form of Eq. (4.1), $\mathbf{L}y = f(x)$. Here the matrix $\mathbf{A}$ plays the role of the differential operator $\mathbf{L}$, the vector $\mathbf{x}$ plays the rule of the function y, and the vector $\mathbf{b}$ plays the role of the forcing function $f(x)$. Although this chapter will consider a variety of topics within linear algebra, nearly all relate in some sense to Eq. (7.1).

As we shall see, use of the "=" in Eq. (7.1) is not always appropriate. To explicitly indicate the dimension of the matrices and vectors, we sometimes write Eq. (7.1) in the expanded form

$$\mathbf{A}_{N \times M} \cdot \mathbf{x}_{M \times 1} = \mathbf{b}_{N \times 1}, \tag{7.2}$$

where $N, M \in \mathbb{N}$ are positive integers. If $N = M$, the matrix is square, and solution techniques are straightforward when $\det \mathbf{A} \neq 0$. For such cases, we are fully justified in using an "=" in Eq. (7.1). However, for $N \neq M$ or when $\det \mathbf{A} = 0$, which arises often in physical problems, the issues are not as straightforward. In some cases we find an infinite number of solutions; in others we find none. However, in what does form *the* paradigm problem of linear algebra, by relaxing our equality constraint, we can always select one or more values of $\mathbf{x}$ to minimize the residual $||\mathbf{A} \cdot \mathbf{x} - \mathbf{b}||_2$. And of those values that minimize the residual, one of them has minimum norm. More formally, we can pose the paradigm problem of linear algebra as finding nonunique $\mathbf{x}_p$ and unique $\hat{\mathbf{x}}$ that are defined as

$$\mathbf{x}_p = \{ \mathbf{x} \mid ||\mathbf{A} \cdot \mathbf{x} - \mathbf{b}||_2 \to \min \}, \tag{7.3}$$

$$\hat{\mathbf{x}} = \{ \mathbf{x}_p \mid ||\mathbf{x}_p||_2 \to \min \}. \tag{7.4}$$

We adapt the notion of the residual, introduced Section 6.11, so as to take the residual $\mathbf{r}$ to be

$$\mathbf{r} = \mathbf{A} \cdot \mathbf{x} - \mathbf{b}. \tag{7.5}$$

In general, we will seek any and all $\mathbf{x}$ that minimize $||\mathbf{r}||_2$; from those, we will sometimes seek the unique $\hat{\mathbf{x}}$. Thus, while Eq. (7.1) is easily justified as an important problem in linear algebra, the fully general Eqs. (7.3,7.4) are better described as the paradigm problem.

7.2 Matrix Fundamentals and Operations

We denote a matrix of size $N \times M$ as

$$\mathbf{A}_{N \times M} = \begin{pmatrix} a_{11} & a_{12} & \cdots & a_{1m} & \cdots & a_{1M} \\ a_{21} & a_{22} & \cdots & a_{2m} & \cdots & a_{2M} \\ \vdots & \vdots & \ddots & \vdots & \vdots & \vdots \\ a_{n1} & a_{n2} & \cdots & a_{nm} & \cdots & a_{nM} \\ \vdots & \vdots & \vdots & \vdots & \ddots & \vdots \\ a_{N1} & a_{N2} & \cdots & a_{Nm} & \cdots & a_{NM} \end{pmatrix}. \tag{7.6}$$

Here a_{ij} is the scalar element occupying the ith row and j^{th} column of $\mathbf{A}_{N \times M}$.

7.2.1 Determinant and Rank

We can take the determinant of a square matrix $\mathbf{A}_{N \times N}$, written $\det \mathbf{A}$. Details of computation of determinants are found in any standard linear algebra reference and are not repeated here. Properties of the determinant include the following:

- The quantity $\det \mathbf{A}_{N \times N}$ is equal to the volume of a parallelepiped in N-dimensional space whose edges are formed by the rows of $\mathbf{A}$.
- If all elements of a row (or column) are multiplied by a scalar, the determinant is also similarly multiplied.
- The elementary operation of subtracting a multiple of one row from another leaves the determinant unchanged.

- If two rows (or columns) of a matrix are interchanged, the sign of the determinant changes.

A *singular* matrix is one whose determinant is zero. The *rank* of a matrix is the size r of the largest square nonsingular matrix that can be formed by deleting rows and columns.

While the determinant is useful to some ends in linear algebra, most of the common problems are better solved without using the determinant at all. It is useful in solving linear systems of equations of small dimension but becomes too cumbersome relative to other methods for commonly encountered large systems of linear algebraic equations. Although it can be used to find the rank, there are also other, more efficient means to calculate this. Furthermore, though a zero value for the determinant almost always has significance, other values are less important. Some matrices that are particularly ill-conditioned for certain problems often have a determinant that gives no clue as to difficulties that may arise.

7.2.2 Matrix Addition

Addition of matrices can be defined as

$$\mathbf{A}_{N \times M} + \mathbf{B}_{N \times M} = \mathbf{C}_{N \times M}, \tag{7.7}$$

where the elements of $\mathbf{C}$ are obtained by adding the corresponding elements of $\mathbf{A}$ and $\mathbf{B}$. Multiplication of a matrix by a scalar α can be defined as

$$\alpha \mathbf{A}_{N \times M} = \mathbf{B}_{N \times M}, \tag{7.8}$$

where the elements of $\mathbf{B}$ are the corresponding elements of $\mathbf{A}$ multiplied by α.

It can be shown that the set of all $N \times M$ matrices is a vector space. We will also refer to an $N \times 1$ matrix as an N-dimensional column vector. Likewise a $1 \times M$ matrix will be called an M-dimensional row vector. Unless otherwise stated, vectors are assumed to be column vectors. In this sense, the inner product of two vectors $\mathbf{x}_{N \times 1}$ and $\mathbf{y}_{N \times 1}$ is $\langle \mathbf{x}, \mathbf{y} \rangle = \bar{\mathbf{x}}^T \cdot \mathbf{y}$.

7.2.3 Column, Row, and Left and Right Null Spaces

The M *column vectors* $\mathbf{c}_m \in \mathbb{C}^N$, $m = 1, 2, \ldots, M$, of the matrix $\mathbf{A}_{N \times M}$ are each of the columns of $\mathbf{A}$. The *column space* is the subspace of $\mathbb{C}^M$ spanned by the column vectors. The N *row vectors* $\mathbf{r}_n \in \mathbb{C}^M$, $n = 1, 2, \ldots, N$, of the same matrix are each of the rows. The *row space* is the subspace of $\mathbb{C}^N$ spanned by the row vectors. The column space vectors and the row space vectors span spaces of the same dimension. Consequently, the column space and row space have the same dimension. The *right null space* is the set of all vectors $\mathbf{x}_{M \times 1} \in \mathbb{C}^M$ for which $\mathbf{A}_{N \times M} \cdot \mathbf{x}_{M \times 1} = \mathbf{0}_{N \times 1}$. The *left null space* is the set of all vectors $\mathbf{y}_{N \times 1} \in \mathbb{C}^N$ for which $\bar{\mathbf{y}}_{N \times 1}^T \cdot \mathbf{A}_{N \times M} = \bar{\mathbf{y}}_{1 \times N} \cdot \mathbf{A}_{N \times M} = \mathbf{0}_{1 \times M}$. The column, row, right null, and left null spaces are also known as the *four fundamental subspaces*.

If we have $\mathbf{A}_{N \times M} : \mathbb{C}^M \to \mathbb{C}^N$, and recall that the rank of $\mathbf{A}$ is r, then we have the following important results:

- The column space of $\mathbf{A}_{N \times M}$ has dimension r, $(r \leq M)$.
- The left null space of $\mathbf{A}_{N \times M}$ has dimension $N - r$.
- The row space of $\mathbf{A}_{N \times M}$ has dimension r, $(r \leq N)$.
- The right null space of $\mathbf{A}_{N \times M}$ has dimension $M - r$.

Given any subspace $\mathbb{S}$, the space of all vectors orthogonal to $\mathbb{S}$ is defined as the *orthogonal complement* of $\mathbb{S}$. Then it can be shown that the right null space is the orthogonal complement of the row space, and the left null space is the orthogonal complement of the column space. This coupled with the four points just delineated form the main ingredients for the *fundamental theorem of linear algebra*.

We also can show

$$\mathbb{C}^N = \text{column space} \oplus \text{left null space} \tag{7.9}$$

$$\mathbb{C}^M = \text{row space} \oplus \text{right null space}. \tag{7.10}$$

See Section 6.3 for the notation $\oplus$ for the direct sum. Also

- Any vector $\mathbf{x} \in \mathbb{C}^M$ can be written as a linear combination of vectors in the row space and the right null space.
- Any M-dimensional vector $\mathbf{x}$ which is in the right null space of $\mathbf{A}$ is orthogonal to any M-dimensional vector in the row space. This comes directly from the definition of the right null space $\{\mathbf{x} \mid \mathbf{A} \cdot \mathbf{x} = \mathbf{0}\}$.
- Any vector $\mathbf{y} \in \mathbb{C}^N$ can be written as the sum of vectors in the column space and the left null space.
- Any N-dimensional vector $\mathbf{y}$ that is in the left null space of $\mathbf{A}$ is orthogonal to any N-dimensional vector in the column space. This comes directly from the definition of the left null space $\{\mathbf{y} \mid \bar{\mathbf{y}}^T \cdot \mathbf{A} = \mathbf{0}^T\}$.

EXAMPLE 7.1

Find the column and row spaces of

$$\mathbf{A} = \begin{pmatrix} 1 & 0 & 1 \\ 0 & 1 & 2 \end{pmatrix} \tag{7.11}$$

and their dimensions.

Restricting ourselves to real vectors, we note first that in the equation $\mathbf{A} \cdot \mathbf{x} = \mathbf{b}$, $\mathbf{A}$ is an operator that maps three-dimensional real vectors $\mathbf{x}$ into vectors $\mathbf{b}$ which are elements of a two-dimensional real space, that is,

$$\mathbf{A} : \mathbb{R}^3 \to \mathbb{R}^2. \tag{7.12}$$

The column vectors are

$$\mathbf{c}_1 = \begin{pmatrix} 1 \\ 0 \end{pmatrix}, \tag{7.13}$$

$$\mathbf{c}_2 = \begin{pmatrix} 0 \\ 1 \end{pmatrix}, \tag{7.14}$$

$$\mathbf{c}_3 = \begin{pmatrix} 1 \\ 2 \end{pmatrix}. \tag{7.15}$$

The column space consists of the vectors $\alpha_1 \mathbf{c}_1 + \alpha_2 \mathbf{c}_2 + \alpha_3 \mathbf{c}_3$, where the αs are any scalars. Because only two of the $\mathbf{c}_i$s are linearly independent, the dimension of the column space is also 2. We can see this by looking at the subdeterminant

$$\det \begin{pmatrix} 1 & 0 \\ 0 & 1 \end{pmatrix} = 1, \tag{7.16}$$

which indicates the rank, $r = 2$. Note the following:

- $\mathbf{c}_1 + 2\mathbf{c}_2 = \mathbf{c}_3$,
- the three column vectors thus lie in a single two-dimensional plane,
- the three column vectors are thus said to span a two-dimensional subspace of $\mathbb{R}^3$.

The two row vectors are

$$\mathbf{r}_1 = \begin{pmatrix} 1 & 0 & 1 \end{pmatrix}, \tag{7.17}$$

$$\mathbf{r}_2 = \begin{pmatrix} 0 & 1 & 2 \end{pmatrix}. \tag{7.18}$$

The row space consists of the vectors $\beta_1 \mathbf{r}_1 + \beta_2 \mathbf{r}_2$, where the βs are any scalars. Because the two $\mathbf{r}_i$s are linearly independent, the dimension of the row space is also 2. That is, the two row vectors are both three-dimensional but span a two-dimensional subspace. We note, for instance, if $\mathbf{x} = (1, 2, 1)^T$, that $\mathbf{A} \cdot \mathbf{x} = \mathbf{b}$ gives

$$\begin{pmatrix} 1 & 0 & 1 \\ 0 & 1 & 2 \end{pmatrix} \begin{pmatrix} 1 \\ 2 \\ 1 \end{pmatrix} = \begin{pmatrix} 2 \\ 4 \end{pmatrix}, \tag{7.19}$$

So

$$\mathbf{b} = 1\mathbf{c}_1 + 2\mathbf{c}_2 + 1\mathbf{c}_3. \tag{7.20}$$

That is, $\mathbf{b}$ is a linear combination of the column space vectors and thus lies in the column space of $\mathbf{A}$. We note for this problem that because an arbitrary $\mathbf{b}$ is two-dimensional and the dimension of the column space is 2, we can represent an arbitrary $\mathbf{b}$ as some linear combination of the column space vectors. For example, we can also say that $\mathbf{b} = 2\mathbf{c}_1 + 4\mathbf{c}_2$. We also note that $\mathbf{x}$ in general *does not* lie in the row space of $\mathbf{A}$, because $\mathbf{x}$ is an arbitrary three-dimensional vector, and we only have enough row vectors to span a two-dimensional subspace (i.e., a plane embedded in a three-dimensional space). However, as seen in Section 7.3, $\mathbf{x}$ does lie in the space defined by the combination of the row space of $\mathbf{A}$ and the *right null space* of $\mathbf{A}$ (the set of vectors $\mathbf{x}$ for which $\mathbf{A} \cdot \mathbf{x} = \mathbf{0}$). In special cases, $\mathbf{x}$ will in fact lie in the row space of $\mathbf{A}$.

7.2.4 Matrix Multiplication

Multiplication of matrices $\mathbf{A}$ and $\mathbf{B}$ can be defined if they are of the proper sizes. Thus

$$\mathbf{A}_{N \times L} \cdot \mathbf{B}_{L \times M} = \mathbf{C}_{N \times M}. \tag{7.21}$$

It may be better to say here that $\mathbf{A}$ is a linear operator that operates on elements that are in a space of dimension $L \times M$ to generate elements that are in a space of dimension $N \times M$; that is, $\mathbf{A} : \mathbb{R}^L \times \mathbb{R}^M \to \mathbb{R}^N \times \mathbb{R}^M$.

EXAMPLE 7.2

Consider the matrix operator

$$\mathbf{A} = \begin{pmatrix} 1 & 2 & 1 \\ -3 & 3 & 1 \end{pmatrix}, \tag{7.22}$$

which operates on 3×4 matrices, that is,

$$\mathbf{A} : \mathbb{R}^3 \times \mathbb{R}^4 \to \mathbb{R}^2 \times \mathbb{R}^4, \tag{7.23}$$

and show how it acts on another matrix.

We can use $\mathbf{A}$ to operate on a 3×4 matrix, as follows:

$$\begin{pmatrix} 1 & 2 & 1 \\ -3 & 3 & 1 \end{pmatrix} \begin{pmatrix} 1 & 0 & 3 & -2 \\ 2 & -4 & 1 & 3 \\ -1 & 4 & 0 & 2 \end{pmatrix} = \begin{pmatrix} 4 & -4 & 5 & 6 \\ 2 & -8 & -6 & 17 \end{pmatrix}. \qquad (7.24)$$

The operation does not exist if the order of the matrices is reversed.

A vector operating on a vector can yield a scalar or a matrix, depending on the order of operation.

EXAMPLE 7.3

Determine $\mathbf{A}_{1\times3} \cdot \mathbf{B}_{3\times1}$ and $\mathbf{B}_{3\times1} \cdot \mathbf{A}_{1\times3}$, where

$$\mathbf{A}_{1\times3} = \mathbf{a}^T = (2 \quad 3 \quad 1), \qquad (7.25)$$

$$\mathbf{B}_{3\times1} = \mathbf{b} = \begin{pmatrix} 3 \\ -2 \\ 5 \end{pmatrix}. \qquad (7.26)$$

Expanding, we have

$$\mathbf{A}_{1\times3} \cdot \mathbf{B}_{3\times1} = \mathbf{a}^T \cdot \mathbf{b} = (2 \quad 3 \quad 1) \begin{pmatrix} 3 \\ -2 \\ 5 \end{pmatrix} = (2)(3) + (3)(-2) + (1)(5) = 5. \qquad (7.27)$$

This is the ordinary inner product $\langle \mathbf{a}, \mathbf{b} \rangle$. The commutation of this operation, however, yields a matrix:

$$\mathbf{B}_{3\times1} \cdot \mathbf{A}_{1\times3} = \mathbf{b}\mathbf{a}^T = \begin{pmatrix} 3 \\ -2 \\ 5 \end{pmatrix} (2 \quad 3 \quad 1) = \begin{pmatrix} (3)(2) & (3)(3) & (3)(1) \\ (-2)(2) & (-2)(3) & (-2)(1) \\ (5)(2) & (5)(3) & (5)(1) \end{pmatrix}, \qquad (7.28)$$

$$= \begin{pmatrix} 6 & 9 & 3 \\ -4 & -6 & -2 \\ 10 & 15 & 5 \end{pmatrix}. \qquad (7.29)$$

This is the *dyadic product* of the two vectors. For vectors (lowercase notation), the dyadic product usually is not characterized by the "dot" operator that we use for the inner product of vectors.

A special case is that of a *square* matrix $\mathbf{A}_{N\times N}$ of size N. For square matrices of the same size, both $\mathbf{A} \cdot \mathbf{B}$ and $\mathbf{B} \cdot \mathbf{A}$ exist. Although $\mathbf{A} \cdot \mathbf{B}$ and $\mathbf{B} \cdot \mathbf{A}$ both yield $N \times N$ matrices, the actual value of the two products may be different. In what follows, we will often, but not always, assume that we are dealing with square matrices. Properties of matrices include the following:

1. $(\mathbf{A} \cdot \mathbf{B}) \cdot \mathbf{C} = \mathbf{A} \cdot (\mathbf{B} \cdot \mathbf{C})$ (associative)
2. $\mathbf{A} \cdot (\mathbf{B} + \mathbf{C}) = \mathbf{A} \cdot \mathbf{B} + \mathbf{A} \cdot \mathbf{C}$ (distributive)
3. $(\mathbf{A} + \mathbf{B}) \cdot \mathbf{C} = \mathbf{A} \cdot \mathbf{C} + \mathbf{B} \cdot \mathbf{C}$ (distributive)
4. $\mathbf{A} \cdot \mathbf{B} \neq \mathbf{B} \cdot \mathbf{A}$ in general (not commutative)
5. $\det \mathbf{A} \cdot \mathbf{B} = (\det \mathbf{A})(\det \mathbf{B})$

7.2.5 Definitions and Properties

We have the following commonly used definitions and properties.

Identity

The *identity* matrix $\mathbf{I}$ is a square diagonal matrix with 1 on the main diagonal and 0 elsewhere. With this definition, we get

$$\mathbf{A}_{N \times M} \cdot \mathbf{I}_{M \times M} = \mathbf{A}_{N \times M}, \tag{7.30}$$

$$\mathbf{I}_{N \times N} \cdot \mathbf{A}_{N \times M} = \mathbf{A}_{N \times M}. \tag{7.31}$$

More compactly, we could say

$$\mathbf{A} \cdot \mathbf{I} = \mathbf{I} \cdot \mathbf{A} = \mathbf{A}, \tag{7.32}$$

where the unsubscripted identity matrix is understood to be square with the correct dimension for matrix multiplication.

Nilpotent

A square matrix $\mathbf{A}$ is *nilpotent* if there exists a positive integer n for which $\mathbf{A}^n = \mathbf{0}$.

Idempotent

A square matrix $\mathbf{A}$ is called *idempotent* if $\mathbf{A} \cdot \mathbf{A} = \mathbf{A}$. The identity matrix $\mathbf{I}$ is idempotent. Projection matrices $\mathbf{P}$ (see Eq. (6.191)), are idempotent. All idempotent matrices that are not the identity matrix are singular. The trace of an idempotent matrix gives its rank. More generally, a function f is idempotent if $f(f(x)) = f(x)$. As an example, the absolute value function is idempotent because $\text{abs}(\text{abs}(x)) = \text{abs}(x)$.

Involutary

A square matrix $\mathbf{A}$ is called *involutary* if it equals its own inverse: $\mathbf{A} \cdot \mathbf{A} = \mathbf{I}$. A function is involutary if $f(f(x)) = x$. As an example, $f(x) = x$ is involutary.

Diagonal

A *diagonal* matrix $\mathbf{D}$ has nonzero terms only along its main diagonal. The sum and product of diagonal matrices are also diagonal. The determinant of a diagonal matrix is the product of all diagonal elements.

Transpose

Here we expand on the earlier discussion of Section 2.1.5. The *transpose* $\mathbf{A}^T$ of a matrix $\mathbf{A}$ is an operation in which the nondiagonal terms are reflected about the diagonal. In terms of scalar components of $\mathbf{A}$, we can say $a_{ij}^T = a_{ji}$. For any matrix $\mathbf{A}_{N \times M}$, we find that $\mathbf{A} \cdot \mathbf{A}^T$ and $\mathbf{A}^T \cdot \mathbf{A}$ are square matrices of size N and M, respectively. Properties of the transpose include the following:

1. $\det \mathbf{A} = \det \mathbf{A}^T$
2. $(\mathbf{A}_{N \times M} \cdot \mathbf{B}_{M \times N})^T = \mathbf{B}^T \cdot \mathbf{A}^T$
3. $(\mathbf{A}_{N \times N} \cdot \mathbf{x}_{N \times 1})^T \cdot \mathbf{y}_{N \times 1} = \mathbf{x}^T \cdot \mathbf{A}^T \cdot \mathbf{y} = \mathbf{x}^T \cdot (\mathbf{A}^T \cdot \mathbf{y})$
4. $(\mathbf{A}_{N \times M} + \mathbf{B}_{N \times M})^T = \mathbf{A}^T + \mathbf{B}^T$

Symmetry, Antisymmetry, and Asymmetry

To reiterate the earlier discussion of Section 2.1.5, a *symmetric* matrix is one for which $\mathbf{A}^T = \mathbf{A}$. An *antisymmetric* or *skew-symmetric* matrix is one for which $\mathbf{A}^T = -\mathbf{A}$.

Any matrix $\mathbf{A}$ can be written as

$$\mathbf{A} = \frac{1}{2}(\mathbf{A} + \mathbf{A}^T) + \frac{1}{2}(\mathbf{A} - \mathbf{A}^T), \tag{7.33}$$

where $(1/2)(\mathbf{A} + \mathbf{A}^T)$ is symmetric and $(1/2)(\mathbf{A} - \mathbf{A}^T)$ is antisymmetric. An *asymmetric* matrix is neither symmetric nor antisymmetric.

Triangular
A lower (or upper) triangular matrix is one in which all entries above (or below) the main diagonal are zero. Lower triangular matrices are often denoted by $\mathbf{L}$, and upper triangular matrices by either $\mathbf{U}$ or $\mathbf{R}$. We will generally use $\mathbf{U}$.

Positive Definite
A *positive definite* matrix $\mathbf{A}$ is a matrix for which $\mathbf{x}^T \cdot \mathbf{A} \cdot \mathbf{x} > 0$ for all nonzero vectors $\mathbf{x}$. A positive definite matrix has real, positive eigenvalues. Every positive definite matrix $\mathbf{A}$ can be written as $\mathbf{A} = \mathbf{U}^T \cdot \mathbf{U}$, where $\mathbf{U}$ is an upper triangular matrix. This is a Cholesky[1] decomposition, which is described further in Section 7.9.2.

Permutation
A *permutation* matrix $\mathbf{P}$ is a square matrix composed of zeros and a single one in each column. None of the ones occurs in the same row. It effects a row exchange when it operates on a general matrix $\mathbf{A}$. It is never singular and is in fact its own inverse, $\mathbf{P} = \mathbf{P}^{-1}$, so $\mathbf{P} \cdot \mathbf{P} = \mathbf{I}$. Also, $||\mathbf{P}||_2 = 1$ and $|\det \mathbf{P}| = 1$. However, we can have $\det \mathbf{P} = \pm 1$, so it can be either a rotation or a reflection. The permutation matrix $\mathbf{P}$ is not to be confused with a projection matrix $\mathbf{P}$, which is usually denoted in the same way. The context should be clear as to which matrix is intended.

EXAMPLE 7.4

Find the permutation matrix $\mathbf{P}$ that effects the exchange of the first and second rows of $\mathbf{A}$, where

$$\mathbf{A} = \begin{pmatrix} 1 & 3 & 5 & 7 \\ 2 & 3 & 1 & 2 \\ 3 & 1 & 3 & 2 \end{pmatrix}. \tag{7.34}$$

To construct $\mathbf{P}$, we begin with a 3×3 identity matrix $\mathbf{I}$. For a first and second row exchange, we replace the ones in the $(1, 1)$ and $(2, 2)$ slot with zeros, then replace the zeros in the $(1, 2)$ and $(2, 1)$ slot with ones. Thus

$$\mathbf{P} = \begin{pmatrix} 0 & 1 & 0 \\ 1 & 0 & 0 \\ 0 & 0 & 1 \end{pmatrix}, \tag{7.35}$$

yielding

$$\mathbf{P} \cdot \mathbf{A} = \begin{pmatrix} 0 & 1 & 0 \\ 1 & 0 & 0 \\ 0 & 0 & 1 \end{pmatrix} \begin{pmatrix} 1 & 3 & 5 & 7 \\ 2 & 3 & 1 & 2 \\ 3 & 1 & 3 & 2 \end{pmatrix} = \begin{pmatrix} 2 & 3 & 1 & 2 \\ 1 & 3 & 5 & 7 \\ 3 & 1 & 3 & 2 \end{pmatrix}. \tag{7.36}$$

---

[1] André-Louis Cholesky, 1875–1918, French mathematician and military officer.

EXAMPLE 7.5

Find the rank and right null space of

$$A = \begin{pmatrix} 1 & 0 & 1 \\ 5 & 4 & 9 \\ 2 & 4 & 6 \end{pmatrix}. \tag{7.37}$$

The rank of A is not three because

$$\det A = 0. \tag{7.38}$$

Because

$$\begin{vmatrix} 1 & 0 \\ 5 & 4 \end{vmatrix} \neq 0, \tag{7.39}$$

the rank of A is 2.

Let

$$x = \begin{pmatrix} x_1 \\ x_2 \\ x_3 \end{pmatrix} \tag{7.40}$$

belong to the right null space of A. Then

$$x_1 + x_3 = 0, \tag{7.41}$$

$$5x_1 + 4x_2 + 9x_3 = 0, \tag{7.42}$$

$$2x_1 + 4x_2 + 6x_3 = 0. \tag{7.43}$$

One strategy to solve singular systems is to take one of the variables to be a known parameter and see if the resulting system can be solved. If the resulting system remains singular, take a second variable to be a second parameter. This ad hoc method will later be made systematic in Section 7.9.3.

So here take $x_1 = t$, and consider the first two equations, which gives

$$\begin{pmatrix} 0 & 1 \\ 4 & 9 \end{pmatrix} \begin{pmatrix} x_2 \\ x_3 \end{pmatrix} = \begin{pmatrix} -t \\ -5t \end{pmatrix}. \tag{7.44}$$

Solving, we find $x_2 = t, x_3 = -t$. So,

$$x = \begin{pmatrix} x_1 \\ x_2 \\ x_3 \end{pmatrix} = \begin{pmatrix} t \\ t \\ -t \end{pmatrix} = t \begin{pmatrix} 1 \\ 1 \\ -1 \end{pmatrix}, \qquad t \in \mathbb{R}^1. \tag{7.45}$$

Therefore, the right null space is the straight line in $\mathbb{R}^3$ that passes through $(0, 0, 0)^T$ and $(1, 1, -1)^T$.

Inverse

Here we specialize discussion from Section 6.7.3 to matrices. For a matrix inverse, we have

Definition: A matrix A has an inverse A^{-1} if $A \cdot A^{-1} = A^{-1} \cdot A = I$.

One can prove the following:

Theorem: A unique inverse exists if the matrix is nonsingular.

Properties of the inverse include the following:

- $(\mathbf{A} \cdot \mathbf{B})^{-1} = \mathbf{B}^{-1} \cdot \mathbf{A}^{-1}$,
- $(\mathbf{A}^{-1})^T = (\mathbf{A}^T)^{-1}$,
- $\det(\mathbf{A}^{-1}) = (\det \mathbf{A})^{-1}$.

If a_{ij} and a_{ij}^{-1} are the elements of $\mathbf{A}$ and $\mathbf{A}^{-1}$, and we define the *cofactor* as

$$c_{ij} = (-1)^{i+j} m_{ij}, \tag{7.46}$$

where the *minor*, m_{ij}, is the determinant of the matrix obtained by canceling the jth row and i^{th} column, the inverse is

$$a_{ij}^{-1} = \frac{c_{ij}}{\det \mathbf{A}}. \tag{7.47}$$

The inverse of a diagonal matrix is also diagonal, but with the reciprocals of the original diagonal elements.

EXAMPLE 7.6

Find the inverse of

$$\mathbf{A} = \begin{pmatrix} 1 & 1 \\ -1 & 1 \end{pmatrix}. \tag{7.48}$$

Omitting the details of cofactors and minors, the inverse is

$$\mathbf{A}^{-1} = \begin{pmatrix} \frac{1}{2} & -\frac{1}{2} \\ \frac{1}{2} & \frac{1}{2} \end{pmatrix}. \tag{7.49}$$

We can confirm that $\mathbf{A} \cdot \mathbf{A}^{-1} = \mathbf{A}^{-1} \cdot \mathbf{A} = \mathbf{I}$.

Similar Matrices

Matrices $\mathbf{A}$ and $\mathbf{B}$ are *similar* if there exists a nonsingular matrix $\mathbf{S}$ such that $\mathbf{B} = \mathbf{S}^{-1} \cdot \mathbf{A} \cdot \mathbf{S}$. Similar matrices have the same determinant, eigenvalues, multiplicities, and eigenvectors.

7.3 Systems of Equations

In general, for matrices that are not necessarily square, the equation $\mathbf{A}_{N \times M} \cdot \mathbf{x}_{M \times 1} = \mathbf{b}_{N \times 1}$ is solvable iff $\mathbf{b}$ can be expressed as combinations of the columns of $\mathbf{A}$. Problems in which $M < N$ are *overconstrained*; in special cases, those in which $\mathbf{b}$ is in the column space of $\mathbf{A}$, a unique solution $\mathbf{x}$ exists. However, in general, no solution $\mathbf{x}$ exists; nevertheless, one can find an $\mathbf{x}$ that will minimize $||\mathbf{A} \cdot \mathbf{x} - \mathbf{b}||_2$. This is closely related to what is known as the method of least squares. Problems in which $M > N$ are generally *underconstrained* and have an infinite number of solutions $\mathbf{x}$ that will satisfy the original equation. Problems for which $M = N$ (square matrices) have a unique solution $\mathbf{x}$ when the rank r of $\mathbf{A}$ is equal to N. If $r < N$, the problem is underconstrained.

7.3.1 Overconstrained

Let us first consider an example of an overconstrained system.

EXAMPLE 7.7

For $\mathbf{x} \in \mathbb{R}^2$, $\mathbf{b} \in \mathbb{R}^3$, $\mathbf{A} : \mathbb{R}^2 \rightarrow \mathbb{R}^3$, identify the best solutions to

$$\begin{pmatrix} 1 & 2 \\ 1 & 0 \\ 1 & 1 \end{pmatrix} \begin{pmatrix} x_1 \\ x_2 \end{pmatrix} = \begin{pmatrix} 0 \\ 1 \\ 3 \end{pmatrix}. \tag{7.50}$$

Here it turns out that $\mathbf{b} = (0, 1, 3)^T$ is not in the column space of $\mathbf{A}$, and there is no solution $\mathbf{x}$ for which $\mathbf{A} \cdot \mathbf{x} = \mathbf{b}$! The column space is a plane defined by two vectors; the vector $\mathbf{b}$ does not happen to lie in the plane defined by the column space. However, we can find a vector $\mathbf{x} = \mathbf{x}_p$, where $\mathbf{x}_p$ can be shown to minimize the Euclidean norm of the residual $||\mathbf{A} \cdot \mathbf{x}_p - \mathbf{b}||_2$. This is achieved by the following procedure, the same employed earlier in Section 6.5, in which we operate on both vectors $\mathbf{A} \cdot \mathbf{x}_p$ and $\mathbf{b}$ by the operator $\mathbf{A}^T$ so as to map both vectors *into the same space*, namely, the row space of $\mathbf{A}$. Once the vectors are in the same space, a unique inversion is possible:

$$\mathbf{A} \cdot \mathbf{x}_p \approx \mathbf{b}, \tag{7.51}$$

$$\mathbf{A}^T \cdot \mathbf{A} \cdot \mathbf{x}_p = \mathbf{A}^T \cdot \mathbf{b}, \tag{7.52}$$

$$\mathbf{x}_p = (\mathbf{A}^T \cdot \mathbf{A})^{-1} \cdot \mathbf{A}^T \cdot \mathbf{b}. \tag{7.53}$$

These operations are, numerically,

$$\begin{pmatrix} 1 & 1 & 1 \\ 2 & 0 & 1 \end{pmatrix} \begin{pmatrix} 1 & 2 \\ 1 & 0 \\ 1 & 1 \end{pmatrix} \begin{pmatrix} x_1 \\ x_2 \end{pmatrix} = \begin{pmatrix} 1 & 1 & 1 \\ 2 & 0 & 1 \end{pmatrix} \begin{pmatrix} 0 \\ 1 \\ 3 \end{pmatrix}, \tag{7.54}$$

$$\begin{pmatrix} 3 & 3 \\ 3 & 5 \end{pmatrix} \begin{pmatrix} x_1 \\ x_2 \end{pmatrix} = \begin{pmatrix} 4 \\ 3 \end{pmatrix}, \tag{7.55}$$

$$\begin{pmatrix} x_1 \\ x_2 \end{pmatrix} = \begin{pmatrix} \frac{11}{6} \\ -\frac{1}{2} \end{pmatrix}. \tag{7.56}$$

The resulting $\mathbf{x}_p$ *will not satisfy* $\mathbf{A} \cdot \mathbf{x}_p = \mathbf{b}$. We can define the difference of $\mathbf{A} \cdot \mathbf{x}_p$ and $\mathbf{b}$ as the residual vector (see Eq. (7.5)), $\mathbf{r} = \mathbf{A} \cdot \mathbf{x}_p - \mathbf{b}$. In fact, $||\mathbf{r}||_2 = ||\mathbf{A} \cdot \mathbf{x}_p - \mathbf{b}||_2 = 2.0412$. If we tried any nearby $\mathbf{x}$, say, $\mathbf{x} = (2, -3/5)^T$, $||\mathbf{A} \cdot \mathbf{x} - \mathbf{b}||_2 = 2.0494 > 2.0412$. Because the problem is linear, this minimum is global; if we take $\mathbf{x} = (10, -24)^T$, then $||\mathbf{A} \cdot \mathbf{x} - \mathbf{b}||_2 = 42.5911 > 2.0412$. Though we have not proved it, our $\mathbf{x}_p$ is the unique vector that minimizes the Euclidean norm of the residual.

Further manipulation shows that we can write our solution as a combination of vectors in the row space of $\mathbf{A}$. As the dimension of the right null space of $\mathbf{A}$ is zero, there is no possible contribution from the right null space vectors. We thus form an arbitrary linear combination of two linearly independent row space vectors, written here as column vectors, and find the coefficients α_1 and α_2 for our $\mathbf{x}_p$:

$$\begin{pmatrix} \frac{11}{6} \\ -\frac{1}{2} \end{pmatrix} = \alpha_1 \begin{pmatrix} 1 \\ 2 \end{pmatrix} + \alpha_2 \begin{pmatrix} 1 \\ 0 \end{pmatrix}, \tag{7.57}$$

$$\begin{pmatrix} \frac{11}{6} \\ -\frac{1}{2} \end{pmatrix} = \begin{pmatrix} 1 & 1 \\ 2 & 0 \end{pmatrix} \begin{pmatrix} \alpha_1 \\ \alpha_2 \end{pmatrix}, \tag{7.58}$$

$$\begin{pmatrix} \alpha_1 \\ \alpha_2 \end{pmatrix} = \begin{pmatrix} -\frac{1}{4} \\ \frac{25}{12} \end{pmatrix}. \tag{7.59}$$

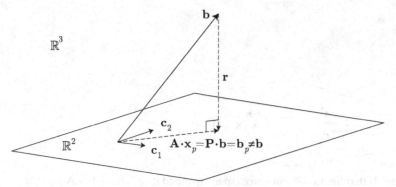

Figure 7.1. Plot for b that lies outside of column space (space spanned by c_1 and c_2) of **A**.

So

$$\begin{pmatrix} x_1 \\ x_2 \end{pmatrix} = \underbrace{-\frac{1}{4}\begin{pmatrix} 1 \\ 2 \end{pmatrix} + \frac{25}{12}\begin{pmatrix} 1 \\ 0 \end{pmatrix}}_{\text{linear combination of row space vectors}}. \tag{7.60}$$

We could also have chosen to expand in terms of the other row space vector $(1,1)^T$, because any two of the three row space vectors span the space $\mathbb{R}^2$.

The vector $\mathbf{A} \cdot \mathbf{x}_p$ actually represents the *projection* of b onto the subspace spanned by the column vectors (i.e., the column space). Call the projected vector $\mathbf{b}_p$:

$$\mathbf{b}_p = \mathbf{A} \cdot \mathbf{x}_p = \underbrace{\mathbf{A} \cdot (\mathbf{A}^T \cdot \mathbf{A})^{-1} \cdot \mathbf{A}^T}_{\text{projection matrix, } \mathbf{P}} \cdot \mathbf{b}. \tag{7.61}$$

For this example, $\mathbf{b}_p = (5/6, 11/6, 4/3)^T$. We can think of $\mathbf{b}_p$ as the shadow cast by b onto the column space. Here, following Eq. (6.191), we have the projection matrix **P** as

$$\mathbf{P} = \mathbf{A} \cdot (\mathbf{A}^T \cdot \mathbf{A})^{-1} \cdot \mathbf{A}^T. \tag{7.62}$$

A sketch of this system is shown in Figure 7.1. Here we indicate what might represent this example in which the column space of **A** does not span the entire space $\mathbb{R}^3$, and for which b lies outside of the column space of **A**. In such a case, $\|\mathbf{A} \cdot \mathbf{x}_p - \mathbf{b}\|_2 > 0$. We have **A** as a matrix that maps two-dimensional vectors **x** into three-dimensional vectors b. Our space is $\mathbb{R}^3$, and embedded within that space are two column vectors c_1 and c_2 that span a column space $\mathbb{R}^2$, which is represented by a plane within a three-dimensional volume. Because b lies outside the column space, there exists no unique vector **x** for which $\mathbf{A} \cdot \mathbf{x} = \mathbf{b}$.

EXAMPLE 7.8

For $\mathbf{x} \in \mathbb{R}^2$, $\mathbf{b} \in \mathbb{R}^3$, $\mathbf{A} : \mathbb{R}^2 \to \mathbb{R}^3$, identify the best solutions to

$$\begin{pmatrix} 1 & 2 \\ 1 & 0 \\ 1 & 1 \end{pmatrix} \begin{pmatrix} x_1 \\ x_2 \end{pmatrix} = \begin{pmatrix} 5 \\ 1 \\ 3 \end{pmatrix}. \tag{7.63}$$

The column space of **A** is spanned by the two column vectors

$$c_1 = \begin{pmatrix} 1 \\ 1 \\ 1 \end{pmatrix}, \qquad c_2 = \begin{pmatrix} 2 \\ 0 \\ 1 \end{pmatrix}. \tag{7.64}$$

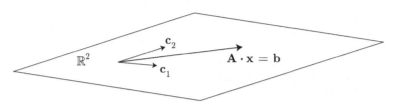

$\mathbb{R}^3$

Figure 7.2. Plot for b that lies in column space (space spanned by c_1 and c_2) of $\mathbf{A}$.

Our equation can also be cast in the form that makes the contribution of the column vectors obvious:

$$x_1 \begin{pmatrix} 1 \\ 1 \\ 1 \end{pmatrix} + x_2 \begin{pmatrix} 2 \\ 0 \\ 1 \end{pmatrix} = \begin{pmatrix} 5 \\ 1 \\ 3 \end{pmatrix}. \tag{7.65}$$

Here we have the unusual case that $\mathbf{b} = (5, 1, 3)^T$ is in the column space of $\mathbf{A}$ (in fact, $\mathbf{b} = \mathbf{c}_1 + 2\mathbf{c}_2$), and we have a unique solution of

$$\mathbf{x} = \begin{pmatrix} 1 \\ 2 \end{pmatrix}. \tag{7.66}$$

In most cases, however, it is not obvious that b lies in the column space. We can still operate on both sides by $\mathbf{A}^T$ and solve, which reveals the correct result:

$$\begin{pmatrix} 1 & 1 & 1 \\ 2 & 0 & 1 \end{pmatrix} \begin{pmatrix} 1 & 2 \\ 1 & 0 \\ 1 & 1 \end{pmatrix} \begin{pmatrix} x_1 \\ x_2 \end{pmatrix} = \begin{pmatrix} 1 & 1 & 1 \\ 2 & 0 & 1 \end{pmatrix} \begin{pmatrix} 5 \\ 1 \\ 3 \end{pmatrix}, \tag{7.67}$$

$$\begin{pmatrix} 3 & 3 \\ 3 & 5 \end{pmatrix} \begin{pmatrix} x_1 \\ x_2 \end{pmatrix} = \begin{pmatrix} 9 \\ 13 \end{pmatrix}, \tag{7.68}$$

$$\begin{pmatrix} x_1 \\ x_2 \end{pmatrix} = \begin{pmatrix} 1 \\ 2 \end{pmatrix}. \tag{7.69}$$

A quick check of the residual shows that, in fact, $\mathbf{r} = \mathbf{A} \cdot \mathbf{x}_p - \mathbf{b} = \mathbf{0}$. So, we have an exact solution for which $\mathbf{x} = \mathbf{x}_p$.

The solution vector $\mathbf{x}$ lies entirely in the row space of $\mathbf{A}$; here it is identically the first row vector $\mathbf{r}_1 = (1, 2)^T$. Also, here the column space is a two-dimensional subspace, in this case, a plane defined by the two column vectors, embedded within a three-dimensional space. The operator $\mathbf{A}$ maps arbitrary two-dimensional vectors $\mathbf{x}$ into the three-dimensional b; however, these b vectors are confined to a two-dimensional subspace within the greater three-dimensional space. Consequently, we cannot always expect to find a vector $\mathbf{x}$ for arbitrary b!

A sketch of this system is shown in Figure 7.2. Here we sketch what might represent this example in which the column space of $\mathbf{A}$ does not span the entire space $\mathbb{R}^3$, but for which b lies in the column space of $\mathbf{A}$. In such a case, $\|\mathbf{A} \cdot \mathbf{x} - \mathbf{b}\|_2 = 0$. We have $\mathbf{A}$ as a matrix that maps two-dimensional vectors $\mathbf{x}$ into three-dimensional vectors b. Our space is $\mathbb{R}^3$, and embedded within that space are two column vectors $\mathbf{c}_1$ and $\mathbf{c}_2$ that span a column space $\mathbb{R}^2$, which is represented by a plane within a three-dimensional volume. Because b in this example happens to lie in the column space, there exists a unique vector $\mathbf{x}$ for which $\mathbf{A} \cdot \mathbf{x} = \mathbf{b}$.

7.3.2 Underconstrained

EXAMPLE 7.9

For $\mathbf{x} \in \mathbb{R}^3$, $\mathbf{b} \in \mathbb{R}^2$, $\mathbf{A} : \mathbb{R}^3 \to \mathbb{R}^2$, identify the best solutions to

$$\begin{pmatrix} 1 & 1 & 1 \\ 2 & 0 & 1 \end{pmatrix} \begin{pmatrix} x_1 \\ x_2 \\ x_3 \end{pmatrix} = \begin{pmatrix} 1 \\ 3 \end{pmatrix}. \tag{7.70}$$

In this case operating on both sides by the transpose is not useful because $(\mathbf{A}^T \cdot \mathbf{A})^{-1}$ does not exist. We take an alternate strategy. Certainly $\mathbf{b} = (1,3)^T$ lies in the column space of $\mathbf{A}$, because for example, $\mathbf{b} = 0(1,2)^T - 2(1,0)^T + 3(1,1)^T$. Setting $x_1 = t$, where t is an arbitrary number, let us solve for x_2 and x_3:

$$\begin{pmatrix} 1 & 1 & 1 \\ 2 & 0 & 1 \end{pmatrix} \begin{pmatrix} t \\ x_2 \\ x_3 \end{pmatrix} = \begin{pmatrix} 1 \\ 3 \end{pmatrix}, \tag{7.71}$$

$$\begin{pmatrix} 1 & 1 \\ 0 & 1 \end{pmatrix} \begin{pmatrix} x_2 \\ x_3 \end{pmatrix} = \begin{pmatrix} 1-t \\ 3-2t \end{pmatrix}. \tag{7.72}$$

Inversion gives

$$\begin{pmatrix} x_2 \\ x_3 \end{pmatrix} = \begin{pmatrix} -2+t \\ 3-2t \end{pmatrix}, \tag{7.73}$$

so

$$\begin{pmatrix} x_1 \\ x_2 \\ x_3 \end{pmatrix} = \begin{pmatrix} t \\ -2+t \\ 3-2t \end{pmatrix} = \begin{pmatrix} 0 \\ -2 \\ 3 \end{pmatrix} + t \underbrace{\begin{pmatrix} 1 \\ 1 \\ -2 \end{pmatrix}}_{\text{right null space}}, \qquad t \in \mathbb{R}^1. \tag{7.74}$$

A useful way to think of problems such as this which are underdetermined is that *the matrix* $\mathbf{A}$ *maps the additive combination of a unique vector from the row space of* $\mathbf{A}$ *plus an arbitrary vector from the right null space of* $\mathbf{A}$ *into the vector* $\mathbf{b}$. Here the vector $(1,1,-2)^T$ is in the right null space; however, the vector $(0,-2,3)^T$ has components in both the right null space and the row space. Let us extract the parts of $(0,-2,3)^T$ which are in each space. Because the row space and right null space are linearly independent, they form a basis, and we can say

$$\begin{pmatrix} 0 \\ -2 \\ 3 \end{pmatrix} = a_1 \underbrace{\begin{pmatrix} 1 \\ 1 \\ 1 \end{pmatrix} + a_2 \begin{pmatrix} 2 \\ 0 \\ 1 \end{pmatrix}}_{\text{row space}} + a_3 \underbrace{\begin{pmatrix} 1 \\ 1 \\ -2 \end{pmatrix}}_{\text{right null space}}. \tag{7.75}$$

In matrix form, we then find

$$\begin{pmatrix} 0 \\ -2 \\ 3 \end{pmatrix} = \underbrace{\begin{pmatrix} 1 & 2 & 1 \\ 1 & 0 & 1 \\ 1 & 1 & -2 \end{pmatrix}}_{\text{invertible}} \begin{pmatrix} a_1 \\ a_2 \\ a_3 \end{pmatrix}. \tag{7.76}$$

The coefficient matrix is nonsingular and thus invertible. Solving, we get

$$\begin{pmatrix} a_1 \\ a_2 \\ a_3 \end{pmatrix} = \begin{pmatrix} -\frac{2}{3} \\ 1 \\ -\frac{4}{3} \end{pmatrix}. \tag{7.77}$$

So **x** can be rewritten as

$$\mathbf{x} = \underbrace{-\frac{2}{3}\begin{pmatrix}1\\1\\1\end{pmatrix} + \begin{pmatrix}2\\0\\1\end{pmatrix}}_{\text{row space}} + \underbrace{\left(t - \frac{4}{3}\right)\begin{pmatrix}1\\1\\-2\end{pmatrix}}_{\text{right null space}}, \qquad t \in \mathbb{R}^1. \tag{7.78}$$

The first two terms in the right-hand side of Eq. (7.78) are the unique linear combination of the row space vectors, while the third term is from the right null space. As by definition, **A** maps any vector from the right null space into the zero element, it makes no contribution to forming **b**; hence, one can allow for an arbitrary constant. Note the analogy here with solutions to inhomogeneous differential equations. The right null space vector can be thought of as a solution to the homogeneous equation, and the terms with the row space vectors can be thought of as particular solutions.

We can also write the solution **x** in matrix form. The matrix is composed of three column vectors, which are the original two row space vectors and the right null space vector, which together form a basis in $\mathbb{R}^3$:

$$\mathbf{x} = \begin{pmatrix}1 & 2 & 1\\1 & 0 & 1\\1 & 1 & -2\end{pmatrix}\begin{pmatrix}-\frac{2}{3}\\1\\t - \frac{4}{3}\end{pmatrix}, \qquad t \in \mathbb{R}^1. \tag{7.79}$$

While the right null space vector is orthogonal to both row space vectors, the row space vectors are not orthogonal to themselves, so this basis is not orthogonal. Leaving out the calculational details, we can use the Gram-Schmidt procedure to cast the solution on an orthonormal basis:

$$\mathbf{x} = \underbrace{\frac{1}{\sqrt{3}}\begin{pmatrix}\frac{1}{\sqrt{3}}\\\frac{1}{\sqrt{3}}\\\frac{1}{\sqrt{3}}\end{pmatrix} + \sqrt{2}\begin{pmatrix}\frac{1}{\sqrt{2}}\\-\frac{1}{\sqrt{2}}\\0\end{pmatrix}}_{\text{row space}} + \underbrace{\sqrt{6}\left(t - \frac{4}{3}\right)\begin{pmatrix}\frac{1}{\sqrt{6}}\\\frac{1}{\sqrt{6}}\\-\sqrt{\frac{2}{3}}\end{pmatrix}}_{\text{right null space}}, \qquad t \in \mathbb{R}^1. \tag{7.80}$$

The first two terms are in the row space now represented on an orthonormal basis, the third is in the right null space. In matrix form, we can say that

$$\mathbf{x} = \begin{pmatrix}\frac{1}{\sqrt{3}} & \frac{1}{\sqrt{2}} & \frac{1}{\sqrt{6}}\\\frac{1}{\sqrt{3}} & -\frac{1}{\sqrt{2}} & \frac{1}{\sqrt{6}}\\\frac{1}{\sqrt{3}} & 0 & -\sqrt{\frac{2}{3}}\end{pmatrix}\begin{pmatrix}\frac{1}{\sqrt{3}}\\\sqrt{2}\\\sqrt{6}\left(t - \frac{4}{3}\right)\end{pmatrix}, \qquad t \in \mathbb{R}^1. \tag{7.81}$$

Of course, there are other orthonormal bases on which the system can be cast.

We see that the minimum length of the vector **x** occurs when $t = 4/3$, that is, when **x** is entirely in the row space. In such a case, we have

$$\min||\mathbf{x}||_2 = \sqrt{\left(\frac{1}{\sqrt{3}}\right)^2 + \left(\sqrt{2}\right)^2} = \sqrt{\frac{7}{3}}. \tag{7.82}$$

Lastly, note that here, we achieved a reasonable answer by setting $x_1 = t$ at the outset. We could have achieved an equivalent result by starting with $x_2 = t$, or $x_3 = t$. This will not work in all problems, as discussed in Section 7.9.3 on *row echelon form*.

7.3.3 Simultaneously Over- and Underconstrained

Some systems of equations are both over- and underconstrained simultaneously. This often happens when the rank r of the matrix is less than both N and M, the matrix dimensions. Such matrices are known as *rank-deficient*.

EXAMPLE 7.10

For $\mathbf{x} \in \mathbb{R}^4$, $\mathbf{b} \in \mathbb{R}^3$, $\mathbf{A} : \mathbb{R}^4 \to \mathbb{R}^3$, identify the best solutions to

$$\begin{pmatrix} 1 & 2 & 0 & 4 \\ 3 & 2 & -1 & 3 \\ -1 & 2 & 1 & 5 \end{pmatrix} \begin{pmatrix} x_1 \\ x_2 \\ x_3 \\ x_4 \end{pmatrix} = \begin{pmatrix} 1 \\ 3 \\ 2 \end{pmatrix}. \tag{7.83}$$

Using elementary row operations to perform Gaussian elimination (see Section A.3), gives rise to the equivalent system:

$$\begin{pmatrix} 1 & 0 & -1/2 & -1/2 \\ 0 & 1 & 1/4 & 9/4 \\ 0 & 0 & 0 & 0 \end{pmatrix} \begin{pmatrix} x_1 \\ x_2 \\ x_3 \\ x_4 \end{pmatrix} = \begin{pmatrix} 0 \\ 0 \\ 1 \end{pmatrix}. \tag{7.84}$$

We immediately see that there is a problem in the last equation, which purports $0 = 1$! What is actually happening is that $\mathbf{A}$ is not full rank $r = 3$ but actually has $r = 2$. It is thus rank-deficient, and vectors $\mathbf{x} \in \mathbb{R}^4$ are mapped into a two-dimensional subspace. So, we do not expect to find any solution to this problem, because our vector $\mathbf{b}$ is an arbitrary three-dimensional vector that most likely does not lie in the two-dimensional subspace. We can, however, find an $\mathbf{x}$ that minimizes the Euclidean norm of the residual. We return to the original equation and operate on a both sides with $\mathbf{A}^T$ to form $\mathbf{A}^T \cdot \mathbf{A} \cdot \mathbf{x} = \mathbf{A}^T \cdot \mathbf{b}$.

$$\begin{pmatrix} 1 & 3 & -1 \\ 2 & 2 & 2 \\ 0 & -1 & 1 \\ 4 & 3 & 5 \end{pmatrix} \begin{pmatrix} 1 & 2 & 0 & 4 \\ 3 & 2 & -1 & 3 \\ -1 & 2 & 1 & 5 \end{pmatrix} \begin{pmatrix} x_1 \\ x_2 \\ x_3 \\ x_4 \end{pmatrix} = \begin{pmatrix} 1 & 3 & -1 \\ 2 & 2 & 2 \\ 0 & -1 & 1 \\ 4 & 3 & 5 \end{pmatrix} \begin{pmatrix} 1 \\ 3 \\ 2 \end{pmatrix}, \tag{7.85}$$

$$\begin{pmatrix} 11 & 6 & -4 & 8 \\ 6 & 12 & 0 & 24 \\ -4 & 0 & 2 & 2 \\ 8 & 24 & 2 & 50 \end{pmatrix} \begin{pmatrix} x_1 \\ x_2 \\ x_3 \\ x_4 \end{pmatrix} = \begin{pmatrix} 8 \\ 12 \\ -1 \\ 23 \end{pmatrix}. \tag{7.86}$$

This operation has mapped both sides of the equation into the same space, namely, the column space of $\mathbf{A}^T$, which is also the row space of $\mathbf{A}$. Because the rank of $\mathbf{A}$ is $r = 2$, the dimension of the row space is also 2, and now the vectors on both sides of the equation have been mapped into the same plane. Again using row operations to perform Gaussian elimination gives rise to

$$\begin{pmatrix} 1 & 0 & -1/2 & -1/2 \\ 0 & 1 & 1/4 & 9/4 \\ 0 & 0 & 0 & 0 \\ 0 & 0 & 0 & 0 \end{pmatrix} \begin{pmatrix} x_1 \\ x_2 \\ x_3 \\ x_4 \end{pmatrix} = \begin{pmatrix} 1/4 \\ 7/8 \\ 0 \\ 0 \end{pmatrix}. \tag{7.87}$$

This suggests that here x_3 and x_4 are arbitrary, so we set $x_3 = s$, $x_4 = t$ and, treating s and t as known quantities, reduce the system to the following:

$$\begin{pmatrix} 1 & 0 \\ 0 & 1 \end{pmatrix} \begin{pmatrix} x_1 \\ x_2 \end{pmatrix} = \begin{pmatrix} 1/4 + s/2 + t/2 \\ 7/8 - s/4 - 9t/4 \end{pmatrix}, \tag{7.88}$$

so

$$\begin{pmatrix} x_1 \\ x_2 \\ x_3 \\ x_4 \end{pmatrix} = \begin{pmatrix} 1/4 \\ 7/8 \\ 0 \\ 0 \end{pmatrix} + s \begin{pmatrix} 1/2 \\ -1/4 \\ 1 \\ 0 \end{pmatrix} + t \begin{pmatrix} 1/2 \\ -9/4 \\ 0 \\ 1 \end{pmatrix}. \tag{7.89}$$

The vectors that are multiplied by s and t are in the right null space of $\mathbf{A}$. The vector $(1/4, 7/8, 0, 0)^T$ is not entirely in the row space of $\mathbf{A}$; it has components in both the row space and right null space. We can thus decompose this vector into a linear combination of row space vectors and right null space vectors using the procedure studied previously, solving the following equation for the coefficients $a_1, \ldots, a_4$, which are the coefficients of the row and right null space vectors:

$$\begin{pmatrix} 1/4 \\ 7/8 \\ 0 \\ 0 \end{pmatrix} = \begin{pmatrix} 1 & 3 & 1/2 & 1/2 \\ 2 & 2 & -1/4 & -9/4 \\ 0 & -1 & 1 & 0 \\ 4 & 3 & 0 & 1 \end{pmatrix} \begin{pmatrix} a_1 \\ a_2 \\ a_3 \\ a_4 \end{pmatrix}. \tag{7.90}$$

Solving, we get

$$\begin{pmatrix} a_1 \\ a_2 \\ a_3 \\ a_4 \end{pmatrix} = \begin{pmatrix} -3/244 \\ 29/244 \\ 29/244 \\ -75/244 \end{pmatrix}. \tag{7.91}$$

So we can recast the solution approximation as

$$\begin{pmatrix} x_1 \\ x_2 \\ x_3 \\ x_4 \end{pmatrix} = \underbrace{-\frac{3}{244} \begin{pmatrix} 1 \\ 2 \\ 0 \\ 4 \end{pmatrix} + \frac{29}{244} \begin{pmatrix} 3 \\ 2 \\ -1 \\ 3 \end{pmatrix}}_{\text{row space}} + \underbrace{\left(s + \frac{29}{244} \right) \begin{pmatrix} 1/2 \\ -1/4 \\ 1 \\ 0 \end{pmatrix} + \left(t - \frac{75}{244} \right) \begin{pmatrix} 1/2 \\ -9/4 \\ 0 \\ 1 \end{pmatrix}}_{\text{right null space}}. \tag{7.92}$$

This choice of $\mathbf{x}$ guarantees that we minimize $||\mathbf{A} \cdot \mathbf{x} - \mathbf{b}||_2$, which in this case is 1.22474. So there are no vectors $\mathbf{x}$ that satisfy the original equation $\mathbf{A} \cdot \mathbf{x} = \mathbf{b}$, but there are a doubly infinite number of vectors $\mathbf{x}$ that can minimize the Euclidean norm of the residual.

We can choose special values of s and t such that we minimize $||\mathbf{x}||_2$ while maintaining $||\mathbf{A} \cdot \mathbf{x} - \mathbf{b}||_2$ at its global minimum. This is done simply by forcing the magnitude of the right null space vectors to zero, so we choose $s = -29/244$, $t = 75/244$, giving

$$\begin{pmatrix} x_1 \\ x_2 \\ x_3 \\ x_4 \end{pmatrix} = \underbrace{-\frac{3}{244} \begin{pmatrix} 1 \\ 2 \\ 0 \\ 4 \end{pmatrix} + \frac{29}{244} \begin{pmatrix} 3 \\ 2 \\ -1 \\ 3 \end{pmatrix}}_{\text{row space}} = \begin{pmatrix} 21/61 \\ 13/61 \\ -29/244 \\ 75/244 \end{pmatrix}. \tag{7.93}$$

This vector has $||\mathbf{x}||_2 = 0.522055$.

7.3.4 Square

A set of N linear algebraic equations in N unknowns can be represented as

$$\mathbf{A}_{N \times N} \cdot \mathbf{x}_{N \times 1} = \mathbf{b}_{N \times 1}. \tag{7.94}$$

One can prove the following:

Theorem: (Cramer's rule) If $\mathbf{A}$ is of dimension $N \times N$ and $\det \mathbf{A} \neq 0$, the unique solution of the ith component of $\mathbf{x}$ for the equation of the form of Eq. (7.94) is

$$x_i = \frac{\det \mathbf{A}_i}{\det \mathbf{A}}, \tag{7.95}$$

where $\mathbf{A}_i$ is the matrix obtained by replacing the ith column of $\mathbf{A}$ by $\mathbf{b}$.

See also Section A.2. Although generally valid, Cramer's rule is most useful for low-dimensional systems. For large systems, Gaussian elimination is a more efficient technique. If $\det \mathbf{A} = 0$, there is either no solution or an infinite number of solutions. In the case where there are no solutions, one can still find an $\mathbf{x}$ that minimizes the normed residual $\|\mathbf{A} \cdot \mathbf{x} - \mathbf{b}\|_2$.

EXAMPLE 7.11

For $\mathbf{A} \colon \mathbb{R}^2 \to \mathbb{R}^2$, solve for $\mathbf{x}$ in $\mathbf{A} \cdot \mathbf{x} = \mathbf{b}$:

$$\begin{pmatrix} 1 & 2 \\ 3 & 2 \end{pmatrix} \begin{pmatrix} x_1 \\ x_2 \end{pmatrix} = \begin{pmatrix} 4 \\ 5 \end{pmatrix}. \tag{7.96}$$

By Cramer's rule,

$$x_1 = \frac{\begin{vmatrix} 4 & 2 \\ 5 & 2 \end{vmatrix}}{\begin{vmatrix} 1 & 2 \\ 3 & 2 \end{vmatrix}} = \frac{-2}{-4} = \frac{1}{2}, \tag{7.97}$$

$$x_2 = \frac{\begin{vmatrix} 1 & 4 \\ 3 & 5 \end{vmatrix}}{\begin{vmatrix} 1 & 2 \\ 3 & 2 \end{vmatrix}} = \frac{-7}{-4} = \frac{7}{4}. \tag{7.98}$$

So

$$\mathbf{x} = \begin{pmatrix} \frac{1}{2} \\ \frac{7}{4} \end{pmatrix}. \tag{7.99}$$

We get the same result by Gaussian elimination. Subtracting three times the first row from the second yields

$$\begin{pmatrix} 1 & 2 \\ 0 & -4 \end{pmatrix} \begin{pmatrix} x_1 \\ x_2 \end{pmatrix} = \begin{pmatrix} 4 \\ -7 \end{pmatrix}. \tag{7.100}$$

Thus, $x_2 = 7/4$. Back substitution into the first equation then gives $x_1 = 1/2$.

EXAMPLE 7.12

With $\mathbf{A} \colon \mathbb{R}^2 \to \mathbb{R}^2$, find the most general $\mathbf{x}$ that best satisfies $\mathbf{A} \cdot \mathbf{x} = \mathbf{b}$ for

$$\begin{pmatrix} 1 & 2 \\ 3 & 6 \end{pmatrix} \begin{pmatrix} x_1 \\ x_2 \end{pmatrix} = \begin{pmatrix} 2 \\ 0 \end{pmatrix}. \tag{7.101}$$

Obviously, there is no unique solution to this system because the determinant of the coefficient matrix is zero. The rank of $\mathbf{A}$ is 1, so in actuality, $\mathbf{A}$ maps vectors from $\mathbb{R}^2$ into a one-dimensional subspace, $\mathbb{R}^1$. For a general $\mathbf{b}$, which does not lie in the one-dimensional subspace, we can find the best solution $\mathbf{x}$ by first multiplying both sides by $\mathbf{A}^T$:

$$\begin{pmatrix} 1 & 3 \\ 2 & 6 \end{pmatrix} \begin{pmatrix} 1 & 2 \\ 3 & 6 \end{pmatrix} \begin{pmatrix} x_1 \\ x_2 \end{pmatrix} = \begin{pmatrix} 1 & 3 \\ 2 & 6 \end{pmatrix} \begin{pmatrix} 2 \\ 0 \end{pmatrix}, \tag{7.102}$$

$$\begin{pmatrix} 10 & 20 \\ 20 & 40 \end{pmatrix} \begin{pmatrix} x_1 \\ x_2 \end{pmatrix} = \begin{pmatrix} 2 \\ 4 \end{pmatrix}. \tag{7.103}$$

This operation maps both sides of the equation into the column space of $\mathbf{A}^T$, which is the row space of $\mathbf{A}$, which has dimension 1. Because the vectors are now in the same space, a solution can be found. Using row reductions to perform Gaussian elimination, we get

$$\begin{pmatrix} 1 & 2 \\ 0 & 0 \end{pmatrix} \begin{pmatrix} x_1 \\ x_2 \end{pmatrix} = \begin{pmatrix} 1/5 \\ 0 \end{pmatrix}. \tag{7.104}$$

We set $x_2 = t$, where t is any arbitrary real number, and solve to get

$$\begin{pmatrix} x_1 \\ x_2 \end{pmatrix} = \begin{pmatrix} 1/5 \\ 0 \end{pmatrix} + t \begin{pmatrix} -2 \\ 1 \end{pmatrix}. \tag{7.105}$$

The vector that t multiplies, $(-2, 1)^T$, is in the right null space of $\mathbf{A}$. We can recast the vector $(1/5, 0)^T$ in terms of a linear combination of the row space vector $(1, 2)^T$ and the right null space vector to get the final form of the solution:

$$\begin{pmatrix} x_1 \\ x_2 \end{pmatrix} = \underbrace{\frac{1}{25} \begin{pmatrix} 1 \\ 2 \end{pmatrix}}_{\text{row space}} + \underbrace{\left(t - \frac{2}{25} \right) \begin{pmatrix} -2 \\ 1 \end{pmatrix}}_{\text{right null space}}. \tag{7.106}$$

This choice of $\mathbf{x}$ guarantees that the Euclidean norm of the residual $||\mathbf{A} \cdot \mathbf{x} - \mathbf{b}||_2$ is minimized. In this case the Euclidean norm of the residual is 1.89737. The vector $\mathbf{x}$ with the smallest norm that minimizes $||\mathbf{A} \cdot \mathbf{x} - \mathbf{b}||_2$ is found by setting the magnitude of the right null space contribution to zero, so we can take $t = 2/25$, giving

$$\begin{pmatrix} x_1 \\ x_2 \end{pmatrix} = \underbrace{\frac{1}{25} \begin{pmatrix} 1 \\ 2 \end{pmatrix}}_{\text{row space}}. \tag{7.107}$$

This gives rise to $||\mathbf{x}||_2 = 0.0894427$.

7.3.5 Fredholm Alternative

The Fredholm alternative, introduced in Section 4.6 for differential equations, applied to linear algebra is as follows. For a given $N \times N$ matrix $\mathbf{A}$, nonzero scalar constant $\lambda \in \mathbb{C}^1$, and $N \times 1$ vector $\mathbf{b}$, there exists either

- a unique $\mathbf{x}$ such that $\mathbf{A} \cdot \mathbf{x} - \lambda \mathbf{I} \cdot \mathbf{x} = \mathbf{b}$, or
- a nonunique $\mathbf{x}$ such that $\mathbf{A} \cdot \mathbf{x} - \lambda \mathbf{I} \cdot \mathbf{x} = \mathbf{0}$.

EXAMPLE 7.13

Examine the Fredholm alternative for

$$\mathbf{A} = \begin{pmatrix} 2 & 0 \\ -1 & 1 \end{pmatrix}, \qquad \mathbf{b} = \begin{pmatrix} 1 \\ 0 \end{pmatrix}. \tag{7.108}$$

The equation $\mathbf{A} \cdot \mathbf{x} - \lambda \mathbf{I} \cdot \mathbf{x} = \mathbf{b}$ is here

$$\begin{pmatrix} 2-\lambda & 0 \\ -1 & 1-\lambda \end{pmatrix} \begin{pmatrix} x_1 \\ x_2 \end{pmatrix} = \begin{pmatrix} 1 \\ 0 \end{pmatrix}. \tag{7.109}$$

For $\lambda \neq 1, 2$, the unique inverse exists, and we find

$$\begin{pmatrix} x_1 \\ x_2 \end{pmatrix} = \begin{pmatrix} \frac{1-\lambda}{\lambda^2 - 3\lambda + 2} \\ \frac{1}{\lambda^2 - 3\lambda + 2} \end{pmatrix}. \tag{7.110}$$

When $\lambda = 1, 2$, the inverse does not exist. Let us then look for $\mathbf{x}$ such that $\mathbf{A} \cdot \mathbf{x} - \lambda \mathbf{I} \cdot \mathbf{x} = 0$. First consider $\lambda = 1$:

$$\mathbf{A} \cdot \mathbf{x} - \lambda \mathbf{I} \cdot \mathbf{x} = \mathbf{0}, \tag{7.111}$$

$$\begin{pmatrix} 1 & 0 \\ -1 & 0 \end{pmatrix} \begin{pmatrix} x_1 \\ x_2 \end{pmatrix} = \begin{pmatrix} 0 \\ 0 \end{pmatrix}. \tag{7.112}$$

By inspection, we see that

$$\mathbf{x} = t \begin{pmatrix} 0 \\ 1 \end{pmatrix}, \qquad t \in \mathbb{R}^1. \tag{7.113}$$

As long as $t \neq 0$, we have found $\mathbf{x}$ that satisfies the Fredholm alternative.

Next consider $\lambda = 2$:

$$\mathbf{A} \cdot \mathbf{x} - \lambda \mathbf{I} \cdot \mathbf{x} = \mathbf{0}, \tag{7.114}$$

$$\begin{pmatrix} 0 & 0 \\ -1 & -1 \end{pmatrix} \begin{pmatrix} x_1 \\ x_2 \end{pmatrix} = \begin{pmatrix} 0 \\ 0 \end{pmatrix}. \tag{7.115}$$

By inspection, we see that

$$\mathbf{x} = t \begin{pmatrix} 1 \\ -1 \end{pmatrix}, \qquad t \in \mathbb{R}^1. \tag{7.116}$$

As long as $t \neq 0$, we have again found $\mathbf{x}$ that satisfies the Fredholm alternative.

EXAMPLE 7.14

Examine the Fredholm alternative for

$$\mathbf{A} = \begin{pmatrix} 2 & 0 \\ -1 & 1 \end{pmatrix}, \qquad \lambda = 1, \qquad \mathbf{b} = \begin{pmatrix} 1 \\ -1 \end{pmatrix}. \tag{7.117}$$

The equation $\mathbf{A} \cdot \mathbf{x} - \lambda \mathbf{I} \cdot \mathbf{x} = \mathbf{b}$ becomes

$$\begin{pmatrix} 1 & 0 \\ -1 & 0 \end{pmatrix} \begin{pmatrix} x_1 \\ x_2 \end{pmatrix} = \begin{pmatrix} 1 \\ -1 \end{pmatrix}. \tag{7.118}$$

Because $\mathbf{b}$ is in the column space of $\mathbf{A} - \mathbf{I}$, we can solve to obtain the nonunique solution that is the sum of a unique row space vector and a nonunique right null space vector:

$$\mathbf{x} = \begin{pmatrix} 1 \\ 0 \end{pmatrix} + t \begin{pmatrix} 0 \\ 1 \end{pmatrix}, \qquad t \in \mathbb{R}^1. \tag{7.119}$$

So there is no unique solution to $\mathbf{A} \cdot \mathbf{x} - \lambda \mathbf{I} \cdot \mathbf{x} = \mathbf{b}$; there is a nonunique solution, however, to this, as well as to $\mathbf{A} \cdot \mathbf{x} - \lambda \mathbf{I} \cdot \mathbf{x} = \mathbf{0}$.

7.4 Eigenvalues and Eigenvectors

7.4.1 Ordinary

Much of the general discussion of eigenvectors and eigenvalues has been covered in Chapter 6 (see especially Section 6.8), and will not be repeated here. A few new concepts are introduced, and some old ones are reinforced.

First, we recall that when one refers to eigenvectors, one typically is referring to the *right* eigenvectors that arise from $\mathbf{A} \cdot \mathbf{e} = \lambda \mathbf{I} \cdot \mathbf{e}$; if no distinction is made, it can be assumed that it is the right set that is being discussed. Though it does not arise as often, there are occasions when one requires the *left* eigenvectors that arise from $\bar{\mathbf{e}}^T \cdot \mathbf{A} = \bar{\mathbf{e}}^T \cdot \mathbf{I}\lambda$. Some important properties and definitions involving eigenvalues are listed next:

- If the matrix $\mathbf{A}$ is self-adjoint, it can be shown that it has the same left and right eigenvectors.
- If $\mathbf{A}$ is not self-adjoint, it has different left and right eigenvectors. The eigenvalues are the same for both left and right eigenvectors of the same operator, whether or not the system is self-adjoint.
- The polynomial equation that arises in the eigenvalue problem is the *characteristic equation* of the matrix.
- The *Cayley-Hamilton*[2] theorem states that a matrix satisfies its own characteristic equation.
- If a matrix is triangular, then its eigenvalues are its diagonal terms.
- Eigenvalues of $\mathbf{A} \cdot \mathbf{A} = \mathbf{A}^2$ are the square of the eigenvalues of $\mathbf{A}$.
- With "eig" denoting a set of eigenvalues,

$$\mathrm{eig}\,(\mathbf{I} + \alpha \mathbf{A}) = 1 + \alpha\,\mathrm{eig}\,\mathbf{A}. \tag{7.120}$$

- Every eigenvector of $\mathbf{A}$ is also an eigenvector of $\mathbf{A}^2$.
- A matrix $\mathbf{A}$ has *spectral radius*, $\rho(\mathbf{A})$, defined as the largest of the absolute values of its eigenvalues:

$$\rho(\mathbf{A}) \equiv \max_n(|\lambda_n|). \tag{7.121}$$

- Recall from Eq. (6.377) that a matrix $\mathbf{A}$ has a *spectral norm*, $||\mathbf{A}||_2$, where

$$||\mathbf{A}||_2 = \sqrt{\max_i(\kappa_i)}, \tag{7.122}$$

  where for real-valued $\mathbf{A}$, κ_i is an eigenvalue of $\mathbf{A}^T \cdot \mathbf{A}$. In general $\rho(\mathbf{A}) \neq ||\mathbf{A}||_2$.
- If $\mathbf{A}$ is square and invertible,

$$\frac{1}{||\mathbf{A}^{-1}||_2} = \sqrt{\min_i(\kappa_i)}, \tag{7.123}$$

  where for real-valued $\mathbf{A}$, κ_i is an eigenvalue of $\mathbf{A}^T \cdot \mathbf{A}$.
- If $\mathbf{A}$ is self-adjoint, $\rho(\mathbf{A}) = ||\mathbf{A}||_2$.
- The *condition number* c of a square matrix $\mathbf{A}$ is

$$c = ||\mathbf{A}||_2 ||\mathbf{A}^{-1}||_2. \tag{7.124}$$

---

[2] Arthur Cayley, 1821–1895, English mathematician, and William Rowan Hamilton, 1805–1865, Anglo-Irish mathematician.

From the definition of the norm, Eq. (7.122), it is easy to show that

$$c = \sqrt{\frac{\max_i(\kappa_i)}{\min_i(\kappa_i)}},$$
(7.125)

where κ_i is again any of the eigenvalues of $\mathbf{A}^T \cdot \mathbf{A}$. For self-adjoint matrices, we get the simpler

$$c = \frac{\max_i |\lambda_i|}{\min_i |\lambda_i|}.$$
(7.126)

- In general, Gelfand's[3] formula holds:

$$\rho(\mathbf{A}) = \lim_{k \to \infty} ||\mathbf{A}^k||^{1/k}.$$
(7.127)

The norm here holds for any matrix norm, including our spectral norm. This notion was discussed earlier in Section 6.9.
- The *trace* of a matrix, introduced on Section 2.1.5, is the sum of the terms on the leading diagonal.
- The trace of a $N \times N$ matrix is the sum of its N eigenvalues.
- The product of the N eigenvalues is the determinant of the matrix.

EXAMPLE 7.15

Demonstrate the theorems and definitions just described for

$$\mathbf{A} = \begin{pmatrix} 0 & 1 & -2 \\ 2 & 1 & 0 \\ 4 & -2 & 5 \end{pmatrix}.$$
(7.128)

The characteristic equation is

$$\lambda^3 - 6\lambda^2 + 11\lambda - 6 = 0.$$
(7.129)

The Cayley-Hamilton theorem is easily verified by direct substitution:

$$\begin{pmatrix} 0 & 1 & -2 \\ 2 & 1 & 0 \\ 4 & -2 & 5 \end{pmatrix} \begin{pmatrix} 0 & 1 & -2 \\ 2 & 1 & 0 \\ 4 & -2 & 5 \end{pmatrix} \begin{pmatrix} 0 & 1 & -2 \\ 2 & 1 & 0 \\ 4 & -2 & 5 \end{pmatrix} - 6 \begin{pmatrix} 0 & 1 & -2 \\ 2 & 1 & 0 \\ 4 & -2 & 5 \end{pmatrix} \begin{pmatrix} 0 & 1 & -2 \\ 2 & 1 & 0 \\ 4 & -2 & 5 \end{pmatrix}$$

$$+ 11 \begin{pmatrix} 0 & 1 & -2 \\ 2 & 1 & 0 \\ 4 & -2 & 5 \end{pmatrix} - 6 \begin{pmatrix} 1 & 0 & 0 \\ 0 & 1 & 0 \\ 0 & 0 & 1 \end{pmatrix} = \begin{pmatrix} 0 & 0 & 0 \\ 0 & 0 & 0 \\ 0 & 0 & 0 \end{pmatrix},$$
(7.130)

$$\begin{pmatrix} -30 & 19 & -38 \\ -10 & 13 & -24 \\ 52 & -26 & 53 \end{pmatrix} + \begin{pmatrix} 36 & -30 & 60 \\ -12 & -18 & 24 \\ -96 & 48 & -102 \end{pmatrix} + \begin{pmatrix} 0 & 11 & -22 \\ 22 & 11 & 0 \\ 44 & -22 & 55 \end{pmatrix} + \begin{pmatrix} -6 & 0 & 0 \\ 0 & -6 & 0 \\ 0 & 0 & -6 \end{pmatrix}$$

$$= \begin{pmatrix} 0 & 0 & 0 \\ 0 & 0 & 0 \\ 0 & 0 & 0 \end{pmatrix},$$
(7.131)

shows that

$$\mathbf{A}^3 - 6\mathbf{A}^2 + 11\mathbf{A} - 6\mathbf{I} = 0.$$
(7.132)

---

[3] Israel Gelfand, 1913–2009, Soviet mathematician.

Considering the traditional right eigenvalue problem, $\mathbf{A} \cdot \mathbf{e} = \lambda \mathbf{I} \cdot \mathbf{e}$, it is easily shown that the eigenvalues and (right) eigenvectors for this system are

$$\lambda_1 = 1, \qquad \mathbf{e}_1 = \begin{pmatrix} 0 \\ 2 \\ 1 \end{pmatrix}, \tag{7.133}$$

$$\lambda_2 = 2, \qquad \mathbf{e}_2 = \begin{pmatrix} \frac{1}{2} \\ 1 \\ 0 \end{pmatrix}, \tag{7.134}$$

$$\lambda_3 = 3, \qquad \mathbf{e}_3 = \begin{pmatrix} -1 \\ -1 \\ 1 \end{pmatrix}. \tag{7.135}$$

One notes that while the eigenvectors do form a basis in $\mathbb{R}^3$, they are not orthogonal; this is a consequence of the matrix not being self-adjoint (or, more specifically, it being asymmetric). The spectral radius is $\rho(\mathbf{A}) = 3$. Now

$$\mathbf{A}^2 = \mathbf{A} \cdot \mathbf{A} = \begin{pmatrix} 0 & 1 & -2 \\ 2 & 1 & 0 \\ 4 & -2 & 5 \end{pmatrix} \begin{pmatrix} 0 & 1 & -2 \\ 2 & 1 & 0 \\ 4 & -2 & 5 \end{pmatrix} = \begin{pmatrix} -6 & 5 & -10 \\ 2 & 3 & -4 \\ 16 & -8 & 17 \end{pmatrix}. \tag{7.136}$$

It is easily shown that the eigenvalues for $\mathbf{A}^2$ are $1, 4, 9$, precisely the squares of the eigenvalues of $\mathbf{A}$.

The trace is

$$\mathrm{tr}(\mathbf{A}) = 0 + 1 + 5 = 6. \tag{7.137}$$

This is the equal to the sum of the eigenvalues

$$\sum_{i=1}^{3} \lambda_i = 1 + 2 + 3 = 6. \tag{7.138}$$

Note also that

$$\det \mathbf{A} = 6 = \lambda_1 \lambda_2 \lambda_3 = (1)(2)(3) = 6. \tag{7.139}$$

All the eigenvalues are positive; *however, the matrix is not positive definite!* For instance, if $\mathbf{x} = (-1, 1, 1)^T$, $\mathbf{x}^T \cdot \mathbf{A} \cdot \mathbf{x} = -1$. We might ask about the positive definiteness of the symmetric part of $\mathbf{A}$, $\mathbf{A}_s = (\mathbf{A} + \mathbf{A}^T)/2$:

$$\mathbf{A}_s = \begin{pmatrix} 0 & \frac{3}{2} & 1 \\ \frac{3}{2} & 1 & -1 \\ 1 & -1 & 5 \end{pmatrix}. \tag{7.140}$$

In this case, $\mathbf{A}_s$ has real eigenvalues, both positive and negative, $\lambda_1 = 5.32, \lambda_2 = -1.39, \lambda_3 = 2.07$. Because of the presence of a negative eigenvalue in the symmetric part of $\mathbf{A}$, we can conclude that both $\mathbf{A}$ and $\mathbf{A}_s$ are not positive definite.

We also note that for real-valued problems $\mathbf{x} \in \mathbb{R}^N$, $\mathbf{A} \in \mathbb{R}^{N \times N}$, the antisymmetric part of a matrix can never be positive definite by the following argument. We can say $\mathbf{x}^T \cdot \mathbf{A} \cdot \mathbf{x} = \mathbf{x}^T \cdot (\mathbf{A}_s + \mathbf{A}_a) \cdot \mathbf{x}$. Then one has $\mathbf{x}^T \cdot \mathbf{A}_a \cdot \mathbf{x} = 0$ for all $\mathbf{x}$ because the tensor inner product of the real antisymmetric $\mathbf{A}_a$ with the symmetric $\mathbf{x}^T$ and $\mathbf{x}$ is identically zero; see Eq. (2.96). So to test the positive definiteness of a real $\mathbf{A}$, it suffices to consider the positive definiteness of its symmetric part: $\mathbf{x}^T \cdot \mathbf{A}_s \cdot \mathbf{x} \geq 0$.

For complex-valued problems, $\mathbf{x} \in \mathbb{C}^N$, $\mathbf{A} \in \mathbb{C}^{N \times N}$, it is not as simple. Recalling that the eigenvalues of an antisymmetric matrix $\mathbf{A}_a$ are purely imaginary, we have, if $\mathbf{x}$ is an eigenvector of $\mathbf{A}_a$, that $\mathbf{x}^H \cdot \mathbf{A}_a \cdot \mathbf{x} = \mathbf{x}^H \cdot (\lambda)\mathbf{x} = \mathbf{x}^H \cdot (i\lambda_I)\mathbf{x} = i\lambda_I \mathbf{x}^H \cdot \mathbf{x} = i\lambda_I \|\mathbf{x}\|_2^2$, where $\lambda_I \in \mathbb{R}^1$. Hence, whenever the vector $\mathbf{x}$ is an eigenvector of $\mathbf{A}_a$, the quantity $\mathbf{x}^H \cdot \mathbf{A}_a \cdot \mathbf{x}$ is a pure imaginary number.

We can also easily solve the left eigenvalue problem, $\mathbf{e}_L^H \cdot \mathbf{A} = \lambda \mathbf{e}_L^H \cdot \mathbf{I}$:

$$\lambda_1 = 1, \quad \mathbf{e}_{(L1)} = \begin{pmatrix} 2 \\ -1 \\ 1 \end{pmatrix}, \tag{7.141}$$

$$\lambda_2 = 2, \quad \mathbf{e}_{L2} = \begin{pmatrix} -3 \\ 1 \\ -2 \end{pmatrix}, \tag{7.142}$$

$$\lambda_3 = 3, \quad \mathbf{e}_{L3} = \begin{pmatrix} 2 \\ -1 \\ 2 \end{pmatrix}. \tag{7.143}$$

We see eigenvalues are the same, but the left and right eigenvectors are different.

We find $||\mathbf{A}||_2$ by considering eigenvalues of $\mathbf{A}^T \cdot \mathbf{A}$, the real variable version of that described in Eq. (6.377):

$$\mathbf{A}^T \cdot \mathbf{A} = \begin{pmatrix} 0 & 2 & 4 \\ 1 & 1 & -2 \\ -2 & 0 & 5 \end{pmatrix} \begin{pmatrix} 0 & 1 & -2 \\ 2 & 1 & 0 \\ 4 & -2 & 5 \end{pmatrix}, \tag{7.144}$$

$$= \begin{pmatrix} 20 & -6 & 20 \\ -6 & 6 & -12 \\ 20 & -12 & 29 \end{pmatrix}. \tag{7.145}$$

This matrix has eigenvalues $\kappa = 49.017, 5.858, 0.125$. The spectral norm is the square root of the largest, giving

$$||\mathbf{A}||_2 = \sqrt{49.017} = 7.00122. \tag{7.146}$$

The eigenvector of $\mathbf{A}^T \cdot \mathbf{A}$ corresponding to $\kappa = 49.017$ is $\hat{\mathbf{e}}_1 = (0.5829, -0.2927, 0.7579)^T$. When we compute the quantity associated with the norm of an operator, we find this vector maps to the norm:

$$\frac{||\mathbf{A} \cdot \hat{\mathbf{e}}_1||_2}{||\hat{\mathbf{e}}_1||_2} = \frac{\left|\left| \begin{pmatrix} 0 & 1 & -2 \\ 2 & 1 & 0 \\ 4 & -2 & 5 \end{pmatrix} \begin{pmatrix} 0.5829 \\ -0.2927 \\ 0.7579 \end{pmatrix} \right|\right|_2}{\left|\left| \begin{pmatrix} 0.582944 \\ -0.292744 \\ 0.757943 \end{pmatrix} \right|\right|_2} = \frac{\left|\left| \begin{pmatrix} -1.80863 \\ 0.873144 \\ 6.70698 \end{pmatrix} \right|\right|_2}{1} = 7.00122.$$

$$\tag{7.147}$$

Had we chosen the eigenvector associated with the eigenvalue of largest magnitude, $\mathbf{e}_3 = (-1, -1, 1)^T$, we would have found $||\mathbf{A} \cdot \mathbf{e}_3||_2/||\mathbf{e}_3||_2 = 3$, the spectral radius. Obviously, this is not the maximum of this operation and thus cannot be a norm.

By Eq. (7.123), we can associate the smallest eigenvalue of $\mathbf{A}^T \cdot \mathbf{A}$, $\kappa = 0.125$, with $||\mathbf{A}^{-1}||_2$ via

$$\frac{1}{||\mathbf{A}^{-1}||_2} = \sqrt{\min_i(\kappa_i)} = \sqrt{0.125} = 0.354, \tag{7.148}$$

$$||\mathbf{A}^{-1}||_2 = \frac{1}{0.354} = 2.8241. \tag{7.149}$$

We can easily verify Gelfand's theorem by direct calculation of $||\mathbf{A}^k||_2^{1/k}$ for various k. We find the following.

k	$||\mathbf{A}^k||_2^{1/k}$
1	7.00122
2	5.27011
3	4.61257
4	4.26334
5	4.03796
10	3.52993
100	3.04984
1000	3.00494
∞	3

As $k \to \infty$, $||\mathbf{A}^k||_2^{1/k}$ approaches the spectral radius $\rho(\mathbf{A}) = 3$.

7.4.2 Generalized in the Second Sense

In Section 6.8, we studied generalized eigenvectors in the first sense. Here we consider a distinct problem that leads to *generalized eigenvalues and eigenvectors in the second sense*. Consider the problem

$$\mathbf{A} \cdot \mathbf{e} = \lambda \mathbf{B} \cdot \mathbf{e}, \tag{7.150}$$

where $\mathbf{A}$ and $\mathbf{B}$ are square matrices, possibly singular, of dimension $N \times N$, $\mathbf{e}$ is a generalized eigenvector in the second sense, and λ is a generalized eigenvalue. If $\mathbf{B}$ were not singular, we could form $(\mathbf{B}^{-1} \cdot \mathbf{A}) \cdot \mathbf{e} = \lambda \mathbf{I} \cdot \mathbf{e}$, which amounts to an ordinary eigenvalue problem. But let us assume that the inverses do not exist. Then Eq. (7.150) can be recast as

$$(\mathbf{A} - \lambda \mathbf{B}) \cdot \mathbf{e} = \mathbf{0}. \tag{7.151}$$

For nontrivial solutions, we simply require

$$\det(\mathbf{A} - \lambda \mathbf{B}) = 0 \tag{7.152}$$

and analyze in a similar manner.

EXAMPLE 7.16

Find the generalized eigenvalues and eigenvectors in the second sense for

$$\underbrace{\begin{pmatrix} 1 & 2 \\ 2 & 1 \end{pmatrix}}_{\mathbf{A}} \mathbf{e} = \lambda \underbrace{\begin{pmatrix} 1 & 0 \\ 1 & 0 \end{pmatrix}}_{\mathbf{B}} \mathbf{e}. \tag{7.153}$$

Here $\mathbf{B}$ is obviously singular. We rewrite as

$$\begin{pmatrix} 1 - \lambda & 2 \\ 2 - \lambda & 1 \end{pmatrix} \begin{pmatrix} e_1 \\ e_2 \end{pmatrix} = \begin{pmatrix} 0 \\ 0 \end{pmatrix}. \tag{7.154}$$

For a nontrivial solution, we require

$$\begin{vmatrix} 1 - \lambda & 2 \\ 2 - \lambda & 1 \end{vmatrix} = 0, \tag{7.155}$$

which gives

$$1 - \lambda - 2(2 - \lambda) = 0, \tag{7.156}$$

$$1 - \lambda - 4 + 2\lambda = 0, \tag{7.157}$$

$$\lambda = 3. \tag{7.158}$$

For **e**, we require

$$\begin{pmatrix} 1 - 3 & 2 \\ 2 - 3 & 1 \end{pmatrix} \begin{pmatrix} e_1 \\ e_2 \end{pmatrix} = \begin{pmatrix} 0 \\ 0 \end{pmatrix}, \tag{7.159}$$

$$\begin{pmatrix} -2 & 2 \\ -1 & 1 \end{pmatrix} \begin{pmatrix} e_1 \\ e_2 \end{pmatrix} = \begin{pmatrix} 0 \\ 0 \end{pmatrix}. \tag{7.160}$$

By inspection, the generalized eigenvector in the second sense,

$$\mathbf{e} = \begin{pmatrix} e_1 \\ e_2 \end{pmatrix} = \alpha \begin{pmatrix} 1 \\ 1 \end{pmatrix}, \tag{7.161}$$

satisfies Eq. (7.153) when $\lambda = 3$, and α is any scalar.

7.5 Matrices as Linear Mappings

By considering a matrix as an operator that effects a linear mapping and applying it to a specific geometry, one can better envision the action of the matrix. Here, then, we envision the alibi approach to the transformation: a vector in one space is transformed by an operator into a different vector in the alibi space. This is demonstrated in the following examples.

EXAMPLE 7.17

Examine how the matrix

$$\mathbf{A} = \begin{pmatrix} 0 & -1 \\ 1 & -1 \end{pmatrix} \tag{7.162}$$

acts on vectors **x**, including those that form a unit square with vertices as $A : (0,0), B : (1,0), C : (1,1), D : (0,1)$.

The original square has area $a = 1$. Each of the vertices map under the linear homogeneous transformation to

$$\underbrace{\begin{pmatrix} 0 & -1 \\ 1 & -1 \end{pmatrix} \underbrace{\begin{pmatrix} 0 \\ 0 \end{pmatrix}}_{A} = \underbrace{\begin{pmatrix} 0 \\ 0 \end{pmatrix}}_{\overline{A}}, \quad \begin{pmatrix} 0 & -1 \\ 1 & -1 \end{pmatrix} \underbrace{\begin{pmatrix} 1 \\ 0 \end{pmatrix}}_{B} = \underbrace{\begin{pmatrix} 0 \\ 1 \end{pmatrix}}_{\overline{B}}, \tag{7.163}$$

$$\begin{pmatrix} 0 & -1 \\ 1 & -1 \end{pmatrix} \underbrace{\begin{pmatrix} 1 \\ 1 \end{pmatrix}}_{C} = \underbrace{\begin{pmatrix} -1 \\ 0 \end{pmatrix}}_{\overline{C}}, \quad \begin{pmatrix} 0 & -1 \\ 1 & -1 \end{pmatrix} \underbrace{\begin{pmatrix} 0 \\ 1 \end{pmatrix}}_{D} = \underbrace{\begin{pmatrix} -1 \\ -1 \end{pmatrix}}_{\overline{D}}. \tag{7.164}$$

In the mapped space, the square has transformed to a parallelogram. This is plotted in Figure 7.3. Here the alibi approach to the mapping is evident. We keep the coordinate axes fixed in Figure 7.3 and rotate and stretch the vectors, instead of keeping the vectors fixed and rotating the axes, as would have been done in an alias transformation. Now we have

$$\det \mathbf{A} = (0)(-1) - (1)(-1) = 1. \tag{7.165}$$

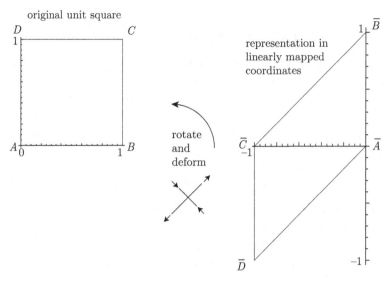

Figure 7.3.   Unit square transforming via stretching and rotation under a linear area- and orientation-preserving alibi mapping.

Thus, the mapping is both orientation- and area-preserving. The orientation-preserving feature is obvious by inspecting the locations of the points A, B, C, and D in both configurations shown in Figure 7.3. We easily calculate the area in the mapped space by combining the areas of two triangles which form the parallelogram:

$$a = \frac{1}{2}(1)(1) + \frac{1}{2}(1)(1) = 1. \tag{7.166}$$

Thus, the area is preserved. The eigenvalues of $\mathbf{A}$ are $-(1/2) \pm \sqrt{3/2}i$, both of which have magnitude of unity. Thus, the spectral radius $\rho(\mathbf{A}) = 1$. However, the spectral norm of $\mathbf{A}$ is nonunity, because

$$\mathbf{A}^T \cdot \mathbf{A} = \begin{pmatrix} 0 & 1 \\ -1 & -1 \end{pmatrix} \begin{pmatrix} 0 & -1 \\ 1 & -1 \end{pmatrix} = \begin{pmatrix} 1 & -1 \\ -1 & 2 \end{pmatrix}, \tag{7.167}$$

which has eigenvalues

$$\kappa = \frac{1}{2}(3 \pm \sqrt{5}). \tag{7.168}$$

The spectral norm is the square root of the maximum eigenvalue of $\mathbf{A}^T \cdot \mathbf{A}$, which is

$$\|\mathbf{A}\|_2 = \sqrt{\frac{1}{2}(3 + \sqrt{5})} = 1.61803. \tag{7.169}$$

It will later be shown (Section 7.9.4), that the action of $\mathbf{A}$ on the unit square can be decomposed into a deformation and a rotation. Both are evident in Figure 7.3.

7.6 Complex Matrices

If $\mathbf{x}$ and $\mathbf{y}$ are complex vectors, we know that their inner product involves the conjugate transpose, also known as the Hermitian transpose, and denote it by a superscript H, as introduced Section 6.3.1. Thus, we define the inner product as

$$\langle \mathbf{x}, \mathbf{y} \rangle = \overline{\mathbf{x}}^T \cdot \mathbf{y} = \mathbf{x}^H \cdot \mathbf{y}. \tag{7.170}$$

Then the norm is given by

$$||\mathbf{x}||_2 = +\sqrt{\mathbf{x}^H \cdot \mathbf{x}}. \qquad (7.171)$$

EXAMPLE 7.18

If

$$\mathbf{x} = \begin{pmatrix} 1+i \\ 3-2i \\ 2 \\ -3i \end{pmatrix}, \qquad (7.172)$$

find $||\mathbf{x}||_2$.

$$||\mathbf{x}||_2 = +\sqrt{\mathbf{x}^H \cdot \mathbf{x}} = + \sqrt{(1-i, 3+2i, 2, +3i) \begin{pmatrix} 1+i \\ 3-2i \\ 2 \\ -3i \end{pmatrix}} = +\sqrt{2+13+4+9} = 2\sqrt{7}.$$

$$(7.173)$$

EXAMPLE 7.19

If

$$\mathbf{x} = \begin{pmatrix} 1+i \\ -2+3i \\ 2-i \end{pmatrix}, \qquad (7.174)$$

$$\mathbf{y} = \begin{pmatrix} 3 \\ 4-2i \\ 3+3i \end{pmatrix}, \qquad (7.175)$$

find $\langle \mathbf{x}, \mathbf{y} \rangle$.

$$\langle \mathbf{x}, \mathbf{y} \rangle = \mathbf{x}^H \cdot \mathbf{y}, \qquad (7.176)$$

$$= (1-i, -2-3i, 2+i) \begin{pmatrix} 3 \\ 4-2i \\ 3+3i \end{pmatrix}, \qquad (7.177)$$

$$= (3-3i) + (-14-8i) + (3+9i), \qquad (7.178)$$

$$= -8 - 2i. \qquad (7.179)$$

Recall also the conjugate or Hermitian transpose of a matrix $\mathbf{A}$ is $\mathbf{A}^H$, given by the transpose of the matrix with each element being replaced by its conjugate:

$$\mathbf{A}^H = \bar{\mathbf{A}}^T. \qquad (7.180)$$

As the Hermitian transpose is the adjoint operator corresponding to a given complex matrix, we can apply an earlier proved theorem for linear operators (Section 6.8), to deduce that the eigenvalues of a complex matrix are the complex conjugates of the Hermitian transpose of that matrix.

The Hermitian transpose is distinguished from a matrix which is Hermitian as follows. A Hermitian matrix is one that is equal to its conjugate transpose. So a matrix that equals its Hermitian transpose is Hermitian. A skew-Hermitian matrix is the negative of its Hermitian transpose. A Hermitian matrix is self-adjoint.

Properties are as follows:

- $\mathbf{x}^H \cdot \mathbf{A} \cdot \mathbf{x}$ is real if $\mathbf{A}$ is Hermitian.
- The eigenvalues of a Hermitian matrix are real.
- The eigenvectors of a Hermitian matrix that correspond to different eigenvalues are orthogonal to each other.
- The determinant of a Hermitian matrix is real.
- The spectral radius of a Hermitian matrix is equal to its spectral norm, $\rho(\mathbf{A}) = \|\mathbf{A}\|_2$.
- If $\mathbf{A}$ is skew-Hermitian, then $i\mathbf{A}$ is Hermitian, and vice versa.

The diagonal elements of a Hermitian matrix must be real as they must be unchanged by conjugation.

EXAMPLE 7.20

Examine features associated with $\mathbf{A}$ and $\mathbf{A} \cdot \mathbf{x} = \mathbf{b}$, where $\mathbf{A} : \mathbb{C}^3 \to \mathbb{C}^3$ with $\mathbf{A}$ the Hermitian matrix and $\mathbf{x}$ the complex vector:

$$\mathbf{A} = \begin{pmatrix} 1 & 2-i & 3 \\ 2+i & -3 & 2i \\ 3 & -2i & 4 \end{pmatrix}, \qquad \mathbf{x} = \begin{pmatrix} 3+2i \\ -1 \\ 2-i \end{pmatrix}. \tag{7.181}$$

First, we have

$$\mathbf{b} = \mathbf{A} \cdot \mathbf{x} = \begin{pmatrix} 1 & 2-i & 3 \\ 2+i & -3 & 2i \\ 3 & -2i & 4 \end{pmatrix} \begin{pmatrix} 3+2i \\ -1 \\ 2-i \end{pmatrix} = \begin{pmatrix} 7 \\ 9+11i \\ 17+4i \end{pmatrix}. \tag{7.182}$$

Now, demonstrate that the properties of Hermitian matrices hold for this case. First

$$\mathbf{x}^H \cdot \mathbf{A} \cdot \mathbf{x} = \begin{pmatrix} 3-2i & -1 & 2+i \end{pmatrix} \begin{pmatrix} 1 & 2-i & 3 \\ 2+i & -3 & 2i \\ 3 & -2i & 4 \end{pmatrix} \begin{pmatrix} 3+2i \\ -1 \\ 2-i \end{pmatrix} = 42 \in \mathbb{R}^1. \tag{7.183}$$

The eigenvalues and (right, same as left here) eigenvectors are

$$\lambda_1 = 6.51907, \qquad \mathbf{e}_1 = \begin{pmatrix} 0.525248 \\ 0.132451 + 0.223964i \\ 0.803339 - 0.105159i \end{pmatrix}, \tag{7.184}$$

$$\lambda_2 = -0.104237, \qquad \mathbf{e}_2 = \begin{pmatrix} -0.745909 \\ -0.385446 + 0.0890195i \\ 0.501844 - 0.187828i \end{pmatrix}, \tag{7.185}$$

$$\lambda_3 = -4.41484, \qquad \mathbf{e}_3 = \begin{pmatrix} 0.409554 \\ -0.871868 - 0.125103i \\ -0.116278 - 0.207222i \end{pmatrix}. \tag{7.186}$$

By inspection $\rho(\mathbf{A}) = 6.51907$. Because $\mathbf{A}$ is Hermitian, we also have $\|\mathbf{A}\|_2 = \rho(\mathbf{A}) = 6.51907$. We find this by first finding the eigenvalues of $\mathbf{A}^H \cdot \mathbf{A}$, which are 42.4983, 19.4908, and 0.010865. The square roots of these are 6.51907, 4.41484, and 0.104237; the spectral norm is the maximum, 6.51907.

Check for orthogonality between two of the eigenvectors, for example, e_1, e_2:

$$\langle e_1, e_2 \rangle = e_1^H \cdot e_2, \tag{7.187}$$

$$= \begin{pmatrix} 0.525248 & 0.132451 - 0.223964i & 0.803339 + 0.105159i \end{pmatrix} \begin{pmatrix} -0.745909 \\ -0.385446 + 0.0890195i \\ 0.501844 - 0.187828i \end{pmatrix}, \tag{7.188}$$

$$= 0 + 0i. \tag{7.189}$$

The same holds for other eigenvectors. It can then be shown that

$$\det \mathbf{A} = 3, \tag{7.190}$$

which is also equal to the product of the eigenvalues. This also tells us that $\mathbf{A}$ is not volume-preserving, but it is orientation-preserving.

Lastly,

$$i\mathbf{A} = \begin{pmatrix} i & 1 + 2i & 3i \\ -1 + 2i & -3i & -2 \\ 3i & 2 & 4i \end{pmatrix} \tag{7.191}$$

is skew-symmetric. It is easily shown that the eigenvalues of $i\mathbf{A}$ are

$$\lambda_1 = 6.51907i, \qquad \lambda_2 = -0.104237i, \qquad \lambda_3 = -4.41484i. \tag{7.192}$$

The eigenvalues of this matrix are just those of the previous multiplied by i.

7.7 Orthogonal and Unitary Matrices

Expanding on a topic introduced on Section 1.6.1, discussed on Section 2.1.2, and briefly discussed in Example 6.46, a set of N N-dimensional orthonormal vectors $\{e_1, e_2, \ldots, e_N\}$ can be formed into a matrix

$$\mathbf{Q} = \begin{pmatrix} \vdots & \vdots & & \vdots \\ e_1 & e_2 & \cdots & e_N \\ \vdots & \vdots & & \vdots \end{pmatrix}. \tag{7.193}$$

If the orthonormal vectors are real, $\mathbf{Q}$ is known as an *orthogonal* matrix. If they are complex, $\mathbf{Q}$ is known as a *unitary* matrix. All orthogonal matrices are also unitary. Many texts will use the label $\mathbf{U}$ to denote a unitary matrix. We do not, because we generally reserve $\mathbf{U}$ for upper triangular matrices; moreover, the properties of each are sufficiently similar that use of a single symbol is warranted.

Properties of unitary matrices include the following:

1. $\mathbf{Q}^H = \mathbf{Q}^{-1}$, and both are unitary.
2. $\mathbf{Q}^H \cdot \mathbf{Q} = \mathbf{Q} \cdot \mathbf{Q}^H = \mathbf{I}$.
3. $||\mathbf{Q}||_2 = 1$, when the domain and range of $\mathbf{Q}$ are in Hilbert spaces.
4. $||\mathbf{Q} \cdot \mathbf{x}||_2 = ||\mathbf{x}||_2$, where $\mathbf{x}$ is a vector.
5. $(\mathbf{Q} \cdot \mathbf{x})^H \cdot (\mathbf{Q} \cdot \mathbf{y}) = \mathbf{x}^H \cdot \mathbf{y}$, where $\mathbf{x}$ and $\mathbf{y}$ are vectors; see Section 6.7.
6. Eigenvalues of $\mathbf{Q}$ have $|\lambda_i| = 1, \lambda_i \in \mathbb{C}^1$, thus, $\rho(\mathbf{Q}) = 1$.
7. Eigenvectors of $\mathbf{Q}$ corresponding to different eigenvalues are orthogonal.
8. $|\det \mathbf{Q}| = 1$.

If $\mathbf{Q}$ is real, we may replace the Hermitian transpose H by the ordinary transpose T. Geometrically, an orthogonal matrix is an operator that transforms but does not stretch a vector. For an orthogonal matrix to be a rotation, which is orientation-preserving, we must have $\det \mathbf{Q} = 1$. If $\mathbf{Q}$ is complex and $\det \mathbf{Q} = 1$, one is tempted to say the unitary matrix operating on a vector induces a pure rotation in a complex space; however, the notion of an angle of rotation in complex space is elusive. Rotation matrices, reflection matrices, and permutation matrices are all unitary matrices. Recall that permutation matrices can also be reflection or rotation matrices.

Let us consider some examples involving orthogonal matrices.

EXAMPLE 7.21

Find the orthogonal matrix corresponding to the self-adjoint

$$\mathbf{A} = \begin{pmatrix} 2 & 1 \\ 1 & 2 \end{pmatrix}. \tag{7.194}$$

The normalized eigenvectors are $(1/\sqrt{2}, 1/\sqrt{2})^T$ and $(-1/\sqrt{2}, 1/\sqrt{2})^T$. The orthogonal matrix is thus

$$\mathbf{Q} = \begin{pmatrix} \frac{1}{\sqrt{2}} & -\frac{1}{\sqrt{2}} \\ \frac{1}{\sqrt{2}} & \frac{1}{\sqrt{2}} \end{pmatrix}. \tag{7.195}$$

In the sense of Eq. (2.54), we can say

$$\mathbf{Q} = \begin{pmatrix} \cos\frac{\pi}{4} & -\sin\frac{\pi}{4} \\ \sin\frac{\pi}{4} & \cos\frac{\pi}{4} \end{pmatrix}, \tag{7.196}$$

and the angle of rotation of the coordinate axes is $\alpha = \pi/4$. We calculate the eigenvalues of $\mathbf{Q}$ to be $\lambda = (1 \pm i)/\sqrt{2}$, which in exponential form becomes $\lambda = \exp(\pm i\pi/4)$, and so we see the rotation angle is embedded within the argument of the polar representation of the eigenvalues. We also see $|\lambda| = 1$. Note that $\mathbf{Q}$ is *not* symmetric. Also, note that $\det \mathbf{Q} = 1$, so this orthogonal matrix is also a rotation matrix.

If $\boldsymbol{\xi}$ is an unrotated Cartesian vector, and our transformation to a rotated frame is $\boldsymbol{\xi} = \mathbf{Q} \cdot \mathbf{x}$, so that $\mathbf{x} = \mathbf{Q}^T \cdot \boldsymbol{\xi}$, we see that the Cartesian unit vector $\boldsymbol{\xi} = (1, 0)^T$ is represented in the rotated coordinate system by

$$\mathbf{x} = \underbrace{\begin{pmatrix} \frac{1}{\sqrt{2}} & \frac{1}{\sqrt{2}} \\ -\frac{1}{\sqrt{2}} & \frac{1}{\sqrt{2}} \end{pmatrix}}_{\mathbf{Q}^T} \begin{pmatrix} 1 \\ 0 \end{pmatrix} = \begin{pmatrix} \frac{1}{\sqrt{2}} \\ -\frac{1}{\sqrt{2}} \end{pmatrix}. \tag{7.197}$$

Thus, the counterclockwise rotation of the axes through angle $\alpha = \pi/4$ gives the Cartesian unit vector $(1, 0)^T$ a new representation of $(1/\sqrt{2}, -1/\sqrt{2})^T$. We see that the other Cartesian unit vector $\boldsymbol{\xi} = (0, 1)^T$ is represented in the rotated coordinate system by

$$\mathbf{x} = \underbrace{\begin{pmatrix} \frac{1}{\sqrt{2}} & \frac{1}{\sqrt{2}} \\ -\frac{1}{\sqrt{2}} & \frac{1}{\sqrt{2}} \end{pmatrix}}_{\mathbf{Q}^T} \begin{pmatrix} 0 \\ 1 \end{pmatrix} = \begin{pmatrix} \frac{1}{\sqrt{2}} \\ \frac{1}{\sqrt{2}} \end{pmatrix}. \tag{7.198}$$

Had $\det \mathbf{Q} = -1$, the transformation would have been nonorientation-preserving.

EXAMPLE 7.22

Analyze the three-dimensional orthogonal matrix

$$\mathbf{Q} = \begin{pmatrix} \frac{1}{\sqrt{3}} & \frac{1}{\sqrt{2}} & \frac{1}{\sqrt{6}} \\ \frac{1}{\sqrt{3}} & -\frac{1}{\sqrt{2}} & \frac{1}{\sqrt{6}} \\ \frac{1}{\sqrt{3}} & 0 & -\sqrt{\frac{2}{3}} \end{pmatrix}. \tag{7.199}$$

Direct calculation reveals $\|\mathbf{Q}\|_2 = 1$, $\det \mathbf{Q} = 1$, and $\mathbf{Q}^T = \mathbf{Q}^{-1}$, so the matrix is a volume- and orientation-preserving rotation matrix. It can also be shown to have a set of eigenvalues and eigenvectors of

$$\lambda_1 = 1, \quad \mathbf{e}_1 = \begin{pmatrix} 0.886452 \\ 0.36718 \\ 0.281747 \end{pmatrix}, \tag{7.200}$$

$$\lambda_2 = \exp(2.9092i), \quad \mathbf{e}_2 = \begin{pmatrix} -0.18406 + 0.27060i \\ -0.076240 - 0.653281i \\ 0.678461 \end{pmatrix}, \tag{7.201}$$

$$\lambda_3 = \exp(-2.9092i), \quad \mathbf{e}_3 = \begin{pmatrix} -0.18406 - 0.27060i \\ -0.076240 + 0.653281i \\ 0.678461 \end{pmatrix}. \tag{7.202}$$

As expected, each eigenvalue has $|\lambda| = 1$. It can be shown that the eigenvector $\mathbf{e}_1$ which is associated with the real eigenvalue, $\lambda_1 = 1$, is aligned with the so-called Euler axis, that is, the axis in three-space about which the rotation occurs. The remaining two eigenvalues are of the form $\exp(\pm\alpha i)$, where α is the angle of rotation about the Euler axis. For this example, we have $\alpha = 2.9092$.

EXAMPLE 7.23

Examine the composite action of three rotations on a vector $\mathbf{x}$:

$$\mathbf{Q}_1 \cdot \mathbf{Q}_2 \cdot \mathbf{Q}_3 \cdot \mathbf{x}$$

$$= \underbrace{\begin{pmatrix} 1 & 0 & 0 \\ 0 & \cos\alpha_1 & -\sin\alpha_1 \\ 0 & \sin\alpha_1 & \cos\alpha_1 \end{pmatrix}}_{\mathbf{Q}_1} \underbrace{\begin{pmatrix} \cos\alpha_2 & 0 & \sin\alpha_2 \\ 0 & 1 & 0 \\ -\sin\alpha_2 & 0 & \cos\alpha_2 \end{pmatrix}}_{\mathbf{Q}_2} \underbrace{\begin{pmatrix} \cos\alpha_3 & -\sin\alpha_3 & 0 \\ \sin\alpha_3 & \cos\alpha_3 & 0 \\ 0 & 0 & 1 \end{pmatrix}}_{\mathbf{Q}_3} \cdot \mathbf{x}. \tag{7.203}$$

It is easy to verify that $\|\mathbf{Q}_1\|_2 = \|\mathbf{Q}_2\|_2 = \|\mathbf{Q}_3\|_2 = 1$, $\det \mathbf{Q}_1 = \det \mathbf{Q}_2 = \det \mathbf{Q}_3 = 1$, so each is a rotation. For $\mathbf{Q}_3$, we find eigenvalues of $\lambda = 1$, $\cos\alpha_3 \pm i\sin\alpha_3$. These can be rewritten as $\lambda = 1, e^{\pm\alpha_3 i}$. The eigenvector associated with the eigenvalue of 1 is $(0, 0, 1)$. Thus, we can consider $\mathbf{Q}_3$ to effect a rotation of α_3 about the 3-axis. Similarly, $\mathbf{Q}_2$ effects a rotation of α_2 about the 2-axis, and $\mathbf{Q}_1$ effects a rotation of α_1 about the 1-axis.

So the action of the combination of rotations on a vector $\mathbf{x}$ is an initial rotation of α_3 about the 3-axis: $\mathbf{Q}_3 \cdot \mathbf{x}$. This vector is then rotated through α_2 about the 2-axis: $\mathbf{Q}_2 \cdot (\mathbf{Q}_3 \cdot \mathbf{x})$. Finally, there is a rotation through α_1 about the 1-axis: $\mathbf{Q}_1 \cdot (\mathbf{Q}_2 \cdot (\mathbf{Q}_3 \cdot \mathbf{x}))$. This is called a 3-2-1 rotation through the so-called *Euler angles* of α_3, α_2, and α_1. Because in general matrix multiplication does not commute, the result will depend on the order of application of the rotations, for example,

$$\mathbf{Q}_1 \cdot \mathbf{Q}_2 \cdot \mathbf{Q}_3 \cdot \mathbf{x} \neq \mathbf{Q}_2 \cdot \mathbf{Q}_1 \cdot \mathbf{Q}_3 \cdot \mathbf{x}, \quad \forall \mathbf{x} \in \mathbb{R}^3, \quad \mathbf{Q}_1, \mathbf{Q}_2, \mathbf{Q}_3 \in \mathbb{R}^3 \times \mathbb{R}^3. \tag{7.204}$$

In contrast, it is not difficult to show that rotations in two dimensions do commute:

$$\mathbf{Q}_1 \cdot \mathbf{Q}_2 \cdot \mathbf{Q}_3 \cdot \mathbf{x} = \mathbf{Q}_2 \cdot \mathbf{Q}_1 \cdot \mathbf{Q}_3 \cdot \mathbf{x}, \quad \forall \mathbf{x} \in \mathbb{R}^2, \quad \mathbf{Q}_1, \mathbf{Q}_2, \mathbf{Q}_3 \in \mathbb{R}^2 \times \mathbb{R}^2. \tag{7.205}$$

Let us next consider examples for unitary matrices.

EXAMPLE 7.24

Demonstrate that the unitary matrix

$$\mathbf{Q} = \begin{pmatrix} \frac{1+i}{\sqrt{3}} & \frac{1-2i}{\sqrt{15}} \\ \frac{1}{\sqrt{3}} & \frac{1+3i}{\sqrt{15}} \end{pmatrix} \tag{7.206}$$

has common properties associated with unitary matrices.

The column vectors are easily seen to be normal. They are also orthogonal:

$$\left(\frac{1-i}{\sqrt{3}}, \frac{1}{\sqrt{3}} \right) \begin{pmatrix} \frac{1-2i}{\sqrt{15}} \\ \frac{1+3i}{\sqrt{15}} \end{pmatrix} = 0 + 0i. \tag{7.207}$$

The matrix itself is not Hermitian. Still, its Hermitian transpose exists:

$$\mathbf{Q}^H = \begin{pmatrix} \frac{1-i}{\sqrt{3}} & \frac{1}{\sqrt{3}} \\ \frac{1+2i}{\sqrt{15}} & \frac{1-3i}{\sqrt{15}} \end{pmatrix}. \tag{7.208}$$

It is then easily verified that

$$\mathbf{Q}^{-1} = \mathbf{Q}^H, \tag{7.209}$$

$$\mathbf{Q} \cdot \mathbf{Q}^H = \mathbf{Q}^H \cdot \mathbf{Q} = \mathbf{I}. \tag{7.210}$$

The eigensystem is

$$\lambda_1 = -0.0986232 + 0.995125i, \quad \mathbf{e}_1 = \begin{pmatrix} 0.688191 - 0.425325i \\ 0.587785 \end{pmatrix}, \tag{7.211}$$

$$\lambda_2 = 0.934172 + 0.356822i, \quad \mathbf{e}_2 = \begin{pmatrix} -0.306358 + 0.501633i \\ -0.721676 - 0.36564i \end{pmatrix}. \tag{7.212}$$

It is easily verified that the eigenvectors are orthogonal and the eigenvalues have magnitude of unity. We find $\det \mathbf{Q} = (1 + 2i)/\sqrt{5}$, which yields $|\det \mathbf{Q}| = 1$. Also, $||\mathbf{Q}||_2 = 1$.

7.7.1 Givens Rotation

A *Givens*[4] *rotation* matrix $\mathbf{G}$ is an $N \times N$ rotation matrix that can be constructed by taking an $N \times N$ identity matrix $\mathbf{I}$, selecting two distinct values for i and j and replacing the ii, jj, ij, and ji entries of $\mathbf{I}$ with

$$g_{jj} = c, \quad g_{ji} = -s \quad g_{ij} = s, \quad g_{ii} = c, \tag{7.213}$$

while additionally requiring that $c^2 + s^2 = 1$.

EXAMPLE 7.25

Develop a 5×5 Givens rotation matrix with $i = 2$ and $j = 4$, $c = 3/5$ and $s = 4/5$.

We first note that

$$c^2 + s^2 = \left(\frac{3}{5} \right)^2 + \left(\frac{4}{5} \right)^2 = 1, \tag{7.214}$$

[4] James Wallace Givens Jr., 1910–1993, American mathematician and computer scientist.

as required. Then the desired Givens rotation matrix $\mathbf{G}$ must have entries $g_{44} = 3/5$, $g_{42} = -4/5$, $g_{24} = 4/5$, $g_{44} = 3/5$. The matrix is

$$\mathbf{G} = \begin{pmatrix} 1 & 0 & 0 & 0 & 0 \\ 0 & \frac{3}{5} & 0 & \frac{4}{5} & 0 \\ 0 & 0 & 1 & 0 & 0 \\ 0 & -\frac{4}{5} & 0 & \frac{3}{5} & 0 \\ 0 & 0 & 0 & 0 & 1 \end{pmatrix}. \tag{7.215}$$

It is easily verified that $\det \mathbf{G} = 1$ and $||\mathbf{G}||_2 = 1$; thus, $\mathbf{G}$ is a rotation matrix.

7.7.2 Householder Reflection

We may wish to transform a vector $\mathbf{x}$ of length N into its image $\overline{\mathbf{x}}$, also of length N, by application of an $N \times N$ matrix $\mathbf{H}$ via $\mathbf{H} \cdot \mathbf{x} = \overline{\mathbf{x}}$ such that the magnitude of $\overline{\mathbf{x}}$ is the same as that of $\mathbf{x}$, but $\overline{\mathbf{x}}$ may have as many as $N - 1$ zeros in its representation. This end can be achieved by application of an appropriate *Householder*[5] *reflection matrix*. There may be nonunique $\mathbf{H}$ and $\overline{\mathbf{x}}$ that achieve this goal. Here the superposed line segment does not connote the complex conjugate; instead, it is a representation of $\mathbf{x}$ in a transformed coordinate system.

In general terms, a Householder reflection matrix $\mathbf{H}$ can be formed via

$$\mathbf{H} = \mathbf{I} - 2\mathbf{u}\mathbf{u}^T, \tag{7.216}$$

as long as $\mathbf{u}$ is any unit vector of length N. We note that

$$\mathbf{H}^T \cdot \mathbf{H} = (\mathbf{I} - 2\mathbf{u}\mathbf{u}^T)^T \cdot (\mathbf{I} - 2\mathbf{u}\mathbf{u}^T). \tag{7.217}$$

Because both $\mathbf{H}$ and $\mathbf{u}\mathbf{u}^T$ are symmetric, we have

$$\mathbf{H}^T \cdot \mathbf{H} = (\mathbf{I} - 2\mathbf{u}\mathbf{u}^T) \cdot (\mathbf{I} - 2\mathbf{u}\mathbf{u}^T), \tag{7.218}$$

$$= \mathbf{I} \cdot \mathbf{I} - 4\mathbf{u}\mathbf{u}^T + 4\mathbf{u}\mathbf{u}^T = \mathbf{I}. \tag{7.219}$$

Because $\mathbf{H}^T \cdot \mathbf{H} = \mathbf{I}$, $\mathbf{H}$ is an orthogonal matrix with $||\mathbf{H}||_2 = 1$. The matrix $\mathbf{H}$ is a reflection matrix because its action on $\mathbf{u}$ yields the additive inverse of $\mathbf{u}$:

$$\mathbf{H} \cdot \mathbf{u} = (\mathbf{I} - 2\mathbf{u}\mathbf{u}^T) \cdot \mathbf{u}, \tag{7.220}$$

$$= \mathbf{u} - 2\mathbf{u} \underbrace{(\mathbf{u}^T \cdot \mathbf{u})}_{=1}, \tag{7.221}$$

$$= -\mathbf{u}. \tag{7.222}$$

So $\mathbf{H}$ reflects $\mathbf{u}$ about the plane for which $\mathbf{u}$ is the normal. As a consequence, we will have $\det \mathbf{H} = -1$. Furthermore, Householder matrices are Hermitian because $\mathbf{H} = \mathbf{H}^T$; that property combined with orthogonality renders Householder matrices to also be involutary so that $\mathbf{H} \cdot \mathbf{H} = \mathbf{I}$.

To achieve our desired end, it can be shown that $\mathbf{u}$ should be selected so that

$$\mathbf{u} = \frac{\mathbf{x} - \overline{\mathbf{x}}}{||\mathbf{x} - \overline{\mathbf{x}}||_2}. \tag{7.223}$$

---

[5] Alston Scott Householder, 1904–1993, American mathematician.

EXAMPLE 7.26

Given the vector $\mathbf{x} = (4, 3)^T$, find a Householder matrix such that the transformed vector $\overline{\mathbf{x}}$ has a zero in its second entry.

We have

$$\|\mathbf{x}\|_2 = \sqrt{4^2 + 3^2} = 5, \qquad (7.224)$$

so for $\overline{\mathbf{x}}$ to have the same magnitude as $\mathbf{x}$ and a zero in its second entry, we consider

$$\overline{\mathbf{x}} = \begin{pmatrix} 5 \\ 0 \end{pmatrix}. \qquad (7.225)$$

So

$$\mathbf{x} - \overline{\mathbf{x}} = \begin{pmatrix} 4 \\ 3 \end{pmatrix} - \begin{pmatrix} 5 \\ 0 \end{pmatrix} = \begin{pmatrix} -1 \\ 3 \end{pmatrix}. \qquad (7.226)$$

Therefore,

$$\mathbf{u} = \frac{\mathbf{x} - \overline{\mathbf{x}}}{\|\mathbf{x} - \overline{\mathbf{x}}\|_2} = \begin{pmatrix} \frac{-1}{\sqrt{10}} \\ \frac{3}{\sqrt{10}} \end{pmatrix}, \qquad (7.227)$$

and

$$\mathbf{H} = \mathbf{I} - 2\mathbf{u}\mathbf{u}^T, \qquad (7.228)$$

$$= \begin{pmatrix} 1 & 0 \\ 0 & 1 \end{pmatrix} - 2 \left(-\frac{1}{\sqrt{10}} \quad \frac{3}{\sqrt{10}} \right) \begin{pmatrix} -\frac{1}{\sqrt{10}} \\ \frac{3}{\sqrt{10}} \end{pmatrix}, \qquad (7.229)$$

$$= \begin{pmatrix} 1 & 0 \\ 0 & 1 \end{pmatrix} - 2 \begin{pmatrix} \frac{1}{10} & \frac{-3}{10} \\ \frac{-3}{10} & \frac{9}{10} \end{pmatrix}, \qquad (7.230)$$

$$= \begin{pmatrix} \frac{4}{5} & \frac{3}{5} \\ \frac{3}{5} & \frac{-4}{5} \end{pmatrix}. \qquad (7.231)$$

It is easily verified that $\mathbf{H}$ has all the remarkable properties of Householder matrices: (1) it is symmetric, so it is its own transpose: $\mathbf{H} = \mathbf{H}^T$; (2) it is involutary, so it is its own inverse: $\mathbf{H} \cdot \mathbf{H} = \mathbf{I}$; (3) it is orthogonal, so its inverse is its transpose: $\mathbf{H}^{-1} = \mathbf{H}^T$; (4) its norm is unity: $\|\mathbf{H}\|_2 = 1$; and (5) it is a reflection, so $\det \mathbf{H} = -1$. Moreover, one also verifies that

$$\mathbf{H} \cdot \mathbf{x} = \begin{pmatrix} \frac{4}{5} & \frac{3}{5} \\ \frac{3}{5} & \frac{-4}{5} \end{pmatrix} \begin{pmatrix} 4 \\ 3 \end{pmatrix} = \begin{pmatrix} 5 \\ 0 \end{pmatrix} = \overline{\mathbf{x}}, \qquad (7.232)$$

as intended. We sketch the various relevant configurations for this problem in Figure 7.4. As depicted, the Householder reflection selects a special angle of reflection which brings the vector $\mathbf{x}$ onto one of the transformed coordinate axes. We also could have considered $\overline{\mathbf{x}} = (-5, 0)^T$ and found similar results.

EXAMPLE 7.27

Given the vector $\mathbf{x} = (1, 2, 2)^T$, find Householder matrices such that the transformed vector $\overline{\mathbf{x}}$ has (1) two zeros in its last two entries and (2) one zero in its last entry.

We have

$$\|\mathbf{x}\|_2 = \sqrt{1^2 + 2^2 + 2^2} = 3. \qquad (7.233)$$

For two zeros in the last entries, we must have $\overline{\mathbf{x}} = (3, 0, 0)^T$ for $\mathbf{x}$ and $\overline{\mathbf{x}}$ to have the same magnitude. Then we get $\mathbf{u} = (\mathbf{x} - \overline{\mathbf{x}})/\|\mathbf{x} - \overline{\mathbf{x}}\|_2 = (-1/\sqrt{3}, 1/\sqrt{3}, 1/\sqrt{3})^T$. This yields a

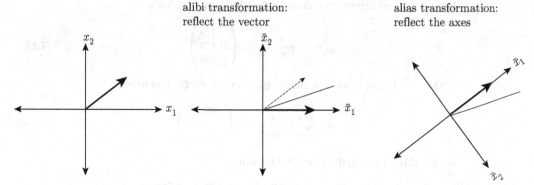

Figure 7.4. Alibi and alias Householder reflection transformations of the vector $\mathbf{x} = (4, 3)^T$ about an axis inclined at $(1/2) \tan^{-1}(3/4) = 18.43°$.

Householder matrix of

$$\mathbf{H} = \begin{pmatrix} \frac{1}{3} & \frac{2}{3} & \frac{2}{3} \\ \frac{2}{3} & \frac{1}{3} & -\frac{2}{3} \\ \frac{2}{3} & -\frac{2}{3} & \frac{1}{3} \end{pmatrix}. \tag{7.234}$$

It is easily verified that $||\mathbf{H}||_2 = 1$, $\det \mathbf{H} = -1$ and $\mathbf{H} \cdot \mathbf{x} = \bar{\mathbf{x}}$.

For one zero in the last entry of $\bar{\mathbf{x}}$, we have an infinite number of choices, of the form $\bar{\mathbf{x}} = (3\cos\theta, 3\sin\theta, 0)^T$. Let us arbitrarily select $\cos\theta = 3/5$ and $\sin\theta = 4/5$, so that $\bar{\mathbf{x}} = (9/5, 12/5, 0)^T$. Then we get $\mathbf{u} = (\mathbf{x} - \bar{\mathbf{x}})/||\mathbf{x} - \bar{\mathbf{x}}||_2 = (-\sqrt{2/15}, -1/\sqrt{30}, \sqrt{5/6})^T$. This yields a Householder matrix of

$$\mathbf{H} = \begin{pmatrix} \frac{11}{15} & -\frac{2}{15} & \frac{2}{3} \\ -\frac{2}{15} & \frac{14}{15} & \frac{1}{3} \\ \frac{2}{3} & \frac{1}{3} & -\frac{2}{3} \end{pmatrix}. \tag{7.235}$$

It is easily verified that $||\mathbf{H}||_2 = 1$, $\det \mathbf{H} = -1$ and $\mathbf{H} \cdot \mathbf{x} = \bar{\mathbf{x}}$.

As demonstrated in the next example, Householder reflections combined with the iteration algorithm of Section 6.9 are a useful strategy to find all the eigenvalues and eigenvectors of a Hermitian matrix. This can be effective for large matrices.

EXAMPLE 7.28
Use iteration combined with appropriate Householder reflection to identify the eigenvalues and eigenvectors of the self-adjoint

$$\mathbf{A} = \begin{pmatrix} 1 & 2 & 3 \\ 2 & 4 & 1 \\ 3 & 1 & 5 \end{pmatrix}. \tag{7.236}$$

We first choose a seed value of $\mathbf{x} = (1, 1, 1)^T$. We then iterate $M = 100$ times forming $\mathbf{x}_M = \mathbf{A}^M \cdot \mathbf{x}$. The Rayleigh quotient gives the extreme eigenvalue estimate

$$\lambda_1 = \frac{\mathbf{x}_M^T \cdot \mathbf{A} \cdot \mathbf{x}_M}{\mathbf{x}_M^T \cdot \mathbf{x}_M} = 7.5896. \tag{7.237}$$

The normalized eigenvector associated with this eigenvalue is

$$\mathbf{e}_1 = \frac{\mathbf{x}_M}{||\mathbf{x}_M||_2} = \begin{pmatrix} 0.479 \\ 0.473 \\ 0.739 \end{pmatrix}. \tag{7.238}$$

With $\bar{\mathbf{e}}_1 = (1, 0, 0)^T$, we then choose the vector for reflection to be

$$\mathbf{u} = \frac{\mathbf{e}_1 - \bar{\mathbf{e}}_1}{||\mathbf{e}_1 - \bar{\mathbf{e}}_1||_2} = \begin{pmatrix} -0.510 \\ 0.464 \\ 0.724 \end{pmatrix}. \tag{7.239}$$

We use this $\mathbf{u}$ to form the Householder matrix

$$\mathbf{H} = \mathbf{I} - \mathbf{u}\mathbf{u}^T = \begin{pmatrix} 0.480 & 0.473 & 0.739 \\ 0.473 & 0.569 & -0.672 \\ 0.739 & -0.672 & -0.0494 \end{pmatrix}. \tag{7.240}$$

We then use $\mathbf{H}$ to operate on $\mathbf{A}$ to obtain

$$\hat{\mathbf{A}} = \mathbf{H}^T \cdot \mathbf{A} \cdot \mathbf{H} = \begin{pmatrix} 7.59 & 0 & 0 \\ 0 & 2.18 & -1.95 \\ 0 & -1.95 & 0.227 \end{pmatrix}. \tag{7.241}$$

Note that λ_1 is now isolated in the upper right corner of $\hat{\mathbf{A}}$. We now remove the first row and column from $\hat{\mathbf{A}}$ and define

$$\tilde{\mathbf{A}} = \begin{pmatrix} 2.18 & -1.95 \\ -1.95 & 0.227 \end{pmatrix}. \tag{7.242}$$

With a seed value of $\mathbf{x} = (1, 1)^T$, we then iterate $M = 100$ times, forming again $\mathbf{x}_M = \tilde{\mathbf{A}}^M \cdot \mathbf{x}$. The Rayleigh quotient gives an estimate of the extreme eigenvalue of $\tilde{\mathbf{A}}$:

$$\lambda_2 = \frac{\mathbf{x}_M^T \cdot \tilde{\mathbf{A}} \cdot \mathbf{x}_M}{\mathbf{x}_M^T \cdot \mathbf{x}_M} = 3.38385. \tag{7.243}$$

We repeat this process for the next eigenvalue. Omitting the details, we find

$$\lambda_3 = -0.973443. \tag{7.244}$$

7.8 Discrete Fourier Transforms

It is a common practice in experimental and theoretical science and engineering to decompose a function or a signal into its Fourier modes. The amplitude of these modes is often a useful description of the function. A Fourier transform (FT) is a linear integral operator that operates on continuous functions and yields results from which amplitudes of each frequency component can be determined. Its discrete analog is the discrete Fourier transform (DFT). The DFT is a matrix that operates on a vector of data to yield a vector of transformed data. There exists a popular, albeit complicated, algorithm to compute the DFT, known as the fast Fourier transform (FFT). This will not be studied here; instead, a simpler and slower method is presented, which will be informally known as a slow Fourier transform (SFT). This discussion simply presents the algorithm for the SFT and demonstrates its use by example.

The FT $\mathbf{Y}(\kappa)$ of a function $y(x)$ is defined as

$$\mathbf{Y}(\kappa) = \hat{\mathbf{Y}}[y(x)] = \int_{-\infty}^{\infty} y(x)e^{-(2\pi i)\kappa x}\, dx, \tag{7.245}$$

and the inverse FT is defined as

$$y(x) = \hat{\mathbf{Y}}^{-1}[\mathbf{Y}(\kappa)] = \int_{-\infty}^{\infty} \mathbf{Y}(\kappa)e^{(2\pi i)\kappa x}\, d\kappa. \tag{7.246}$$

Here κ is the wavenumber and is the reciprocal of the wavelength. The FT has a discrete analog. The connection between the two is often not transparent in the literature. With some effort a connection *can* be made at the expense of diverging from one school's notation to the other's. Here we will be satisfied with a form that demonstrates the analogs between the continuous and discrete transform but will not be completely linked. To make the connection, one can construct a discrete approximation to the integral of the FT and, with some effort, arrive at an equivalent result.

For the DFT, consider a function $y(x)$, $x \in [x_{min}, x_{max}]$, $x \in \mathbb{R}^1, y \in \mathbb{R}^1$. Now discretize the domain into N uniformly distributed points so that every x_j is mapped to a y_j for $j = 0, \ldots, N-1$. Here we comply with the traditional, yet idiosyncratic, limits on j, which are found in many texts on DFT. This offsets standard vector and matrix numbering schemes by 1, and so care must be exercised in implementing these algorithms with common software. We seek a discrete analog of the continuous inverse FT of the form

$$y_j = \frac{1}{\sqrt{N}} \sum_{k=0}^{N-1} c_k \exp\left((2\pi i)k \left(\frac{N-1}{N} \right) \left(\frac{x_j - x_{min}}{x_{max} - x_{min}} \right) \right), \qquad j = 0, \ldots, N-1. \tag{7.247}$$

Here k plays the role of κ, and c_k plays the role of $\mathbf{Y}(\kappa)$. For uniformly spaced x_j, one has

$$j = (N-1) \left(\frac{x_j - x_{min}}{x_{max} - x_{min}} \right), \tag{7.248}$$

so that we then seek

$$y_j = \frac{1}{\sqrt{N}} \sum_{k=0}^{N-1} c_k \exp\left((2\pi i)\frac{kj}{N} \right), \qquad j = 0, \ldots, N-1. \tag{7.249}$$

Now consider the equation

$$z^N = 1, \qquad z \in \mathbb{C}^1. \tag{7.250}$$

This equation has N distinct roots

$$z = e^{2\pi i \frac{j}{N}}, \qquad j = 0, \ldots, N-1. \tag{7.251}$$

Taking for convenience

$$w \equiv e^{2\pi i/N}, \tag{7.252}$$

one sees that the N roots are also described by $w^0, w^1, w^2, \ldots, w^{N-1}$. Now define the following matrix:

$$\mathbf{Q} = \frac{1}{\sqrt{N}} \begin{pmatrix} 1 & 1 & 1 & \cdots & 1 \\ 1 & w & w^2 & \cdots & w^{N-1} \\ 1 & w^2 & w^4 & \cdots & w^{2(N-1)} \\ \vdots & \vdots & \vdots & \vdots & \vdots \\ 1 & w^{N-1} & w^{2(N-1)} & \cdots & w^{(N-1)^2} \end{pmatrix}. \tag{7.253}$$

It is easy to demonstrate for arbitrary N that $\mathbf{Q}$ is unitary, that is, $\mathbf{Q}^H \cdot \mathbf{Q} = \mathbf{I}$. Because of this, it is known that $\mathbf{Q}^H = \mathbf{Q}^{-1}$, that $||\mathbf{Q}||_2 = 1$, that the eigenvalues of $\mathbf{Q}$ have magnitude of unity, and that the column vectors of $\mathbf{Q}$ are orthonormal. Note that $\mathbf{Q}$ is not Hermitian. Also, note that many texts omit the factor $1/\sqrt{N}$ in the definition of $\mathbf{Q}$; this is not a major problem but does render $\mathbf{Q}$ to be nonunitary.

Now given a vector $\mathbf{y} = y_j$, $j = 0, \ldots, N-1$, the DFT is defined as the following mapping:

$$\mathbf{c} = \mathbf{Q}^H \cdot \mathbf{y}. \tag{7.254}$$

The inverse transform is trivial owing to the unitary nature of $\mathbf{Q}$:

$$\mathbf{Q} \cdot \mathbf{c} = \mathbf{Q} \cdot \mathbf{Q}^H \cdot \mathbf{y}, \tag{7.255}$$

$$= \mathbf{Q} \cdot \mathbf{Q}^{-1} \cdot \mathbf{y}, \tag{7.256}$$

$$= \mathbf{I} \cdot \mathbf{y}, \tag{7.257}$$

$$\mathbf{y} = \mathbf{Q} \cdot \mathbf{c}. \tag{7.258}$$

Because our $\mathbf{Q}$ is unitary, it preserves the length of vectors. Thus, it induces a Parseval's equation (see Section 6.6):

$$||\mathbf{y}||_2 = ||\mathbf{c}||_2. \tag{7.259}$$

EXAMPLE 7.29

Find a five-term DFT of the function

$$y = x^2, \qquad x \in [0, 4]. \tag{7.260}$$

Take then for $N = 5$ a set of uniformly distributed points in the domain and their image in the range

$$x_0 = 0, \quad x_1 = 1, \quad x_2 = 2, \quad x_3 = 3, \quad x_4 = 4, \tag{7.261}$$

$$y_0 = 0, \quad y_1 = 1, \quad y_2 = 4, \quad y_3 = 9, \quad y_4 = 16. \tag{7.262}$$

Now for $N = 5$, one has

$$w = e^{2\pi i/5} = \underbrace{\left(\frac{1}{4}(-1 + \sqrt{5}) \right)}_{=\Re(w)} + \underbrace{\left(\frac{1}{2}\sqrt{\frac{1}{2}(5 + \sqrt{5})} \right)}_{=\Im(w)} i = 0.309 + 0.951i. \tag{7.263}$$

Here the notation $\Re(w)$ means the real part of w; likewise $\Im(w)$ means the imaginary part of w. The five distinct roots of $z^5 = 1$ are

$$z^{(0)} = w^0 = 1, \tag{7.264}$$

$$z^{(1)} = w^1 = 0.309 + 0.951i, \tag{7.265}$$

$$z^{(2)} = w^2 = -0.809 + 0.588i, \tag{7.266}$$

$$z^{(3)} = w^3 = -0.809 - 0.588i, \tag{7.267}$$

$$z^{(4)} = w^4 = 0.309 - 0.951i. \tag{7.268}$$

The matrix $\mathbf{Q}$ is then

$$\mathbf{Q} = \frac{1}{\sqrt{5}} \begin{pmatrix} 1 & 1 & 1 & 1 & 1 \\ 1 & w & w^2 & w^3 & w^4 \\ 1 & w^2 & w^4 & w^6 & w^8 \\ 1 & w^3 & w^6 & w^9 & w^{12} \\ 1 & w^4 & w^8 & w^{12} & w^{16} \end{pmatrix} = \frac{1}{\sqrt{5}} \begin{pmatrix} 1 & 1 & 1 & 1 & 1 \\ 1 & w & w^2 & w^3 & w^4 \\ 1 & w^2 & w^4 & w^1 & w^3 \\ 1 & w^3 & w^1 & w^4 & w^2 \\ 1 & w^4 & w^3 & w^2 & w^1 \end{pmatrix}, \tag{7.269}$$

$$= \frac{1}{\sqrt{5}} \begin{pmatrix} 1 & 1 & 1 & 1 & 1 \\ 1 & 0.309 + 0.951i & -0.809 + 0.588i & -0.809 - 0.588i & 0.309 - 0.951i \\ 1 & -0.809 + 0.588i & 0.309 - 0.951i & 0.309 + 0.951i & -0.809 - 0.588i \\ 1 & -0.809 - 0.588i & 0.309 + 0.951i & 0.309 - 0.951i & -0.809 + 0.588i \\ 1 & 0.309 - 0.951i & -0.809 - 0.588i & -0.809 + 0.588i & 0.309 + 0.951i \end{pmatrix}. \tag{7.270}$$

Now $\mathbf{c} = \mathbf{Q}^H \cdot \mathbf{y}$, so

$$\begin{pmatrix} c_0 \\ c_1 \\ c_2 \\ c_3 \\ c_4 \end{pmatrix} =$$

$$\frac{1}{\sqrt{5}} \begin{pmatrix} 1 & 1 & 1 & 1 & 1 \\ 1 & 0.309 - 0.951i & -0.809 - 0.588i & -0.809 + 0.588i & 0.309 + 0.951i \\ 1 & -0.809 - 0.588i & 0.309 + 0.951i & 0.309 - 0.951i & -0.809 + 0.588i \\ 1 & -0.809 + 0.588i & 0.309 - 0.951i & 0.309 + 0.951i & -0.809 - 0.588i \\ 1 & 0.309 + 0.951i & -0.809 + 0.588i & -0.809 - 0.588i & 0.309 - 0.951i \end{pmatrix} \begin{pmatrix} 0 \\ 1 \\ 4 \\ 9 \\ 16 \end{pmatrix},$$

$$= \begin{pmatrix} 13.416 \\ -2.354 + 7.694i \\ -4.354 + 1.816i \\ -4.354 - 1.816i \\ -2.354 - 7.694i \end{pmatrix}. \tag{7.271}$$

One is often interested in the magnitude of the components of $\mathbf{c}$, which gives a measure of the so-called energy associated with each Fourier mode. So one calculates a vector of the magnitude of each component as

$$\begin{pmatrix} \sqrt{c_0 \overline{c_0}} \\ \sqrt{c_1 \overline{c_1}} \\ \sqrt{c_2 \overline{c_2}} \\ \sqrt{c_3 \overline{c_3}} \\ \sqrt{c_4 \overline{c_4}} \end{pmatrix} = \begin{pmatrix} |c_0| \\ |c_1| \\ |c_2| \\ |c_3| \\ |c_4| \end{pmatrix} = \begin{pmatrix} 13.4164 \\ 8.0463 \\ 4.7178 \\ 4.7178 \\ 8.0463 \end{pmatrix}. \tag{7.272}$$

Now owing to a phenomenon known as *aliasing*, explained in detail in standard texts, the values of c_k which have significance are the first half $c_k, k = 0, \ldots, N/2$. Here

$$\|\mathbf{y}\|_2 = \|\mathbf{c}\|_2 = \sqrt{354} = 18.8149. \tag{7.273}$$

By construction,

$$y_0 = \frac{1}{\sqrt{5}} \left(c_0 + c_1 + c_2 + c_3 + c_4 \right), \tag{7.274}$$

$$y_1 = \frac{1}{\sqrt{5}} \left(c_0 + c_1 e^{2\pi i/5} + c_2 e^{4\pi i/5} + c_3 e^{6\pi i/5} + c_4 e^{8\pi i/5} \right), \tag{7.275}$$

$$y_2 = \frac{1}{\sqrt{5}} \left(c_0 + c_1 e^{4\pi i/5} + c_2 e^{8\pi i/5} + c_3 e^{12\pi i/5} + c_4 e^{16\pi i/5} \right), \tag{7.276}$$

$$y_3 = \frac{1}{\sqrt{5}} \left(c_0 + c_1 e^{6\pi i/5} + c_2 e^{12\pi i/5} + c_3 e^{18\pi i/5} + c_4 e^{24\pi i/5} \right), \tag{7.277}$$

$$y_4 = \frac{1}{\sqrt{5}} \left(c_0 + c_1 e^{8\pi i/5} + c_2 e^{16\pi i/5} + c_3 e^{24\pi i/5} + c_4 e^{32\pi i/5} \right). \tag{7.278}$$

In general, it is seen that y_j can be described by

$$y_j = \frac{1}{\sqrt{N}} \sum_{k=0}^{N-1} c_k \exp\left((2\pi i)\frac{kj}{N} \right), \qquad j = 0, \dots, N-1. \tag{7.279}$$

Realizing now that for a uniform discretization, such as done here, that

$$\Delta x = \frac{x_{max} - x_{min}}{N-1}, \tag{7.280}$$

and that

$$x_j = j\Delta x + x_{min}, \qquad j = 0, \dots, N-1, \tag{7.281}$$

one has

$$x_j = j\left(\frac{x_{max} - x_{min}}{N-1} \right) + x_{min}, \qquad j = 0, \dots, N-1. \tag{7.282}$$

Solving for j, one gets

$$j = (N-1)\left(\frac{x_j - x_{min}}{x_{max} - x_{min}} \right), \tag{7.283}$$

so that y_j can be expressed as a Fourier-type expansion in terms of x_j as

$$y_j = \frac{1}{\sqrt{N}} \sum_{k=1}^{N} c_k \exp\left((2\pi i)k \left(\frac{N-1}{N} \right) \left(\frac{x_j - x_{min}}{x_{max} - x_{min}} \right) \right), \qquad j = 0, \dots, N-1. \tag{7.284}$$

Here, the wavenumber of mode k, κ_k, is seen to be

$$\kappa_k = k\frac{N-1}{N}. \tag{7.285}$$

And as $N \to \infty$, one has

$$\kappa_k \approx k. \tag{7.286}$$

A real utility of the DFT is seen in its ability to select amplitudes of modes of signals at certain frequencies as demonstrated in the following example.

EXAMPLE 7.30

Find the DFT of the signal for $x \in [0, 3]$

$$y(x) = 10\sin\left((2\pi)\frac{2x}{3} \right) + 2\sin\left((2\pi)\frac{10x}{3} \right) + \sin\left((2\pi)\frac{100x}{3} \right). \tag{7.287}$$

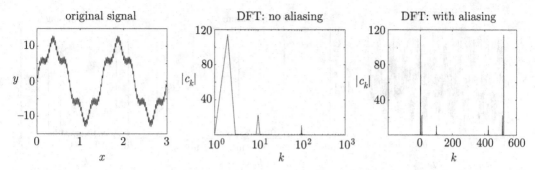

Figure 7.5. Plot of a three-term sinusoid $y(x)$ and its discrete Fourier transform for $N = 523$. The first DFT is plotted from $k = 0, \ldots, N/2$ and thus represents the original signal well. The second DFT is plotted from $k = 0, \ldots, N - 1$ and exhibits aliasing effects at high k.

Rescaling the domain so as to take $x \in [0, 3]$ into $\tilde{x} \in [0, 1]$ via the transformation $\tilde{x} = x/3$, one has

$$y(\tilde{x}) = 10 \sin\left((2\pi)2\tilde{x}\right) + 2\sin\left((2\pi)10\tilde{x}\right) + \sin\left((2\pi)100\tilde{x}\right). \tag{7.288}$$

To capture the high wavenumber components of the signal, one must have a sufficiently large value of N. In the transformed domain, the smallest wavelength is $\lambda = 1/100 = 0.01$. So for a domain length of unity, one needs *at least* $N = 100$ sampling points. In fact, let us choose to take more points, $N = 523$. There is no problem in choosing an unusual number of points for this so-called slow Fourier transform. If an FFT were attempted, one would have to choose integral powers of 2 as the number of points.

A plot of the function $y(x)$ and two versions of its DFT, $|c_k|$ versus k, is given in Figure 7.5. Note that $|c_k|$ has its peaks at $k = 2$, $k = 10$, and $k = 100$, equal to the wavenumbers of the generating sine functions, $\kappa_1 = 2$, $\kappa_2 = 10$, and $\kappa_3 = 100$. To avoid the confusing, and nonphysical, aliasing effect, only half the $|c_k|$ values have been plotted the first DFT of Figure 7.5. The second DFT here plots all values of $|c_k|$ and thus exhibits aliasing for large k.

EXAMPLE 7.31

Now take the DFT of a signal that is corrupted by so-called white, or random, noise. The signal here is given in $x \in [0, 1]$ by

$$y(x) = \sin\left((2\pi)10x\right) + \sin\left((2\pi)100x\right) + f_{rand}[-1, 1](x). \tag{7.289}$$

Here $f_{rand}[-1, 1](x)$ returns a random number between -1 and 1 for any value of x. A plot of the function $y(x)$ and two versions of its 607 point DFT, $|c_k|$ versus k, is given in Figure 7.6. In the raw data plotted in Figure 7.6, it is difficult to distinguish the signal from the random noise. But on examination of the accompanying DFT plot, it is clear that there are unambiguous components of the signal that peak at $k = 10$ and $k = 100$, which indicates there is a strong component of the signal with $\kappa = 10$ and $\kappa = 100$. Once again, to avoid the confusing, and nonphysical, aliasing effect, only half the $|c_k|$ values have been plotted in the first DFT of Figure 7.6. The second DFT gives all values of $|c_k|$ and exhibits aliasing.

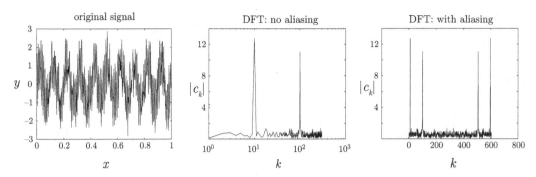

Figure 7.6. Plot of $y(x)$, a two-term sinusoid accompanied by random noise, and its discrete Fourier transform for $N = 607$ points. The first DFT is plotted from $k = 0, \ldots, N/2$ and thus represents the original signal well. The second DFT is plotted from $k = 0, \ldots, N - 1$ and exhibits aliasing effects at high k.

7.9 Matrix Decompositions

One of the most important tasks, especially in the numerical solution of algebraic and differential equations, is decomposing general matrices into simpler components. A brief discussion will be given here of some of the more important decompositions. It is noted that many popular software programs have algorithms that routinely calculate these decompositions.

7.9.1 L · D · U

Probably the most important technique in solving linear systems of algebraic equations of the form $\mathbf{A} \cdot \mathbf{x} = \mathbf{b}$ uses the decomposition

$$\mathbf{A} = \mathbf{P}^{-1} \cdot \mathbf{L} \cdot \mathbf{D} \cdot \mathbf{U}, \tag{7.290}$$

where $\mathbf{A}$ is a square matrix,[6] $\mathbf{P}$ is a never-singular permutation matrix, $\mathbf{L}$ is a lower triangular matrix, $\mathbf{D}$ is a diagonal matrix, and $\mathbf{U}$ is an upper triangular matrix. In other contexts $\mathbf{R}$ is sometimes used for an upper triangular matrix, and $\mathbf{P}$ is sometimes used for a projection matrix. The decomposition is often called either an $\mathbf{L} \cdot \mathbf{D} \cdot \mathbf{U}$ or more simply an $\mathbf{L} \cdot \mathbf{U}$ decomposition. The method amounts to a form of Gaussian elimination, by which all terms can be found. The permutation matrix is necessary in case row exchanges are necessary in the Gaussian elimination.

A common numerical algorithm to solve for $\mathbf{x}$ in $\mathbf{A} \cdot \mathbf{x} = \mathbf{b}$ is as follows.

- Factor $\mathbf{A}$ into $\mathbf{P}^{-1} \cdot \mathbf{L} \cdot \mathbf{D} \cdot \mathbf{U}$ so that $\mathbf{A} \cdot \mathbf{x} = \mathbf{b}$ becomes

$$\underbrace{\mathbf{P}^{-1} \cdot \mathbf{L} \cdot \mathbf{D} \cdot \mathbf{U}}_{\mathbf{A}} \cdot \mathbf{x} = \mathbf{b}. \tag{7.291}$$

- Operate on both sides of Eq. (7.291) with $\left(\mathbf{P}^{-1} \cdot \mathbf{L} \cdot \mathbf{D}\right)^{-1}$ to get

$$\mathbf{U} \cdot \mathbf{x} = \left(\mathbf{P}^{-1} \cdot \mathbf{L} \cdot \mathbf{D}\right)^{-1} \cdot \mathbf{b}. \tag{7.292}$$

---

[6] If $\mathbf{A}$ is not square, there is an equivalent decomposition, known as *row echelon form*, to be discussed in Section 7.9.3.

- Solve next for the new variable **c** in the new equation

$$\mathbf{P}^{-1} \cdot \mathbf{L} \cdot \mathbf{D} \cdot \mathbf{c} = \mathbf{b}, \tag{7.293}$$

so

$$\mathbf{c} = \left(\mathbf{P}^{-1} \cdot \mathbf{L} \cdot \mathbf{D}\right)^{-1} \cdot \mathbf{b}. \tag{7.294}$$

The triangular form of $\mathbf{L} \cdot \mathbf{D}$ renders the inversion of $\mathbf{P}^{-1} \cdot \mathbf{L} \cdot \mathbf{D}$ to be more computationally efficient than inversion of an arbitrary square matrix.

- Substitute **c** from Eq. (7.294) into Eq. (7.292), the modified version of the original equation, to get

$$\mathbf{U} \cdot \mathbf{x} = \mathbf{c}, \tag{7.295}$$

so

$$\mathbf{x} = \mathbf{U}^{-1} \cdot \mathbf{c}. \tag{7.296}$$

Again because $\mathbf{U}$ is triangular, the inversion is computationally efficient.

EXAMPLE 7.32

Find the $\mathbf{L} \cdot \mathbf{D} \cdot \mathbf{U}$ decomposition of the matrix:

$$\mathbf{A} = \begin{pmatrix} -5 & 4 & 9 \\ -22 & 14 & 18 \\ 16 & -8 & -6 \end{pmatrix}. \tag{7.297}$$

The process is essentially a series of row operations, which is the essence of Gaussian elimination. First we operate to transform the -22 and 16 in the first column into zeros. Crucial in this step is the necessity of the term in the 1,1 slot, known as the *pivot*, to be nonzero. If it is zero, a row exchange will be necessary, mandating a permutation matrix that is not the identity matrix. In this case, there are no such problems. We multiply the first row by $22/5$ and subtract from the second row, then multiply the first row by $-16/5$ and subtract from the third row. The factors $22/5$ and $-16/5$ will go in the 2,1 and 3,1 slots of the matrix $\mathbf{L}$. The diagonal of $\mathbf{L}$ always is filled with ones. This row operation yields

$$\mathbf{A} = \begin{pmatrix} -5 & 4 & 9 \\ -22 & 14 & 18 \\ 16 & -8 & -6 \end{pmatrix} = \begin{pmatrix} 1 & 0 & 0 \\ 22/5 & 1 & 0 \\ -16/5 & 0 & 1 \end{pmatrix} \begin{pmatrix} -5 & 4 & 9 \\ 0 & -18/5 & -108/5 \\ 0 & 24/5 & 114/5 \end{pmatrix}. \tag{7.298}$$

Now multiplying the new second row by $-4/3$, subtracting this from the third row, and depositing the factor $-4/3$ into 3,2 slot of the matrix $\mathbf{L}$, we get

$$\mathbf{A} = \begin{pmatrix} -5 & 4 & 9 \\ -22 & 14 & 18 \\ 16 & -8 & -6 \end{pmatrix} = \underbrace{\begin{pmatrix} 1 & 0 & 0 \\ 22/5 & 1 & 0 \\ -16/5 & -4/3 & 1 \end{pmatrix}}_{\mathbf{L}} \underbrace{\begin{pmatrix} -5 & 4 & 9 \\ 0 & -18/5 & -108/5 \\ 0 & 0 & -6 \end{pmatrix}}_{\mathbf{U}}. \tag{7.299}$$

The form given in Eq. (7.299) is often described as the $\mathbf{L} \cdot \mathbf{U}$ decomposition of $\mathbf{A}$. We can force the diagonal terms of the upper triangular matrix to unity by extracting a diagonal matrix D to form the $L \cdot D \cdot U$ decomposition:

$$A = \begin{pmatrix} -5 & 4 & 9 \\ -22 & 14 & 18 \\ 16 & -8 & -6 \end{pmatrix}$$

$$= \underbrace{\begin{pmatrix} 1 & 0 & 0 \\ 22/5 & 1 & 0 \\ -16/5 & -4/3 & 1 \end{pmatrix}}_{L} \underbrace{\begin{pmatrix} -5 & 0 & 0 \\ 0 & -18/5 & 0 \\ 0 & 0 & -6 \end{pmatrix}}_{D} \underbrace{\begin{pmatrix} 1 & -4/5 & -9/5 \\ 0 & 1 & 6 \\ 0 & 0 & 1 \end{pmatrix}}_{U}. \quad (7.300)$$

Note that D *does not* contain the eigenvalues of A. Also, because there were no row exchanges necessary $P = P^{-1} = I$, and it has not been included.

EXAMPLE 7.33

Find the $L \cdot D \cdot U$ decomposition of the matrix A:

$$A = \begin{pmatrix} 0 & 1 & 2 \\ 1 & 1 & 1 \\ 1 & 0 & 0 \end{pmatrix}. \quad (7.301)$$

There is a zero in the pivot, so a row exchange is necessary:

$$P \cdot A = \begin{pmatrix} 0 & 0 & 1 \\ 0 & 1 & 0 \\ 1 & 0 & 0 \end{pmatrix} \begin{pmatrix} 0 & 1 & 2 \\ 1 & 1 & 1 \\ 1 & 0 & 0 \end{pmatrix} = \begin{pmatrix} 1 & 0 & 0 \\ 1 & 1 & 1 \\ 0 & 1 & 2 \end{pmatrix}. \quad (7.302)$$

Performing Gaussian elimination by subtracting 1 times the first row from the second and depositing the 1 in the 2,1 slot of L, we get

$$P \cdot A = L \cdot U \rightarrow \begin{pmatrix} 0 & 0 & 1 \\ 0 & 1 & 0 \\ 1 & 0 & 0 \end{pmatrix} \begin{pmatrix} 0 & 1 & 2 \\ 1 & 1 & 1 \\ 1 & 0 & 0 \end{pmatrix} = \begin{pmatrix} 1 & 0 & 0 \\ 1 & 1 & 0 \\ 0 & 0 & 1 \end{pmatrix} \begin{pmatrix} 1 & 0 & 0 \\ 0 & 1 & 1 \\ 0 & 1 & 2 \end{pmatrix}. \quad (7.303)$$

Now subtracting 1 times the second row, and depositing the 1 in the 3,2 slot of L, we find

$$P \cdot A = L \cdot U \rightarrow \begin{pmatrix} 0 & 0 & 1 \\ 0 & 1 & 0 \\ 1 & 0 & 0 \end{pmatrix} \begin{pmatrix} 0 & 1 & 2 \\ 1 & 1 & 1 \\ 1 & 0 & 0 \end{pmatrix} = \begin{pmatrix} 1 & 0 & 0 \\ 1 & 1 & 0 \\ 0 & 1 & 1 \end{pmatrix} \begin{pmatrix} 1 & 0 & 0 \\ 0 & 1 & 1 \\ 0 & 0 & 1 \end{pmatrix}. \quad (7.304)$$

Now U already has ones on the diagonal, so the diagonal matrix D is simply the identity matrix. Using this and inverting P, which is P itself (!), we get the final decomposition:

$$A = P^{-1} \cdot L \cdot D \cdot U = \underbrace{\begin{pmatrix} 0 & 0 & 1 \\ 0 & 1 & 0 \\ 1 & 0 & 0 \end{pmatrix}}_{P^{-1}} \underbrace{\begin{pmatrix} 1 & 0 & 0 \\ 1 & 1 & 0 \\ 0 & 1 & 1 \end{pmatrix}}_{L} \underbrace{\begin{pmatrix} 1 & 0 & 0 \\ 0 & 1 & 0 \\ 0 & 0 & 1 \end{pmatrix}}_{D} \underbrace{\begin{pmatrix} 1 & 0 & 0 \\ 0 & 1 & 1 \\ 0 & 0 & 1 \end{pmatrix}}_{U}. \quad (7.305)$$

7.9.2 Cholesky

If A is a Hermitian positive definite matrix, we can define a Cholesky decomposition, briefly introduced on Section 7.2.5. Because A must be positive definite, it must be square. The Cholesky decomposition is as follows:

$$\mathbf{A} = \mathbf{U}^H \cdot \mathbf{U}. \tag{7.306}$$

Here $\mathbf{U}$ is an upper triangular matrix. One might think of $\mathbf{U}$ as the rough equivalent of the square root of the positive definite $\mathbf{A}$. We also have the related decomposition:

$$\mathbf{A} = \hat{\mathbf{U}}^H \cdot \mathbf{D} \cdot \hat{\mathbf{U}}, \tag{7.307}$$

where $\hat{\mathbf{U}}$ is upper triangular with a value of unity on its diagonal, and $\mathbf{D}$ is diagonal.

If we define a lower triangular $\mathbf{L}$ as $\mathbf{L} = \mathbf{U}^H$, the Cholesky decomposition can be rewritten as

$$\mathbf{A} = \mathbf{L} \cdot \mathbf{L}^H. \tag{7.308}$$

There also exists an analogous decomposition

$$\mathbf{A} = \hat{\mathbf{L}} \cdot \mathbf{D} \cdot \hat{\mathbf{L}}^H. \tag{7.309}$$

These definitions hold as well for real $\mathbf{A}$; in such cases, we can simply replace the Hermitian transpose by the ordinary transpose.

EXAMPLE 7.34

The Cholesky decomposition of a Hermitian matrix $\mathbf{A}$ is as follows:

$$\mathbf{A} = \begin{pmatrix} 5 & 4i \\ -4i & 5 \end{pmatrix} = \mathbf{U}^H \cdot \mathbf{U} = \underbrace{\begin{pmatrix} \sqrt{5} & 0 \\ -\frac{4i}{\sqrt{5}} & \frac{3}{\sqrt{5}} \end{pmatrix}}_{\mathbf{U}^H} \underbrace{\begin{pmatrix} \sqrt{5} & \frac{4i}{\sqrt{5}} \\ 0 & \frac{3}{\sqrt{5}} \end{pmatrix}}_{\mathbf{U}}. \tag{7.310}$$

The eigenvalues of $\mathbf{A}$ are $\lambda = 1, \lambda = 9$, so the matrix is indeed positive definite.

We can also write in alternative form

$$\mathbf{A} = \begin{pmatrix} 5 & 4i \\ -4i & 5 \end{pmatrix} = \hat{\mathbf{U}}^H \cdot \mathbf{D} \cdot \hat{\mathbf{U}} = \underbrace{\begin{pmatrix} 1 & 0 \\ -\frac{4i}{5} & 1 \end{pmatrix}}_{\hat{\mathbf{U}}^H} \underbrace{\begin{pmatrix} 5 & 0 \\ 0 & \frac{9}{5} \end{pmatrix}}_{\mathbf{D}} \underbrace{\begin{pmatrix} 1 & \frac{4i}{5} \\ 0 & 1 \end{pmatrix}}_{\hat{\mathbf{U}}}. \tag{7.311}$$

Here $\mathbf{D}$ does not contain the eigenvalues of $\mathbf{A}$.

7.9.3 Row Echelon Form

When $\mathbf{A}$ is not square, we can still use Gaussian elimination to cast the matrix in *row echelon form*:

$$\mathbf{A} = \mathbf{P}^{-1} \cdot \mathbf{L} \cdot \mathbf{D} \cdot \mathbf{U}. \tag{7.312}$$

Again $\mathbf{P}$ is a never-singular permutation matrix, $\mathbf{L}$ is lower triangular and square, $\mathbf{D}$ is diagonal and square, and $\mathbf{U}$ is upper triangular and rectangular and of the same dimension as $\mathbf{A}$. The strategy is to use row operations in such a fashion that ones or zeros appear on the diagonal.

EXAMPLE 7.35

Determine the row echelon form of the nonsquare matrix,

$$\mathbf{A} = \begin{pmatrix} 1 & -3 & 2 \\ 2 & 0 & 3 \end{pmatrix}. \tag{7.313}$$

We take 2 times the first row and subtract the result from the second row. The scalar 2 is deposited in the 2,1 slot in the **L** matrix. So

$$\mathbf{A} = \begin{pmatrix} 1 & -3 & 2 \\ 2 & 0 & 3 \end{pmatrix} = \underbrace{\begin{pmatrix} 1 & 0 \\ 2 & 1 \end{pmatrix}}_{\mathbf{L}} \underbrace{\begin{pmatrix} 1 & -3 & 2 \\ 0 & 6 & -1 \end{pmatrix}}_{\mathbf{U}}. \tag{7.314}$$

Again, Eq. (7.314) is also known as an $\mathbf{L} \cdot \mathbf{U}$ decomposition and is often as useful as the $\mathbf{L} \cdot \mathbf{D} \cdot \mathbf{U}$ decomposition. There is no row exchange so the permutation matrix and its inverse are the identity matrix. We extract a 1 and 6 to form the diagonal matrix **D**, so the final form is

$$\mathbf{A} = \mathbf{P}^{-1} \cdot \mathbf{L} \cdot \mathbf{D} \cdot \mathbf{U} = \underbrace{\begin{pmatrix} 1 & 0 \\ 0 & 1 \end{pmatrix}}_{\mathbf{P}^{-1}} \underbrace{\begin{pmatrix} 1 & 0 \\ 2 & 1 \end{pmatrix}}_{\mathbf{L}} \underbrace{\begin{pmatrix} 1 & 0 \\ 0 & 6 \end{pmatrix}}_{\mathbf{D}} \underbrace{\begin{pmatrix} 1 & -3 & 2 \\ 0 & 1 & -\frac{1}{6} \end{pmatrix}}_{\mathbf{U}}. \tag{7.315}$$

Row echelon form is an especially useful form for underconstrained systems, as illustrated in the following example.

EXAMPLE 7.36

Find best estimates for the unknown **x** in the equation $\mathbf{A} \cdot \mathbf{x} = \mathbf{b}$, where $\mathbf{A} : \mathbb{R}^5 \to \mathbb{R}^3$ is known and **b** is left general, but considered to be known:

$$\begin{pmatrix} 2 & 1 & -1 & 1 & 2 \\ 4 & 2 & -2 & 1 & 0 \\ -2 & -1 & 1 & -2 & -6 \end{pmatrix} \begin{pmatrix} x_1 \\ x_2 \\ x_3 \\ x_4 \\ x_5 \end{pmatrix} = \begin{pmatrix} b_1 \\ b_2 \\ b_3 \end{pmatrix}. \tag{7.316}$$

We perform Gaussian elimination row operations on the second and third rows to get zeros in the first column:

$$\begin{pmatrix} 2 & 1 & -1 & 1 & 2 \\ 0 & 0 & 0 & -1 & -4 \\ 0 & 0 & 0 & -1 & -4 \end{pmatrix} \begin{pmatrix} x_1 \\ x_2 \\ x_3 \\ x_4 \\ x_5 \end{pmatrix} = \begin{pmatrix} b_1 \\ -2b_1 + b_2 \\ b_1 + b_3 \end{pmatrix}. \tag{7.317}$$

The next round of Gaussian elimination works on the third row and yields

$$\begin{pmatrix} 2 & 1 & -1 & 1 & 2 \\ 0 & 0 & 0 & -1 & -4 \\ 0 & 0 & 0 & 0 & 0 \end{pmatrix} \begin{pmatrix} x_1 \\ x_2 \\ x_3 \\ x_4 \\ x_5 \end{pmatrix} = \begin{pmatrix} b_1 \\ -2b_1 + b_2 \\ 3b_1 - b_2 + b_3 \end{pmatrix}. \tag{7.318}$$

The reduced third equation gives

$$0 = 3b_1 - b_2 + b_3. \tag{7.319}$$

This is the equation of a plane in $\mathbb{R}^3$. Thus, arbitrary $\mathbf{b} \in \mathbb{R}^3$ *will not* satisfy the original equation. Said another way, the operator **A** maps arbitrary five-dimensional vectors **x** into a two-dimensional subspace of a three-dimensional vector space. The rank of **A** is 2. Thus, the dimension of both the row space and the column space is 2; the dimension of the right null space is 3, and the dimension of the left null space is 1.

We also note that there are two nontrivial equations remaining. The first nonzero elements from the left of each row are known as the pivots. The number of pivots is equal to the rank of the matrix. Variables that correspond to each pivot are known as *basic variables*. Variables with no pivot are known as *free variables*. Here the basic variables are x_1 and x_4, while the free variables are x_2, x_3, and x_5.

Now enforcing the constraint $3b_1 - b_2 + b_3 = 0$, without which there will be *no* exact solution, we can set each free variable to an arbitrary value and then solve the resulting square system. Take $x_2 = r$, $x_3 = s$, $x_5 = t$, where here r, s, and t are arbitrary real scalar constants. So

$$\begin{pmatrix} 2 & 1 & -1 & 1 & 2 \\ 0 & 0 & 0 & -1 & -4 \\ 0 & 0 & 0 & 0 & 0 \end{pmatrix} \begin{pmatrix} x_1 \\ r \\ s \\ x_4 \\ t \end{pmatrix} = \begin{pmatrix} b_1 \\ -2b_1 + b_2 \\ 0 \end{pmatrix}, \tag{7.320}$$

which gives

$$\begin{pmatrix} 2 & 1 \\ 0 & -1 \end{pmatrix} \begin{pmatrix} x_1 \\ x_4 \end{pmatrix} = \begin{pmatrix} b_1 - r + s - 2t \\ -2b_1 + b_2 + 4t \end{pmatrix}, \tag{7.321}$$

yielding

$$x_4 = 2b_1 - b_2 - 4t, \tag{7.322}$$

$$x_1 = \frac{1}{2}(-b_1 + b_2 - r + s + 2t). \tag{7.323}$$

Thus

$$\mathbf{x} = \begin{pmatrix} x_1 \\ x_2 \\ x_3 \\ x_4 \\ x_5 \end{pmatrix} = \begin{pmatrix} \frac{1}{2}(-b_1 + b_2 - r + s + 2t) \\ r \\ s \\ 2b_1 - b_2 - 4t \\ t \end{pmatrix}, \tag{7.324}$$

$$= \begin{pmatrix} \frac{1}{2}(-b_1 + b_2) \\ 0 \\ 0 \\ 2b_1 - b_2 \\ 0 \end{pmatrix} + r \begin{pmatrix} -\frac{1}{2} \\ 1 \\ 0 \\ 0 \\ 0 \end{pmatrix} + s \begin{pmatrix} \frac{1}{2} \\ 0 \\ 1 \\ 0 \\ 0 \end{pmatrix} + t \begin{pmatrix} 1 \\ 0 \\ 0 \\ -4 \\ 1 \end{pmatrix}, r, s, t \in \mathbb{R}^1. \tag{7.325}$$

The coefficients r, s, and t multiply the three right null space vectors. These, in combination with two independent row space vectors, form a basis for any vector $\mathbf{x}$. Thus, we can again cast the solution as a particular solution that is a unique combination of independent row space vectors and a nonunique combination of the right null space vectors (the homogeneous solution):

$$\mathbf{x} = \begin{pmatrix} x_1 \\ x_2 \\ x_3 \\ x_4 \\ x_5 \end{pmatrix} = \underbrace{\frac{25b_1 - 13b_2}{106} \begin{pmatrix} 2 \\ 1 \\ -1 \\ 1 \\ 2 \end{pmatrix} + \frac{-13b_1 + 11b_2}{106} \begin{pmatrix} 4 \\ 2 \\ -2 \\ 1 \\ 0 \end{pmatrix}}_{\text{row space}} + \underbrace{\hat{r} \begin{pmatrix} -\frac{1}{2} \\ 1 \\ 0 \\ 0 \\ 0 \end{pmatrix} + \hat{s} \begin{pmatrix} \frac{1}{2} \\ 0 \\ 1 \\ 0 \\ 0 \end{pmatrix} + \hat{t} \begin{pmatrix} 1 \\ 0 \\ 0 \\ -4 \\ 1 \end{pmatrix}}_{\text{right null space}}. \tag{7.326}$$

In matrix form, we can say that

$$
\mathbf{x} = \begin{pmatrix} x_1 \\ x_2 \\ x_3 \\ x_4 \\ x_5 \end{pmatrix} = \begin{pmatrix} 2 & 4 & -\frac{1}{2} & \frac{1}{2} & 1 \\ 1 & 2 & 1 & 0 & 0 \\ -1 & -2 & 0 & 1 & 0 \\ 1 & 1 & 0 & 0 & -4 \\ 2 & 0 & 0 & 0 & 1 \end{pmatrix} \begin{pmatrix} \frac{25b_1 - 13b_2}{106} \\ \frac{-13b_1 + 11b_2}{106} \\ \hat{r} \\ \hat{s} \\ \hat{t} \end{pmatrix} . \tag{7.327}
$$

Here we have taken $\hat{r} = r + (b_1 - 9b_2)/106$, $\hat{s} = s + (-b_1 + 9b_2)/106$, and $\hat{t} = (-30b_1 + 26b_2)/106$; as they are arbitrary constants multiplying vectors in the right null space, the relationship to b_1 and b_2 is actually unimportant. As before, although the null space basis vectors are orthogonal to the row space basis vectors, the entire matrix is not orthogonal. The Gram-Schmidt procedure could be used to cast the solution on either an orthogonal or orthonormal basis.

It is also noted that we have effectively found the $\mathbf{L} \cdot \mathbf{U}$ decomposition of $\mathbf{A}$. The terms in $\mathbf{L}$ are from the Gaussian elimination, and we have already $\mathbf{U}$:

$$
\mathbf{A} = \mathbf{L} \cdot \mathbf{U} \rightarrow \underbrace{\begin{pmatrix} 2 & 1 & -1 & 1 & 2 \\ 4 & 2 & -2 & 1 & 0 \\ -2 & -1 & 1 & -2 & -6 \end{pmatrix}}_{\mathbf{A}} = \underbrace{\begin{pmatrix} 1 & 0 & 0 \\ 2 & 1 & 0 \\ -1 & 1 & 1 \end{pmatrix}}_{\mathbf{L}} \underbrace{\begin{pmatrix} 2 & 1 & -1 & 1 & 2 \\ 0 & 0 & 0 & -1 & -4 \\ 0 & 0 & 0 & 0 & 0 \end{pmatrix}}_{\mathbf{U}} . \tag{7.328}
$$

The $\mathbf{L} \cdot \mathbf{D} \cdot \mathbf{U}$ decomposition is

$$
\underbrace{\begin{pmatrix} 2 & 1 & -1 & 1 & 2 \\ 4 & 2 & -2 & 1 & 0 \\ -2 & -1 & 1 & -2 & -6 \end{pmatrix}}_{\mathbf{A}} = \underbrace{\begin{pmatrix} 1 & 0 & 0 \\ 2 & 1 & 0 \\ -1 & 1 & 1 \end{pmatrix}}_{\mathbf{L}} \underbrace{\begin{pmatrix} 2 & 0 & 0 \\ 0 & -1 & 0 \\ 0 & 0 & 0 \end{pmatrix}}_{\mathbf{D}} \underbrace{\begin{pmatrix} 1 & \frac{1}{2} & -\frac{1}{2} & \frac{1}{2} & 1 \\ 0 & 0 & 0 & 1 & 4 \\ 0 & 0 & 0 & 0 & 0 \end{pmatrix}}_{\mathbf{U}} . \tag{7.329}
$$

There were no row exchanges, so in effect the permutation matrix $\mathbf{P}$ is the identity matrix, and there is no need to include it.

Lastly, we note that a more robust alternative to the method shown here would be to *first* apply the $\mathbf{A}^T$ operator to both sides of the equation so to map both sides into the column space of $\mathbf{A}$. Then there would be no need to restrict $\mathbf{b}$ so that it lies in the column space. Our results are then interpreted as giving us only a projection of $\mathbf{x}$. Taking $\mathbf{A}^T \cdot \mathbf{A} \cdot \mathbf{x} = \mathbf{A}^T \cdot \mathbf{b}$ and then casting the result into row echelon form gives

$$
\begin{pmatrix} 1 & 1/2 & -1/2 & 0 & -1 \\ 0 & 0 & 0 & 1 & 4 \\ 0 & 0 & 0 & 0 & 0 \\ 0 & 0 & 0 & 0 & 0 \\ 0 & 0 & 0 & 0 & 0 \end{pmatrix} \begin{pmatrix} x_1 \\ x_2 \\ x_3 \\ x_4 \\ x_5 \end{pmatrix} = \begin{pmatrix} (1/22)(b_1 + 7b_2 + 4b_3) \\ (1/11)(b_1 - 4b_2 - 7b_3) \\ 0 \\ 0 \\ 0 \end{pmatrix} . \tag{7.330}
$$

This suggests we take $x_2 = r$, $x_3 = s$, and $x_5 = t$ and solve so to get

$$
\begin{pmatrix} x_1 \\ x_2 \\ x_3 \\ x_4 \\ x_5 \end{pmatrix} = \begin{pmatrix} (1/22)(b_1 + 7b_2 + 4b_3) \\ 0 \\ 0 \\ (1/11)(b_1 - 4b_2 - 7b_3) \\ 0 \end{pmatrix} + r \begin{pmatrix} -1/2 \\ 1 \\ 0 \\ 0 \\ 0 \end{pmatrix} + s \begin{pmatrix} 1/2 \\ 0 \\ 1 \\ 0 \\ 0 \end{pmatrix} + t \begin{pmatrix} 1 \\ 0 \\ 0 \\ -4 \\ 1 \end{pmatrix} . \tag{7.331}
$$

We could go on to cast this in terms of combinations of row vectors and right null space vectors, but will not do so here. It is reiterated that this result is valid for arbitrary $\mathbf{b}$, but that it only represents a solution which minimizes $\|\mathbf{A} \cdot \mathbf{x} - \mathbf{b}\|_2$.

7.9.4 $Q \cdot U$

The $Q \cdot U$ decomposition allows us to formulate a matrix as the product of an orthogonal (unitary if complex) matrix Q and an upper triangular matrix U, of the same dimension as A. The decomposition is more commonly described as a $Q \cdot R$ decomposition, but we are generally describing upper triangular matrices by U. We seek Q and U such that

$$A = Q \cdot U. \tag{7.332}$$

We omit details of the algorithm by which the decomposition is constructed. The matrix A can be square or rectangular. It can be thought of as a deformation due to U followed by a volume-preserving rotation or reflection due to Q.

EXAMPLE 7.37

The $Q \cdot U$ decomposition of the matrix we considered in a previous example, Example 7.32, is as follows:

$$A = \underbrace{\begin{pmatrix} -5 & 4 & 9 \\ -22 & 14 & 18 \\ 16 & -8 & -6 \end{pmatrix}}_{A} = Q \cdot U, \tag{7.333}$$

$$= \underbrace{\begin{pmatrix} -0.1808 & -0.4982 & 0.8480 \\ -0.7954 & -0.4331 & -0.4240 \\ 0.5785 & -0.7512 & -0.3180 \end{pmatrix}}_{Q} \underbrace{\begin{pmatrix} 27.6586 & -16.4867 & -19.4153 \\ 0 & -2.0465 & -7.7722 \\ 0 & 0 & 1.9080 \end{pmatrix}}_{U}. \tag{7.334}$$

Note that $\det Q = 1$, so it is volume- and orientation-preserving. Noting further that $||Q||_2 = 1$, we deduce that $||U||_2 = ||A||_2 = 37.9423$. Also, recalling how matrices can be thought of as transformations, we see how to think of A as a stretching (U) followed by rotation (Q).

By changing the sign on appropriate columns of Q, U can always be constructed such that it has positive elements on its diagonal. In this example we can change the sign on all the elements of the second column of Q and recover

$$A = \underbrace{\begin{pmatrix} -5 & 4 & 9 \\ -22 & 14 & 18 \\ 16 & -8 & -6 \end{pmatrix}}_{A} = Q \cdot U$$

$$= \underbrace{\begin{pmatrix} -0.1808 & 0.4982 & 0.8480 \\ -0.7954 & 0.4331 & -0.4240 \\ 0.5785 & 0.7512 & -0.3180 \end{pmatrix}}_{Q} \underbrace{\begin{pmatrix} 27.6586 & -16.4867 & -19.4153 \\ 0 & 2.0465 & 7.7722 \\ 0 & 0 & 1.9080 \end{pmatrix}}_{U}. \tag{7.335}$$

Of course this sign change induces $\det Q = -1$, so it is no longer a rotation.

EXAMPLE 7.38

Find the $Q \cdot U$ decomposition for our nonsquare matrix from Example 7.35

$$A = \begin{pmatrix} 1 & -3 & 2 \\ 2 & 0 & 3 \end{pmatrix}. \tag{7.336}$$

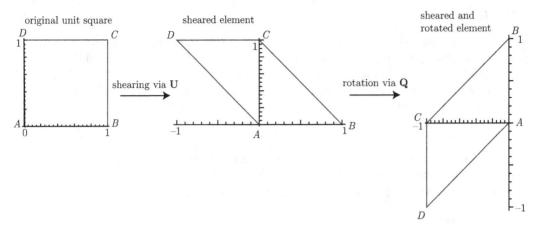

Figure 7.7. Unit square transforming via explicit shear stretching ($\mathbf{U}$), and rotation ($\mathbf{Q}$) under a linear area- and orientation-preserving alibi mapping.

The decomposition is

$$\mathbf{A} = \underbrace{\begin{pmatrix} 0.4472 & -0.8944 \\ 0.8944 & 0.4472 \end{pmatrix}}_{\mathbf{Q}} \cdot \underbrace{\begin{pmatrix} 2.2361 & -1.3416 & 3.577 \\ 0 & 2.6833 & -0.4472 \end{pmatrix}}_{\mathbf{U}}. \tag{7.337}$$

Once again $\det \mathbf{Q} = 1$, so $\mathbf{Q}$ effects an area- and orientation-preserving mapping. It is easy to show that $||\mathbf{A}||_2 = ||\mathbf{U}||_2 = 4.63849$.

EXAMPLE 7.39

Give a geometric interpretation of the $\mathbf{Q} \cdot \mathbf{U}$ decomposition in the context of the discussion surrounding the transformation of a unit square by the matrix $\mathbf{A}$ considered earlier in Example 7.17,

$$\mathbf{A} = \begin{pmatrix} 0 & -1 \\ 1 & -1 \end{pmatrix}. \tag{7.338}$$

The decomposition is

$$\mathbf{A} = \underbrace{\begin{pmatrix} 0 & -1 \\ 1 & 0 \end{pmatrix}}_{\mathbf{Q}} \underbrace{\begin{pmatrix} 1 & -1 \\ 0 & 1 \end{pmatrix}}_{\mathbf{U}}. \tag{7.339}$$

Now $\det \mathbf{A} = 1$. Moreover, $\det \mathbf{Q} = 1$ and $\det \mathbf{U} = 1$, so both of these matrices preserve area and orientation. As usual, $||\mathbf{Q}||_2 = 1$, so its operation preserves the lengths of vectors. The deformation is embodied in $\mathbf{U}$, which has $||\mathbf{U}||_2 = ||\mathbf{A}||_2 = 1.61803$. Decomposing the transformation of the unit square depicted in Figure 7.3 by first applying $\mathbf{U}$ to each of the vertices, and then applying $\mathbf{Q}$ to each of the stretched vertices, we see that $\mathbf{U}$ effects an area- and orientation-preserving shear deformation, and $\mathbf{Q}$ effects a counterclockwise rotation of $\pi/2$. This is depicted in Figure 7.7.

The $\mathbf{Q} \cdot \mathbf{U}$ decomposition can be shown to be closely related to the Gram-Schmidt orthogonalization process. It is also useful in increasing the efficiency of estimating $\mathbf{x}$ for $\mathbf{A} \cdot \mathbf{x} \approx \mathbf{b}$ when the system is overconstrained; that is, $\mathbf{b}$ is not in the column space of $\mathbf{A}$. If we as usual operate on both sides, as follows,

$$\mathbf{A} \cdot \mathbf{x} \approx \mathbf{b}, \tag{7.340}$$

$$\mathbf{A}^T \cdot \mathbf{A} \cdot \mathbf{x} = \mathbf{A}^T \cdot \mathbf{b}, \qquad \mathbf{A} = \mathbf{Q} \cdot \mathbf{U}, \tag{7.341}$$

$$(\mathbf{Q} \cdot \mathbf{U})^T \cdot \mathbf{Q} \cdot \mathbf{U} \cdot \mathbf{x} = (\mathbf{Q} \cdot \mathbf{U})^T \cdot \mathbf{b}, \tag{7.342}$$

$$\mathbf{U}^T \cdot \mathbf{Q}^T \cdot \mathbf{Q} \cdot \mathbf{U} \cdot \mathbf{x} = \mathbf{U}^T \cdot \mathbf{Q}^T \cdot \mathbf{b}, \tag{7.343}$$

$$\mathbf{U}^T \cdot \mathbf{Q}^{-1} \cdot \mathbf{Q} \cdot \mathbf{U} \cdot \mathbf{x} = \mathbf{U}^T \cdot \mathbf{Q}^T \cdot \mathbf{b}, \tag{7.344}$$

$$\mathbf{U}^T \cdot \mathbf{U} \cdot \mathbf{x} = \mathbf{U}^T \cdot \mathbf{Q}^T \cdot \mathbf{b}, \tag{7.345}$$

$$\mathbf{x} = (\mathbf{U}^T \cdot \mathbf{U})^{-1} \cdot \mathbf{U}^T \cdot \mathbf{Q}^T \cdot \mathbf{b}, \tag{7.346}$$

$$\mathbf{Q} \cdot \mathbf{U} \cdot \mathbf{x} = \mathbf{Q} \cdot \left(\mathbf{U} \cdot (\mathbf{U}^T \cdot \mathbf{U})^{-1} \cdot \mathbf{U}^T \right) \cdot \mathbf{Q}^T \cdot \mathbf{b}, \tag{7.347}$$

$$\mathbf{A} \cdot \mathbf{x} = \underbrace{\mathbf{Q} \cdot \left(\mathbf{U} \cdot (\mathbf{U}^T \cdot \mathbf{U})^{-1} \cdot \mathbf{U}^T \right) \cdot \mathbf{Q}^T}_{\mathbf{P}} \cdot \mathbf{b}. \tag{7.348}$$

When rectangular, $\mathbf{U}$ has no zeros on its diagonal, $\mathbf{U} \cdot (\mathbf{U}^T \cdot \mathbf{U})^{-1} \cdot \mathbf{U}^T$ has all zeros, except for r ones on the diagonal, where r is the rank of $\mathbf{U}$. This makes solution of overconstrained problems particularly simple. We note lastly that $\mathbf{Q} \cdot \mathbf{U} \cdot (\mathbf{U}^T \cdot \mathbf{U})^{-1} \cdot \mathbf{U}^T \cdot \mathbf{Q}^T = \mathbf{P}$, a projection matrix, defined first in Eq. (6.191) and to be discussed in Section 7.10.

7.9.5 Diagonalization

Casting a matrix into a form in which all (or sometimes most) of its off-diagonal elements have zero value has its most important application in solving systems of differential equations (as seen in Section 9.5), but also in other scenarios. For many cases, we can decompose a square matrix $\mathbf{A}$ into the form

$$\mathbf{A} = \mathbf{S} \cdot \mathbf{\Lambda} \cdot \mathbf{S}^{-1}, \tag{7.349}$$

where $\mathbf{S}$ is a nonsingular matrix and $\mathbf{\Lambda}$ is a diagonal matrix. To diagonalize a square matrix $\mathbf{A}$, we must find $\mathbf{S}$, a diagonalizing matrix, such that $\mathbf{S}^{-1} \cdot \mathbf{A} \cdot \mathbf{S}$ is diagonal. Not all matrices are diagonalizable. By inversion, we can also say

$$\mathbf{\Lambda} = \mathbf{S}^{-1} \cdot \mathbf{A} \cdot \mathbf{S}. \tag{7.350}$$

Considering $\mathbf{A}$ to be the original matrix, we have subjected it to a general linear transformation, which in general stretches and rotates, to arrive at $\mathbf{\Lambda}$; this transformation has the same form as that previously considered in Eq. (6.345). One can define the *algebraic multiplicity* of an eigenvalue as the number of times it occurs and the *geometric multiplicity* of an eigenvalue as the number of eigenvectors it has. One can prove the following:

Theorem: A matrix with distinct eigenvalues can be diagonalized, but the diagonalizing matrix is not unique.

Theorem: Nonzero eigenvectors corresponding to different eigenvalues are linearly independent.

Theorem: If $\mathbf{A}$ is an $N \times N$ matrix with N linearly independent right eigenvectors $\{\mathbf{e}_1, \mathbf{e}_2, \ldots, \mathbf{e}_n, \ldots, \mathbf{e}_N\}$ corresponding to eigenvalues $\{\lambda_1, \lambda_2, \ldots, \lambda_n, \ldots, \lambda_N\}$ (not necessarily distinct), then the $N \times N$ matrix $\mathbf{S}$ whose columns are populated by the eigenvectors of $\mathbf{A}$,

$$\mathbf{S} = \begin{pmatrix} \vdots & \vdots & \ldots & \vdots & \ldots & \vdots \\ \mathbf{e}_1 & \mathbf{e}_2 & \ldots & \mathbf{e}_n & \ldots & \mathbf{e}_N \\ \vdots & \vdots & \ldots & \vdots & \ldots & \vdots \end{pmatrix}, \tag{7.351}$$

makes

$$\mathbf{S}^{-1} \cdot \mathbf{A} \cdot \mathbf{S} = \mathbf{\Lambda}, \tag{7.352}$$

where

$$\mathbf{\Lambda} = \begin{pmatrix} \lambda_1 & 0 & \ldots & \ldots & \ldots & 0 \\ 0 & \lambda_2 & \ldots & \ldots & \ldots & \ldots \\ \vdots & \vdots & \ddots & \vdots & \vdots & \vdots \\ \vdots & \vdots & \vdots & \lambda_n & \vdots & \vdots \\ \vdots & \vdots & \vdots & \vdots & \ddots & 0 \\ 0 & \ldots & \ldots & \ldots & 0 & \lambda_N \end{pmatrix} \tag{7.353}$$

is a diagonal matrix of eigenvalues. The matrices $\mathbf{A}$ and $\mathbf{\Lambda}$ are similar.

Let us see if this recipe works when we fill the columns of $\mathbf{S}$ with the eigenvectors. First operate on Eq. (7.352) with $\mathbf{S}$ to arrive at a more general version of the eigenvalue problem:

$$\mathbf{A} \cdot \mathbf{S} = \mathbf{S} \cdot \mathbf{\Lambda}, \tag{7.354}$$

$$\underbrace{\begin{pmatrix} a_{11} & \ldots & a_{1N} \\ \vdots & \ddots & \vdots \\ a_{N1} & \ldots & a_{NN} \end{pmatrix}}_{=\mathbf{A}} \underbrace{\begin{pmatrix} \vdots & \ldots & \vdots \\ \mathbf{e}_1 & \ldots & \mathbf{e}_N \\ \vdots & \ldots & \vdots \end{pmatrix}}_{=\mathbf{S}} = \underbrace{\begin{pmatrix} \vdots & \ldots & \vdots \\ \mathbf{e}_1 & \ldots & \mathbf{e}_N \\ \vdots & \ldots & \vdots \end{pmatrix}}_{=\mathbf{S}} \underbrace{\begin{pmatrix} \lambda_1 & \ldots & 0 \\ \vdots & \ddots & \vdots \\ 0 & \ldots & \lambda_N \end{pmatrix}}_{=\mathbf{\Lambda}}, \tag{7.355}$$

$$= \underbrace{\begin{pmatrix} \vdots & \ldots & \vdots \\ \lambda_1 \mathbf{e}_1 & \ldots & \lambda_N \mathbf{e}_N \\ \vdots & \ldots & \vdots \end{pmatrix}}_{=\mathbf{S} \cdot \mathbf{\Lambda}}. \tag{7.356}$$

This induces N eigenvalue problems for each eigenvector:

$$\mathbf{A} \cdot \mathbf{e}_1 = \lambda_1 \mathbf{I} \cdot \mathbf{e}_1, \tag{7.357}$$

$$\mathbf{A} \cdot \mathbf{e}_2 = \lambda_2 \mathbf{I} \cdot \mathbf{e}_2, \tag{7.358}$$

$$\vdots \quad \vdots$$

$$\mathbf{A} \cdot \mathbf{e}_N = \lambda_N \mathbf{I} \cdot \mathbf{e}_N. \tag{7.359}$$

The effect of postmultiplication of both sides of Eq. (7.354) by $\mathbf{S}^{-1}$ is:

$$\mathbf{A} \cdot \mathbf{S} \cdot \mathbf{S}^{-1} = \mathbf{S} \cdot \mathbf{\Lambda} \cdot \mathbf{S}^{-1} \tag{7.360}$$

$$\mathbf{A} = \mathbf{S} \cdot \mathbf{\Lambda} \cdot \mathbf{S}^{-1}. \tag{7.361}$$

EXAMPLE 7.40

Diagonalize the matrix considered in a previous example, Example 7.32:

$$\mathbf{A} = \begin{pmatrix} -5 & 4 & 9 \\ -22 & 14 & 18 \\ 16 & -8 & -6 \end{pmatrix}, \tag{7.362}$$

and check.

The eigenvalue-eigenvector pairs are

$$\lambda_1 = -6, \quad \mathbf{e}_1 = \begin{pmatrix} -1 \\ -2 \\ 1 \end{pmatrix}, \tag{7.363}$$

$$\lambda_2 = 3, \quad \mathbf{e}_2 = \begin{pmatrix} 1 \\ 2 \\ 0 \end{pmatrix}, \tag{7.364}$$

$$\lambda_3 = 6, \quad \mathbf{e}_3 = \begin{pmatrix} 2 \\ 1 \\ 2 \end{pmatrix}. \tag{7.365}$$

$$\tag{7.366}$$

Then

$$\mathbf{S} = \begin{pmatrix} \vdots & \vdots & \vdots \\ \mathbf{e}_1 & \mathbf{e}_2 & \mathbf{e}_3 \\ \vdots & \vdots & \vdots \end{pmatrix} = \begin{pmatrix} -1 & 1 & 2 \\ -2 & 2 & 1 \\ 1 & 0 & 2 \end{pmatrix}. \tag{7.367}$$

The inverse is

$$\mathbf{S}^{-1} = \begin{pmatrix} -\frac{4}{3} & \frac{2}{3} & 1 \\ -\frac{5}{3} & \frac{4}{3} & 1 \\ \frac{2}{3} & -\frac{1}{3} & 0 \end{pmatrix}. \tag{7.368}$$

Thus,

$$\mathbf{A} \cdot \mathbf{S} = \begin{pmatrix} 6 & 3 & 12 \\ 12 & 6 & 6 \\ -6 & 0 & 12 \end{pmatrix}, \tag{7.369}$$

and

$$\mathbf{\Lambda} = \mathbf{S}^{-1} \cdot \mathbf{A} \cdot \mathbf{S} = \begin{pmatrix} -6 & 0 & 0 \\ 0 & 3 & 0 \\ 0 & 0 & 6 \end{pmatrix}. \tag{7.370}$$

Let us also note the complementary decomposition of $\mathbf{A}$:

$$\mathbf{A} = \mathbf{S} \cdot \mathbf{\Lambda} \cdot \mathbf{S}^{-1} = \underbrace{\begin{pmatrix} -1 & 1 & 2 \\ -2 & 2 & 1 \\ 1 & 0 & 2 \end{pmatrix}}_{\mathbf{S}} \underbrace{\begin{pmatrix} -6 & 0 & 0 \\ 0 & 3 & 0 \\ 0 & 0 & 6 \end{pmatrix}}_{\mathbf{\Lambda}} \underbrace{\begin{pmatrix} -\frac{4}{3} & \frac{2}{3} & 1 \\ -\frac{5}{3} & \frac{4}{3} & 1 \\ \frac{2}{3} & -\frac{1}{3} & 0 \end{pmatrix}}_{\mathbf{S}^{-1}} = \underbrace{\begin{pmatrix} -5 & 4 & 9 \\ -22 & 14 & 18 \\ 16 & -8 & -6 \end{pmatrix}}_{\mathbf{A}}. \tag{7.371}$$

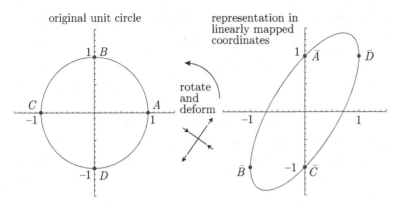

Figure 7.8. Unit circle transforming via stretching and rotation under a linear area- and orientation-preserving alibi mapping.

Because the matrix is not symmetric, the eigenvectors are not orthogonal, for example, $\mathbf{e}_1^T \cdot \mathbf{e}_2 = -5$.

If $\mathbf{A}$ is symmetric (Hermitian), then its eigenvectors must be orthogonal; thus, it is possible to normalize the eigenvectors so that the matrix $\mathbf{S}$ is in fact orthogonal (unitary if complex). Thus, for symmetric $\mathbf{A}$, we can have

$$\mathbf{A} = \mathbf{Q} \cdot \mathbf{\Lambda} \cdot \mathbf{Q}^{-1}. \tag{7.372}$$

Because $\mathbf{Q}^{-1} = \mathbf{Q}^T$, we have

$$\mathbf{A} = \mathbf{Q} \cdot \mathbf{\Lambda} \cdot \mathbf{Q}^T. \tag{7.373}$$

Geometrically, the action of a symmetric $\mathbf{A}$ on a geometric entity can be considered as volume-preserving rotation or reflection via $\mathbf{Q}^T$, followed by a stretching due to $\mathbf{\Lambda}$, completed by another volume-preserving rotation or reflection via $\mathbf{Q}$, which acts opposite to the effect of $\mathbf{Q}^T$. Note also that with $\mathbf{A} \cdot \mathbf{S} = \mathbf{S} \cdot \mathbf{\Lambda}$, the column vectors of $\mathbf{S}$ (which are the right eigenvectors of $\mathbf{A}$) form a basis in $\mathbb{C}^N$.

EXAMPLE 7.41

Demonstrate how the matrix of Example 7.17

$$\mathbf{A} = \begin{pmatrix} 0 & -1 \\ 1 & -1 \end{pmatrix}, \tag{7.374}$$

acts on vectors $\mathbf{x}$ which form the unit circle in the untransformed space:

$$\mathbf{x} = \begin{pmatrix} \cos\theta \\ \sin\theta \end{pmatrix}, \qquad \theta \in [0, 2\pi]. \tag{7.375}$$

In the alibi space, we have

$$\bar{\mathbf{x}} = \mathbf{A} \cdot \mathbf{x} = \begin{pmatrix} 0 & -1 \\ 1 & -1 \end{pmatrix} \begin{pmatrix} \cos\theta \\ \sin\theta \end{pmatrix} = \begin{pmatrix} -\sin\theta \\ \cos\theta - \sin\theta \end{pmatrix}. \tag{7.376}$$

We recognize here that the overline notation denotes the transformed space, not the complex conjugate. Our mapping of the unit circle is plotted in Figure 7.8. The figure represents both the rotation and the deformation of the mapping.

Now, a detailed analysis reveals that the transformed curve is actually an ellipse with the following properties:

- It is centered at the origin.
- Points on the ellipse shown in Figure 7.8, $\overline{A}$, $\overline{B}$, $\overline{C}$, and $\overline{D}$, are images of points on the circle, A, B, C, and D.
- Its major and minor axes are aligned with the eigenvectors of $\mathbf{A}^{-1^T} \cdot \mathbf{A}^{-1}$; because $\mathbf{A}^{-1^T} \cdot \mathbf{A}^{-1}$ is symmetric, the major and minor axes are guaranteed to be orthogonal.
- The lengths of its major and minor axes are given by

$$\text{major axis} = \frac{1}{\sqrt{\lambda_1}}, \qquad \text{minor axis} = \frac{1}{\sqrt{\lambda_2}}, \qquad (7.377)$$

where λ_1 and λ_2 are the eigenvalues, guaranteed real and positive, of $\mathbf{A}^{-1^T} \cdot \mathbf{A}^{-1}$. For this example, we find that the major and minor axes lengths are given by

$$\frac{1}{\sqrt{\lambda_1}} = \sqrt{\frac{2}{3 - \sqrt{5}}} = 1.61803, \qquad (7.378)$$

$$\frac{1}{\sqrt{\lambda_2}} = \sqrt{\frac{2}{3 + \sqrt{5}}} = 0.61803. \qquad (7.379)$$

- The area a of the ellipse is the product of π and the major and minor axes, which here is

$$a = \pi \frac{1}{\sqrt{\lambda_1}} \frac{1}{\sqrt{\lambda_2}} = \pi \sqrt{\frac{2}{3 - \sqrt{5}}} \sqrt{\frac{2}{3 + \sqrt{5}}} = \pi. \qquad (7.380)$$

The area of the original circle of radius unity is $a = \pi r^2 = \pi$. Our result confirms the transformation is area-preserving.

- It is described by the equation

$$\left(\frac{0.851\overline{x}_1 - 0.526\overline{x}_2}{0.618034} \right)^2 + \left(\frac{0.526\overline{x}_1 + 0.851\overline{x}_2}{1.61803} \right)^2 = 1. \qquad (7.381)$$

We arrive at these results via the following analysis:

$$\overline{\mathbf{x}} = \mathbf{A} \cdot \mathbf{x}, \qquad (7.382)$$

$$\mathbf{A}^{-1} \cdot \overline{\mathbf{x}} = \mathbf{x}, \qquad (7.383)$$

$$\mathbf{A}^{-1^T} \cdot \mathbf{A}^{-1} \cdot \overline{\mathbf{x}} = \mathbf{A}^{-1^T} \cdot \mathbf{x}. \qquad (7.384)$$

Now, $\mathbf{A}^{-1^T} \cdot \mathbf{A}^{-1}$ is symmetric, so it has the decomposition

$$\mathbf{A}^{-1^T} \cdot \mathbf{A}^{-1} = \mathbf{Q} \cdot \mathbf{\Lambda} \cdot \mathbf{Q}^T. \qquad (7.385)$$

Thus, our mapping can be written as

$$\mathbf{Q} \cdot \mathbf{\Lambda} \cdot \mathbf{Q}^T \cdot \overline{\mathbf{x}} = \mathbf{A}^{-1^T} \cdot \mathbf{x}, \qquad (7.386)$$

$$\overline{\mathbf{x}}^T \cdot \mathbf{Q} \cdot \mathbf{\Lambda} \cdot \mathbf{Q}^T \cdot \overline{\mathbf{x}} = \overline{\mathbf{x}}^T \cdot \mathbf{A}^{-1^T} \cdot \mathbf{x}, \qquad (7.387)$$

$$\overline{\mathbf{x}}^T \cdot \mathbf{Q} \cdot \mathbf{\Lambda} \cdot \mathbf{Q}^T \cdot \overline{\mathbf{x}} = \left(\mathbf{A}^{-1} \cdot \overline{\mathbf{x}} \right)^T \cdot \mathbf{x}, \qquad (7.388)$$

$$\left(\mathbf{Q}^T \cdot \overline{\mathbf{x}} \right)^T \cdot \mathbf{\Lambda} \cdot \left(\mathbf{Q}^T \cdot \overline{\mathbf{x}} \right) = \mathbf{x}^T \cdot \mathbf{x}. \qquad (7.389)$$

Now on our unit circle which defines $\mathbf{x}$, we have $\mathbf{x}^T \cdot \mathbf{x} = 1$, so the ellipse is defined by

$$\left(\mathbf{Q}^T \cdot \overline{\mathbf{x}} \right)^T \cdot \mathbf{\Lambda} \cdot \left(\mathbf{Q}^T \cdot \overline{\mathbf{x}} \right) = 1. \qquad (7.390)$$

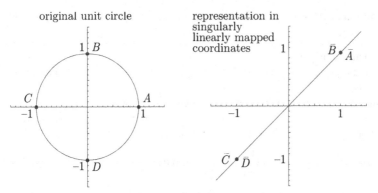

Figure 7.9. Unit circle transforming under a singular alibi mapping.

Here $\mathbf{Q}$ effects a rotation of axes, and $\mathbf{\Lambda}$ effects the stretching.
If $\mathbf{A}$ had been singular, for example,

$$\mathbf{A} = \begin{pmatrix} 1 & 1 \\ 1 & 1 \end{pmatrix}, \tag{7.391}$$

the action of $\mathbf{A}$ on the unit circle would have been to map it into a line inclined at $\pi/4$. As such a mapping takes many to one, it is a surjection. The action on the unit circle is depicted in Figure 7.9.

EXAMPLE 7.42

Determine the action of the matrix

$$\mathbf{A} = \begin{pmatrix} 2 & 1 \\ 1 & 1 \end{pmatrix} \tag{7.392}$$

on a unit square in terms of the diagonal decomposition of $\mathbf{A}$.

We first note that $\det \mathbf{A} = 1$, so it preserves volumes and orientations. We easily calculate that $\|\mathbf{A}\|_2 = 3/2 + \sqrt{5}/2 = 2.61803$, so it has the potential to stretch a vector. It is symmetric, so it has real eigenvalues, which are $\lambda = 3/2 \pm \sqrt{5}/2$. Its spectral radius is thus $\rho(\mathbf{A}) = 3/2 + \sqrt{5}/2$, which is equal to its spectral norm. Its eigenvectors are orthogonal, so they can be orthonormalized to form an orthogonal matrix. After detailed calculation, one finds the diagonal decomposition to be

$$\mathbf{A} = \underbrace{\begin{pmatrix} \sqrt{\frac{5+\sqrt{5}}{10}} & -\sqrt{\frac{2}{5+\sqrt{5}}} \\ \sqrt{\frac{2}{5+\sqrt{5}}} & \sqrt{\frac{5+\sqrt{5}}{10}} \end{pmatrix}}_{\mathbf{Q}} \underbrace{\begin{pmatrix} \frac{3+\sqrt{5}}{2} & 0 \\ 0 & \frac{3-\sqrt{5}}{2} \end{pmatrix}}_{\mathbf{\Lambda}} \underbrace{\begin{pmatrix} \sqrt{\frac{5+\sqrt{5}}{10}} & \sqrt{\frac{2}{5+\sqrt{5}}} \\ -\sqrt{\frac{2}{5+\sqrt{5}}} & \sqrt{\frac{5+\sqrt{5}}{10}} \end{pmatrix}}_{\mathbf{Q}^T}. \tag{7.393}$$

The action of this composition of matrix operations on a unit square is depicted in Figure 7.10. The first rotation is induced by $\mathbf{Q}^T$ and is clockwise through an angle of $\pi/5.67511 = 31.717°$. This is followed by an eigenstretching of $\mathbf{\Lambda}$. The action is completed by a rotation induced by $\mathbf{Q}$. The second rotation reverses the angle of the first in a counterclockwise rotation of $\pi/5.67511 = 31.717°$.

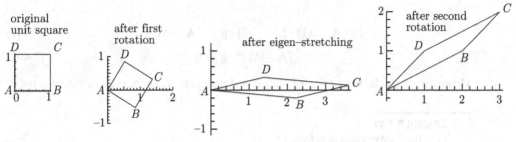

Figure 7.10. Unit square transforming via rotation, stretching, and rotation of the diagonalization decomposition under a linear area- and orientation-preserving alibi mapping.

7.9.6 Jordan Canonical Form

A square matrix $\mathbf{A}$ without a sufficient number of linearly independent eigenvectors can still be decomposed into a near-diagonal form:

$$\mathbf{A} = \mathbf{S} \cdot \mathbf{J} \cdot \mathbf{S}^{-1}. \tag{7.394}$$

This form is known as the Jordan[7] (upper) canonical form in which the near-diagonal matrix $\mathbf{J}$,

$$\mathbf{J} = \mathbf{S}^{-1} \cdot \mathbf{A} \cdot \mathbf{S}, \tag{7.395}$$

has zeros everywhere, except for eigenvalues along the principal diagonal and unity above the missing eigenvectors. The form is sometimes called a Jordan normal form.

   Consider the eigenvalue λ of algebraic multiplicity $N - L + 1$ of the matrix $\mathbf{A}_{N \times N}$. Then

$$(\mathbf{A} - \lambda \mathbf{I}) \cdot \mathbf{e} = 0 \tag{7.396}$$

gives some linearly independent eigenvectors $\mathbf{e}_1, \mathbf{e}_2, \ldots, \mathbf{e}_L$. If $L = N$, the algebraic multiplicity is unity, and the matrix can be diagonalized. If, however, $L < N$ we need $N - L$ more linearly independent vectors. These are the *eigenvectors generalized in the first sense*, first introduced in Section 6.8. One can be obtained from

$$(\mathbf{A} - \lambda \mathbf{I}) \cdot \mathbf{g}_1 = \mathbf{e}, \tag{7.397}$$

and others from

$$(\mathbf{A} - \lambda \mathbf{I}) \cdot \mathbf{g}_{j+1} = \mathbf{g}_j \quad \text{for} \quad j = 1, 2, \ldots, N - L - 1. \tag{7.398}$$

This procedure is continued until N linearly independent eigenvectors and generalized eigenvectors are obtained, which is the most that we can have in $\mathbb{R}^N$. Then

$$\mathbf{S} = \begin{pmatrix} \vdots & \cdots & \vdots & \vdots & \cdots & \vdots \\ \mathbf{e}_1 & \cdots & \mathbf{e}_L & \mathbf{g}_1 & \cdots & \mathbf{g}_{N-L} \\ \vdots & \cdots & \vdots & \vdots & \cdots & \vdots \end{pmatrix} \tag{7.399}$$

gives $\mathbf{S}^{-1} \cdot \mathbf{A} \cdot \mathbf{S} = \mathbf{J}$, where $\mathbf{J}$ is of the Jordan canonical form.

   Notice that $\mathbf{g}_n$ also satisfies $(\mathbf{A} - \lambda \mathbf{I})^n \cdot \mathbf{g}_n = \mathbf{0}$. For example, if

$$(\mathbf{A} - \lambda \mathbf{I}) \cdot \mathbf{g} = \mathbf{e}, \tag{7.400}$$

---

[7] Marie Ennemond Camille Jordan, 1838–1922, French mathematician.

then

$$(\mathbf{A} - \lambda\mathbf{I}) \cdot (\mathbf{A} - \lambda\mathbf{I}) \cdot \mathbf{g} = (\mathbf{A} - \lambda\mathbf{I}) \cdot \mathbf{e}, \tag{7.401}$$

$$(\mathbf{A} - \lambda\mathbf{I})^2 \cdot \mathbf{g} = \mathbf{0}. \tag{7.402}$$

However, any solution of Eq. (7.402) is not necessarily a generalized eigenvector.

EXAMPLE 7.43

Find the Jordan canonical form of

$$\mathbf{A} = \begin{pmatrix} 4 & 1 & 3 \\ 0 & 4 & 1 \\ 0 & 0 & 4 \end{pmatrix}. \tag{7.403}$$

The eigenvalues are $\lambda = 4$ with multiplicity 3. For this value,

$$(\mathbf{A} - \lambda\mathbf{I}) = \begin{pmatrix} 0 & 1 & 3 \\ 0 & 0 & 1 \\ 0 & 0 & 0 \end{pmatrix}. \tag{7.404}$$

The eigenvectors are obtained from $(\mathbf{A} - \lambda\mathbf{I}) \cdot \mathbf{e}_1 = \mathbf{0}$, which gives $x_2 + 3x_3 = 0, x_3 = 0$. The most general form of the eigenvector is

$$\mathbf{e}_1 = \begin{pmatrix} a \\ 0 \\ 0 \end{pmatrix}. \tag{7.405}$$

Only one eigenvector can be obtained from this eigenvalue. To get a generalized eigenvector, we take $(\mathbf{A} - \lambda\mathbf{I}) \cdot \mathbf{g}_1 = \mathbf{e}_1$, which gives $x_2 + 3x_3 = a, x_3 = 0$, so that

$$\mathbf{g}_1 = \begin{pmatrix} b \\ a \\ 0 \end{pmatrix}. \tag{7.406}$$

Another generalized eigenvector can be similarly obtained from $(\mathbf{A} - \lambda\mathbf{I}) \cdot \mathbf{g}_2 = \mathbf{g}_1$, so that $x_2 + 3x_3 = b, x_3 = a$. Thus, we get

$$\mathbf{g}_2 = \begin{pmatrix} c \\ b - 3a \\ a \end{pmatrix}. \tag{7.407}$$

From the eigenvector and generalized eigenvectors

$$\mathbf{S} = \begin{pmatrix} \vdots & \vdots & \vdots \\ \mathbf{e}_1 & \mathbf{g}_1 & \mathbf{g}_2 \\ \vdots & \vdots & \vdots \end{pmatrix} = \begin{pmatrix} a & b & c \\ 0 & a & b - 3a \\ 0 & 0 & a \end{pmatrix}, \tag{7.408}$$

and

$$\mathbf{S}^{-1} = \begin{pmatrix} \frac{1}{a} & -\frac{b}{a^2} & \frac{-b^2 + 3ba + ca}{a^3} \\ 0 & \frac{1}{a} & \frac{-b + 3a}{a^2} \\ 0 & 0 & \frac{1}{a} \end{pmatrix}. \tag{7.409}$$

The Jordan canonical form is

$$\mathbf{J} = \mathbf{S}^{-1} \cdot \mathbf{A} \cdot \mathbf{S} = \begin{pmatrix} 4 & 1 & 0 \\ 0 & 4 & 1 \\ 0 & 0 & 4 \end{pmatrix}. \tag{7.410}$$

In Eq. (7.409), a, b, and c are any constants. Choosing $a = 1$, $b = c = 0$, for example, simplifies the algebra giving

$$\mathbf{S} = \begin{pmatrix} 1 & 0 & 0 \\ 0 & 1 & -3 \\ 0 & 0 & 1 \end{pmatrix},$$ (7.411)

and

$$\mathbf{S}^{-1} = \begin{pmatrix} 1 & 0 & 0 \\ 0 & 1 & 3 \\ 0 & 0 & 1 \end{pmatrix}.$$ (7.412)

7.9.7 Schur

The Schur[8] decomposition of a square matrix $\mathbf{A}$ is as follows:

$$\mathbf{A} = \mathbf{Q} \cdot \mathbf{U} \cdot \mathbf{Q}^T.$$ (7.413)

Here $\mathbf{Q}$ is an orthogonal (unitary if complex) matrix, and $\mathbf{U}$ is upper triangular, with the eigenvalues this time along the diagonal. We omit details of the numerical algorithm to calculate the Schur decomposition.

EXAMPLE 7.44

The Schur decomposition of the matrix we diagonalized in a previous example, Example 7.32, is as follows:

$$\underbrace{\mathbf{A} = \begin{pmatrix} -5 & 4 & 9 \\ -22 & 14 & 18 \\ 16 & -8 & -6 \end{pmatrix}}_{\mathbf{A}} = \mathbf{Q} \cdot \mathbf{U} \cdot \mathbf{Q}^T =$$ (7.414)

$$\underbrace{\begin{pmatrix} -0.4082 & 0.1826 & 0.8944 \\ -0.8165 & 0.3651 & -0.4472 \\ 0.4082 & 0.9129 & 0 \end{pmatrix}}_{\mathbf{Q}} \underbrace{\begin{pmatrix} -6 & -20.1246 & 31.0376 \\ 0 & 3 & 5.7155 \\ 0 & 0 & 6 \end{pmatrix}}_{\mathbf{U}} \underbrace{\begin{pmatrix} -0.4082 & -0.8165 & 0.4082 \\ 0.1826 & 0.3651 & 0.9129 \\ 0.8944 & -0.4472 & 0 \end{pmatrix}}_{\mathbf{Q}^T}.$$ (7.415)

This decomposition was achieved with numerical software. This particular $\mathbf{Q}$ has det $\mathbf{Q} = -1$, so if it were used in a coordinate transformation, it would be volume-preserving but not orientation-preserving. Because the Schur decomposition is nonunique, it could be easily recalculated if one also wanted to preserve orientation.

EXAMPLE 7.45

The Schur decomposition of the matrix considered in Example 7.17 is as follows:

$$\mathbf{A} = \begin{pmatrix} 0 & -1 \\ 1 & -1 \end{pmatrix},$$ (7.416)

$$= \underbrace{\begin{pmatrix} \frac{1+\sqrt{3}i}{2\sqrt{2}} & -\frac{1}{\sqrt{2}} \\ \frac{1}{\sqrt{2}} & \frac{1-\sqrt{3}i}{2\sqrt{2}} \end{pmatrix}}_{\mathbf{Q}} \underbrace{\begin{pmatrix} \frac{-1+\sqrt{3}i}{2} & \frac{-1+\sqrt{3}i}{2} \\ 0 & \frac{-1-\sqrt{3}i}{2} \end{pmatrix}}_{\mathbf{U}} \underbrace{\begin{pmatrix} \frac{1-\sqrt{3}i}{2\sqrt{2}} & \frac{1}{\sqrt{2}} \\ -\frac{1}{\sqrt{2}} & \frac{1+\sqrt{3}i}{2\sqrt{2}} \end{pmatrix}}_{\mathbf{Q}^H}.$$ (7.417)

---

[8] Issai Schur, 1875–1941, Belrussian-born German-based mathematician.

This is a nonunique decomposition. Unusually, the form given here is exact; most require numerical approximation. Because $\mathbf{U}$ has the eigenvalues of $\mathbf{A}$ on its diagonal, we must consider unitary matrices. When this is recomposed, we recover the original real $\mathbf{A}$. Once again, we have $||\mathbf{U}||_2 = ||\mathbf{A}||_2 = 1.61803$. Here $\det \mathbf{Q} = \det \mathbf{Q}^H = 1$, so both are area- and orientation-preserving.

We can imagine the operation of $\mathbf{A}$ on a real vector $\mathbf{x}$ as an initial rotation into the complex plane effected by application of $\mathbf{Q}^H$: $\mathbf{x}' = \mathbf{Q}^H \cdot \mathbf{x}$. This is followed by a stretching effected by $\mathbf{U}$ so that $\mathbf{x}'' = \mathbf{U} \cdot \mathbf{x}'$. Application of $\mathbf{Q}$ rotates back into the real plane: $\mathbf{x}''' = \mathbf{Q} \cdot \mathbf{x}''$. The composite effect is $\mathbf{x}''' = \mathbf{Q} \cdot \mathbf{U} \cdot \mathbf{Q}^H \cdot \mathbf{x} = \mathbf{A} \cdot \mathbf{x}$.

If $\mathbf{A}$ is symmetric, then the upper triangular matrix $\mathbf{U}$ reduces to the diagonal matrix with eigenvalues on the diagonal, $\mathbf{\Lambda}$; the Schur decomposition is in this case simply $\mathbf{A} = \mathbf{Q} \cdot \mathbf{\Lambda} \cdot \mathbf{Q}^T$.

7.9.8 Singular Value

The singular value decomposition (SVD) can be used for square or nonsquare matrices and is the most general form of diagonalization. Any complex matrix $\mathbf{A}_{N \times M}$ can be factored into the SVD,

$$\mathbf{A}_{N \times M} = \mathbf{Q}_{N \times N} \cdot \mathbf{\Sigma}_{N \times M} \cdot \mathbf{Q}_{M \times M}^H, \tag{7.418}$$

where $\mathbf{Q}_{N \times N}$ and $\mathbf{Q}_{M \times M}^H$ are orthogonal (unitary, if complex) matrices, and $\mathbf{\Sigma}$ has positive numbers σ_i, $(i = 1, 2, \ldots, r)$ in the first r positions on the main diagonal, and zero everywhere else. It turns out that r is the rank of $\mathbf{A}_{N \times M}$. The columns of $\mathbf{Q}_{N \times N}$ are the eigenvectors of $\mathbf{A}_{N \times M} \cdot \mathbf{A}_{N \times M}^H$. The columns of $\mathbf{Q}_{M \times M}$ are the eigenvectors of $\mathbf{A}_{N \times M}^H \cdot \mathbf{A}_{N \times M}$. The values σ_i, $(i = 1, 2, \ldots, r) \in \mathbb{R}^1$ are called the *singular values* of $\mathbf{A}$. They are analogous to eigenvalues and are in fact the positive square roots of the eigenvalues of $\mathbf{A}_{N \times M} \cdot \mathbf{A}_{N \times M}^H$ or $\mathbf{A}_{N \times M}^H \cdot \mathbf{A}_{N \times M}$. Because the matrix from which the eigenvalues are drawn is Hermitian, the eigenvalues, and thus the singular values, are guaranteed real. If $\mathbf{A}$ itself is square and Hermitian, the absolute value of the eigenvalues of $\mathbf{A}$ will equal its singular values. If $\mathbf{A}$ is square and non-Hermitian, there is no simple relation between its eigenvalues and singular values.

The column vectors of $\mathbf{Q}_{N \times N}$ and $\mathbf{Q}_{M \times M}$ are even more than orthonormal: they also must be chosen in such a way that $\mathbf{A}_{N \times M} \cdot \mathbf{Q}_{M \times M}$ is a scalar multiple of $\mathbf{Q}_{N \times N}$. This comes directly from post-multiplying the general form of the singular value decomposition, Eq. (7.418), by $\mathbf{Q}_{M \times M}$. We find $\mathbf{A}_{N \times M} \cdot \mathbf{Q}_{M \times M} = \mathbf{Q}_{N \times N} \cdot \mathbf{\Sigma}_{N \times M}$. So in fact a more robust way of computing the singular value decomposition is to first compute one of the orthogonal matrices, and then compute the other orthogonal matrix with which the first one is consistent.

EXAMPLE 7.46

Find the singular value decomposition of the matrix from Example 7.35,

$$\mathbf{A}_{2 \times 3} = \begin{pmatrix} 1 & -3 & 2 \\ 2 & 0 & 3 \end{pmatrix}. \tag{7.419}$$

The matrix is real so we do not need to consider the conjugate transpose; we will retain the notation for generality though here the ordinary transpose would suffice. First consider $\mathbf{A} \cdot \mathbf{A}^H$:

$$\mathbf{A} \cdot \mathbf{A}^H = \underbrace{\begin{pmatrix} 1 & -3 & 2 \\ 2 & 0 & 3 \end{pmatrix}}_{\mathbf{A}} \underbrace{\begin{pmatrix} 1 & 2 \\ -3 & 0 \\ 2 & 3 \end{pmatrix}}_{\mathbf{A}^H} = \begin{pmatrix} 14 & 8 \\ 8 & 13 \end{pmatrix}. \tag{7.420}$$

The diagonal eigenvalue matrix and corresponding orthogonal matrix composed of the normalized eigenvectors in the columns are

$$\mathbf{\Lambda}_{2\times 2} = \begin{pmatrix} 21.5156 & 0 \\ 0 & 5.48439 \end{pmatrix}, \quad \mathbf{Q}_{2\times 2} = \begin{pmatrix} 0.728827 & -0.684698 \\ 0.684698 & 0.728827 \end{pmatrix}. \tag{7.421}$$

Next we consider $\mathbf{A}^H \cdot \mathbf{A}$:

$$\mathbf{A}^H \cdot \mathbf{A} = \underbrace{\begin{pmatrix} 1 & 2 \\ -3 & 0 \\ 2 & 3 \end{pmatrix}}_{\mathbf{A}^H} \underbrace{\begin{pmatrix} 1 & -3 & 2 \\ 2 & 0 & 3 \end{pmatrix}}_{\mathbf{A}} = \begin{pmatrix} 5 & -3 & 8 \\ -3 & 9 & -6 \\ 8 & -6 & 13 \end{pmatrix}. \tag{7.422}$$

The diagonal eigenvalue matrix and corresponding orthogonal matrix composed of the normalized eigenvectors in the columns are

$$\mathbf{\Lambda}_{3\times 3} = \begin{pmatrix} 21.52 & 0 & 0 \\ 0 & 5.484 & 0 \\ 0 & 0 & 0 \end{pmatrix}, \quad \mathbf{Q}_{3\times 3} = \begin{pmatrix} 0.4524 & 0.3301 & -0.8285 \\ -0.4714 & 0.8771 & 0.09206 \\ 0.7571 & 0.3489 & 0.5523 \end{pmatrix}. \tag{7.423}$$

We take

$$\mathbf{\Sigma}_{2\times 3} = \begin{pmatrix} \sqrt{21.52} & 0 & 0 \\ 0 & \sqrt{5.484} & 0 \end{pmatrix} = \begin{pmatrix} 4.639 & 0 & 0 \\ 0 & 2.342 & 0 \end{pmatrix} \tag{7.424}$$

and can easily verify that

$$\mathbf{Q}_{2\times 2} \cdot \mathbf{\Sigma}_{2\times 3} \cdot \mathbf{Q}_{3\times 3}^H$$

$$= \underbrace{\begin{pmatrix} 0.7288 & -0.6847 \\ 0.6847 & 0.7288 \end{pmatrix}}_{\mathbf{Q}_{2\times 2}} \underbrace{\begin{pmatrix} 4.639 & 0 & 0 \\ 0 & 2.342 & 0 \end{pmatrix}}_{\mathbf{\Sigma}_{2\times 3}} \underbrace{\begin{pmatrix} 0.4524 & -0.4714 & 0.7571 \\ 0.3301 & 0.8771 & 0.3489 \\ -0.8285 & 0.09206 & 0.5523 \end{pmatrix}}_{\mathbf{Q}_{3\times 3}^H}, \tag{7.425}$$

$$= \begin{pmatrix} 1 & -3 & 2 \\ 2 & 0 & 3 \end{pmatrix} = \mathbf{A}_{2\times 3}. \tag{7.426}$$

The singular values here are $\sigma_1 = 4.639$, $\sigma_2 = 2.342$. As an aside, both $\det \mathbf{Q}_{2\times 2} = 1$ and $\det \mathbf{Q}_{3\times 3} = 1$, so they are orientation-preserving.

Let us see how we can get another singular value decomposition of the same matrix. Here we will employ the more robust technique of computing the decomposition. The orthogonal matrices $\mathbf{Q}_{3\times 3}$ and $\mathbf{Q}_{2\times 2}$ are not unique as one can multiply any row or column by -1 and still maintain orthonormality. For example, instead of the value found earlier, let us presume that we found

$$\mathbf{Q}_{3\times 3} = \begin{pmatrix} -0.4524 & 0.3301 & -0.8285 \\ 0.4714 & 0.8771 & 0.09206 \\ -0.7571 & 0.3489 & 0.5523 \end{pmatrix}. \tag{7.427}$$

Here the first column of the original $\mathbf{Q}_{3\times 3}$ has been multiplied by -1. If we used this new $\mathbf{Q}_{3\times 3}$ in conjunction with the previously found matrices to form $\mathbf{Q}_{2\times 2} \cdot \mathbf{A}_{2\times 3} \cdot \mathbf{Q}_{3\times 3}^H$, we would not recover $\mathbf{A}_{2\times3}$. The more robust way is to take

$$\mathbf{A}_{2\times3} = \mathbf{Q}_{2\times2} \cdot \boldsymbol{\Sigma}_{2\times3} \cdot \mathbf{Q}^H_{3\times3}, \tag{7.428}$$

$$\mathbf{A}_{2\times3} \cdot \mathbf{Q}_{3\times3} = \mathbf{Q}_{2\times2} \cdot \boldsymbol{\Sigma}_{2\times3}, \tag{7.429}$$

$$\underbrace{\begin{pmatrix} 1 & -3 & 2 \\ 2 & 0 & 3 \end{pmatrix}}_{\mathbf{A}_{2\times3}} \underbrace{\begin{pmatrix} -0.4524 & 0.3301 & -0.8285 \\ 0.4714 & 0.8771 & 0.09206 \\ -0.7571 & 0.3489 & 0.5523 \end{pmatrix}}_{\mathbf{Q}_{3\times3}} = \underbrace{\begin{pmatrix} q_{11} & q_{12} \\ q_{21} & q_{22} \end{pmatrix}}_{\mathbf{Q}_{2\times2}} \underbrace{\begin{pmatrix} 4.639 & 0 & 0 \\ 0 & 2.342 & 0 \end{pmatrix}}_{\boldsymbol{\Sigma}_{2\times3}},$$

$$\tag{7.430}$$

$$\begin{pmatrix} -3.381 & -1.603 & 0 \\ -3.176 & 1.707 & 0 \end{pmatrix} = \begin{pmatrix} 4.639q_{11} & 2.342q_{12} & 0 \\ 4.639q_{21} & 2.342q_{22} & 0 \end{pmatrix}. \tag{7.431}$$

Solving for q_{ij}, we find that

$$\mathbf{Q}_{2\times2} = \begin{pmatrix} -0.7288 & -0.6847 \\ -0.6847 & 0.7288 \end{pmatrix}. \tag{7.432}$$

It is easily seen that this version of $\mathbf{Q}_{2\times2}$ differs from the first version by a sign change in the first column. Direct substitution shows that the new decomposition also recovers $\mathbf{A}_{2\times3}$:

$$\mathbf{Q}_{2\times2} \cdot \boldsymbol{\Sigma}_{2\times3} \cdot \mathbf{Q}^H_{3\times3}$$

$$= \underbrace{\begin{pmatrix} -0.7288 & -0.6847 \\ -0.6847 & 0.7288 \end{pmatrix}}_{\mathbf{Q}_{2\times2}} \underbrace{\begin{pmatrix} 4.639 & 0 & 0 \\ 0 & 2.342 & 0 \end{pmatrix}}_{\boldsymbol{\Sigma}_{2\times3}} \underbrace{\begin{pmatrix} -0.4524 & 0.4714 & -0.7571 \\ 0.3301 & 0.8771 & 0.3489 \\ -0.8285 & 0.09206 & 0.5523 \end{pmatrix}}_{\mathbf{Q}^H_{3\times3}} \tag{7.433}$$

$$= \begin{pmatrix} 1 & -3 & 2 \\ 2 & 0 & 3 \end{pmatrix} = \mathbf{A}_{2\times3}. \tag{7.434}$$

Both of the orthogonal matrices $\mathbf{Q}$ used in this section have determinant of -1, so they do not preserve orientation.

EXAMPLE 7.47

The singular value decomposition of another matrix considered in earlier examples (e.g., Example 7.17), is as follows:

$$\mathbf{A} = \begin{pmatrix} 0 & -1 \\ 1 & -1 \end{pmatrix}, \tag{7.435}$$

$$= \underbrace{\begin{pmatrix} \sqrt{\frac{2}{5+\sqrt{5}}} & -\sqrt{\frac{2}{5-\sqrt{5}}} \\ \sqrt{\frac{2}{5-\sqrt{5}}} & \sqrt{\frac{2}{5+\sqrt{5}}} \end{pmatrix}}_{\mathbf{Q}_2} \underbrace{\begin{pmatrix} \sqrt{\frac{1}{2}(3+\sqrt{5})} & 0 \\ 0 & \sqrt{\frac{1}{2}(3-\sqrt{5})} \end{pmatrix}}_{\boldsymbol{\Sigma}} \underbrace{\begin{pmatrix} \sqrt{\frac{2}{5+\sqrt{5}}} & -\sqrt{\frac{2}{5-\sqrt{5}}} \\ \sqrt{\frac{2}{5-\sqrt{5}}} & \sqrt{\frac{2}{5+\sqrt{5}}} \end{pmatrix}}_{\mathbf{Q}^T_1}. \tag{7.436}$$

The singular value decomposition here is $\mathbf{A} = \mathbf{Q}_2 \cdot \boldsymbol{\Sigma} \cdot \mathbf{Q}^T_1$. All matrices are 2×2, because $\mathbf{A}$ is square of dimension 2×2. Interestingly, $\mathbf{Q}_2 = \mathbf{Q}^T_1$. Both induce a counterclockwise rotation of $\alpha = \arcsin\sqrt{2/(5 - \sqrt{5})} = \pi/3.0884 = 58.2°$. We also have $\det \mathbf{Q}_1 = \det \mathbf{Q}_2 = \|\mathbf{Q}_2\|_2 = \|\mathbf{Q}_1\|_2 = 1$. Thus, both are pure rotations. By inspection, $\|\boldsymbol{\Sigma}\|_2 = \|\mathbf{A}\|_2 = \sqrt{(3 + \sqrt{5})/2} = 1.61803$.

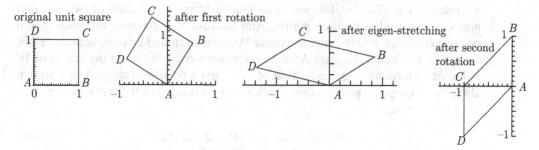

Figure 7.11. Unit square transforming via rotation, stretching, and rotation of the singular value decomposition under a linear area- and orientation-preserving alibi mapping.

The action of this composition of matrix operations on a unit square is depicted in Figure 7.11. The first rotation is induced by Q_1^T. This is followed by an eigenstretching of Σ. The action is completed by a rotation induced by Q_2.

It is also easily shown that the singular values of a square Hermitian matrix are identical to the eigenvalues of that matrix. The singular values of a square non-Hermitian matrix are not, in general, the eigenvalues of that matrix.

7.9.9 Polar

A square invertible, potentially non-Hermitian matrix A with $\det A > 0$ has the following two polar decompositions:

$$A = Q \cdot W, \tag{7.437}$$

$$A = V \cdot Q. \tag{7.438}$$

We take Q to be an orthogonal (or unitary) rotation matrix, and both W and V are symmetric (Hermitian) positive definite square matrices. Equation (7.437) is known as the right polar decomposition, and Eq. (7.438) is known as the left polar decomposition. Both are easily derived from the singular value decomposition.

Let us take the square, non-Hermitian A to have the following singular value decomposition:

$$A = Q_2 \cdot \Sigma \cdot Q_1^T. \tag{7.439}$$

For convenience, we construct both Q_1 and Q_2 so they are rotation matrices. Because A is square, it has the same dimensions as Q_1, Σ, and Q_2. When A is asymmetric, $Q_1 \neq Q_2$. Now, recalling that $Q_1^T \cdot Q_1 = I$, we can recast Eq. (7.439) as

$$A = Q_2 \cdot \underbrace{Q_1^T \cdot Q_1}_{I} \cdot \Sigma \cdot Q_1^T, \tag{7.440}$$

$$= \underbrace{Q_2 \cdot Q_1^T}_{=Q} \cdot \underbrace{Q_1 \cdot \Sigma \cdot Q_1^T}_{=W}. \tag{7.441}$$

If we take

$$Q = Q_2 \cdot Q_1^T, \tag{7.442}$$

$$W = Q_1 \cdot \Sigma \cdot Q_1^T, \tag{7.443}$$

we recover Eq. (7.437). Because $\mathbf{Q}$ is formed from the product of two rotation matrices, it is itself a rotation matrix. And because $\mathbf{\Sigma}$ is diagonal with guaranteed positive entries, it and the similar matrix $\mathbf{W}$ are guaranteed positive definite. Thus, the action of a general matrix $\mathbf{A}$ on a vector is the composition of the action of $\mathbf{W}$ followed by a rotation $\mathbf{Q}$. The action of $\mathbf{W}$ is first a rotation, followed by a stretch along the orthogonal principal axes, followed by a reverse of the first rotation.

Further note that

$$\mathbf{W} \cdot \mathbf{W} = \mathbf{Q}_1 \cdot \mathbf{\Sigma} \cdot \underbrace{\mathbf{Q}_1^T \cdot \mathbf{Q}_1}_{=\mathbf{I}} \cdot \mathbf{\Sigma} \cdot \mathbf{Q}_1^T, \tag{7.444}$$

$$= \mathbf{Q}_1 \cdot \mathbf{\Sigma} \cdot \mathbf{\Sigma} \cdot \mathbf{Q}_1^T. \tag{7.445}$$

Noting that $\mathbf{\Sigma} \cdot \mathbf{\Sigma}$ is diagonal, the right side of Eq. (7.445) is nothing more than the diagonalization of $\mathbf{A}^T \cdot \mathbf{A}$; thus,

$$\mathbf{W} \cdot \mathbf{W} = \mathbf{A}^T \cdot \mathbf{A} \equiv \mathbf{C}. \tag{7.446}$$

We extend the notion of a square root to matrix operators which are positive semi-definite such as $\mathbf{A}^T \cdot \mathbf{A}$, which allows us to derive from Eq. (7.446) the equation

$$\mathbf{W} = \sqrt{\mathbf{A}^T \cdot \mathbf{A}}. \tag{7.447}$$

When then matrix $\mathbf{A}$ is associated with coordinate transformations, the matrix $\mathbf{C}$ is known as the *right Cauchy-Green tensor*. It is the equivalent of the metric tensor defined in Eq. (1.257).

The analysis is similar for the left polar decomposition. Recalling that $\mathbf{Q}_2^T \cdot \mathbf{Q}_2 = \mathbf{I}$, we recast Eq. (7.439) as

$$\mathbf{A} = \mathbf{Q}_2 \cdot \mathbf{\Sigma} \cdot \underbrace{\mathbf{Q}_2^T \cdot \mathbf{Q}_2}_{\mathbf{I}} \cdot \mathbf{Q}_1^T, \tag{7.448}$$

$$= \underbrace{\mathbf{Q}_2 \cdot \mathbf{\Sigma} \cdot \mathbf{Q}_2^T}_{=\mathbf{V}} \cdot \underbrace{\mathbf{Q}_2 \cdot \mathbf{Q}_1^T}_{=\mathbf{Q}}. \tag{7.449}$$

If we take

$$\mathbf{V} = \mathbf{Q}_2 \cdot \mathbf{\Sigma} \cdot \mathbf{Q}_2^T \tag{7.450}$$

and $\mathbf{Q}$ as in Eq. (7.442), we recover the left polar decomposition of Eq. (7.438). Similarly, it is easy to show that

$$\mathbf{V} \cdot \mathbf{V} = \mathbf{A} \cdot \mathbf{A}^T \equiv \mathbf{B}. \tag{7.451}$$

When $\mathbf{A}$ is associated with coordinate transformations, the matrix $\mathbf{B}$ is known as the *left Cauchy-Green tensor* and can be related to the metric tensor. We also see that

$$\mathbf{V} = \sqrt{\mathbf{A} \cdot \mathbf{A}^T}. \tag{7.452}$$

When $\mathbf{A}$ is symmetric, $\mathbf{Q}_1 = \mathbf{Q}_2$, $\mathbf{Q} = \mathbf{I}$, and $\mathbf{W} = \mathbf{V}$; that is, there is no distinction between the left and right polar decompositions. If $\det \mathbf{A} < 0$, a similar decomposition can be made; however, the matrix $\mathbf{Q}$ may additionally require reflection, $\det \mathbf{Q} = -1$, and is not a pure rotation. For complex $\mathbf{A}$, similar decompositions can also be made, but the geometric interpretation is more difficult.

The *Green–St. Venant*[9] tensor $\mathbf{E}$ is defined as

$$\mathbf{E} = \frac{1}{2}\left(\mathbf{A}^T \cdot \mathbf{A} - \mathbf{I}\right). \tag{7.453}$$

If $\mathbf{A}$ is a pure rotation $\mathbf{Q}$, then $\mathbf{E} = \mathbf{0}$.

EXAMPLE 7.48

Find the right and left polar decompositions of the matrix considered in the related Example 7.47:

$$\mathbf{A} = \begin{pmatrix} 0 & -1 \\ 1 & -1 \end{pmatrix}. \tag{7.454}$$

Here we have a square, invertible, asymmetric matrix with $\det \mathbf{A} = 1$, so its action is area- and orientation-preserving. The relevant singular value decomposition has been identified in the earlier example. For the right polar decomposition, we can then easily form the rotation matrix

$$\mathbf{Q} = \mathbf{Q}_2 \cdot \mathbf{Q}_1^T = \begin{pmatrix} -\frac{1}{\sqrt{5}} & -\frac{2}{\sqrt{5}} \\ \frac{2}{\sqrt{5}} & -\frac{1}{\sqrt{5}} \end{pmatrix}. \tag{7.455}$$

We can also find

$$\mathbf{W} = \mathbf{Q}_1 \cdot \boldsymbol{\Sigma} \cdot \mathbf{Q}_1^T = \begin{pmatrix} \frac{2}{\sqrt{5}} & -\frac{1}{\sqrt{5}} \\ -\frac{1}{\sqrt{5}} & \frac{3}{\sqrt{5}} \end{pmatrix}. \tag{7.456}$$

The right Cauchy-Green tensor is

$$\mathbf{C} = \mathbf{W} \cdot \mathbf{W} = \mathbf{A}^T \cdot \mathbf{A} = \begin{pmatrix} 1 & -1 \\ -1 & 2 \end{pmatrix}. \tag{7.457}$$

It is easily verified that $\mathbf{Q} \cdot \mathbf{W} = \mathbf{A}$. It is also easily verified that $||\mathbf{Q}||_2 = 1$ and $||\mathbf{W}||_2 = ||\mathbf{A}||_2 = ||\boldsymbol{\Sigma}||_2 = \sqrt{(3 + \sqrt{5})/2}$.
One can also find

$$\mathbf{V} = \mathbf{Q}_2 \cdot \boldsymbol{\Sigma} \cdot \mathbf{Q}_2^T = \begin{pmatrix} \frac{2}{\sqrt{5}} & \frac{1}{\sqrt{5}} \\ \frac{1}{\sqrt{5}} & \frac{3}{\sqrt{5}} \end{pmatrix}. \tag{7.458}$$

The left Cauchy-Green tensor is

$$\mathbf{B} = \mathbf{V} \cdot \mathbf{V} = \mathbf{A} \cdot \mathbf{A}^T = \begin{pmatrix} 1 & 1 \\ 1 & 2 \end{pmatrix}. \tag{7.459}$$

It is easily confirmed that $\mathbf{V} \cdot \mathbf{Q} = \mathbf{A}$. It is also easily verified that $||\mathbf{Q}||_2 = 1$ and $||\mathbf{V}||_2 = ||\mathbf{A}||_2 = \sqrt{(3 + \sqrt{5})/2}$.
Lastly, we see the Green–St. Venant tensor is

$$\mathbf{E} = \frac{1}{2}\left(\underbrace{\begin{pmatrix} 0 & 1 \\ -1 & -1 \end{pmatrix}}_{\mathbf{A}^T} \underbrace{\begin{pmatrix} 0 & -1 \\ 1 & -1 \end{pmatrix}}_{\mathbf{A}} - \underbrace{\begin{pmatrix} 1 & 0 \\ 0 & 1 \end{pmatrix}}_{\mathbf{I}} \right) = \begin{pmatrix} 0 & -\frac{1}{2} \\ -\frac{1}{2} & \frac{1}{2} \end{pmatrix}. \tag{7.460}$$

It is a symmetric tensor.

---

[9] Adhémar Jean Claude Barré de Saint-Venant, 1797–1886, French mechanician.

EXAMPLE 7.49

Find the left polar decomposition of the 1×1 matrix

$$\mathbf{A} = (1+i).\tag{7.461}$$

We can by inspection rewrite this as

$$\mathbf{A} = (\sqrt{2}e^{i\pi/4}) = \underbrace{(\sqrt{2})}_{\mathbf{V}}\underbrace{(e^{i\pi/4})}_{\mathbf{Q}}.\tag{7.462}$$

One can see the analogy between the polar decomposition of a complex scalar and that of a matrix. Clearly $\mathbf{V}$ is positive definite and Hermitian, and $\mathbf{Q}$ is a unitary matrix.

7.9.10 Hessenberg

A square matrix $\mathbf{A}$ can be decomposed into Hessenberg[10] form

$$\mathbf{A} = \mathbf{Q} \cdot \mathbf{H} \cdot \mathbf{Q}^T,\tag{7.463}$$

where $\mathbf{Q}$ is an orthogonal (or unitary) matrix and $\mathbf{H}$ has zeros below the first subdiagonal. When $\mathbf{A}$ is Hermitian, $\mathbf{Q}$ is tridiagonal, which is easy to invert numerically. Also, $\mathbf{H}$ has the same eigenvalues as $\mathbf{A}$. Here the $\mathbf{H}$ of the Hessenberg form is not to be confused with the Hessian matrix, which often is denoted by the same symbol; see Eq. (1.148).

EXAMPLE 7.50

The Hessenberg form of our example square matrix $\mathbf{A}$ from Example 7.32 is

$$\mathbf{A} = \begin{pmatrix} -5 & 4 & 9 \\ -22 & 14 & 18 \\ 16 & -8 & -6 \end{pmatrix} = \mathbf{Q} \cdot \mathbf{H} \cdot \mathbf{Q}^T,\tag{7.464}$$

$$= \underbrace{\begin{pmatrix} 1 & 0 & 0 \\ 0 & -0.8087 & 0.5882 \\ 0 & 0.5882 & 0.8087 \end{pmatrix}}_{\mathbf{Q}} \underbrace{\begin{pmatrix} -5 & 2.0586 & 9.6313 \\ 27.2029 & 2.3243 & -24.0451 \\ 0 & 1.9459 & 5.6757 \end{pmatrix}}_{\mathbf{H}} \underbrace{\begin{pmatrix} 1 & 0 & 0 \\ 0 & -0.8087 & 0.5882 \\ 0 & 0.5882 & 0.8087 \end{pmatrix}}_{\mathbf{Q}^T}.$$
$$\tag{7.465}$$

The matrix $\mathbf{Q}$ found here has determinant of -1; it could be easily recalculated to arrive at an orientation-preserving value of $+1$.

7.10 Projection Matrix

Here we consider a topic discussed earlier in a broader context, the projection matrix defined in Eq. (6.191). The vector $\mathbf{A} \cdot \mathbf{x}$ belongs to the column space of $\mathbf{A}$. Here $\mathbf{A}$ is not necessarily square. Consider the equation $\mathbf{A} \cdot \mathbf{x} = \mathbf{b}$, where $\mathbf{A}$ and $\mathbf{b}$ are given. If the given vector $\mathbf{b}$ does not lie in the column space of $\mathbf{A}$, the equation cannot be solved for $\mathbf{x}$. Still, we would like to find $\mathbf{x}_p$ such that

$$\mathbf{A} \cdot \mathbf{x}_p = \mathbf{b}_p,\tag{7.466}$$

[10] Karl Hessenberg, 1904–1959, German mathematician and engineer.

which does lie in the column space of $\mathbf{A}$, such that $\mathbf{b}_p$ is the projection of $\mathbf{b}$ onto the column space. The residual vector from Eq. (7.5) is also expressed as

$$\mathbf{r} = \mathbf{b}_p - \mathbf{b}. \tag{7.467}$$

For a projection, this residual should be orthogonal to all vectors $\mathbf{A} \cdot \mathbf{z}$ that belong to the column space, where the components of $\mathbf{z}$ are arbitrary. Enforcing this condition, we get

$$0 = (\mathbf{A} \cdot \mathbf{z})^T \cdot \mathbf{r}, \tag{7.468}$$

$$= (\mathbf{A} \cdot \mathbf{z})^T \cdot \underbrace{(\mathbf{b}_p - \mathbf{b})}_{\mathbf{r}}, \tag{7.469}$$

$$= \mathbf{z}^T \cdot \mathbf{A}^T \cdot \underbrace{(\mathbf{A} \cdot \mathbf{x}_p - \mathbf{b})}_{\mathbf{b}_p}, \tag{7.470}$$

$$= \mathbf{z}^T \cdot (\mathbf{A}^T \cdot \mathbf{A} \cdot \mathbf{x}_p - \mathbf{A}^T \cdot \mathbf{b}). \tag{7.471}$$

Because $\mathbf{z}$ is an arbitrary vector,

$$\mathbf{A}^T \cdot \mathbf{A} \cdot \mathbf{x}_p - \mathbf{A}^T \cdot \mathbf{b} = 0, \tag{7.472}$$

from which

$$\mathbf{A}^T \cdot \mathbf{A} \cdot \mathbf{x}_p = \mathbf{A}^T \cdot \mathbf{b}, \tag{7.473}$$

$$\mathbf{x}_p = (\mathbf{A}^T \cdot \mathbf{A})^{-1} \cdot \mathbf{A}^T \cdot \mathbf{b}, \tag{7.474}$$

$$\mathbf{A} \cdot \mathbf{x}_p = \mathbf{A} \cdot (\mathbf{A}^T \cdot \mathbf{A})^{-1} \cdot \mathbf{A}^T \cdot \mathbf{b}, \tag{7.475}$$

$$\mathbf{b}_p = \underbrace{\mathbf{A} \cdot (\mathbf{A}^T \cdot \mathbf{A})^{-1} \cdot \mathbf{A}^T}_{\equiv \mathbf{P}} \cdot \mathbf{b}. \tag{7.476}$$

This is equivalent to that given in Eq. (6.191). The projection matrix $\mathbf{P}$ defined by $\mathbf{b}_p = \mathbf{P} \cdot \mathbf{b}$ is

$$\mathbf{P} = \mathbf{A} \cdot (\mathbf{A}^T \cdot \mathbf{A})^{-1} \cdot \mathbf{A}^T. \tag{7.477}$$

The projection matrix for an operator $\mathbf{A}$, when operating on an arbitrary vector $\mathbf{b}$, yields the projection of $\mathbf{b}$ onto the column space of $\mathbf{A}$. Many vectors $\mathbf{b}$ could have the same projection onto the column space of $\mathbf{A}$. It can be shown that an $N \times N$ matrix $\mathbf{P}$ is a projection matrix iff $\mathbf{P} \cdot \mathbf{P} = \mathbf{P}$. Because of this, the projection matrix is idempotent: $\mathbf{P} \cdot \mathbf{x} = \mathbf{P} \cdot \mathbf{P} \cdot \mathbf{x} = \cdots = \mathbf{P}^n \cdot \mathbf{x}$. Moreover, the rank of $\mathbf{P}$ is its trace.

EXAMPLE 7.51

Determine and analyze the projection matrix associated with projecting a vector $\mathbf{b} \in \mathbb{R}^3$ onto the two-dimensional space spanned by the basis vectors $(1, 2, 3)^T$ and $(1, 1, 1)^T$.

We form the matrix $\mathbf{A}$ by populating its columns with the basis vectors. So

$$\mathbf{A} = \begin{pmatrix} 1 & 1 \\ 2 & 1 \\ 3 & 1 \end{pmatrix}. \tag{7.478}$$

Then we find the projection matrix $\mathbf{P}$ via Eq. (7.477):

$$\mathbf{P} = \begin{pmatrix} 1 & 1 \\ 2 & 1 \\ 3 & 1 \end{pmatrix} \left(\begin{pmatrix} 1 & 2 & 3 \\ 1 & 1 & 1 \end{pmatrix} \begin{pmatrix} 1 & 1 \\ 2 & 1 \\ 3 & 1 \end{pmatrix} \right)^{-1} \begin{pmatrix} 1 & 2 & 3 \\ 1 & 1 & 1 \end{pmatrix} = \begin{pmatrix} \frac{5}{6} & \frac{1}{3} & -\frac{1}{6} \\ \frac{1}{3} & \frac{1}{3} & \frac{1}{3} \\ -\frac{1}{6} & \frac{1}{3} & \frac{5}{6} \end{pmatrix}. \tag{7.479}$$

By inspection **P** is self-adjoint, thus it is guaranteed to possess real eigenvalues, which are $\lambda = 1, 1, 0$. There is one nonzero eigenvalue for each of the two linearly independent basis vectors that form **A**. It is easily shown that $||\mathbf{P}||_2 = 1$, $\rho(\mathbf{P}) = 1$, and $\det \mathbf{P} = 0$. Thus, **P** is singular. This is because it maps vectors in three-space to two-space. The rank of **P** is 2 as is its trace. As required of all projection matrices, $\mathbf{P} \cdot \mathbf{P} = \mathbf{P}$:

$$
\begin{pmatrix} \frac{5}{6} & \frac{1}{3} & -\frac{1}{6} \\ \frac{1}{3} & \frac{1}{3} & \frac{1}{3} \\ -\frac{1}{6} & \frac{1}{3} & \frac{5}{6} \end{pmatrix} \begin{pmatrix} \frac{5}{6} & \frac{1}{3} & -\frac{1}{6} \\ \frac{1}{3} & \frac{1}{3} & \frac{1}{3} \\ -\frac{1}{6} & \frac{1}{3} & \frac{5}{6} \end{pmatrix} = \begin{pmatrix} \frac{5}{6} & \frac{1}{3} & -\frac{1}{6} \\ \frac{1}{3} & \frac{1}{3} & \frac{1}{3} \\ -\frac{1}{6} & \frac{1}{3} & \frac{5}{6} \end{pmatrix}.
\tag{7.480}
$$

That is to say, **P** is idempotent.

It is easily shown that the singular value decomposition of **P** is equivalent to a diagonalization, giving

$$
\mathbf{P} = \mathbf{Q} \cdot \mathbf{\Lambda} \cdot \mathbf{Q}^T = \begin{pmatrix} \frac{1}{\sqrt{2}} & \frac{1}{\sqrt{3}} & \frac{1}{\sqrt{6}} \\ 0 & \frac{1}{\sqrt{3}} & -\sqrt{\frac{2}{3}} \\ -\frac{1}{\sqrt{2}} & \frac{1}{\sqrt{3}} & \frac{1}{\sqrt{6}} \end{pmatrix} \begin{pmatrix} 1 & 0 & 0 \\ 0 & 1 & 0 \\ 0 & 0 & 0 \end{pmatrix} \begin{pmatrix} \frac{1}{\sqrt{2}} & 0 & -\frac{1}{\sqrt{2}} \\ \frac{1}{\sqrt{3}} & \frac{1}{\sqrt{3}} & \frac{1}{\sqrt{3}} \\ \frac{1}{\sqrt{6}} & -\sqrt{\frac{2}{3}} & \frac{1}{\sqrt{6}} \end{pmatrix}.
\tag{7.481}
$$

The matrix **Q** has $||\mathbf{Q}||_2 = 1$ and $\det \mathbf{Q} = 1$, so it is a true rotation. Thus, when **P** is applied to a vector **b** to obtain $\mathbf{b}_p$, we can consider **b** to be first rotated into the configuration aligned with the two basis vectors via application of $\mathbf{Q}^T$. Then in this configuration, one of the modes of **b** is suppressed via application of $\mathbf{\Lambda}$, while the other two modes are preserved. The result is returned to its original configuration via application of **Q**, which precisely provides a counterrotation to $\mathbf{Q}^T$. Also, the decomposition is equivalent to that previously discussed in Section 7.9.4; here $\mathbf{\Lambda} = \mathbf{U} \cdot \left(\mathbf{U}^T \cdot \mathbf{U} \right)^{-1} \cdot \mathbf{U}^T$, where **U** is as was defined there. Specifically, **P** has rank $r = 2$ and that $\mathbf{\Lambda}$ has $r = 2$ values of unity on its diagonal.

7.11 Least Squares

One important application of projection matrices is the method of least squares. This method is often used to fit data to a given functional form. The form is most often in terms of polynomials, but there is absolutely no restriction; trigonometric functions, logarithmic functions, or Bessel functions can all serve as well. Now if one has, say, 10 data points, one can in principle find a ninth-order polynomial that will pass through all the data points. Often, especially when there is much experimental error in the data, such a function may be subject to wild oscillations, which are unwarranted by the underlying physics, and thus is not useful as a predictive tool. In such cases, it may be more useful to choose a lower-order curve that does not exactly pass through all experimental points but that does minimize the residual.

In this method, one

- examines the data,
- makes a nonunique judgment of what the functional form might be,
- substitutes each data point into the assumed form so as to form an overconstrained system of linear equations,
- uses the technique associated with projection matrices to solve for the coefficients that best represent the given data.

7.11.1 Unweighted

This is the most common method used when one has equal confidence in all the data.

EXAMPLE 7.52

Find the best straight line to approximate the measured data relating x to t:

$$
\begin{array}{c|c}
t & x \\
\hline
0 & 5 \\
1 & 7 \\
2 & 10 \\
3 & 12 \\
6 & 15 \\
\end{array}
\tag{7.482}
$$

A straight line fit will have the form

$$x = a_0 + a_1 t, \tag{7.483}$$

where a_0 and a_1 are the terms to be determined. Substituting each data point to the assumed form, we get five equations in two unknowns:

$$5 = a_0 + 0a_1, \tag{7.484}$$

$$7 = a_0 + 1a_1, \tag{7.485}$$

$$10 = a_0 + 2a_1, \tag{7.486}$$

$$12 = a_0 + 3a_1, \tag{7.487}$$

$$15 = a_0 + 6a_1. \tag{7.488}$$

Rearranging, we get

$$
\begin{pmatrix}
1 & 0 \\
1 & 1 \\
1 & 2 \\
1 & 3 \\
1 & 6
\end{pmatrix}
\begin{pmatrix} a_0 \\ a_1 \end{pmatrix}
=
\begin{pmatrix}
5 \\
7 \\
10 \\
12 \\
15
\end{pmatrix}.
\tag{7.489}
$$

This is of the form $\mathbf{A} \cdot \mathbf{a} = \mathbf{b}$. We then find that

$$\mathbf{a} = \left(\mathbf{A}^T \cdot \mathbf{A} \right)^{-1} \cdot \mathbf{A}^T \cdot \mathbf{b}. \tag{7.490}$$

Substituting, we find that

$$
\underbrace{\begin{pmatrix} a_0 \\ a_1 \end{pmatrix}}_{\mathbf{a}}
=
\left(
\underbrace{\begin{pmatrix} 1 & 1 & 1 & 1 & 1 \\ 0 & 1 & 2 & 3 & 6 \end{pmatrix}}_{\mathbf{A}^T}
\underbrace{\begin{pmatrix} 1 & 0 \\ 1 & 1 \\ 1 & 2 \\ 1 & 3 \\ 1 & 6 \end{pmatrix}}_{\mathbf{A}}
\right)^{-1}
\underbrace{\begin{pmatrix} 1 & 1 & 1 & 1 & 1 \\ 0 & 1 & 2 & 3 & 6 \end{pmatrix}}_{\mathbf{A}^T}
\underbrace{\begin{pmatrix} 5 \\ 7 \\ 10 \\ 12 \\ 15 \end{pmatrix}}_{\mathbf{b}}
=
\begin{pmatrix} 5.7925 \\ 1.6698 \end{pmatrix}.
\tag{7.491}
$$

So the best fit estimate is

$$x = 5.7925 + 1.6698t. \tag{7.492}$$

The Euclidean norm of the residual is $\|\mathbf{A} \cdot \mathbf{a} - \mathbf{b}\|_2 = 1.9206$. This represents the ℓ_2 residual of the prediction. A plot of the raw data and the best fit straight line is shown in Figure 7.12.

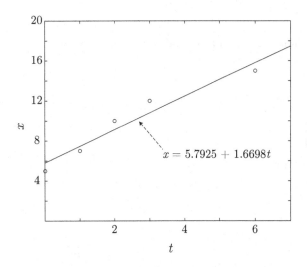

Figure 7.12. Plot of x, t data and best least squares straight line fit.

7.11.2 Weighted

If one has more confidence in some data points than others, one can define a weighting function to give more priority to those particular data points.

EXAMPLE 7.53

Find the best straight line fit for the data in the previous example. Now, however, assume that we have five times the confidence in the accuracy of the final two data points, relative to the other points. Define a square weighting matrix $\mathbf{W}$:

$$\mathbf{W} = \begin{pmatrix} 1 & 0 & 0 & 0 & 0 \\ 0 & 1 & 0 & 0 & 0 \\ 0 & 0 & 1 & 0 & 0 \\ 0 & 0 & 0 & 5 & 0 \\ 0 & 0 & 0 & 0 & 5 \end{pmatrix}. \tag{7.493}$$

Now we perform the following operations:

$$\mathbf{A} \cdot \mathbf{a} = \mathbf{b}, \tag{7.494}$$

$$\mathbf{W} \cdot \mathbf{A} \cdot \mathbf{a} = \mathbf{W} \cdot \mathbf{b}, \tag{7.495}$$

$$(\mathbf{W} \cdot \mathbf{A})^T \cdot \mathbf{W} \cdot \mathbf{A} \cdot \mathbf{a} = (\mathbf{W} \cdot \mathbf{A})^T \cdot \mathbf{W} \cdot \mathbf{b}, \tag{7.496}$$

$$\mathbf{a} = \left((\mathbf{W} \cdot \mathbf{A})^T \cdot \mathbf{W} \cdot \mathbf{A} \right)^{-1} (\mathbf{W} \cdot \mathbf{A})^T \cdot \mathbf{W} \cdot \mathbf{b}. \tag{7.497}$$

With values of $\mathbf{W}$ from Eq. (7.493), direct substitution leads to

$$\mathbf{a} = \begin{pmatrix} a_0 \\ a_1 \end{pmatrix} = \begin{pmatrix} 8.0008 \\ 1.1972 \end{pmatrix}. \tag{7.498}$$

So the best weighted least squares fit is

$$x = 8.0008 + 1.1972\, t. \tag{7.499}$$

A plot of the raw data and the best fit straight line is shown in Figure 7.13.

When the measurements are independent and equally reliable, $\mathbf{W}$ is the identity matrix. If the measurements are independent but not equally reliable, $\mathbf{W}$ is at most

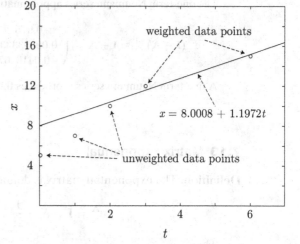

Figure 7.13. Plot of x, t data and best weighted least squares straight line fit.

diagonal. If the measurements are not independent, then nonzero terms can appear off the diagonal in $\mathbf{W}$. It is often advantageous, for instance, in problems in which one wants to control a process in real time, to give priority to recent data estimates over old data estimates and to continually employ a least squares technique to estimate future system behavior. The previous example does just that. A prominent fast algorithm for such problems is known as a *Kálmán*[11] *filter*.

7.12 Neumann Series

Matrices can often be treated in a similar fashion as scalars. For instance, for $x \in \mathbb{R}^1$, we have the well-known geometric series

$$\frac{1}{1-x} = 1 + x + x^2 + \cdots, \qquad |x| < 1. \tag{7.500}$$

Similarly, for a matrix, we have the *Neumann series*

$$(\mathbf{I} - \mathbf{A})^{-1} = \mathbf{I} + \mathbf{A} + \mathbf{A}^2 + \cdots, \qquad ||\mathbf{A}||_2 < 1. \tag{7.501}$$

To achieve a given error of approximation, fewer terms are needed as $||\mathbf{A}|| \to 0$.

EXAMPLE 7.54

Consider the matrix

$$\mathbf{A} = \begin{pmatrix} -\frac{1}{40} & \frac{1}{30} & \frac{7}{120} \\ \frac{1}{120} & \frac{1}{30} & \frac{3}{40} \\ -\frac{1}{24} & -\frac{1}{120} & -\frac{1}{60} \end{pmatrix}. \tag{7.502}$$

Find an approximation to $(\mathbf{I} - \mathbf{A})^{-1}$.

We first note that $||\mathbf{A}||_2 = 0.107586$, so we expect the Neumann series to converge in a few terms. Direct calculation reveals that

$$(\mathbf{I} - \mathbf{A})^{-1} = \begin{pmatrix} 0.973509 & 0.0330667 & 0.0582964 \\ 0.00529342 & 1.03401 & 0.0765828 \\ -0.0399413 & -0.00983064 & 0.98059 \end{pmatrix}. \tag{7.503}$$

---

[11] Rudolf Emil Kálmán, 1930–, Hungarian/American electrical engineer.

The one-term Neumann series approximation is good:

$$(\mathbf{I} - \mathbf{A})^{-1} \approx \mathbf{I} + \mathbf{A} = \begin{pmatrix} 0.975 & 0.0333333 & 0.0583333 \\ 0.00833333 & 1.03333 & 0.075 \\ -0.0416667 & -0.00833333 & 0.983333 \end{pmatrix}. \tag{7.504}$$

A five-term Neumann series reproduces the results to the original precision.

7.13 Matrix Exponential

Definition: The exponential matrix is defined as

$$e^{\mathbf{A}} = \mathbf{I} + \mathbf{A} + \frac{1}{2!}\mathbf{A}^2 + \frac{1}{3!}\mathbf{A}^3 + \cdots. \tag{7.505}$$

Thus

$$e^{\mathbf{A}t} = \mathbf{I} + \mathbf{A}t + \frac{1}{2!}\mathbf{A}^2 t^2 + \frac{1}{3!}\mathbf{A}^3 t^3 + \cdots, \tag{7.506}$$

$$\frac{d}{dt}\left(e^{\mathbf{A}t}\right) = \mathbf{A} + \mathbf{A}^2 t + \frac{1}{2!}\mathbf{A}^3 t^2 + \dots, \tag{7.507}$$

$$= \mathbf{A} \cdot \underbrace{\left(\mathbf{I} + \mathbf{A}t + \frac{1}{2!}\mathbf{A}^2 t^2 + \frac{1}{3!}\mathbf{A}^3 t^3 + \dots\right)}_{=e^{\mathbf{A}t}}, \tag{7.508}$$

$$= \mathbf{A} \cdot e^{\mathbf{A}t}. \tag{7.509}$$

Properties of the matrix exponential include

$$e^{a\mathbf{I}} = e^a \mathbf{I}, \tag{7.510}$$

$$(e^{\mathbf{A}})^{-1} = e^{-\mathbf{A}}, \tag{7.511}$$

$$e^{\mathbf{A}(t+s)} = e^{\mathbf{A}t} e^{\mathbf{A}s}. \tag{7.512}$$

But $e^{\mathbf{A}+\mathbf{B}} = e^{\mathbf{A}} e^{\mathbf{B}}$ only if $\mathbf{A} \cdot \mathbf{B} = \mathbf{B} \cdot \mathbf{A}$. Thus, $e^{t\mathbf{I}+s\mathbf{A}} = e^t e^{s\mathbf{A}}$.

EXAMPLE 7.55

Find $e^{\mathbf{A}t}$ if

$$\mathbf{A} = \begin{pmatrix} a & 1 & 0 \\ 0 & a & 1 \\ 0 & 0 & a \end{pmatrix}. \tag{7.513}$$

We have

$$\mathbf{A} = a\mathbf{I} + \mathbf{B}, \tag{7.514}$$

where

$$\mathbf{B} = \begin{pmatrix} 0 & 1 & 0 \\ 0 & 0 & 1 \\ 0 & 0 & 0 \end{pmatrix}. \tag{7.515}$$

Thus

$$\mathbf{B}^2 = \begin{pmatrix} 0 & 0 & 1 \\ 0 & 0 & 0 \\ 0 & 0 & 0 \end{pmatrix}, \tag{7.516}$$

$$\mathbf{B}^3 = \begin{pmatrix} 0 & 0 & 0 \\ 0 & 0 & 0 \\ 0 & 0 & 0 \end{pmatrix}, \tag{7.517}$$

$$\vdots$$

$$\mathbf{B}^n = \begin{pmatrix} 0 & 0 & 0 \\ 0 & 0 & 0 \\ 0 & 0 & 0 \end{pmatrix}, \quad n \geq 4. \tag{7.518}$$

Furthermore,

$$\mathbf{I} \cdot \mathbf{B} = \mathbf{B} \cdot \mathbf{I} = \mathbf{B}. \tag{7.519}$$

Thus

$$e^{\mathbf{A}t} = e^{(a\mathbf{I}+\mathbf{B})t}, \tag{7.520}$$

$$= e^{at\mathbf{I}} \cdot e^{\mathbf{B}t}, \tag{7.521}$$

$$= \underbrace{\left(\mathbf{I} + at\mathbf{I} + \frac{1}{2!}a^2t^2\mathbf{I}^2 + \frac{1}{3!}a^3t^3\mathbf{I}^3 + \cdots \right)}_{=e^{at\mathbf{I}}=e^{at}\mathbf{I}} \cdot \left(\mathbf{I} + \mathbf{B}t + \frac{1}{2!}\mathbf{B}^2t^2 + \overbrace{\frac{1}{3!}\mathbf{B}^3t^3 + \cdots}^{=0} \right),$$
$$\underbrace{\phantom{\left(\mathbf{I} + \mathbf{B}t + \frac{1}{2!}\mathbf{B}^2t^2 \right)}}_{=e^{\mathbf{B}t}} \tag{7.522}$$

$$= e^{at}\mathbf{I} \cdot \left(\mathbf{I} + \mathbf{B}t + \mathbf{B}^2\frac{t^2}{2} \right), \tag{7.523}$$

$$= e^{at} \begin{pmatrix} 1 & t & \frac{t^2}{2} \\ 0 & 1 & t \\ 0 & 0 & 1 \end{pmatrix}. \tag{7.524}$$

If $\mathbf{A}$ can be diagonalized, the calculation is simplified. Then

$$e^{\mathbf{A}t} = e^{\mathbf{S}\cdot\mathbf{\Lambda}\cdot\mathbf{S}^{-1}t} = \mathbf{I} + \mathbf{S}\cdot\mathbf{\Lambda}\cdot\mathbf{S}^{-1}t + \cdots + \frac{1}{N!}\left(\mathbf{S}\cdot\mathbf{\Lambda}\cdot\mathbf{S}^{-1}t \right)^N + \cdots. \tag{7.525}$$

Noting that

$$\left(\mathbf{S}\cdot\mathbf{\Lambda}\cdot\mathbf{S}^{-1} \right)^2 = \mathbf{S}\cdot\mathbf{\Lambda}\cdot\mathbf{S}^{-1}\cdot\mathbf{S}\cdot\mathbf{\Lambda}\cdot\mathbf{S}^{-1} = \mathbf{S}\cdot\mathbf{\Lambda}^2\cdot\mathbf{S}^{-1}, \tag{7.526}$$

$$\left(\mathbf{S}\cdot\mathbf{\Lambda}\cdot\mathbf{S}^{-1} \right)^N = \mathbf{S}\cdot\mathbf{\Lambda}\cdot\mathbf{S}^{-1}\cdots\cdots\mathbf{S}\cdot\mathbf{\Lambda}\cdot\mathbf{S}^{-1} = \mathbf{S}\cdot\mathbf{\Lambda}^N\cdot\mathbf{S}^{-1}, \tag{7.527}$$

the original expression reduces to

$$e^{\mathbf{A}t} = \mathbf{S}\cdot\left(\mathbf{I} + \mathbf{\Lambda}t + \cdots + \frac{1}{N!}\left(\mathbf{\Lambda}^Nt^N \right) + \cdots \right)\cdot\mathbf{S}^{-1}, \tag{7.528}$$

$$= \mathbf{S}\cdot e^{\mathbf{\Lambda}t}\cdot\mathbf{S}^{-1}. \tag{7.529}$$

7.14 Quadratic Form

At times one may be given a polynomial equation for which one wants to determine conditions under which the expression is positive. For example, if we have

$$f(\xi_1, \xi_2, \xi_3) = 18\xi_1^2 - 16\xi_1\xi_2 + 5\xi_2^2 + 12\xi_1\xi_3 - 4\xi_2\xi_3 + 6\xi_3^2, \tag{7.530}$$

it is not obvious whether or not there exist (ξ_1, ξ_2, ξ_3) which will give positive or negative values of f. However, it is easily verified that f can be rewritten as

$$f(\xi_1, \xi_2, \xi_3) = 2(\xi_1 - \xi_2 + \xi_3)^2 + 3(2\xi_1 - \xi_2)^2 + 4(\xi_1 + \xi_3)^2. \tag{7.531}$$

So in this case, $f \geq 0$ for all (ξ_1, ξ_2, ξ_3). How to demonstrate positivity (or nonpositivity) of such expressions is the topic of this section. A *quadratic form* is an expression

$$f(\xi_1, \cdots, \xi_N) = \sum_{j=1}^{N} \sum_{i=1}^{N} a_{ij}\xi_i\xi_j, \tag{7.532}$$

where $\{a_{ij}\}$ is a real, symmetric matrix that we will also call $\mathbf{A}$. The surface represented by the equation $\sum_{j=1}^{N} \sum_{i=1}^{N} a_{ij}\xi_i\xi_j =$ constant is a *quadric* surface. With the coefficient matrix defined, we can represent f as

$$f = \boldsymbol{\xi}^T \cdot \mathbf{A} \cdot \boldsymbol{\xi}. \tag{7.533}$$

Now, by Eq. (7.372), $\mathbf{A}$ can be decomposed as $\mathbf{Q} \cdot \mathbf{\Lambda} \cdot \mathbf{Q}^{-1}$, where $\mathbf{Q}$ is the orthogonal matrix populated by the orthonormalized eigenvectors of $\mathbf{A}$, and $\mathbf{\Lambda}$ is the corresponding diagonal matrix of eigenvalues. Thus, Eq. (7.533) becomes

$$f = \boldsymbol{\xi}^T \cdot \underbrace{\mathbf{Q} \cdot \mathbf{\Lambda} \cdot \mathbf{Q}^{-1}}_{\mathbf{A}} \cdot \boldsymbol{\xi}. \tag{7.534}$$

Because $\mathbf{Q}$ is orthogonal, $\mathbf{Q}^T = \mathbf{Q}^{-1}$, and we find

$$f = \boldsymbol{\xi}^T \cdot \mathbf{Q} \cdot \mathbf{\Lambda} \cdot \mathbf{Q}^T \cdot \boldsymbol{\xi}. \tag{7.535}$$

Now, define $\mathbf{x}$ so that $\mathbf{x} = \mathbf{Q}^T \cdot \boldsymbol{\xi} = \mathbf{Q}^{-1} \cdot \boldsymbol{\xi}$. Consequently, $\boldsymbol{\xi} = \mathbf{Q} \cdot \mathbf{x}$. Thus, Eq. (7.535) becomes

$$f = (\mathbf{Q} \cdot \mathbf{x})^T \cdot \mathbf{Q} \cdot \mathbf{\Lambda} \cdot \mathbf{x}, \tag{7.536}$$

$$= \mathbf{x}^T \cdot \mathbf{Q}^T \cdot \mathbf{Q} \cdot \mathbf{\Lambda} \cdot \mathbf{x}, \tag{7.537}$$

$$= \mathbf{x}^T \cdot \mathbf{Q}^{-1} \cdot \mathbf{Q} \cdot \mathbf{\Lambda} \cdot \mathbf{x}, \tag{7.538}$$

$$= \mathbf{x}^T \cdot \mathbf{\Lambda} \cdot \mathbf{x}. \tag{7.539}$$

This *standard form* of a quadratic form is one in which the cross-product terms (i.e., $\xi_i\xi_j$, $i \neq j$) do not appear. One has the following:

Principal Axis Theorem: If $\mathbf{Q}$ is the orthogonal matrix and $\lambda_1, \cdots, \lambda_N$ the eigenvalues corresponding to $\{a_{ij}\}$, a change in coordinates,

$$\begin{pmatrix} \xi_1 \\ \vdots \\ \xi_N \end{pmatrix} = \mathbf{Q} \cdot \begin{pmatrix} x_1 \\ \vdots \\ x_N \end{pmatrix}, \tag{7.540}$$

will reduce the quadratic form, Eq. (7.532), to its standard quadratic form

$$f(x_1, \ldots, x_N) = \lambda_1 x_1^2 + \lambda_2 x_2^2 + \cdots + \lambda_N x_N^2. \tag{7.541}$$

It is perhaps better to consider this as an alias rather than an alibi transformation.

EXAMPLE 7.56

Change

$$f(\xi_1, \xi_2) = 2\xi_1^2 + 2\xi_1\xi_2 + 2\xi_2^2 \tag{7.542}$$

to a standard quadratic form.

For $N = 2$, Eq. (7.532) becomes

$$f(\xi_1, \xi_2) = a_{11}\xi_1^2 + (a_{12} + a_{21})\xi_1\xi_2 + a_{22}\xi_2^2. \tag{7.543}$$

We choose $\{a_{ij}\}$ such that the matrix is symmetric. This gives us

$$a_{11} = 2, \tag{7.544}$$

$$a_{12} = 1, \tag{7.545}$$

$$a_{21} = 1, \tag{7.546}$$

$$a_{22} = 2. \tag{7.547}$$

So we get

$$\mathbf{A} = \begin{pmatrix} 2 & 1 \\ 1 & 2 \end{pmatrix}. \tag{7.548}$$

The eigenvalues of $\mathbf{A}$ are $\lambda = 1, \lambda = 3$. The orthogonal matrix corresponding to $\mathbf{A}$ is

$$\mathbf{Q} = \begin{pmatrix} \frac{1}{\sqrt{2}} & \frac{1}{\sqrt{2}} \\ -\frac{1}{\sqrt{2}} & \frac{1}{\sqrt{2}} \end{pmatrix}, \qquad \mathbf{Q}^{-1} = \mathbf{Q}^T = \begin{pmatrix} \frac{1}{\sqrt{2}} & -\frac{1}{\sqrt{2}} \\ \frac{1}{\sqrt{2}} & \frac{1}{\sqrt{2}} \end{pmatrix}. \tag{7.549}$$

The transformation $\boldsymbol{\xi} = \mathbf{Q} \cdot \mathbf{x}$ is

$$\xi_1 = \frac{1}{\sqrt{2}}(x_1 + x_2), \tag{7.550}$$

$$\xi_2 = \frac{1}{\sqrt{2}}(-x_1 + x_2). \tag{7.551}$$

We have $\det \mathbf{Q} = 1$, so the transformation is orientation-preserving. The inverse transformation $\mathbf{x} = \mathbf{Q}^{-1} \cdot \boldsymbol{\xi} = \mathbf{Q}^T \cdot \boldsymbol{\xi}$ is

$$x_1 = \frac{1}{\sqrt{2}}(\xi_1 - \xi_2), \tag{7.552}$$

$$x_2 = \frac{1}{\sqrt{2}}(\xi_1 + \xi_2). \tag{7.553}$$

Using Eqs. (7.550,7.551) to eliminate ξ_1 and ξ_2 in Eq. (7.542), we get a result in the form of Eq. (7.541):

$$f(x_1, x_2) = x_1^2 + 3x_2^2. \tag{7.554}$$

In terms of the original variables, we get

$$f(\xi_1, \xi_2) = \frac{1}{2}(\xi_1 - \xi_2)^2 + \frac{3}{2}(\xi_1 + \xi_2)^2. \tag{7.555}$$

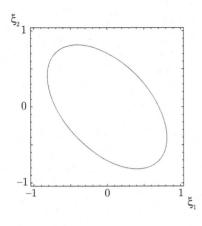

Figure 7.14. Quadric associated with $f(\xi_1, \xi_2) = (\xi_1 - \xi_2)^2/2 + 3(\xi_1 + \xi_2)^2/2 = 1$.

For this two-variable problem, a quadric "surface" is actually an ellipse in the (ξ_1, ξ_2) plane. We give a plot of the quadric for $f(\xi_1, \xi_2) = 1$ in Figure 7.14.

EXAMPLE 7.57

Change

$$f(\xi_1, \xi_2, \xi_3) = 18\xi_1^2 - 16\xi_1\xi_2 + 5\xi_2^2 + 12\xi_1\xi_3 - 4\xi_2\xi_3 + 6\xi_3^2 \tag{7.556}$$

to a standard quadratic form.

For $N = 3$, Eq. (7.532) becomes

$$f(\xi_1, \xi_2, \xi_3) = (\xi_1 \quad \xi_2 \quad \xi_3) \begin{pmatrix} 18 & -8 & 6 \\ -8 & 5 & -2 \\ 6 & -2 & 6 \end{pmatrix} \begin{pmatrix} \xi_1 \\ \xi_2 \\ \xi_3 \end{pmatrix} = \boldsymbol{\xi}^T \cdot \mathbf{A} \cdot \boldsymbol{\xi}. \tag{7.557}$$

The eigenvalues of $\mathbf{A}$ are $\lambda_1 = 1, \lambda_2 = 4, \lambda_3 = 24$. The orthogonal matrix corresponding to $\mathbf{A}$ is

$$\mathbf{Q} = \begin{pmatrix} -\frac{4}{\sqrt{69}} & -\frac{1}{\sqrt{30}} & \frac{13}{\sqrt{230}} \\ -\frac{7}{\sqrt{69}} & \sqrt{\frac{2}{15}} & -3\sqrt{\frac{2}{115}} \\ \frac{2}{\sqrt{69}} & \sqrt{\frac{5}{6}} & \sqrt{\frac{5}{46}} \end{pmatrix}, \qquad \mathbf{Q}^{-1} = \mathbf{Q}^T = \begin{pmatrix} -\frac{4}{\sqrt{69}} & -\frac{7}{\sqrt{69}} & \frac{2}{\sqrt{69}} \\ -\frac{1}{\sqrt{30}} & \sqrt{\frac{2}{15}} & \sqrt{\frac{5}{6}} \\ \frac{13}{\sqrt{230}} & -3\sqrt{\frac{2}{115}} & \sqrt{\frac{5}{46}} \end{pmatrix}. \tag{7.558}$$

For this nonunique choice of $\mathbf{Q}$, we note that $\det \mathbf{Q} = -1$, so it is a reflection. The inverse transformation $\mathbf{x} = \mathbf{Q}^{-1} \cdot \boldsymbol{\xi} = \mathbf{Q}^T \cdot \boldsymbol{\xi}$ is

$$x_1 = \frac{-4}{\sqrt{69}}\xi_1 - \frac{7}{\sqrt{69}}\xi_2 + \frac{2}{\sqrt{69}}\xi_3, \tag{7.559}$$

$$x_2 = -\frac{1}{\sqrt{30}}\xi_1 + \sqrt{\frac{2}{15}}\xi_2 + \sqrt{\frac{5}{6}}\xi_3, \tag{7.560}$$

$$x_3 = \frac{13}{\sqrt{230}}\xi_1 - 3\sqrt{\frac{2}{115}}\xi_2 + \sqrt{\frac{5}{46}}\xi_3. \tag{7.561}$$

Directly imposing, then, the standard quadratic form of Eq. (7.541) onto Eq. (7.556), we get

$$f(x_1, x_2, x_3) = x_1^2 + 4x_2^2 + 24x_3^2. \tag{7.562}$$

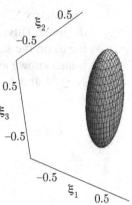

Figure 7.15. Quadric associated with $f(\xi_1, \xi_2, \xi_3) = 2(\xi_1 - \xi_2 + \xi_3)^2 + 3(2\xi_1 - \xi_2)^2 + 4(\xi_1 + \xi_3)^2 = 1$.

In terms of the original variables, we get

$$f(\xi_1, \xi_2, \xi_3) = \left(\frac{-4}{\sqrt{69}}\xi_1 - \frac{7}{\sqrt{69}}\xi_2 + \frac{2}{\sqrt{69}}\xi_3 \right)^2$$

$$+ 4 \left(-\frac{1}{\sqrt{30}}\xi_1 + \sqrt{\frac{2}{15}}\xi_2 + \sqrt{\frac{5}{6}}\xi_3 \right)^2$$

$$+ 24 \left(\frac{13}{\sqrt{230}}\xi_1 - 3\sqrt{\frac{2}{115}}\xi_2 + \sqrt{\frac{5}{46}}\xi_3 \right)^2. \tag{7.563}$$

It is clear that $f(\xi_1, \xi_2, \xi_3)$ is positive definite. Moreover, by performing the multiplications, it is easily seen that the original form is recovered. Further manipulation would also show that

$$f(\xi_1, \xi_2, \xi_3) = 2(\xi_1 - \xi_2 + \xi_3)^2 + 3(2\xi_1 - \xi_2)^2 + 4(\xi_1 + \xi_3)^2, \tag{7.564}$$

so we see the particular quadratic form is not unique. For this three-variable problem, a quadric surface is an ellipsoid in the (ξ_1, ξ_2, ξ_3) volume. We give a plot of the quadric for $f(\xi_1, \xi_2, \xi_3) = 1$ in Figure 7.15.

7.15 Moore-Penrose Pseudoinverse

We seek the Moore-Penrose[12] pseudoinverse $\mathbf{A}_{M \times N}^+$ such that the following four conditions are satisfied:

$$\mathbf{A}_{N \times M} \cdot \mathbf{A}_{M \times N}^+ \cdot \mathbf{A}_{N \times M} = \mathbf{A}_{N \times M}, \tag{7.565}$$

$$\mathbf{A}_{M \times N}^+ \cdot \mathbf{A}_{N \times M} \cdot \mathbf{A}_{M \times N}^+ = \mathbf{A}_{M \times N}^+, \tag{7.566}$$

$$\left(\mathbf{A}_{N \times M} \cdot \mathbf{A}_{M \times N}^+ \right)^H = \mathbf{A}_{N \times M} \cdot \mathbf{A}_{M \times N}^+, \tag{7.567}$$

$$\left(\mathbf{A}_{M \times N}^+ \cdot \mathbf{A}_{N \times M} \right)^H = \mathbf{A}_{M \times N}^+ \cdot \mathbf{A}_{N \times M}. \tag{7.568}$$

This will be achieved if we define

$$\mathbf{A}_{M \times N}^+ = \mathbf{Q}_{M \times M} \cdot \mathbf{\Sigma}_{M \times N}^+ \cdot \mathbf{Q}_{N \times N}^H. \tag{7.569}$$

---

[12] Eliakim Hastings Moore, 1862–1932, American mathematician, and Sir Roger Penrose, 1931–, English mathematician. The operation is also credited to Arne Bjerhammar, 1917–2011, Swedish geodesist.

The matrix $\boldsymbol{\Sigma}^+$ is $M \times N$ with σ_i^{-1}, $(i = 1, 2, \ldots)$ in the first r positions on the main diagonal. This is closely related to the $N \times M$ matrix $\boldsymbol{\Sigma}$, defined in Section 7.9.8, having σ_i on the same diagonal positions. The Moore-Penrose pseudoinverse, $\mathbf{A}_{M \times N}^+$, is also known as the *generalized inverse*. This is because in the special case in which $N \leq M$ and $N = r$, it can be shown that

$$\mathbf{A}_{N \times M} \cdot \mathbf{A}_{M \times N}^+ = \mathbf{I}_{N \times N}. \tag{7.570}$$

Let us check this with our definitions for the case when $N \leq M$, $N = r$:

$$\mathbf{A}_{N \times M} \cdot \mathbf{A}_{M \times N}^+ = \left(\mathbf{Q}_{N \times N} \cdot \boldsymbol{\Sigma}_{N \times M} \cdot \mathbf{Q}_{M \times M}^H \right) \cdot \left(\mathbf{Q}_{M \times M} \cdot \boldsymbol{\Sigma}_{M \times N}^+ \cdot \mathbf{Q}_{N \times N}^H \right), \tag{7.571}$$

$$= \mathbf{Q}_{N \times N} \cdot \boldsymbol{\Sigma}_{N \times M} \cdot \mathbf{Q}_{M \times M}^{-1} \cdot \mathbf{Q}_{M \times M} \cdot \boldsymbol{\Sigma}_{M \times N}^+ \cdot \mathbf{Q}_{N \times N}^H, \tag{7.572}$$

$$= \mathbf{Q}_{N \times N} \cdot \boldsymbol{\Sigma}_{N \times M} \cdot \boldsymbol{\Sigma}_{M \times N}^+ \cdot \mathbf{Q}_{N \times N}^H, \tag{7.573}$$

$$= \mathbf{Q}_{N \times N} \cdot \mathbf{I}_{N \times N} \cdot \mathbf{Q}_{N \times N}^H, \tag{7.574}$$

$$= \mathbf{Q}_{N \times N} \cdot \mathbf{Q}_{N \times N}^H, \tag{7.575}$$

$$= \mathbf{Q}_{N \times N} \cdot \mathbf{Q}_{N \times N}^{-1}, \tag{7.576}$$

$$= \mathbf{I}_{N \times N}. \tag{7.577}$$

We note for this special case that precisely because of the way we defined $\boldsymbol{\Sigma}^+$, $\boldsymbol{\Sigma}_{N \times M} \cdot \boldsymbol{\Sigma}_{M \times N}^+ = \mathbf{I}_{N \times N}$. When $N > M$, $\boldsymbol{\Sigma}_{N \times M} \cdot \boldsymbol{\Sigma}_{M \times N}^+$ yields a matrix with r ones on the diagonal and zeros elsewhere.

One also has the analogs to Eqs. (7.122, 7.123):

$$||\mathbf{A}||_2 = \max_i \sigma_i, \tag{7.578}$$

$$\frac{1}{||\mathbf{A}^+||_2} = \min_i \sigma_i. \tag{7.579}$$

EXAMPLE 7.58

Find the Moore-Penrose pseudoinverse, $\mathbf{A}_{3 \times 2}^+$, of $\mathbf{A}_{2 \times 3}$ from the matrix of a previous example (Example 7.35):

$$\mathbf{A}_{2 \times 3} = \begin{pmatrix} 1 & -3 & 2 \\ 2 & 0 & 3 \end{pmatrix}. \tag{7.580}$$

$$\mathbf{A}_{3 \times 2}^+ = \mathbf{Q}_{3 \times 3} \cdot \boldsymbol{\Sigma}_{3 \times 2}^+ \cdot \mathbf{Q}_{2 \times 2}^H, \tag{7.581}$$

$$\mathbf{A}_{3 \times 2}^+ =$$

$$\begin{pmatrix} 0.452350 & 0.330059 & -0.828517 \\ -0.471378 & 0.877114 & 0.0920575 \\ 0.757088 & 0.348902 & 0.552345 \end{pmatrix} \begin{pmatrix} \frac{1}{4.6385} & 0 \\ 0 & \frac{1}{2.3419} \\ 0 & 0 \end{pmatrix} \begin{pmatrix} 0.728827 & 0.684698 \\ -0.684698 & 0.728827 \end{pmatrix}, \tag{7.582}$$

$$\mathbf{A}_{3 \times 2}^+ = \begin{pmatrix} -0.0254237 & 0.169492 \\ -0.330508 & 0.20339 \\ 0.0169492 & 0.220339 \end{pmatrix}. \tag{7.583}$$

Note that

$$\mathbf{A}_{2 \times 3} \cdot \mathbf{A}_{3 \times 2}^+ = \begin{pmatrix} 1 & -3 & 2 \\ 2 & 0 & 3 \end{pmatrix} \begin{pmatrix} -0.0254237 & 0.169492 \\ -0.330508 & 0.20339 \\ 0.0169492 & 0.220339 \end{pmatrix} = \begin{pmatrix} 1 & 0 \\ 0 & 1 \end{pmatrix}. \tag{7.584}$$

Both $\mathbf{Q}$ matrices have a determinant of $+1$ and are thus volume- and orientation-preserving.

EXAMPLE 7.59

Use the Moore-Penrose pseudoinverse to solve the problem $\mathbf{A} \cdot \mathbf{x} = \mathbf{b}$ studied in an earlier example (Example 7.12):

$$\begin{pmatrix} 1 & 2 \\ 3 & 6 \end{pmatrix} \begin{pmatrix} x_1 \\ x_2 \end{pmatrix} = \begin{pmatrix} 2 \\ 0 \end{pmatrix}. \tag{7.585}$$

We first seek the singular value decomposition of $\mathbf{A}$, $\mathbf{A} = \mathbf{Q}_2 \cdot \mathbf{\Sigma} \cdot \mathbf{Q}_1^H$. Now

$$\mathbf{A}^H \cdot \mathbf{A} = \begin{pmatrix} 1 & 3 \\ 2 & 6 \end{pmatrix} \begin{pmatrix} 1 & 2 \\ 3 & 6 \end{pmatrix} = \begin{pmatrix} 10 & 20 \\ 20 & 40 \end{pmatrix}. \tag{7.586}$$

The eigensystem with normalized eigenvectors corresponding to $\mathbf{A}^H \cdot \mathbf{A}$ is

$$\lambda_1 = 50, \qquad \mathbf{e}_1 = \begin{pmatrix} \frac{1}{\sqrt{5}} \\ \frac{2}{\sqrt{5}} \end{pmatrix}, \tag{7.587}$$

$$\lambda_2 = 0, \qquad \mathbf{e}_2 = \begin{pmatrix} -\frac{2}{\sqrt{5}} \\ \frac{1}{\sqrt{5}} \end{pmatrix}, \tag{7.588}$$

so

$$\mathbf{Q}_1 = \begin{pmatrix} \frac{1}{\sqrt{5}} & -\frac{2}{\sqrt{5}} \\ \frac{2}{\sqrt{5}} & \frac{1}{\sqrt{5}} \end{pmatrix}, \tag{7.589}$$

$$\mathbf{\Sigma} = \begin{pmatrix} \sqrt{50} & 0 \\ 0 & 0 \end{pmatrix} = \begin{pmatrix} 5\sqrt{2} & 0 \\ 0 & 0 \end{pmatrix}, \tag{7.590}$$

so taking $\mathbf{A} \cdot \mathbf{Q}_1 = \mathbf{Q}_2 \cdot \mathbf{\Sigma}$ gives

$$\underbrace{\begin{pmatrix} 1 & 2 \\ 3 & 6 \end{pmatrix}}_{\mathbf{A}} \underbrace{\begin{pmatrix} \frac{1}{\sqrt{5}} & -\frac{2}{\sqrt{5}} \\ \frac{2}{\sqrt{5}} & \frac{1}{\sqrt{5}} \end{pmatrix}}_{=\mathbf{Q}_1} = \underbrace{\begin{pmatrix} q_{11} & q_{12} \\ q_{21} & q_{22} \end{pmatrix}}_{\mathbf{Q}_2} \underbrace{\begin{pmatrix} 5\sqrt{2} & 0 \\ 0 & 0 \end{pmatrix}}_{\mathbf{\Sigma}}, \tag{7.591}$$

$$\sqrt{5} \begin{pmatrix} 1 & 0 \\ 3 & 0 \end{pmatrix} = \begin{pmatrix} 5\sqrt{2}q_{11} & 0 \\ 5\sqrt{2}q_{21} & 0 \end{pmatrix}. \tag{7.592}$$

Solving, we get

$$\begin{pmatrix} q_{11} \\ q_{21} \end{pmatrix} = \begin{pmatrix} \frac{1}{\sqrt{5}} \\ \frac{3}{\sqrt{10}} \end{pmatrix}. \tag{7.593}$$

Imposing orthonormality to find q_{12} and q_{22}, we get

$$\begin{pmatrix} q_{12} \\ q_{22} \end{pmatrix} = \begin{pmatrix} \frac{3}{\sqrt{10}} \\ -\frac{1}{\sqrt{10}} \end{pmatrix}, \tag{7.594}$$

so

$$\mathbf{Q}_2 = \begin{pmatrix} \frac{1}{\sqrt{10}} & \frac{3}{\sqrt{10}} \\ \frac{3}{\sqrt{10}} & -\frac{1}{\sqrt{10}} \end{pmatrix}, \tag{7.595}$$

and

$$\mathbf{A} = \mathbf{Q}_2 \cdot \mathbf{\Sigma} \cdot \mathbf{Q}_1^H = \underbrace{\begin{pmatrix} \frac{1}{\sqrt{10}} & \frac{3}{\sqrt{10}} \\ \frac{3}{\sqrt{10}} & -\frac{1}{\sqrt{10}} \end{pmatrix}}_{\mathbf{Q}_2} \underbrace{\begin{pmatrix} 5\sqrt{2} & 0 \\ 0 & 0 \end{pmatrix}}_{\mathbf{\Sigma}} \underbrace{\begin{pmatrix} \frac{1}{\sqrt{5}} & \frac{2}{\sqrt{5}} \\ -\frac{2}{\sqrt{5}} & \frac{1}{\sqrt{5}} \end{pmatrix}}_{\mathbf{Q}_1^H} = \begin{pmatrix} 1 & 2 \\ 3 & 6 \end{pmatrix}. \tag{7.596}$$

As an aside, note that $\mathbf{Q}_1$ is orientation-preserving, while $\mathbf{Q}_2$ is not, though this property is not important for this analysis.

We will need $\boldsymbol{\Sigma}^+$, which is easily calculated by taking the inverse of each diagonal term of $\boldsymbol{\Sigma}$:

$$\boldsymbol{\Sigma}^+ = \begin{pmatrix} \frac{1}{5\sqrt{2}} & 0 \\ 0 & 0 \end{pmatrix}. \tag{7.597}$$

Now the Moore-Penrose pseudoinverse is

$$\mathbf{A}^+ = \mathbf{Q}_1 \cdot \boldsymbol{\Sigma}^+ \cdot \mathbf{Q}_2^H = \underbrace{\begin{pmatrix} \frac{1}{\sqrt{5}} & -\frac{2}{\sqrt{5}} \\ \frac{2}{\sqrt{5}} & \frac{1}{\sqrt{5}} \end{pmatrix}}_{\mathbf{Q}_1} \underbrace{\begin{pmatrix} \frac{1}{5\sqrt{2}} & 0 \\ 0 & 0 \end{pmatrix}}_{\boldsymbol{\Sigma}^+} \underbrace{\begin{pmatrix} \frac{1}{\sqrt{10}} & \frac{3}{\sqrt{10}} \\ \frac{3}{\sqrt{10}} & -\frac{1}{\sqrt{10}} \end{pmatrix}}_{\mathbf{Q}_2^H} = \begin{pmatrix} \frac{1}{50} & \frac{3}{50} \\ \frac{2}{50} & \frac{6}{50} \end{pmatrix}. \tag{7.598}$$

Direct multiplication shows that $\mathbf{A} \cdot \mathbf{A}^+ \neq \mathbf{I}$. This is a consequence of $\mathbf{A}$ not being a full rank matrix. However, the four Moore-Penrose conditions are satisfied: $\mathbf{A} \cdot \mathbf{A}^+ \cdot \mathbf{A} = \mathbf{A}$, $\mathbf{A}^+ \cdot \mathbf{A} \cdot \mathbf{A}^+ = \mathbf{A}^+$, $(\mathbf{A} \cdot \mathbf{A}^+)^H = \mathbf{A} \cdot \mathbf{A}^+$, and $(\mathbf{A}^+ \cdot \mathbf{A})^H = \mathbf{A}^+ \cdot \mathbf{A}$.

Lastly, applying the Moore-Penrose pseudoinverse operator to the vector b to form $\mathbf{x} = \mathbf{A}^+ \cdot \mathbf{b}$, we get

$$\mathbf{x} = \mathbf{A}^+ \cdot \mathbf{b} = \underbrace{\begin{pmatrix} \frac{1}{50} & \frac{3}{50} \\ \frac{2}{50} & \frac{6}{50} \end{pmatrix}}_{\mathbf{A}^+} \underbrace{\begin{pmatrix} 2 \\ 0 \end{pmatrix}}_{\mathbf{b}} = \begin{pmatrix} \frac{1}{25} \\ \frac{2}{25} \end{pmatrix}. \tag{7.599}$$

We see that the Moore-Penrose operator acting on b has yielded an x vector that is in the row space of $\mathbf{A}$. As there is no right null space component, it is the minimum length vector that minimizes the residual $\|\mathbf{A} \cdot \mathbf{x} - \mathbf{b}\|_2$. It is fully consistent with the solution we found using Gaussian elimination in (Example 7.12).

EXERCISES

1. Determine the action of the matrix

$$\mathbf{A} = \begin{pmatrix} 3 & 1 \\ 2 & 2 \end{pmatrix}$$

on the unit square with vertices at $(0,0)$, $(1,0)$, $(1,1)$, and $(0,1)$. Give a plot of the original unit square and its image following the alibi mapping. Also decompose $\mathbf{A}$ under (a) the $\mathbf{Q} \cdot \mathbf{U}$ decomposition and (b) the singular value decomposition, and for each decomposition, plot the series of mappings under the action of each component of the decomposition.

2. Find the nth power of $\begin{pmatrix} 1 & 3 \\ 3 & 1 \end{pmatrix}$.

3. Show that the following matrix is nilpotent, and find its matrix exponential:

$$\begin{pmatrix} 0 & 0 & 1 & 0 & 0 \\ 0 & 1 & -1 & 0 & 0 \\ 0 & 1 & -1 & 0 & 0 \\ 0 & 2 & -2 & 0 & 0 \\ 0 & 0 & -1 & 0 & 0 \end{pmatrix}.$$

4. Find the inverse of

$$\begin{pmatrix} 1/4 & 1/2 & 3/4 \\ 3/4 & 1/2 & 1/4 \\ 1/4 & 1/2 & 1/2 \end{pmatrix}.$$

5. Find the most general set of x_i that best satisfies
   (a)

$$\begin{pmatrix} 1 & 0 & 1 \\ 0 & 1 & 0 \\ 1 & 0 & -1 \\ 0 & 1 & 0 \\ 1 & 0 & 0 \end{pmatrix} \begin{pmatrix} x_1 \\ x_2 \\ x_3 \end{pmatrix} = \begin{pmatrix} 1 \\ 3 \\ 3 \\ 3 \\ 2 \end{pmatrix}.$$

   (b)

$$\begin{pmatrix} 1 & 1 & 0 & 0 \\ 3 & 3 & 0 & 7 \\ 0 & 1 & 1 & 0 \end{pmatrix} \begin{pmatrix} x_1 \\ x_2 \\ x_3 \\ x_4 \end{pmatrix} = \begin{pmatrix} 9 \\ 20 \\ 2 \end{pmatrix}.$$

   (c)

$$\begin{pmatrix} 1 & 3 & 4 & 5 & 0 \\ 0 & 1 & 1 & 1 & 0 \\ 1 & 3 & 5 & 6 & 7 \\ 2 & 6 & 10 & 12 & 14 \end{pmatrix} \begin{pmatrix} x_1 \\ x_2 \\ x_3 \\ x_4 \\ x_5 \end{pmatrix} = \begin{pmatrix} 1 \\ 3 \\ 0 \\ 1 \end{pmatrix}.$$

6. If $\mathbf{A} : \ell_2^2 \to \ell_2^2$, find $||\mathbf{A}||_2$ when

$$\mathbf{A} = \begin{pmatrix} 1 & -1 \\ 1 & 1 \end{pmatrix}.$$

   Also find its inverse and adjoint.

7. Check that the matrix $\mathbf{A}$, where

$$\mathbf{A} = \begin{pmatrix} 5 & -2 & -2 \\ -2 & 5 & -2 \\ -2 & -2 & 5 \end{pmatrix},$$

   satisfies the following equivalent definitions for a real positive definite symmetric matrix: (a) For $\mathbf{x} \neq 0$, $\mathbf{x}^T \cdot \mathbf{A} \cdot \mathbf{x} > 0$. Hint: Show that the left side can be written as $(2x_1 - x_2)^2 + (2x_2 - x_3)^2 + (2x_3 - x_1)^2$ where $\mathbf{x} = (x_1, x_2, x_3)^T$. (b) All eigenvalues are positive. (c) Upper left submatrices (i.e., starting from the upper left corner, submatrices of size 1×1, 2×2, etc.) have positive determinants.

8. For the following values of $\mathbf{A}$ and $\mathbf{b}$, find the unique value of $\mathbf{x}$ that minimizes $||\mathbf{A} \cdot \mathbf{x} - \mathbf{b}||_2$ and simultaneously minimizes $||\mathbf{x}||_2$:

   (a) $\mathbf{A} = \begin{pmatrix} 0 & 0 \\ 0 & 0 \end{pmatrix}$, $\mathbf{b} = \begin{pmatrix} 1 \\ 2 \end{pmatrix}$.

   (b) $\mathbf{A} = \begin{pmatrix} 1 & 2 \\ 2 & 2 \end{pmatrix}$, $\mathbf{b} = \begin{pmatrix} 1 \\ 2 \end{pmatrix}$.

   (c) $\mathbf{A} = \begin{pmatrix} 1 & 2 \\ 1 & 2 \end{pmatrix}$, $\mathbf{b} = \begin{pmatrix} 1 \\ 2 \end{pmatrix}$.

(d) $\mathbf{A} = \begin{pmatrix} 1+i & 2+i \\ 1 & 2 \end{pmatrix}, \mathbf{b} = \begin{pmatrix} 1 \\ 2 \end{pmatrix}.$

9. For the following values of $\mathbf{A}$ and $\mathbf{b}$, find the unique value of $\mathbf{x}$ which minimizes $||\mathbf{A} \cdot \mathbf{x} - \mathbf{b}||_2$ and simultaneously minimizes $||\mathbf{x}||_2$:

(a) $\mathbf{A} = \begin{pmatrix} 1 & 2 & 3 \\ 4 & 5 & 6 \end{pmatrix}, \mathbf{b} = \begin{pmatrix} 3 \\ 2 \end{pmatrix}.$

(b) $\mathbf{A} = \begin{pmatrix} 1 & 2 \\ 3 & 4 \\ 5 & 6 \end{pmatrix}, \mathbf{b} = \begin{pmatrix} 3 \\ 2 \\ 1 \end{pmatrix}.$

(c) $\mathbf{A} = \begin{pmatrix} 1 & 2 & 3 \\ 4 & 5 & 9 \\ 7 & 8 & 15 \\ 1 & 2 & 3 \end{pmatrix}, \mathbf{b} = \begin{pmatrix} 0 \\ 1 \\ 2 \\ 3 \end{pmatrix}.$

(d) $\mathbf{A} = \begin{pmatrix} i & 2i & 3i \\ 4 & 5 & 9 \\ 7 & 8 & 15 \\ 1 & 2 & 3 \end{pmatrix}, \mathbf{b} = \begin{pmatrix} 0 \\ 1 \\ 2 \\ 3 \end{pmatrix}.$

10. Examine the Fredholm alternative for

(a) $\mathbf{A} = \begin{pmatrix} 0 & 0 \\ 0 & 0 \end{pmatrix}, \mathbf{b} = \begin{pmatrix} 1 \\ 2 \end{pmatrix},$

(b) $\mathbf{A} = \begin{pmatrix} 1 & 2 \\ 2 & 2 \end{pmatrix}, \mathbf{b} = \begin{pmatrix} 1 \\ 2 \end{pmatrix},$

(c) $\mathbf{A} = \begin{pmatrix} 1 & 2 \\ 1 & 2 \end{pmatrix}, \mathbf{b} = \begin{pmatrix} 1 \\ 2 \end{pmatrix},$

(d) $\mathbf{A} = \begin{pmatrix} 1+i & 2+i \\ 1 & 2 \end{pmatrix}, \mathbf{b} = \begin{pmatrix} 1 \\ 2 \end{pmatrix}.$

11. Show that the left and right eigenvalues of $\mathbf{A}$ are the same but the eigenvectors are different, where

$$\mathbf{A} = \begin{pmatrix} 5 & 6 & 3 \\ 3 & 1 & 4 \\ 8 & 6 & 8 \end{pmatrix}.$$

Show that $\mathbf{l}_i^T \cdot \mathbf{r}_j = 0$ where $\mathbf{l}_i$ and $\mathbf{r}_j$ are the left and right eigenvectors corresponding to different eigenvalues λ_i and λ_j. Find also the eigenvalues and eigenvectors of $\mathbf{A} \cdot \mathbf{A}^T$.

12. Find two normalized eigenvectors (including, if needed, a generalized eigenvector) of

$$\begin{pmatrix} 0 & 1 \\ -1 & -2 \end{pmatrix}.$$

13. Find the eigenvectors of unit length of

$$\begin{pmatrix} 1 & -2 \\ 3 & -4 \end{pmatrix}.$$

Are they orthogonal?

14. Find the eigenvalues and inverse of

$$\begin{pmatrix} \frac{1}{2} & 1 & 0 \\ 0 & 1 & 1 \\ 0 & 0 & 2 \end{pmatrix}.$$

15. Find the eigenvectors and generalized eigenvectors of

$$\mathbf{A} = \begin{pmatrix} 1 & 1 & 1 & 1 \\ 0 & 1 & 1 & 1 \\ 0 & 0 & 0 & 1 \\ 0 & 0 & 0 & 0 \end{pmatrix}.$$

16. Show that the eigenvectors and generalized eigenvectors of

$$\begin{pmatrix} 1 & 1 & 2 & 0 \\ 0 & 1 & 3 & 0 \\ 0 & 0 & 2 & 2 \\ 0 & 0 & 0 & 1 \end{pmatrix}$$

span $\mathbb{R}^4$.

17. (a) Show that the characteristic equation for any 3×3 matrix $\mathbf{A}$ can be written as

$$p(\lambda) = \det \mathbf{A} - \frac{1}{2}\left((\operatorname{tr}\mathbf{A})^2 - \operatorname{tr}(\mathbf{A}^2)\right)\lambda + (\operatorname{tr}\mathbf{A})\lambda^2 - \lambda^3.$$

   (b) Thus, write the characteristic equation for an $N \times N$ matrix.
   (c) Using the Cayley-Hamilton theorem, show that the inverse of any invertible matrix $\mathbf{A}$ is

$$\mathbf{A}^{-1} = \frac{1}{\det \mathbf{A}}\left((-1)^{n-1}\mathbf{A}^{n-1} + (-1)^{n-2}(\operatorname{tr}\mathbf{A})\,\mathbf{A}^{n-2} + \ldots\right).$$

18. Determine the action of a matrix on the set of points forming a square of unit area with vertices at $(1/2, 1/2)$, $(-1/2, 1/2)$, $(-1/2, -1/2)$, $(1/2, -1/2)$. Give a plot of the original square and its mapped image if the matrix is

   (a) $\begin{pmatrix} 2 & 1 \\ 0 & 1 \end{pmatrix}$,

   (b) $\begin{pmatrix} 2 & 1 \\ 0 & -1 \end{pmatrix}$,

   (c) $\begin{pmatrix} 2 & 1 \\ 2 & 1 \end{pmatrix}$.

Interpret your results in relation to the area- and orientation-preserving properties of the matrix.

19. Determine the action of a matrix on the set of points forming a circle of unit area with center at the origin. Give a plot of the original circle and its mapped image if the matrix is

   (a) $\begin{pmatrix} 2 & 1 \\ 0 & 1 \end{pmatrix}$,

   (b) $\begin{pmatrix} 2 & 1 \\ 0 & -1 \end{pmatrix}$,

   (c) $\begin{pmatrix} 2 & 1 \\ 2 & 1 \end{pmatrix}$.

Interpret your results in relation to the area- and orientation-preserving properties of the matrix.

20. For the complex matrices, **A**, find eigenvectors, find eigenvalues, demonstrate whether the eigenvectors are orthogonal, find the matrix **S** such that $\mathbf{S}^{-1} \cdot \mathbf{A} \cdot \mathbf{S}$ is of Jordan form, and find the singular value decomposition if

    (a)

$$\mathbf{A} = \begin{pmatrix} 2+i & 2 \\ 2 & 1 \end{pmatrix},$$

    (b)

$$\mathbf{A} = \begin{pmatrix} 2 & 4i & 2+i \\ -4i & 1 & 3 \\ 2-i & 3 & -2 \end{pmatrix}.$$

21. (a) Starting with a vector in the direction $(1, 2, 0)^T$, use the Gram-Schmidt procedure to find a set of orthonormal vectors in $\mathbb{R}^3$. Using these vectors, construct (b) an orthogonal matrix **Q**, and then find (c) the angles between **x** and $\mathbf{Q} \cdot \mathbf{x}$, where **x** is $(1, 0, 0)^T$, $(0, 1, 0)^T$ and $(0, 0, 1)^T$, respectively.

22. Show that the matrix

$$\frac{1}{2} \begin{pmatrix} 1+i & 1-i \\ 1-i & 1+i \end{pmatrix}$$

    is unitary.

23. Find a five term DFT of the function $y = x$ with $x \in [0, 4]$.

24. Write **A** in row echelon form, where

$$\mathbf{A} = \begin{pmatrix} 0 & 0 & 1 & 0 \\ 2 & -2 & 0 & 0 \\ 1 & 0 & 1 & 2 \end{pmatrix}.$$

25. Diagonalize, or find the Jordan canonical form of, the matrix

$$\begin{pmatrix} 0 & 1 \\ -1 & -2 \end{pmatrix}.$$

26. Diagonalize or reduce to Jordan canonical form

$$\mathbf{A} = \begin{pmatrix} 5 & 2 & -1 \\ 0 & 5 & 1 \\ 0 & 0 & 5 \end{pmatrix}.$$

27. Decompose **A** into $\mathbf{S} \cdot \mathbf{J} \cdot \mathbf{S}^{-1}$, $\mathbf{P}^{-1} \cdot \mathbf{L} \cdot \mathbf{D} \cdot \mathbf{U}$, $\mathbf{Q} \cdot \mathbf{U}$, Schur form, and Hessenberg form:

$$\mathbf{A} = \begin{pmatrix} 0 & 1 & 0 & 1 \\ 1 & 0 & 1 & 0 \\ 0 & 1 & 0 & 1 \\ 1 & 0 & 1 & 0 \end{pmatrix}.$$

28. Find the matrix **S** that will convert the following to Jordan canonical form:

(a) $\begin{pmatrix} 6 & -1 & -3 & 1 \\ -1 & 6 & 1 & -3 \\ -3 & 1 & 6 & -1 \\ 1 & -3 & -1 & 6 \end{pmatrix}$,

(b) $\begin{pmatrix} 8 & -2 & -2 & 0 \\ 0 & 6 & 2 & -4 \\ -2 & 0 & 8 & -2 \\ 2 & -4 & 0 & 6 \end{pmatrix}$,

and show the Jordan canonical form.

29. If

$$A = \begin{pmatrix} 5 & 4 \\ 1 & 2 \end{pmatrix},$$

find a matrix **S** such that $S^{-1} \cdot A \cdot S$ is a diagonal matrix. Show by multiplication that it is indeed diagonal.

30. Determine if **A** and **B** are similar, where

$$A = \begin{pmatrix} 6 & 2 \\ -2 & 1 \end{pmatrix},$$

$$B = \begin{pmatrix} 8 & 6 \\ -3 & -1 \end{pmatrix}.$$

31. Find the eigenvalues, eigenvectors, and the matrix **S** such that $S^{-1} \cdot A \cdot S$ is diagonal or of Jordan form, where **A** is

(a) $\begin{pmatrix} 5 & 0 & 0 \\ 1 & 0 & 1 \\ 0 & 0 & -2 \end{pmatrix}$,

(b) $\begin{pmatrix} -2 & 0 & 2 \\ 2 & 1 & 0 \\ 0 & 0 & -2i \end{pmatrix}$,

(c) $\begin{pmatrix} 3 & 0 & -1 \\ -1 & 2 & 2i \\ 1 & 0 & 1+i \end{pmatrix}$.

Put each of the matrices in $L \cdot D \cdot U$ form, in $Q \cdot U$ form, and in Schur form.

32. Find the Schur and Cholesky decompositions of **A**, where

$$A = \begin{pmatrix} 0 & 0 & 0 & 0 \\ 0 & 1 & -3 & 0 \\ 0 & -3 & 1 & 0 \\ 0 & 0 & 0 & 0 \end{pmatrix}.$$

33. Let

$$A = \begin{pmatrix} 1 & 1 & 2 \\ 0 & 1 & 1 \\ 0 & 0 & 1 \end{pmatrix}.$$

Find **S** such that $S^{-1} \cdot A \cdot S = J$, where **J** is of the Jordan form. Show by multiplication that $A \cdot S = S \cdot J$.

34. For

$$\mathbf{A} = \begin{pmatrix} 8 & 5 & -2 & -1 \\ 6 & 8 & -2 & 8 \\ -1 & 2 & 0 & 1 \end{pmatrix},$$

(a) find the $\mathbf{P}^{-1} \cdot \mathbf{L} \cdot \mathbf{D} \cdot \mathbf{U}$ decomposition, (b) find the singular values and the singular value decomposition.

35. For each of the following matrices $\mathbf{A}$, find the left and right polar decompositions, the corresponding left and right Cauchy-Green tensors, and the Green–St. Venant tensor:

(a)

$$\mathbf{A} = \begin{pmatrix} 1 & -3 \\ 1 & 2 \end{pmatrix},$$

(b)

$$\mathbf{A} = \begin{pmatrix} 1+i & -3 \\ 1 & 2 \end{pmatrix},$$

(c)

$$\mathbf{A} = \begin{pmatrix} 1 & 2 & 3 \\ 3 & 2 & 1 \\ 1 & 3 & 2 \end{pmatrix}.$$

36. A Sylvester[13] equation takes the form

$$\mathbf{A} \cdot \mathbf{X} + \mathbf{X} \cdot \mathbf{B} = \mathbf{C},$$

where $\mathbf{A}$, $\mathbf{B}$, and $\mathbf{C}$ are known matrices of dimension $N \times N$, and $\mathbf{X}$ is an unknown matrix of the same dimension. Solutions exist when $\mathbf{A}$ and $\mathbf{B}$ have no common eigenvalues. Solve the following Sylvester equations for $\mathbf{X}$ arising from the following sets of matrices:

(a)

$$\mathbf{A} = \begin{pmatrix} 1 & 0 \\ 0 & 2 \end{pmatrix}, \quad \mathbf{B} = \begin{pmatrix} 1 & 1 \\ 1 & 2 \end{pmatrix}, \quad \mathbf{C} = \begin{pmatrix} 1 & 2 \\ 3 & 4 \end{pmatrix},$$

(b)

$$\mathbf{A} = \begin{pmatrix} 3 & 1 \\ 2 & 3 \end{pmatrix}, \quad \mathbf{B} = \begin{pmatrix} 1 & 2 \\ 2 & 4 \end{pmatrix}, \quad \mathbf{C} = \begin{pmatrix} 1 & 0 \\ 0 & 0 \end{pmatrix}.$$

37. Consider square matrices $\mathbf{A}$ and $\mathbf{B}$. The commutator $[\mathbf{A}, \mathbf{B}]$ is defined as

$$[\mathbf{A}, \mathbf{B}] = \mathbf{A} \cdot \mathbf{B} - \mathbf{B} \cdot \mathbf{A}$$

and gives an indication as to how well matrix multiplication commutes between $\mathbf{A}$ and $\mathbf{B}$. Find $[\mathbf{A}, \mathbf{B}]$ if

(a) $\mathbf{A} = \begin{pmatrix} 1 & 0 \\ -1 & 2 \end{pmatrix}, \mathbf{B} = \begin{pmatrix} -1 & 0 \\ -3 & 2 \end{pmatrix},$

(b) $\mathbf{A} = \begin{pmatrix} 1 & 0 \\ -1 & 2 \end{pmatrix}, \mathbf{B} = \begin{pmatrix} -1 & 1 \\ -3 & 2 \end{pmatrix}.$

---

[13] James Joseph Sylvester, 1814–1897, English mathematician.

Find a general condition under which $\mathbf{A} \cdot \mathbf{B} = \mathbf{B} \cdot \mathbf{A}$.

38. Determine if

$$\mathbf{A} \cdot \mathbf{B} = 0$$

implies that either $\mathbf{A}$ or $\mathbf{B}$ *must* be 0. Show relevant examples.

39. Find the projection matrix onto the space spanned by $(1, 2, 3)^T$ and $(2, 3, 5)^T$. Find the projection of the vector $(7, 8, 9)^T$ onto this space.

40. Find the constants for a quadratic least squares formula

$$y(x) = a + bx + cx^2,$$

and find the best fit for the following measured data:

$\mathbf{x} = (-3, -2.5, -2, -1.5, -1, -0.5, 0, 0.5, 1, 1.5, 2, 2.5, 3)^T$

$\mathbf{y} = (28.58, 24.05, 13.24, 10.29, 3.27, 2.65, 1.21, 3.81, 5.99, 12.39, 17.02, 27.16, 34.42)^T.$

41. Find the $\mathbf{x}$ with smallest $||\mathbf{x}||_2$ that minimizes $||\mathbf{A} \cdot \mathbf{x} - \mathbf{b}||_2$ for

$$\mathbf{A} = \begin{pmatrix} 1 & 0 & 3 \\ 2 & -1 & 3 \\ 3 & -1 & 5 \end{pmatrix}, \qquad \mathbf{b} = \begin{pmatrix} 1 \\ 0 \\ 1 \end{pmatrix}.$$

42. Find the most general $\mathbf{x}$ that minimizes $||\mathbf{A} \cdot \mathbf{x} - \mathbf{b}||_2$ for

$$\mathbf{A} = \begin{pmatrix} 1 & 0 \\ 2 & -1 \\ 3 & -2 \end{pmatrix}, \qquad \mathbf{b} = \begin{pmatrix} 1 \\ 0 \\ 1 \end{pmatrix}.$$

43. Find $\mathbf{x}$ with the smallest $||\mathbf{x}||_2$ that minimizes $||\mathbf{A} \cdot \mathbf{x} - \mathbf{b}||_2$ for

$$\mathbf{A} = \begin{pmatrix} 1 & 0 & 1 & 4 \\ 1 & 0 & 2 & -1 \\ 2 & 1 & 3 & -2 \end{pmatrix}, \qquad \mathbf{b} = \begin{pmatrix} 2 \\ 1 \\ -3 \end{pmatrix}.$$

44. Find the $\mathbf{x}$ with minimum $||\mathbf{x}||_2$ that minimizes $||\mathbf{A} \cdot \mathbf{x} - \mathbf{b}||_2$ in the following problems. For each, find the right null space and show that the most general solution vector can be represented as a linear combination of a unique vector in the row space of $\mathbf{A}$ plus an arbitrary scalar multiple of the right null space of $\mathbf{A}$:

(a)

$$\mathbf{A} = \begin{pmatrix} -4 & 1 & 0 \\ 2 & 0 & 0 \\ -2 & 1 & 0 \end{pmatrix}, \qquad \mathbf{b} = \begin{pmatrix} 1 \\ 3 \\ 2 \end{pmatrix},$$

(b)

$$\mathbf{A} = \begin{pmatrix} 1 & 3 & 2 & 5 & 6 \\ 7 & 2 & 1 & -4 & 5 \\ 1 & 4 & 2 & 13 & 7 \end{pmatrix}, \qquad \mathbf{b} = \begin{pmatrix} 1 \\ 4 \\ 1 \end{pmatrix}.$$

45. An experiment yields the following data:

t	x
0.00	1.001
0.10	1.089
0.23	1.240
0.70	1.654
0.90	1.738
1.50	2.120
2.65	1.412
3.00	1.301

We have 15 times as much confidence in the first four data points than we do in all the others. Find the least squares best fit coefficients a, b, and c if the assumed functional form is

(a) $x = a + bt + ct^2$
(b) $x = a + b\sin t + c\sin 2t$.

Plot on a single graph the data points and the two best fit estimates. Which best fit estimate has the smallest least squares residual?

46. Use a one-, two-, and three-term Neumann series to approximate $(\mathbf{I} - \mathbf{A})^{-1}$ when

(a) $\mathbf{A} = \begin{pmatrix} 1/10 & 1/20 \\ 1/100 & 1/50 \end{pmatrix}$,

(b) $\mathbf{A} = \begin{pmatrix} i/10 & (1+i)/20 \\ i/100 & i/50 \end{pmatrix}$.

Compare the approximation with the exact solution.

47. Prove that

(a) $e^{a\mathbf{I}} = e^a\mathbf{I}$,
(b) $(e^{\mathbf{A}})^{-1} = e^{-\mathbf{A}}$,
(c) $e^{\mathbf{A}(t+s)} = e^{\mathbf{A}t}\,e^{\mathbf{A}s}$.

48. Find $e^{\mathbf{A}}$ if

$$\mathbf{A} = \begin{pmatrix} 1 & 1 & 1 \\ 0 & 3 & 2 \\ 0 & 0 & 5 \end{pmatrix}.$$

49. Find $\exp \begin{pmatrix} 0 & 0 & i \\ 0 & 1 & 0 \\ 1 & 0 & 0 \end{pmatrix}$.

50. Show that

$$e^{\mathbf{A}} = \begin{pmatrix} \cos(1) & \sin(1) \\ -\sin(1) & \cos(1) \end{pmatrix},$$

if

$$\mathbf{A} = \begin{pmatrix} 0 & 1 \\ -1 & 0 \end{pmatrix}.$$

51. Find the standard quadratic form for

$$f(x_1, x_2) = 5.5x_1^2 - 3x_1x_2 + 5.5x_2^2.$$

52. Reduce $4x^2 + 4y^2 + 2z^2 - 4xy + 4yz + 4zx$ to standard quadratic form.

53. Show that the function

$$f(x, y, z) = x^2 + y^2 + z^2 + yz - zx - xy$$

is always nonnegative.

54. Is the quadratic form

$$f(x_1, x_2, x_3) = 4x_1^2 + 2x_1x_2 + 4x_1x_3$$

positive definite?

55. Find the Moore-Penrose pseudoinverse of

(a) $\begin{pmatrix} 2 & -1 & 0 \\ -3 & 2 & 5 \end{pmatrix}$,

(b) $\begin{pmatrix} 1 & 1 & 1 \\ 2 & 2 & 2 \\ 3 & 3 & 3 \end{pmatrix}$,

(c) $\begin{pmatrix} 1 & 1 \\ 2 & 2 \\ 3 & 3 \\ i & 1 \end{pmatrix}$.

8 Linear Integral Equations

In this chapter we introduce an important, though often less emphasized, topic: *integral equations*. We have already touched on related topics in solving for the Green's function in Section 4.3.3 and in an example within Section 6.7.2. Integral equations, and their cousins the integro-differential equations, often arise naturally in engineering problems where nonlocal effects are significant, that is, when what is happening at a given point in space-time is affected by the past or by points at a distance, or by both. They may arise in such areas as radiation heat transfer and statistical mechanics. They also arise in problems involving the Green's functions of linear operators, which may originate from a wide variety of problems in engineering such as heat transfer, elasticity, or electromagnetics. Our focus is on linear integral equations, though one could extend to the nonlinear theory if desired. Our previous study of linear equations of the sort $\mathbf{L}y = f$ has mainly addressed cases where $\mathbf{L}$ is either a linear differential operator or a matrix. Here we take it to be an integral. We are then able to apply standard notions from eigenvalue analysis to aid in the interpretation of the solutions to such equations. When the integral operator is discretized, integral equations can be approximated as linear algebra problems, and all of the methods of Chapter 7 are valid.

8.1 Definitions

We consider integral equations that take the form

$$h(x)y(x) = f(x) + \lambda \int_a^b K(x,s)y(s)\,ds. \tag{8.1}$$

Such an equation is linear in $y(x)$, the unknown dependent variable for which we seek a solution. Here $K(x,s)$, the so-called *kernel*, is known, $h(x)$ and $f(x)$ are known functions, and λ is a constant parameter. We could rewrite Eq. (8.1) as

$$\underbrace{\left(h(x)\,(\cdot)|_{s=x} - \lambda \int_a^b K(x,s)\,(\cdot)\,ds \right)}_{\mathbf{L}} y(s) = f(x), \tag{8.2}$$

so that it takes the explicit form $\mathbf{L}y = f$. Here $(\cdot)$ is a placeholder for the operand. If $f(x) = 0$, our integral equation is *homogeneous*. When a and b are fixed constants,

Eq. (8.1) is called a *Fredholm equation*. If the upper limit is instead the variable x, we have a *Volterra*[1] *equation*:

$$h(x)y(x) = f(x) + \lambda \int_a^x K(x,s)y(s)\,ds. \tag{8.3}$$

A Fredholm equation whose kernel has the property $K(x,s) = 0$ for $s > x$ is in fact a Volterra equation. If one or both of the limits is infinite, the equation is known as a *singular integral equation*, for example,

$$h(x)y(x) = f(x) + \lambda \int_a^\infty K(x,s)y(s)\,ds. \tag{8.4}$$

If $h(x) = 0$, we have what is known as a *Fredholm equation of the first kind*:

$$0 = f(x) + \lambda \int_a^b K(x,s)y(s)\,ds. \tag{8.5}$$

Here we can expect difficulties in solving for $y(s)$ if, for a given x, $K(x,s)$ takes on a value of zero or near zero for $s \in [a,b]$. That is because when $K(x,s) = 0$, it maps all $y(s)$ into zero, rendering the solution nonunique. The closer $K(x,s)$ is to zero, the more challenging it is to estimate $y(s)$.

If $h(x) = 1$, we have a *Fredholm equation of the second kind*:

$$y(x) = f(x) + \lambda \int_a^b K(x,s)y(s)\,ds. \tag{8.6}$$

Equations of this kind have a more straightforward solution than those of the first kind.

8.2 Homogeneous Fredholm Equations

Let us here consider homogeneous Fredholm equations, that is, those with $f(x) = 0$.

8.2.1 First Kind

A homogeneous Fredholm equation of the first kind takes the form

$$0 = \int_a^b K(x,s)y(s)\,ds. \tag{8.7}$$

Solutions to Eq. (8.7) are functions $y(s)$ that lie in the null space of the linear integral operator. Certainly $y(s) = 0$ satisfies, but there may be other nontrivial solutions, based on the nature of the kernel $K(x,s)$. Certainly, for a given x, if there are points or regions where $K(x,s) = 0$ in $s \in [a,b]$, one would expect nontrivial and nonunique $y(s)$ to exist that would still satisfy Eq. (8.7). Also, if $K(x,s)$ oscillates appropriately about zero for $s \in [a,b]$, one may find nontrivial and nonunique $y(s)$.

EXAMPLE 8.1

Find solutions $y(x)$ to the homogeneous Fredholm equation of the first kind:

$$0 = \int_0^1 xsy(s)\,ds. \tag{8.8}$$

---

[1] Vito Volterra, 1860–1940, Italian mathematician.

Assuming $x \neq 0$, we can factor to say

$$0 = \int_0^1 sy(s) \, ds. \tag{8.9}$$

Certainly solutions for y are nonunique. For example, any function that is odd and symmetric about $s = 1/2$ and scaled by s will satisfy, for example,

$$y(x) = C \frac{\sin(2n\pi x)}{x}, \qquad C \in \mathbb{R}^1, \; n \in \mathbb{Q}^1. \tag{8.10}$$

The piecewise function

$$y(x) = \begin{cases} C, & x = 0, \\ 0, & x \in (0, 1], \end{cases} \tag{8.11}$$

also satisfies, where $C \in \mathbb{R}^1$.

8.2.2 Second Kind

A homogeneous Fredholm equation of the second kind takes the form

$$y(x) = \lambda \int_a^b K(x, s) y(s) \, ds. \tag{8.12}$$

Obviously, when $y(s) = 0$, Eq. (8.12) is satisfied. But we might expect that there exist nontrivial eigenfunctions and corresponding eigenvalues that also satisfy Eq. (8.12). This is because Eq. (8.12) takes the form of $(1/\lambda)y = \mathbf{L}y$, where $\mathbf{L}$ is the linear integral operator. In the theory of integral equations, it is more traditional to have the eigenvalue λ play the role of the reciprocal of the usual eigenvalue.

Separable Kernel
In the special case in which the kernel is what is known as a *separable kernel* or *degenerate kernel* with the form

$$K(x, s) = \sum_{i=1}^{N} \phi_i(x) \psi_i(s), \tag{8.13}$$

significant simplification arises. We then substitute into Eq. (8.12) to get

$$y(x) = \lambda \int_a^b \left(\sum_{i=1}^{N} \phi_i(x) \psi_i(s) \right) y(s) \, ds, \tag{8.14}$$

$$= \lambda \sum_{i=1}^{N} \phi_i(x) \underbrace{\int_a^b \psi_i(s) y(s) \, ds}_{c_i}. \tag{8.15}$$

Then we define the constants c_i, $i = 1, \ldots, N$, as

$$c_i = \int_a^b \psi_i(s) y(s) \, ds, \qquad i = 1, \ldots, N, \tag{8.16}$$

and find

$$y(x) = \lambda \sum_{i=1}^{N} c_i \phi_i(x). \tag{8.17}$$

We get the constants c_i by substituting Eq. (8.17) into Eq. (8.16):

$$c_i = \int_a^b \psi_i(s) \lambda \sum_{j=1}^N c_j \phi_j(s) \, ds, \qquad (8.18)$$

$$= \lambda \sum_{j=1}^N c_j \underbrace{\int_a^b \psi_i(s) \phi_j(s) \, ds}_{B_{ij}}. \qquad (8.19)$$

Defining the constant matrix B_{ij} as $B_{ij} = \int_a^b \psi_i(s)\phi_j(s) \, ds$, we then have

$$c_i = \lambda \sum_{j=1}^N B_{ij} c_j. \qquad (8.20)$$

In Gibbs notation, we would say

$$\mathbf{c} = \lambda \mathbf{B} \cdot \mathbf{c}, \qquad (8.21)$$

$$\mathbf{0} = (\lambda \mathbf{B} - \mathbf{I}) \cdot \mathbf{c}. \qquad (8.22)$$

This is an eigenvalue problem for $\mathbf{c}$. Here the reciprocal of the traditional eigenvalues of $\mathbf{B}$ give the values of λ, and the eigenvectors are the associated values of $\mathbf{c}$.

EXAMPLE 8.2

Find the eigenvalues and eigenfunctions for the homogeneous Fredholm equation of the second kind with the degenerate kernel, $K(x, s) = xs$, on the domain $x \in [0, 1]$:

$$y(x) = \lambda \int_0^1 xs y(s) \, ds. \qquad (8.23)$$

The equation simplifies to

$$y(x) = \lambda x \int_0^1 s y(s) \, ds. \qquad (8.24)$$

Take then

$$c = \int_0^1 s y(s) \, ds, \qquad (8.25)$$

so that

$$y(x) = \lambda x c. \qquad (8.26)$$

Thus,

$$c = \int_0^1 s \lambda s c \, ds, \qquad (8.27)$$

$$1 = \lambda \int_0^1 s^2 \, ds, \qquad (8.28)$$

$$= \lambda \left. \frac{s^3}{3} \right|_0^1, \qquad (8.29)$$

$$= \lambda \left(\frac{1}{3} \right), \qquad (8.30)$$

$$\lambda = 3. \qquad (8.31)$$

Thus, there is a single eigenfunction $y = x$ associated with a single eigenvalue $\lambda = 3$. Any constant multiplied by the eigenfunction is also an eigenfunction.

_____

Nonseparable Kernel

For many problems, the kernel is not separable, and we must resort to numerical methods. Let us consider Eq. (8.12) with $a = 0, b = 1$:

$$y(x) = \lambda \int_0^1 K(x, s) y(s) \, ds. \tag{8.32}$$

Now, although there are many sophisticated numerical methods to evaluate the integral in Eq. (8.32), it is easiest to convey our ideas via the simplest method: the rectangular rule with evenly spaced intervals. Let us distribute N points uniformly in $x \in [0, 1]$ so that $x_i = (i - 1)/(N - 1)$, $i = 1, \ldots, N$. We form the same distribution for $s \in [0, 1]$ with $s_j = (j - 1)/(N - 1)$, $j = 1, \ldots, N$. For a given $x = x_i$, this distribution defines $N - 1$ rectangles of width $\Delta s = 1/(N - 1)$ and of height $K(x_i, s_j) \equiv K_{ij}$. We can think of K_{ij} as a matrix of dimension $(N - 1) \times (N - 1)$. We can estimate the integral by adding the areas of all of the individual rectangles. By the nature of the rectangular rule, this method has a small asymmetry that ignores the influence of the function values at $i = j = N$. In the limit of large N, this is not a problem. Next, let $y(x_i) \equiv y_i$, $i = 1, \ldots, N - 1$, and $y(s_j) \equiv y_j$, $j = 1, \ldots, N - 1$, and write Eq. (8.32) in a discrete approximation as

$$y_i = \lambda \sum_{j=1}^{N-1} K_{ij} y_j \Delta s. \tag{8.33}$$

In vector form, we could say

$$\mathbf{y} = \lambda \mathbf{K} \cdot \mathbf{y} \Delta s, \tag{8.34}$$

$$0 = \left(\mathbf{K} - \frac{1}{\lambda \Delta s} \mathbf{I} \right) \cdot \mathbf{y}, \tag{8.35}$$

$$= (\mathbf{K} - \sigma \mathbf{I}) \cdot \mathbf{y}. \tag{8.36}$$

Obviously, this is a eigenvalue problem in linear algebra. The eigenvalues of $\mathbf{K}$, $\sigma_i = 1/(\lambda_i \Delta s)$, $i = 1, \ldots, N - 1$, approximate the eigenvalues of $\mathbf{L} = \int_0^1 K(x, s)(\cdot) \, ds$, and the eigenvectors are approximations to the eigenfunctions of $\mathbf{L}$.

_____

EXAMPLE 8.3

Find numerical approximations of the first nine eigenvalues and eigenfunctions of the homogeneous Fredholm equation of the second kind:

$$y(x) = \lambda \int_0^1 \sin(10xs) y(s) \, ds. \tag{8.37}$$

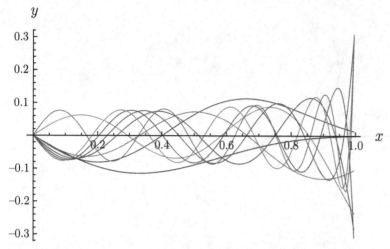

Figure 8.1. First nine eigenfunctions for $y(x) = \lambda \int_0^1 \sin(10xs)y(s)\, ds$.

Discretization leads to a matrix equation in the form of Eq. (8.34). For display purposes only, we examine a coarse discretization of $N = 6$. In this case, our discrete equation is

$$
\underbrace{\begin{pmatrix} y_1 \\ y_2 \\ y_3 \\ y_4 \\ y_5 \end{pmatrix}}_{\mathbf{y}} = \lambda \underbrace{\begin{pmatrix} 0 & 0 & 0 & 0 & 0 \\ 0 & 0.389 & 0.717 & 0.932 & 1.000 \\ 0 & 0.717 & 1.000 & 0.675 & -0.058 \\ 0 & 0.932 & 0.675 & -0.443 & -0.996 \\ 0 & 1.000 & -0.058 & -0.996 & 0.117 \end{pmatrix}}_{\mathbf{K}} \underbrace{\begin{pmatrix} y_1 \\ y_2 \\ y_3 \\ y_4 \\ y_5 \end{pmatrix}}_{\mathbf{y}} \underbrace{\left(\frac{1}{5} \right)}_{\Delta s}. \tag{8.38}
$$

Obviously, $\mathbf{K}$ is not full rank because of the row and column of zeros. In fact it has a rank of 4. The zeros exist because $K(x, s) = 0$ for both $x = 0$ and $s = 0$. This, however, poses no issues for computing the eigenvalues and eigenvectors. However, $N = 6$ is too small to resolve either the eigenvalues or eigenfunctions of the underlying continuous operator. Choosing $N = 201$ points gives acceptable resolution for the first nine eigenvalues, which are

$$\lambda_1 = 2.523, \qquad \lambda_2 = -2.526, \qquad \lambda_3 = 2.792, \tag{8.39}$$

$$\lambda_4 = -7.749, \qquad \lambda_5 = 72.867, \qquad \lambda_6 = -1225.2, \tag{8.40}$$

$$\lambda_7 = 3.014 \times 10^4, \qquad \lambda_8 = -1.011 \times 10^6, \qquad \lambda_9 = 4.417 \times 10^7. \tag{8.41}$$

The corresponding eigenfunctions are plotted in Figure 8.1.

EXAMPLE 8.4

Find numerical approximations of the first six eigenvalues and eigenfunctions of the homogeneous Fredholm equation of the second kind:

$$y(x) = \lambda \int_0^1 g(x, s)y(s)\, ds, \tag{8.42}$$

where

$$g(x, s) = \begin{cases} x(s - 1), & x \leq s, \\ s(x - 1), & x \geq s. \end{cases} \tag{8.43}$$

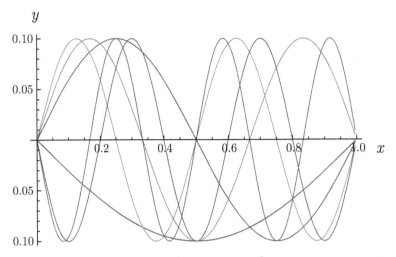

Figure 8.2. First six eigenfunctions for $y(x) = \lambda \int_0^1 g(x,s)y(s)\,ds$, where $g(x,s)$ is the Green's function for $d^2y/dx^2 = f(x)$, $y(0) = y(1) = 0$.

This kernel is the Green's function for a problem studied earlier, Eqs. (4.168,4.169), $d^2y/dx^2 = f(x)$ with $y(0) = y(1) = 0$. The Green's function solution is $y(x) = \int_0^1 g(x,s)f(s)\,ds$. For our example problem, we have $f(s) = \lambda y(s)$; thus, we are also solving the eigenvalue problem $d^2y/dx^2 = \lambda y$.

Choosing $N = 201$ points gives acceptable resolution for the first six eigenvalues, which are

$$\lambda_1 = -9.869, \qquad \lambda_2 = -39.48, \qquad \lambda_3 = -88.81, \tag{8.44}$$

$$\lambda_4 = -157.9, \qquad \lambda_5 = -246.6, \qquad \lambda_6 = -355.0. \tag{8.45}$$

These compare well with the known eigenvalues of $\lambda = -n^2\pi^2$, $n = 1, 2, \ldots$:

$$\lambda_1 = -9.870, \qquad \lambda_2 = -39.48, \qquad \lambda_3 = -88.83, \tag{8.46}$$

$$\lambda_4 = -157.9, \qquad \lambda_5 = -246.7, \qquad \lambda_6 = -355.3. \tag{8.47}$$

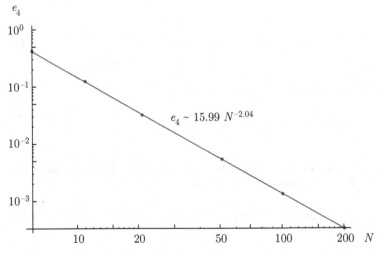

Figure 8.3. Convergence of the relative error in approximation of $\lambda_4 = -16\pi^2$ for $y(x) = \lambda \int_0^1 g(x,s)y(s)\,ds$, where $g(x,s)$ is the Green's function for $d^2y/dx^2 = f(x)$, $y(0) = y(1) = 0$.

The corresponding eigenfunctions are plotted in Figure 8.2. The eigenfunctions appear to approximate well the known eigenfunctions $\sin(n\pi x)$, $n = 1, 2, \ldots$.

We can gain some understanding of the accuracy of our method by studying how its error converges as the number of terms in the approximation increases. There are many choices as to how to evaluate the error. Here let us choose one of the eigenvalues, say, λ_4; others could have been chosen. We know the exact value is $\lambda_4 = -16\pi^2$. Let us take the relative error to be

$$e_4 = \frac{|\lambda_{4N} + 16\pi^2|}{16\pi^2}, \tag{8.48}$$

where λ_{4N} here is understood to be the numerical approximation to λ_4, which is a function of N. Figure 8.3 shows the convergence, which is well approximated by the curve fit $e_4 \approx 15.99 N^{-2.04}$.

8.3 Inhomogeneous Fredholm Equations

Inhomogeneous integral equations can also be studied, and we do so here.

8.3.1 First Kind

EXAMPLE 8.5

Consider solutions $y(x)$ to the inhomogeneous Fredholm equation of the first kind:

$$0 = x + \int_0^1 \sin(10xs)y(s)\, ds. \tag{8.49}$$

Here we have $f(x) = x$, $\lambda = 1$, and $K(x, s) = \sin(10xs)$. For a given value of x, we have $K(x, s) = 0$, when $s = 0$, and so we expect a nonunique solution for y.

Let us once again solve this by discretization techniques identical to previous examples. In short,

$$0 = f(x) + \int_0^1 K(x, s)y(s)\, ds, \tag{8.50}$$

leads to the matrix equation

$$0 = \mathbf{f} + \mathbf{K} \cdot \mathbf{y}\Delta s, \tag{8.51}$$

where $\mathbf{f}$ is a vector of length $N - 1$ containing the values of $f(x_i)$, $i = 1, \ldots, N - 1$, $\mathbf{K}$ is a matrix of dimension $(N - 1) \times (N - 1)$ populated by values of $K(x_i, s_j)$, $i = 1, \ldots, N - 1$, $j = 1, \ldots, N - 1$, and $\mathbf{y}$ is a vector of length $N - 1$ containing the unknown values of $y(x_j)$, $j = 1, \ldots, N - 1$.

When we evaluate the rank of $\mathbf{K}$, we find for $K(x, s) = \sin(xs)$ that the rank of the discrete $\mathbf{K}$ is $r = N - 2$. This is because $K(x, s)$ evaluates to zero at $x = 0$ and $s = 0$. Thus, the right null space is of dimension unity. Now, we have no guarantee that $\mathbf{f}$ lies in the column space of $\mathbf{K}$, so the best we can imagine is that there exists a unique solution $\mathbf{y}$ that minimizes $\|\mathbf{f} + \mathbf{K} \cdot \mathbf{y}\Delta s\|_2$ that itself has no components in the null space of $\mathbf{K}$, so that $\mathbf{y}$ itself is of minimum "length." So, we say our best $\mathbf{y}$ is

$$\mathbf{y} = -\frac{1}{\Delta s}\mathbf{K}^+ \cdot \mathbf{f}, \tag{8.52}$$

where $\mathbf{K}^+$ is the Moore-Penrose pseudoinverse of $\mathbf{K}$.

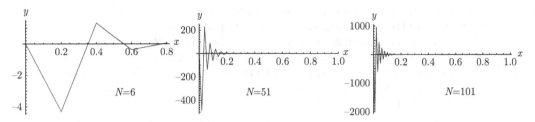

Figure 8.4. Approximations $\mathbf{y} \approx y(x)$ that have minimum norm while best satisfying the discrete Fredholm equation of the first kind $\mathbf{0} = \mathbf{f} + \mathbf{K} \cdot \mathbf{y}\Delta s$, modeling the continuous $0 = x + \int_0^1 \sin(xs)y(s)\, ds$.

Letting $N = 6$ gives rise to the matrix equation

$$
\underbrace{\begin{pmatrix} 0 \\ 0 \\ 0 \\ 0 \\ 0 \end{pmatrix}}_{\mathbf{0}} = \underbrace{\begin{pmatrix} 0 \\ 0.2 \\ 0.4 \\ 0.6 \\ 0.8 \end{pmatrix}}_{\mathbf{f}} + \underbrace{\begin{pmatrix} 0 & 0 & 0 & 0 & 0 \\ 0 & 0.389 & 0.717 & 0.932 & 1.000 \\ 0 & 0.717 & 1.000 & 0.675 & -0.058 \\ 0 & 0.932 & 0.675 & -0.443 & 0.996 \\ 0 & 1.000 & -0.058 & 0.996 & 0.117 \end{pmatrix}}_{\mathbf{K}} \underbrace{\begin{pmatrix} y_1 \\ y_2 \\ y_2 \\ y_4 \\ y_5 \end{pmatrix}}_{\mathbf{y}} \underbrace{\left(\tfrac{1}{5} \right)}_{\Delta s}. \tag{8.53}
$$

Solving for the $\mathbf{y}$ of minimum length that minimizes $\|\mathbf{f} + \mathbf{K} \cdot \mathbf{y}\Delta s\|_2$, we find

$$
\mathbf{y} = \begin{pmatrix} 0 \\ -4.29 \\ 1.32 \\ -0.361 \\ 0.057 \end{pmatrix}. \tag{8.54}
$$

We see by inspection that the vector $(1, 0, 0, 0, 0)^T$ lies in the right null space of $\mathbf{K}$. So $\mathbf{K}$ operating on any scalar multiple, α, of this null space vector maps into zero and does not contribute to the error. So the following set of solution vectors $\mathbf{y}$ all have the same error in approximation:

$$
\mathbf{y} = \begin{pmatrix} \alpha \\ -4.29 \\ 1.32 \\ -0.361 \\ 0.057 \end{pmatrix}, \qquad \alpha \in \mathbb{R}^1. \tag{8.55}
$$

We also find the error to be, for $N = 6$,

$$
\|\mathbf{f} + \mathbf{K} \cdot \mathbf{y}\Delta s\|_2 = 0. \tag{8.56}
$$

Because the error is zero, we have selected a function $f(x) = x$ whose discrete approximation lies in the column space of $\mathbf{K}$; for more general functions, this will not be the case. This is a consequence of our selected function, $f(x) = x$, evaluating to zero at $x = 0$. For example, for $f(x) = x + 2$, we would find $\|\mathbf{f} + \mathbf{K} \cdot \mathbf{y}\Delta s\|_2 = f(0) = 2$, with all of the error at $x = 0$ and none at other points in the domain.

This seems to be a rational way to approximate the best continuous $y(x)$ to satisfy the continuous integral equation. However, as N increases, we find the approximation $\mathbf{y}$ does not converge to a finite well-behaved function, as displayed for $N = 6, 51, 101$ in Figure 8.4. This lack of convergence is likely related to the ill-conditioned nature of $\mathbf{K}$. For $N = 6$, the condition number c, which is the ratio of the largest and smallest singular values (see Eq. (7.125)), is $c = 45$; for $N = 51$, we find $c = 10^{137}$; for $N = 101, c = 10^{232}$. This ill-conditioned behavior is typical for Fredholm equations of the first kind. Although the function itself does not converge with increasing N, the error $\|\mathbf{f} + \mathbf{K} \cdot \mathbf{y}\Delta s\|_2$ remains zero for all N for $f(x) = x$ (or any other f that has $f(0) = 0$).

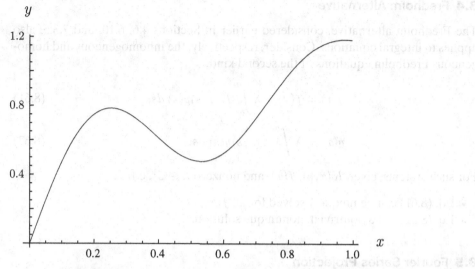

Figure 8.5. The function $y(x)$ that solves the Fredholm equation of the second kind $y(x) = x + \int_0^1 \sin(10xs)y(s)\,ds$.

8.3.2 Second Kind

EXAMPLE 8.6

Identify the solution $y(x)$ to the inhomogeneous Fredholm equation of the second kind:

$$y(x) = x + \int_0^1 \sin(10xs)y(s)\,ds. \tag{8.57}$$

Again, we have $f(x) = x$, $\lambda = 1$, and $K(x, s) = \sin(10xs)$. Let us once again solve this by discretization techniques identical to previous examples. In short,

$$y(x) = f(x) + \int_0^1 K(x, s)y(s)\,ds, \tag{8.58}$$

leads to the matrix equation

$$\mathbf{y} = \mathbf{f} + \mathbf{K} \cdot \mathbf{y}\Delta s, \tag{8.59}$$

where $\mathbf{f}$ is a vector of length $N - 1$ containing the values of $f(x_i)$, $i = 1, \ldots, N - 1$, $\mathbf{K}$ is a matrix of dimension $(N - 1) \times (N - 1)$ populated by values of $K(x_i, s_j)$, $i = 1, \ldots, N - 1$, $j = 1, \ldots, N - 1$, and $\mathbf{y}$ is a vector of length $N - 1$ containing the unknown values of $y(x_j)$, $j = 1, \ldots, N - 1$.

Solving for $\mathbf{y}$, we find

$$\mathbf{y} = (\mathbf{I} - \mathbf{K}\Delta s)^{-1} \cdot \mathbf{f}. \tag{8.60}$$

The matrix $\mathbf{I} - \mathbf{K}\Delta s$ is not singular, and thus we find a unique solution. The only error in this solution is that associated with the discrete nature of the approximation. This discretization error approaches zero as N becomes large. The converged solution is plotted in Figure 8.5. In contrast to Fredholm equations of the first kind, those of the second kind generally have unambiguous solution.

8.4 Fredholm Alternative

The Fredholm alternative, considered earlier in Sections 4.6, 6.10, and 7.3.5, also applies to integral equations. Consider, respectively, the inhomogeneous and homogeneous Fredholm equations of the second kind,

$$y(x) = f(x) + \lambda \int_a^b K(x,s)y(s)\,ds, \tag{8.61}$$

$$y(x) = \lambda \int_a^b K(x,s)y(s)\,ds. \tag{8.62}$$

For such systems, given $K(x,s)$, $f(x)$, and nonzero $\lambda \in \mathbb{C}^1$, either

- Eq. (8.61) can be uniquely solved for all $f(x)$, or
- Eq. (8.62) has a nontrivial, nonunique solution.

8.5 Fourier Series Projection

Analogous to the method considered first in Section 4.5, we can use the eigenfunctions of the linear integral operator as a basis on which to project a general function. This then yields a Fourier series approximation of the general function.

First let us take the inner product to be defined in a typical fashion for functions $u, v \in \mathbb{L}_2[0,1]$:

$$\langle u,v \rangle = \int_0^1 u(s)v(s)\,ds. \tag{8.63}$$

If $u(s)$ and $v(s)$ are sampled at N uniformly spaced points in the domain $s \in [0,1]$, with $s_1 = 0$, $s_N = 1$, $\Delta s = 1/(N-1)$, the inner product can be approximated by what amounts to the rectangular method of numerical integration:

$$\langle u,v \rangle \approx \sum_{n=1}^{N-1} u_n v_n \Delta s. \tag{8.64}$$

Then, if we consider u_n, v_n, to be the components of vectors $\mathbf{u}$ and $\mathbf{v}$, each of length $N-1$, we can cast the inner product as

$$\langle u,v \rangle \approx (\mathbf{u}^T \cdot \mathbf{v})\Delta s, \tag{8.65}$$

$$\approx \left(\mathbf{u}\sqrt{\Delta s}\right)^T \cdot \left(\mathbf{v}\sqrt{\Delta s}\right). \tag{8.66}$$

The functions u and v are orthogonal if $\langle u,v \rangle = 0$ when $u \neq v$. The norm of a function is, as usual,

$$\|u\|_2 = \sqrt{\langle u,u \rangle} = \sqrt{\int_0^1 u^2(s)\,ds}. \tag{8.67}$$

In the discrete approximation, we have

$$\|u\|_2 \approx \sqrt{(\mathbf{u}^T \cdot \mathbf{u})\Delta s}, \tag{8.68}$$

$$\approx \sqrt{\left(\mathbf{u}\sqrt{\Delta s}\right)^T \cdot \left(\mathbf{u}\sqrt{\Delta s}\right)}. \tag{8.69}$$

Now consider the integral equation defining our eigenfunctions $y(x)$, Eq. (8.32):

$$y(x) = \lambda \int_0^1 K(x, s) y(s) \, ds. \tag{8.70}$$

We restrict attention to problems where $K(x, s) = K(s, x)$. With this, the integral operator is self-adjoint, and the eigenfunctions are thus guaranteed to be orthogonal, as has been shown in Section 6.7.2. Consequently, we are dealing with a problem from Hilbert-Schmidt theory. Discretization, as before, leads to Eq. (8.36):

$$0 = (\mathbf{K} - \sigma \mathbf{I}) \cdot \mathbf{y}. \tag{8.71}$$

Because $K(x, s) = K(s, x)$, its discrete form gives $K_{ij} = K_{ji}$. Thus, $\mathbf{K} = \mathbf{K}^T$, and the discrete operator is self-adjoint. We find a set of $N - 1$ eigenvectors, each of length $N - 1$, $\mathbf{y}_i$, $i = 1, \ldots, N - 1$.

Now if $y_i(x)$ is the eigenfunction, we can define a corresponding orthonormal eigenfunction $\varphi_i(x)$ by scaling $y_i(x)$ by its norm:

$$\varphi_i(x) = \frac{y_i(x)}{||y_i||_2} = \frac{y_i(x)}{\sqrt{\int_0^1 y_i^2(s) \, ds}}. \tag{8.72}$$

The discrete analog is

$$\phi_i = \frac{\mathbf{y}_i}{\sqrt{(\mathbf{y}_i^T \cdot \mathbf{y}_i) \Delta s}}, \tag{8.73}$$

$$= \frac{\mathbf{y}_i}{\sqrt{(\mathbf{y}_i \sqrt{\Delta s})^T \cdot (\mathbf{y}_i \sqrt{\Delta s})}}. \tag{8.74}$$

Now for an M-term Fourier series, we take the discrete version of Eq. (4.504) to approximate $f(x)$ by

$$f(x) \approx f(x_j) = \mathbf{f}_p^T = \sum_{i=1}^{M} \alpha_i \varphi_i(x_j) = \boldsymbol{\alpha}^T \cdot \boldsymbol{\Phi}. \tag{8.75}$$

Here $\mathbf{f}_p$ is an $(N - 1) \times 1$ vector containing the projection of f, $\boldsymbol{\Phi}$ is a matrix of dimension $M \times (N - 1)$ with each row populated by an eigenvector ϕ_i:

$$\boldsymbol{\Phi} = \begin{pmatrix} \cdots & \phi_1 & \cdots \\ \cdots & \phi_2 & \cdots \\ & \vdots & \\ \cdots & \phi_M & \cdots \end{pmatrix}. \tag{8.76}$$

So if $M = 4$, we would have the approximation

$$f(x) = f(x_j) = \mathbf{f}_p = \alpha_1 \phi_1 + \alpha_2 \phi_2 + \alpha_3 \phi_3 + \alpha_4 \phi_4. \tag{8.77}$$

However, we need an expression for the Fourier coefficients $\boldsymbol{\alpha}$. Now in the continuous limit, Eq. (4.511) holds that $\alpha_i = \int_0^1 f(s)\varphi_i(s)\,ds$. The discrete analog of this is

$$\boldsymbol{\alpha}^T = (\mathbf{f}^T \cdot \boldsymbol{\Phi}^T)\Delta s, \tag{8.78}$$

$$= (\mathbf{f}\sqrt{\Delta s})^T \cdot (\boldsymbol{\Phi}\sqrt{\Delta s})^T, \tag{8.79}$$

$$\boldsymbol{\alpha} = (\boldsymbol{\Phi}\sqrt{\Delta s}) \cdot (\mathbf{f}\sqrt{\Delta s}), \tag{8.80}$$

$$\frac{\boldsymbol{\alpha}}{\sqrt{\Delta s}} = (\boldsymbol{\Phi}\sqrt{\Delta s}) \cdot \mathbf{f}. \tag{8.81}$$

The vector $\mathbf{f}$ is of length $N-1$ and contains the values of $f(x)$ evaluated at each x_i. When $M = N-1$, the matrix $\boldsymbol{\Phi}\sqrt{\Delta s}$ is square and, moreover, orthogonal. Thus, its norm is unity, and its transpose is its inverse. When square, it can always be constructed such that its determinant is unity, thus rendering it to be a rotation. In this case, $\mathbf{f}$ is rotated by $\boldsymbol{\Phi}\sqrt{\Delta s}$ to form $\boldsymbol{\alpha}/\sqrt{\Delta s}$.

We could also represent Eq. (8.75) as

$$\mathbf{f}_p^T = \frac{\boldsymbol{\alpha}^T}{\sqrt{\Delta s}} \cdot \left(\boldsymbol{\Phi}\sqrt{\Delta s}\right), \tag{8.82}$$

$$\mathbf{f}_p = \left(\boldsymbol{\Phi}^T \sqrt{\Delta s}\right) \cdot \frac{\boldsymbol{\alpha}}{\sqrt{\Delta s}}. \tag{8.83}$$

Using Eq. (8.81) to eliminate $\boldsymbol{\alpha}/\sqrt{\Delta s}$ in Eq. (8.83), we can say

$$\mathbf{f}_p = \underbrace{\left(\boldsymbol{\Phi}^T \sqrt{\Delta s}\right) \cdot \left(\boldsymbol{\Phi}\sqrt{\Delta s}\right)}_{\mathbf{P}} \cdot \mathbf{f}. \tag{8.84}$$

The matrix $\left(\boldsymbol{\Phi}^T \sqrt{\Delta s}\right) \cdot \left(\boldsymbol{\Phi}\sqrt{\Delta s}\right)$ is a projection matrix $\mathbf{P}$:

$$\mathbf{P} = \left(\boldsymbol{\Phi}^T \sqrt{\Delta s}\right) \cdot \left(\boldsymbol{\Phi}\sqrt{\Delta s}\right). \tag{8.85}$$

The matrix $\mathbf{P}$ has dimension $(N-1) \times (N-1)$ and is of rank M. It has M eigenvalues of unity and $N-1-M$ eigenvalues that are zero. If $\boldsymbol{\Phi}$ is square, $\mathbf{P}$ becomes the identity matrix $\mathbf{I}$, rendering $\mathbf{f}_p = \mathbf{f}$, and no information is lost. That is, the approximation at each of the $N-1$ points is exact. Still, if the underlying function $f(x)$ has fine-scale structures, one must take $N-1$ to be sufficiently large to capture those structures.

EXAMPLE 8.7

Find a Fourier series approximation for the function $f(x) = 1 - x^2$, $x \in [0, 1]$, where the basis functions are the orthonormalized eigenfunctions of the integral equation

$$y(x) = \lambda \int_0^1 \sin(10xs)y(s)\,ds. \tag{8.86}$$

We have found the unnormalized eigenfunction approximation $\mathbf{y}$ in an earlier example by solving the discrete equation

$$\mathbf{0} = (\mathbf{K} - \sigma\mathbf{I}) \cdot \mathbf{y}. \tag{8.87}$$

Here $\mathbf{K}$ is of dimension $(N-1) \times (N-1)$, is populated by $\sin(10 x_i s_j)$, $i, j = 1, \ldots, N-1$, and is obviously symmetric.

Let us first select a coarse approximation with $N = 6$. Thus, $\Delta s = 1/(N-1) = 1/5$. This yields the same $\mathbf{K}$ we saw earlier in Eq. (8.38):

$$\mathbf{K} = \begin{pmatrix} 0 & 0 & 0 & 0 & 0 \\ 0 & 0.389 & 0.717 & 0.932 & 1.000 \\ 0 & 0.717 & 1.000 & 0.675 & -0.058 \\ 0 & 0.932 & 0.675 & -0.443 & -0.996 \\ 0 & 1.000 & -0.058 & -0.996 & 0.117 \end{pmatrix}. \tag{8.88}$$

We then find the eigenvectors of $\mathbf{K}$ and use them to construct the matrix $\boldsymbol{\Phi}$. For completeness, we present $\boldsymbol{\Phi}$ for the case where $M = N - 1 = 5$:

$$\boldsymbol{\Phi}_{5\times5} = \begin{pmatrix} \cdots & \phi_1 & \cdots \\ \cdots & \phi_2 & \cdots \\ \cdots & \phi_3 & \cdots \\ \cdots & \phi_4 & \cdots \\ \cdots & \phi_5 & \cdots \end{pmatrix} = \begin{pmatrix} 0 & 1.3 & 1.6 & 0.86 & 0.21 \\ 0 & 1.1 & 0.051 & -1.5 & -1.2 \\ 0 & 0.88 & -0.52 & -0.86 & 1.8 \\ 0 & -1.1 & 1.5 & -1.1 & 0.44 \\ 2.2 & 0 & 0 & 0 & 0 \end{pmatrix}. \tag{8.89}$$

Now, let us consider an $M = 3$-term Fourier series approximation. Then we will restrict attention to the first three eigenfunctions and consider $\boldsymbol{\Phi}$ to be a matrix of dimension $M \times (N-1) = 3 \times 5$:

$$\boldsymbol{\Phi} = \begin{pmatrix} \cdots & \phi_1 & \cdots \\ \cdots & \phi_2 & \cdots \\ \cdots & \phi_3 & \cdots \end{pmatrix} = \begin{pmatrix} 0 & 1.3 & 1.6 & 0.86 & 0.21 \\ 0 & 1.1 & 0.051 & -1.5 & -1.2 \\ 0 & 0.88 & -0.52 & -0.86 & 1.8 \end{pmatrix}. \tag{8.90}$$

Now we consider the value of $f(x)$ at each of the $N-1$ sample points given in the vector $\mathbf{x}$:

$$\mathbf{x} = \begin{pmatrix} 0 \\ \frac{1}{5} \\ \frac{2}{5} \\ \frac{3}{5} \\ \frac{4}{5} \end{pmatrix} = \begin{pmatrix} 0 \\ 0.2 \\ 0.4 \\ 0.6 \\ 0.8 \end{pmatrix}. \tag{8.91}$$

At each point $f(x)$ gives us the vector $\mathbf{f}$, of length $N-1$:

$$\mathbf{f} = \begin{pmatrix} 1 \\ \frac{24}{25} \\ \frac{21}{25} \\ \frac{16}{25} \\ \frac{9}{25} \end{pmatrix} = \begin{pmatrix} 1.00 \\ 0.96 \\ 0.84 \\ 0.64 \\ 0.36 \end{pmatrix}. \tag{8.92}$$

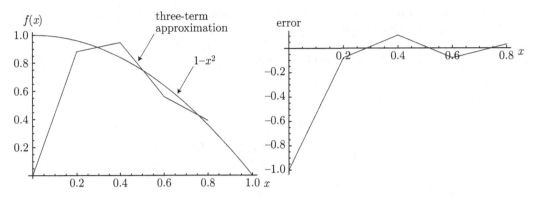

Figure 8.6. $M = 3$-term Fourier approximation of $f(x) = 1 - x^2$, $x \in [0, 1]$ where the basis functions are eigenfunctions of the $N - 1 = 5$-term discretization of the integral operator with the symmetric kernel $K(x, s) = \sin(10xs)$, along with the error distribution.

We can find the projected value $\mathbf{f}_p$ by direct application of Eq. (8.84):

$$
\mathbf{f}_p = \underbrace{\left(\underbrace{\begin{pmatrix} 0 & 0 & 0 \\ 1.3 & 1.1 & 0.88 \\ 1.6 & 0.051 & -0.52 \\ 0.86 & -1.5 & -0.86 \\ 0.21 & -1.2 & 1.8 \end{pmatrix}}_{\mathbf{\Phi}^T} \underbrace{\sqrt{\frac{1}{5}}}_{\sqrt{\Delta s}} \right) \left(\underbrace{\begin{pmatrix} 0 & 1.3 & 1.6 & 0.86 & 0.21 \\ 0 & 1.1 & 0.051 & -1.5 & -1.2 \\ 0 & 0.88 & -0.52 & -0.86 & 1.8 \end{pmatrix}}_{\mathbf{\Phi}} \underbrace{\sqrt{\frac{1}{5}}}_{\sqrt{\Delta s}} \right)}_{\mathbf{P}} \underbrace{\begin{pmatrix} 1.00 \\ 0.96 \\ 0.84 \\ 0.64 \\ 0.36 \end{pmatrix}}_{\mathbf{f}},
$$

$$
\tag{8.93}
$$

$$
= \begin{pmatrix} 0 \\ 0.88 \\ 0.95 \\ 0.56 \\ 0.39 \end{pmatrix}. \tag{8.94}
$$

A plot of the $M = 3$-term approximation for $N - 1 = 5$ superposed onto the exact solution, and in a separate plot, the error distribution, is shown in Figure 8.6. We see the approximation is generally a good one even with only three terms. At $x = 0$, the approximation is bad because all the selected basis functions evaluate to zero there, while the function evaluates to unity.

The Fourier coefficients $\boldsymbol{\alpha}$ are found from Eq. (8.80) and are given by

$$
\boldsymbol{\alpha} = \left(\begin{pmatrix} 0 & 1.3 & 1.6 & 0.86 & 0.21 \\ 0 & 1.1 & 0.051 & -1.5 & -1.2 \\ 0 & 0.88 & -0.52 & -0.86 & 1.8 \end{pmatrix} \sqrt{\frac{1}{5}} \right) \left(\begin{pmatrix} 1.00 \\ 0.96 \\ 0.84 \\ 0.64 \\ 0.36 \end{pmatrix} \sqrt{\frac{1}{5}} \right) = \begin{pmatrix} 0.64 \\ -0.061 \\ 0.10 \end{pmatrix}.
$$

$$
\tag{8.95}
$$

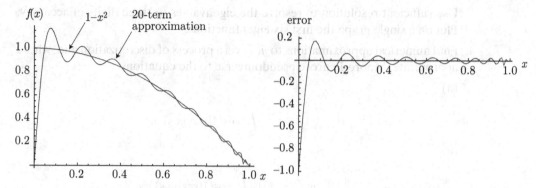

Figure 8.7. $M = 20$-term Fourier approximation of $f(x) = 1 - x^2$, $x \in [0, 1]$ where the basis functions are eigenfunctions of the $N - 1 = 100$-term discretization of the integral operator with the symmetric kernel $K(x, s) = \sin(10xs)$, along with the error distribution.

So the Fourier series is

$$
\mathbf{f}_p = \alpha_1 \boldsymbol{\phi}_1 + \alpha_2 \boldsymbol{\phi}_2 + \alpha_3 \boldsymbol{\phi}_3 = 0.64 \begin{pmatrix} 0 \\ 1.3 \\ 1.6 \\ 0.86 \\ 0.21 \end{pmatrix} - 0.061 \begin{pmatrix} 0 \\ 1.1 \\ 0.051 \\ -1.5 \\ -1.2 \end{pmatrix} + 0.10 \begin{pmatrix} 0 \\ 0.88 \\ -0.52 \\ -0.86 \\ 1.8 \end{pmatrix} = \begin{pmatrix} 0 \\ 0.88 \\ 0.95 \\ 0.56 \\ 0.39 \end{pmatrix}.
$$

(8.96)

If we increase $N - 1$, while holding M fixed, our basis functions become smoother, but the error remains roughly the same. If we increase M while holding $N - 1$ fixed, we can reduce the error; we achieve no error when $M = N - 1$. Let us examine a case where $N - 1 = 100$, so the basis functions are much smoother, and $M = 20$, so the error is reduced. A plot of the $M = 20$-term approximation for $N - 1 = 100$ superposed onto the exact solution, and in a separate plot, the error distribution, is shown in Figure 8.7.

EXERCISES

1. Solve the Volterra equation

$$
a + \int_0^t e^{bs} u(s) \, ds = a e^{bt}.
$$

Hint: Differentiate.

2. Find any and all eigenvalues λ and associated eigenfunctions y that satisfy

$$
y(x) = \lambda \int_0^1 \frac{x}{s} y(s) \, ds.
$$

3. Find a numerical approximation to the first six eigenvalues and eigenfunctions of

$$
y(x) = \lambda \int_0^1 \cos(10xs) y(s) \, ds.
$$

Use sufficient resolution to resolve the eigenvalues to three digits of accuracy. Plot on a single graph the first six eigenfunctions.

4. Find numerical approximations to $y(x)$ via a process of discretization and, where appropriate, Moore-Penrose pseudoinverse to the equations

(a)

$$0 = x + \int_0^1 \cos(10xs)y(s)\, ds,$$

(b)

$$y(x) = x + \int_0^1 \cos(10xs)y(s)\, ds.$$

In each, demonstrate whether the solution converges as the discretization is made finer.

5. Find any and all solutions, $y(x)$, that best satisfy

(a) $y(x) = \int_0^1 y(s)\, ds,$
(b) $y(x) = x + \int_0^1 y(s)\, ds,$
(c) $y(x) = \int_0^1 x^2 s^2 y(s)\, ds,$
(d) $y(x) = x^2 + \int_0^1 x^2 s^2 y(s)\, ds.$

6. Using the eigenfunctions $y_i(x)$ of the equation

$$y(x) = \lambda \int_0^1 e^{xs} y(s)\, ds,$$

approximate the following functions $f(x)$ for $x \in [0, 1]$ in 10-term expansions of the form

$$f(x) = \sum_{i=1}^{10} \alpha_i y_i(x);$$

(a) $f(x) = x,$
(b) $f(x) = \sin(\pi x).$

The eigenfunctions will need to be estimated by numerical approximation.

9 Dynamical Systems

In this chapter, we consider the time-like evolution of sets of state variables, a subject often called *dynamical systems*. We begin with a brief consideration of discrete dynamical systems known as *iterated maps*, which are a nonlinear extension of the difference equations posed in Section 4.9 or the finite difference method posed in Example 4.24. We also show some of the striking geometries that result from iterated maps known as *fractals*. Generally, however, we are concerned with systems that can be described by sets of ordinary differential equations, both linear and nonlinear. In that solution to nonlinear differential equations is usually done in discrete form, there is a strong connection to iterated maps. And we shall see that discrete and continuous dynamical systems share many features.

This final chapter appropriately coalesces many topics studied earlier: discrete systems, ordinary differential equations, perturbation analysis, linear algebra, and geometry. Its use in modeling physical engineering systems is widespread and comes in two main classes: (1) systems that are modeled by coupled ordinary differential equations and (2) systems that are modeled by one or more partial differential equations. Systems of the first type often arise in time-dependent problems with spatial homogeneity in which discrete entities interact. Systems of the second type generally involve time-evolution of interacting systems with spatial inhomogeneity and reduce to large systems of ordinary differential equations following either (1) discretization of one or more independent variables or (2) projection of the dependent variable onto a finite basis in function space. The chapter closes with examples that draw problems from the domain of partial differential equations into the domain of ordinary differential equations and thus demonstrates how the methods of this book can be brought to bear on this critical area of mathematics of engineering systems.

9.1 Iterated Maps

Similar to the nonlinear difference equation introduced by Eq. (4.610), a map $f_n :$ $\mathbb{R}^N \to \mathbb{R}^N$ can be iterated to give a dynamical system of the form

$$x_n^{k+1} = f_n(x_1^k, x_2^k, \ldots, x_N^k), \qquad n = 1, \ldots, N, \quad k = 0, 1, 2, \ldots. \tag{9.1}$$

This dynamical system is obviously not a differential equation. Also, k is not an exponent; instead, it is an index that indicates the local iteration number. Given an initial point x_n^0, $(n = 1, \ldots, N)$ in $\mathbb{R}^N$, a series of images $x_n^1, x_n^2, x_n^3, \ldots$ can be found as $k = 0, 1, 2, \ldots$. The map is dissipative or conservative according to whether the diameter of a set is larger than that of its image or the same, respectively, that is, if the determinant of the Jacobian matrix, $\det \partial f_n / \partial x_j \leq 1$.

The point $x_i = \overline{x}_i$ is a *fixed point*[1] of the map if it maps to itself, that is, if

$$\overline{x}_n = f_n(\overline{x}_1, \overline{x}_2, \ldots, \overline{x}_N), \qquad n = 1, \ldots, N. \tag{9.2}$$

The fixed point $\overline{x}_n$ is linearly unstable if a small perturbation from it leads the images farther and farther away. Otherwise, it is stable. A special case of this is asymptotic stability wherein the image returns arbitrarily close to the fixed point.

A linear map can be written as $x_i^{k+1} = \sum_{j=1}^{N} A_{ij} x_j^k$, $(i = 1, 2, \ldots)$ or $\mathbf{x}^{k+1} = \mathbf{A} \cdot \mathbf{x}^k$. The origin $\mathbf{x} = \mathbf{0}$ is a fixed point of this map. If $||\mathbf{A}|| > 1$, then $||\mathbf{x}^{k+1}|| > ||\mathbf{x}^k||$, and the map is unstable. Otherwise, it is stable.

EXAMPLE 9.1

Examine the linear stability of the fixed points of the logistics map, popularized by May,[2]

$$x^{k+1} = rx^k(1 - x^k), \tag{9.3}$$

and study its nonlinear behavior for large k away from fixed points for $r \in [0, 4]$.

We take $r \in [0, 4]$ so that $x^k \in [0, 1]$ maps onto $x^{k+1} \in [0, 1]$. That is, the mapping is onto itself. The fixed points are solutions of

$$\overline{x} = r\overline{x}(1 - \overline{x}), \tag{9.4}$$

which are

$$\overline{x} = 0, \qquad \overline{x} = 1 - \frac{1}{r}. \tag{9.5}$$

Consider now the mapping. For an initial seed x^0, we generate a series of x^k. For example, if we take $r = 0.4$ and $x^0 = 0.3$, we get

$$x^0 = 0.3, \tag{9.6}$$

$$x^1 = 0.4(0.3)(1 - 0.3) = 0.084, \tag{9.7}$$

$$x^2 = 0.4(0.084)(1 - 0.084) = 0.0307776, \tag{9.8}$$

$$x^3 = 0.4(0.0307776)(1 - 0.0307776) = 0.0119321, \tag{9.9}$$

$$x^4 = 0.4(0.0119321)(1 - 0.0119321) = 0.0047159, \tag{9.10}$$

$$x^5 = 0.4(0.0047159)(1 - 0.0047159) = 0.00187747, \tag{9.11}$$

$$\vdots$$

$$x^\infty = 0. \tag{9.12}$$

---

[1] In this chapter, the superposed line indicates a fixed point or equilibrium; it has no relation to our earlier usage of this notation for either a transformed coordinate system or a complex conjugate. The context of the discussion should clarify any questions.

[2] Robert McCredie May, 1938–, Australian-Anglo ecologist.

For this value of r, the solution approaches the fixed point of 0. Consider $r = 4/3$ and $x^0 = 0.3$:

$$x^0 = 0.3, \tag{9.13}$$

$$x^1 = (4/3)(0.3)(1 - 0.3) = 0.28, \tag{9.14}$$

$$x^2 = (4/3)(0.28)(1 - 0.28) = 0.2688, \tag{9.15}$$

$$x^3 = (4/3)(0.2688)(1 - 0.2688) = 0.262062, \tag{9.16}$$

$$x^4 = (4/3)(0.262062)(1 - 0.262062) = 0.257847, \tag{9.17}$$

$$x^5 = (4/3)(0.257847)(1 - 0.257847) = 0.255149, \tag{9.18}$$

$$\vdots$$

$$x^\infty = 0.250 = 1 - \frac{1}{r}. \tag{9.19}$$

In this case, the solution was attracted to the alternate fixed point.

To analyze the stability of each fixed point, we give it a small perturbation $\tilde{x}$. Thus, $\overline{x} + \tilde{x}$ is mapped to $\overline{x} + \tilde{\tilde{x}}$, where

$$\overline{x} + \tilde{\tilde{x}} = r(\overline{x} + \tilde{x})(1 - \overline{x} - \tilde{x}) = r(\overline{x} - \overline{x}^2 + \tilde{x} - 2\overline{x}\tilde{x} + \tilde{x}^2). \tag{9.20}$$

Neglecting small terms, we get

$$\overline{x} + \tilde{\tilde{x}} = r(\overline{x} - \overline{x}^2 + \tilde{x} - 2\overline{x}\tilde{x}) = r\overline{x}(1 - \overline{x}) + r\tilde{x}(1 - 2\overline{x}). \tag{9.21}$$

Simplifying, we get

$$\tilde{\tilde{x}} = r\tilde{x}(1 - 2\overline{x}). \tag{9.22}$$

A fixed point is stable if $|\tilde{\tilde{x}}/\tilde{x}| \leq 1$. This indicates that the perturbation is decaying. Now consider each fixed point in turn.

$\overline{x} = 0$:

$$\tilde{\tilde{x}} = r\tilde{x}(1 - 2(0)), \tag{9.23}$$

$$= r\tilde{x}, \tag{9.24}$$

$$\left| \frac{\tilde{\tilde{x}}}{\tilde{x}} \right| = r. \tag{9.25}$$

This is stable if $r < 1$.

$\overline{x} = 1 - 1/r$:

$$\tilde{\tilde{x}} = r\tilde{x}\left(1 - 2\left(1 - \frac{1}{r}\right)\right), \tag{9.26}$$

$$= (2 - r)\tilde{x}, \tag{9.27}$$

$$\left| \frac{\tilde{\tilde{x}}}{\tilde{x}} \right| = |2 - r|. \tag{9.28}$$

This is unstable for $r < 1$, stable for $1 \leq r \leq 3$, and unstable for $r > 3$.

Examine what happens to the map for $r > 3$. Consider $r = 3.2$ and $x^0 = 0.3$:

$$x^0 = 0.3, \tag{9.29}$$

$$x^1 = 3.2(0.3)(1 - 0.3) = 0.672, \tag{9.30}$$

$$x^2 = 3.2(0.672)(1 - 0.672) = 0.705331, \tag{9.31}$$

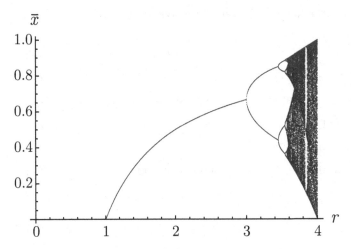

Figure 9.1. Bifurcation diagram of $\overline{x} = \lim_{k\to\infty} x^k$ as a function of r for the logistics map, $x^{k+1} = rx^k(1 - x^k)$ for $r \in [0, 4]$.

$$x^3 = 3.2(0.705331)(1 - 0.705331) = 0.665085, \tag{9.32}$$

$$x^4 = 3.2(0.665085)(1 - 0.665085) = 0.71279, \tag{9.33}$$

$$x^5 = 3.2(0.71279)(1 - 0.71279) = 0.655105, \tag{9.34}$$

$$x^6 = 3.2(0.655105)(1 - 0.655105) = 0.723016, \tag{9.35}$$

$$x^7 = 3.2(0.723016)(1 - 0.723016) = 0.640845, \tag{9.36}$$

$$x^8 = 3.2(0.640845)(1 - 0.640845) = 0.736521, \tag{9.37}$$

$$\vdots$$

$$x^{\infty - 1} = 0.799455, \tag{9.38}$$

$$x^{\infty} = 0.513045. \tag{9.39}$$

This system is said to have *bifurcated*. It oscillates between two points, never going to a fixed point. The two points about which it oscillates are constant for this value of r. For greater values of r, the system moves between $4, 8, 16, \ldots$ points. Such is the essence of bifurcation phenomena. The term refers to a qualitative change in the behavior of a dynamical system in response to a small change in a parameter. A plot, known as a *bifurcation diagram*, of the equilibrium values $\overline{x}$ as a function of r is given in Figure 9.1.

Other maps that have been studied are the

- Hénon[3] map:

$$x_{k+1} = y_k + 1 - ax_k^2, \tag{9.40}$$

$$y_{k+1} = bx_k. \tag{9.41}$$

For $a = 1.3, b = 0.34$, the attractor is periodic, while for $a = 1.4, b = 0.34$, the map has what is known as a strange attractor, to be studied in Section 9.11.3.

---

[3] Michel Hénon, 1931–, French mathematician and astronomer.

Figure 9.2. Cantor set.

- Dissipative standard map:

$$x_{k+1} = x_k + y_{k+1} \bmod 2\pi, \tag{9.42}$$

$$y_{k+1} = \lambda y_k + k \sin x_k. \tag{9.43}$$

If $\lambda = 1$, the map is area preserving.

9.2 Fractals

We next take up the unusual geometries that have been described as *fractal*. These are objects that are not smooth but arise frequently in the solution of dynamical systems. The best-known have their origins in iterated maps. A fractal can be loosely defined as a geometrical shape in which the parts are in some way similar to the whole. This self-similarity may be exact; that is, a piece of the fractal, if magnified, may look exactly like the whole fractal.

Before discussing examples, we need to put forward a working definition of dimension. Though there are many definitions in current use, we present here the Hausdorff-Besicovitch[4] dimension D. If N_ϵ is the number of "boxes" of side length ϵ needed to cover an object, then

$$D = \lim_{\epsilon \to 0} \frac{\ln N_\epsilon}{\ln(1/\epsilon)}. \tag{9.44}$$

We can check that this definition corresponds to the common geometrical shapes.

1. Point: $N_\epsilon = 1, D = 0$ because $D = \lim_{\epsilon \to 0} \dfrac{\ln 1}{-\ln \epsilon} = 0$

2. Line of length ℓ: $N_\epsilon = \ell/\epsilon, D = 1$ because $D = \lim_{\epsilon \to 0} \dfrac{\ln(\ell/\epsilon)}{-\ln \epsilon} = \dfrac{\ln \ell - \ln \epsilon}{-\ln \epsilon} = 1$

3. Surface of size ℓ^2: $N_\epsilon = (\ell/\epsilon)^2, D = 2$ because $D = \lim_{\epsilon \to 0} \dfrac{\ln(\ell^2/\epsilon^2)}{-\ln \epsilon} = \dfrac{2\ln \ell - 2\ln \epsilon}{-\ln \epsilon} = 2$

4. Volume of size ℓ^3: $N_\epsilon = (\ell/\epsilon)^3, D = 3$ because $D = \lim_{\epsilon \to 0} \dfrac{\ln(\ell^3/\epsilon^3)}{-\ln \epsilon} = \dfrac{3\ln \ell - 3\ln \epsilon}{-\ln \epsilon} = 3$

A fractal has a dimension that is not an integer. Many physical objects are fractal-like in that they are fractal within a range of length scales. Coastlines are among the geographical features that are of this shape. If there are N_ϵ units of a measuring stick of length ϵ, the measured length of the coastline will be of the power-law form $\epsilon N_\epsilon = \epsilon^{1-D}$, where D is the dimension.

9.2.1 Cantor Set

We describe here the Cantor[5] set. Consider the line corresponding to $k = 0$ in Figure 9.2. Take away the middle third to leave the two portions; this is shown

---

[4] Felix Hausdorff, 1868–1942, German mathematician, and Abram Samoilovitch Besicovitch, 1891–1970, Russian mathematician.

[5] Georg Ferdinand Ludwig Philipp Cantor, 1845–1918, Russian-born, German-based mathematician.

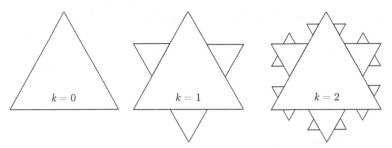

Figure 9.3. Koch curve.

as $k = 1$. Repeat the process to get $k = 2, 3, \ldots$. If $k \to \infty$, what is left is called the Cantor set. Let us take the length of the line segment to be unity when $k = 0$. Because $N_\epsilon = 2^k$ and $\epsilon = 1/3^k$, the dimension of the Cantor set is

$$D = \lim_{\epsilon \to 0} \frac{\ln N_\epsilon}{\ln(1/\epsilon)} = \lim_{k \to \infty} \frac{\ln 2^k}{\ln 3^k} = \frac{k \ln 2}{k \ln 3} = \frac{\ln 2}{\ln 3} = 0.6309 \ldots. \qquad (9.45)$$

It can be seen that the endpoints of the removed intervals are never removed; it can be shown the Cantor set contains an infinite number of points, and it is an uncountable set.

9.2.2 Koch Curve

We next describe a Koch[6] curve. Here we start with an equilateral triangle shown in Figure 9.3 as $k = 0$. Each side of the original triangle has unit length. The middle third of each side of the triangle is removed, and two sides of a triangle are drawn on that. This is shown as $k = 1$. The process is continued, and in the limit gives a continuous, closed curve that is nowhere smooth. Because $N_\epsilon = 3 \times 4^k$ and $\epsilon = 1/3^k$, the dimension of the Koch curve is

$$D = \lim_{\epsilon \to 0} \frac{\ln N_\epsilon}{\ln(1/\epsilon)} = \lim_{k \to \infty} \frac{\ln(3)4^k}{\ln 3^k} = \lim_{k \to \infty} \frac{\ln 3 + k \ln 4}{k \ln 3} = \frac{\ln 4}{\ln 3} = 1.261 \ldots. \qquad (9.46)$$

The limit curve itself has infinite length, it is nowhere differentiable, and it surrounds a finite area.

9.2.3 Menger Sponge

An example of a fractal that is an iterate of an object that starts in three-dimensional space is a "Menger sponge."[7] A Menger sponge is depicted in Figure 9.4. It is constructed by iteration. One starts with an ordinary cube. Then, every face of the cube is divided into nine equal squares. These surface lines are projected inward so as to divide the original cube into 27 packed cubes. The cube in the middle of each face is removed, as is the cube at the inner center. This process is then repeated ad infinitum for each of the cubes that remain.

---

[6] Niels Fabian Helge von Koch, 1870–1924, Swedish mathematician.
[7] Karl Menger, 1902–1985, Austrian and American mathematician.

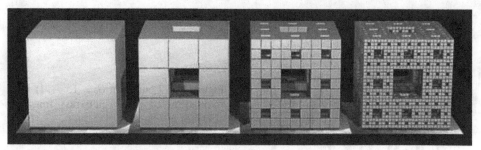

Figure 9.4. Menger sponge. Image from `wikimedia commons`.

9.2.4 Weierstrass Function

For a, b, $t \in \mathbb{R}^1$, $W : \mathbb{R}^1 \to \mathbb{R}^1$, the Weierstrass function is

$$W(t) = \sum_{k=1}^{\infty} a^k \cos b^k t, \qquad (9.47)$$

where $a \in (0, 1)$, b is a positive odd integer and $ab > 1 + 3\pi/2$. It is everywhere continuous but nowhere differentiable! Both require some effort to prove. A Weierstrass function, for $a = 1/2$, $b = 13$, truncated at four terms, is plotted in Figure 9.5. For these values of a and b, its fractal character can be seen when one recognizes that cosine waves of ever-higher frequency and ever-lower amplitude are superposed onto low-frequency cosine waves. For $a = 1/2$ and $b = 13$, the explicit form of $W(t)$ is

$$W(t) = \frac{1}{2} \cos(13t) + \frac{1}{4} \cos(169t) + \frac{1}{8} \cos(2197t) + \frac{1}{16} \cos(28561t) + \cdots . \qquad (9.48)$$

9.2.5 Mandelbrot and Julia Sets

For $z \in \mathbb{C}^1$, $c \in \mathbb{C}^1$, the Mandelbrot[8] set is the set of all c for which

$$z_{k+1} = z_k^2 + c \qquad (9.49)$$

stays bounded as $k \to \infty$, when $z_0 = 0$. The boundaries of this set are fractal. A Mandelbrot set is sketched in Figure 9.6a. Associated with each c for the Mandelbrot set is a Julia[9] set, the set of complex initial seeds z_0 that allows $z_{k+1} = z_k^2 + c$ to converge for fixed complex c. A Julia set for $c = -1 + 0i$ is plotted in Figure 9.6b.

9.3 Introduction to Differential Systems

We turn now to a main topic of relevance to the mathematics of engineering systems: nonlinear systems of ordinary differential equations that are continuous in a time-like variable. We introduce this expansive subject by first considering some paradigm problems that will illustrate the solution techniques.

---

[8] Benoît Mandelbrot, 1924–2010, Polish-born mathematician based mainly in France.
[9] Gaston Maurice Julia, 1893–1978, Algerian-born French mathematician.

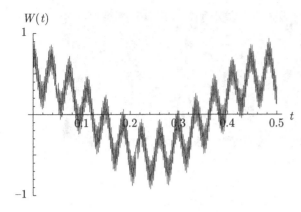

Figure 9.5. Four-term $(k = 1, \ldots, 4)$ approximation to the Weierstrass function, $W(t)$ for $a = 1/2, b = 13$.

The fundamental procedure for analyzing a system of nonlinear ordinary differential equations is to

- cast the system into a standard form,
- identify the equilibria (fixed points) of the system,
- if possible, linearize the system about its equilibria,
- if linearizable, ascertain the stability of the linearized system to small disturbances,
- if not linearizable, attempt to ascertain the stability of the nonlinear system near its equilibria,
- if possible, solve the full nonlinear system, usually by discrete numerical approximation via an iterated map.

9.3.1 Autonomous Example

First consider a simple example of what is known as an *autonomous* system. An autonomous system of ordinary differential equations can be written in the form

$$\frac{d\mathbf{x}}{dt} = \mathbf{f}(\mathbf{x}). \tag{9.50}$$

Notice that the independent variable t does not appear explicitly in the inhomogeneous term $\mathbf{f}(\mathbf{x})$.

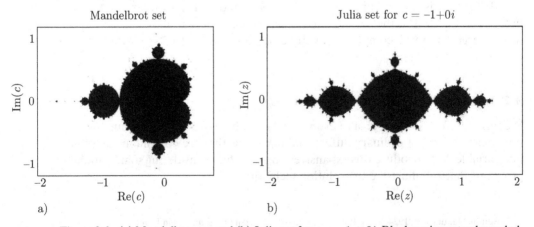

Figure 9.6. (a) Mandelbrot set and (b) Julia set for $c = -1 + 0i$. Black regions stay bounded.

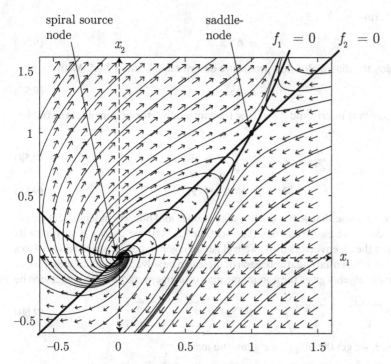

Figure 9.7. Phase plane for Eqs. (9.51–9.52), $dx_1/dt = x_2 - x_1^2$, $dx_2/dt = x_2 - x_1$, along with equilibrium points $(0,0)$ and $(1,1)$, separatrices $x_2 - x_1^2 = 0$, $x_2 - x_1 = 0$, solution trajectories, and corresponding vector field.

EXAMPLE 9.2

For $\mathbf{x} \in \mathbb{R}^2, t \in \mathbb{R}^1, \mathbf{f} : \mathbb{R}^2 \to \mathbb{R}^2$, explore features of the system

$$\frac{dx_1}{dt} = x_2 - x_1^2 = f_1(x_1, x_2), \tag{9.51}$$

$$\frac{dx_2}{dt} = x_2 - x_1 = f_2(x_1, x_2). \tag{9.52}$$

The curves defined in the (x_1, x_2) plane by $f_1 = 0$ and $f_2 = 0$ are useful in determining both the fixed points (found at the intersection of f_1 and f_2) and the behavior of the solution. In fact one can sketch trajectories of paths in this phase space by inspection in many cases. The loci of points where $f_1 = 0$ and $f_2 = 0$ are plotted in Figure 9.7. The zeros, which are fixed-point equilibria, are found at $(x_1, x_2)^T = (0, 0)^T$, $(1, 1)^T$. Linearize about both points by neglecting quadratic and higher powers of deviations from the fixed points to find the local behavior of the solution near these points. Near $(0, 0)^T$, the linearization is

$$\frac{dx_1}{dt} = x_2, \tag{9.53}$$

$$\frac{dx_2}{dt} = x_2 - x_1, \tag{9.54}$$

or

$$\frac{d}{dt} \begin{pmatrix} x_1 \\ x_2 \end{pmatrix} = \begin{pmatrix} 0 & 1 \\ -1 & 1 \end{pmatrix} \begin{pmatrix} x_1 \\ x_2 \end{pmatrix}. \tag{9.55}$$

This is of the form

$$\frac{d\mathbf{x}}{dt} = \mathbf{A} \cdot \mathbf{x}. \tag{9.56}$$

We next employ the diagonalization procedure described in Section 7.9.5. With

$$\mathbf{S} \cdot \mathbf{z} \equiv \mathbf{x}, \tag{9.57}$$

where $\mathbf{S}$ is a constant matrix and $\mathbf{z}$ is a new transformed vector of dependent variables, we get

$$\frac{d}{dt}(\mathbf{S} \cdot \mathbf{z}) = \mathbf{S} \cdot \frac{d\mathbf{z}}{dt} = \mathbf{A} \cdot \mathbf{S} \cdot \mathbf{z}, \tag{9.58}$$

$$\frac{d\mathbf{z}}{dt} = \mathbf{S}^{-1} \cdot \mathbf{A} \cdot \mathbf{S} \cdot \mathbf{z}. \tag{9.59}$$

At this point we assume that $\mathbf{A}$ has distinct eigenvalues and linearly independent eigenvectors; other cases are easily handled (see Section 9.5.2). If we choose $\mathbf{S}$ such that its columns contain the eigenvectors of $\mathbf{A}$, we will get a diagonal matrix, which will lead to a set of uncoupled differential equations; each of these can be solved individually. So for our $\mathbf{A}$, standard linear algebra gives the nonunique matrix of eigenvectors and its inverse to be

$$\mathbf{S} = \begin{pmatrix} \frac{1}{2} + \frac{\sqrt{3}}{2}i & \frac{1}{2} - \frac{\sqrt{3}}{2}i \\ 1 & 1 \end{pmatrix}, \qquad \mathbf{S}^{-1} = \begin{pmatrix} \frac{i}{\sqrt{3}} & \frac{1}{2} + \frac{\sqrt{3}}{6}i \\ -\frac{i}{\sqrt{3}} & \frac{1}{2} - \frac{\sqrt{3}}{6}i \end{pmatrix}. \tag{9.60}$$

With this choice, we get the diagonal eigenvalue matrix

$$\mathbf{\Lambda} = \mathbf{S}^{-1} \cdot \mathbf{A} \cdot \mathbf{S} = \begin{pmatrix} \frac{1}{2} - \frac{\sqrt{3}}{2}i & 0 \\ 0 & \frac{1}{2} + \frac{\sqrt{3}}{2}i \end{pmatrix}. \tag{9.61}$$

Thus,

$$\frac{d\mathbf{z}}{dt} = \mathbf{\Lambda} \cdot \mathbf{z}. \tag{9.62}$$

So we get two uncoupled equations for $\mathbf{z}$:

$$\frac{dz_1}{dt} = \underbrace{\left(\frac{1}{2} - \frac{\sqrt{3}}{2}i\right)}_{=\lambda_1} z_1, \tag{9.63}$$

$$\frac{dz_2}{dt} = \underbrace{\left(\frac{1}{2} + \frac{\sqrt{3}}{2}i\right)}_{=\lambda_2} z_2, \tag{9.64}$$

which have solutions

$$z_1 = c_1 \exp\left(\left(\frac{1}{2} - \frac{\sqrt{3}}{2}i\right)t\right), \tag{9.65}$$

$$z_2 = c_2 \exp\left(\left(\frac{1}{2} + \frac{\sqrt{3}}{2}i\right)t\right). \tag{9.66}$$

Then we form $\mathbf{x}$ by taking $\mathbf{x} = \mathbf{S} \cdot \mathbf{z}$ so that

$$x_1 = \left(\frac{1}{2} + \frac{\sqrt{3}}{2}i\right) \underbrace{c_1 \exp\left(\left(\frac{1}{2} - \frac{\sqrt{3}}{2}i\right)t\right)}_{=z_1} + \left(\frac{1}{2} - \frac{\sqrt{3}}{2}i\right) \underbrace{c_2 \exp\left(\left(\frac{1}{2} + \frac{\sqrt{3}}{2}i\right)t\right)}_{=z_2}, \tag{9.67}$$

$$x_2 = \underbrace{c_1 \exp\left(\left(\frac{1}{2} - \frac{\sqrt{3}}{2}i\right)t\right)}_{=z_1} + \underbrace{c_2 \exp\left(\left(\frac{1}{2} + \frac{\sqrt{3}}{2}i\right)t\right)}_{=z_2}. \tag{9.68}$$

Because there is a positive real coefficient in the exponential terms, both x_1 and x_2 grow exponentially. The imaginary component indicates that this is an oscillatory growth. Hence, there is no tendency for a solution which is initially close to $(0,0)^T$ to remain there. So the fixed point is *unstable*.

Consider the next fixed point near $(1,1)^T$. First define a new set of local variables whose values are small in the neighborhood of the equilibrium:

$$\tilde{x}_1 = x_1 - 1, \tag{9.69}$$

$$\tilde{x}_2 = x_2 - 1. \tag{9.70}$$

Then

$$\frac{dx_1}{dt} = \frac{d\tilde{x}_1}{dt} = (\tilde{x}_2 + 1) - (\tilde{x}_1 + 1)^2, \tag{9.71}$$

$$\frac{dx_2}{dt} = \frac{d\tilde{x}_2}{dt} = (\tilde{x}_2 + 1) - (\tilde{x}_1 + 1). \tag{9.72}$$

Expanding, we get

$$\frac{d\tilde{x}_1}{dt} = (\tilde{x}_2 + 1) - \tilde{x}_1^2 - 2\tilde{x}_1 - 1, \tag{9.73}$$

$$\frac{d\tilde{x}_2}{dt} = (\tilde{x}_2 + 1) - (\tilde{x}_1 + 1). \tag{9.74}$$

Linearizing about $(\tilde{x}_1, \tilde{x}_2)^T = (0,0)^T$, we find

$$\frac{d\tilde{x}_1}{dt} = \tilde{x}_2 - 2\tilde{x}_1, \tag{9.75}$$

$$\frac{d\tilde{x}_2}{dt} = \tilde{x}_2 - \tilde{x}_1, \tag{9.76}$$

or

$$\frac{d}{dt}\begin{pmatrix} \tilde{x}_1 \\ \tilde{x}_2 \end{pmatrix} = \begin{pmatrix} -2 & 1 \\ -1 & 1 \end{pmatrix}\begin{pmatrix} \tilde{x}_1 \\ \tilde{x}_2 \end{pmatrix}. \tag{9.77}$$

Going through an essentially identical exercise as before shows the eigenvalues to be

$$\lambda_1 = -\frac{1}{2} + \frac{\sqrt{5}}{2} > 0, \tag{9.78}$$

$$\lambda_2 = -\frac{1}{2} - \frac{\sqrt{5}}{2} < 0, \tag{9.79}$$

which in itself shows the solution to be unstable because there is a positive eigenvalue. After the usual linear algebra and back transformations, one obtains the local solution:

$$x_1 = 1 + c_1\left(\frac{3-\sqrt{5}}{2}\right)\exp\left(\left(-\frac{1}{2} + \frac{\sqrt{5}}{2}\right)t\right) + c_2\left(\frac{3+\sqrt{5}}{2}\right)\exp\left(\left(-\frac{1}{2} - \frac{\sqrt{5}}{2}\right)t\right), \tag{9.80}$$

$$x_2 = 1 + c_1\exp\left(\left(-\frac{1}{2} + \frac{\sqrt{5}}{2}\right)t\right) + c_2\exp\left(\left(-\frac{1}{2} - \frac{\sqrt{5}}{2}\right)t\right). \tag{9.81}$$

Although this solution is generally unstable, if one has the special case in which $c_1 = 0$, the fixed point is stable. Such is characteristic of a *saddle point*.

As an interesting aside, we can use Eq. (2.372) to calculate the curvature field for this system. With the notation of the present section, the curvature field is given by

$$\kappa = \frac{\sqrt{(\mathbf{f}^T \cdot \mathbf{F} \cdot \mathbf{F}^T \cdot \mathbf{f})(\mathbf{f}^T \cdot \mathbf{f}) - (\mathbf{f}^T \cdot \mathbf{F}^T \cdot \mathbf{f})^2}}{(\mathbf{f}^T \cdot \mathbf{f})^{3/2}}, \tag{9.82}$$

where $\mathbf{F}$, the gradient of the vector field $\mathbf{f}$, is given by the analog of Eq. (2.371):

$$\mathbf{F} = \nabla \mathbf{f}^T. \tag{9.83}$$

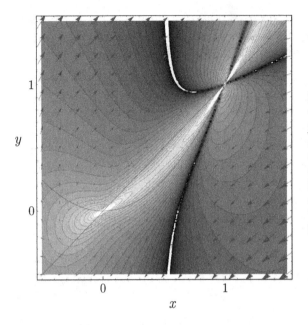

Figure 9.8. Contours with shading of $\ln \kappa$, where κ is trajectory curvature for trajectories of solutions to Eqs. (9.51–9.52), $dx_1/dt = x_2 - x_1^2$, $dx_2/dt = x_2 - x_1$. Separatrices $x_2 - x_1^2 = 0$ and $x_2 - x_1 = 0$ are also plotted. Light shading corresponds to large trajectory curvature; dark shading corresponds to small trajectory curvature. The vector field is associated with the solution trajectories.

So with

$$\mathbf{f} = \begin{pmatrix} f_1(x_1, x_2) \\ f_2(x_1, x_2) \end{pmatrix} = \begin{pmatrix} x_2 - x_1^2 \\ x_2 - x_1 \end{pmatrix}, \qquad \mathbf{F} = \begin{pmatrix} \frac{\partial f_1}{\partial x_1} & \frac{\partial f_2}{\partial x_1} \\ \frac{\partial f_1}{\partial x_2} & \frac{\partial f_2}{\partial x_2} \end{pmatrix} = \begin{pmatrix} -2x_1 & -1 \\ 1 & 1 \end{pmatrix}, \quad (9.84)$$

detailed calculation reveals that

$$\kappa = \frac{\sqrt{(-x_1^2 + x_1^3 + x_1^4 + x_1 x_2 - x_1^2 x_2 - 2x_1^3 x_2 - x_2^2 + 2x_1 x_2^2)^2}}{(x_1^2 + x_1^4 - 2x_1 x_2 - 2x_1^2 x_2 + 2x_2^2)^{3/2}}. \qquad (9.85)$$

A plot of the curvature field is shown in Figure 9.8. Because κ varies over orders of magnitude, the contours are for $\ln \kappa$ to more easily visualize the variation. Regions of high curvature are noted near both critical points and in the regions between the curves $x_2 = x_1$ and $x_2 = x_1^2$ for $x_1 \in [0, 1]$. Comparison with Figure 9.7 reveals consistency.

9.3.2 Nonautonomous Example

Next, consider a more complicated example. Among other things, the system as originally cast will be *nonautonomous* in that the independent variable t appears explicitly. Some operations will be necessary to cast the system in standard autonomous form.

EXAMPLE 9.3

For $\mathbf{x} \in \mathbb{R}^2, t \in \mathbb{R}^1, \mathbf{f} : \mathbb{R}^2 \times \mathbb{R}^1 \to \mathbb{R}^2$, analyze

$$t\frac{dx_1}{dt} + x_2 x_1 \frac{dx_2}{dt} = x_1 + t \qquad = f_1(x_1, x_2, t), \qquad (9.86)$$

$$x_1 \frac{dx_1}{dt} + x_2^2 \frac{dx_2}{dt} = x_1 t \qquad = f_2(x_1, x_2, t), \qquad (9.87)$$

$$x_1(0) = x_{10}, \qquad x_2(0) = x_{20}. \qquad (9.88)$$

First, let us transform the system into an autonomous form. We will render t to be a dependent variable and introduce a new independent variable s and, in so doing, raise the order of the system from 2 to 3. Let

$$\frac{dt}{ds} = 1, \qquad t(0) = 0, \tag{9.89}$$

and further let $y_1 = x_1, y_2 = x_2, y_3 = t$. Then with $s \in \mathbb{R}^1, \mathbf{y} \in \mathbb{R}^3, \mathbf{g} : \mathbb{R}^3 \to \mathbb{R}^3$,

$$y_3 \frac{dy_1}{ds} + y_2 y_1 \frac{dy_2}{ds} = y_1 + y_3 \qquad = g_1(y_1, y_2, y_3), \tag{9.90}$$

$$y_1 \frac{dy_1}{ds} + y_2^2 \frac{dy_2}{ds} = y_1 y_3 \qquad = g_2(y_1, y_2, y_3), \tag{9.91}$$

$$\frac{dy_3}{ds} = 1 \qquad = g_3(y_1, y_2, y_3), \tag{9.92}$$

$$y_1(0) = y_{10}, \qquad y_2(0) = y_{20}, \qquad y_3(0) = 0. \tag{9.93}$$

In matrix form, we have

$$\begin{pmatrix} y_3 & y_2 y_1 & 0 \\ y_1 & y_2^2 & 0 \\ 0 & 0 & 1 \end{pmatrix} \begin{pmatrix} \frac{dy_1}{ds} \\ \frac{dy_2}{ds} \\ \frac{dy_3}{ds} \end{pmatrix} = \begin{pmatrix} y_1 + y_3 \\ y_1 y_3 \\ 1 \end{pmatrix}. \tag{9.94}$$

Inverting the coefficient matrix, we obtain the following equation, which is in autonomous form:

$$\frac{d}{ds} \begin{pmatrix} y_1 \\ y_2 \\ y_3 \end{pmatrix} = \begin{pmatrix} \frac{y_1 y_2 - y_1^2 y_3 + y_2 y_3}{y_2 y_3 - y_1^2} \\ \frac{y_1(y_3^2 - y_1 - y_3)}{y_2(y_2 y_3 - y_1^2)} \\ 1 \end{pmatrix} = \begin{pmatrix} h_1(y_1, y_2, y_3) \\ h_2(y_1, y_2, y_3) \\ h_3(y_1, y_2, y_3) \end{pmatrix}. \tag{9.95}$$

There are potential singularities at $y_2 = 0$ and $y_2 y_3 = y_1^2$. Under such conditions, the determinant of the coefficient matrix is zero, and dy_i/ds is not uniquely determined. One way to address the potential singularities is by defining a new independent variable $u \in \mathbb{R}^1$ via the equation

$$\frac{ds}{du} = y_2 \left(y_2 y_3 - y_1^2 \right). \tag{9.96}$$

The system, Eq. (9.95), then transforms to

$$\frac{d}{du} \begin{pmatrix} y_1 \\ y_2 \\ y_3 \end{pmatrix} = \begin{pmatrix} y_2 \left(y_1 y_2 - y_1^2 y_3 + y_2 y_3 \right) \\ y_1 \left(y_3^2 - y_1 - y_3 \right) \\ y_2 \left(y_2 y_3 - y_1^2 \right) \end{pmatrix} = \begin{pmatrix} p_1(y_1, y_2, y_3) \\ p_2(y_1, y_2, y_3) \\ p_3(y_1, y_2, y_3) \end{pmatrix}. \tag{9.97}$$

This equation actually has an infinite number of fixed points, all of which lie on a line in the three-dimensional phase volume. The line is given parametrically by $(y_1, y_2, y_3)^T = (0, 0, v)^T$, $v \in \mathbb{R}^1$. Here v is just a parameter used in describing the line of fixed points. However, it turns out in this case that the Taylor series expansions yield no linear contribution near any of the fixed points, so we do not get to use the standard local linearized analysis technique! The problem has an essential nonlinear character, even near fixed points. More potent methods would need to be employed, but the example demonstrates the principle. Figure 9.9 gives a numerically obtained solution for $y_1(u), y_2(u), y_3(u)$ along with a trajectory in (y_1, y_2, y_3) space when $y_1(0) = 1, y_2(0) = -1, y_3(0) = 0$. This corresponds to $x_1(t = 0) = 1, x_2(t = 0) = -1$.

While the solutions are single-valued functions of the variable u, they are not single-valued in t after the transformation back to $x_1(t), x_2(t)$ is effected. Also, while it appears there are points ($u = 0.38, u = 0.84, u = 1.07$) where the derivatives

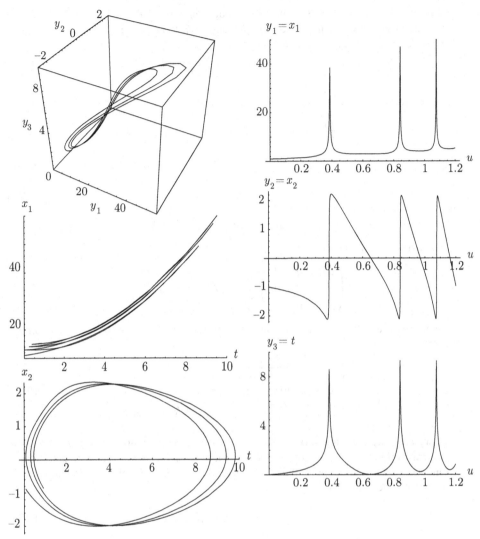

Figure 9.9. Solutions for the initial conditions, $y_1(0) = 1$, $y_2(0) = -1$, $y_3(0) = 0$, for the second paradigm example of Section 9.3.2: trajectory in phase volume (y_1, y_2, y_3); also $y_1(u), y_2(u), y_3(u)$ and $x_1(t)$, $x_2(t)$. Here $y_1 = x_1$, $y_2 = x_2$, $y_3 = t$.

$dy_1/du, dy_2/du, dy_3/du$ become unbounded, closer inspection reveals that they are simply points of steep, but bounded, derivatives. However, at points where the slope $dy_3/du = dt/du$ changes sign, the derivatives dx_1/dt and dx_2/dt formally are infinite, as is reflected in the cyclic behavior exhibited in the plots of x_1 versus t or x_2 versus t.

9.3.3 General Approach

Consider $\mathbf{x} \in \mathbb{R}^N, t \in \mathbb{R}^1, \hat{\mathbf{f}} : \mathbb{R}^N \times \mathbb{R}^N \times \mathbb{R}^1 \to \mathbb{R}^N$. A general nonlinear system of differential-algebraic equations takes on the form

$$\hat{\mathbf{f}}\left(\frac{d\mathbf{x}}{dt}, \mathbf{x}, t\right) = \mathbf{0}. \tag{9.98}$$

Such general problems can be challenging. Let us here restrict to a form that is quasi-linear in the time derivatives. Thus, consider $\mathbf{x} \in \mathbb{R}^N, t \in \mathbb{R}^1, \mathbf{A} : \mathbb{R}^N \times \mathbb{R}^1 \to \mathbb{R}^N \times \mathbb{R}^N, \mathbf{f} : \mathbb{R}^N \times \mathbb{R}^1 \to \mathbb{R}^N$. At this stage, we allow $\mathbf{A}$ to be potentially singular. Then the nonlinear problem, albeit one whose derivatives $d\mathbf{x}/dt$ appear only in linear combinations, of the form

$$\mathbf{A}(\mathbf{x}, t) \cdot \frac{d\mathbf{x}}{dt} = \mathbf{f}(\mathbf{x}, t), \qquad \mathbf{x}(0) = \mathbf{x}_0, \tag{9.99}$$

can be reduced to autonomous form in the following manner. With

$$\mathbf{x} = \begin{pmatrix} x_1 \\ \vdots \\ x_N \end{pmatrix}, \ \mathbf{A}(\mathbf{x}, t) = \begin{pmatrix} a_{11}(\mathbf{x}, t) & \cdots & a_{1N}(\mathbf{x}, t) \\ \vdots & \ddots & \vdots \\ a_{N1}(\mathbf{x}, t) & \cdots & a_{NN}(\mathbf{x}, t) \end{pmatrix}, \ \mathbf{f}(\mathbf{x}, t) = \begin{pmatrix} f_1(x_1, \ldots, x_N, t) \\ \vdots \\ f_N(x_1, \ldots, x_N, t) \end{pmatrix},$$

$$\tag{9.100}$$

define $s \in \mathbb{R}^1$ such that

$$\frac{dt}{ds} = 1, \qquad t(0) = 0. \tag{9.101}$$

Then define $\mathbf{y} \in \mathbb{R}^{N+1}, \mathbf{B} : \mathbb{R}^{N+1} \to \mathbb{R}^{N+1} \times \mathbb{R}^{N+1}, \mathbf{g} : \mathbb{R}^{N+1} \to \mathbb{R}^{N+1}$, such that along with $s \in \mathbb{R}^1$,

$$\mathbf{y} = \begin{pmatrix} y_1 \\ \vdots \\ y_N \\ y_{N+1} \end{pmatrix} = \begin{pmatrix} x_1 \\ \vdots \\ x_N \\ t \end{pmatrix}, \tag{9.102}$$

$$\mathbf{B}(\mathbf{y}) = \begin{pmatrix} a_{11}(\mathbf{y}) & \cdots & a_{1N}(\mathbf{y}) & 0 \\ \vdots & \ddots & \vdots & \vdots \\ a_{N1}(\mathbf{y}) & \cdots & a_{NN}(\mathbf{y}) & 0 \\ 0 & \cdots & 0 & 1 \end{pmatrix}, \tag{9.103}$$

$$\mathbf{g}(\mathbf{y}) = \begin{pmatrix} g_1(y_1, \ldots, y_{N+1}) \\ \vdots \\ g_N(y_1, \ldots, y_{N+1}) \\ g_{N+1}(y_1, \ldots, y_{N+1}) \end{pmatrix} = \begin{pmatrix} f_1(x_1, \ldots, x_N, t) \\ \vdots \\ f_N(x_1, \ldots, x_N, t) \\ 1 \end{pmatrix}. \tag{9.104}$$

Equation (9.99) then transforms to

$$\mathbf{B}(\mathbf{y}) \cdot \frac{d\mathbf{y}}{ds} = \mathbf{g}(\mathbf{y}). \tag{9.105}$$

By forming $\mathbf{B}^{-1}$, assuming $\mathbf{B}$ is nonsingular, Eq. (9.105) can be written as

$$\frac{d\mathbf{y}}{ds} = \mathbf{B}^{-1}(\mathbf{y}) \cdot \mathbf{g}(\mathbf{y}), \tag{9.106}$$

or by taking $\mathbf{B}^{-1}(\mathbf{y}) \cdot \mathbf{g}(\mathbf{y}) \equiv \mathbf{h}(\mathbf{y})$, we get an autonomous form, with $s \in \mathbb{R}^1, \mathbf{y} \in \mathbb{R}^{N+1}, \mathbf{h} : \mathbb{R}^{N+1} \to \mathbb{R}^{N+1}$:

$$\frac{d\mathbf{y}}{ds} = \mathbf{h}(\mathbf{y}). \tag{9.107}$$

The new inhomogeneous function $\mathbf{h}(\mathbf{y})$ cannot be formed when $\mathbf{B}(\mathbf{y})$ is singular, and thus noninvertible. At such singular points, we cannot form a linearly independent set of $d\mathbf{y}/ds$, and the system is better considered as a set of differential-algebraic equations, described in the upcoming Section 9.7. If the source of the singularity can be identified, a singularity-free autonomous set of equations can often be written. For example, suppose $\mathbf{h}$ can be rewritten as

$$\mathbf{h}(\mathbf{y}) = \frac{\mathbf{p}(\mathbf{y})}{q(\mathbf{y})}, \tag{9.108}$$

where $\mathbf{p}$ and q have no singularities. A likely choice would be $q(\mathbf{y}) = \det \mathbf{B}(\mathbf{y})$. Then we can remove the singularity by introducing the new independent variable $u \in \mathbb{R}^1$ such that

$$\frac{ds}{du} = q(\mathbf{y}). \tag{9.109}$$

Using the chain rule, the system then becomes

$$\frac{d\mathbf{y}}{ds} = \frac{\mathbf{p}(\mathbf{y})}{q(\mathbf{y})}, \tag{9.110}$$

$$\frac{ds}{du}\frac{d\mathbf{y}}{ds} = q(\mathbf{y})\frac{\mathbf{p}(\mathbf{y})}{q(\mathbf{y})}, \tag{9.111}$$

$$\frac{d\mathbf{y}}{du} = \mathbf{p}(\mathbf{y}), \tag{9.112}$$

which has no singularities.

Casting ordinary differential equation systems in autonomous form is the starting point for most problems and most theoretical development. The task from here generally proceeds as follows:

- Find all the zeros, or fixed points, of $\mathbf{h}$. This is an algebra problem, which can be topologically difficult for nonlinear problems. Recognize that fixed points may be isolated or part of a continuum.
- If $\mathbf{h}$ has any singularities, redefine variables in the manner demonstrated to remove them.
- If possible, linearize $\mathbf{h}$ (or its equivalent) about each of its zeros.
- Perform a local analysis of the system of differential equations near zeros.
- If the system is linear, an eigenvalue analysis is sufficient to reveal stability; for nonlinear systems, the situation is not always straightforward.

9.4 High-Order Scalar Differential Equations

An equation with $x \in \mathbb{R}^1, t \in \mathbb{R}^1, a : \mathbb{R}^1 \times \mathbb{R}^1 \to \mathbb{R}^N, f : \mathbb{R}^1 \to \mathbb{R}^1$ of the form

$$\frac{d^N x}{dt^N} + a_N(x,t)\frac{d^{N-1}x}{dt^{N-1}} + \cdots + a_2(x,t)\frac{dx}{dt} + a_1(x,t)x = f(t) \tag{9.113}$$

can be expressed as a system of $N+1$ first-order autonomous equations. Such a formulation may be especially convenient when one is trying to solve such an equation with numerical methods. Let $x = y_1, dx/dt = y_2, \ldots, d^{N-1}x/dt^{N-1} = y_N, t = y_{N+1}$. Then, with $y \in \mathbb{R}^{N+1}, s = t \in \mathbb{R}^1, a : \mathbb{R}^1 \times \mathbb{R}^1 \to \mathbb{R}^N, f : \mathbb{R}^1 \to \mathbb{R}^1$, we get

$$\frac{dy_1}{ds} = y_2, \tag{9.114}$$

$$\frac{dy_2}{ds} = y_3, \tag{9.115}$$

$$\vdots$$

$$\frac{dy_{N-1}}{ds} = y_N, \tag{9.116}$$

$$\frac{dy_N}{ds} = -a_N(y_1, y_{N+1})y_N - a_{N-1}(y_1, y_{N+1})y_{N-1} - \cdots - a_1(y_1, y_{N+1})y_1$$

$$+ f(y_{N+1}), \tag{9.117}$$

$$\frac{dy_{N+1}}{ds} = 1. \tag{9.118}$$

EXAMPLE 9.4

For $x \in \mathbb{R}^1, t \in \mathbb{R}^1$, formulate the forced Airy equation:

$$\frac{d^2x}{dt^2} + tx = \sin t, \qquad x(0) = 0, \qquad \left.\frac{dx}{dt}\right|_{t=0} = 0, \tag{9.119}$$

as an autonomous system of first-order equations and explore features of its solution.

Here $a_2(x, t) = 0, a_1(x, t) = t, f(t) = \sin t$. In (x, t) space, this is a linear system with nonconstant coefficients. Now this differential equation with homogeneous boundary conditions and nonzero forcing has no simple analytical solution. With effort, a solution in terms of Airy functions is in principle possible (see Section A.7.5). It can certainly be solved numerically; most solution techniques require recasting as an autonomous system

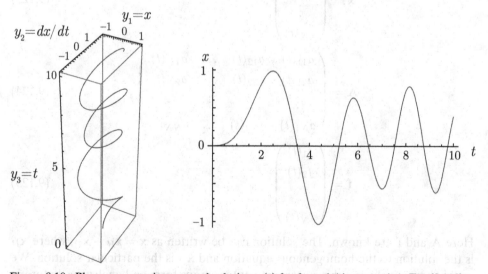

Figure 9.10. Phase space trajectory and solution $x(t)$ for forced Airy equation, Eq. (9.119).

of first-order equations. To achieve this, with $y \in \mathbb{R}^3, s \in \mathbb{R}^1$, consider

$$x = y_1, \qquad \frac{dx}{dt} = y_2, \qquad t = s = y_3. \tag{9.120}$$

Then $d/dt = d/ds$, and Eq. (9.119) transforms to

$$\frac{d}{ds}\begin{pmatrix} y_1 \\ y_2 \\ y_3 \end{pmatrix} = \begin{pmatrix} y_2 \\ -y_3 y_1 + \sin y_3 \\ 1 \end{pmatrix} = \begin{pmatrix} h_1(y_1, y_2, y_3) \\ h_2(y_1, y_2, y_3) \\ h_3(y_1, y_2, y_3) \end{pmatrix}, \qquad \begin{pmatrix} y_1(0) \\ y_2(0) \\ y_3(0) \end{pmatrix} = \begin{pmatrix} 0 \\ 0 \\ 0 \end{pmatrix}. \tag{9.121}$$

In this transformed space, the presence of the terms $-y_3 y_1$ and $\sin y_3$ renders the system nonlinear. This system has no equilibrium point as there exists no y_1, y_2, y_3 for which $h_1 = h_2 = h_3 = 0$. Once the numerical solution is obtained, one transforms back to (x, t) space. Figure 9.10 gives the trajectory in the (y_1, y_2, y_3) phase space and a plot of the corresponding solution $x(t)$ for $t \in [0, 10]$.

9.5 Linear Systems

Here we consider dynamical systems that are linear differential equations. For a linear system, the coefficients $a_N, \ldots, a_2, a_1$ in Eq. (9.113) are independent of x.

9.5.1 Inhomogeneous with Variable Coefficients

In general, for $\mathbf{x} \in \mathbb{R}^N, t \in \mathbb{R}^1, \mathbf{A} : \mathbb{R}^1 \to \mathbb{R}^N \times \mathbb{R}^N, \mathbf{f} : \mathbb{R}^1 \to \mathbb{R}^N$, any linear system may be written in matrix form as

$$\frac{d\mathbf{x}}{dt} = \mathbf{A}(t) \cdot \mathbf{x} + \mathbf{f}(t), \tag{9.122}$$

where

$$\mathbf{x} = \begin{pmatrix} x_1(t) \\ x_2(t) \\ \vdots \\ x_N(t) \end{pmatrix}, \tag{9.123}$$

$$\mathbf{A} = \begin{pmatrix} a_{11}(t) & a_{12}(t) & \cdots & a_{1N}(t) \\ a_{21}(t) & a_{22}(t) & \cdots & a_{2N}(t) \\ \vdots & \vdots & \vdots & \vdots \\ a_{N1}(t) & a_{N2}(t) & \cdots & a_{NN}(t) \end{pmatrix}, \tag{9.124}$$

$$\mathbf{f} = \begin{pmatrix} f_1(t) \\ f_2(t) \\ \vdots \\ f_N(t) \end{pmatrix}. \tag{9.125}$$

Here $\mathbf{A}$ and $\mathbf{f}$ are known. The solution can be written as $\mathbf{x} = \mathbf{x}_H + \mathbf{x}_P$, where $\mathbf{x}_H$ is the solution to the homogeneous equation and $\mathbf{x}_P$ is the particular solution. We could alternatively cast this system into autonomous form and analyze as before in Section 9.3.3.

9.5.2 Homogeneous with Constant Coefficients

Here we expand on analysis first introduced in Section 4.2.1. For $\mathbf{x} \in \mathbb{R}^N, t \in \mathbb{R}^1, \mathbf{A} \in \mathbb{R}^N \times \mathbb{R}^N$, consider the solution of the homogeneous equation

$$\frac{d\mathbf{x}}{dt} = \mathbf{A} \cdot \mathbf{x}, \tag{9.126}$$

where $\mathbf{A}$ is a matrix of constants. In this case, there is only one fixed point, namely, the null vector:

$$\mathbf{x} = \mathbf{0}. \tag{9.127}$$

We have already solved this problem via one common method in Section 9.3.1. As this problem is of foundational importance, it is useful to understand its solution from another perspective. First, set

$$\mathbf{x} = \mathbf{e} e^{\lambda t}, \tag{9.128}$$

where $\mathbf{e} \in \mathbb{R}^N$ is a constant vector and λ is a constant scalar. Substituting into Eq. (9.126), we get

$$\lambda \mathbf{e} e^{\lambda t} = \mathbf{A} \cdot \mathbf{e} e^{\lambda t}, \tag{9.129}$$

$$\lambda \mathbf{e} = \mathbf{A} \cdot \mathbf{e}. \tag{9.130}$$

This is an eigenvalue problem where λ is an eigenvalue and $\mathbf{e}$ is an eigenvector. Details of the solution depend on whether $\mathbf{A}$ has a complete set of N linearly independent eigenvectors. We examine both cases in full next.

N Eigenvectors

Here we will assume that there is a full set of eigenvectors. The eigenvalues may or may not be distinct and may or may not be zero. If $\mathbf{e}_1, \mathbf{e}_2, \ldots, \mathbf{e}_N$, are the eigenvectors corresponding to eigenvalues $\lambda_1, \lambda_2, \ldots, \lambda_N$, then

$$\mathbf{x} = \sum_{n=1}^{N} c_n \mathbf{e}_n e^{\lambda_n t} \tag{9.131}$$

is the general solution, where $c_1, c_2, \ldots, c_N$, are arbitrary constants.

EXAMPLE 9.5

For $\mathbf{x} \in \mathbb{R}^3, t \in \mathbb{R}^1, \mathbf{A} \in \mathbb{R}^3 \times \mathbb{R}^3$, solve $d\mathbf{x}/dt = \mathbf{A} \cdot \mathbf{x}$, where

$$\mathbf{A} = \begin{pmatrix} 1 & -1 & 4 \\ 3 & 2 & -1 \\ 2 & 1 & -1 \end{pmatrix}. \tag{9.132}$$

The eigenvalues and eigenvectors are

$$\lambda_1 = 1, \qquad \mathbf{e}_1 = \begin{pmatrix} -1 \\ 4 \\ 1 \end{pmatrix}, \tag{9.133}$$

$$\lambda_2 = 3, \qquad \mathbf{e}_2 = \begin{pmatrix} 1 \\ 2 \\ 1 \end{pmatrix}, \tag{9.134}$$

$$\lambda_3 = -2, \qquad \mathbf{e}_3 = \begin{pmatrix} -1 \\ 1 \\ 1 \end{pmatrix}. \tag{9.135}$$

Thus, the solution is

$$\mathbf{x} = c_1 \begin{pmatrix} -1 \\ 4 \\ 1 \end{pmatrix} e^t + c_2 \begin{pmatrix} 1 \\ 2 \\ 1 \end{pmatrix} e^{3t} + c_3 \begin{pmatrix} -1 \\ 1 \\ 1 \end{pmatrix} e^{-2t}. \tag{9.136}$$

Expanding, we get

$$x_1(t) = -c_1 e^t + c_2 e^{3t} - c_3 e^{-2t}, \tag{9.137}$$

$$x_2(t) = 4c_1 e^t + 2c_2 e^{3t} + c_3 e^{-2t}, \tag{9.138}$$

$$x_3(t) = c_1 e^t + c_2 e^{3t} + c_3 e^{-2t}. \tag{9.139}$$

Each dependent variable $x_i(t)$ has dependence on each eigenmode. This is the essence of the coupled nature of the system.

EXAMPLE 9.6

For $\mathbf{x} \in \mathbb{R}^3, t \in \mathbb{R}^1, \mathbf{A} \in \mathbb{R}^3 \times \mathbb{R}^3$, solve $d\mathbf{x}/dt = \mathbf{A} \cdot \mathbf{x}$, where

$$\mathbf{A} = \begin{pmatrix} 1 & 1 & -1 \\ 0 & 1 & 0 \\ 0 & 1 & 0 \end{pmatrix}. \tag{9.140}$$

Note that $\mathbf{A}$ has two identical rows and is thus singular, the eigenvalues and eigenvectors are

$$\lambda_1 = 1, \qquad \mathbf{e}_1 = \begin{pmatrix} 0 \\ 1 \\ 1 \end{pmatrix}, \tag{9.141}$$

$$\lambda_2 = 1, \qquad \mathbf{e}_2 = \begin{pmatrix} 1 \\ 0 \\ 0 \end{pmatrix}, \tag{9.142}$$

$$\lambda_3 = 0, \qquad \mathbf{e}_3 = \begin{pmatrix} 1 \\ 0 \\ 1 \end{pmatrix}. \tag{9.143}$$

Regardless of the singular nature of $\mathbf{A}$, its eigenvectors are linearly independent, so our present method applies. Here we have both repeated eigenvalues, $\lambda_1 = \lambda_2 = 1$, and because of the singular nature of $\mathbf{A}$, eigenvalues which are zero, $\lambda_3 = 0$, none of which poses any difficulty. The solution is

$$\mathbf{x} = c_1 \begin{pmatrix} 0 \\ 1 \\ 1 \end{pmatrix} e^t + c_2 \begin{pmatrix} 1 \\ 0 \\ 0 \end{pmatrix} e^t + c_3 \begin{pmatrix} 1 \\ 0 \\ 1 \end{pmatrix} e^0. \tag{9.144}$$

Expanding, we get

$$x_1(t) = c_2 e^t + c_3, \tag{9.145}$$

$$x_2(t) = c_1 e^t, \tag{9.146}$$

$$x_3(t) = c_1 e^t + c_3. \tag{9.147}$$

EXAMPLE 9.7

For $\mathbf{x} \in \mathbb{R}^3, t \in \mathbb{R}^1, \mathbf{A} \in \mathbb{R}^3 \times \mathbb{R}^3$, solve $d\mathbf{x}/dt = \mathbf{A} \cdot \mathbf{x}$, where

$$\mathbf{A} = \begin{pmatrix} 2 & -1 & -1 \\ 2 & 1 & -1 \\ 0 & -1 & 1 \end{pmatrix}. \tag{9.148}$$

The eigenvalues and eigenvectors are

$$\lambda_1 = 2, \quad \mathbf{e}_1 = \begin{pmatrix} 0 \\ 1 \\ -1 \end{pmatrix}, \tag{9.149}$$

$$\lambda_2 = 1+i, \quad \mathbf{e}_2 = \begin{pmatrix} 1 \\ -i \\ 1 \end{pmatrix}, \tag{9.150}$$

$$\lambda_3 = 1-i, \quad \mathbf{e}_3 = \begin{pmatrix} 1 \\ i \\ 1 \end{pmatrix}. \tag{9.151}$$

Thus, the solution is

$$\mathbf{x} = c_1 \begin{pmatrix} 0 \\ 1 \\ -1 \end{pmatrix} e^{2t} + c_2 \begin{pmatrix} 1 \\ -i \\ 1 \end{pmatrix} e^{(1+i)t} + c_3 \begin{pmatrix} 1 \\ i \\ 1 \end{pmatrix} e^{(1-i)t}, \tag{9.152}$$

$$= c_1 \begin{pmatrix} 0 \\ 1 \\ -1 \end{pmatrix} e^{2t} + c_2' \begin{pmatrix} \cos t \\ \sin t \\ \cos t \end{pmatrix} e^t + c_3' \begin{pmatrix} \sin t \\ -\cos t \\ \sin t \end{pmatrix} e^t, \tag{9.153}$$

where $c_2' = c_2 + c_3, c_3' = i(c_2 - c_3)$.

< N Eigenvectors

Here we present procedures for the more difficult case in which the matrix $\mathbf{A}$ does not possess a complete set of linearly independent eigenvectors.

One solution of $d\mathbf{x}/dt = \mathbf{A} \cdot \mathbf{x}$ is $\mathbf{x} = e^{\mathbf{A}t} \cdot \mathbf{e}$, where $\mathbf{e}$ is a constant vector. If $\mathbf{e}_1, \mathbf{e}_2, \ldots, \mathbf{e}_K, K < N$, are linearly independent vectors, then $\mathbf{x}_k = e^{\mathbf{A}t} \cdot \mathbf{e}_k$, $k = 1, \ldots, K$, are linearly independent solutions. We would like to choose $\mathbf{e}_k$, $k = 1, 2, \ldots, K$, such that each $e^{\mathbf{A}t} \cdot \mathbf{e}_k$ is a series with a finite number of terms. This can be done in the following manner. Because

$$e^{\mathbf{A}t} \cdot \mathbf{e} = e^{\lambda \mathbf{I}t} \cdot e^{(\mathbf{A}-\lambda\mathbf{I})t} \cdot \mathbf{e}, \tag{9.154}$$

$$= e^{\lambda t} \mathbf{I} \cdot e^{(\mathbf{A}-\lambda\mathbf{I})t} \cdot \mathbf{e}, \tag{9.155}$$

$$= e^{\lambda t} e^{(\mathbf{A}-\lambda\mathbf{I})t} \cdot \mathbf{e}, \tag{9.156}$$

$$= e^{\lambda t} \left(\mathbf{I} + (\mathbf{A} - \lambda\mathbf{I})t + \left(\frac{1}{2!}\right)(\mathbf{A} - \lambda\mathbf{I})^2 t^2 + \cdots \right) \cdot \mathbf{e}, \tag{9.157}$$

the series will be finite if

$$(\mathbf{A} - \lambda\mathbf{I})^k \cdot \mathbf{e} = 0 \tag{9.158}$$

for some positive integer k.

SUMMARY OF METHOD. The procedure to find $\mathbf{x}_n$, $n = 1, 2, \ldots, N$, the N linearly independent solutions of

$$\frac{d\mathbf{x}}{dt} = \mathbf{A} \cdot \mathbf{x}, \tag{9.159}$$

where $\mathbf{A}$ is a constant, is the following. First find all eigenvalues λ_n, $n = 1, \ldots, N$, and as many eigenvectors $\mathbf{e}_k$, $i = 1, 2, \ldots, K$, as possible.

1. If $K = N$, the N linearly independent solutions are $\mathbf{x}_n = e^{\lambda_n t} \mathbf{e}_n$.
2. If $K < N$, there are only K linearly independent solutions of the type $\mathbf{x}_k = e^{\lambda_k t} \mathbf{e}_k$. To find additional solutions corresponding to a multiple eigenvalue λ, find all linearly independent $\mathbf{g}$ such that $(\mathbf{A} - \lambda\mathbf{I})^2 \cdot \mathbf{g} = 0$, but $(\mathbf{A} - \lambda\mathbf{I}) \cdot \mathbf{g} \neq 0$. Notice that generalized eigenvectors will satisfy the requirement, though there are other solutions. For each such $\mathbf{g}$, we have

$$e^{\mathbf{A}t} \cdot \mathbf{g} = e^{\lambda t} \left(\mathbf{g} + t(\mathbf{A} - \lambda\mathbf{I}) \cdot \mathbf{g} \right), \tag{9.160}$$

   which is a solution.
3. If more solutions are needed, then find all linearly independent $\mathbf{g}$ for which $(\mathbf{A} - \lambda\mathbf{I})^3 \cdot \mathbf{g} = 0$, but $(\mathbf{A} - \lambda\mathbf{I})^2 \cdot \mathbf{g} \neq 0$. The corresponding solution is

$$e^{\mathbf{A}t} \cdot \mathbf{g} = e^{\lambda t} \left(\mathbf{g} + t(\mathbf{A} - \lambda\mathbf{I}) \cdot \mathbf{g} + \frac{t^2}{2}(\mathbf{A} - \lambda\mathbf{I})^2 \cdot \mathbf{g} \right). \tag{9.161}$$

4. Continue until N linearly independent solutions have been found.

A linear combination of the N linearly independent solutions

$$\mathbf{x} = \sum_{n=1}^{N} c_n \mathbf{x}_n \tag{9.162}$$

is the general solution, where $c_1, c_2, \ldots, c_N$, are arbitrary constants.

ALTERNATIVE METHOD. As an alternative to the method just described, which is easily seen to be equivalent, we can use the Jordan canonical form (see Section 7.9.6) in a straightforward way to arrive at the solution. Recall that the Jordan form exists for all matrices. We begin with

$$\frac{d\mathbf{x}}{dt} = \mathbf{A} \cdot \mathbf{x}. \tag{9.163}$$

Then we use the Jordan decomposition, Eq. (7.394), $\mathbf{A} = \mathbf{S} \cdot \mathbf{J} \cdot \mathbf{S}^{-1}$, to write

$$\frac{d\mathbf{x}}{dt} = \underbrace{\mathbf{S} \cdot \mathbf{J} \cdot \mathbf{S}^{-1}}_{=\mathbf{A}} \cdot \mathbf{x}. \tag{9.164}$$

If we apply the matrix operator $\mathbf{S}^{-1}$, which is a constant, to both sides, we get

$$\frac{d}{dt} \left(\underbrace{\mathbf{S}^{-1} \cdot \mathbf{x}}_{\mathbf{z}} \right) = \mathbf{J} \cdot \underbrace{\mathbf{S}^{-1} \cdot \mathbf{x}}_{\mathbf{z}}. \tag{9.165}$$

Now taking $\mathbf{z} \equiv \mathbf{S}^{-1} \cdot \mathbf{x}$, we get

$$\frac{d\mathbf{z}}{dt} = \mathbf{J} \cdot \mathbf{z}. \tag{9.166}$$

We then solve each equation one by one, starting with the last equation, $dz_N/dt = \lambda_N z_N$, and proceeding to the first. In the process of solving these equations sequentially, there will be influence from each off-diagonal term that will give rise to a secular term in the solution. Once $\mathbf{z}$ is determined, we solve for $\mathbf{x}$ by taking $\mathbf{x} = \mathbf{S} \cdot \mathbf{z}$.

This method of course works in the common case in which the matrix $\mathbf{J}$ is diagonal; that is, it applies for cases in which there are N differential equations and N ordinary eigenvectors.

EXAMPLE 9.8

For $\mathbf{x} \in \mathbb{R}^3, t \in \mathbb{R}^1, \mathbf{A} \in \mathbb{R}^3 \times \mathbb{R}^3$, find the general solution of

$$\frac{d\mathbf{x}}{dt} = \mathbf{A} \cdot \mathbf{x}, \tag{9.167}$$

where

$$\mathbf{A} = \begin{pmatrix} 4 & 1 & 3 \\ 0 & 4 & 1 \\ 0 & 0 & 4 \end{pmatrix}. \tag{9.168}$$

$\mathbf{A}$ has an eigenvalue $\lambda = 4$ with multiplicity 3. The only eigenvector is

$$\mathbf{e} = \begin{pmatrix} 1 \\ 0 \\ 0 \end{pmatrix}, \tag{9.169}$$

which gives a solution

$$e^{4t} \begin{pmatrix} 1 \\ 0 \\ 0 \end{pmatrix}. \tag{9.170}$$

A generalized eigenvector in the first sense is

$$\mathbf{g}_1 = \begin{pmatrix} 0 \\ 1 \\ 0 \end{pmatrix}, \tag{9.171}$$

which leads to the solution

$$e^{4t} \left(\mathbf{g}_1 + t(\mathbf{A} - \lambda\mathbf{I}) \cdot \mathbf{g}_1 \right) = e^{4t} \left(\begin{pmatrix} 0 \\ 1 \\ 0 \end{pmatrix} + t \begin{pmatrix} 0 & 1 & 3 \\ 0 & 0 & 1 \\ 0 & 0 & 0 \end{pmatrix} \begin{pmatrix} 0 \\ 1 \\ 0 \end{pmatrix} \right), \tag{9.172}$$

$$= e^{4t} \begin{pmatrix} t \\ 1 \\ 0 \end{pmatrix}. \tag{9.173}$$

Another generalized eigenvector in the first sense,

$$\mathbf{g}_2 = \begin{pmatrix} 0 \\ -3 \\ 1 \end{pmatrix}, \tag{9.174}$$

gives the solution

$$e^{4t} \left(\mathbf{g}_2 + t(\mathbf{A} - \lambda\mathbf{I}) \cdot \mathbf{g}_2 + \frac{t^2}{2}(\mathbf{A} - \lambda\mathbf{I})^2 \cdot \mathbf{g}_2 \right)$$

$$= e^{4t} \left(\begin{pmatrix} 0 \\ -3 \\ 1 \end{pmatrix} + t \begin{pmatrix} 0 & 1 & 3 \\ 0 & 0 & 1 \\ 0 & 0 & 0 \end{pmatrix} \begin{pmatrix} 0 \\ -3 \\ 1 \end{pmatrix} + \frac{t^2}{2} \begin{pmatrix} 0 & 0 & 1 \\ 0 & 0 & 0 \\ 0 & 0 & 0 \end{pmatrix} \begin{pmatrix} 0 \\ -3 \\ 1 \end{pmatrix} \right), \tag{9.175}$$

$$= e^{4t} \begin{pmatrix} \frac{t^2}{2} \\ -3 + t \\ 1 \end{pmatrix}. \tag{9.176}$$

The general solution is

$$\mathbf{x} = c_1 e^{4t} \begin{pmatrix} 1 \\ 0 \\ 0 \end{pmatrix} + c_2 e^{4t} \begin{pmatrix} t \\ 1 \\ 0 \end{pmatrix} + c_3 e^{4t} \begin{pmatrix} \frac{t^2}{2} \\ -3 + t \\ 1 \end{pmatrix}, \tag{9.177}$$

where c_1, c_2, c_3 are arbitrary constants.

ALTERNATIVE METHOD. Alternatively, we can simply use the Jordan decomposition to form the solution. When we form the matrix $\mathbf{S}$ from the eigenvectors and generalized eigenvectors, we have

$$\mathbf{S} = \begin{pmatrix} \vdots & \vdots & \vdots \\ \mathbf{e} & \mathbf{g}_1 & \mathbf{g}_2 \\ \vdots & \vdots & \vdots \end{pmatrix} = \begin{pmatrix} 1 & 0 & 0 \\ 0 & 1 & -3 \\ 0 & 0 & 1 \end{pmatrix}. \tag{9.178}$$

We then get

$$\mathbf{S}^{-1} = \begin{pmatrix} 1 & 0 & 0 \\ 0 & 1 & 3 \\ 0 & 0 & 1 \end{pmatrix}. \tag{9.179}$$

Thus,

$$\mathbf{J} = \mathbf{S}^{-1} \cdot \mathbf{A} \cdot \mathbf{S} = \begin{pmatrix} 4 & 1 & 0 \\ 0 & 4 & 1 \\ 0 & 0 & 4 \end{pmatrix}. \tag{9.180}$$

Now with $\mathbf{z} = \mathbf{S}^{-1} \cdot \mathbf{x}$, we solve $d\mathbf{z}/dt = \mathbf{J} \cdot \mathbf{z}$,

$$\frac{d}{dt} \begin{pmatrix} z_1 \\ z_2 \\ z_3 \end{pmatrix} = \begin{pmatrix} 4 & 1 & 0 \\ 0 & 4 & 1 \\ 0 & 0 & 4 \end{pmatrix} \begin{pmatrix} z_1 \\ z_2 \\ z_3 \end{pmatrix}. \tag{9.181}$$

The final equation is uncoupled; solving $dz_3/dt = 4z_3$, we get

$$z_3(t) = c_3 e^{4t}. \tag{9.182}$$

Now consider the second equation,

$$\frac{dz_2}{dt} = 4z_2 + z_3. \tag{9.183}$$

Using our solution for z_3, we get

$$\frac{dz_2}{dt} = 4z_2 + c_3 e^{4t}. \tag{9.184}$$

Solving, we get

$$z_2(t) = c_2 e^{4t} + c_3 t e^{4t}. \tag{9.185}$$

Now consider the first equation,

$$\frac{dz_1}{dt} = 4z_1 + z_2. \tag{9.186}$$

Using our solution for z_2, we find

$$\frac{dz_1}{dt} = 4z_1 + c_2 e^{4t} + c_3 t e^{4t}. \tag{9.187}$$

Solving, we see that

$$z_1(t) = c_1 e^{4t} + \frac{1}{2} t e^{4t} \left(2c_2 + t c_3 \right), \tag{9.188}$$

so we have

$$z(t) = \begin{pmatrix} c_1 e^{4t} + \frac{1}{2} t e^{4t} (2c_2 + tc_3) \\ c_2 e^{4t} + c_3 t e^{4t} \\ c_3 e^{4t} \end{pmatrix}. \tag{9.189}$$

Then for $x = S \cdot z$, we recover

$$x = c_1 e^{4t} \begin{pmatrix} 1 \\ 0 \\ 0 \end{pmatrix} + c_2 e^{4t} \begin{pmatrix} t \\ 1 \\ 0 \end{pmatrix} + c_3 e^{4t} \begin{pmatrix} \frac{t^2}{2} \\ -3 + t \\ 1 \end{pmatrix}, \tag{9.190}$$

which is identical to our earlier result.

Fundamental Matrix

If x_n, $n = 1, \ldots, N$, are linearly independent solutions of $dx/dt = A \cdot x$, then

$$\Omega = \begin{pmatrix} \vdots & \vdots & \cdots & \vdots \\ x_1 & x_2 & \cdots & x_N \\ \vdots & \vdots & \cdots & \vdots \end{pmatrix} \tag{9.191}$$

is called a *fundamental matrix*. It is not unique. The general solution is

$$x = \Omega \cdot c, \tag{9.192}$$

where

$$c = \begin{pmatrix} c_1 \\ \vdots \\ c_N \end{pmatrix}. \tag{9.193}$$

EXAMPLE 9.9

Find the fundamental matrix of the previous example problem.

The fundamental matrix is

$$\Omega = e^{4t} \begin{pmatrix} 1 & t & \frac{t^2}{2} \\ 0 & 1 & -3 + t \\ 0 & 0 & 1 \end{pmatrix}, \tag{9.194}$$

so that

$$x = \Omega \cdot c = e^{4t} \begin{pmatrix} 1 & t & \frac{t^2}{2} \\ 0 & 1 & -3 + t \\ 0 & 0 & 1 \end{pmatrix} \begin{pmatrix} c_1 \\ c_2 \\ c_3 \end{pmatrix}. \tag{9.195}$$

Discretization

Most common numerical methods for obtaining approximate solutions to systems of ordinary differential equations reduce to solving linear algebra problems, often as an iterated map. Let us examine in the next example our linear homogeneous differential system $dx/dt = A \cdot x$ in the context of a common numerical method that relies on discretization to obtain an approximate solution. In so doing, we shall see more connections between differential equations and linear algebra.

EXAMPLE 9.10

Examine the linear homogeneous system $dx/dt = \mathbf{A} \cdot \mathbf{x}$ in terms of an explicit finite difference approximation, and give a geometric interpretation of the combined action of the differential and matrix operator on $\mathbf{x}$.

As introduced in Example 4.24, a first-order explicit finite difference approximation to the differential equation takes the form

$$\frac{\mathbf{x}^{k+1} - \mathbf{x}^k}{\Delta t} = \mathbf{A} \cdot \mathbf{x}^k, \tag{9.196}$$

$$\mathbf{x}^{k+1} = \mathbf{x}^k + \Delta t \mathbf{A} \cdot \mathbf{x}^k, \tag{9.197}$$

$$= (\mathbf{I} + \Delta t \mathbf{A}) \cdot \mathbf{x}^k. \tag{9.198}$$

This method is sometimes known as the forward Euler method. We can also define $\mathbf{B} = \mathbf{I} + \Delta t \mathbf{A}$ and rewrite Eq. (9.198) as

$$\mathbf{x}^{k+1} = \mathbf{B} \cdot \mathbf{x}^k. \tag{9.199}$$

Let us decompose $\mathbf{A}$ into symmetric and antisymmetric parts,

$$\mathbf{A}_s = \frac{\mathbf{A} + \mathbf{A}^H}{2}, \tag{9.200}$$

$$\mathbf{A}_a = \frac{\mathbf{A} - \mathbf{A}^H}{2}, \tag{9.201}$$

so that $\mathbf{A} = \mathbf{A}_s + \mathbf{A}_a$. Then Eq. (9.198) becomes

$$\mathbf{x}^{k+1} = (\mathbf{I} + \Delta t \mathbf{A}_s + \Delta t \mathbf{A}_a) \cdot \mathbf{x}^k. \tag{9.202}$$

Now because $\mathbf{A}_s$ is symmetric, it can be diagonally decomposed as

$$\mathbf{A}_s = \mathbf{Q} \cdot \mathbf{\Lambda}_s \cdot \mathbf{Q}^H, \tag{9.203}$$

where $\mathbf{Q}$ is a unitary matrix, which we will restrict to be a rotation matrix, and $\mathbf{\Lambda}_s$ is a diagonal matrix with the guaranteed real eigenvalues of $\mathbf{A}_s$ on its diagonal. It can also be shown that the antisymmetric $\mathbf{A}_a$ has a related decomposition,

$$\mathbf{A}_a = \hat{\mathbf{Q}} \cdot \mathbf{\Lambda}_a \cdot \hat{\mathbf{Q}}^H, \tag{9.204}$$

where $\hat{\mathbf{Q}}$ is a unitary matrix and $\mathbf{\Lambda}_a$ is a diagonal matrix with the purely imaginary eigenvalues of $\mathbf{A}_a$ on its diagonal. Substituting Eqs. (9.203,9.204) into Eq. (9.202), we get

$$\mathbf{x}^{k+1} = \left(\mathbf{I} + \Delta t \underbrace{\mathbf{Q} \cdot \mathbf{\Lambda}_s \cdot \mathbf{Q}^H}_{\mathbf{A}_s} + \Delta t \underbrace{\hat{\mathbf{Q}} \cdot \mathbf{\Lambda}_a \cdot \hat{\mathbf{Q}}^H}_{\mathbf{A}_a} \right) \cdot \mathbf{x}^k. \tag{9.205}$$

Now because $\mathbf{Q} \cdot \mathbf{Q}^H = \mathbf{I} = \mathbf{Q} \cdot \mathbf{I} \cdot \mathbf{Q}^H$, we can operate on the first and third terms of Eq. (9.205) to get

$$\mathbf{x}^{k+1} = \left(\mathbf{Q} \cdot \mathbf{I} \cdot \mathbf{Q}^H + \Delta t \mathbf{Q} \cdot \mathbf{\Lambda}_s \cdot \mathbf{Q}^H + \Delta t \mathbf{Q} \cdot \mathbf{Q}^H \cdot \hat{\mathbf{Q}} \cdot \mathbf{\Lambda}_a \cdot \hat{\mathbf{Q}}^H \cdot \mathbf{Q} \cdot \mathbf{Q}^H \right) \cdot \mathbf{x}^k, \tag{9.206}$$

$$= \mathbf{Q} \cdot \left(\mathbf{I} + \Delta t \mathbf{\Lambda}_s + \Delta t \mathbf{Q}^H \cdot \hat{\mathbf{Q}} \cdot \mathbf{\Lambda}_a \cdot \hat{\mathbf{Q}}^H \cdot \mathbf{Q} \right) \cdot \mathbf{Q}^H \cdot \mathbf{x}^k, \tag{9.207}$$

$$\mathbf{Q}^H \cdot \mathbf{x}^{k+1} = \underbrace{\mathbf{Q}^H \cdot \mathbf{Q}}_{=\mathbf{I}} \cdot \left(\mathbf{I} + \Delta t \mathbf{\Lambda}_s + \Delta t \mathbf{Q}^H \cdot \hat{\mathbf{Q}} \cdot \mathbf{\Lambda}_a \cdot \hat{\mathbf{Q}}^H \cdot \mathbf{Q} \right) \cdot \mathbf{Q}^H \cdot \mathbf{x}^k. \tag{9.208}$$

Now, let us define a rotated coordinate system as $\hat{\mathbf{x}} = \mathbf{Q}^H \cdot \mathbf{x}$, so that Eq. (9.208) becomes

$$\hat{\mathbf{x}}^{k+1} = \left(\mathbf{I} + \Delta t \mathbf{\Lambda}_s + \Delta t \mathbf{Q}^H \cdot \hat{\mathbf{Q}} \cdot \mathbf{\Lambda}_a \cdot \hat{\mathbf{Q}}^H \cdot \mathbf{Q} \right) \cdot \hat{\mathbf{x}}^k, \tag{9.209}$$

$$= \left(\mathbf{I} + \underbrace{\Delta t \mathbf{\Lambda}_s}_{\text{stretching}} + \underbrace{\Delta t (\mathbf{Q}^H \cdot \hat{\mathbf{Q}}) \cdot \mathbf{\Lambda}_a \cdot (\mathbf{Q}^H \cdot \hat{\mathbf{Q}})^H}_{\text{rotation}} \right) \cdot \hat{\mathbf{x}}^k. \tag{9.210}$$

This rotated coordinate system is aligned with the principal axes of deformation associated with $\mathbf{A}_s$. The new value, $\hat{\mathbf{x}}^{k+1}$, is composed of the sum of three terms: (1) the old value, due to the action of $\mathbf{I}$, (2) a stretching along the coordinate axes by the term $\Delta t \mathbf{\Lambda}_s$, and (3) a rotation, normal to the coordinate axes by the term $\Delta t (\mathbf{Q}^H \cdot \hat{\mathbf{Q}}) \cdot \mathbf{\Lambda}_a \cdot (\mathbf{Q}^H \cdot \hat{\mathbf{Q}})^H$. Because both $\mathbf{Q}$ and $\hat{\mathbf{Q}}$ have a norm of unity, it is the magnitude of the eigenvalues, along with Δt that determines the amount of stretching and rotation that occurs. Although $\mathbf{\Lambda}_a$ and $\hat{\mathbf{Q}}$ have imaginary components, when combined together, they yield a real result, if the original system is real.

We can get a similar interpretation using the polar decomposition of Section 7.9.9. Let us impose the polar decomposition of Eq. (7.437) to Eq. (9.198) to get

$$\mathbf{x}^{k+1} = (\mathbf{I} + \Delta t \mathbf{Q} \cdot \mathbf{W}) \cdot \mathbf{x}^k. \tag{9.211}$$

Now employ similar operations as done earlier:

$$\mathbf{x}^{k+1} = (\mathbf{Q} \cdot \mathbf{I} \cdot \mathbf{Q}^H + \Delta t \mathbf{Q} \cdot \mathbf{W} \cdot \mathbf{Q} \cdot \mathbf{Q}^H) \cdot \mathbf{x}^k, \tag{9.212}$$

$$= \mathbf{Q} \cdot (\mathbf{I} + \Delta t \mathbf{W} \cdot \mathbf{Q}) \cdot \mathbf{Q}^H \cdot \mathbf{x}^k, \tag{9.213}$$

$$\mathbf{Q}^H \cdot \mathbf{x}^{k+1} = \mathbf{Q}^H \cdot \mathbf{Q} \cdot (\mathbf{I} + \Delta t \mathbf{W} \cdot \mathbf{Q}) \cdot \mathbf{Q}^H \cdot \mathbf{x}^k, \tag{9.214}$$

$$\hat{\mathbf{x}}^{k+1} = (\mathbf{I} + \Delta t \mathbf{W} \cdot \mathbf{Q}) \cdot \hat{\mathbf{x}}^k. \tag{9.215}$$

Here the stretching and rotation, segregated additively in the earlier decomposition, has been restructured.

Now let us imagine that the real parts of the eigenvalues of $\mathbf{A}$ are all less than or equal to zero, thus insuring that the equilibrium $\mathbf{x} = \mathbf{0}$ of the original system $d\mathbf{x}/dt = \mathbf{A} \cdot \mathbf{x}$ is stable. What can we say about the stability of our discrete first-order finite difference approximation? Certainly for stability of the corresponding linear mapping, Eq. (9.199), we must insist that

$$||\mathbf{B}||_2 = ||\mathbf{I} + \Delta t \mathbf{A}||_2 \le 1. \tag{9.216}$$

This requires that

$$\rho \left((\mathbf{I} + \Delta t \mathbf{A})^H \cdot (\mathbf{I} + \Delta t \mathbf{A}) \right) \le 1, \tag{9.217}$$

where ρ is the spectral radius, that is the maximum eigenvalue of the matrix in question. Expanding, we get

$$\rho \left((\mathbf{I}^H + \Delta t \mathbf{A}^H) \cdot (\mathbf{I} + \Delta t \mathbf{A}) \right) \le 1, \tag{9.218}$$

$$\rho \left(\mathbf{I} + \Delta t \mathbf{A}^H + \Delta t \mathbf{A} + \Delta t^2 \mathbf{A}^H \cdot \mathbf{A} \right) \le 1, \tag{9.219}$$

$$\rho \left(\mathbf{I} + 2 \Delta t \mathbf{A}_s + \Delta t^2 \mathbf{A}^H \cdot \mathbf{A} \right) \le 1. \tag{9.220}$$

Expanding on the definition of the spectral radius, we find

$$\max \left| \text{eig} \left(\mathbf{I} + 2 \Delta t \mathbf{A}_s + \Delta t^2 \mathbf{A}^H \cdot \mathbf{A} \right) \right| \le 1. \tag{9.221}$$

Recalling Eq. (7.120), we reduce our condition to

$$\max |1 + \text{eig} \left(2\Delta t \mathbf{A}_s + \Delta t^2 \mathbf{A}^H \cdot \mathbf{A} \right)| \le 1. \tag{9.222}$$

If $\mathbf{A}_s = \mathbf{0}$, it is impossible to satisfy the stability criterion because the eigenvalues of $\mathbf{A}^H \cdot \mathbf{A}$ must be real and positive. Thus, the forward Euler method is unconditionally unstable for all Δt in the case in which $\mathbf{A}$ is purely antisymmetric.

Let us consider a particular $\mathbf{A}$ and examine the consequences of several choices. Let

$$\mathbf{A} = \begin{pmatrix} -1 & 10 \\ -10 & -2 \end{pmatrix}. \tag{9.223}$$

This matrix has eigenvalues of $\lambda_{1,2} = -3/2 \pm \sqrt{399}i/2 = -1.5 \pm 9.987i$, so it is describing an underdamped oscillatory solution with the time scale of amplitude decay longer than that of oscillation. The matrix is well conditioned with a condition number of $c = ||\mathbf{A}||_2 ||\mathbf{A}^{-1}||_2 = (205 + \sqrt{409})/204 = 1.104$. Taking $\mathbf{x}(0) = (1, 1)^T$, solution to $d\mathbf{x}/dt = \mathbf{A} \cdot \mathbf{x}$ yields

$$x_1 = e^{-3t/2} \left(\cos\left(\frac{\sqrt{399}t}{2} \right) + \frac{\sqrt{399}}{19} \sin\left(\frac{\sqrt{399}t}{2} \right) \right), \tag{9.224}$$

$$x_2 = e^{-3t/2} \left(\cos\left(\frac{\sqrt{399}t}{2} \right) - \frac{\sqrt{399}}{19} \sin\left(\frac{\sqrt{399}t}{2} \right) \right). \tag{9.225}$$

The exact solution is obviously stable in that both x_1 and x_2 approach the origin as $t \to \infty$. Moreover, $||\mathbf{x}||_2$ is well behaved, decreasing smoothly from a maximum of $\sqrt{2}$ at $t = 0$ to zero at $t \to \infty$.

Now it is easy to show the stability criterion for this particular first-order explicit numerical method applied to the scalar equation $dx/dt = \lambda x$ is $\Delta t \le 2/|\lambda|$ for $\lambda \in \mathbb{R}^1$. While one might be tempted to extrapolate this to our system by taking $\Delta t \le 2/|\lambda|_{max}$ where λ is a potentially complex eigenvalue of $\mathbf{A}$, this can induce instabilities in the numerical approximation of the exact solution. Here the eigenvalues of $\mathbf{A}$ both have the same magnitude, yielding $|\lambda_{1,2}| = \sqrt{102} = 10.0995$, giving rise to an extrapolated apparent stability criterion of $\Delta t \le 0.19803$. In contrast, Eq. (9.216) or (9.222) induces the more stringent correct criterion, which arises from solving

$$||\mathbf{B}||_2 = ||\mathbf{I} + \Delta t \mathbf{A}||_2 = \frac{\sqrt{205\Delta t^2 + \left(\sqrt{409\Delta t^2 - 12\Delta t + 4} - 6 \right) \Delta t + 2}}{\sqrt{2}} = 1 \tag{9.226}$$

for the stability boundary of Δt. Identification of the boundary yields

$$\Delta t \le 0.0196078 \tag{9.227}$$

as the condition for stability.

Let us examine the ability of the discrete approximation, iterative application of Eq. (9.198), to capture the essence of the known stable exact solution for three values: $\Delta t = 0.01, 0.03, 0.05$. All three have $\Delta t < 2/|\lambda|_{max}$. And all three fall below the scale of oscillation of the exact solution, $2/\sqrt{399} = 0.100$. But only $\Delta t = 0.01$ is below the more stringent criterion of Eq. (9.227) based on the spectral norm. Plots of $x_1(t)$ and its discrete approximations are shown in Figure 9.11. Obviously for $\Delta t = 0.01$, the discrete approximation is both stable and moderately accurate. Smaller Δt would enhance the accuracy. The two larger values of Δt are both unstable. The instability is induced by the nearly antisymmetric, *nonnormal* nature of the matrix $\mathbf{A}$. Additional discussion is given by Trefethen and Embree (2005).

We close by noting that there exist matrices $\mathbf{A}$ for which no Δt exists to stabilize the forward Euler method. For example, in a problem to be studied soon with an alternate

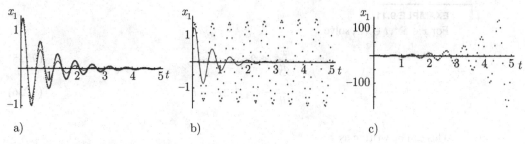

Figure 9.11. Exact stable solution for $x_1(t)$ and three approximations using a first-order explicit discretization with Δt of (a) 0.01, (b) 0.03, and (c) 0.05.

stable numerical method (see Example 9.14) with

$$\mathbf{A} = \begin{pmatrix} 0 & -1 \\ 1 & -1 \end{pmatrix}, \tag{9.228}$$

we have a stable exact solution because the eigenvalues of $\mathbf{A}$ are $-1/2 \pm \sqrt{3}i/2$. But

$$||\mathbf{B}||_2 = ||\mathbf{I} + \Delta t \mathbf{A}||_2 = \frac{\sqrt{3\Delta t^2 + \left(\sqrt{5\Delta t^2 - 4\Delta t + 4} - 2\right)\Delta t + 2}}{\sqrt{2}} \tag{9.229}$$

is greater than unity for all $\Delta t \neq 0$, and unity only if $\Delta t = 0$; thus, any nontrivial forward Euler approximation for this case is unconditionally unstable.

9.5.3 Inhomogeneous with Constant Coefficients

If $\mathbf{A}$ is a constant matrix that is diagonalizable, the system of differential equations represented by

$$\frac{d\mathbf{x}}{dt} = \mathbf{A} \cdot \mathbf{x} + \mathbf{f}(t) \tag{9.230}$$

can be decoupled into a set of scalar equations, each of which is in terms of a single dependent variable. From Eq. (7.350), let $\mathbf{S}$ be such that $\mathbf{S}^{-1} \cdot \mathbf{A} \cdot \mathbf{S} = \Lambda$, where Λ is a diagonal matrix of eigenvalues. Taking $\mathbf{x} = \mathbf{S} \cdot \mathbf{z}$, we get

$$\frac{d(\mathbf{S} \cdot \mathbf{z})}{dt} = \mathbf{A} \cdot \mathbf{S} \cdot \mathbf{z} + \mathbf{f}(t), \tag{9.231}$$

$$\mathbf{S} \cdot \frac{d\mathbf{z}}{dt} = \mathbf{A} \cdot \mathbf{S} \cdot \mathbf{z} + \mathbf{f}(t). \tag{9.232}$$

Applying $\mathbf{S}^{-1}$ to both sides,

$$\frac{d\mathbf{z}}{dt} = \underbrace{\mathbf{S}^{-1} \cdot \mathbf{A} \cdot \mathbf{S}}_{=\Lambda} \cdot \mathbf{z} + \mathbf{S}^{-1} \cdot \mathbf{f}(t), \tag{9.233}$$

$$= \Lambda \cdot \mathbf{z} + \mathbf{g}(t), \tag{9.234}$$

where $\Lambda = \mathbf{S}^{-1} \cdot \mathbf{A} \cdot \mathbf{S}$ and $\mathbf{g}(t) = \mathbf{S}^{-1} \cdot \mathbf{f}(t)$. This is the decoupled form of the original equation.

EXAMPLE 9.11

For $x \in \mathbb{R}^2, t \in \mathbb{R}^1$, solve

$$\frac{dx_1}{dt} = 2x_1 + x_2 + 1, \tag{9.235}$$

$$\frac{dx_2}{dt} = x_1 + 2x_2 + t. \tag{9.236}$$

This can be written as

$$\frac{d}{dt}\begin{pmatrix} x_1 \\ x_2 \end{pmatrix} = \begin{pmatrix} 2 & 1 \\ 1 & 2 \end{pmatrix}\begin{pmatrix} x_1 \\ x_2 \end{pmatrix} + \begin{pmatrix} 1 \\ t \end{pmatrix}. \tag{9.237}$$

We have

$$\mathbf{S} = \begin{pmatrix} 1 & 1 \\ -1 & 1 \end{pmatrix}, \qquad \mathbf{S}^{-1} = \begin{pmatrix} \frac{1}{2} & -\frac{1}{2} \\ \frac{1}{2} & \frac{1}{2} \end{pmatrix}, \qquad \mathbf{\Lambda} = \begin{pmatrix} 1 & 0 \\ 0 & 3 \end{pmatrix}, \tag{9.238}$$

so that

$$\frac{d}{dt}\begin{pmatrix} z_1 \\ z_2 \end{pmatrix} = \begin{pmatrix} 1 & 0 \\ 0 & 3 \end{pmatrix}\begin{pmatrix} z_1 \\ z_2 \end{pmatrix} + \frac{1}{2}\begin{pmatrix} 1-t \\ 1+t \end{pmatrix}. \tag{9.239}$$

The solution is

$$z_1 = ae^t + \frac{t}{2}, \tag{9.240}$$

$$z_2 = be^{3t} - \frac{2}{9} - \frac{t}{6}, \tag{9.241}$$

which, using $x_1 = z_1 + z_2$ and $x_2 = -z_1 + z_2$, transforms to

$$x_1 = ae^t + be^{3t} - \frac{2}{9} + \frac{t}{3}, \tag{9.242}$$

$$x_2 = -ae^t + be^{3t} - \frac{2}{9} - \frac{2t}{3}. \tag{9.243}$$

EXAMPLE 9.12

Solve the system

$$\frac{d\mathbf{x}}{dt} = \mathbf{A} \cdot (\mathbf{x} - \mathbf{x}_0) + \mathbf{b}, \qquad \mathbf{x}(t_0) = \mathbf{x}_0. \tag{9.244}$$

Such a system arises naturally when one linearizes a nonlinear system of the form $d\mathbf{x}/dt = \mathbf{f}(\mathbf{x})$ about a point $\mathbf{x} = \mathbf{x}_0$. Here, then, $\mathbf{A}$ is the Jacobian matrix $\partial \mathbf{f}/\partial \mathbf{x}|_{\mathbf{x}=\mathbf{x}_0}$. The system is in equilibrium when

$$\mathbf{A} \cdot (\mathbf{x} - \mathbf{x}_0) = -\mathbf{b}, \tag{9.245}$$

or

$$\mathbf{x} = \mathbf{x}_0 - \mathbf{A}^{-1} \cdot \mathbf{b}. \tag{9.246}$$

Furthermore, note that if $\mathbf{b} = \mathbf{0}$, the initial condition $\mathbf{x} = \mathbf{x}_0$ is also an equilibrium condition and is the unique solution to the differential equation.

First define a new dependent variable $\mathbf{z}$:

$$\mathbf{z} \equiv \mathbf{x} - \mathbf{x}_0 + \mathbf{A}^{-1} \cdot \mathbf{b}, \tag{9.247}$$

which represents a deviation from equilibrium. So we have

$$\mathbf{x} = \mathbf{z} + \mathbf{x}_0 - \mathbf{A}^{-1} \cdot \mathbf{b}. \tag{9.248}$$

At $t = t_0$, we then get

$$\mathbf{z}(t_0) = \mathbf{A}^{-1} \cdot \mathbf{b}. \tag{9.249}$$

Then substitute into the original differential equation system to get

$$\frac{d}{dt} \left(\mathbf{z} + \mathbf{x}_0 - \mathbf{A}^{-1} \cdot \mathbf{b} \right) = \mathbf{A} \cdot \left(\mathbf{z} - \mathbf{A}^{-1} \cdot \mathbf{b} \right) + \mathbf{b}, \qquad \mathbf{z}(t_0) = \mathbf{A}^{-1} \cdot \mathbf{b}, \tag{9.250}$$

$$\frac{d\mathbf{z}}{dt} = \mathbf{A} \cdot \mathbf{z}, \qquad \mathbf{z}(t_0) = \mathbf{A}^{-1} \cdot \mathbf{b}. \tag{9.251}$$

Now assume that the Jacobian is fully diagonalizable so that we can take $\mathbf{A} = \mathbf{S} \cdot \mathbf{\Lambda} \cdot \mathbf{S}^{-1}$. Thus, we have

$$\frac{d\mathbf{z}}{dt} = \mathbf{S} \cdot \mathbf{\Lambda} \cdot \mathbf{S}^{-1} \cdot \mathbf{z}, \qquad \mathbf{z}(t_0) = \mathbf{A}^{-1} \cdot \mathbf{b}. \tag{9.252}$$

Take now

$$\mathbf{w} \equiv \mathbf{S}^{-1} \cdot \mathbf{z}, \qquad \mathbf{z} = \mathbf{S} \cdot \mathbf{w}, \tag{9.253}$$

so that the differential equations and initial conditions become

$$\frac{d}{dt} \left(\mathbf{S} \cdot \mathbf{w} \right) = \mathbf{S} \cdot \mathbf{\Lambda} \cdot \mathbf{w}, \qquad \mathbf{S} \cdot \mathbf{w}(t_0) = \mathbf{A}^{-1} \cdot \mathbf{b}. \tag{9.254}$$

Because $\mathbf{S}$ and $\mathbf{S}^{-1}$ are constant, we can apply the operator $\mathbf{S}^{-1}$ to both sides of the differential equations and initial conditions to get

$$\mathbf{S}^{-1} \cdot \frac{d}{dt} \left(\mathbf{S} \cdot \mathbf{w} \right) = \mathbf{S}^{-1} \cdot \mathbf{S} \cdot \mathbf{\Lambda} \cdot \mathbf{w}, \qquad \mathbf{S}^{-1} \cdot \mathbf{S} \cdot \mathbf{w}(t_0) = \mathbf{S}^{-1} \cdot \mathbf{A}^{-1} \cdot \mathbf{b}, \tag{9.255}$$

$$\frac{d}{dt} \left(\mathbf{S}^{-1} \cdot \mathbf{S} \cdot \mathbf{w} \right) = \mathbf{I} \cdot \mathbf{\Lambda} \cdot \mathbf{w}, \qquad \mathbf{I} \cdot \mathbf{w}(t_0) = \mathbf{S}^{-1} \cdot \mathbf{A}^{-1} \cdot \mathbf{b}, \tag{9.256}$$

$$\frac{d\mathbf{w}}{dt} = \mathbf{\Lambda} \cdot \mathbf{w}, \qquad \mathbf{w}(t_0) = \mathbf{S}^{-1} \cdot \mathbf{A}^{-1} \cdot \mathbf{b}. \tag{9.257}$$

This is in diagonal form and has solution

$$\mathbf{w}(t) = e^{\mathbf{\Lambda}(t-t_0)} \cdot \mathbf{S}^{-1} \cdot \mathbf{A}^{-1} \cdot \mathbf{b}. \tag{9.258}$$

In terms of $\mathbf{z}$, the solution is

$$\mathbf{z}(t) = \mathbf{S} \cdot e^{\mathbf{\Lambda}(t-t_0)} \cdot \mathbf{S}^{-1} \cdot \mathbf{A}^{-1} \cdot \mathbf{b}. \tag{9.259}$$

Then using the definition of $\mathbf{z}$, one can write the solution in terms of the original $\mathbf{x}$ as

$$\mathbf{x}(t) = \mathbf{x}_0 + \left(\mathbf{S} \cdot e^{\mathbf{\Lambda}(t-t_0)} \cdot \mathbf{S}^{-1} - \mathbf{I} \right) \cdot \mathbf{A}^{-1} \cdot \mathbf{b}. \tag{9.260}$$

The scales of evolution with t are determined by $\mathbf{\Lambda}$; in particular, the scales of each mode, τ_i, are $\tau_i = 1/\lambda_i$, where λ_i is an entry in $\mathbf{\Lambda}$. The constant vector $\mathbf{b}$ plays a secondary role in determining the scales of t. Lastly, one infers from the discussion of the matrix exponential, Eq. (7.529), that $e^{\mathbf{A}(t-t_0)} = \mathbf{S} \cdot e^{\mathbf{\Lambda}(t-t_0)} \cdot \mathbf{S}^{-1}$, so we get the final form of

$$\mathbf{x}(t) = \mathbf{x}_0 + \left(e^{\mathbf{A}(t-t_0)} - \mathbf{I} \right) \cdot \mathbf{A}^{-1} \cdot \mathbf{b}. \tag{9.261}$$

The method of undetermined coefficients for systems is similar to that presented for scalar equations in Section 4.3.1.

EXAMPLE 9.13

For $\mathbf{x} \in \mathbb{R}^3, t \in \mathbb{R}^1, \mathbf{A} \in \mathbb{R}^3 \times \mathbb{R}^3, \mathbf{f} : \mathbb{R}^1 \to \mathbb{R}^3$, solve $d\mathbf{x}/dt = \mathbf{A} \cdot \mathbf{x} + \mathbf{f}(t)$ with

$$
\mathbf{A} = \begin{pmatrix} 4 & 1 & 3 \\ 0 & 4 & 1 \\ 0 & 0 & 4 \end{pmatrix}, \qquad \mathbf{f} = \begin{pmatrix} 3e^t \\ 0 \\ 0 \end{pmatrix}. \tag{9.262}
$$

The homogeneous part of this problem has been solved in Example 9.8. Let the particular solution be

$$
\mathbf{x}_P = \mathbf{c}e^t. \tag{9.263}
$$

Substituting into the equation, we get

$$
\mathbf{c}e^t = \mathbf{A} \cdot \mathbf{c}e^t + \begin{pmatrix} 3 \\ 0 \\ 0 \end{pmatrix} e^t. \tag{9.264}
$$

We can cancel the exponential to get

$$
(\mathbf{I} - \mathbf{A}) \cdot \mathbf{c} = \begin{pmatrix} 3 \\ 0 \\ 0 \end{pmatrix}, \tag{9.265}
$$

which can be solved to get

$$
\mathbf{c} = \begin{pmatrix} -1 \\ 0 \\ 0 \end{pmatrix}. \tag{9.266}
$$

Therefore,

$$
\mathbf{x} = \mathbf{x}_H + \begin{pmatrix} -1 \\ 0 \\ 0 \end{pmatrix} e^t. \tag{9.267}
$$

The method must be modified if $\mathbf{f} = \mathbf{c}e^{\lambda t}$, where λ is an eigenvalue of $\mathbf{A}$. Then the particular solution must be of the form $\mathbf{x}_P = (\mathbf{c}_0 + t\mathbf{c}_1 + t^2\mathbf{c}_2 + \cdots)e^{\lambda t}$, where the series is finite, and we take as many terms as necessary. The method of variation of parameters can also be used and follows the general procedure explained in Section 4.3.2.

9.6 Nonlinear Systems

Nonlinear systems can be difficult to solve. Even for algebraic systems, general solutions do not exist for polynomial equations of arbitrary degree. Nonlinear differential equations, both ordinary and partial, admit analytical solutions only in special cases. Because these equations are common in engineering applications, many techniques for approximate numerical and analytical solutions have been developed. Our purpose in this section is more restricted; we revisit in more detail concepts such as equilibrium and linearization first discussed in Section 9.3, introduce some standard nomenclature, and explore some of the limited analytical tools available to study the stability of nonlinear systems. We continue discussion of nonlinear systems beyond this section for most of the remainder of the book.

9.6.1 Definitions

With $x \in \mathbb{R}^N, t \in \mathbb{R}^1, f : \mathbb{R}^N \to \mathbb{R}^N$, consider a system of N nonlinear first-order ordinary differential equations,

$$\frac{dx_n}{dt} = f_n(x_1, x_2, \ldots, x_N), \qquad n = 1, \ldots, N, \tag{9.268}$$

where t is time and f_n is a vector field. The vector field f_n is said to have *structural stability* if any vector field g_n near f_n is topologically equivalent (see Section 6.7), to f_n. The system is *autonomous* because f_n is not a function of t. The coordinates $x_1, x_2, \ldots, x_N$, form a *phase* or *state* space. The divergence of the vector field, div $f_n = \sum_{n=1}^{N} \partial f_n / \partial x_n$, indicates the change of a given volume of initial conditions in phase space. If the divergence is zero, the volume remains constant, and the system is said to be *conservative*. If the divergence is negative, the volume shrinks with t, and the system is *dissipative*. The volume in a dissipative system eventually goes to zero. This final state to which some initial set of points in phase space goes is called an *attractor*. Attractors may be points, closed curves, tori, or fractals (strange). A given dynamical system may have several attractors that coexist. Each attractor has its own *basin of attraction* in $\mathbb{R}^N$; initial conditions that lie within this basin tend to that particular attractor.

The steady state solutions $x_n = \overline{x}_n$ of Eq. (9.268) are called *critical* (or *fixed*, *singular*, or *stationary*) points. Thus, by definition,

$$f_n(\overline{x}_1, \overline{x}_2, \ldots, \overline{x}_N) = 0, \qquad n = 1, \ldots, N, \tag{9.269}$$

which is an algebraic, potentially transcendental set of equations. The dynamics of the system are analyzed by studying the stability of the critical point. For this we perturb the system so that

$$x_n = \overline{x}_n + \tilde{x}_n, \tag{9.270}$$

where the $\sim$ denotes a perturbation. If $||\tilde{x}_n||$ is bounded for $t \to \infty$, the critical point is said to be *stable*; otherwise, it is *unstable*. As a special case, if $||\tilde{x}_n|| \to 0$ as $t \to \infty$, the critical point is *asymptotically* stable. If $||\tilde{x}_n||$ remains at a nonzero constant as $t \to \infty$, the critical point is *neutrally stable*. These equilibria may be located at finite points within the phase space or points at infinity.

Dynamical systems are often thought to evolve on *manifolds*. Loosely defined we often think of manifolds as some topological surface of given dimension embedded within a space of equal or higher dimension. Most two-dimensional surfaces and one-dimensional curves may be manifolds within a three-dimensional space, for example. Solutions of dynamical systems typically evolve on one-dimensional manifolds embedded within an N-dimensional space. These one-dimensional manifolds are called *trajectories* or orbits. Trajectories are examples of so-called *invariant manifolds*, sets of points which, when an initial point on the invariant manifold is acted on by the dynamical system, evolve so as to stay on the invariant manifold.

EXAMPLE 9.14

Evaluate some of the properties of dynamical systems for the degenerate case of a linear system:

$$\frac{d}{dt} \begin{pmatrix} x_1 \\ x_2 \end{pmatrix} = \begin{pmatrix} 0 & -1 \\ 1 & -1 \end{pmatrix} \begin{pmatrix} x_1 \\ x_2 \end{pmatrix}. \tag{9.271}$$

This is of the form $d\mathbf{x}/dt = \mathbf{A} \cdot \mathbf{x}$. This particular alibi mapping $\mathbf{f} = \mathbf{A} \cdot \mathbf{x}$ was studied in an earlier example in Section 7.5, and the differential system itself was briefly considered on Example 9.10. Here $f_1 = -x_2$ and $f_2 = x_1 - x_2$ define a vector field in phase space. Its divergence is

$$\operatorname{div} \mathbf{f} = \frac{\partial f_1}{\partial x_1} + \frac{\partial f_2}{\partial x_2} = 0 - 1 = -1, \tag{9.272}$$

so the system is dissipative; that is, a volume composed of a set of points shrinks as t increases. In this case the equilibrium state, $f_i = 0$, exists at a unique point, the origin, $x_1 = \overline{x}_1 = 0$, $x_2 = \overline{x}_2 = 0$. The eigenvalues of $\mathbf{A} = \partial f_i / \partial x_j$ are $-1/2 \pm \sqrt{3}i/2$. Thus, $\rho(\mathbf{A}) = |-1/2 \pm \sqrt{3}i/2| = 1$, the equilibrium is stable, and the basin of attraction is the entire (x_1, x_2) plane.

Note that $\det \mathbf{A} = 1$, and thus the mapping $\mathbf{A} \cdot \mathbf{x}$ is volume- and orientation-preserving. We also find from Eq. (7.169) that $||\mathbf{A}||_2 = \sqrt{(3 + \sqrt{5})/2} = 1.61803$, so $\mathbf{A}$ operating on $\mathbf{x}$ can lengthen $\mathbf{x}$. This would seem to contradict the dissipative nature of the dynamical system, which is volume shrinking! However, one must realize that the mapping of a vector $\mathbf{x}$ by the dynamical system is more complicated.

Returning to the definition of the derivative, the dynamical system can also be approximated, using the so-called first-order *implicit finite difference* method, as

$$\lim_{\Delta t \to 0} \begin{pmatrix} \frac{x_1^{k+1} - x_1^k}{\Delta t} \\ \frac{x_2^{k+1} - x_2^k}{\Delta t} \end{pmatrix} = \lim_{\Delta t \to 0} \begin{pmatrix} 0 & -1 \\ 1 & -1 \end{pmatrix} \begin{pmatrix} x_1^{k+1} \\ x_2^{k+1} \end{pmatrix}. \tag{9.273}$$

This method is also known as the *backward Euler method*. Had the right-hand side been evaluated at k instead of $k + 1$, the finite difference formulation would be known as explicit, such as that studied in Section 9.5.2. We have selected the implicit formulation so as to maintain the proper dissipative property of the continuous system, which for this problem would not be obtained with a first-order explicit scheme. We demand that $\lim_{\Delta t \to 0} x_i^k = x_i, i = 1, 2$. We focus on small finite Δt, though our analysis allows for large Δt as well, and rearrange Eq. (9.273) to get the iterated map

$$\begin{pmatrix} x_1^{k+1} \\ x_2^{k+1} \end{pmatrix} = \begin{pmatrix} 0 & -\Delta t \\ \Delta t & -\Delta t \end{pmatrix} \begin{pmatrix} x_1^{k+1} \\ x_2^{k+1} \end{pmatrix} + \begin{pmatrix} x_1^k \\ x_2^k \end{pmatrix}, \tag{9.274}$$

$$\begin{pmatrix} 1 & \Delta t \\ -\Delta t & 1 + \Delta t \end{pmatrix} \begin{pmatrix} x_1^{k+1} \\ x_2^{k+1} \end{pmatrix} = \begin{pmatrix} x_1^k \\ x_2^k \end{pmatrix}, \tag{9.275}$$

$$\begin{pmatrix} x_1^{k+1} \\ x_2^{k+1} \end{pmatrix} = \underbrace{\begin{pmatrix} \frac{1 + \Delta t}{1 + \Delta t + \Delta t^2} & \frac{-\Delta t}{1 + \Delta t + \Delta t^2} \\ \frac{\Delta t}{1 + \Delta t + \Delta t^2} & \frac{1}{1 + \Delta t + \Delta t^2} \end{pmatrix}}_{=\mathbf{B}} \begin{pmatrix} x_1^k \\ x_2^k \end{pmatrix}. \tag{9.276}$$

So our dynamical system, for finite Δt, is appropriately considered as an iterated map of the form

$$\mathbf{x}^{k+1} = \mathbf{B} \cdot \mathbf{x}^k, \tag{9.277}$$

where

$$\mathbf{B} = \begin{pmatrix} \frac{1 + \Delta t}{1 + \Delta t + \Delta t^2} & \frac{-\Delta t}{1 + \Delta t + \Delta t^2} \\ \frac{\Delta t}{1 + \Delta t + \Delta t^2} & \frac{1}{1 + \Delta t + \Delta t^2} \end{pmatrix}. \tag{9.278}$$

The matrix $\mathbf{B}$ has

$$\det \mathbf{B} = \frac{1}{1 + \Delta t + \Delta t^2}. \tag{9.279}$$

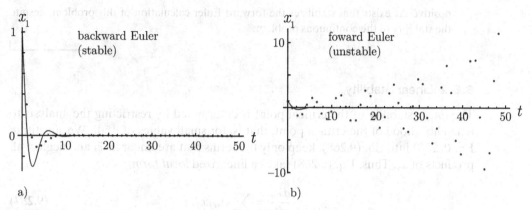

a)   b)

Figure 9.12. Predictions of $x_1(t)$ that solve Eq. (9.271) along with $x_1(0) = x_2(0) = 1$ from a (a) first-order implicit backward Euler method and (b) first-order explicit forward Euler method, each with $\Delta t = 1.1$ along with the exact solution.

For $\Delta t > 0$, $\det \mathbf{B} \in (0, 1)$, indicating an orientation-preserving shrinking of the volume element, consistent with $\operatorname{div} \mathbf{f} < 0$. The eigenvalues of $\mathbf{B}$ are

$$\frac{1 + \frac{\Delta t}{2} \pm \frac{\sqrt{3}}{2} i}{1 + \Delta t + \Delta t^2}, \tag{9.280}$$

which for small Δt expand as

$$1 - \left(1 \pm \sqrt{3} i\right) \frac{\Delta t}{2} + \cdots . \tag{9.281}$$

More importantly, the spectral norm of $\mathbf{B}$ is the square root of the largest eigenvalue of $\mathbf{B} \cdot \mathbf{B}^T$. Detailed calculation reveals this, and its series expansion in two limits, to be

$$\|\mathbf{B}\|_2 = \frac{1 + \Delta t + \frac{3}{2} \Delta t^2 + \Delta t \sqrt{1 + \Delta t + \frac{5}{4} \Delta t^2}}{1 + 2\Delta t + 3\Delta t^2 + 2\Delta t^3 + \Delta t^4}, \tag{9.282}$$

$$\lim_{\Delta t \to 0} \|\mathbf{B}\|_2 = 1 - \frac{\Delta t^2}{2} + \cdots , \tag{9.283}$$

$$\lim_{\Delta t \to \infty} \|\mathbf{B}\|_2 = \sqrt{\frac{3 + \sqrt{5}}{2}} \frac{1}{\Delta t} = \frac{\|\mathbf{A}\|_2}{\Delta t}. \tag{9.284}$$

In both limits of Δt, we see that $\|\mathbf{B}\|_2 < 1$; this can be shown to hold for all Δt. It takes on a value of unity only for $\Delta t = 0$. Then, because $\|\mathbf{B}\|_2 \leq 1$, $\forall \, \Delta t$, the action of $\mathbf{B}$ on any $\mathbf{x}$ is to diminish its norm; thus, the system is dissipative. Now $\mathbf{B}$ has a nonzero antisymmetric part, which is typically associated with rotation. One could show via a variety of decompositions that the action of $\mathbf{B}$ on a vector is to compress and rotate it.

For $x_1(0) = 1$, $x_2(0) = 1$, the exact solution is

$$x_1(t) = e^{-t/2} \left(\cos\left(\frac{\sqrt{3}t}{2}\right) - \frac{\sqrt{3}}{3} \sin\left(\frac{\sqrt{3}t}{2}\right) \right), \tag{9.285}$$

$$x_2(t) = e^{-t/2} \left(\cos\left(\frac{\sqrt{3}t}{2}\right) + \frac{\sqrt{3}}{3} \sin\left(\frac{\sqrt{3}t}{2}\right) \right). \tag{9.286}$$

It is obviously stable and oscillatory. With $\Delta t = 1.1$, we plot approximate discrete solutions for $x_1(t)$ from the backward and forward Euler methods along with the exact solution in Figure 9.12. The implicit and explicit methods yield stable and unstable predictions, respectively, for these conditions. As noted earlier in Example 9.10, no real positive Δt exists that stabilizes the forward Euler calculation of this problem, despite the stability of the continuous problem.

_____

9.6.2 Linear Stability

The *linear* stability of the critical point is determined by restricting the analysis to a neighborhood of the critical point, that is, for small values of $||\tilde{x}_i||$. We substitute Eq. (9.270) into Eq. (9.268), keep only the terms that are linear in $\tilde{x}_i$ and neglect all products of $\tilde{x}_i$. Thus, Eq. (9.268) takes a linearized *local form*

$$\frac{d\tilde{x}_n}{dt} = \sum_{j=1}^{N} A_{nj}\tilde{x}_j. \tag{9.287}$$

Another way of obtaining the same result is to expand the vector field in a Taylor series around $x_j = \bar{x}_j$ so that

$$f_n(x_j) = \sum_{j=1}^{N} \left.\frac{\partial f_n}{\partial x_j}\right|_{x_j=\bar{x}_j} \tilde{x}_j + \cdots, \tag{9.288}$$

in which the higher order terms have been neglected. Thus, in Eq. (9.287),

$$A_{nj} = \left.\frac{\partial f_n}{\partial x_j}\right|_{x_j=\bar{x}_j} \tag{9.289}$$

is the Jacobian of f_n evaluated at the critical point. In matrix form, the linearized equation for the perturbation $\tilde{\mathbf{x}}$ is

$$\frac{d\tilde{\mathbf{x}}}{dt} = \mathbf{A} \cdot \tilde{\mathbf{x}}. \tag{9.290}$$

The real parts of the eigenvalues of $\mathbf{A}$ determine the linear stability of the critical point $\tilde{\mathbf{x}} = \mathbf{0}$ and the behavior of the solution near it:

- If all eigenvalues have negative real parts, the critical point is asymptotically stable.
- If *at least one* eigenvalue has a positive real part, the critical point is unstable.
- If all eigenvalues have real parts less than or equal to zero, and some have zero real parts, then the critical point is stable if $\mathbf{A}$ has k linearly independent eigenvectors for each eigenvalue of multiplicity k. Otherwise, it is unstable.

The following are some terms used in classifying critical points according to the real and imaginary parts of the eigenvalues of $\mathbf{A}$.

Classification	*Eigenvalues*
Hyperbolic	Nonzero real part
Saddle	Some real parts negative, others positive
Stable node or sink	All real parts negative
ordinary sink	All real parts negative, imaginary parts zero
spiral sink	All real parts negative, imaginary parts nonzero
Unstable node or source	All real parts positive
ordinary source	All real parts positive, imaginary parts zero
spiral source	All real parts positive, imaginary parts nonzero
Center	All purely imaginary and nonzero

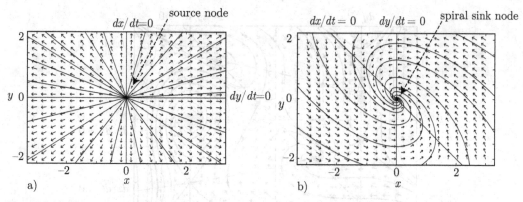

Figure 9.13. Phase plane for (a) ordinary source node; $dx/dt = x$, $dy/dt = y$, and (b) spiral sink node, $dx/dx = -x - y$, $dy/dt = x$.

Figures 9.13 and 9.14 show examples of phase planes, vector fields, and trajectories for linear homogeneous systems that describe ordinary source, spiral sink, center, and saddle-nodes, respectively. We often associate the special trajectories evident in Figure 9.14b for saddle-nodes with so-called *stable* and *unstable manifolds*. The pair of trajectories pointing toward the saddle equilibrium form the stable manifold, and the pair pointing away form the unstable manifold. Figure 9.15 gives a phase plane, vector field, and trajectories for a complicated nonlinear system with many nodes present. Here the nodes are spiral nodes and saddle-nodes.

9.6.3 Heteroclinic and Homoclinic Trajectories

All trajectories not on a limit cycle reach an equilibrium when one integrates either forward or backward in the independent variable. Sometimes that equilibrium is at infinity; sometimes it is finite. A trajectory that traverses from the neighborhood of one equilibrium point to another distinct equilibrium is known as a *heteroclinic trajectory*. A trajectory that traverses from the neighborhood of one equilibrium, moving initially away from that equilibrium, but ultimately returning to the same point is known as a *homoclinic trajectory*. We give examples of both next.

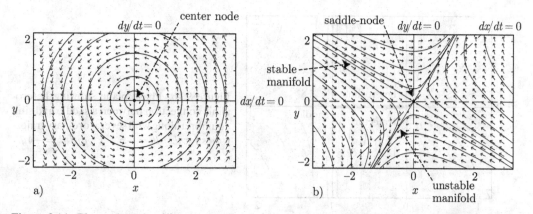

Figure 9.14. Phase plane for (a) center node; $dx/dt = -y$, $dy/dt = x$, and (b) saddle-node, $dx/dt = y - x$, $dy/dt = x$.

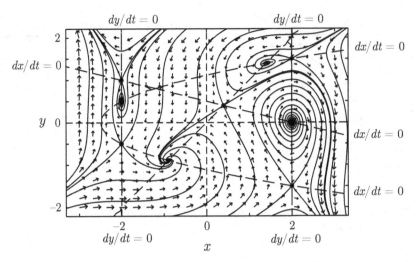

Figure 9.15. Phase plane for system with many nodes of varying nature, $dx/dt = (y + x/4 - 1/2)$, $dy/dt = (y - x)(x - 2)(x + 2)$.

EXAMPLE 9.15

Examine heteroclinic trajectories linking the equilibria of the dynamical system:

$$\frac{dx}{dt} = -x + x^2, \tag{9.291}$$

$$\frac{dy}{dt} = -4y. \tag{9.292}$$

We take t to represent time. As an aside, this nonlinear system has an exact solution. One can divide one by the other to get $dy/dx = 4y/(x - x^2)$, which can be solved by separation of variables to yield $y(x) = Cx^4/(1 - x)^4$, and solutions can be plotted for various C. In terms of our more general dynamical systems theory, it is easy to show there are two finite equilibria $(\bar{x}, \bar{y})^T = (0, 0)^T$ and $(1, 0)^T$. Linearization reveals that the first is a sink and the second is a saddle. The unstable manifold of the saddle has a heteroclinic connection to the sink. That heteroclinic connection is given parametrically by $x(s) = s$, $y(s) = 0$ for $s \in [1, 0]$.

Figure 9.16 gives a phase plane, vector field, and trajectories for this system. The heteroclinic trajectory lies along the x axis; from the unstable manifold of the saddle

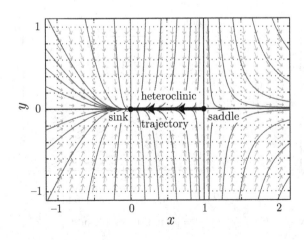

Figure 9.16. Phase plane showing heteroclinic connection between a saddle and sink equilibria along with several ordinary trajectories for $dx/dt = -x + x^2$, $dy/dt = -4y$.

at $(1,0)^T$, it moves into the sink at $(0,0)^T$. The heteroclinic trajectory is an invariant manifold because it is a trajectory. Visually, one might be inclined to describe it additionally as an *attractive* invariant manifold, as nearby trajectories appear to collapse onto it. Moreover, the time constant associated with the approach to the attracting manifold is $\tau = 1/4$, much smaller than the time constant associated with motion on the attracting manifold, which, near the sink, is $\tau = 1$. For this system, the slow evolution modes are confined to the attracting manifold at $y = 0$. In a physical system, one would be more likely to be able to observe the slow dynamics relative to the fast dynamics.

The notion of how to determine the attractiveness of a heteroclinic trajectory is not straightforward, as can be seen in the following example.

EXAMPLE 9.16

Consider the example of Mengers,[10]

$$\frac{dx_1}{dt} = \frac{1}{20}(1 - x_1^2), \tag{9.293}$$

$$\frac{dx_2}{dt} = -2x_2 - \frac{35}{16}x_2 + 2(1 - x_1^2)x_3, \tag{9.294}$$

$$\frac{dx_3}{dt} = x_2 + x_3, \tag{9.295}$$

and identify heteroclinic trajectories and their attractiveness. Here we can consider t to represent time.

There are only two finite equilibria for this system, a saddle at $(-1,0,0)^T$ and a sink at $(1,0,0)$. Because the first equation is uncoupled from the second two and is sufficiently simple, it can be integrated exactly to form $x_1 = \tanh(t/20)$. This, coupled with $x_2 = 0$ and $x_3 = 0$, satisfies all differential equations and connects the equilibria, so the x_1 axis for $x_1 \in [-1, 1]$ is the heteroclinic trajectory. One then asks if nearby trajectories are attracted to it. This can be answered by a local geometry-based analysis. Our system is of the form $d\mathbf{x}/dt = \mathbf{f}(\mathbf{x})$. Let us consider its behavior in the neighborhood of a generic point $\mathbf{x}_0$, which is on the heteroclinic trajectory but far from equilibrium. We then locally linearize our system as

$$\frac{d}{dt}(\mathbf{x} - \mathbf{x}_0) = \underbrace{\mathbf{f}(\mathbf{x}_0)}_{\text{translation}} + \underbrace{\mathbf{J}_s|_{\mathbf{x}_0} \cdot (\mathbf{x} - \mathbf{x}_0)}_{\text{stretch}} + \underbrace{\mathbf{J}_a|_{\mathbf{x}_0} \cdot (\mathbf{x} - \mathbf{x}_0)}_{\text{rotation}} + \cdots. \tag{9.296}$$

Here we have defined the local Jacobian matrix $\mathbf{J}$ as well as its symmetric ($\mathbf{J}_s$) and antisymmetric ($\mathbf{J}_a$) parts:

$$\mathbf{J} = \frac{\partial \mathbf{f}}{\partial \mathbf{x}} = \mathbf{J}_s + \mathbf{J}_a,$$

$$\mathbf{J}_s = \frac{\mathbf{J} + \mathbf{J}^T}{2}, \quad \mathbf{J}_a = \frac{\mathbf{J} - \mathbf{J}^T}{2}.$$

The symmetry of $\mathbf{J}_s$ allows definition of a real orthonormal basis. For this three-dimensional system, the dual vector $\boldsymbol{\omega}$ of the antisymmetric $\mathbf{J}_a$ defines the axis of rotation, and its magnitude ω describes the rotation rate. Now the relative volumetric stretching rate is given by $\operatorname{tr}\mathbf{J} = \operatorname{tr}\mathbf{J}_s = \operatorname{div}\mathbf{f}$. And it is not difficult to show that the linear stretching rate σ associated with any direction with unit normal $\boldsymbol{\alpha}$ is $\sigma = \boldsymbol{\alpha}^T \cdot \mathbf{J}_s \cdot \boldsymbol{\alpha}$.

---

[10] J. D. Mengers, "Slow Invariant Manifolds for Reaction-Diffusion Systems," Ph.D. dissertation, University of Notre Dame, Notre Dame, IN, 2012.

For our system, we have

$$
\mathbf{J} = \begin{pmatrix} -\frac{x_1}{10} & 0 & 0 \\ -4x_1x_3 & -2 & -\frac{35}{16} + 2(1-x_1^2) \\ 0 & 1 & 1 \end{pmatrix}.
\tag{9.297}
$$

We see that the relative volumetric stretch rate is

$$
\operatorname{tr}\mathbf{J} = -1 - \frac{x_1}{10}.
\tag{9.298}
$$

Because on the heteroclinic trajectory $x_1 \in [-1,1]$, we always have a locally shrinking volume on that trajectory. Now by inspection, the unit tangent vector to the heteroclinic trajectory is $\boldsymbol{\alpha}_t = (1,0,0)^T$. So the tangential stretching rate on the heteroclinic trajectory is

$$
\sigma_t = \boldsymbol{\alpha}_t^T \cdot \mathbf{J} \cdot \boldsymbol{\alpha}_t = -\frac{x_1}{10}.
\tag{9.299}
$$

So near the saddle we have $\sigma_t = 1/10$, and near the sink we have $\sigma_t = -1/10$. Now we are concerned with stretching in directions normal to the heteroclinic trajectory. Certainly two unit normal vectors are $\boldsymbol{\alpha}_{n1} = (0,1,0)^T$ and $\boldsymbol{\alpha}_{n2} = (0,0,1)^T$. But there are also infinitely many other unit normals. A detailed optimization calculation reveals, however, that if we (1) form the 3×2 matrix $\mathbf{Q}_n$ with $\boldsymbol{\alpha}_{n1}$ and $\boldsymbol{\alpha}_{n2}$ in its columns,

$$
\mathbf{Q}_n = \begin{pmatrix} 0 & 0 \\ 1 & 0 \\ 0 & 1 \end{pmatrix},
\tag{9.300}
$$

and (2) form the 2×2 Jacobians $\mathbf{J}_{ns}$ and $\mathbf{J}_{na}$ associated with the plane normal to the heteroclinic trajectory,

$$
\mathbf{J}_{ns} = \mathbf{Q}_n^T \cdot \mathbf{J}_s \cdot \mathbf{Q}_n, \qquad \mathbf{J}_{na} = \mathbf{Q}_n^T \cdot \mathbf{J}_a \cdot \mathbf{Q}_n,
\tag{9.301}
$$

that (a) the eigenvalues of $\mathbf{J}_{ns}$ give the extreme values of the normal stretching rates σ_{n1} and σ_{n2}, and the normalized eigenvectors give the associated directions for extreme normal stretching, and (b) the magnitude of extremal rotation in the hyperplane normal to $\boldsymbol{\alpha}_t$ is given by $\omega = ||\mathbf{J}_{na}||_2$. On the heteroclinic trajectory, we find

$$
\mathbf{J}_s = \begin{pmatrix} -\frac{x_1}{10} & 0 & 0 \\ 0 & -2 & -\frac{19}{32} + 1 - x_1^2 \\ 0 & -\frac{19}{32} + 1 - x_1^2 & 1 \end{pmatrix}.
\tag{9.302}
$$

The reduced Jacobian associated with dynamics in the normal plane is

$$
\mathbf{J}_{ns} = \mathbf{Q}_n^T \cdot \mathbf{J}_s \cdot \mathbf{Q}_n = \begin{pmatrix} -2 & -\frac{19}{32} + 1 - x_1^2 \\ -\frac{19}{32} + 1 - x_1^2 & 1 \end{pmatrix}.
\tag{9.303}
$$

Its eigenvalues give the extremal normal stretching rates, which are

$$
\sigma_{n,1,2} = -\frac{1}{2} \pm \frac{\sqrt{2473 - 832x_1^2 + 1024x_1^4}}{32}.
\tag{9.304}
$$

For $x_1 \in [-1,1]$, we have $\sigma_{n,1} \approx 1$ and $\sigma_{n,2} \approx -2$. *Because of the presence of a positive normal stretching rate, one cannot guarantee that trajectories are attracted to the heteroclinic trajectory, even though volume of nearby points is shrinking.* Positive normal stretching does not guarantee divergence from the heteroclinic trajectory; it permits it. Rotation can orient a collection of nearby points into regions where there is either positive or negative normal stretching. There are two possibilities for the heteroclinic trajectory to be attracting: either (1) all normal stretching rates are negative or (2) the rotation rate is sufficiently fast and the overall system is dissipative so that the integrated effect is relaxation to the heteroclinic trajectory. For the heteroclinic trajectory to have the

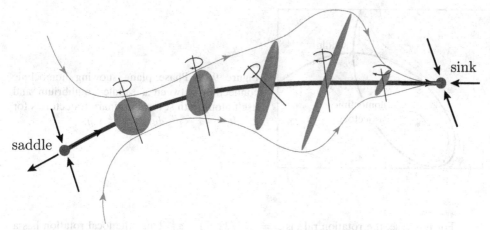

Figure 9.17. Sketch of a dissipative phase volume showing heteroclinic connection between a saddle and sink equilibria along with the evolution of a set of points initially configured as a sphere as they move into regions with some positive normal stretching rates.

additional property of being restricted to the slow dynamics, we must additionally require that the smallest normal stretching rate be larger than the tangential stretching rate.

We illustrate these notions in the sketch of Figure 9.17. Here we imagine a sphere of points as initial conditions near the saddle. We imagine that the system is dissipative so that the volume shrinks as the sphere moves. While the overall volume shrinks, one of the normal stretching rates is positive, admitting divergence of nearby trajectories from the heteroclinic trajectory. Rotation orients the volume into a region where negative normal stretching brings all points ultimately to the sink.

For our system, families of trajectories are shown in Figure 9.18a, and it is seen that there is divergence from the heteroclinic trajectory. This must be attributed to some points experiencing positive normal stretching away from the heteroclinic trajectory.

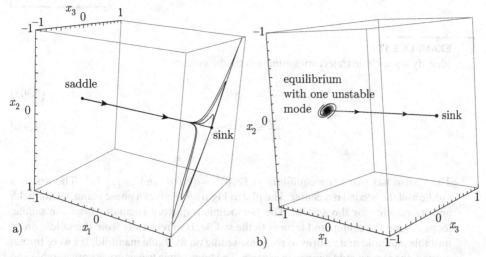

Figure 9.18. Plots of trajectories near the heteroclinic connection between equilibria with one unstable mode and a sink illustrating (a) divergence of nearby trajectories due to positive normal stretching with insufficiently rapid rotation and (b) convergence of nearby trajectories in the presence of positive normal stretching with sufficiently rapid rotation.

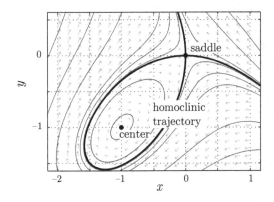

Figure 9.19. Phase plane showing homoclinic connection between a saddle equilibrium and itself along with several ordinary trajectories for $dx/dt = -x - y^2$, $dy/dt = x^2 + y$.

For this case, the rotation rate is $\omega = -51/32 + 1 - x_1^2$. Thus, the local rotation has a magnitude of near-unity near the heteroclinic orbit, and the time scales of rotation are close to the time scales of normal stretching.

We can modify the system to include more rotation. For instance, replacing Eq. (9.295) by $dx_3/dt = 10x_2 + x_3$ introduces a sufficient amount of rotation to render the heteroclinic trajectory to be attractive to nearby trajectories. Detailed analysis reveals that this small change (1) does not change the location of the two equilibria, (2) does not change the heteroclinic trajectory connecting the two equilibria, (3) modifies the dynamics near each equilibrium such that both have two stable oscillatory modes, with the equilibrium at $(-1, 0, 0)^T$ also containing a third unstable mode and that at $(1, 0, 0)^T$ containing a third stable mode, (4) does not change that the system has a negative volumetric stretch rate on the heteroclinic trajectory, (5) does not change that a positive normal stretching mode exists on the heteroclinic trajectory, and (6) enhances the rotation such that the heteroclinic trajectory is locally attractive. This is illustrated in Figure 9.18b.

Had the local Jacobian been purely symmetric, interpretation would have been much easier. It is the effect of a nonzero antisymmetric part of $\mathbf{J}$ that induces the geometrical complexities of rotation. Such systems are often known as *nonnormal dynamical systems*.

EXAMPLE 9.17

Identify homoclinic trajectories admitted by the system

$$\frac{dx}{dt} = -x - y^2, \tag{9.305}$$

$$\frac{dy}{dt} = x^2 + y. \tag{9.306}$$

The system has two finite equilibria at $(\overline{x}, \overline{y})^T = (0, 0)^T$ and $(-1, -1)^T$. The first is a saddle and the second is a center. The plot in Figure 9.19 gives a phase plane, vector field, and trajectories for this system. The homoclinic trajectory emanates from the saddle, loops around the center, and returns to the saddle. It moves away from the saddle on its unstable manifold and returns to the same saddle on its stable manifold. If t were turned around, the same would happen in reverse. The homoclinic trajectory is both a stable and unstable manifold! Trajectories originating within the homoclinic orbit are limit cycles about the center equilibrium point.

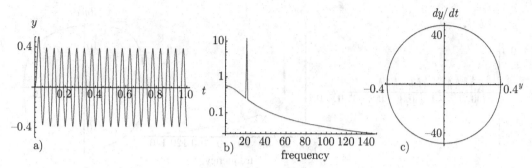

Figure 9.20. Response $y(t)$ of a linear forced mass-spring-damper system: (a) y versus t, (b) DFT of $y(t)$, (c) phase portrait for $t \in [1/2, 1]$.

9.6.4 Nonlinear Forced Mass-Spring-Damper

Let us briefly consider an important nonlinear system in engineering: a sinusoidally forced mass-spring-damper with nonlinear effects. In the linear limit, exact methods are available from Section 4.8. As studied there, such systems exhibit decaying transient output characterized by the natural frequency of the unforced system coupled with a long time oscillation at the driving frequency. Introduction of nonlinearity introduces new oscillatory phenomena to this output. The oscillatory behavior is compactly analyzed by use of the discrete Fourier transform (DFT) of Section 7.8, as considered in the following two examples.

EXAMPLE 9.18

Use the DFT to analyze the output of the forced mass-spring-damper system

$$m\frac{d^2y}{dt^2} + b\frac{dy}{dt}\left(1 + \beta\left(\frac{dy}{dt}\right)^2\right) + ky = F\sin(2\pi\nu t), \qquad y(0) = 0, \qquad \dot{y}(0) = 0,$$
(9.307)

for $m = 1$, $b = 200$, $k = 10000$, $F = 10000$, $\nu = 20$. Here distance is y and time is t. Consider the linear case when $\beta = 0$ and the nonlinear case when $\beta = 10^{-2}$. All of these variables and parameters have an appropriate set of physical units, not given here, when one is considering this to represent a physical problem.

We omit details as to actual solution of the differential equation. The linear system may be addressed by the methods of Chapter 4; the nonlinear system requires numerical solution. A simplistic estimate of the natural frequency of the system from Section 4.8 is $\sqrt{k/m} = 100$. Here, however, we will study a time sufficiently long so that initial transients which oscillate at that frequency have decayed. We thus expect the long time dynamics to reflect the character of the driving force, driven here at a lower frequency, $\nu = 20$, than the natural frequency.

Results for the linear system, $\beta = 0$, are shown in Figure 9.20. After the rapid decay of the initial transient, the system exhibits purely sinusoidal oscillation at a single frequency, that of the driver, $\nu = 20$. This is clear from the DFT, which has a single spike at $\nu = 20$. The phase portrait, plotted for $t \in [1/2, 1]$ so as to remove the initial transient, shows that when properly scaled, y versus dy/dt has a limit cycle that is a circle, consistent with simple harmonic motion of a linear oscillator.

Results for the nonlinear system, $\beta = 10^{-2}$, are shown in Figure 9.21. Relative to the linear system, the long time amplitude has decayed from roughly 0.4 to around 0.15; this is a consequence of increased damping. The behavior appears sinusoidal but actually

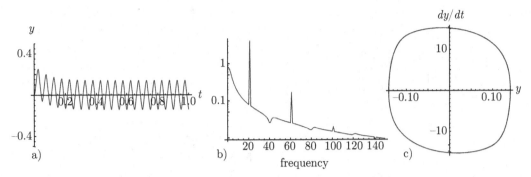

Figure 9.21. Response $y(t)$ of a nonlinear forced mass-spring-damper system: (a) y versus t, (b) DFT of $y(t)$, (c) phase portrait for $t \in [1/2, 1]$.

has small deviations. The DFT reveals the nonlinear system to exhibit a similar peak at $\nu = 20$; however, new peaks are realized at odd integer multiples of the driving frequency, $60, 100, \ldots$. Moreover, small minima are realized at even integer multiples, $40, 80, \ldots$. The input signal at a single frequency has been modulated by the nonlinear system such that its output has finite amplitude modes at discrete frequencies. In informal terms, one might summarize this important concept as

- *linear systems: one frequency in, one frequency out; the output is proportional to the input,*
- *nonlinear systems: one frequency in, many frequencies out; the output is not proportional to the input.*

In this case, the amplitude of the modes decays as frequency increases. For both systems, it is likely the nonzero amplitude of the DFT away from the peaks can be attributed to the discrete nature of many of the approximations. The phase portrait reveals that y versus dy/dt has a limit cycle which resembles a circle but is slightly warped; thus, the motion has small deviation from simple harmonic motion.

EXAMPLE 9.19

Use the DFT to analyze the output of the nonlinear forced mass-spring-damper system

$$m\frac{d^2y}{dt^2} + b\frac{dy}{dt} + k(1 + \kappa y^2)y = F\sin(2\pi\nu t), \qquad y(0) = 0, \qquad \dot{y}(0) = 0, \quad (9.308)$$

for $m = 1, b = 200, k = 10000, F = 10000, \nu = 20, \kappa = 5000$.

This nonlinear equation is a forced Duffing equation with linear damping. Again, we omit details as to actual solution of the differential equation. The linear version with $\kappa = 0$ has been considered in the previous example; the nonlinear system again requires numerical solution. Results are shown in Figure 9.22. Relative to the linear system, the long time amplitude has decayed from roughly 0.4 to around 0.07; this is a consequence of increased spring stiffness. The behavior appears to have a fundamental oscillation with a frequency of $\nu = 20$. However, there are higher frequency modes in the output $y(t)$. This is verified by examination of the DFT, which shows a main peak at $\nu = 20$, but other peaks with nearly the same magnitude at $60, 100, 140, 180, 220, 260, \ldots$. The phase portrait reveals a limit cycle with large deviation from circularity; this is not close to simple harmonic motion. So even more so than the previous example, we see that for nonlinear systems, the output may be far from proportional to the input.

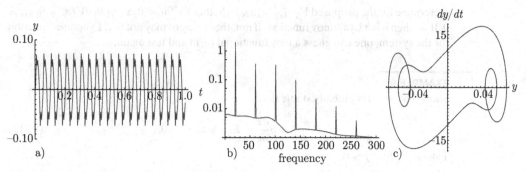

Figure 9.22. Response $y(t)$ of a nonlinear forced mass-spring-damper system: (a) y versus t, (b) DFT of $y(t)$, (c) phase portrait for $t \in [1/2, 1]$.

9.6.5 Lyapunov Functions

This section describes one of the few methods available that can allow one to make definitive statements about the stability of critical points of nonlinear systems. It only works for some systems.

For $x \in \mathbb{R}^N, t \in \mathbb{R}^1, f : \mathbb{R}^N \to \mathbb{R}^N$, consider the system of differential equations

$$\frac{dx_n}{dt} = f_n(x_1, x_2, \ldots, x_N), \qquad n = 1, 2, \ldots, N, \tag{9.309}$$

with $x_n = 0$ as a critical point. If there exists a $V(x_1, x_2, \ldots, x_N) : \mathbb{R}^N \to \mathbb{R}^1$ such that

- V is positive definite,
- dV/dt is negative semidefinite.

then the equilibrium point of the differential equations, $x_i = 0$, is globally stable to all perturbations, large or small. The function $V(x_1, x_2, \ldots, x_N)$ is called a Lyapunov[11] function.

Although one cannot always find a Lyapunov function for a given system of differential equations, we can pose a trial-and-error method to seek a Lyapunov function given a set of autonomous ordinary differential equations. Although the method lacks robustness, it is always straightforward to guess a functional form for a Lyapunov function and test whether the proposed function satisfies the criteria. The method is as follows.

1. Choose a test function $V(x_1, \ldots, x_N)$. The function should be chosen to be strictly positive for $x_n \neq 0$ and zero for $x_n = 0$.
2. Calculate

$$\frac{dV}{dt} = \frac{\partial V}{\partial x_1}\frac{dx_1}{dt} + \frac{\partial V}{\partial x_2}\frac{dx_2}{dt} + \cdots + \frac{\partial V}{\partial x_N}\frac{dx_N}{dt}, \tag{9.310}$$

$$= \frac{\partial V}{\partial x_1} f_1(x_1, \ldots, x_N) + \frac{\partial V}{\partial x_2} f_2(x_1, \ldots, x_N) + \cdots + \frac{\partial V}{\partial x_N} f_N(x_1, \ldots, x_N). \tag{9.311}$$

It is in this step where the differential equations actually enter into the calculation.

---

[11] Alexandr Mikhailovich Lyapunov, 1857–1918, Russian mathematician.

3. Determine for the proposed $V(x_1, \ldots, x_N)$ whether $dV/dt < 0, x_n \neq 0; dV/dt = 0, x_n = 0$. If so, then it is a Lyapunov function. If not, there may or may not be a Lyapunov function for the system; one can guess a new functional form and test again.

EXAMPLE 9.20

Show that $x = 0$ is globally stable if

$$m\frac{d^2x}{dt^2} + b\frac{dx}{dt} + k_1 x + k_2 x^3 = 0. \tag{9.312}$$

Take $m, b, k_1, k_2 > 0$.

This system models the motion of a mass-spring-damper system when the spring is non-linear; it is a type of Duffing equation similar to that considered in the previous example. Here, however, there is no forcing. Breaking the original second-order differential equation into two first-order equations, we get

$$\frac{dx}{dt} = y, \tag{9.313}$$

$$\frac{dy}{dt} = -\frac{b}{m}y - \frac{k_1}{m}x - \frac{k_2}{m}x^3. \tag{9.314}$$

Here x represents the position and y represents the velocity. Let us guess that the Lyapunov function has the form

$$V(x, y) = a_1 x^2 + a_2 y^2 + a_3 x^4, \qquad a_1, a_2, a_3 > 0. \tag{9.315}$$

Note that $V(x, y) \geq 0$ and that $V(0, 0) = 0$. Then

$$\frac{dV}{dt} = \frac{\partial V}{\partial x}\frac{dx}{dt} + \frac{\partial V}{\partial y}\frac{dy}{dt}, \tag{9.316}$$

$$= 2a_1 x\frac{dx}{dt} + 4a_3 x^3\frac{dx}{dt} + 2a_2 y\frac{dy}{dt}, \tag{9.317}$$

$$= (2a_1 x + 4a_3 x^3)y + 2a_2 y\left(-\frac{b}{m}y - \frac{k_1}{m}x - \frac{k_2}{m}x^3\right), \tag{9.318}$$

$$= 2\left(a_1 - \frac{a_2 k_1}{m}\right)xy + 2\left(2a_3 - \frac{a_2 k_2}{m}\right)x^3 y - \frac{2a_2}{m}by^2. \tag{9.319}$$

If we choose $a_2 = m/2$, $a_1 = 1/2k_1$, $a_3 = k_2/4$, then the coefficients on xy and $x^3 y$ in the expression for dV/dt are identically zero, and we get

$$\frac{dV}{dt} = -by^2, \tag{9.320}$$

which for $b > 0$ is negative $\forall y \neq 0$ and zero for $y = 0$. Furthermore, with these choices of a, b, c, the Lyapunov function itself is

$$V = \frac{1}{2}k_1 x^2 + \frac{1}{4}k_2 x^4 + \frac{1}{2}my^2 \geq 0. \tag{9.321}$$

Checking, we see

$$\frac{dV}{dt} = k_1 x\frac{dx}{dt} + k_2 x^3\frac{dx}{dt} + my\frac{dy}{dt}, \tag{9.322}$$

$$= k_1 xy + k_2 x^3 y + my\left(-\frac{b}{m}y - \frac{k_1}{m}x - \frac{k_2}{m}x^3\right), \tag{9.323}$$

$$= k_1 xy + k_2 x^3 y - by^2 - k_1 xy - k_2 x^3 y, \tag{9.324}$$

$$= -by^2 \leq 0. \tag{9.325}$$

Thus, V is a Lyapunov function, and $x = y = 0$ is globally stable. Actually, in this case, $V = (\text{kinetic energy} + \text{potential energy})$, where kinetic energy $= (1/2)my^2$, and potential energy $= (1/2)k_1x^2 + (1/4)k_2x^4$. Note that $V(x, y)$ is just an algebraic function of the system's state variables. When we take the time derivative of V, we are forced to invoke our original system, which defines the differential equations. Because V is strictly positive or zero for all x, y, and because it is decaying for all time, it is necessarily implied that $V \to 0$, hence $x, y \to 0$.

9.6.6 Hamiltonian Systems

Closely related to the Lyapunov function is the Hamiltonian, which exists for some systems that are nondissipative, that is, those systems for which $dV/dt = 0$. In such a case, we define the Hamiltonian H to be the Lyapunov function $H = V$ with $dH/dt \equiv 0$. For such systems, we integrate once to find that $H(x_i, y_i)$ must be a constant for all x_i, y_i. Such systems are said to be *conservative*.

With $x \in \mathbb{R}^N, y \in \mathbb{R}^N, t \in \mathbb{R}^1, f : \mathbb{R}^{2N} \to \mathbb{R}^N, g : \mathbb{R}^{2N} \to \mathbb{R}^N$. We say a system of equations of the form

$$\frac{dx_n}{dt} = f_n(x_1, \ldots, x_N, y_1, \ldots, y_N), \qquad \frac{dy_n}{dt} = g_n(x_1, \ldots, x_N, y_1, \ldots, y_N),$$

$$n = 1, \ldots, N, \tag{9.326}$$

is *Hamiltonian* if we can find a function $H(x_n, y_n) : \mathbb{R}^N \times \mathbb{R}^N \to \mathbb{R}^1$ such that

$$\frac{dH}{dt} = \frac{\partial H}{\partial x_n} \frac{dx_n}{dt} + \frac{\partial H}{\partial y_n} \frac{dy_n}{dt} = 0, \tag{9.327}$$

$$= \frac{\partial H}{\partial x_n} f_n(x_1, \ldots, x_N, y_1, \ldots, y_N) + \frac{\partial H}{\partial y_n} g_n(x_1, \ldots, x_N, y_1, \ldots, y_N) = 0. \tag{9.328}$$

This differential equation can at times be solved directly by the method of separation of variables in which we assume a specific functional form for $H(x_i, y_i)$.

Alternatively, we can also determine H by demanding that

$$\frac{\partial H}{\partial y_n} = \frac{dx_n}{dt}, \qquad \frac{\partial H}{\partial x_n} = -\frac{dy_n}{dt}. \tag{9.329}$$

Substituting from the original differential equations, we are led to equations for $H(x_i, y_i)$:

$$\frac{\partial H}{\partial y_i} = f_i(x_1, \ldots, x_N, y_1, \ldots, y_N), \qquad \frac{\partial H}{\partial x_i} = -g_i(x_1, \ldots, x_N, y_1, \ldots, y_N). \tag{9.330}$$

EXAMPLE 9.21

Find the Hamiltonian for a linear mass-spring system:

$$m\frac{d^2x}{dt^2} + kx = 0, \qquad x(0) = x_0, \qquad \left.\frac{dx}{dt}\right|_0 = \dot{x}_0. \tag{9.331}$$

Taking $dx/dt = y$ to reduce this to a system of two first-order equations, we have

$$\frac{dx}{dt} = f(x, y) = y, \qquad x(0) = x_0, \tag{9.332}$$

$$\frac{dy}{dt} = g(x, y) = -\frac{k}{m}x, \qquad y(0) = y_0. \tag{9.333}$$

For this system, $N = 1$.

We seek $H(x, y)$ such that $dH/dt = 0$. That is,

$$\frac{dH}{dt} = \frac{\partial H}{\partial x}\frac{dx}{dt} + \frac{\partial H}{\partial y}\frac{dy}{dt} = 0. \tag{9.334}$$

Substituting from the given system of differential equations, we have

$$\frac{\partial H}{\partial x}y + \frac{\partial H}{\partial y}\left(-\frac{k}{m}x\right) = 0. \tag{9.335}$$

As with all partial differential equations, one has to transform to a system of ordinary differential equations to solve. Here we will take the approach of the method of separation of variables and assume a solution of the form

$$H(x, y) = A(x) + B(y), \tag{9.336}$$

where A and B are functions to be determined. With this assumption, we get

$$y\frac{dA}{dx} - \frac{k}{m}x\frac{dB}{dy} = 0. \tag{9.337}$$

Rearranging, we get

$$\frac{1}{x}\frac{dA}{dx} = \frac{k}{my}\frac{dB}{dy}. \tag{9.338}$$

Now the term on the left is a function of x only, and the term on the right is a function of y only. The only way this can be generally valid is if both terms are equal to the same constant, which we take to be C. Hence,

$$\frac{1}{x}\frac{dA}{dx} = \frac{k}{my}\frac{dB}{dy} = C, \tag{9.339}$$

from which we get two ordinary differential equations:

$$\frac{dA}{dx} = Cx, \qquad \frac{dB}{dy} = \frac{Cm}{k}y. \tag{9.340}$$

The solution is

$$A(x) = \frac{1}{2}Cx^2 + K_1, \qquad B(y) = \frac{1}{2}\frac{Cm}{k}y^2 + K_2. \tag{9.341}$$

A general solution is

$$H(x, y) = \frac{1}{2}C\left(x^2 + \frac{m}{k}y^2\right) + K_1 + K_2. \tag{9.342}$$

Although this general solution is perfectly valid, we can obtain a common physical interpretation by taking $C = k, K_1 + K_2 = 0$. With these choices, the Hamiltonian becomes

$$H(x, y) = \frac{1}{2}kx^2 + \frac{1}{2}my^2. \tag{9.343}$$

The first term represents the potential energy of the spring, and the second term represents the kinetic energy. Because by definition $dH/dt = 0$, this system conserves its mechanical energy. Verifying the properties of a Hamiltonian, we see

$$\frac{dH}{dt} = \frac{\partial H}{\partial x}\frac{dx}{dt} + \frac{\partial H}{\partial y}\frac{dy}{dt}, \tag{9.344}$$

$$= kxy + my\left(-\frac{k}{m}x\right), \tag{9.345}$$

$$= 0. \tag{9.346}$$

Because this system has $dH/dt = 0$, then $H(x, y)$ must be constant for all t, including $t = 0$, when the initial conditions apply. So

$$H(x(t), y(t)) = H(x(0), y(0)) = \frac{1}{2} \left(kx_0^2 + my_0^2 \right). \tag{9.347}$$

Thus, the system has the integral

$$\frac{1}{2} \left(kx^2 + my^2 \right) = \frac{1}{2} \left(kx_0^2 + my_0^2 \right). \tag{9.348}$$

We can take an alternate solution approach by consideration of Eq. (9.330) as applied to this problem:

$$\frac{\partial H}{\partial y} = f = y, \qquad \frac{\partial H}{\partial x} = -g = \frac{k}{m} x. \tag{9.349}$$

Integrating the first of these, we get

$$H(x, y) = \frac{1}{2} y^2 + F(x). \tag{9.350}$$

Differentiating with respect to x, we get

$$\frac{\partial H}{\partial x} = \frac{dF}{dx}, \tag{9.351}$$

and this must be

$$\frac{dF}{dx} = \frac{k}{m} x. \tag{9.352}$$

So

$$F(x) = \frac{k}{2m} x^2 + K. \tag{9.353}$$

Thus,

$$H(x, y) = \frac{1}{2m} \left(kx^2 + my^2 \right) + K. \tag{9.354}$$

We can choose $K = 0$, and because $dH/dt = 0$, we have H as a constant that is set by the initial conditions, thus giving

$$\frac{1}{2m} \left(kx^2 + my^2 \right) = \frac{1}{2m} \left(kx_0^2 + my_0^2 \right), \tag{9.355}$$

which gives identical information as does Eq. (9.348).

9.7 Differential-Algebraic Systems

Many dynamical systems are better considered as differential-algebraic systems of equations of the general form given in Eq. (9.98). There is a rich theory for such systems, which we will not fully exploit here. Instead, we shall consider briefly certain types of linear and nonlinear differential-algebraic systems.

9.7.1 Linear Homogeneous

Consider the system of homogeneous differential-algebraic equations of the form

$$\mathbf{B} \cdot \frac{d\mathbf{x}}{dt} = \mathbf{A} \cdot \mathbf{x}. \tag{9.356}$$

Here $\mathbf{A}$ and $\mathbf{B}$ are constant matrices, and we take $\mathbf{B}$ to be singular; thus, it cannot be inverted. We will assume $\mathbf{A}$ is invertible. There is an apparent equilibrium when $\mathbf{x} = \mathbf{0}$, but the singularity of $\mathbf{B}$ gives us concern that this may not always hold. In any case, we can assume solutions of the type $\mathbf{x} = \mathbf{e}e^{\lambda t}$, where $\mathbf{e}$ is a constant vector and λ is a constant scalar, and substitute into Eq. (9.356) to get

$$\mathbf{B} \cdot \mathbf{e}\lambda e^{\lambda t} = \mathbf{A} \cdot \mathbf{e}e^{\lambda t}, \tag{9.357}$$

$$\mathbf{B} \cdot \mathbf{e}\lambda = \mathbf{A} \cdot \mathbf{e}, \tag{9.358}$$

$$(\mathbf{A} - \lambda\mathbf{B}) \cdot \mathbf{e} = \mathbf{0}. \tag{9.359}$$

Equation (9.359) is a generalized eigenvalue problem in the second sense, as considered in Section 7.4.2. It may be solved for $\mathbf{e}$ and λ; the solution for $\mathbf{e}$ is nonunique.

EXAMPLE 9.22

Solve the linear homogeneous differential-algebraic system

$$\frac{dx_1}{dt} + 2\frac{dx_2}{dt} = x_1 + x_2, \tag{9.360}$$

$$0 = 2x_1 - x_2. \tag{9.361}$$

Although this problem is simple enough to directly eliminate x_2 in favor of x_1, other problems are not that simple, so let us illustrate the general method. In matrix form, we can say

$$\begin{pmatrix} 1 & 2 \\ 0 & 0 \end{pmatrix} \begin{pmatrix} \frac{dx_1}{dt} \\ \frac{dx_2}{dt} \end{pmatrix} = \begin{pmatrix} 1 & 1 \\ 2 & -1 \end{pmatrix} \begin{pmatrix} x_1 \\ x_2 \end{pmatrix}. \tag{9.362}$$

Taking $x_1 = e_1 e^{\lambda t}$ and $x_2 = e_2 e^{\lambda t}$ gives

$$\lambda \begin{pmatrix} 1 & 2 \\ 0 & 0 \end{pmatrix} \begin{pmatrix} e_1 \\ e_2 \end{pmatrix} e^{\lambda t} = \begin{pmatrix} 1 & 1 \\ 2 & -1 \end{pmatrix} \begin{pmatrix} e_1 \\ e_2 \end{pmatrix} e^{\lambda t}, \tag{9.363}$$

$$\begin{pmatrix} \lambda & 2\lambda \\ 0 & 0 \end{pmatrix} \begin{pmatrix} e_1 \\ e_2 \end{pmatrix} = \begin{pmatrix} 1 & 1 \\ 2 & -1 \end{pmatrix} \begin{pmatrix} e_1 \\ e_2 \end{pmatrix}, \tag{9.364}$$

$$\begin{pmatrix} 1-\lambda & 1-2\lambda \\ 2 & -1 \end{pmatrix} \begin{pmatrix} e_1 \\ e_2 \end{pmatrix} = \begin{pmatrix} 0 \\ 0 \end{pmatrix}. \tag{9.365}$$

The determinant of the coefficient matrix must be zero, giving

$$-(1-\lambda) - 2(1-2\lambda) = 0, \tag{9.366}$$

$$-1 + \lambda - 2 + 4\lambda = 0, \tag{9.367}$$

$$\lambda = \frac{3}{5}. \tag{9.368}$$

With this generalized eigenvalue, our generalized eigenvectors in the second sense are found via

$$\begin{pmatrix} 1-\frac{3}{5} & 1-2\left(\frac{3}{5}\right) \\ 2 & -1 \end{pmatrix} \begin{pmatrix} e_1 \\ e_2 \end{pmatrix} = \begin{pmatrix} 0 \\ 0 \end{pmatrix}, \tag{9.369}$$

$$\begin{pmatrix} \frac{2}{5} & -\frac{1}{5} \\ 2 & -1 \end{pmatrix} \begin{pmatrix} e_1 \\ e_2 \end{pmatrix} = \begin{pmatrix} 0 \\ 0 \end{pmatrix}. \tag{9.370}$$

By inspection, the nonunique solution must be of the form

$$\begin{pmatrix} e_1 \\ e_2 \end{pmatrix} = C_1 \begin{pmatrix} 1 \\ 2 \end{pmatrix}. \tag{9.371}$$

So the general solution is

$$\begin{pmatrix} x_1 \\ x_2 \end{pmatrix} = \begin{pmatrix} C_1 e^{3t/5} \\ 2C_1 e^{3t/5} \end{pmatrix}. \tag{9.372}$$

There is only one arbitrary constant for this system.

A less desirable approach to differential-algebraic systems is to differentiate the constraint. This requires care in that an initial condition must be imposed that is consistent with the original constraint. Applying this method to our example problem gives rise to the system

$$\frac{dx_1}{dt} + 2\frac{dx_2}{dt} = x_1 + x_2, \tag{9.373}$$

$$2\frac{dx_1}{dt} - \frac{dx_2}{dt} = 0. \tag{9.374}$$

In matrix form, this gives

$$\begin{pmatrix} 1 & 2 \\ 2 & -1 \end{pmatrix} \begin{pmatrix} \frac{dx_1}{dt} \\ \frac{dx_2}{dt} \end{pmatrix} = \begin{pmatrix} 1 & 1 \\ 0 & 0 \end{pmatrix} \begin{pmatrix} x_1 \\ x_2 \end{pmatrix}, \tag{9.375}$$

$$\begin{pmatrix} \frac{dx_1}{dt} \\ \frac{dx_2}{dt} \end{pmatrix} = \begin{pmatrix} \frac{1}{5} & \frac{1}{5} \\ \frac{2}{5} & \frac{2}{5} \end{pmatrix} \begin{pmatrix} x_1 \\ x_2 \end{pmatrix}. \tag{9.376}$$

The eigenvalues of the coefficient matrix are $\lambda = 0$ and $\lambda = 3/5$. Whenever one finds an eigenvalue of zero in a dynamical system, there is actually a hidden algebraic constraint within the system. Diagonalization allows us to write the system as

$$\begin{pmatrix} \frac{dx_1}{dt} \\ \frac{dx_2}{dt} \end{pmatrix} = \begin{pmatrix} \frac{2}{3} & \frac{2}{3} \\ -\frac{2}{3} & \frac{1}{3} \end{pmatrix}^{-1} \begin{pmatrix} \frac{3}{5} & 0 \\ 0 & 0 \end{pmatrix} \begin{pmatrix} \frac{2}{3} & \frac{2}{3} \\ -\frac{2}{3} & \frac{1}{3} \end{pmatrix} \begin{pmatrix} x_1 \\ x_2 \end{pmatrix}, \tag{9.377}$$

$$\begin{pmatrix} \frac{2}{3} & \frac{2}{3} \\ -\frac{2}{3} & \frac{1}{3} \end{pmatrix} \begin{pmatrix} \frac{dx_1}{dt} \\ \frac{dx_2}{dt} \end{pmatrix} = \begin{pmatrix} \frac{3}{5} & 0 \\ 0 & 0 \end{pmatrix} \begin{pmatrix} \frac{2}{3} & \frac{2}{3} \\ -\frac{2}{3} & \frac{1}{3} \end{pmatrix} \begin{pmatrix} x_1 \\ x_2 \end{pmatrix}. \tag{9.378}$$

Regrouping, we can say

$$\frac{d}{dt}(x_1 + x_2) = \frac{3}{5}(x_1 + x_2), \tag{9.379}$$

$$\frac{d}{dt}(-2x_1 + x_2) = 0. \tag{9.380}$$

Solving gives

$$x_1 + x_2 = C_1 e^{3t/5}, \tag{9.381}$$

$$-2x_1 + x_2 = C_2. \tag{9.382}$$

So the problem with the differentiated constraint yields two arbitrary constants. For consistency with the original formulation, we must take $C_2 = 0$, thus $x_2 = 2x_1$. Thus,

$$x_1 = \frac{1}{3} C_1 e^{3t/5}, \tag{9.383}$$

$$x_2 = \frac{2}{3} C_1 e^{3t/5}. \tag{9.384}$$

Because C_1 is arbitrary, this is fully consistent with our previous solution.

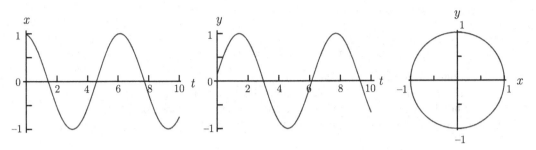

Figure 9.23. Solution to the differential-algebraic system of Eq. (9.388).

9.7.2 Nonlinear

Let us consider two simple nonlinear examples for differential-algebraic equation systems.

EXAMPLE 9.23

Solve

$$\frac{dx}{dt} = -y, \tag{9.385}$$

$$x^2 + y^2 = 1, \tag{9.386}$$

$$x(0) = 0.99. \tag{9.387}$$

The system is nonlinear because of the nonlinear constraint. However, we can also view this system as a Hamiltonian system for a linear oscillator. The nonlinear constraint is the Hamiltonian. We recognize that if we differentiate the nonlinear constraint, the system of nonlinear differential-algebraic equations reduces to a linear system of differential equations, $dx/dt = -y$, $dy/dt = x$, which is that of a linear oscillator.

Formulated as a differential-algebraic system, we can say

$$\begin{pmatrix} 1 & 0 \\ 0 & 0 \end{pmatrix} \begin{pmatrix} \frac{dx}{dt} \\ \frac{dy}{dt} \end{pmatrix} = \begin{pmatrix} -y \\ x^2 + y^2 - 1 \end{pmatrix}, \qquad x(0) = 0.99. \tag{9.388}$$

We might imagine an equilibrium to be located at $(x, y) = (\pm 1, 0)$. Certainly at such a point, $dx/dt = 0$, and the constraint is satisfied. However, at such a point, $dy/dt \neq 0$, so it is not a true equilibrium. Linearization near $(\pm 1, 0)$ would induce another generalized eigenvalue problem in the second sense. For the full problem, the form presented is suitable for numerical integration by many appropriate differential-algebraic software packages. We do so and find the result plotted in Figure 9.23. For this system, what is seen to be a pseudo-equilibrium at $(x, y)^T = (\pm 1, 0)^T$ is realized periodically. The point is not a formal equilibrium, because it does not remain there as $t \to \infty$. We also see that the trajectory in the (x, y) plane is confined to the unit circle, as required by the constraint.

EXAMPLE 9.24

Solve

$$\frac{dx}{dt} = y^2 + xy, \tag{9.389}$$

$$2x^2 + y^2 = 1, \tag{9.390}$$

$$x(0) = 0. \tag{9.391}$$

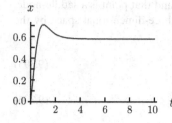
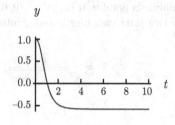
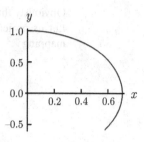

Figure 9.24. Solution to the differential-algebraic system of Eq. (9.392).

Formulated as a differential-algebraic system, we can say

$$\begin{pmatrix} 1 & 0 \\ 0 & 0 \end{pmatrix} \begin{pmatrix} \frac{dx}{dt} \\ \frac{dy}{dt} \end{pmatrix} = \begin{pmatrix} y^2 + xy \\ 2x^2 + y^2 - 1 \end{pmatrix}, \qquad x(0) = 0. \tag{9.392}$$

We could linearize near the potential equilibria, located at $(x, y)^T = (\pm 1/\sqrt{3}, \mp 1/\sqrt{3})^T$, $(\pm\sqrt{1/2}, 0)^T$. This would induce another generalized eigenvalue problem in the second sense. For the full problem, the form presented is suitable for numerical integration by many appropriate differential-algebraic software packages. We do so and find the result plotted in Figure 9.24. For this system, a true equilibrium at $(x, y)^T = (1/\sqrt{3}, -1/\sqrt{3})^T$ is realized. We also see that the trajectory in the (x, y) plane is confined to the ellipse, as required by the constraint.

9.8 Fixed Points at Infinity

Often in dynamical systems, there are additional fixed points, not readily seen in finite phase space. These fixed points are actually at infinity, and such points can play a role in determining the dynamics of a system as well as aiding in finding basins of attraction. Fixed points at infinity can be studied in a variety of ways. One method involves the so-called *Poincaré sphere*. Another method uses what is called *projective space*.

9.8.1 Poincaré Sphere

For two-dimensional dynamical systems, a good way to address the dynamics in the far field is to transform the doubly-infinite phase plane onto the surface of a sphere with radius unity. The projection will be such that points at infinity are mapped onto the equator of the sphere. One can then view the sphere from the north pole and see more clearly how the dynamics develop on the surface of the sphere.

EXAMPLE 9.25

Using the Poincaré sphere, find the global dynamics, including at infinity, for the simple system

$$\frac{dx}{dt} = x, \tag{9.393}$$

$$\frac{dy}{dt} = -y. \tag{9.394}$$

Obviously the equilibrium point is at $(x, y)^T = (0, 0)^T$, and that point is a saddle-node. Let us project the two state variables x and y into a three-dimensional space by the mapping $\mathbb{R}^2 \to \mathbb{R}^3$:

$$X = \frac{x}{\sqrt{1 + x^2 + y^2}}, \tag{9.395}$$

$$Y = \frac{y}{\sqrt{1 + x^2 + y^2}}, \tag{9.396}$$

$$Z = \frac{1}{\sqrt{1 + x^2 + y^2}}. \tag{9.397}$$

We actually could alternatively analyze this system with a closely related mapping from $\mathbb{R}^2 \to \mathbb{R}^2$, but this makes some of the analysis less geometrically transparent. Note that

$$\lim_{x \to \infty} X = 1 \quad \forall \, y < \infty, \tag{9.398}$$

$$\lim_{y \to \infty} Y = 1 \quad \forall \, x < \infty. \tag{9.399}$$

Furthermore, if both x and y go to infinity, say, on the line $y = mx$, then

$$\lim_{x \to \infty, y = mx} X = \frac{1}{\sqrt{m^2 + 1}}, \tag{9.400}$$

$$\lim_{x \to \infty, y = mx} Y = \frac{m}{\sqrt{m^2 + 1}}, \tag{9.401}$$

$$\lim_{x \to \infty, y = mx} X^2 + Y^2 = 1. \tag{9.402}$$

So points at infinity are mapping onto a unit circle in (X, Y) space. Also, going into the saddle-node at $(x, y)^T = (0, 0)^T$ along the same line gives

$$\lim_{x \to 0, y = mx} X = x + \cdots, \tag{9.403}$$

$$\lim_{x \to 0, y = mx} Y = y + \cdots. \tag{9.404}$$

So the original and transformed spaces have the same essential behavior near the finite equilibrium point. Last, note that

$$X^2 + Y^2 + Z^2 = \frac{x^2 + y^2 + 1}{1 + x^2 + y^2} = 1. \tag{9.405}$$

Thus, in fact, the mapping takes one onto a unit sphere in (X, Y, Z) space. The surface $X^2 + Y^2 + Z^2 = 1$ is called the Poincaré sphere. One can actually view this in the same way one does an actual map of the surface of the earth. Just as a Mercator[12] projection map is a representation of the spherical surface of the earth projected onto a flat surface, the original (x, y) phase space is a planar representation of the surface of the Poincaré sphere.

Let us find the inverse transformation. By inspection, it is seen that

$$x = \frac{X}{Z}, \tag{9.406}$$

$$y = \frac{Y}{Z}. \tag{9.407}$$

---

[12] Geradus Mercator, 1512–1594, (b. Gerard de Kremer) Flemish cartographer.

Now apply the transformation, Eqs. (9.406,9.407) to our dynamical system, Eqs. (9.393,9.394):

$$\underbrace{\frac{d}{dt}\left(\frac{X}{Z}\right)}_{dx/dt} = \underbrace{\frac{X}{Z}}_{x}, \tag{9.408}$$

$$\underbrace{\frac{d}{dt}\left(\frac{Y}{Z}\right)}_{dy/dt} = \underbrace{-\frac{Y}{Z}}_{-y}. \tag{9.409}$$

Expand using the quotient rule to get

$$\frac{1}{Z}\frac{dX}{dt} - \frac{X}{Z^2}\frac{dZ}{dt} = \frac{X}{Z}, \tag{9.410}$$

$$\frac{1}{Z}\frac{dY}{dt} - \frac{Y}{Z^2}\frac{dZ}{dt} = -\frac{Y}{Z}. \tag{9.411}$$

Now on the unit sphere $X^2 + Y^2 + Z^2 = 1$, we must have

$$2X\,dX + 2Y\,dY + 2Z\,dZ = 0, \tag{9.412}$$

so dividing by dt and solving for dZ/dt, we must have

$$\frac{dZ}{dt} = -\frac{X}{Z}\frac{dX}{dt} - \frac{Y}{Z}\frac{dY}{dt}. \tag{9.413}$$

Using Eq. (9.413) to eliminate dZ/dt in Eqs. (9.410,9.411), our dynamical system can be written as

$$\frac{1}{Z}\frac{dX}{dt} - \frac{X}{Z^2}\underbrace{\left(-\frac{X}{Z}\frac{dX}{dt} - \frac{Y}{Z}\frac{dY}{dt}\right)}_{dZ/dt} = \frac{X}{Z}, \tag{9.414}$$

$$\frac{1}{Z}\frac{dY}{dt} - \frac{Y}{Z^2}\underbrace{\left(-\frac{X}{Z}\frac{dX}{dt} - \frac{Y}{Z}\frac{dY}{dt}\right)}_{dZ/dt} = -\frac{Y}{Z}. \tag{9.415}$$

Multiply Eqs. (9.414,9.415) by Z^3 to get

$$Z^2\frac{dX}{dt} + X\left(X\frac{dX}{dt} + Y\frac{dY}{dt}\right) = Z^2 X, \tag{9.416}$$

$$Z^2\frac{dY}{dt} + Y\left(X\frac{dX}{dt} + Y\frac{dY}{dt}\right) = -Z^2 Y. \tag{9.417}$$

Regroup to find

$$(X^2 + Z^2)\frac{dX}{dt} + XY\frac{dY}{dt} = Z^2 X, \tag{9.418}$$

$$XY\frac{dX}{dt} + (Y^2 + Z^2)\frac{dY}{dt} = -Z^2 Y. \tag{9.419}$$

Now, eliminate Z by demanding $X^2 + Y^2 + Z^2 = 1$ to get

$$(1 - Y^2)\frac{dX}{dt} + XY\frac{dY}{dt} = (1 - X^2 - Y^2)X, \tag{9.420}$$

$$XY\frac{dX}{dt} + (1 - X^2)\frac{dY}{dt} = -(1 - X^2 - Y^2)Y. \tag{9.421}$$

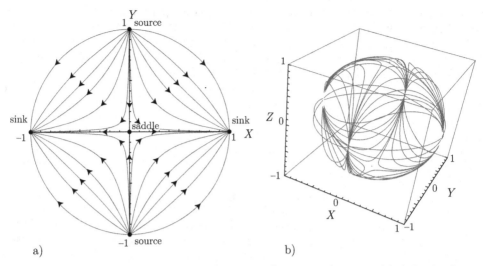

Figure 9.25. Global phase portraits of the system $dx/dt = x, dy/dt = -y$: (a) projection from the Poincaré sphere onto the (X, Y) plane, (b) full projection onto the Poincaré sphere in (X, Y, Z) space.

Solve this quasi-linear system for dX/dt and dY/dt to arrive at

$$\frac{dX}{dt} = X - X^3 + XY^2, \tag{9.422}$$

$$\frac{dY}{dt} = -Y + Y^3 - X^2Y. \tag{9.423}$$

The five equilibrium points, and their stability, for this system are easily verified to be

$$(X, Y)^T = (0, 0)^T, \qquad \text{saddle}, \tag{9.424}$$

$$(X, Y)^T = (1, 0)^T, \qquad \text{sink}, \tag{9.425}$$

$$(X, Y)^T = (-1, 0)^T, \qquad \text{sink}, \tag{9.426}$$

$$(X, Y)^T = (0, 1)^T, \qquad \text{source}, \tag{9.427}$$

$$(X, Y)^T = (0, -1)^T, \qquad \text{source}. \tag{9.428}$$

In this space, four new equilibria have appeared. As we are also confined to the Poincaré sphere on which $X^2 + Y^2 + Z^2 = 1$, we can also see that each of the new equilibria has $Z = 0$; that is, the new equilibrium points lie on the equator of the Poincaré sphere. Transforming back to the original space, we find the equilibria are at

$$(x, y)^T = (0, 0)^T, \qquad \text{saddle}, \tag{9.429}$$

$$(x, y)^T = (\infty, 0)^T, \qquad \text{sink}, \tag{9.430}$$

$$(x, y)^T = (-\infty, 0), \qquad \text{sink}, \tag{9.431}$$

$$(x, y)^T = (0, \infty)^T, \qquad \text{source}, \tag{9.432}$$

$$(x, y)^T = (0, -\infty)^T, \qquad \text{source}. \tag{9.433}$$

Phase portraits showing several trajectories projected into (X, Y) and (X, Y, Z) space are shown in Figure 9.25. Figure 9.25a represents the Poincaré sphere from above the north pole; Figure 9.25b depicts the entire Poincaré sphere. On the sphere itself, there are some additional complexities due to so-called antipodal equilibrium points. In this example, both the north pole and the south pole are saddle equilibria, when the entire sphere is considered. For more general problems, one must realize that this projection induces pairs of equilibria, and that usually only one member of each pair needs to be considered in detail.

Additionally, one notes in the global phase portraits three interesting features for two-dimensional phase spaces:

- except at critical points, individual trajectories never cross each other,
- all trajectories connect one critical point to another,
- it formally requires $t \to \infty$ to reach a critical point.

In the untransformed (x, y) space, we were not certain of the fate of trajectories as they moved far from the saddle equilibrium; in the transformed space, we see that there are heteroclinic connections to points at infinity. Any trajectory is also an invariant manifold. Certain of these manifolds are attracting invariant manifolds in that nearby trajectories are attracted to them. The line $Y = 0$, and so $y = 0$, represents an attracting invariant manifold for this system. A finite initial condition can only approach two fixed points at infinity. But the curve representing points at infinity, $Z = 0$, is an invariant manifold. Except for trajectories that originate at the two source points, what happens at infinity stays at infinity.

9.8.2 Projective Space

When extended to higher dimension, the Poincaré sphere approach becomes unwieldy. A more efficient approach is provided by projective space. This approach does not have the graphical appeal of the Poincaré sphere.

EXAMPLE 9.26

Using projective space, find the global dynamics, including at infinity, for the same simple system

$$\frac{dx}{dt} = x, \tag{9.434}$$

$$\frac{dy}{dt} = -y. \tag{9.435}$$

Again, it is obvious that the equilibrium point is at $(x, y)^T = (0, 0)^T$, and that point is a saddle-node. Let us project the two state variables x and y into a new two-dimensional space by the mapping $\mathbb{R}^2 \to \mathbb{R}^2$:

$$X = \frac{1}{x}, \tag{9.436}$$

$$Y = \frac{y}{x}. \tag{9.437}$$

Along the line $y = mx$, as $x \to \infty$, we get $X \to 0, Y \to m$. So for $x \neq 0$, a point at infinity in (x, y) space maps to a finite point in (X, Y) space. By inspection, the inverse mapping is

$$x = \frac{1}{X}, \tag{9.438}$$

$$y = \frac{Y}{X}. \tag{9.439}$$

Under this transformation, Eqs. (9.434–9.435) become

$$\frac{d}{dt}\left(\frac{1}{X}\right) = \frac{1}{X},\tag{9.440}$$

$$\frac{d}{dt}\left(\frac{Y}{X}\right) = -\frac{Y}{X}.\tag{9.441}$$

Expanding, we find

$$-\frac{1}{X^2}\frac{dX}{dt} = \frac{1}{X},\tag{9.442}$$

$$\frac{1}{X}\frac{dY}{dt} - \frac{Y}{X^2}\frac{dX}{dt} = -\frac{Y}{X}.\tag{9.443}$$

Simplifying gives

$$\frac{dX}{dt} = -X,\tag{9.444}$$

$$X\frac{dY}{dt} - Y\frac{dX}{dt} = -XY.\tag{9.445}$$

Solving for the derivatives, the system reduces to

$$\frac{dX}{dt} = -X,\tag{9.446}$$

$$\frac{dY}{dt} = -2Y.\tag{9.447}$$

By inspection, there is a sink at $(X, Y)^T = (0, 0)^T$. At such a point, the inverse mapping tells us $x \to \pm\infty$ depending on whether X is positive or negative, and y is indeterminate. If we approach $(X, Y)^T = (0, 0)^T$ along the line $Y = mX$, then y approaches the finite number m. This is consistent with trajectories being swept away from the origin toward $x \to \pm\infty$ in the original phase space, indicating an attraction at $x \to \pm\infty$. But it does not account for the trajectories emanating from $y \to \pm\infty$. This is because the transformation selected obscured this root.

To recover it, we can consider the alternate transformation $\hat{X} = x/y$, $\hat{Y} = 1/y$. Doing so leads to the system $d\hat{X}/dt = 2\hat{X}$, $d\hat{Y}/dt = \hat{Y}$, which has a source at $(\hat{X}, \hat{Y})^T = (0, 0)^T$, which is consistent with the source-like behavior in the original (x, y) space as $y \to \pm\infty$. This transformation, however, obscures the sink-like behavior at $x \to \pm\infty$.

To capture both points at infinity, we can consider a nondegenerate transformation, of which there are infinitely many. One is $\tilde{X} = 1/(x + y)$, $\tilde{Y} = (x - y)/(x + y)$. Doing so leads to the system $d\tilde{X}/dt = -\tilde{X}\tilde{Y}$, $d\tilde{Y}/dt = 1 - \tilde{Y}^2$. This system has two roots, a source at $(\tilde{X}, \tilde{Y})^T = (0, -1)^T$ and a sink at $(\tilde{X}, \tilde{Y})^T = (0, 1)^T$. The source corresponds to $y \to \pm\infty$. The sink corresponds to $x \to \pm\infty$.

9.9 Bifurcations

We study here for continuous systems the phenomenon of bifurcation. This was introduced for discrete systems in Section 9.1. Dynamical systems representing some physical problem frequently have parameters associated with them. Thus, for $x \in \mathbb{R}^N, t \in \mathbb{R}^1, r \in \mathbb{R}^1, f : \mathbb{R}^N \to \mathbb{R}^N$, we can write the continuous analog of Eq. (9.1),

$$\frac{dx_n}{dt} = f_n(x_1, x_2, \ldots, x_N; r), \qquad n = 1, \ldots, N,\tag{9.448}$$

where we have explicitly included r as a parameter; we consider t to represent time. The theory can easily be extended if there is more than one parameter. We would like to consider the changes in the long time ($t \to \infty$) behavior of solutions as the real number r, called the *bifurcation parameter*, is varied. The nature of the critical point may change as r is varied; other critical points may appear or disappear, or its stability may change. This is a bifurcation, and the r at which it happens is the bifurcation point. The study of the solutions and bifurcations of the steady state falls under *singularity theory*.

Let us look at some of the bifurcations obtained for different vector fields. Most of the examples will be one-dimensional, that is, $x \in \mathbb{R}^1, r \in \mathbb{R}^1, f : \mathbb{R}^1 \to \mathbb{R}^1$:

$$\frac{dx}{dt} = f(x; r). \tag{9.449}$$

Even though this can be solved exactly in many cases, we assume that such a solution is not available so that techniques of analysis can be developed for more complicated systems.

9.9.1 Pitchfork

For $x \in \mathbb{R}^1, t \in \mathbb{R}^1, r \in \mathbb{R}^1, r_0 \in \mathbb{R}^1$, consider

$$\frac{dx}{dt} = -x(x^2 - (r - r_0)). \tag{9.450}$$

The critical points are $\bar{x} = 0$, and $\bar{x} = \pm\sqrt{r - r_0}$. We see that $r = r_0$ is a *bifurcation point*; for $r < r_0$, there is only one critical point, whereas for $r > r_0$, there are three.

Linearizing around the critical point $\bar{x} = 0$, we get

$$\frac{d\tilde{x}}{dt} = (r - r_0)\tilde{x}. \tag{9.451}$$

This has solution

$$\tilde{x}(t) = \tilde{x}(0) \exp\left((r - r_0)t\right). \tag{9.452}$$

For $r < r_0$, the critical point is asymptotically stable; for $r > r_0$, it is unstable.

Notice that the function $V(x) = x^2$ satisfies the following conditions: $V > 0$ for $x \neq 0, V = 0$ for $x = 0$, and $dV/dt = (dV/dx)(dx/dt) = -2x^2(x^2 - (r - r_0)) \le 0$ for $r < r_0$. Thus, $V(x)$ is a Lyapunov function and $\bar{x} = 0$ is globally stable for all perturbations, large or small, as long as $r < r_0$.

Now let us examine the critical point $\bar{x} = \sqrt{r - r_0}$ that exists only for $r > r_0$. Putting $x = \bar{x} + \tilde{x}$, the right side of Eq. (9.450) becomes

$$f(x) = -\left(\sqrt{r - r_0} + \tilde{x}\right)\left(\left(\sqrt{r - r_0} + \tilde{x}\right)^2 - (r - r_0)\right). \tag{9.453}$$

Linearizing for small $\tilde{x}$, we get

$$\frac{d\tilde{x}}{dt} = -2(r - r_0)\tilde{x}. \tag{9.454}$$

This has solution

$$\tilde{x}(t) = \tilde{x}(0) \exp\left(-2(r - r_0)t\right). \tag{9.455}$$

For $r > r_0$, this critical point is stable. The other critical point $\bar{x} = -\sqrt{r - r_0}$ is also found to be stable for $r > r_0$. The results are summarized in the bifurcation diagram

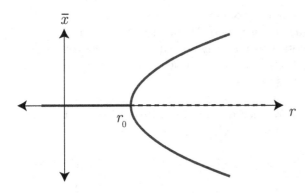

Figure 9.26. Sketch of a pitchfork bifurcation. Heavy lines are stable equilibria; dashed lines are unstable equilibria.

sketched in Figure 9.26. At the bifurcation point, $r = r_0$, we have

$$\frac{dx}{dt} = -x^3. \tag{9.456}$$

This equation has a critical point at $x = 0$, but its linearization does not yield any significant information. We must do a nonlinear analysis to determine the stability of the critical point. In this case, it is straightforward. One can easily observe that any initial condition on either side of the origin will move toward the origin. However, one can also solve directly, applying an initial condition, to obtain

$$x(t) = \pm \frac{x(0)}{\sqrt{1 + 2x(0)^2 t}}, \tag{9.457}$$

$$\lim_{t \to \infty} x(t) = 0. \tag{9.458}$$

Because the system approaches the critical point as $t \to \infty$ for all values of $x(0)$, the critical point $x = 0$ is unconditionally stable.

9.9.2 Transcritical

For $x \in \mathbb{R}^1, t \in \mathbb{R}^1, r \in \mathbb{R}^1, r_0 \in \mathbb{R}^1$, consider

$$\frac{dx}{dt} = -x(x - (r - r_0)). \tag{9.459}$$

The critical points are $\bar{x} = 0$ and $\bar{x} = r - r_0$. The bifurcation occurs at $r = r_0$. Once again the linear stability of the solutions can be determined. Near $\bar{x} = 0$, the linearization is

$$\frac{d\tilde{x}}{dt} = (r - r_0)\tilde{x}, \tag{9.460}$$

which has solution

$$\tilde{x}(t) = \tilde{x}(0) \exp\left((r - r_0)t\right). \tag{9.461}$$

This critical point is stable for $r < r_0$. Near $\bar{x} = r - r_0$, we take $\tilde{x} = x - (r - r_0)$. The resulting linearization is

$$\frac{d\tilde{x}}{dt} = -(r - r_0)\tilde{x}, \tag{9.462}$$

which has solution

$$\tilde{x}(t) = \tilde{x}(0) \exp\left(-(r - r_0)t\right). \tag{9.463}$$

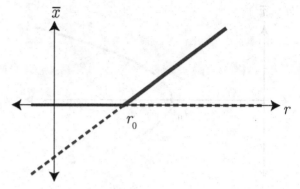

Figure 9.27. Sketch of a transcritical bifurcation. Heavy lines are stable equilibria; dashed lines are unstable equilibria.

This critical point is stable for $r > r_0$.

At the bifurcation point, $r = r_0$, there is no linearization, and the system becomes

$$\frac{dx}{dt} = -x^2, \tag{9.464}$$

which has solution

$$x(t) = \frac{x(0)}{1 + x(0)t}. \tag{9.465}$$

Here the asymptotic stability depends on the initial condition! For $x(0) \geq 0$, the critical point at $x = 0$ is stable. For $x(0) < 0$, there is a blowup phenomenon at $t = -1/x(0)$. The results are summarized in the bifurcation diagram sketched in Figure 9.27.

9.9.3 Saddle-Node

For $x \in \mathbb{R}^1, t \in \mathbb{R}^1, r \in \mathbb{R}^1, r_0 \in \mathbb{R}^1$, consider

$$\frac{dx}{dt} = -x^2 + (r - r_0). \tag{9.466}$$

The critical points are $\bar{x} = \pm\sqrt{r - r_0}$. Taking $\tilde{x} = x \mp \sqrt{r - r_0}$ and linearizing, we obtain

$$\frac{d\tilde{x}}{dt} = \mp 2\sqrt{r - r_0}\tilde{x}, \tag{9.467}$$

which has solution

$$\tilde{x}(t) = \tilde{x}(0) \exp\left(\mp\sqrt{r - r_0}t\right). \tag{9.468}$$

For $r > r_0$, the root $x = +\sqrt{r - r_0}$ is asymptotically stable. The root $x = -\sqrt{r - r_0}$ is asymptotically unstable.

At the point, $r = r_0$, there is no linearization, and the system becomes

$$\frac{dx}{dt} = -x^2, \tag{9.469}$$

which has solution

$$x(t) = \frac{x(0)}{1 + x(0)t}. \tag{9.470}$$

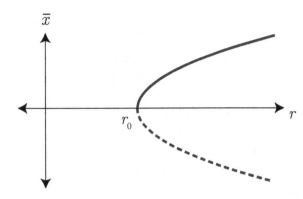

Figure 9.28. Sketch of saddle-node bifurcation. Heavy lines are stable equilibria; dashed lines are unstable equilibria.

Here the asymptotic stability again depends on the initial condition. For $x(0) \geq 0$, the critical point at $x = 0$ is stable. For $x(0) < 0$, there is a blowup phenomenon at $t = -1/x(0)$. The results are summarized in the bifurcation diagram sketched in Figure 9.28.

9.9.4 Hopf

Let us consider a two-dimensional vector field with potentially oscillatory behavior resulting from what is known as a Hopf[13] bifurcation.

EXAMPLE 9.27

With $x, y, t, r, r_0 \in \mathbb{R}^1$, take

$$\frac{dx}{dt} = (r - r_0)x - y - x(x^2 + y^2), \tag{9.471}$$

$$\frac{dy}{dt} = x + (r - r_0)y - y(x^2 + y^2). \tag{9.472}$$

The origin $(0,0)^T$ is a critical point. The linearized perturbation equations are

$$\frac{d}{dt}\begin{pmatrix} \tilde{x} \\ \tilde{y} \end{pmatrix} = \begin{pmatrix} r - r_0 & -1 \\ 1 & r - r_0 \end{pmatrix}\begin{pmatrix} \tilde{x} \\ \tilde{y} \end{pmatrix}. \tag{9.473}$$

The eigenvalues λ of the coefficient matrix are $\lambda = (r - r_0) \pm i$. For $r < r_0$, the real part is negative, and the origin is stable. At $r = r_0$, there is a Hopf bifurcation as the eigenvalues cross the imaginary axis of the complex plane as r is changed. For $r > r_0$, a periodic orbit in the (x, y) phase plane appears. The linearized analysis will not give the amplitude of the motion. Writing the given equation in polar coordinates (ρ, θ) yields

$$\frac{d\rho}{dt} = \rho(r - r_0) - \rho^3, \tag{9.474}$$

$$\frac{d\theta}{dt} = 1. \tag{9.475}$$

This system exhibits a pitchfork bifurcation in the amplitude of the oscillation ρ.

---

[13] Eberhard Frederich Ferdinand Hopf, 1902–1983, Austrian-born German mathematician.

9.10 Projection of Partial Differential Equations

One of the most important topics in engineering is the solution of partial differential equations, an expansive topic that we only briefly broach in this penultimate section. We largely avoid the details of this rich and varied subject matter but simply take the opportunity here to illustrate through an example some important principles as they relate to linear and nonlinear dynamical systems. Those principles are as follows:

- Solution of a partial differential equation can always be cast in terms of solving an infinite set of ordinary differential equations.
- Approximate solution of a partial differential equation can be cast in terms of solving a finite set of ordinary differential equations.
- A linear partial differential equation induces an uncoupled system of linear ordinary differential equations.
- A nonlinear partial differential equation induces a coupled set of nonlinear ordinary differential equations.

There are many viable methods to represent a partial differential equation as system of ordinary differential equations. Among them are methods in which one or more dependent and independent variables are discretized; important examples are the finite difference and finite element methods, which will not be considered here. Another key method involves projecting the dependent variable onto a set of basis functions and truncating this infinite series. We illustrate such a process here with an example involving a projection incorporating the method of weighted residuals. Our method will be spectral in the sense described in Section 6.11 in that the basis functions will be orthonormal and the method will be Galerkin.

EXAMPLE 9.28

Convert the nonlinear partial differential equation and initial and boundary conditions,

$$\frac{\partial T}{\partial t} = \frac{\partial}{\partial x}\left((1+\epsilon T)\frac{\partial T}{\partial x}\right), \qquad T(x,0) = x - x^2, \quad T(0,t) = T(1,t) = 0, \quad (9.476)$$

to a system of ordinary differential equations using a Galerkin projection method and find a two-term approximation.

This well-known equation, sometimes called the *heat equation*, can be shown to describe the time evolution of a spatial temperature field in a one-dimensional geometry with material properties that have weak temperature dependency when $0 < \epsilon \ll 1$. The boundary conditions are homogeneous, and the initial condition is symmetric about $x = 1/2$. We can think of T as temperature, x as distance, and t as time, all of which have been suitably scaled. For $\epsilon = 0$, the material properties are constant, and the equation is linear; otherwise, the material properties are temperature-dependent, and the equation is nonlinear because of the product $T \partial T / \partial x$. We can use the product rule to rewrite Eq. (9.476) as

$$\frac{\partial T}{\partial t} = \frac{\partial^2 T}{\partial x^2} + \epsilon T \frac{\partial^2 T}{\partial x^2} + \epsilon \left(\frac{\partial T}{\partial x}\right)^2. \qquad (9.477)$$

Now let us assume that $T(x,t)$ can be approximated in an N-term series by

$$T(x,t) = \sum_{n=1}^{N} \alpha_n(t)\varphi_n(x). \qquad (9.478)$$

We presume the exact solution is approached as $N \to \infty$. We can consider $\alpha_n(t)$ to be a set of N time-dependent amplitudes that modulate each spatial basis function,

$\varphi_n(x)$. For convenience, we will insist that the spatial basis functions satisfy the spatial boundary conditions $\varphi_n(0) = \varphi_n(1) = 0$ as well as an orthonormality condition for $x \in [0, 1]$: $\langle \varphi_n, \varphi_m \rangle = \delta_{nm}$. At the initial state, we have

$$T(x, 0) = x - x^2 = \sum_{n=1}^{N} \alpha_n(0)\varphi_n(x). \tag{9.479}$$

The terms $\alpha_n(0)$ are simply the constants in the Fourier series expansion (see Eq. (6.224)), of $x - x^2$:

$$\alpha_n(0) = \langle \varphi_n, (x - x^2) \rangle. \tag{9.480}$$

The partial differential equation expands as

$$\underbrace{\sum_{n=1}^{N} \frac{d\alpha_n}{dt}\varphi_n(x)}_{\partial T/\partial t} = \underbrace{\sum_{n=1}^{N} \alpha_n(t)\frac{d^2\varphi_n}{dx^2}}_{\partial^2 T/\partial x^2} + \epsilon \underbrace{\left(\sum_{n=1}^{N} \alpha_n(t)\varphi_n(x)\right)}_{T} \underbrace{\left(\sum_{n=1}^{N} \alpha_n(t)\frac{d^2\varphi_n}{dx^2}\right)}_{\partial^2 T/\partial x^2}$$

$$+ \epsilon \underbrace{\left(\sum_{n=1}^{N} \alpha_n(t)\frac{d\varphi_n}{dx}\right)^2}_{(\partial T/\partial x)^2}. \tag{9.481}$$

We change one of the dummy indices in each of the nonlinear terms from n to m and rearrange to find

$$\sum_{n=1}^{N} \frac{d\alpha_n}{dt}\varphi_n(x) = \sum_{n=1}^{N} \alpha_n(t)\frac{d^2\varphi_n}{dx^2} + \epsilon \sum_{n=1}^{N}\sum_{m=1}^{N} \alpha_n(t)\alpha_m(t)\left(\varphi_n(x)\frac{d^2\varphi_m}{dx^2} + \frac{d\varphi_n}{dx}\frac{d\varphi_m}{dx}\right). \tag{9.482}$$

Next, for the Galerkin procedure, one selects the weighting functions $\psi_l(x)$ to be the basis functions $\varphi_l(x)$ and takes the inner product of the equation with the weighting functions, yielding

$$\left\langle \varphi_l(x), \sum_{n=1}^{N} \frac{d\alpha_n}{dt}\varphi_n(x) \right\rangle$$

$$= \left\langle \varphi_l(x), \sum_{n=1}^{N} \alpha_n(t)\frac{d^2\varphi_n}{dx^2} + \epsilon \sum_{n=1}^{N}\sum_{m=1}^{N} \alpha_n(t)\alpha_m(t)\left(\varphi_n(x)\frac{d^2\varphi_m}{dx^2} + \frac{d\varphi_n}{dx}\frac{d\varphi_m}{dx}\right)\right\rangle. \tag{9.483}$$

Because of the orthonormality of the basis functions, the left side has obvious simplifications, yielding

$$\frac{d\alpha_l}{dt} = \left\langle \varphi_l(x), \sum_{n=1}^{N} \alpha_n(t)\frac{d^2\varphi_n}{dx^2} + \epsilon \sum_{n=1}^{N}\sum_{m=1}^{N} \alpha_n(t)\alpha_m(t)\left(\varphi_n(x)\frac{d^2\varphi_m}{dx^2} + \frac{d\varphi_n}{dx}\frac{d\varphi_m}{dx}\right)\right\rangle. \tag{9.484}$$

The right side can also be simplified via a complicated set of integration by parts and application of boundary conditions. If we further select $\varphi_n(x)$ to be an eigenfunction of d^2/dx^2, the first term on the right side will simplify considerably, though this choice is not required. In any case, this all serves to remove the explicit dependency on x, thus yielding a system of N ordinary differential equations of the form

$$\frac{d\boldsymbol{\alpha}}{dt} = \mathbf{f}(\boldsymbol{\alpha}), \tag{9.485}$$

where $\boldsymbol{\alpha}$ is a vector of length N and $\mathbf{f}$ is in general a nonlinear function of $\boldsymbol{\alpha}$.

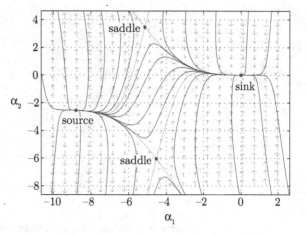

Figure 9.29. Phase plane dynamics of $N = 2$ amplitudes of spatial modes of solution to a weakly nonlinear heat equation.

We select our orthonormal basis functions as the eigenfunctions of d^2/dx^2 that also satisfy the appropriate boundary conditions,

$$\varphi_n(x) = \sqrt{2}\sin((2n-1)\pi x), \qquad n = 1, \dots, N. \qquad (9.486)$$

Because of the symmetry of our system about $x = 1/2$, it can be shown that only odd multiples of πx are present in the trigonometric sin approximation. Had we chosen an initial condition without such symmetry, we would have required both even and odd powers. We then apply the necessary Fourier expansion to find $\alpha_n(0)$, perform a detailed analysis of all of the necessary inner products, select $N = 2$, and arrive at the following nonlinear system of ordinary differential equations for the evolution of the time-dependent amplitudes:

$$\frac{d\alpha_1}{dt} = -\pi^2\alpha_1 + \sqrt{2}\pi\epsilon\left(-\frac{4}{3}\alpha_1^2 + \frac{8}{15}\alpha_2\alpha_1 - \frac{36}{35}\alpha_2^2\right), \qquad \alpha_1(0) = \frac{4\sqrt{2}}{\pi^3}, \qquad (9.487)$$

$$\frac{d\alpha_2}{dt} = -9\pi^2\alpha_2 + \sqrt{2}\pi\epsilon\left(\frac{12}{5}\alpha_1^2 - \frac{648}{35}\alpha_2\alpha_1 - 4\alpha_2^2\right), \qquad \alpha_2(0) = \frac{4\sqrt{2}}{27\pi^3}. \qquad (9.488)$$

By inspection the system has a equilibrium when $(\alpha_1, \alpha_2) = (0, 0)$. And by inspection that equilibrium is a sink in phase space. But calculation reveals there are three other equilibria that must be considered. The character of those roots can be revealed by examination of the eigenvalues of the local Jacobian matrix obtained by linearization near those roots. For $\epsilon = 1/5$, the three additional roots and their character are found to be

$$(\alpha_1, \alpha_2) = (-4.53, -6.04), \qquad \text{saddle}, \qquad (9.489)$$

$$(\alpha_1, \alpha_2) = (-8.78, -2.54), \qquad \text{source}, \qquad (9.490)$$

$$(\alpha_1, \alpha_2) = (-5.15, 3.46), \qquad \text{saddle}. \qquad (9.491)$$

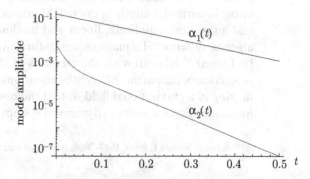

Figure 9.30. Evolution of $N = 2$ amplitudes of spatial modes of solution to a weakly nonlinear heat equation.

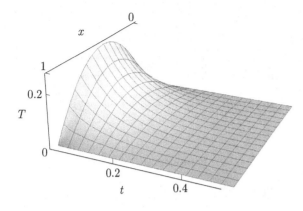

Figure 9.31. $T(x,t)$ from $N = 2$ term Galerkin projection for a weakly nonlinear heat equation.

When $\epsilon = 0$, and because we selected our basis functions to be the eigenfunctions of d^2/dx^2, we see the system is linear and uncoupled with exact solution

$$\alpha_1(t) = \frac{4\sqrt{2}}{\pi^3} e^{-\pi^2 t}, \tag{9.492}$$

$$\alpha_2(t) = \frac{4\sqrt{2}}{27\pi^3} e^{-9\pi^2 t}. \tag{9.493}$$

Thus, for $\epsilon = 0$, the two-term approximation is

$$T(x,t) \approx \frac{4\sqrt{2}}{\pi^3} e^{-\pi^2 t} \sin(\pi x) + \frac{4\sqrt{2}}{27\pi^3} e^{-9\pi^2 t} \sin(3\pi x). \tag{9.494}$$

For $\epsilon \neq 0$, numerical solution is required. We do so for $\epsilon = 1/5$ and plot the phase plane dynamics in Figure 9.29 for arbitrary initial conditions. Many initial conditions lead one to the finite sink at $(0,0)$. It is likely that the dynamics are also influenced by equilibria at infinity, not shown here. One can show that the solutions in the neighborhood of the sink are the most relevant to the underlying physical problem.

We plot results of $\alpha_1(t), \alpha_2(t)$ for our initial conditions in Figure 9.30. We see the first mode has significantly more amplitude than the second mode. Both modes are decaying rapidly to the sink at $(0,0)$. The $N = 2$ solution with full time and space dependency is

$$T(x,t) \approx \alpha_1(t) \sin(\pi x) + \alpha_2(t) \sin(3\pi x) \tag{9.495}$$

and is plotted in Figure 9.31.

9.11 Lorenz Equations

We close with a section presenting a celebrated and influential problem whose solution is best illuminated with a cornucopia of important tools and ideas of this book: linearized analysis in the neighborhood of equilibria, nonlinear analysis near and away from equilibria, linear and nonlinear transformations, geometry, linear algebra, differential equations, and bifurcation theory. The problem is that studied by Lorenz,[14] who drew on the methods of the previous section to use a Galerkin projection of the partial differential equations describing the motion of a heated fluid moving in a gravitational field so that the essence of the motion was captured by a low-dimensional nonlinear dynamical system. Omitting details of the reduction, for

[14] Edward Norton Lorenz, 1917–2008, American meteorologist.

independent variable $t \in \mathbb{R}^1$, thought of as time, dependent variables $(x, y, z)^T \in \mathbb{R}^3$, and parameters $\sigma, r, b \in \mathbb{R}^1$, $\sigma > 0$, $r > 0$, $b > 0$, the Lorenz equations are

$$\frac{dx}{dt} = \sigma(y - x), \tag{9.496}$$

$$\frac{dy}{dt} = rx - y - xz, \tag{9.497}$$

$$\frac{dz}{dt} = -bz + xy. \tag{9.498}$$

The bifurcation parameter will be taken to be r.

9.11.1 Linear Stability

The critical points are obtained from

$$\overline{y} - \overline{x} = 0, \tag{9.499}$$

$$r\overline{x} - \overline{y} - \overline{x}\,\overline{z} = 0, \tag{9.500}$$

$$-b\overline{z} + \overline{x}\,\overline{y} = 0, \tag{9.501}$$

which gives

$$\begin{pmatrix} \overline{x} \\ \overline{y} \\ \overline{z} \end{pmatrix} = \begin{pmatrix} 0 \\ 0 \\ 0 \end{pmatrix}, \begin{pmatrix} \sqrt{b(r-1)} \\ \sqrt{b(r-1)} \\ r-1 \end{pmatrix}, \begin{pmatrix} -\sqrt{b(r-1)} \\ -\sqrt{b(r-1)} \\ r-1 \end{pmatrix}. \tag{9.502}$$

When $r = 1$, there is only one critical point at the origin. For more general r, a linear stability analysis of each of the three critical points follows.

- $\overline{x} = \overline{y} = \overline{z} = 0$. Small perturbations around this point give

$$\frac{d}{dt} \begin{pmatrix} \tilde{x} \\ \tilde{y} \\ \tilde{z} \end{pmatrix} = \begin{pmatrix} -\sigma & \sigma & 0 \\ r & -1 & 0 \\ 0 & 0 & -b \end{pmatrix} \begin{pmatrix} \tilde{x} \\ \tilde{y} \\ \tilde{z} \end{pmatrix}. \tag{9.503}$$

The characteristic equation is

$$(\lambda + b)(\lambda^2 + \lambda(\sigma + 1) - \sigma(r - 1)) = 0, \tag{9.504}$$

from which we get the eigenvalues

$$\lambda = -b, \qquad \lambda = \frac{1}{2}\left(-(1 + \sigma) \pm \sqrt{(1 + \sigma)^2 - 4\sigma(1 - r)}\right). \tag{9.505}$$

For $0 < r < 1$, the eigenvalues are real and negative, because $(1 + \sigma)^2 > 4\sigma(1 - r)$. At $r = 1$, there is a pitchfork bifurcation with one zero eigenvalue. For $r > 1$, the origin becomes unstable.

- $\overline{x} = \overline{y} = \sqrt{b(r - 1)}$, $\overline{z} = r - 1$. We need $r \geq 1$ for a real solution. Small perturbations give

$$\frac{d}{dt} \begin{pmatrix} \tilde{x} \\ \tilde{y} \\ \tilde{z} \end{pmatrix} = \begin{pmatrix} -\sigma & \sigma & 0 \\ 1 & -1 & -\sqrt{b(r-1)} \\ \sqrt{b(r-1)} & \sqrt{b(r-1)} & -b \end{pmatrix} \begin{pmatrix} \tilde{x} \\ \tilde{y} \\ \tilde{z} \end{pmatrix}. \tag{9.506}$$

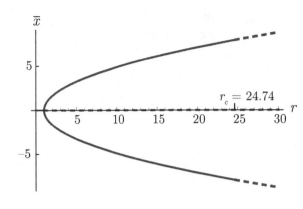

Figure 9.32. Bifurcation diagram for Lorenz equations, Eqs. (9.496–9.498), with $\sigma = 10, b = 8/3$.

The characteristic equation is

$$\lambda^3 + (\sigma + b + 1)\lambda^2 + (\sigma + r)b\lambda + 2\sigma b(r - 1) = 0. \tag{9.507}$$

This system is difficult to fully analyze. Detailed analysis reveals of a critical value of r:

$$r = r_c = \frac{\sigma(\sigma + b + 3)}{\sigma - b - 1}. \tag{9.508}$$

At $r = r_c$ the characteristic equation, Eq. (9.507), can be factored to give the eigenvalues

$$\lambda = -(\sigma + b + 1), \qquad \lambda = \pm i\sqrt{\frac{2b\sigma(\sigma + 1)}{\sigma - b - 1}}. \tag{9.509}$$

If $\sigma > b + 1$, two of the eigenvalues are purely imaginary, and this corresponds to a Hopf bifurcation. The periodic solution that is created at this value of r can be shown to be unstable so that the bifurcation is what is known as *subcritical*.

∘ If $r = r_c$ and $\sigma < b + 1$, one can find all real eigenvalues, including at least one positive eigenvalue, that tells us this is unstable.

∘ We also find instability if $r > r_c$. If $r > r_c$ and $\sigma > b + 1$, we can find one negative real eigenvalue and two complex eigenvalues with positive real parts; hence, this is unstable. If $r > r_c$, and $\sigma < b + 1$, we can find three real eigenvalues, with at least one positive; this is unstable.

∘ For $1 < r < r_c$ and $\sigma < b + 1$, we find three real eigenvalues, one of which is positive; this is unstable.

∘ For stability, we can take

$$1 < r < r_c, \quad \text{and} \quad \sigma > b + 1. \tag{9.510}$$

In this case, we can find one negative real eigenvalue and two eigenvalues (which could be real or complex) with negative real parts; hence, this is stable.

• $\bar{x} = \bar{y} = -\sqrt{b(r - 1)}, \bar{z} = r - 1$. Analysis of this critical point is essentially identical to that of the previous point.

For a particular case, these results are summarized in the bifurcation diagram of Figure 9.32. Shown here are results when $\sigma = 10$, $b = 8/3$. For these values, Eq. (9.508) tells us $r_c = 24.74$. Also, $\sigma > b + 1$. For real equilibria, we need $r > 0$. The equilibrium at the origin is stable for $r \in [0, 1]$ and unstable for $r > 1$; the instability is denoted by the dashed line. At $r = 1$, there is a pitchfork bifurcation, and two new real equilibria are available. These are both linearly stable for $r \in [1, r_c]$. For

$r \in [1, 1.34562]$, the eigenvalues are both real and negative. For $r \in [1.134562, r_c]$, two of the eigenvalues become complex, but all three have negative real parts, so local linear stability is maintained. For $r > r_c$, all three equilibria are unstable and indicated by dashed lines. As an aside, because of nonlinear effects, some initial conditions yield trajectories that do not relax to a stable equilibrium for $r < r_c$. It can be shown, for example, if $x(0) = y(0) = z(0) = 1$, $r = 24 < r_c$ gives rise to a trajectory that never reaches either of the linearly stable critical points.

9.11.2 Nonlinear Stability: Center Manifold Projection

The nonlinear stability of dynamical systems is a broad and difficult subject with no universal approach guaranteed to yield results. Here we consider one strategy that sometimes works, the so-called *center manifold projection*. It is a procedure for obtaining the nonlinear behavior near an eigenvalue with zero real part. As an example, we look at the Lorenz system at the bifurcation point $r = 1$. Recall that when $r = 1$, the Lorenz equations have a single equilibrium at the origin. Linearization of the Lorenz equations near the equilibrium point at $(0, 0, 0)$ gives rise to a system of the form $dx/dt = \mathbf{A} \cdot \mathbf{x}$, where

$$\mathbf{A} = \begin{pmatrix} -\sigma & \sigma & 0 \\ 1 & -1 & 0 \\ 0 & 0 & -b \end{pmatrix}. \tag{9.511}$$

The matrix $\mathbf{A}$ has eigenvalues and eigenvectors

$$\lambda_1 = 0, \quad \mathbf{e}_1 = \begin{pmatrix} 1 \\ 1 \\ 0 \end{pmatrix}, \tag{9.512}$$

$$\lambda_2 = -(\sigma + 1), \quad \mathbf{e}_2 = \begin{pmatrix} -\sigma \\ 1 \\ 0 \end{pmatrix}, \tag{9.513}$$

$$\lambda_3 = -b, \quad \mathbf{e}_3 = \begin{pmatrix} 0 \\ 0 \\ 1 \end{pmatrix}. \tag{9.514}$$

The fact that $\lambda_1 = 0$ suggests that there is a local algebraic dependency between at least two of the state variables, and that locally, the system behaves as a differential-algebraic system, such as studied in Section 9.7.

We use the eigenvectors as a basis to define new coordinates (u, v, w) where

$$\begin{pmatrix} x \\ y \\ z \end{pmatrix} = \begin{pmatrix} 1 & -\sigma & 0 \\ 1 & 1 & 0 \\ 0 & 0 & 1 \end{pmatrix} \begin{pmatrix} u \\ v \\ w \end{pmatrix}. \tag{9.515}$$

This linear transformation has a Jacobian whose determinant is $J = 1 + \sigma$; thus, for $\sigma > -1$, it is orientation-preserving. It is volume-preserving only if $\sigma = 0$ or -2. Inversion shows that

$$u = \frac{x + \sigma y}{1 + \sigma}, \tag{9.516}$$

$$v = \frac{y - x}{1 + \sigma}, \tag{9.517}$$

$$w = z. \tag{9.518}$$

In terms of the new variables, the derivatives are expressed as

$$\frac{dx}{dt} = \frac{du}{dt} - \sigma \frac{dv}{dt}, \tag{9.519}$$

$$\frac{dy}{dt} = \frac{du}{dt} + \frac{dv}{dt}, \tag{9.520}$$

$$\frac{dz}{dt} = \frac{dw}{dt}, \tag{9.521}$$

so that original nonlinear Lorenz equations (9.496–9.498) become

$$\frac{du}{dt} - \sigma \frac{dv}{dt} = \sigma(1 + \sigma)v, \tag{9.522}$$

$$\frac{du}{dt} + \frac{dv}{dt} = -(1 + \sigma)v - (u - \sigma v)w, \tag{9.523}$$

$$\frac{dw}{dt} = -bw + (u - \sigma v)(u + v). \tag{9.524}$$

Solving directly for the derivatives so as to place the equations in autonomous form, we get

$$\frac{du}{dt} = 0u - \frac{\sigma}{1 + \sigma}(u - \sigma v)w = \lambda_1 u + \text{nonlinear terms}, \tag{9.525}$$

$$\frac{dv}{dt} = -(1 + \sigma)v - \frac{1}{1 + \sigma}(u - \sigma v)w = \lambda_2 v + \text{nonlinear terms}, \tag{9.526}$$

$$\frac{dw}{dt} = -bw + (u - \sigma v)(u + v) = \lambda_3 w + \text{nonlinear terms}. \tag{9.527}$$

The objective of using the eigenvectors as basis vectors is to change the original system to diagonal form in the linear terms. Notice that the coefficient on each linear term is an eigenvalue. We have thus rotated the system to be aligned with a local set of principal axes, much in the same way as considered in Section 2.1.5. Furthermore, the eigenvalues $\lambda_2 = -(1 + \sigma)$ and $\lambda_3 = -b$ are negative, ensuring that the linear behavior $v = e^{-(1+\sigma)t}$ and $w = e^{-bt}$ takes the solution quickly to zero in these variables.

It would appear then that we are only left with an equation in $u(t)$ for large t. However, if we put $v = w = 0$ in the right side, dv/dt and dw/dt would be zero if it were not for the u^2 term in dw/dt, implying that the dynamics are confined to $v = w = 0$ only if we ignore this term. According to the center manifold theorem, it is possible to find a manifold (called the center manifold) that is tangent to $u = 0$ but is not necessarily the tangent itself, to which the dynamics is indeed confined.

We can get as good an approximation to the center manifold as we want by choosing new variables. Expanding Eq. (9.527), which has the potential problem, we get

$$\frac{dw}{dt} = -bw + u^2 - (\sigma - 1)uv - \sigma v^2. \tag{9.528}$$

Letting

$$\tilde{w} = w - \frac{u^2}{b}, \tag{9.529}$$

so that $-bw + u^2 = -b\tilde{w}$, we can eliminate the potential problem with the derivative of w.

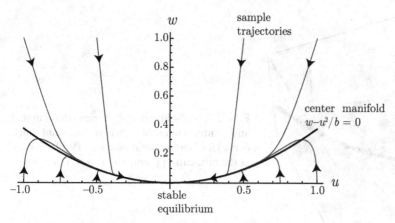

Figure 9.33. Projection onto the (u, w) plane (note $x = u - v$, $y = u + v$, $z = w$) of solution trajectories (thin curves) and center manifold (thick curve) for Lorenz equations, Eqs. (9.496–9.498), at the bifurcation point; $r = 1$, $\sigma = 1$, $b = 8/3$.

In the new variables $(u, v, \tilde{w})$, the full Lorenz equations are written as

$$\frac{du}{dt} = -\frac{\sigma}{1+\sigma}(u - \sigma v)\left(\tilde{w} + \frac{u^2}{b}\right), \tag{9.530}$$

$$\frac{dv}{dt} = -(1+\sigma)v - \frac{1}{1+\sigma}(u - \sigma v)\left(\tilde{w} + \frac{u^2}{b}\right), \tag{9.531}$$

$$\frac{d\tilde{w}}{dt} = -b\tilde{w} - (\sigma - 1)uv - \sigma v^2 + \frac{2\sigma}{b(1+\sigma)}u(u - \sigma v)\left(\tilde{w} + \frac{u^2}{b}\right). \tag{9.532}$$

Once again, the variables v and $\tilde{w}$ go to zero quickly. Formally setting them to zero, we recast Eqs. (9.530–9.532) as

$$\frac{du}{dt} = -\frac{\sigma}{b(1+\sigma)}u^3, \tag{9.533}$$

$$\frac{dv}{dt} = -\frac{1}{b(1+\sigma)}u^3, \tag{9.534}$$

$$\frac{d\tilde{w}}{dt} = \frac{2\sigma}{b^2(1+\sigma)}u^4. \tag{9.535}$$

Here, dv/dt and $d\tilde{w}/dt$ approach zero if u approaches zero. Now the equation for the evolution of u, Eq. (9.533), suggests that this is the case. Simply integrating Eq. (9.533) and applying an initial condition, we get

$$u(t) = \pm(u(0))\sqrt{\frac{b(1+\sigma)}{b(1+\sigma) + 2\sigma(u(0))^2t}}, \tag{9.536}$$

which is asymptotically stable as $t \to \infty$. So, to this level of approximation, the dynamics are confined to the $v = \tilde{w} = 0$ line. The bifurcation at $r = 1$ is said to be *supercritical*. Higher order terms can be included to obtain improved accuracy, if necessary.

We next focus attention on a particular case where the parameters were chosen to be $r = 1$, $\sigma = 1$, and $b = 8/3$. Figure 9.33 gives the projection onto the (u, w) phase

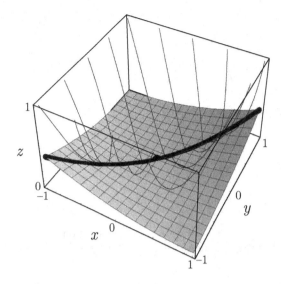

Figure 9.34. Solution trajectories (thin curves) and center manifold (gray surface and thick curve) for Lorenz equations, Eqs. (9.496–9.498), at the bifurcation point; $r = 1$, $\sigma = 1$, $b = 8/3$.

space of several solution trajectories calculated in (u, v, w) phase space for a wide variety of initial conditions along with the center manifold, $\tilde{w} = w - u^2/b = 0$. It is seen that a given solution trajectory indeed approaches the center manifold on its way to the equilibrium point at the origin. The center manifold approximates the solution trajectories well in the neighborhood of the origin. Far from the origin, not shown here, it is not an attracting manifold.

We can gain more insight into the center manifold by transforming back into (x, y, z) space. Figure 9.34 shows in that space several solution trajectories, a representation of the surface that constitutes the center manifold, as well as a curve embedded within the center manifold to which trajectories are further attracted. We can in fact decompose the motion of a trajectory into the following regimes, for the parameters $r = 1$, $\sigma = 1$, $b = 8/3$.

- *Very fast attraction to the two-dimensional center manifold, $\tilde{w} = 0$*: Because $b > \sigma + 1$, for this case, $\tilde{w}$ approaches zero faster than v approaches zero, via exponential decay dictated by Eqs. (9.531, 9.532). So on a time scale of $1/b$, the trajectory first approaches $\tilde{w} = 0$, which means it approaches $w - u^2/b = 0$. Transforming back to (x, y, z) via Eqs. (9.516–9.518), a trajectory thus approaches the surface:

$$z = \frac{1}{b} \underbrace{\left(\frac{x}{1+\sigma} + \frac{\sigma y}{1+\sigma} \right)}_{u}^{2} \Bigg|_{\sigma=1, b=8/3} = \frac{3}{8} \left(\frac{x+y}{2} \right)^2. \tag{9.537}$$

- *Fast attraction to the one-dimensional curve, $v = 0$*: Once on the two-dimensional manifold, the slower time scale relaxation with time constant $1/(\sigma + 1)$ to the curve given by $v = 0$ occurs. When $v = 0$, we also have $x = y$, so this curve takes the parametric form

$$x(s) = s, \tag{9.538}$$

$$y(s) = s, \tag{9.539}$$

$$z(s) = \frac{1}{b} \left(\frac{s}{1+\sigma} + \frac{\sigma s}{1+\sigma} \right)^2 \Bigg|_{\sigma=1, b=8/3} = \frac{3}{8} s^2. \tag{9.540}$$

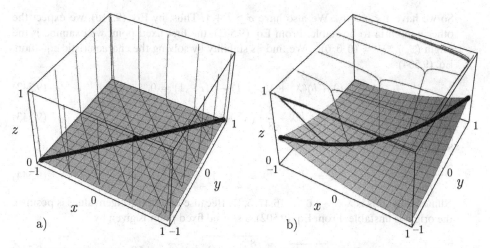

Figure 9.35. Solution trajectories (thin curves) and center manifold (gray surface and thick curve) for Lorenz equations, Eqs. (9.496–9.498), at the bifurcation point: (a) $r = 1$, $\sigma = 1$, $b = 100$, and (b) $r = 1$, $\sigma = 100$, $b = 8/3$.

- *Slow attraction to the zero-dimensional equilibrium point at* $(0, 0, 0)$: This final relaxation brings the system to rest.

For different parameters, this sequence of events is modified, as depicted in Figure 9.35. In Figure 9.35a, we take $r = 1$, $\sigma = 1$, $b = 100$. By Eqs. (9.531,9.532), these parameters induce an even faster relaxation to $\tilde{w} = 0$; as before, this is followed by a fast relaxation to $v = 0$, where $x = y$, and a final slow relaxation to equilibrium. One finds that the center manifold surface $\tilde{w} = 0$ has less curvature and that the trajectories, following an initial nearly vertical descent, have sharp curvature as they relax onto the flatter center manifold, where they again approach equilibrium at the origin.

In Figure 9.35b, we take $r = 1$, $\sigma = 100$, $b = 8/3$. By Eqs. (9.531,9.532), these parameters induce an initial fast relaxation to $v = 0$, where $x = y$. This is followed by a fast relaxation to the center manifold where $\tilde{w} = 0$ and then a slow relaxation to equilibrium at the origin.

9.11.3 Transition to Chaos

By varying the bifurcation parameter r, we can predict what is called a transition to chaos. We illustrate this transition for two sets of parameters for the Lorenz equations. The first will have trajectories that relax to a stable fixed point; the second will have so-called *chaotic* trajectories that relax to what is known as a *strange attractor*.

EXAMPLE 9.29

Examine the solution to the Lorenz equations for conditions: $\sigma = 10$, $r = 10$, $b = 8/3$ with initial conditions $x(0) = y(0) = z(0) = 1$.

We first note that $r > 1$, so we expect the origin to be unstable. Next note from Eq. (9.508) that

$$r_c = \frac{\sigma(\sigma + b + 3)}{\sigma - b - 1} = \frac{10(10 + \frac{8}{3} + 3)}{10 - \frac{8}{3} - 1} = \frac{470}{19} = 24.74. \tag{9.541}$$

So we have $1 < r < r_c$. We also have $\sigma > b + 1$. Thus, by Eq. (9.510), we expect the other equilibria to be stable. From Eq. (9.502), the first fixed point we examine is the origin $(\bar{x}, \bar{y}, \bar{z})^T = (0, 0, 0)^T$. We find its stability by solving the characteristic equation, Eq. (9.504):

$$(\lambda + b)(\lambda^2 + \lambda(\sigma + 1) - \sigma(r - 1) = 0, \tag{9.542}$$

$$\left(\lambda + \frac{8}{3}\right)(\lambda^2 + 11\lambda - 90) = 0. \tag{9.543}$$

Solution gives

$$\lambda = -\frac{8}{3}, \qquad \lambda = \frac{1}{2}\left(-11 \pm \sqrt{481}\right). \tag{9.544}$$

Numerically, this is $\lambda = -2.67, -16.47, 5.47$. Because one of the eigenvalues is positive, the origin is unstable. From Eq. (9.502), a second fixed point is given by

$$\bar{x} = \sqrt{b(r - 1)} = \sqrt{\frac{8}{3}(10 - 1)} = 2\sqrt{6} = 4.90, \tag{9.545}$$

$$\bar{y} = \sqrt{b(r - 1)} = \sqrt{\frac{8}{3}(10 - 1)} = 2\sqrt{6} = 4.90, \tag{9.546}$$

$$\bar{z} = r - 1 = 10 - 1 = 9. \tag{9.547}$$

Consideration of the roots of Eq. (9.507) shows the second fixed point is stable:

$$\lambda^3 + (\sigma + b + 1)\lambda^2 + (\sigma + r)b\lambda + 2\sigma b(r - 1) = 0, \tag{9.548}$$

$$\lambda^3 + \frac{41}{3}\lambda^2 + \frac{160}{3}\lambda + 480 = 0. \tag{9.549}$$

Solution gives

$$\lambda = -12.48, \qquad \lambda = -0.595 \pm 6.17\, i. \tag{9.550}$$

From Eq. (9.502), a third fixed point is given by

$$\bar{x} = -\sqrt{b(r - 1)} = -\sqrt{\frac{8}{3}(10 - 1)} = -2\sqrt{6} = -4.90, \tag{9.551}$$

$$\bar{y} = -\sqrt{b(r - 1)} = -\sqrt{\frac{8}{3}(10 - 1)} = -2\sqrt{6} = -4.90, \tag{9.552}$$

$$\bar{z} = r - 1 = 10 - 1 = 9. \tag{9.553}$$

The stability analysis for this point is essentially identical as that for the second point. The eigenvalues are identical $\lambda = -12.48, -0.595 \pm 6.17i$; thus, the root is linearly stable. Because we have two stable roots, we might expect some initial conditions to induce trajectories to one of the stable roots and other initial conditions to induce trajectories to the other. Figure 9.36 shows the phase space trajectories in (x, y, z) space and the behavior in the time domain, $x(t), y(t), z(t)$. Examination of the solution reveals that for this set of initial conditions, the second equilibrium is attained.

EXAMPLE 9.30

Now consider the conditions: $\sigma = 10$, $r = 28$, $b = 8/3$. Initial conditions remain $x(0) = y(0) = z(0) = 1$.

The analysis is similar to the previous example, except that we have changed the bifurcation parameter r. We first note that $r > 1$, so we expect the origin to be an unstable

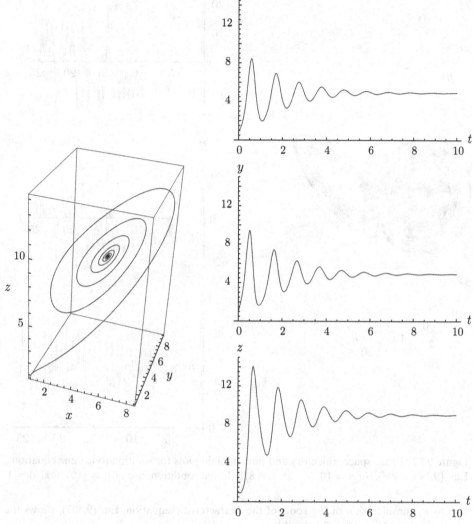

Figure 9.36. Solution to Lorenz equations, Eqs. (9.496–9.498), for $\sigma = 10$, $r = 10$, $b = 8/3$. Initial conditions are $x(0) = y(0) = z(0) = 1$.

equilibrium. We next note from Eq. (9.508) that

$$r_c = \frac{\sigma(\sigma + b + 3)}{\sigma - b - 1} = \frac{10(10 + \frac{8}{3} + 3)}{10 - \frac{8}{3} - 1} = \frac{470}{19} = 24.74 \qquad (9.554)$$

remains unchanged from the previous example. So we have $r > r_c$. Thus, we expect the other equilibria to be unstable as well.

From Eq. (9.502), the origin is again a fixed point, and again it can be shown to be unstable. From Eq. (9.502), the second fixed point is now given by

$$\overline{x} = \sqrt{b(r - 1)} = \sqrt{\frac{8}{3}(28 - 1)} = 8.485, \qquad (9.555)$$

$$\overline{y} = \sqrt{b(r - 1)} = \sqrt{\frac{8}{3}(28 - 1)} = 8.485, \qquad (9.556)$$

$$\overline{z} = r - 1 = 28 - 1 = 27. \qquad (9.557)$$

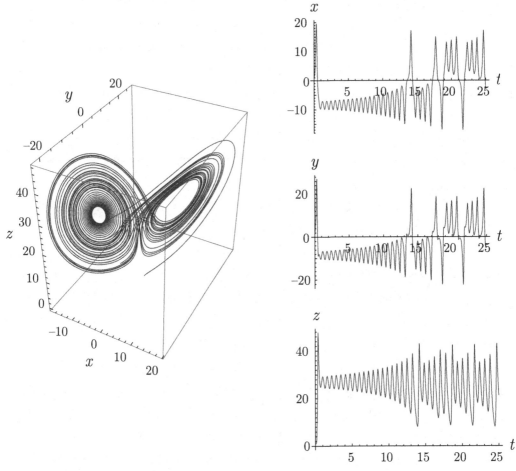

Figure 9.37. Phase space trajectory and time domain plots for solution to Lorenz equations, Eqs. (9.496–9.498), for $\sigma = 10, r = 28, b = 8/3$. Initial conditions are $x(0) = y(0) = z(0) = 1$.

Now, consideration of the roots of the characteristic equation, Eq. (9.507), shows the second fixed point here is unstable:

$$\lambda^3 + (\sigma + b + 1)\lambda^2 + (\sigma + r)b\lambda + 2\sigma b(r - 1) = 0, \tag{9.558}$$

$$\lambda^3 + \frac{41}{3}\lambda^2 + \frac{304}{3}\lambda + 1440 = 0. \tag{9.559}$$

Solution gives

$$\lambda = -13.8546, \qquad \lambda = 0.094 \pm 10.2\,i. \tag{9.560}$$

Moreover, the third fixed point is unstable in exactly the same fashion as the second. The consequence of this is that there is no possibility of achieving an equilibrium as $t \to \infty$. More importantly, numerical solution reveals the solution to approach what is known as a strange attractor of Hausdorff-Besicovitch dimension of about 2.06. Moreover, numerical experimentation would reveal an extreme, exponential sensitivity of the solution trajectories to the initial conditions. That is, a small change in initial conditions would induce a large deviation of a trajectory in a finite time. Such systems are known as chaotic. Figure 9.37 shows the phase space, the strange attractor, and the behavior in the time domain of this system, which has undergone a transition to a chaotic state.

EXERCISES

1. For the logistics map $x_{k+1} = rx_k(1 - x_k); 0 < x_k < 1, 0 < r < 4$, write a short program that determines the value of x as $k \to \infty$. Generate the bifurcation diagram, that is the limiting value of x as a function of r for $0 < r < 4$. If r_i is the ith bifurcation point, that is the value at which the number of fixed points changes, make an estimate of Feigenbaum's[15] constant defined as

$$\delta = \lim_{n \to \infty} \frac{r_{n-1} - r_n}{r_n - r_{n+1}}.$$

2. Generate the bifurcation diagram for the iterated Gaussian map $x_{n+1} = \exp(-\alpha x_n^2) + \beta$, with $\alpha = 4.9$ and $\beta \in [-1, 1]$.

3. For the two-dimensional Hénon map

$$x_{n+1} = y_n + 1 - ax_n^2$$

$$y_{n+1} = bx_n,$$

calculate and plot x_n and y_n versus n with $a = 1.4$ and $b = 0.3$.

4. Consider a straight line between $x = 0$ and $x = l$. Remove the middle half (i.e., the portion between $x = l/4$ and $x = 3l/4$). Repeat the process on the two pieces that are left. Find the dimension of what is left after an infinite number of iterations.

5. Determine the dimension of the Menger sponge.

6. Write the system of equations

$$x\frac{dx}{dt} + xy\frac{dy}{dt} = x - 1$$

$$(x + y)\frac{dx}{dt} + x\frac{dy}{dt} = y + 1$$

in the form

$$\frac{dx}{dt} = f(x, y)$$

$$\frac{dy}{dt} = g(x, y).$$

Plot curves on which $f = 0, g = 0$ in the (x, y) phase plane. Also plot in this plane the vector field defined by the differential equations. With a combination of analysis and numerics, find a path in phase space from one critical point to another critical point. For this path, which is a heteroclinic orbit, plot $x(t)$, $y(t)$ and include the path in the (x, y) phase plane.

7. Reduce

$$\frac{d}{dt}\begin{pmatrix} x_1 \\ x_2 \end{pmatrix} = \begin{pmatrix} a_{11} & a_{12} \\ a_{21} & a_{22} \end{pmatrix}\begin{pmatrix} x_1 \\ x_2 \end{pmatrix},$$

where the as are nonzero, to a single high-order scalar differential equation.

[15] Mitchell Feigenbaum, 1944–, American mathematical physicist.

8. Write the Blasius[16] equation

$$f''' + \frac{1}{2}ff'' = 0$$

as three first-order equations in at least two different ways. Hint: Let $y_1 = f$, $y_2 = f + f'$, $y_3 = f' + f''$.

9. Find the general solution of $dx/dt = \mathbf{A} \cdot \mathbf{x}$ where

$$\mathbf{A} = \begin{pmatrix} 1 & -3 & 1 \\ 2 & -1 & -2 \\ 2 & -3 & 0 \end{pmatrix}.$$

10. Solve $dx/dt = \mathbf{A} \cdot \mathbf{x}$ where

$$\mathbf{A} = \begin{pmatrix} 2 & 1 \\ 0 & 2 \end{pmatrix},$$

using the matrix exponential.

11. Find the solution of $dx/dt = \mathbf{A} \cdot \mathbf{x}$ if

(a)

$$\mathbf{A} = \begin{pmatrix} 1 & 0 & -1 \\ -1 & 2 & 1 \\ 1 & 0 & 1 \end{pmatrix}, \qquad \mathbf{x}(0) = \begin{pmatrix} 0 \\ 0 \\ 1 \end{pmatrix},$$

(b)

$$\mathbf{A} = \begin{pmatrix} 1 & -3 & 2 \\ 0 & -1 & 0 \\ 0 & -1 & -2 \end{pmatrix}, \qquad \mathbf{x}(0) = \begin{pmatrix} 1 \\ 2 \\ 1 \end{pmatrix},$$

(c)

$$\mathbf{A} = \begin{pmatrix} 1 & 0 & 0 \\ 0 & 1 & -1 \\ 0 & 1 & 1 \end{pmatrix}, \qquad \mathbf{x}(0) = \begin{pmatrix} 1 \\ 1 \\ 1 \end{pmatrix},$$

(d)

$$\mathbf{A} = \begin{pmatrix} -3 & 0 & 2 & 0 \\ 0 & -2 & 0 & 0 \\ 0 & 0 & 1 & 1 \\ 0 & 0 & 0 & 0 \end{pmatrix}, \qquad \mathbf{x}(0) = \begin{pmatrix} 1 \\ 1 \\ 1 \\ 1 \end{pmatrix}.$$

12. Solve the system of equations

(a)

$$\dot{x} + \dot{y} = e^t,$$
$$\dot{x} - \dot{y} = e^{-t},$$

(b)

$$\dot{x} + y = 1,$$
$$\dot{y} + 4x = 0,$$

with $x(0) = y(0) = 1$.

---

[16] Paul Richard Heinrich Blasius, 1883–1970, German fluid mechanician.

13. Write in matrix form and solve

$$\frac{dx}{dt} = y + z,$$

$$\frac{dy}{dt} = z + x,$$

$$\frac{dz}{dt} = x + y.$$

14. Solve the system of differential equations

$$\frac{dx}{dt} = -5x + 2y + z,$$

$$\frac{dy}{dt} = -5y + 3z,$$

$$\frac{dz}{dt} = -5z$$

using generalized eigenvectors.

15. Express

$$\frac{dx_1}{dt} + x_1 + \frac{dx_2}{dt} + 3x_2 = 0$$

$$\frac{dx_1}{dt} + 3\frac{dx_2}{dt} + x_2 = 0$$

in the form $dx/dt = \mathbf{A} \cdot \mathbf{x}$ and solve. Plot some solution trajectories in the (x_1, x_2) phase plane as well as the vector field defined by the system of equations.

16. Let

$$\mathbf{A} = \begin{pmatrix} 1 & 1 & 2 \\ 0 & 1 & 1 \\ 0 & 0 & 1 \end{pmatrix}.$$

Solve the equation

$$\frac{d\mathbf{x}}{dt} = \mathbf{A} \cdot \mathbf{x}.$$

Determine the critical points and their stability.

17. Determine if the origin is stable if $dx/dt = \mathbf{A} \cdot \mathbf{x}$, where

$$\mathbf{A} = \begin{pmatrix} 3 & -3 & 0 \\ 0 & -5 & -2 \\ -6 & 0 & -3 \end{pmatrix}.$$

18. For each of the values of $\mathbf{A}$ given as follows, consider $dx/dt = \mathbf{A} \cdot \mathbf{x}$ with $\mathbf{x}(0) = (1, 1)^T$ and (i) find and plot an exact solution, (ii) identify any and all stability restrictions on Δt for the first-order forward Euler method, and (iii) choose a variety of values of Δt and generate forward Euler numerical approximations to the exact solution so as to reveal the accuracy and stability of your approximation for $t \in [0, 5]$:

(a) $\begin{pmatrix} -1 & 0 \\ 0 & -3 \end{pmatrix}$,

(b) $\begin{pmatrix} -1 & 0 \\ 0 & 3 \end{pmatrix}$,

(c) $\begin{pmatrix} -1 & 2 \\ 0 & -3 \end{pmatrix}$,

(d) $\begin{pmatrix} -1 & 2 \\ -2 & -3 \end{pmatrix}$,

(e) $\begin{pmatrix} -1 & 20 \\ 0 & -3 \end{pmatrix}$,

(f) $\begin{pmatrix} -1 & 20 \\ -20 & -3 \end{pmatrix}$,

(g) $\begin{pmatrix} 0 & 20 \\ -20 & 0 \end{pmatrix}$.

19. Repeat the previous problem using the backward Euler method.

20. Find the dynamical system corresponding to the Hamiltonian $H(x, y) = x^2 + 2xy + y^2$, and then solve it.

21. Show that the Hénon-Heiles[17] system

$$\frac{d^2 x}{dt^2} = -x - 2xy$$

$$\frac{d^2 y}{dt^2} = -y + y^2 - x^2$$

is Hamiltonian. Find the Hamiltonian of the system, and determine the stability of the critical point at the origin.

22. Find the real critical point(s) of the nonlinear equation

$$\ddot{x} + \dot{x} + x + x^5 = 0$$

and investigate its (their) *linear* stability.

23. Classify the critical points of

$$\frac{dx}{dt} = x - y - 3$$

$$\frac{dy}{dt} = y - x^2 + 1$$

and analyze their stability. Plot the global (x, y) phase plane, including critical points, vector fields, and trajectories.

24. Classify the critical points of

$$\frac{dx}{dt} = x + y - 2$$

$$\frac{dy}{dt} = 2y - x^2 + 1$$

and analyze their stability.

---

[17] Carl Eugene Heiles, 1939–, American astrophysicist.

25. Find a critical point of the following system and show its local and global stability:

$$\frac{dx}{dt} = (x - 2)\left((y - 1)^2 - 1\right),$$

$$\frac{dy}{dt} = (2 - y)\left((x - 2)^2 + 1\right),$$

$$\frac{dz}{dt} = (4 - z).$$

26. Show that for all initial conditions the solutions of

$$\frac{dx}{dt} = -x + x^2 y - y^2,$$

$$\frac{dy}{dt} = -x^3 + xy - 6z,$$

$$\frac{dz}{dt} = 2y$$

tend to $x = y = z = 0$ as $t \to \infty$.

27. Find the critical point(s) of the dynamical system and investigate its (their) *global* stability:

$$\dot{x} = -x - yz,$$

$$\dot{y} = -y + 2xz,$$

$$\dot{z} = -z - xy.$$

28. Find a Lyapunov function of the form $V = ax^2 + by^2$ to investigate the global stability of the critical point $x = y = 0$ of the system of equations

$$\frac{dx}{dt} = -2x^3 + 3xy^2,$$

$$\frac{dy}{dt} = -x^2 y - y^3.$$

29. Given the dynamical system

$$\frac{dy_1}{dt} = y_2 - y_1^3,$$

$$\frac{dy_2}{dt} = -y_1 + y_2 - y_2^3,$$

determine the stability of the origin to (a) small and (b) large perturbations.

30. The populations x and y of two competing animal species are governed by

$$\frac{dx}{dt} = x - 2xy,$$

$$\frac{dy}{dt} = -y + xy.$$

What are the steady state populations? Is the situation stable?

31. Analyze the linear stability of the critical point of

$$\frac{dx}{dt} = 2y + y^2,$$

$$\frac{dy}{dt} = -r + 2x^2.$$

32. Find a Lyapunov function for the system

$$\frac{dx}{dt} = -x - 2y^2,$$

$$\frac{dy}{dt} = xy - y^3.$$

33. Analyze the local stability of the origin in the following system:

$$\frac{dx}{dt} = -2x + y + 3z + 8y^3,$$

$$\frac{dy}{dt} = -6y - 5z + 2z^3,$$

$$\frac{dz}{dt} = z + x^2 + y^3.$$

34. Show that the origin is linearly stable,

$$\frac{dx}{dt} = (x - by)(x^2 + y^2 - 1)$$

$$\frac{dy}{dt} = (ax + y)(x^2 + y^2 - 1),$$

where $a, b > 0$. Show also that the origin is stable to large perturbations, as long as they satisfy $x^2 + y^2 < 1$.

35. Show that solutions of the system of differential equations

$$\frac{dx}{dt} = -x + y^3 - z^3,$$

$$\frac{dy}{dt} = -y + z^3 - x^3,$$

$$\frac{dz}{dt} = -z + x^3 - y^3$$

eventually approach the origin for all initial conditions.

36. Show that the solutions of

$$\frac{dx}{dt} = y - x^3$$

$$\frac{dy}{dt} = -x - y^3$$

tend to $(0,0)$ as $t \to \infty$.

37. Find the critical point (or points) of the van der Pol equation

$$\frac{d^2x}{dt^2} - a(1 - x^2)\frac{dx}{dt} + x = 0, \ a > 0$$

and determine its (or their) stability to small perturbations. For $a = 1$, plot the $dx/dt, x$ phase plane including critical points, vector fields, and trajectories.

38. Sketch the steady state bifurcation diagram of

$$\frac{dx}{dt} = (x - r)(x + r)((x - 3)^2 + (r - 1)^2 - 1),$$

where r is the bifurcation parameter. Determine the linear stability of each branch; indicate the stable and unstable ones differently on the diagram.

39. Classify the critical point of

$$\frac{d^2 x}{dt^2} + (r - r_0)x = 0.$$

40. Show that $x = 0$ is a stable critical point of the differential equation

$$\frac{dx}{dt} = -\sum_{n=0}^{N} a_n x^{2n+1},$$

where $a_n \geq 0$, $n = 0, 1, \ldots, N$.

41. Find the stability of the critical points of the Duffing equation

$$\frac{d^2 x}{dt^2} + a\frac{dx}{dt} + bx + x^3 = 0$$

for positive and negative values of a and b. Sketch the flow lines.

42. Write the system of equations

$$\dot{x}_1 + \dot{x}_2 + \dot{x}_3 = 6e^t,$$
$$\dot{x}_2 + \dot{x}_3 = 5e^t,$$
$$x_1 + x_2 - x_3 = 0$$

in the form $dx/dt = \mathbf{f}(t)$ by differentiating the algebraic constraint, and solve the three equations. Solve the original system without differentiating the constraint.

43. Write

$$\begin{pmatrix} 1 & 1 \\ 1 & 1 \end{pmatrix} \frac{d}{dt} \begin{pmatrix} x_1 \\ x_2 \end{pmatrix} = \begin{pmatrix} 1 & 1 \\ 1 & 2 \end{pmatrix} \begin{pmatrix} x_1 \\ x_2 \end{pmatrix}$$

as separate differential and algebraic equations.

44. Use the Poincaré sphere to find all critical points, finite and infinite, of the system

$$\frac{dx}{dt} = 2x - 2xy,$$
$$\frac{dy}{dt} = 2y - x^2 + y^2.$$

Plot families of trajectories in the (x, y) phase space and the (X, Y) projection of the Poincaré sphere.

45. Draw the bifurcation diagram of

$$\frac{dx}{dt} = x^3 + x\left((r - 2)^3 - 1\right),$$

where r is the bifurcation parameter, indicating stable and unstable branches.

46. Graph the bifurcation diagrams of the following equations. Determine and indicate the linear stability of each branch:

(a) $\dfrac{dx}{dt} = -\left(\dfrac{1}{x} - r\right)(2x - r),$

(b) $\dfrac{dx}{dt} = -x\left((x - 2)^2 - (r - 1)\right).$

47. Plot a bifurcation diagram for the differential equation

$$\frac{dx}{dt} = (x - 3)(x^2 - r),$$

where r is the bifurcation parameter. Analyze linear stability and indicate stable and unstable branches.

48. Show in parameter space the different possible behaviors of

$$\frac{dx}{dt} = a + x^2 y - 2bx - x$$

$$\frac{dy}{dt} = bx - x^2 y,$$

where $a, b > 0$.

49. Generate the bifurcation diagram showing the stable and unstable steady states of

$$\frac{dx}{dt} = rx(1 - x) - x,$$

where r is the bifurcation parameter.

50. Find and plot all critical points $(\overline{x}, \overline{y})$ of

$$\frac{dx}{dt} = (r - 1)x - 3xy^2 - x^3$$

$$\frac{dy}{dt} = (r - 1)y - 3x^2 y - y^3$$

as functions of r. Determine the stability of $(\overline{x}, \overline{y}) = (0, 0)$ and of *one* postbifurcation branch.

51. Plot the bifurcation diagram and analyze the stability of

$$\frac{dx}{dt} = -x(x^3 - r - 1) - \frac{1}{10},$$

where r is the bifurcation parameter.

52. Draw the bifurcation diagram of

$$\frac{dx}{dt} = (x^2 - 2)^2 - 2(x^2 + 1)(r - 1) + (r - 1)^2,$$

where r is the bifurcation parameter, indicating the stability of each branch.

53. A two-dimensional dynamical system expressed in polar form is

$$\frac{dr}{dt} = r(r - 2)(r - 3)$$

$$\frac{d\theta}{dt} = 2.$$

Find the (a) critical point(s) and (b) periodic solution(s) and (c) analyze their stability.

54. Find the critical points and their stability of the Rössler[18] equations

$$\frac{dx}{dt} = -y - z,$$

$$\frac{dy}{dt} = x + ay,$$

$$\frac{dz}{dt} = b + z(x - c).$$

Plot the (x, y) projection of the numerical solution for $a = b = 0.1$ and values of $c = 4, 6, 8.7, 9, 12.6$ and 13.

55. For the following cases, convert the partial differential equation (which form (a) linear and (b) nonlinear heat equations) and initial and boundary conditions to a system of ordinary differential equations using a Galerkin projection method and find a one-, two-, and three-term approximate solution:

(a)

$$\frac{\partial T}{\partial t} = \frac{\partial^2 T}{\partial x^2}, \qquad T(x, 0) = x - x^3, \qquad T(0, t) = T(1, t) = 0.$$

(b)

$$\frac{\partial T}{\partial t} = \frac{\partial}{\partial x}\left(\left(1 + \frac{T}{10}\right)\frac{\partial T}{\partial x}\right), \qquad T(x, 0) = x - x^3, \quad T(0, t) = T(1, t) = 0.$$

56. Integrate the Lorenz equations numerically, and for each case, plot the trajectory in (x, y, z) phase space and plot $x(t), y(t), z(t)$ for $t \in [0, 50]$.

(a) $b = 8/3, r = 28$, initial conditions $x(0) = 2, y(0) = 1, z(0) = 3$, and (i) $\sigma = 1$, and (ii) $\sigma = 10$. Change the initial condition on x to $x(0) = 2.002$ and plot the difference in the predictions of x versus time for both values of σ.

(b) $\sigma = 10, b = 8/3$, initial conditions $x(0) = 0, y(0) = 1, z(0) = 0$, and (i) $r = 10$, (ii) $r = 24$, and (iii) $r = 28$. Change the initial condition on x to $x(0) = 0.002$ and plot the difference in the predictions of x versus time for all three values of r.

57. A linear control system can be represented by

$$\dot{\mathbf{x}} = \mathbf{A} \cdot \mathbf{x} + \mathbf{B} \cdot \mathbf{u},$$

where $\mathbf{x} \in \mathbb{R}^{n \times 1}$ is the state of the plant and $\mathbf{u} \in \mathbb{R}^{m \times 1}$ is a control signal. For each one of the following cases, determine if there exists a $\mathbf{u}$ such that $\mathbf{x}$ can be taken from an arbitrary $\mathbf{x}$ at $t = 0$ to the origin in a finite time τ and, if possible, suggest a $\mathbf{u}$:

(a) $\mathbf{A} = \begin{pmatrix} 0 & 1 \\ 0 & 0 \end{pmatrix}, \quad \mathbf{B} = \begin{pmatrix} 0 \\ 1 \end{pmatrix},$

(b) $\mathbf{A} = \begin{pmatrix} -1 & 0 \\ 0 & -2 \end{pmatrix}, \quad \mathbf{B} = \begin{pmatrix} 1 \\ 2 \end{pmatrix},$

(c) $\mathbf{A} = \begin{pmatrix} 0 & 1 \\ 0 & 0 \end{pmatrix}, \quad \mathbf{B} = \begin{pmatrix} 1 \\ 0 \end{pmatrix},$

(d) $\mathbf{A} = \begin{pmatrix} 0 & 1 \\ 0 & 0 \end{pmatrix}, \quad \mathbf{B} = \begin{pmatrix} 0 & 1 \\ 1 & 0 \end{pmatrix}.$

[18] Otto Rössler, 1940–, German biochemist.

58. The following equations arise in a natural circulation loop problem,

$$\frac{dx}{dt} = y - x,$$

$$\frac{dy}{dt} = a - zx,$$

$$\frac{dz}{dt} = xy - b,$$

where a and b are nonnegative parameters. Find the critical points and analyze their linear stability. Find numerically the attractors for (a) $a = 2$, $b = 1$, (b) $a = 0.95$, $b = 1$, and (c) $a = 0$, $b = 1$.

59. The motion of a freely spinning object in space is given by

$$\frac{dx}{dt} = yz,$$

$$\frac{dy}{dt} = -2xz,$$

$$\frac{dz}{dt} = xy,$$

where x, y, z represent the angular velocities about the three principal axes. Show that $x^2 + y^2 + z^2$ is a constant. Find the critical points and analyze their linear stability. Check by throwing a nonspherical object (e.g., this book held together with a rubber band) in the air.

60. A bead moves along a smooth circular wire of radius a that is rotating about a vertical axis with constant angular speed ω. Taking gravity and centrifugal forces into account, the motion of the bead is given by

$$a\frac{d^2\theta}{dt^2} = -g\sin\theta + a\omega^2\cos\theta\sin\theta,$$

where θ is the angular position of the bead with respect to the downward vertical position. Find the equilibrium positions and their stability as the parameter $\mu = a\omega^2/g$ is varied.

61. Change the Orr[19]-Sommerfeld[20] equation

$$\frac{\mu}{i\alpha\rho}\left(\frac{d^2}{dx^2} - \alpha^2\right)^2\varphi = (U - c)\left(\frac{d^2}{dx^2} - \alpha^2\right)\varphi - \frac{d^2U}{dx^2}\varphi,$$

which governs two-dimensional instability in parallel flow, to four first-order equations; $\varphi(x)$ and x are the dependent and independent variables, respectively, $U = U(x)$, $i = \sqrt{-1}$, and the rest of the symbols are constant parameters.

62. Calculate numerical solutions of the van der Pol equation

$$\ddot{y} - a(1 - y^2)\dot{y} + k^2y = 0.$$

With t as time, plot (a) the time-dependent behavior y versus t and (b) path in the phase plane $(y, \dot{y})$. Choose small, medium, and large values of a and comment on the effect it has on the solution.

---

[19] William McFadden Orr, 1866–1934, Irish-British mathematician.
[20] Arnold Johannes Wilhelm Sommerfeld, 1868–1951, German physicist.

63. The position (r, ϕ) of a body moving under the gravitational field of a fixed body at the origin is governed by

$$\ddot{r} - r\dot{\phi}^2 = -\frac{k}{r^2},$$

$$2\dot{r}\dot{\phi} + r\ddot{\phi} = 0,$$

where k is a constant. From this show that the radius vector from the fixed to the moving body sweeps out equal areas in equal times (Kepler's[21] second law).

---

[21] Johannes Kepler, 1571–1630, German astronomer.

APPENDIX A

Throughout the book there are instances when the reader may have to consult a favorite undergraduate text. To mitigate this need, the following topics have been assembled and summarized.

A.1 Roots of Polynomial Equations

Here we discuss how to obtain roots of polynomial equations of various orders.

A.1.1 First-Order

The solution to the trivial first-order polynomial

$$x + a_0 = 0, \tag{A.1}$$

where $a_0 \in \mathbb{C}^1$, is by inspection

$$x = -a_0. \tag{A.2}$$

A.1.2 Quadratic

The well-known solutions to the quadratic equation

$$x^2 + a_1 x + a_0 = 0, \tag{A.3}$$

where $a_0, a_1 \in \mathbb{C}^1$, are

$$x = \frac{-a_1 \pm \sqrt{a_1^2 - 4a_0}}{2}. \tag{A.4}$$

EXAMPLE A.1

Find all roots to

$$x^2 + (1 + 3i)x + (-2 + 2i) = 0. \tag{A.5}$$

Here we have $a_0 = -2 + 2i$, $a_1 = 1 + 3i$. Applying Eq. (A.4), we find

$$x = \frac{-1 - 3i \pm \sqrt{(1 + 3i)^2 - 4(-2 + 2i)}}{2}, \tag{A.6}$$

$$= \frac{-1 - 3i \pm \sqrt{-8 + 6i + 8 - 8i}}{2}, \tag{A.7}$$

$$= \frac{-1 - 3i \pm \sqrt{-2i}}{2}, \tag{A.8}$$

$$= \frac{-1 - 3i \pm \sqrt{(1-i)^2}}{2}, \tag{A.9}$$

$$= \frac{-1 - 3i \pm (1-i)}{2}. \tag{A.10}$$

Thus, the two roots are

$$x_1 = -2i, \tag{A.11}$$

$$x_2 = -1 - i. \tag{A.12}$$

Because the coefficients of the original polynomial are complex, it is not required for the complex solutions to be a complex conjugate pair.

A.1.3 Cubic

The solution for cubic equations is not as straightforward, and one will find many approaches in the literature. Here we present a robust algorithm. Consider

$$x^3 + a_2 x^2 + a_1 x + a_0 = 0, \tag{A.13}$$

where we take $a_0, a_1, a_2 \in \mathbb{C}^1$. Let

$$p = a_1 - \frac{a_2^2}{3}, \tag{A.14}$$

$$q = a_0 - \frac{a_1 a_2}{3} + \frac{2a_2^3}{27}. \tag{A.15}$$

Then find three roots, w_1, w_2, w_3, to the following equation:

$$w^3 = -\frac{1}{2}\left(q + \sqrt{q^2 + \frac{4p^3}{27}}\right). \tag{A.16}$$

The right-hand side of Eq. (A.16) is in general a complex constant and can be cast as

$$w^3 = -\frac{1}{2}\left(q + \sqrt{q^2 + \frac{4p^3}{27}}\right) = re^{i\theta}, \tag{A.17}$$

where r and θ are constants determined by standard complex algebra (see Section A.8). Because of periodicity, this is equivalent to

$$w^3 = re^{i(\theta + 2n\pi)}, \qquad n = 0, 1, 2, \ldots. \tag{A.18}$$

Considering only the $n = 0, 1, 2$, cases, as for higher n no new roots are found, we solve to get

$$w_1 = \sqrt[3]{r}\, e^{i\theta/3}, \tag{A.19}$$

$$w_2 = \sqrt[3]{r}\, e^{i(\theta + 2\pi)/3}, \tag{A.20}$$

$$w_3 = \sqrt[3]{r}\, e^{i(\theta + 4\pi)/3}. \tag{A.21}$$

These can easily be recast into Cartesian form. The three roots of the original cubic polynomial are then given by

$$x_1 = w_1 - \frac{p}{3w_1} - \frac{a_2}{3}, \tag{A.22}$$

$$x_2 = w_2 - \frac{p}{3w_2} - \frac{a_2}{3}, \tag{A.23}$$

$$x_3 = w_3 - \frac{p}{3w_3} - \frac{a_2}{3}. \tag{A.24}$$

$$\tag{A.25}$$

EXAMPLE A.2

Find all roots to

$$x^3 + x^2 + x + 1 = 0. \tag{A.26}$$

Here we have $a_0 = a_1 = a_2 = 1$. We find

$$p = 1 - \frac{1^2}{3} = \frac{2}{3}, \tag{A.27}$$

$$q = 1 - \frac{(1)(1)}{3} + \frac{(2)(1)^3}{27} = \frac{20}{27}. \tag{A.28}$$

We then find

$$w^3 = \frac{2}{27} \left(-5 + 3\sqrt{3} \right) = \frac{2}{27} \left(-5 + 3\sqrt{3} \right) e^{i(0+2n\pi)}. \tag{A.29}$$

Noting that $(2(-5 + 3\sqrt{3})/27)^{1/3} = -1/3 + 1/\sqrt{3}$, we find the three roots (one real, two complex) to be, for $n = 0, 1, 2$,

$$w_1 = \left(-\frac{1}{3} + \frac{1}{\sqrt{3}} \right) e^{i0} = \left(-\frac{1}{3} + \frac{1}{\sqrt{3}} \right), \tag{A.30}$$

$$w_2 = \left(-\frac{1}{3} + \frac{1}{\sqrt{3}} \right) e^{i2\pi/3} = \left(-\frac{1}{3} + \frac{1}{\sqrt{3}} \right) \left(-\frac{1}{2} + i\frac{\sqrt{3}}{2} \right), \tag{A.31}$$

$$w_3 = \left(-\frac{1}{3} + \frac{1}{\sqrt{3}} \right) e^{i4\pi/3} = \left(-\frac{1}{3} + \frac{1}{\sqrt{3}} \right) \left(-\frac{1}{2} - i\frac{\sqrt{3}}{2} \right). \tag{A.32}$$

Substituting into $x_i = w_i - p/3/w_i - a_2/3$, $i = 1, 2, 3$, gives the final simple result

$$x_1 = -1, \tag{A.33}$$

$$x_2 = i, \tag{A.34}$$

$$x_3 = -i. \tag{A.35}$$

The solutions can be easily verified by direct substitution into the original cubic equation.

A.1.4 Quartic

We give an algorithm to solve the quartic

$$x^4 + a_3 x^3 + a_2 x^2 + a_1 x + a_0 = 0, \tag{A.36}$$

where $a_0, a_1, a_2, a_3 \in \mathbb{C}^1$.

First find the three roots, y_1, y_2, y_3, of the cubic

$$y^3 - a_2 y^2 + (a_1 a_3 - 4a_0)y - (a_1^2 + a_0 a_3^2 - 4a_0 a_2) = 0. \qquad (A.37)$$

At least one of these roots, y_i, $i = 1, 2, 3$, when substituted into the two quadratic equations

$$x^2 + \left(\frac{a_3}{2} \mp \left(\frac{a_3^2}{4} + y_i - a_2 \right)^{1/2} \right) x + \frac{y_i}{2} \mp \left(\left(\frac{y_i}{2} \right)^2 - a_0 \right)^{1/2} = 0, \qquad (A.38)$$

will yield the appropriate four solutions, x_1, x_2, x_3, x_4, which solve the original quartic. One can simply try each value of y_i, compute the roots for x, and test by substitution into the original equation whether that value of y_i yielded the correct roots for x.

EXAMPLE A.3

Find all roots to

$$x^4 + x^3 + x^2 + x + 1 = 0. \qquad (A.39)$$

Here we have $a_0 = a_1 = a_2 = a_3 = 1$. We first substitute into Eq. (A.37) to get the associated cubic equation

$$y^3 - y^2 - 3y + 2 = 0. \qquad (A.40)$$

We use the procedure of the previous section to find the three roots of Eq. (A.40) to be

$$y_1 = 2, \qquad (A.41)$$

$$y_2 = -\frac{1}{2}(1 + \sqrt{5}), \qquad (A.42)$$

$$y_3 = -\frac{1}{2}(1 - \sqrt{5}). \qquad (A.43)$$

In this particular case, substitution of either y_1, y_2, or y_3 into Eq. (A.38) yields a viable solution. For simplicity, we substitute for y_1, yielding the two quadratic equations

$$x^2 + \left(\frac{1}{2} \pm \frac{\sqrt{5}}{2} \right) x + 1 = 0. \qquad (A.44)$$

These can be solved easily to obtain the four roots

$$x_1 = -(-1)^{3/5} = \frac{-1 + \sqrt{5}}{4} - i\sqrt{\frac{5 + \sqrt{5}}{8}}, \qquad (A.45)$$

$$x_2 = (-1)^{2/5} = \frac{-1 + \sqrt{5}}{4} + i\sqrt{\frac{5 + \sqrt{5}}{8}}, \qquad (A.46)$$

$$x_3 = -(-1)^{1/5} = -\frac{1 + \sqrt{5}}{4} - i\sqrt{\frac{5 - \sqrt{5}}{8}}, \qquad (A.47)$$

$$x_4 = (-1)^{4/5} = -\frac{1 + \sqrt{5}}{4} + i\sqrt{\frac{5 - \sqrt{5}}{8}}. \qquad (A.48)$$

The solutions can be easily verified by direct substitution into the original quartic equation.

A.1.5 Quintic and Higher

The Abel-Ruffini[1] theorem shows that there is no general algebraic formula for the solutions to quintic or higher order polynomials. However, many numerical algorithms exist for finding roots to polynomials of all finite order.

A.2 Cramer's Rule

A set of linear algebraic equations can be written as

$$\mathbf{A} \cdot \mathbf{x} = \mathbf{b}, \tag{A.49}$$

where the matrix $\mathbf{A} \in \mathbb{R}^N \times \mathbb{R}^N$ and vectors $\mathbf{x}, \mathbf{b} \in \mathbb{R}^N$. The solution is

$$x_i = \frac{D_i}{D}, \qquad i = 1, \ldots, N. \tag{A.50}$$

where D is the determinant of the system, that is, $D = |\mathbf{A}|$, and D_i is the determinant of the same matrix but with the ith column replaced by $\mathbf{b}$. The rule is useful in practice for a relatively small N. Notice also that the method works as long as $D \neq 0$; if $D = 0$, the set of equations is linearly dependent, and an infinite number of solutions is possible.

EXAMPLE A.4

Find the solution of

$$\begin{pmatrix} 1 & 2 & 3 \\ 1 & 2 & 1 \\ 3 & 2 & 1 \end{pmatrix} \begin{pmatrix} x_1 \\ x_2 \\ x_3 \end{pmatrix} = \begin{pmatrix} 14 \\ 8 \\ 10 \end{pmatrix}, \tag{A.51}$$

using Cramer's rule.

Because

$$D = \begin{vmatrix} 1 & 2 & 3 \\ 1 & 2 & 1 \\ 3 & 2 & 1 \end{vmatrix} = -8, \tag{A.52}$$

$$D_1 = \begin{vmatrix} 14 & 2 & 3 \\ 8 & 2 & 1 \\ 10 & 2 & 1 \end{vmatrix} = -8, \tag{A.53}$$

$$D_2 = \begin{vmatrix} 1 & 14 & 3 \\ 1 & 8 & 1 \\ 3 & 10 & 1 \end{vmatrix} = -16, \tag{A.54}$$

$$D_3 = \begin{vmatrix} 1 & 2 & 14 \\ 1 & 2 & 8 \\ 3 & 2 & 10 \end{vmatrix} = -24, \tag{A.55}$$

the solution is

$$x_1 = \frac{D_1}{D} = \frac{-8}{-8} = 1, \tag{A.56}$$

$$x_2 = \frac{D_2}{D} = \frac{-16}{-8} = 2, \tag{A.57}$$

$$x_3 = \frac{D_3}{D} = \frac{-24}{-8} = 3. \tag{A.58}$$

---

[1] Paolo Ruffini, 1765–1822, Italian mathematician.

A.3 Gaussian Elimination

Suppose that we have to find $\mathbf{x}$ such that

$$\begin{pmatrix} a_{11} & a_{12} & a_{13} \\ a_{21} & a_{22} & a_{23} \\ a_{31} & a_{32} & a_{33} \end{pmatrix} \begin{pmatrix} x_1 \\ x_2 \\ x_3 \end{pmatrix} = \begin{pmatrix} b_1 \\ b_2 \\ b_3 \end{pmatrix}, \tag{A.59}$$

where the matrix is nonsingular. This is of the form

$$\mathbf{A} \cdot \mathbf{x} = \mathbf{b}, \tag{A.60}$$

and the so-called augmented matrix is

$$\mathbf{B} = \begin{pmatrix} a_{11} & a_{12} & a_{13} & b_1 \\ a_{21} & a_{22} & a_{23} & b_2 \\ a_{31} & a_{32} & a_{33} & b_3 \end{pmatrix}, \tag{A.61}$$

which has the elements of both $\mathbf{A}$ and $\mathbf{b}$.

Multiply the first row by $\alpha_{21} = a_{21}/a_{11}$ (assuming that $a_{11} \neq 0$), and subtract the second row to eliminate a_{21}. This gives us

$$\mathbf{B}' = \begin{pmatrix} a'_{11} & a'_{12} & a'_{13} & b'_1 \\ 0 & a'_{22} & a'_{23} & b'_2 \\ a'_{31} & a'_{32} & a'_{33} & b'_3 \end{pmatrix}, \tag{A.62}$$

where

$$a'_{22} = \frac{a_{21}a_{12}}{a_{11}} - a_{22}, \tag{A.63}$$

$$a'_{23} = \frac{a_{21}a_{13}}{a_{11}} - a_{23}, \tag{A.64}$$

$$b'_2 = \frac{a_{21}b_1}{a_{11}} - b_2, \tag{A.65}$$

and the other elements are the same. Next, multiply the first row by $\alpha_{31} = a'_{31}/a'_{11}$ (with $a'_{11} \neq 0$) and subtract the third row to eliminate a'_{31}. After this, we will have

$$\mathbf{B}'' = \begin{pmatrix} a''_{11} & a''_{12} & a''_{13} & b''_1 \\ 0 & a''_{22} & a''_{23} & b''_2 \\ 0 & a''_{32} & a''_{33} & b''_3 \end{pmatrix}, \tag{A.66}$$

where

$$a''_{32} = \frac{a'_{31}a'_{12}}{a'_{11}} - a'_{32}, \tag{A.67}$$

$$a''_{33} = \frac{a'_{31}a'_{13}}{a'_{11}} - a'_{33}, \tag{A.68}$$

$$b''_3 = \frac{a'_{31}b'_1}{a'_{11}} - b'_3, \tag{A.69}$$

and all the other elements remain the same. Operate on the last two rows in a similar manner using $\alpha_{32} = a''_{32}/a''_{22}$ (for $a''_{22} \neq 0$) to get

$$
\mathbf{B}''' = \begin{pmatrix} a'''_{11} & a'''_{12} & a'''_{13} & b'''_{1} \\ 0 & a'''_{22} & a_{21}a'''_{23} & b'''_{2} \\ 0 & 0 & a'''_{33} & b'''_{3} \end{pmatrix}, \tag{A.70}
$$

where

$$
a'''_{33} = \frac{a''_{32}a''_{23}}{a''_{22}} - a''_{33}, \tag{A.71}
$$

$$
b'''_{3} = \frac{a''_{33}b''_{2}}{a''_{22}} - b''_{3}, \tag{A.72}
$$

with the other elements remaining unchanged.

At this stage the Gaussian elimination procedure enables the values of x_1, x_2, x_3 to be determined by back substitution. Thus, the last row gives $x_3 = b'''_3/a'''_{33}$. This, substituted into the second row, gives x_2, and, when both are substituted into the first row, gives x_1. Gaussian elimination is closely related to the $\mathbf{L} \cdot \mathbf{D} \cdot \mathbf{U}$ decomposition studied in Section 7.9.1.

A.4 Trapezoidal Rule

The integral

$$
I = \int_a^b f(x)\, dx \tag{A.73}
$$

can be approximated by first selecting a set of N points, $x_1, x_2, \ldots, x_N$, not necessarily uniformly spaced, with $x_1 < x_2 <, \ldots, < x_N$, and $x_1 = a$ and $x_N = b$, second evaluating f at each of the points, taking $f_i = f(x_i)$, and then forming the sum

$$
I \approx \frac{1}{2} \sum_{i=2}^{N} (x_i - x_{i-1})(f_i + f_{i-1}). \tag{A.74}
$$

A.5 Trigonometric Relations

The following are some standard relations from trigonometry:

$$
\sin x \sin y = \frac{1}{2}\cos(x - y) - \frac{1}{2}\cos(x + y), \tag{A.75}
$$

$$
\sin x \cos y = \frac{1}{2}\sin(x + y) + \frac{1}{2}\sin(x - y), \tag{A.76}
$$

$$
\cos x \cos y = \frac{1}{2}\cos(x - y) + \frac{1}{2}\cos(x + y), \tag{A.77}
$$

$$
\sin^2 x = \frac{1}{2} - \frac{1}{2}\cos 2x, \tag{A.78}
$$

$$
\sin x \cos x = \frac{1}{2}\sin 2x, \tag{A.79}
$$

$$
\cos^2 x = \frac{1}{2} + \frac{1}{2}\cos 2x, \tag{A.80}
$$

$$\sin^3 x = \frac{3}{4} \sin x - \frac{1}{4} \sin 3x, \tag{A.81}$$

$$\sin^2 x \cos x = \frac{1}{4} \cos x - \frac{1}{4} \cos 3x, \tag{A.82}$$

$$\sin x \cos^2 x = \frac{1}{4} \sin x + \frac{1}{4} \sin 3x, \tag{A.83}$$

$$\cos^3 x = \frac{3}{4} \cos x + \frac{1}{4} \cos 3x, \tag{A.84}$$

$$\sin^4 x = \frac{3}{8} - \frac{1}{2} \cos 2x + \frac{1}{8} \cos 4x, \tag{A.85}$$

$$\sin^3 x \cos x = \frac{1}{4} \sin 2x - \frac{1}{8} \sin 4x, \tag{A.86}$$

$$\sin^2 x \cos^2 x = \frac{1}{8} - \frac{1}{8} \cos 4x, \tag{A.87}$$

$$\sin x \cos^3 x = \frac{1}{4} \sin 2x + \frac{1}{8} \sin 4x, \tag{A.88}$$

$$\cos^4 x = \frac{3}{8} + \frac{1}{2} \cos 2x + \frac{1}{8} \cos 4x, \tag{A.89}$$

$$\sin^5 x = \frac{5}{8} \sin x - \frac{5}{16} \sin 3x + \frac{1}{16} \sin 5x, \tag{A.90}$$

$$\sin^4 x \cos x = \frac{1}{8} \cos x - \frac{3}{16} \cos 3x + \frac{1}{16} \cos 5x, \tag{A.91}$$

$$\sin^3 x \cos^2 x = \frac{1}{8} \sin x + \frac{1}{16} \sin 3x - \frac{1}{16} \sin 5x, \tag{A.92}$$

$$\sin^2 x \cos^3 x = -\frac{1}{8} \cos x - \frac{1}{16} \cos 3x - \frac{1}{16} \cos 5x, \tag{A.93}$$

$$\sin x \cos^4 x = \frac{1}{8} \sin x + \frac{3}{16} \sin 3x + \frac{1}{16} \sin 5x, \tag{A.94}$$

$$\cos^5 x = \frac{5}{8} \cos x + \frac{5}{16} \cos 3x + \frac{1}{16} \cos 5x, \tag{A.95}$$

$$a \sin kx + b \cos kx = \sqrt{a^2 + b^2} \sin \left(kx + \mathrm{Tan}^{-1}(b, a) \right). \tag{A.96}$$

In Eq. (A.96), we have employed a version of the inverse tangent function that is more useful in calculation. It is sometimes known as atan2. It is defined as follows:

$$\mathrm{Tan}^{-1}(b, a) = \begin{cases} \tan^{-1}\left(\frac{b}{a}\right), & a > 0, \\ \tan^{-1}\left(\frac{b}{a}\right) + \pi, & b \geq 0, \ a < 0, \\ \tan^{-1}\left(\frac{b}{a}\right) - \pi, & b < 0, \ a < 0, \\ +\frac{\pi}{2}, & b > 0, \ a = 0, \\ -\frac{\pi}{2}, & b < 0, \ a = 0, \\ \text{undefined}, & b = 0, \ a = 0. \end{cases} \tag{A.97}$$

If $a > 0$, Eq. (A.96) reduces to

$$a \sin kx + b \cos kx = \sqrt{a^2 + b^2} \sin \left(kx + \tan^{-1}\left(\frac{b}{a}\right) \right), \qquad a > 0. \tag{A.98}$$

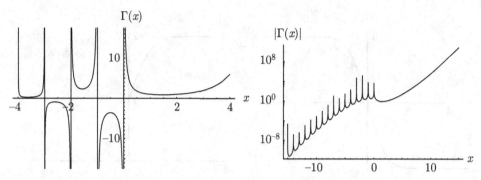

Figure A.1. Gamma function and amplitude of gamma function.

One can also obtain a single formula for Tan^{-1}, which is

$$\mathrm{Tan}^{-1}(b, a) = 2 \tan^{-1} \frac{b}{a + \sqrt{a^2 + b^2}}. \tag{A.99}$$

A.6 Hyperbolic Functions

The hyperbolic functions are defined as follows:

$$\sinh \theta = \frac{e^\theta - e^{-\theta}}{2}, \tag{A.100}$$

$$\cosh \theta = \frac{e^\theta + e^{-\theta}}{2}. \tag{A.101}$$

A.7 Special Functions

We consider a set of special functions that may be encountered in engineering.

A.7.1 Gamma

The gamma function may be thought of as an extension to the factorial function. Recall that the factorial function requires an integer argument. The gamma function admits real arguments; when the argument of the gamma function is an integer, one finds it is directly related to the factorial function. The gamma function is defined by

$$\Gamma(x) = \int_0^\infty e^{-t} t^{x-1} \, dt. \tag{A.102}$$

Generally, we are interested in $x > 0$, but results are available for all x. Some properties are as follows:

1. $\Gamma(1) = 1$
2. $\Gamma(x) = (x - 1)\Gamma(x - 1), x > 1$
3. $\Gamma(x) = (x - 1)(x - 2)\cdots(x - r)\Gamma(x - r), x > r$
4. $\Gamma(n) = (n - 1)!$, where n is a positive integer
5. $\Gamma(x) \sim \sqrt{\frac{2\pi}{x}} x^x e^{-x} \left(1 + \frac{1}{12x} + \frac{1}{288x^2} + \cdots\right)$ (Stirling's approximation)

It can be shown that Stirling's approximation is a divergent series. It is an asymptotic series, but as more terms are added, the solution can actually get worse. A remedy exists, attributed to Bayes.[2] The gamma function and its amplitude are plotted in Figure A.1.

---

[2] Thomas Bayes, 1701–1761, English mathematician and cleric.

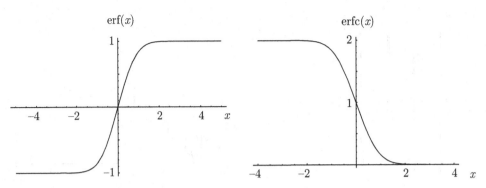

Figure A.2. Error function and error function complement.

A.7.2 Error

The error function is defined by

$$\operatorname{erf}(x) = \frac{2}{\sqrt{\pi}} \int_0^x e^{-\xi^2} \, d\xi, \tag{A.103}$$

and the complementary error function by

$$\operatorname{erfc}(x) = 1 - \operatorname{erf} x. \tag{A.104}$$

The error function and the complementary error function are plotted in Figure A.2. The imaginary error function is defined by

$$\operatorname{erfi}(z) = -i \operatorname{erf}(iz), \tag{A.105}$$

where $z \in \mathbb{C}^1$. For real arguments, $x \in \mathbb{R}^1$, it can be shown that $\operatorname{erfi}(x) = -i \operatorname{erf}(ix) \in \mathbb{R}^1$. The imaginary error function is plotted in Figure A.3 for a real argument, $x \in \mathbb{R}^1$.

A.7.3 Sine, Cosine, and Exponential Integral

The sine integral function is defined by

$$\operatorname{Si}(x) = \int_0^x \frac{\sin \xi}{\xi} \, d\xi, \tag{A.106}$$

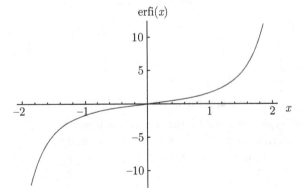

Figure A.3. Imaginary error function, $\operatorname{erfi}(x)$, for real argument, $x \in \mathbb{R}^1$.

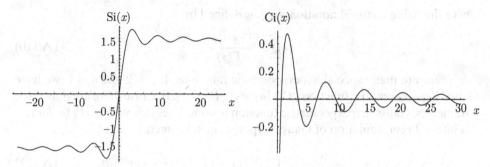

Figure A.4. Sine integral function, $\mathrm{Si}(x)$, and cosine integral function $\mathrm{Ci}(x)$.

and the cosine integral function by

$$\mathrm{Ci}(x) = -\int_x^\infty \frac{\cos\xi}{\xi}\,d\xi. \tag{A.107}$$

The sine integral function is real-valued for $x \in (-\infty, \infty)$. The cosine integral function is real-valued for $x \in [0, \infty)$. We also have $\lim_{x\to 0_+} \mathrm{Ci}(x) \to -\infty$. The cosine integral takes on a value of zero at discrete positive real values and has an amplitude that slowly decays as $x \to \infty$. The sine integral and cosine integral functions are plotted in Figure A.4.

The exponential integral function is defined by

$$\mathrm{Ei}(x) = -\fint_{-x}^\infty \frac{e^{-\xi}}{\xi}\,d\xi = \int_{-\infty}^x \frac{e^\xi}{\xi}\,d\xi. \tag{A.108}$$

The exponential integral function is plotted in Figure A.5. We must use the Cauchy principal value of the integral if $x > 0$.

A.7.4 Hypergeometric

A generalized *hypergeometric function* is defined by

$$_pF_q\left(\{a_1, \ldots, a_p\}, \{b_1, \ldots, b_q\}; x\right) = \sum_{k=1}^\infty \frac{(a_1)_k (a_2)_k \ldots (a_p)_k}{(b_1)_k (b_2)_k \ldots (b_q)_k} \frac{x^k}{k!}, \tag{A.109}$$

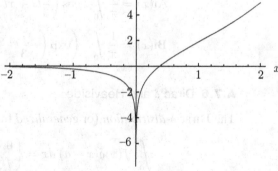

Figure A.5. Exponential integral function, $\mathrm{Ei}(x)$.

where the rising factorial notation, $(s)_k$, is defined by

$$(s)_k \equiv \frac{\Gamma(s+k)}{\Gamma(s)}. \tag{A.110}$$

There are many special hypergeometric functions. If $p = 2$ and $q = 1$, we have Gauss' hypergeometric function $_2F_1(\{a_1, a_2\}, \{b_1\}; x)$. Because there are only three parameters, Gauss' hypergeometric function is sometimes denoted as $_2F_1(a, b, c, x)$. An integral representation of Gauss' hypergeometric function is

$$_2F_1(a, b, c, x) = \frac{\Gamma(c)}{\Gamma(b)\Gamma(c-b)} \int_0^1 t^{b-1}(1-t)^{c-b-1}(1-tx)^{-a} \, dt. \tag{A.111}$$

For special values of parameters, hypergeometric functions can reduce to other functions such as $\tanh^{-1}$.

A.7.5 Airy

The Airy functions $\text{Ai}(x)$ and $\text{Bi}(x)$ are most compactly defined as the two linearly independent solutions to the second-order differential equation known as the Airy equation,

$$\frac{d^2y}{dx^2} - xy = 0, \tag{A.112}$$

yielding solutions of the form

$$y = C_1\text{Ai}(x) + C_2\text{Bi}(x). \tag{A.113}$$

They can be expressed in a variety of other forms. In terms of the so-called confluent hypergeometric limit function $_0F_1$, we have

$$\text{Ai}(x) = \frac{1}{3^{2/3}\Gamma\left(\frac{2}{3}\right)}{_0F_1}\left(\{\}; \left\{\frac{2}{3}\right\}; \frac{1}{9}x^3\right) - \frac{x}{3^{1/3}\Gamma\left(\frac{1}{3}\right)}{_0F_1}\left(\{\}; \left\{\frac{4}{3}\right\}; \frac{1}{9}x^3\right), \tag{A.114}$$

$$\text{Bi}(x) = \frac{1}{3^{1/6}\Gamma\left(\frac{2}{3}\right)}{_0F_1}\left(\{\}; \left\{\frac{2}{3}\right\}; \frac{1}{9}x^3\right) - \frac{3^{1/6}x}{\Gamma\left(\frac{1}{3}\right)}{_0F_1}\left(\{\}; \left\{\frac{4}{3}\right\}; \frac{1}{9}x^3\right). \tag{A.115}$$

The Airy functions are plotted in Figure A.6.

In integral form, the Airy functions are, for $x \in \mathbb{R}^1$,

$$\text{Ai}(x) = \frac{1}{\pi} \int_0^\infty \cos\left(\frac{1}{3}t^3 + xt\right) dt, \tag{A.116}$$

$$\text{Bi}(x) = \frac{1}{\pi} \int_0^\infty \left(\exp\left(-\frac{1}{3}t^3 + xt\right) + \sin\left(\frac{1}{3}t^3 + xt\right)\right) dt. \tag{A.117}$$

A.7.6 Dirac δ and Heaviside

The Dirac δ-*distribution* (or *generalized* function, or simply function) is defined by

$$\int_\alpha^\beta f(x)\delta(x-a) \, dx = \begin{cases} 0, & a \notin [\alpha, \beta], \\ f(a), & a \in [\alpha, \beta]. \end{cases} \tag{A.118}$$

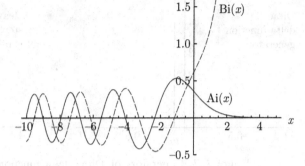

Figure A.6. Airy functions Ai(x) (solid)
and Bi(x) (dashed).

From this it follows that

$$\delta(x-a) = 0, \quad x \neq a, \tag{A.119}$$

$$\int_{-\infty}^{\infty} \delta(x-a)\, dx = 1. \tag{A.120}$$

The δ-distribution may be imagined in a limiting fashion as

$$\delta(x-a) = \lim_{\epsilon \to 0^+} \Delta_\epsilon(x-a), \tag{A.121}$$

where $\Delta_\epsilon(x-a)$ has one of the following forms:

1.

$$\Delta_\epsilon(x-a) = \begin{cases} 0, & x < a - \frac{\epsilon}{2}, \\ \frac{1}{\epsilon}, & a - \frac{\epsilon}{2} \leq x \leq a + \frac{\epsilon}{2}, \\ 0, & x > a + \frac{\epsilon}{2}, \end{cases} \tag{A.122}$$

2.

$$\Delta_\epsilon(x-a) = \frac{\epsilon}{\pi((x-a)^2 + \epsilon^2)}, \tag{A.123}$$

3.

$$\Delta_\epsilon(x-a) = \frac{1}{\sqrt{\pi\epsilon}} e^{-(x-a)^2/\epsilon}. \tag{A.124}$$

The derivative of the function

$$h_\epsilon(x-a) = \begin{cases} 0, & x < a - \frac{\epsilon}{2}, \\ \frac{1}{\epsilon}(x-a) + \frac{1}{2}, & a - \frac{\epsilon}{2} \leq x \leq a + \frac{\epsilon}{2}, \\ 1, & x > a + \frac{\epsilon}{2}, \end{cases} \tag{A.125}$$

is $\Delta_\epsilon(x-a)$ in Eq. (A.122). If we define the *Heaviside* function, $H(x-a)$, as

$$H(x-a) = \lim_{\epsilon \to 0^+} h_\epsilon(x-a), \tag{A.126}$$

then

$$\frac{d}{dx} H(x-a) = \delta(x-a). \tag{A.127}$$

The generator of the Dirac function $\Delta_\epsilon(x-a)$ and the generator of the Heaviside function $h_\epsilon(x-a)$ are plotted for $a = 0$ and $\epsilon = 1/5$ in Figure A.7. As $\epsilon \to 0$, Δ_ϵ has its width decrease and its height increase in such a fashion that its area remains

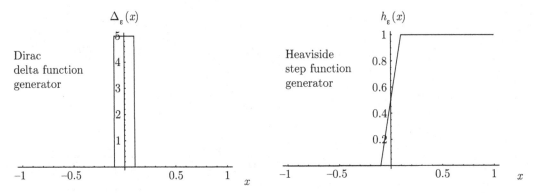

Figure A.7. Generators of Dirac delta function and Heaviside function, $\Delta_\epsilon(x - a)$ and $h_\epsilon(x - a)$, plotted for $a = 0$ and $\epsilon = 1/5$.

constant; simultaneously, h_ϵ has its slope steepen in the region where it jumps from zero to unity as $\epsilon \to 0$.

A.8 Complex Numbers

Here we briefly introduce some basic elements of complex number theory. Recall that the imaginary number i is defined such that

$$i^2 = -1, \qquad i = \sqrt{-1}. \tag{A.128}$$

A.8.1 Euler's Formula

We can get the useful *Euler's formula*, by considering the following Taylor series expansions of common functions about $t = 0$:

$$e^t = 1 + t + \frac{1}{2!}t^2 + \frac{1}{3!}t^3 + \frac{1}{4!}t^4 + \frac{1}{5!}t^5 \cdots, \tag{A.129}$$

$$\sin t = 0 + t + 0\frac{1}{2!}t^2 - \frac{1}{3!}t^3 + 0\frac{1}{4!}t^4 + \frac{1}{5!}t^5 \cdots, \tag{A.130}$$

$$\cos t = 1 + 0t - \frac{1}{2!}t^2 + 0\frac{1}{3!}t^3 + \frac{1}{4!}t^4 + 0\frac{1}{5!}t^5 \cdots. \tag{A.131}$$

With these expansions, now consider the combinations $(\cos t + i \sin t)|_{t=\theta}$ and $e^t|_{t=i\theta}$:

$$\cos \theta + i \sin \theta = 1 + i\theta - \frac{1}{2!}\theta^2 - i\frac{1}{3!}\theta^3 + \frac{1}{4!}\theta^4 + i\frac{1}{5!}\theta^5 + \cdots, \tag{A.132}$$

$$e^{i\theta} = 1 + i\theta + \frac{1}{2!}(i\theta)^2 + \frac{1}{3!}(i\theta)^3 + \frac{1}{4!}(i\theta)^4 + \frac{1}{5!}(i\theta)^5 + \cdots, \tag{A.133}$$

$$= 1 + i\theta - \frac{1}{2!}\theta^2 - i\frac{1}{3!}\theta^3 + \frac{1}{4!}\theta^4 + i\frac{1}{5!}\theta^5 + \cdots. \tag{A.134}$$

As the two series are identical, we have Euler's formula

$$e^{i\theta} = \cos \theta + i \sin \theta. \tag{A.135}$$

Powers of complex numbers can be easily obtained using *de Moivre's* formula:

$$e^{in\theta} = \cos n\theta + i \sin n\theta. \tag{A.136}$$

We also see that

$$e^{-in\theta} = \cos n\theta - i \sin n\theta. \tag{A.137}$$

Combining Eqs. (A.136) and (A.137), we solve for $\sin n\theta$ and $\cos n\theta$, obtaining

$$\sin n\theta = \frac{e^{in\theta} - e^{-in\theta}}{2i}, \tag{A.138}$$

$$\cos n\theta = \frac{e^{in\theta} + e^{-in\theta}}{2}. \tag{A.139}$$

A.8.2 Polar and Cartesian Representations

Now if we take x and y to be real numbers and define the complex number z to be

$$z = x + iy, \tag{A.140}$$

we can multiply and divide by $\sqrt{x^2 + y^2}$ to obtain

$$z = \sqrt{x^2 + y^2} \left(\frac{x}{\sqrt{x^2 + y^2}} + i \frac{y}{\sqrt{x^2 + y^2}} \right). \tag{A.141}$$

Noting the similarities between this and the transformation between Cartesian and polar coordinates suggests we adopt

$$r = \sqrt{x^2 + y^2}, \qquad \cos \theta = \frac{x}{\sqrt{x^2 + y^2}}, \qquad \sin \theta = \frac{y}{\sqrt{x^2 + y^2}}. \tag{A.142}$$

Thus, we have

$$z = r \left(\cos \theta + i \sin \theta \right), \tag{A.143}$$

$$= r e^{i\theta}. \tag{A.144}$$

The polar and Cartesian representations of a complex number z are shown in Figure A.8.

Now we can define the *complex conjugate* $\bar{z}$ as

$$\bar{z} = x - iy, \tag{A.145}$$

$$= \sqrt{x^2 + y^2} \left(\frac{x}{\sqrt{x^2 + y^2}} - i \frac{y}{\sqrt{x^2 + y^2}} \right), \tag{A.146}$$

$$= r \left(\cos \theta - i \sin \theta \right), \tag{A.147}$$

$$= r \left(\cos(-\theta) + i \sin(-\theta) \right), \tag{A.148}$$

$$= r e^{-i\theta}. \tag{A.149}$$

Note now that

$$z\bar{z} = (x + iy)(x - iy) = x^2 + y^2 = |z|^2, \tag{A.150}$$

$$= r e^{i\theta} r e^{-i\theta} = r^2 = |z|^2. \tag{A.151}$$

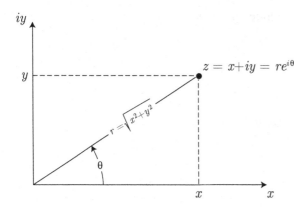

Figure A.8. Polar and Cartesian representations of a complex number z.

We also have

$$\sin \theta = \frac{e^{i\theta} - e^{-i\theta}}{2i}, \tag{A.152}$$

$$\cos \theta = \frac{e^{i\theta} + e^{-i\theta}}{2}. \tag{A.153}$$

EXERCISES

1. Expand each one of the following and then solve.
   (a) $(x - 1)(x - 2) = 0$.
   (b) $(x - 1)(x - 2)(x - 3) = 0$.
   (c) $(x - 1)(x - 2)(x - 3)(x - 4) = 0$.

2. Solve:
   (a) $x^3 - x^2 + x + 1 = 0$.
   (b) $x^4 - 7x^3 + 4x^2 + 12x = 0$.
   (c) $x^4 + 5x^2 + 4 = 0$.

3. For the system of equations

   $$x + y + z = 6,$$
   $$x + 2y + 2z = 11,$$
   $$x + y + 2z = 9,$$

   solve using
   (a) Cramer's rule
   (b) Gaussian elimination.

4. Determine $f(x) = x^2$ for $x = 0$, 0.25, 0.5, 0.75, and 1, and then calculate $\int_0^1 f(x)\, dx$ using the trapezoidal rule and a calculator.

5. Write a computer program to evaluate the integral in Problem 4 by dividing $[0, 1]$ into n parts. Plot the error as a function of n.

6. Using complex numbers, show that
   (a) $\cos^3 x = \frac{1}{4}(\cos 3x + 3 \cos x)$,
   (b) $\sin^3 x = \frac{1}{4}(3 \sin x - \sin 3x)$.

7. Show that

$$\sinh(ix) = i \sin x$$

$$\cosh(ix) = \cos x.$$

8. Determine $\lim_{n\to\infty} z_n$ for $z_n = 3/n + ((n+1)/(n+2))i$.

9. The error function is defined as erf $(x) = (2/\sqrt{\pi}) \int_0^x e^{-u^2} \, du$. Determine its derivative with respect to x.

10. Show that

$$\lim_{\epsilon\to0} \int_\alpha^\beta f(x)\Delta_\epsilon(x-a) \, dx = f(a)$$

for $a \in (\alpha, \beta)$, where Δ_ϵ is defined in Eq. (A.122)

11. Determine

(a) $\sqrt[4]{i}$,

(b) $i^i \sqrt[i]{i}$.

12. Find all complex numbers $z = x + iy$ such that $|z + 2i| = |1 + i|$.

13. Find the three cube roots of unity.

14. Find all the fourth roots of unity.

Bibliography

R. Abraham, J. E. Marsden, and T. Ratiu, *Manifolds, Tensor Analysis, and Applications*, Springer, New York, 1988.

M. Abramowitz and I. A. Stegun, eds., *Handbook of Mathematical Functions*, Dover, New York, 1964.

A. A. Andronov, *Qualitative Theory of Second Order Dynamical Systems*, John Wiley, New York, 1973.

P. J. Antsaklis and A. N. Michel, *Linear Systems*, Birkhäuser, Boston, 1997.

P. J. Antsaklis and A. N. Michel, *A Linear Systems Primer*, Birkhäuser, Boston, 2007.

T. M. Apostol, *Calculus: One-Variable Calculus, with an Introduction to Linear Algebra*, Vol. 1, 2nd ed., John Wiley, New York, 1991a.

T. M. Apostol, *Calculus: Multi-Variable Calculus and Linear Algebra with Applications to Differential Equations and Probability*, Vol. 2, 2nd ed., John Wiley, New York, 1991b.

G. B. Arfken, H. J. Weber, and F. E. Harris, *Mathematical Methods for Physicists*, 7th ed., Academic Press, Waltham, MA, 2012.

R. Aris, *Vectors, Tensors, and the Basic Equations of Fluid Mechanics*, Dover, New York, 1962.

V. I. Arnold, *Ordinary Differential Equations*, MIT Press, Cambridge, MA, 1973.

V. I. Arnold, *Geometrical Methods in the Theory of Ordinary Differential Equations*, Springer, New York, 1983.

D. Arrowsmith and C. M. Place, *Dynamical Systems: Differential Equations, Maps, and Chaotic Behaviour*, Chapman Hall/CRC, Boca Raton, FL, 1992.

N. H. Asmar, *Applied Complex Analysis with Partial Differential Equations*, Prentice Hall, Upper Saddle River, NJ, 2002.

G. I. Barenblatt, *Scaling, Self-Similarity, and Intermediate Asymptotics*, Cambridge University Press, Cambridge, UK, 1996.

R. Bellman and K. L. Cooke, *Differential-Difference Equations*, Academic Press, New York, 1963.

C. M. Bender and S. A. Orszag, *Advanced Mathematical Methods for Scientists and Engineers*, Springer, New York, 1999.

M. L. Boas, *Mathematical Methods in the Physical Sciences*, 3rd ed., John Wiley, New York, 2005.

A. I. Borisenko and I. E. Tarapov, *Vector and Tensor Analysis with Applications*, Dover, New York, 1968.

W. E. Boyce and R. C. DiPrima, *Elementary Differential Equations and Boundary Value Problems*, 10th Edition, John Wiley, New York, 2012.

M. Braun, *Differential Equations and Their Applications*, Springer, New York, 1983.

K. E. Brenan, S. L. Campbell, and L. R. Petzold, *Numerical Solution of Initial-Value Problems in Differential-Algebraic Equations*, SIAM, Philadelphia, 1996.

I. N. Bronshtein and K. A. Semendyayev, *Handbook of Mathematics*, Springer, Berlin, 1998.

C. Canuto, M. Y. Hussaini, A. Quarteroni, and T. A. Zang, *Spectral Methods in Fluid Dynamics*, Springer, New York, 1988.

J. Carr, *Applications of Centre Manifold Theory*, Springer, New York, 1981.

G. F. Carrier and C. E. Pearson, *Ordinary Differential Equations*, SIAM, Philadelphia, 1991.

R. V. Churchill, *Fourier Series and Boundary Value Problems*, McGraw-Hill, New York, 1941.

P. G. Ciarlet, *Introduction to Numerical Linear Algebra and Optimisation*, Cambridge University Press, Cambridge, UK, 1989.

T. B. Co, *Methods of Applied Mathematics for Engineers and Scientists*, Cambridge University Press, Cambridge, UK, 2013.

J. A. Cochran, H. C. Wiser, and B. J. Rice, *Advanced Engineering Mathematics*, 2nd ed., Brooks/Cole, Monterey, CA, 1987.

E. A. Coddington, *An Introduction to Ordinary Differential Equations*, Dover, New York, 1989.

E. A. Coddington and N. Levinson, *Theory of Ordinary Differential Equations*, Krieger, Malabar, FL, 1987.

R. Courant, *Differential and Integral Calculus*, 2 vols., John Wiley, New York, 1988.

R. Courant and D. Hilbert, *Methods of Mathematical Physics*, 2 vols., John Wiley, New York, 1989.

I. Daubechies, *Ten Lectures on Wavelets*, SIAM, Philadelphia, 1992.

L. Debnath and P. Mikusinski, *Introduction to Hilbert Spaces with Applications*, 3rd ed., Elsevier, Amsterdam, 2005.

J. W. Dettman, *Mathematical Methods in Physics and Engineering*, McGraw-Hill, New York, 1962.

P. G. Drazin, *Nonlinear Systems*, Cambridge University Press, Cambridge, UK, 1992.

R. D. Driver, *Ordinary and Delay Differential Equations*, Springer, New York, 1977.

J. Feder, *Fractals*, Plenum Press, New York, 1988.

B. A. Finlayson, *The Method of Weighted Residuals and Variational Principles*, Academic Press, New York, 1972.

C. A. J. Fletcher, *Computational Techniques for Fluid Dynamics*, 2nd ed., Springer, Berlin, 1991.

B. Fornberg, *A Practical Guide to Pseudospectral Methods*, Cambridge University Press, Cambridge, UK, 1998.

B. Friedman, *Principles and Techniques of Applied Mathematics*, Dover, New York, 1956.

I. M. Gelfand and S. V. Fomin, *Calculus of Variations*, Dover, New York, 2000.

J. Gleick, *Chaos*, Viking, New York, 1987.

G. H. Golub and C. F. Van Loan, *Matrix Computations*, 3rd ed., Johns Hopkins University Press, Baltimore, MD, 1996.

S. W. Goode, *An Introduction to Differential Equations and Linear Algebra*, Prentice-Hall, Englewood Cliffs, NJ, 1991.

B. Goodwine, *Engineering Differential Equations: Theory and Applications*, Springer, New York, 2011.

D. Gottlieb and S. A. Orszag, *Numerical Analysis of Spectral Methods: Theory and Applications*, SIAM, Philadelphia, 1977.

M. D. Greenberg, *Foundations of Applied Mathematics*, Prentice-Hall, Englewood Cliffs, NJ, 1978.

M. D. Greenberg, *Advanced Engineering Mathematics*, 2nd ed., Pearson, Upper Saddle River, NJ, 1998.

D. H. Griffel, *Applied Functional Analysis*, Dover, New York, 2002.

J. Guckenheimer and P. H. Holmes, *Nonlinear Oscillations, Dynamical Systems, and Bifurcations of Vector Fields*, Springer, New York, 2002.

J. Hale and H. Koçak, *Dynamics and Bifurcations*, Springer, New York, 1991.

M. T. Heath, *Scientific Computing*, 2nd ed., McGraw-Hill, Boston, 2002.

F. B. Hildebrand, *Advanced Calculus for Applications*, 2nd ed., Prentice-Hall, Englewood Cliffs, NJ, 1976.

E. J. Hinch, *Perturbation Methods*, Cambridge University Press, Cambridge, UK, 1991.

M. W. Hirsch and S. Smale, *Differential Equations, Dynamical Systems, and Linear Algebra*, Academic Press, Boston, 1974.

M. W. Hirsch, S. Smale, and R. L. Devaney, *Differential Equations, Dynamical Systems, and an Introduction to Chaos*, 3rd ed., Academic Press, Waltham, MA, 2013.

M. H. Holmes, *Introduction to Perturbation Methods*, Springer, New York, 1995.

M. H. Holmes, *Introduction to the Foundations of Applied Mathematics*, Springer, New York, 2009.

R. A. Howland, *Intermediate Dynamics: A Linear Algebraic Approach*, Springer, New York, 2006.

J. H. Hubbard and B. B. Hubbard, *Vector Calculus, Linear Algebra, and Differential Forms: A Unified Approach*, 4th ed., Matrix Editions, Ithaca, NY, 2009.

M. Humi and W. Miller, *Second Course in Ordinary Differential Equations for Scientists and Engineers*, Springer, New York, 1988.

A. Iserles, *A First Course in the Numerical Analysis of Differential Equations*, 2nd ed., Cambridge University Press, Cambridge, UK, 2009.

E. T. Jaynes, *Probability Theory: The Logic of Science*, Cambridge University Press, Cambridge, UK, 2003.

H. Jeffreys and B. Jeffreys, *Methods of Mathematical Physics*, 3rd ed., Cambridge University Press, Cambridge, UK, 1972.

D. W. Jordan and P. Smith, *Nonlinear Ordinary Differential Equations: An Introduction for Scientists and Engineers*, 4th ed., Oxford University Press, Oxford, UK, 2007.

P. B. Kahn, *Mathematical Methods for Engineers and Scientists*, Dover, New York, 2004.

W. Kaplan, *Advanced Calculus*, 5th ed., Addison-Wesley, Boston, 2003.

D. C. Kay, *Tensor Calculus*, Schaum's Outline Series, McGraw-Hill, New York, 1988.

J. Kevorkian and J. D. Cole, *Perturbation Methods in Applied Mathematics*, Springer, New York, 1981.

J. Kevorkian and J. D. Cole, *Multiple Scale and Singular Perturbation Methods*, Springer, New York, 1996.

A. N. Kolmogorov and S. V. Fomin, *Elements of the Theory of Functions and Functional Analysis*, Dover, New York, 1999.

L. D. Kovach, *Advanced Engineering Mathematics*, Addison-Wesley, Reading, MA, 1982.

E. Kreyszig, *Introductory Functional Analysis with Applications*, John Wiley, New York, 1978.

E. Kreyszig, *Advanced Engineering Mathematics*, 10th ed., John Wiley, New York, 2011.

C. Lanczos, *The Variational Principles of Mechanics*, 4th ed., Dover, New York, 2000.

P. D. Lax, *Functional Analysis*, John Wiley, New York, 2002.

P. D. Lax, *Linear Algebra and Its Applications*, 2nd ed., John Wiley, Hoboken, NJ, 2007.

J. R. Lee, *Advanced Calculus with Linear Analysis*, Academic Press, New York, 1972.

R. J. LeVeque, *Finite Volume Methods for Hyperbolic Problems*, Cambridge University Press, Cambridge, UK, 2002.

R. J. LeVeque, *Finite Difference Methods for Ordinary and Partial Differential Equations*, SIAM, Philadelphia, 2007.

A. J. Lichtenberg and M. A. Lieberman, *Regular and Chaotic Dynamics*, 2nd ed., Springer, Berlin, 1992.

C. C. Lin and L. A. Segel, *Mathematics Applied to Deterministic Problems in the Natural Sciences*, SIAM, Philadelphia, 1988.

J. D. Logan, *Applied Mathematics*, 4th ed., John Wiley, Hoboken, NJ, 2013.

R. J. Lopez, *Advanced Engineering Mathematics*, Addison Wesley Longman, Boston, 2001.

D. Lovelock and H. Rund, *Tensors, Differential Forms, and Variational Principles*, Dover, New York, 1989.

J. E. Marsden and A. Tromba, *Vector Calculus*, 6th ed., W. H. Freeman, San Francisco, 2011.

J. Mathews and R. L. Walker, *Mathematical Methods of Physics*, Addison-Wesley, Redwood City, CA, 1970.

A. J. McConnell, *Applications of Tensor Analysis*, Dover, New York, 1957.

A. N. Michel and C. J. Herget, *Applied Algebra and Functional Analysis*, Dover, New York, 1981.

R. K. Miller and A. N. Michel, *Ordinary Differential Equations*, Dover, New York, 2007.

P. M. Morse and H. Feshbach, *Methods of Theoretical Physics*, 2 vols., McGraw-Hill, New York, 1953.

J. A. Murdock, *Perturbations, Theory and Methods*, SIAM, Philadelphia, 1987.

G. M. Murphy, *Ordinary Differential Equations and Their Solutions*, Dover, New York, 2011.

J. T. Oden and L. F. Demkowicz, *Applied Functional Analysis*, 2nd ed., CRC, Boca Raton, FL, 2010.

P. V. O'Neil, *Advanced Engineering Mathematics*, 7th ed., Cennage, Stamford, CT, 2012.

L. Perko, *Differential Equations and Dynamical Systems*, 3rd ed., Springer, Berlin, 2006.

W. H. Press, S. A. Teukolsky, W. T. Vetterling, and B. P. Flannery, *Numerical Recipes in Fortran 77*, Cambridge University Press, Cambridge, UK, 1986.

A. Prosperetti, *Advanced Mathematics for Applications*, Cambridge University Press, Cambridge, UK, 2011.

J. N. Reddy, *Applied Functional Analysis and Variational Methods in Engineering*, McGraw-Hill, New York, 1986.

J. N. Reddy and M. L. Rasmussen, *Advanced Engineering Analysis*, John Wiley, New York, 1982.

R. M. Redheffer, *Differential Equations: Theory and Applications*, Jones and Bartlett, Boston, 1991.

R. D. Richtmyer and K. W. Morton, *Difference Methods for Initial-Value Problems*, 2nd ed., Krieger, Malabar, FL, 1994.

F. Riesz and B. Sz.-Nagy, *Functional Analysis*, Dover, New York, 1990.

K. F. Riley, M. P. Hobson, and S. J. Bence, *Mathematical Methods for Physics and Engineering: A Comprehensive Guide*, 3rd ed., Cambridge University Press, Cambridge, UK, 2006.

P. D. Ritger and N. J. Rose, *Differential Equations with Applications*, Dover, New York, 2010.

J. C. Robinson, *Infinite-Dimensional Dynamical Systems*, Cambridge University Press, Cambridge, UK, 2001.

M. Rosenlicht, *Introduction to Analysis*, Dover, New York, 1968.

T. L. Saaty and J. Bram, *Nonlinear Mathematics*, Dover, New York, 2010.

H. Sagan, *Boundary and Eigenvalue Problems in Mathematical Physics*, Dover, New York, 1989.

D. A. Sanchez, R. C. Allen, and W. T. Kyner, *Differential Equations*, Addison-Wesley, Boston, 1988.

H. M. Schey, *Div, Grad, Curl, and All That*, 4th ed., W. W. Norton, London, 2005.

M. J. Schramm, *Introduction to Real Analysis*, Prentice-Hall, Englewood Cliffs, NJ, 1996.

L. A. Segel, *Mathematics Applied to Continuum Mechanics*, Dover, New York, 1987.

D. S. Sivia and J. Skilling, *Data Analysis: A Bayesian Tutorial*, 2nd ed., Oxford University Press, Oxford, UK, 2006.

I. S. Sokolnikoff and R. M. Redheffer, *Mathematics of Physics and Modern Engineering*, 2nd ed., McGraw-Hill, New York, 1966.

G. Stephenson and P. M. Radmore, *Advanced Mathematical Methods for Engineering and Science Students*, Cambridge University Press, Cambridge, UK, 1990.

G. Strang, *Introduction to Applied Mathematics*, Wellesley-Cambridge, Wellesley, MA, 1986.

G. Strang, *Linear Algebra and Its Applications*, 4th ed., Cennage Learning, Stamford, CT, 2005.

G. Strang, *Computational Science and Engineering*, Wellesley-Cambridge, Wellesley, MA, 2007.

S. H. Strogatz, *Nonlinear Dynamics and Chaos with Applications to Physics, Biology, Chemistry, and Engineering*, Westview, Boulder, CO, 2001.

R. Temam, *Infinite-Dimensional Dynamical Systems in Mechanics and Physics*, 2nd ed., Springer, New York, 1997.

G. B. Thomas and R. L. Finney, *Calculus and Analytic Geometry*, 9th ed., Addison-Wesley, Boston, 1995.

L. N. Trefethen and D. Bau, *Numerical Linear Algebra*, SIAM, Philadelphia, 1997.

L. N. Trefethen and M. Embree, *Spectra and Pseudospectra: The Behavior of Nonnormal Matrices and Operators*, Princeton University Press, Princeton, NJ, 2005.

M. Van Dyke, *Perturbation Methods in Fluid Mechanics*, Parabolic Press, Stanford, CA, 1975.

A. Varma and M. Morbinelli, *Mathematical Methods in Chemical Engineering*, Oxford University Press, Oxford, UK, 1997.

S. Wiggins, *Global Bifurcations and Chaos: Analytical Methods*, Springer, New York, 1988.

S. Wiggins, *Introduction to Applied Nonlinear Dynamical Systems and Chaos*, 2nd ed., Springer, New York, 2003.

H. J. Wilcox and L. W. Lamm, *An Introduction to Lebesgue Integration and Fourier Series*, Dover, New York, 2012.

C. R. Wylie and L. C. Barrett, *Advanced Engineering Mathematics*, 6th ed., McGraw-Hill, New York, 1995.

D. Xiu, *Numerical Methods for Stochastic Computations*, Princeton University Press, Princeton, NJ, 2010.

E. Zeidler, *Applied Functional Analysis: Main Principles and Their Applications*, Springer, New York, 1995.

E. Zeidler, *Applied Functional Analysis: Applications to Mathematical Physics*, Springer, New York, 1999.

D. G. Zill and M. R. Cullen, *Advanced Engineering Mathematics*, 4th ed., Jones and Bartlett, Boston, 2009.

Index

Printed in the United States
by Baker & Taylor Publisher Services

Printed in the United States
by Baker & Taylor Publisher Services